The Role of Polyphenols and Bioactive Compounds for Combating Neurodegenerative Diseases

Ali Imran
Government College University, Pakistan

Ghulam Hussain
https://orcid.org/0000-0001-9090-7789
Government College University, Pakistan

IGI Global
Scientific Publishing
Publishing Tomorrow's Research Today

Vice President of Editorial	Melissa Wagner
Director of Acquisitions	Mikaela Felty
Director of Book Development	Jocelynn Hessler
Production Manager	Mike Brehm
Cover Design	Jose Rosado

Published in the United States of America by
IGI Global Scientific Publishing
701 East Chocolate Avenue
Hershey, PA, 17033, USA
Tel: 717-533-8845 | Fax: 717-533-7115
Website: https://www.igi-global.com E-mail: cust@igi-global.com

Library of Congress Cataloging-in-Publication Data

LCCN: 2025941809 (CIP Data Pending)
ISBN13: 9798337341392
Isbn13Softcover: 9798337341408
EISBN13: 9798337341415
British Cataloguing in Publication Data
A Cataloguing in Publication record for this book is available from the British Library.

Table of Contents

Detailed Table of Contents

The gut microbiota is a key component of the complex, bidirectional communication system that connects the central nervous system with the gastrointestinal tract. This chapter examines the way in which dietary bioactive, including short-chain fatty acids, polyphenols, probiotics, and prebiotics, alter the gut microbiota and have an impact on the gut-brain axis's neurological, hormonal, immunological, and metabolic pathways. Vagal and endocrine signaling immunological modulation, and gut barrier integrity maintenance are important mechanisms. The chapter also covers the impact of dysbiosis on mental and neurological illnesses, the role of microbial metabolites in brain function, and the therapeutic potential of certain dietary strategies. The standardization, bioavailability, and personalization of these analyses continue to present difficulties despite encouraging preclinical and clinical results. Future research should concentrate on customized diet based on microbiota profiles and comprehensive clinical trials to verify efficacy and safety.

Neurodegenerative and neurological disorders act as a serious global public health problem. These conditions result in oxidative stress, neuronal loss, neuroinflammation, synaptic dysfunction, and mitochondrial impairment. Polyphenolic compounds such as flavonoids, phenolic acids, stilbenes, and lignans have gained attention due to their antioxidant, antinflammatory, cardioprotective, and neuroprotective properties. The underlying mechanism of neuroprotection of these compounds involves inhibiting the activation of cyclin-dependent kinase-5 and GSK3β, activating the PI3K/AKT pathway, increasing antioxidant enzymes, reducing the generation of oxidative species, inhibiting the NF-κB, and activating the NLRP3 pathway. Additionally, they regulate neurotransmitters (dopamine, glutamate), increase neurotrophic factors to support neuronal growth and plasticity, prevent misfolded proteins, and neurotoxicity for neuronal survival. This chapter will discuss the neuroprotective potential of these polyphenols and the current understanding of their neuroprotective mechanism.

Neurodegenerative diseases like Alzheimer's, Parkinson's, stroke, and multiple sclerosis pose a major global health challenge due to progressive neuronal damage. Flavonoids, abundant in plant-based foods, offer neuroprotective potential through antioxidant, anti-inflammatory, and signaling-modulatory effects. This chapter reviews their structural diversity, mechanisms of action—including regulation of oxidative stress, inflammation, and cell survival—and their role in preclinical and clinical models. It also discusses nanotechnology-based delivery systems that improve flavonoid bioavailability and brain targeting. Despite promising effects, challenges like poor bioavailability and variable clinical outcomes highlight the need for further research and innovative therapeutic strategies.

The rising incidence of Alzheimer's and Parkinson's diseases has prompted the exploration of treatment options beyond conventional medicine. Increasing attention is being given to plant-based polyphenols for their role in modulating gut microbiota to support brain health. This chapter integrates pharmacy, microbiology, and neuroscience to explore polyphenols like curcumin, resveratrol, and quercetin, focusing on their anti-inflammatory and neuroprotective effects. It highlights how gut microbiota influences the activation and effectiveness of these compounds through short-chain fatty acids, neurotransmitters, and immune responses. Challenges such as poor bioavailability are addressed through advanced drug delivery systems including nanoformulations, phytosomes, and colon-targeted drugs. The chapter also evaluates the combined use of phytopharmaceuticals, functional foods, and probiotics in neurodegenerative disease management.

Neurodegeneration is driven by complex interplay of oxidative stress, chronic neuroinflammation, and mitochondrial dysfunction. These interconnected mechanisms not only exacerbate neuronal damage but also challenge the effectiveness of single-target therapies. Polyphenols, naturally occurring compounds, have significant multi-target therapeutic potential to modulate these pathological mechanisms. This chapter gives a comprehensive overview of the therapeutic potentials of polyphenols against neuroprotection. It highlights how polyphenols counteract neuroinflammation by focusing on their roles in limiting glial activation and downregulating pro-inflammatory molecules and enzymes via targeting multiple associated pathways. In addition, polyphenols increase mitochondrial integrity by reducing reactive oxygen species, preserving membrane potential, and promoting biogenesis. By targeting various neurodegenerative hallmarks, polyphenols signify the compelling strategies for disease modification, paving the way for innovative therapeutic interventions in neurodegenerative disorders.

Human microbiota is a diverse and rich ecosystem composed of trillion microorganisms, such as bacteria, fungi, and viruses. These microbial populations reside at different locations of the body including the skin, mouth, and more importantly, gastrointestinal (GI) tract. The gut microbiota is among them and serves the main physiologic homeostasis, supportive metabolic processes, immune-modulatory and defensive functions against the invasion of pathogenic organisms. The term probiotics refers to live microorganisms which when taken in sufficient quantities exert a number of health benefits on the host. They are important to the gut health, as they provide necessary micronutrients, enzymes, and vitamins (this includes B complex and vitamin K that is vital to digestive processes and mucosal integrity). In this chapter, the association of human microbiota with probiotics is highlighted and the use of probiotics to rebalance the microbe in the host is demonstrated as contributing to good health.

Neurodegenerative illnesses afflict hundreds of millions of people globally, resulting in significant healthcare expenses. Conventional treatments have hardly an effect on the disease, which at most are symptomatic, and recent studies have found, that the action of polyphenols and bioactive molecules are beneficial against the disease progression and also in maintaining cognitive function. These phytochemicals abundant in fruits, vegetables, herbs and medicinal plants exhibit a range of beneficial effects which could lead to the transformation of the concept of neuroprotection. Polyphenols such as flavonoids, phenolic acids, and tannins, have gained significant attention due to their potential in antioxidant and anti-inflammatory properties. It can therefore modulate specific signalling pathways implicated in autophagy, amyloid clearance and synaptic plasticity thus, could be potentially beneficial in the treatment of pathologies such Alzheimer's, Parkinson's and Huntington's diseases.

Polyphenols, are a diverse group of plant-derived compounds found in various foods and beverages such as fruits, berries, legumes, tea and red wine. These compounds exhibit complex structures and are known for their potent bioactive properties. Among polyphenols, flavonoids are the most abundant and have demonstrated significant therapeutic potential, especially in combating neurodegenerative disorders like Alzheimer's, Parkinson's, Huntington's diseases and amyotrophic lateral sclerosis. The increasing prevalence of NNDs due to aging populations underscores the urgency of exploring effective treatments. Flavonoids exert neuroprotective effects through multiple mechanisms. These involve antioxidant activity, modulation of neuroinflammation, inhibition of protein aggregation, and mitigation of endoplasmic reticulum stress. They also influence the microbe-gut-brain axis and have shown promise in enhancing cognitive function and memory.

Neurodegenerative disorders (NDs), notably Parkinson's disease (PD), Alzheimer's disease (AD), Huntington's disease (HD), Epilepsy, etc., are marked by progressive neuronal dysfunction, cognitive decline, and sensory-motor impairments. All these NDs result in clinical and biochemical changes, caused by the accumulation of misfolded proteins, such as Amyloid-β and tau in Alzheimer's disease and α-synuclein in PD. There are currently no cures for most of the NDs, and treatments are primarily

Preface

INTRODUCTION

We are pleased to introduce this edited volume, *The Role of Polyphenols and Bioactive Compounds for Combating Neurodegenerative Diseases*, which brings together emerging scientific perspectives at the intersection of neuroscience, nutrition, and bioactive compound research. Neurodegenerative disorders—including Alzheimer's disease, Parkinson's disease, and various forms of cognitive decline—represent some of the most pressing global health challenges of our time. As populations age and the prevalence of these conditions rises, the urgency to identify safe, effective, and accessible interventions has never been greater.

Within this context, naturally occurring compounds such as polyphenols and other bioactives have attracted increasing attention. A growing body of research suggests that these compounds—found widely in fruits, vegetables, teas, herbs, and other plant-derived foods—play significant roles in antioxidant defense, anti-inflammatory modulation, and neuroprotective signaling pathways. Their ability to influence molecular networks implicated in neurodegeneration positions them as promising candidates in both the prevention and management of cognitive disorders.

This book seeks to consolidate and advance current knowledge on these bioactive compounds by illuminating the molecular mechanisms that underpin their neuroprotective effects, presenting cutting-edge evidence from clinical and preclinical studies, and highlighting integrative dietary strategies that support brain health. Through contributions from experts across multiple disciplines, the volume also underscores the potential for functional food development and nutraceutical innovation grounded in rigorous scientific understanding.

We envision this work as a resource that bridges fundamental science with applied practice. By encouraging interdisciplinary dialogue—among neuroscientists, nutrition researchers, food technologists, clinicians, and public health practitioners—we hope to foster continued innovation in the rapidly expanding field of neuro-nutrition. In doing so, we also aim to inspire new research directions that translate natural compounds into meaningful, evidence-based approaches for combating neurodegenerative diseases.

This book is designed for a diverse audience: researchers, academics, graduate students, healthcare professionals, and industry innovators working at the frontiers of food science, neurology, pharmacology, and functional food development. It is our sincere hope that the insights presented here contribute to both scientific progress and public health impact, supporting global efforts to improve brain health through natural, accessible, and sustainable means.

ORGANIZATION OF THE BOOK

Chapter 1: A Cross Talk Between Natural Biomolecules and Neuroglia

This chapter explores the intricate communication between neurons and glial cells and its critical role in maintaining brain health. Disruption of neuron-glia interactions contributes to neurodegenerative disorders such as Alzheimer's and Parkinson's disease. The chapter highlights natural bioactive compounds—including alkaloids, terpenoids, and polyphenols—as modulators of neuroglial function. By exerting antioxidant, anti-inflammatory, and neuroprotective effects, these compounds support cognition, mood, and neurological outcomes, offering complementary approaches to conventional therapies. Medicinal plants and compounds such as cannabinoids and ß-asarone are discussed for their potential to provide safer, effective interventions targeting neuroglia.

Chapter 2: Beyond The Surface—Polyphenols as Cellular Reprogrammers in Neurodegenerative Diseases

Here, we examine the emerging concept of polyphenols not only as antioxidant and anti-inflammatory agents but also as modulators of core cellular processes. Polyphenols influence epigenetic modifications, synaptic plasticity, mitochondrial biogenesis, and cellular senescence, supporting neuronal survival and neurogenesis. By targeting transcription factors and signaling pathways, they represent a paradigm shift from symptomatic relief to modifying disease progression, positioning polyphenols as promising tools for precision medicine approaches in neurodegenerative disorders.

Chapter 3: Bioactive Peptides Produced from Microbial Sources and Their Neuroprotective Effects

This chapter focuses on microbial-derived bioactive peptides, highlighting their neuroprotective potential. Produced via fermentation, these peptides demonstrate antioxidant, anti-inflammatory, antimicrobial, and immunomodulatory properties, and modulate the gut-brain axis. Mechanisms discussed include mitigation of oxidative stress, inhibition of apoptosis, enhancement of mitochondrial function, and promotion of neurotrophic signaling. The chapter emphasizes the promise of microbial peptides in preventing and treating neurodegenerative diseases, offering an overview of sources, production, and future research directions.

Chapter 4: Blueberries and Brainpower—Exploring the Role of Anthocyanins in Cognitive Health

This chapter highlights the cognitive benefits of blueberries, rich in anthocyanins. Anthocyanins reduce neuroinflammation, protect neurons, and enhance memory and focus. Integrating insights from nutrition science and clinical evidence, the chapter connects blueberry consumption to brain-friendly diets such as MIND and DASH, and discusses practical dietary strategies. Variability in individual response due to gut microbiota and genetics is also considered, making the chapter both scientifically grounded and practically applicable for everyday cognitive health.

Chapter 5: Gut-Brain Axis and Microbiome Modulation by Dietary Bioactives

Focusing on the gut-brain axis, this chapter examines how dietary bioactives—including polyphenols, prebiotics, probiotics, and short-chain fatty acids—modulate gut microbiota to influence neurological, hormonal, immunological, and metabolic pathways. Mechanisms such as vagal signaling, immune modulation, and gut barrier maintenance are detailed. The chapter also addresses dysbiosis in mental and neurological disorders, highlighting the therapeutic potential of targeted dietary interventions while acknowledging challenges in standardization, bioavailability, and personalized approaches.

Chapter 6: Mechanisms of Neuroprotection by Flavonoids, Phenolic Acids, Stilbenes, and Lignans

This chapter delves into the neuroprotective mechanisms of various polyphenolic compounds. Flavonoids, phenolic acids, stilbenes, and lignans counter oxidative stress, neuroinflammation, synaptic dysfunction, and mitochondrial impairment. The discussion covers regulation of signaling pathways, neurotransmitters, neurotrophic factors, and protein misfolding. By providing a comprehensive understanding of these mechanisms, the chapter underscores the potential of polyphenols to support neuronal survival and functional recovery in neurodegenerative conditions.

Chapter 7: Neuroprotective Flavonoids—Mechanisms and Delivery Strategies

The neuroprotective benefits of flavonoids are explored in the context of neurodegenerative diseases, including Alzheimer's, Parkinson's, and multiple sclerosis. Beyond their antioxidant and anti-inflammatory properties, flavonoids modulate signaling pathways critical for cell survival. The chapter also addresses challenges in bioavailability and clinical translation, highlighting innovative nanotechnology-based delivery systems to enhance brain targeting and therapeutic efficacy.

Chapter 8: Phytopharmaceuticals and the Gut-Brain Axis in Neurodegenerative Therapy

This chapter integrates insights from pharmacy, microbiology, and neuroscience to explore how plant-derived polyphenols modulate gut microbiota to support brain health. Polyphenols such as curcumin, resveratrol, and quercetin are discussed in the context of anti-inflammatory and neuroprotective effects. The chapter emphasizes gut-mediated modulation of compound efficacy, advanced delivery systems to overcome bioavailability challenges, and combined strategies involving phytopharmaceuticals, functional foods, and probiotics in neurodegenerative disease management.

Chapter 9: Polyphenol-Mediated Neuroprotection Against Traumatic Injury to the Nervous System

Polyphenols demonstrate potent neuroprotective effects in traumatic brain injury, spinal cord injury, and peripheral nerve injuries. By modulating oxidative stress, inflammation, apoptosis, and axonal regeneration, polyphenols support neuronal survival and functional recovery. The chapter presents detailed mechanisms of action, highlighting key compounds such as gallic acid, EGCG, resveratrol, hesperidin, curcumin, and quercetin, and emphasizes their role in mitigating secondary damage and promoting repair.

Chapter 10: Polyphenol-Rich Medicinal Plants—Mechanistic Insights and Therapeutic Potentials Against Alzheimer's Disease

Focusing on Alzheimer's disease, this chapter explores the neuroprotective potential of polyphenol-rich plants like Ginkgo biloba, Curcuma longa, Camellia sinensis, and Withania somnifera. Mechanisms include free radical scavenging, anti-inflammatory action, acetylcholinesterase inhibition, and modulation of Nrf2 and NF-κB pathways. Laboratory, animal, and clinical evidence are discussed, alongside bioavailability challenges and advanced delivery strategies, providing a comprehensive resource on polyphenol-based AD interventions.

Chapter 11: Polyphenols Against Oxidative Stress, Neuroinflammation, and Mitochondrial Dysfunction

This chapter synthesizes the multi-target potential of polyphenols in neuroprotection. By simultaneously mitigating oxidative stress, chronic neuroinflammation, and mitochondrial dysfunction, polyphenols address key mechanisms underlying neurodegeneration. The discussion highlights how polyphenols preserve mitochondrial integrity, regulate glial activation, and modulate multiple signaling pathways, offering an integrated approach to disease modification.

Chapter 12: Probiotics and Human Microbiome—Therapeutic Potentials in Cancer, Cardiovascular Disease, and Cognitive Disorders

Here, the focus is on human microbiota and the therapeutic application of probiotics. The chapter details the composition and physiological roles of gut microbiota and demonstrates how probiotics can rebalance microbial populations to support health, including cognitive function. Emphasis is placed on micronutrient production, immune modulation, and metabolic homeostasis, highlighting probiotics as an adjunct strategy in neuroprotective interventions.

Chapter 13: The Role of Polyphenols and Bioactive Compounds for Combating Neurodegenerative Diseases

This chapter provides an overview of polyphenols and other bioactive compounds as neuroprotective agents. Highlighting their antioxidant and anti-inflammatory properties, the chapter discusses modulation of autophagy, amyloid clearance, and synaptic plasticity, illustrating the potential of these phytochemicals to influence disease progression in Alzheimer's, Parkinson's, and Huntington's diseases. The discussion frames polyphenols as transformative tools in neuroprotection.

Chapter 14: Neurodegenerative Natural Foes—Polyphenols, Especially Flavonoids as Therapeutic Agents

Focusing specifically on flavonoids, this chapter details their multifaceted therapeutic potential against neurodegenerative disorders. Mechanisms include antioxidant activity, modulation of neuroinflammation, inhibition of protein aggregation, mitigation of endoplasmic reticulum stress, and engagement with the gut-brain axis. The chapter emphasizes flavonoids' role in enhancing cognition and memory, highlighting their relevance in addressing age-related neurological challenges.

Chapter 15: Neuroprotective Potential of Bioactive Compounds

The final chapter synthesizes evidence on the neuroprotective potential of polyphenols and other bioactive compounds across major neurodegenerative disorders, including Parkinson's, Alzheimer's, Huntington's disease, and epilepsy. It examines mechanisms such as antioxidant activity, anti-inflammatory effects, and modulation of misfolded proteins, while highlighting bioactive compounds from plant sources like Naringenin, Ashwagandha, Celastrus, Neem, Ginkgo biloba, and Seabuckthorn. The chapter underscores the growing interest in natural compounds as interventions to slow progression, support neuronal survival, and enhance cognitive function.

CONCLUSION

In bringing together the diverse perspectives and research presented in this volume, we hope to offer a comprehensive and integrative understanding of the role of polyphenols and bioactive compounds in promoting brain health and mitigating neurodegenerative disorders. Across sixteen chapters, contributors have highlighted not only the molecular mechanisms and therapeutic potential of these natural compounds, but also their translational relevance—from dietary interventions and functional foods to advanced delivery strategies and precision medicine approaches.

This work underscores the multifaceted nature of neurodegeneration, encompassing oxidative stress, inflammation, mitochondrial dysfunction, protein misfolding, and dysregulated neuroglial and gut-brain signaling. Importantly, the chapters collectively demonstrate how polyphenols, flavonoids, microbial peptides, and other bioactive compounds can modulate these pathways, offering both neuroprotection and the promise of cognitive resilience.

We anticipate that this reference will serve as a valuable resource for researchers, clinicians, nutritionists, and innovators in the fields of neuroscience, pharmacology, and functional foods, while inspiring further interdisciplinary collaboration. Ultimately, our goal is to support the development of safe, effective, and accessible strategies to maintain brain health and address the growing global burden of neurodegenerative diseases.

By weaving together foundational science, emerging clinical insights, and practical applications, this book stands as a testament to the potential of nature's compounds to illuminate new paths in neuroprotection and brain longevity.

Chapter 1
A Cross Talk Between Natural Biomolecules and Neuroglia

Tehreem Iman
https://orcid.org/0009-0002-5992-6791
Government College University, Faisalabad, Pakistan

Rabia Akram
https://orcid.org/0000-0001-9416-6391
Government College University, Faisalabad, Pakistan

Ammara Tehreem
Government College University, Faisalabad, Pakistan

Aliza Asghar
https://orcid.org/0009-0001-9782-8618
Government College University, Faisalabad, Pakistan

Maryam Mustafa
Government College University, Faisalabad, Pakistan

Javed Iqbal
Allied Hospital, Pakistan

ABSTRACT

The human brain's billions of neurons rely on glial cells for repair, maintenance, and protection. Disruption in glia–neuron communication drives chronic neurodegenerative disorders (NDs) such as Alzheimer's and Parkinson's, marked by protein aggregates, neuronal loss, and functional decline. Conventional therapies are costly, have side effects, and often lack efficacy, highlighting the need for novel approaches. Natural bioactive compounds—such as alkaloids (caffeine, berberine), terpenoids (limonene, ginkgolides), and polyphenols (quercetin, resveratrol, curcumin)—exert antioxidant, anti-inflammatory, and neuroprotective effects, thereby modulating neuroglial function. Medicinal plants (Withania somnifera, Curcuma longa, Bacopa monnieri) and agents like cannabinoids, β-asarone, and pergolide show promise in improving cognition, mood, and neurological outcomes. Targeting neuroglia with these compounds offers a complementary, effective, and safer approach to managing CNS disorders.

1 INTRODUCTION

The central nervous system and the peripheral nervous system are the two main topographic divisions of the nervous system. However, from a functional perspective, the two systems work together in a coordinated and integrated manner and are closely connected (Thau et al., 2022). The brain and spinal

DOI: 10.4018/979-8-3373-4139-2.ch001

cord form the central nervous system, while the cranial and peripheral nerves and associated neurological structures comprise the peripheral nervous system (Bazira, 2021).

1.1 Neuroglia Cells: Classification, Function, and Overview

The two primary cell types that make up the brain are the glial and neuronal cells. Both are vital parts of the neurological system. As the nervous system's homeostatic and defensive cells, glial cells include astrocytes, oligodendrocytes (OLs), microglia, and ependymal cells. They are remarkably diverse in shape and function. They possess exceptional flexibility that allows them to adapt throughout life to environmental challenges (Gabriel Ortiz et al., 2016; Hossain et al., 2017; Verkhratsky et al., 2025). Their systemic role is to maintain homeostasis at all levels of the central nervous system, from molecules to organs (Ibrahim et al., 2020; Matejuk & Ransohoff, 2020; Peng et al., 2014). These cells facilitate tissue healing after injury, and their malfunction may contribute to neurodegenerative and mental illnesses (Y. Yu et al., 2024). The majority of cells in the human CNS are astrocytes. They perform homeostatic, structural, and metabolic functions (Riaz et al, 2023), as well as neuroprotective roles such as promoting synapse formation, stabilizing and regulating the blood-brain barrier (BBB), and removing excess neurotransmitters (Benarroch, 2005; Oberheim et al., 2012; Ransom & Ransom, 2012; Vasile et al., 2017; Wei & Morrison, 2023). In the CNS, microglia serve as the main immune cells, responding to damage and infection. OLs are essential for the formation and maintenance of myelin sheaths around neuronal axons, enabling rapid signal transmission. The brain's ventricular system is lined by ependymal cells, which contribute to the circulation and production of cerebrospinal fluid (CSF). They are crucial for maintaining CNS homeostasis, supporting neuronal activity, and responding to pathogenic conditions (Camberos-Barraza et al., 2024; Dar et al., 2024; Quincozes-Santos et al., 2021; Verkhratsky & Semyanov, 2025).

Figure 1. Classification and function of Neuroglial cells of CNS and PNS

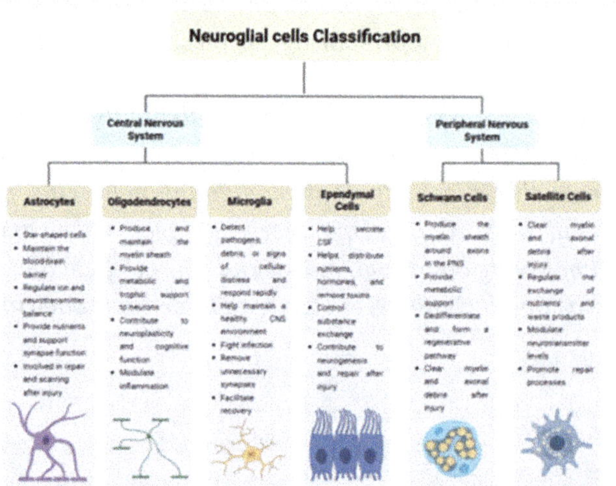

1.2 Neuroglia: Key Players in CNS Homeostasis

The ability of the adult CNS to regenerate after illness or damage is restricted. The primary glial populations in the CNS include astrocytes, OLs, microglia, and ependymal cells. These cells are essential for both the disease process and brain growth and homeostasis (Quan et al., 2022). Neuroglial cells have recently been shown to be critical for immunological surveillance, metabolic support, neural transmission, and cognitive functions, including memory and learning. Additionally, a growing number of neurological and psychiatric problems are linked to their malfunction, ranging from mental ailments like depression and schizophrenia to neurodegenerative diseases like Alzheimer's and Parkinson's diseases (Escartin et al., 2021; Gao et al., 2023a; Khelfaoui et al., 2024).

1.2.1 Astrocytes

The astrocyte is a core glial cell that forms a tripartite synapse with neurons to play a crucial role in the structure and functioning of neural circuits and brain processes. As true secretory cells in the CNS, astrocytes produce ions, neurotransmitters, neurohormones, trophic factors, immunoactive chemicals, and amino acids. They are a component of the gliocrine system, together with microglia (Verkhratsky et al., 2016). As supporting cells, they have long been thought to play important functions in maintaining metabolic homeostasis by producing glycogen and giving neurons energy substrates, facilitating synaptic transmission through ionic balance and neurotransmitter clearance, and defending the CNS. Astrocytes also show experience-dependent plasticity (Pathak & Sriram, 2023; Takano & Soderling, 2021; Verkhratsky & Nedergaard, 2018; Williamson et al., 2024). Additionally, astrocytes support the pathways for fatty acid and amino acid metabolism, water and ion balance, oxidative stress resistance, and anti-inflammatory responses (Morris et al., 2020; Sofroniew, 2020). Hippocampus ensembles of learning-associated astrocytes (LAAs) are critical for memory recall (Williamson et al., 2024).

1.2.2 Microglia

Microglia are the primary immune effector cells that are essential for CNS development, homeostasis, and the body's reaction to brain damage. They are classified as both glia and resident macrophage-like immune cells. Under healthy conditions, microglia communicate with neighboring neurons, astrocytes, and OLs to establish synapses and demonstrate their homeostatic role (Wright-Jin & Gutmann, 2019). When there are pathological disruptions to homeostasis, they react first, just like peripheral immune cells. Similar to peripheral macrophages, T cells, and B cells, microglia must scan their surroundings, phagocytose debris, and release cytokines even under less than ideal energy circumstances (Bernier et al., 2020).

1.2.3 Oligodendrocytes

The myelinating cells of the CNS are called oligodendrocytes (OL), and they are distinguished by their unique development through a multi-step process that involves modifications to cell markers and shape (Chiou & Piao, 2021). The main function of OL is to form the insulating myelin sheaths composed of lipids and myelin proteins around axons in the CNS (Verkhratsky & Semyanov, 2025). This is essential for both axonal energy conservation and the quick saltatory conduction of action potentials along axons (Dar et al., 2024; Kister & Kister, 2022). Historically, OLs are the cells primarily responsible for remyelination, instead of pre-existing mature OLs (Crawford et al., 2016; Zhou et al., 2020). Each OL will myelinate between 20 and 60 axons, with myelin internodes ranging in length from 20 to 200 μm, depending on the CNS area. Because OL cells not only myelinate but also maintain axon metabolism, their development is becoming more and more linked to neurodegenerative illnesses like multiple sclerosis (Jäkel et al., 2019).

1.2.4 Ependymal cells

Ependymal cells are multiciliated neuroepithelial cells that line the cerebral ventricles and spinal cord (Deng et al., 2023). They are found where the brain parenchyma and cerebrospinal fluid (CSF) meet, and they are essential to the development of the brain-CSF barrier. They offer important sensory and mechanical support, allowing chemicals to move between the CSF and parenchyma to preserve its composition and keep harmful compounds out of the CNS. Additionally, ependymal cells can release cytokines and other signaling molecules, which aid in immunological and metabolic control as well as nutritional maintenance. With their strong proliferation ability and multipotent responses to damage, ependymal cells can be triggered in pathological circumstances such as spinal cord injury (SCI) and neurodegenerative illnesses, making them potential endogenous stem cells (Xie & Li, 2024).

2 NATURAL BIOMOLECULES

2.1 Plant-Derived Compounds

Plant-derived compounds are known as phytochemicals. The plants have distinctive tastes, colors, and therapeutic qualities, which are due to these compounds (Esmeeta et al., 2022). These compounds comprise polyphenols, terpenes, flavonoids, alkaloids, and many more (Am & a, 2023). By influencing the action of receptors for the main inhibitory neurotransmitters, bioactive compounds derived from medicinal plants contribute significantly to the preservation of the chemical equilibrium in the brain. Medicinal plants include *Panax ginseng, Valeriana officinalis, Bacopa monniera, Nardostachys jatamansi, Ginkgo biloba,* and *Withania somnifera*. Several traditional therapeutic systems have made extensive use of them due to their neuroprotective, adaptogenic, and psychotropic qualities (Awuchi, 2023; G. P. Kumar et al., 2015).

2.1.1 Polyphenols

Natural products, vegetables, and fruits contain large levels of polyphenols, a class of naturally occurring secondary metabolites (Vauzour, 2012). Their biological functions are diverse and include the modulation of neurogenesis and microglial activation, as well as anti-inflammatory, neuroprotective, and antioxidant properties (Badjakov et al., 2025). Polyphenolic compounds include curcumin, epigallocatechin gallate (EGCG), and resveratrol (grape, peanut, and berry) (Y. Fu et al., 2018; Jalouli et al., 2025; Yoon, 2025).

2.1.2 Flavonoids

Flavonoids are a class of dietary polyphenols derived from plants that have garnered interest due to their pro-cognitive, neuroprotective, antioxidant (removing reactive oxygen species (ROS), and neuro-inflammation-suppressive properties. They also improve brain function and have anti-inflammatory properties (Cichon et al., 2020; Hasan et al., 2023; Jaeger et al., 2018; Vauzour et al., 2008). Beverages, like wine and tea, honey, vegetables, fruits, flowers, stems, roots, bark, and grains, are naturally occurring flavonoids (Panche et al., 2016; Zbieta & Ebas, 2025).

2.1.3 Alkaloids

Alkaloids are compounds found in several plant species that contain nitrogen. They are one of the biggest groups of natural products; they are produced by plants, marine life, amphibians, certain bacteria, and fungi (Kong et al., 2021; Wink, 2000). They have shown promise as preventative measures against chronic inflammation and neurodegenerative diseases (NDDs) (Aryal et al., 2022). They have a broad range of different pharmaceutical tools in the therapeutic field, including antipyretic drugs (like quinine), anticancer drugs (like vincristine), antihypertensive drugs (like reserpine), analgesics (like morphine), antihyperglycemics (e.g., piperine), and antiasthmatics (like ephedrine) (Ng et al., 2015).

2.1.4 Terpenoids

The biggest class of phytochemicals is terpenoids, which are made up of a variety of molecules with wildly different chemical structures (Sharma et al., 2017; W. Yang et al., 2020). Numerous antibacterial, anti-inflammatory, anti-allergic, anti-metastatic, antioxidant, anti-carcinogenic, anti-angiogenesis, neuroprotective, active against various mental diseases, and other beneficial uses that might improve human health have been found in terpenoids (Câmara et al., 2024; Del Prado-Audelo et al., 2021; Mony et al., 2022; Siddiqui et al., 2024; Subedi & Yumnam, 2021; B. Xu et al., 2022).

2.1.5 Tannins

A class of phenolic chemicals known as tannins is found in a wide range of foods, including grains, legume seeds, tree wood and bark, wild herbs, and, most notably, in a variety of fruits and vegetables (Hossain MT et al., 2021). In many medications, medical treatments, and nutraceuticals, tannins are increasingly regarded as an essential ingredient having both internal and external effects. Tannin is used externally to heal wounds and inflammation of the skin, and it may also stop many chronic illnesses before they start. Antioxidant, free radical scavenging, antimicrobial, anti-carcinogenic, anti-mutagenic, antitumor, anti-parasitic, antiviral, anti-inflammatory, antimicrobial, anti-allergic, anthelmintic, immunomodulation, hepato-protective, an inhibitor of the detrimental pro-oxidative enzyme, and anti-nutrient effects are some of the effects that tannin compounds may have on the gastrointestinal tract. Additionally, certain tannins have chemopreventive effects on skin, stomach, prostate, oral, and breast cancer (Cosme et al., 2025; Fraga-Corral et al., 2021; Hossain MT et al., 2021; Ozogul et al., 2025; Smeriglio et al., 2017; Wu et al., 2019).

3 NEUROGLIA IN CNS DISORDERS: MEDIATORS OF DAMAGE AND REPAIR

Deterioration of neuronal structure and function is a hallmark of a wide class of progressive illnesses known as neurodegenerative diseases (NDs), psychiatric disorders, and traumatic brain injuries. Among the most difficult health issues of the twenty-first century are Alzheimer's disease (AD), Parkinson's disease (PD), amyotrophic lateral sclerosis/motor neuron disease (ALS/MND), and multiple sclerosis (MS). Collectively, these conditions affect millions of people globally, and their numbers are expected to increase dramatically by 2050, primarily due to an aging population (Nichols et al., 2019).

3.1 Role of Astrocytes in CNS Disorders

3.1.1 Neurodegenerative Diseases

Astrocytes are the most prevalent glial cells in the CNS. These star-shaped cells play a crucial role in preserving homeostasis, supporting neuronal metabolism, and controlling neurotransmitter levels (Dar et al., 2024; Mulica et al., 2021). They are important for the onset and progression of neurodegenerative diseases (Deng et al., 2024). Reactive astrocytes come in two varieties: A1 astrocytes, which are neurotoxic, and A2 astrocytes, which may be neuroprotective (Chang et al., 2023; Clarke et al., 2018; Liddelow et al., 2017) The most common pathogenic feature of several neurodegenerative illnesses, such as Alzheimer's disease (AD), Parkinson's disease (PD), Huntington's disease (HD), and amyotrophic lateral sclerosis (ALS), is cellular aggregations of misfolded proteins (Aguzzi & O'Connor, 2010). Autophagic mechanisms in the brain, mainly in neurons but also in glial cells, are widely dysfunctional in neurodegenerative illnesses. This leads to a poor clearance of misfolded proteins and consequent buildup of these proteins, e.g., amyloid beta peptide (Aβ) in AD, α-Synuclein (α-Syn) in PD, huntingtin in HD, resulting in neuroinflammation (Guo et al., 2018). Autophagy and neuroinflammation are linked reciprocally, with autophagy influencing inflammation and inflammation being modulated by autophagy (Kim et al., 2024). After neuronal injury, astrocytes become activated. In many neurological illnesses, reactive astrocytes cause the development of neuropathology (Zhou et al., 2019). Reactive astrocytes that surround Aβ plaques help to eliminate Aβ by absorbing it, mostly destroying it via autophagy, and releasing Aβ-degrading proteases into the extracellular space (Frost & Li, 2017; Kim et al., 2024). In PD, astrocytes play a crucial role in the endocytosis of extracellularα-Syn produced by neurons (Tsunemi et al., 2020). As with Aβ, the autophagy-lysosome system is the primary degrader of α-Syn in astrocytes; however, due to reduced autophagic activity, senescent cells have a delayed clearance of α-Syn (Hong et al., 2024). Apart from eliminating harmful clusters, autophagy in astrocytes can regulate the release of pro-inflammatory cytokines and microglia activation. Neuroinflammation and neurodegeneration may be worsened by any dysregulation of autophagy, whether it be through blockage or overactivation. Autophagy can, in short, increase inflammation, control the production of inflammatory markers, and function as an effector of inflammation regulated by cytokines, i.e., TNF-α, interferon (IFN)-γ, interleukin-1β (IL-1β), IL18, IL-4, IL-10, and IL-17 (H. Gan et al., 2024). Reactive astrogliosis's mitochondrial dysfunction is another key factor in the development of CNS disorders. The two most prevalent neurodegenerative illnesses, AD and PD, as well as ageing, have been linked to mitochondrial abnormalities that affect their dynamics, transport, and exchange. There have been reports of impaired mitophagy in ALS, PD, AD, and HD (Martinez-Vicente, 2017; X. Wang & Jia, 2023). Mitophagy is crucial for removing amyloid precursor protein (APP), which causes amyloid plaques (Vaillant-Beuchot et al., 2021). Amyloid precursor protein (APP) cleaves to create the toxic Aβ and the hyperphosphorylated tau protein (pTau), which builds up in the AD brain (De Strooper & Karran, 2016).

Figure 2. Role of astrocyte in health and disease conditions

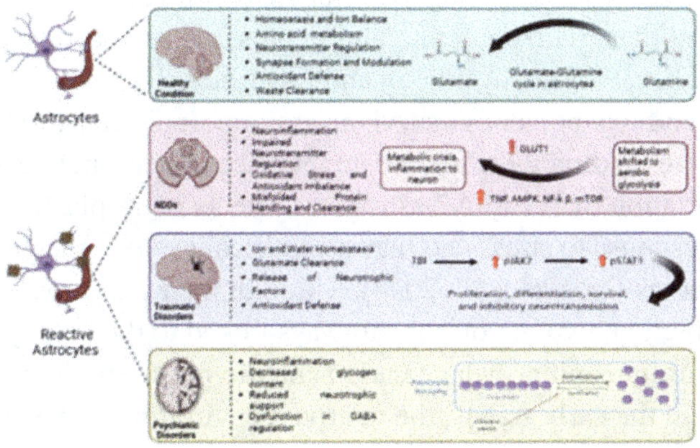

3.1.2 Traumatic CNS Disorders

An acute injury that occurs as a result of violence, falls, athletic activities, warfare, and motor accidents is known as Traumatic Brain Injury (TBI) (Kardos et al., 2017). Such injuries trigger functional and structural changes in the astrocytes (Zhou et al., 2020). Following TBI, astrocytes coordinate neurorestorative mechanisms that facilitate neurological progress and release neuroprotective factors to eliminate cell debris in the impaired brain. On the other hand, astrocytes become dysregulated through injury, leading to the development of cytotoxic mediators that impede the CNS by impairing mechanisms such as apoptosis and neuronal dysfunction (M. Yuan & Wu, 2022). These activated astrocytes act as key modulators of the inflammatory reaction that is initiated by the TBI. They not only trigger early inflammatory reactions after injury but also play a necessary role in the secretion of the inflammasome and activation of astrocytes, which are significant processes that contribute to the TBI pathophysiology. These processes have a considerable impact on the advancement of the secondary insults and persistence (H. Zhang et al., 2025). Astrocytes have a complicated role in the repair of spinal cord injury. Major contributors to the chemical and physical impediment known as the glial scar to axonal regeneration are reactive astrocytes (White & Jakeman, 2008). They regulate the blood-spinal cord barrier and also control spinal blood circulation, the availability of fluids, and homeostasis of ions (Attwell et al., 2010; Takano et al., 2006). During SCI and other pathological conditions, they become activated and polarized. Their phenotypic and functional plasticity play an essential role in regeneration and repair after SCI (Li et al., 2024).

3.1.3 Psychiatric Disorders

It has now been established that glia play several vital physiological roles and that abnormalities in their function may significantly contribute to the symptoms of many psychiatric disorders, even though they were initially believed to be only present in the brain to support neurons in a physical, metabolic, and nutritional capacity (Rawani et al., 2025). Oligodendrocyte or white matter lesions and mental illnesses such as autism, bipolar disorder, schizophrenia, and depression are correlated. Astrocytic alterations impact mood and complex behavior in physiology and functioning in certain disorders (Li et al., 2016). Astrocytes affect neuronal and synaptogenesis survival; they are significant modulators of neuroplasticity. The malfunction in astrocytes is directly related to an increase in neuronal death in depressive illnesses (Peng et al., 2015). The relation of A1 to A2 astrocytes affects neuroplasticity and synaptic function. They influence neuronal response to stress and play a significant role in the pathophysiology of depression (Hao et al., 2020; Puentes-Orozco et al., 2024). Through their participation in Ca2+ signaling and dopamine–glutamate interaction, two important pathways linked to the pathophysiology of schizophrenia, astrocytes control neurotransmitter balance and synaptic plasticity, impacting early susceptibility and disease development. In the early stages, the neuronal environment may already be affected by the genetic variables linked to the "two-hits" model of schizophrenia. Astrocytes play a role in the formation of a modified neural matrix, which leads to excessive synaptic pruning and demyelination, interferes with neuronal communication, and exacerbates the dopamine–glutamate imbalance. Both are primary symptoms of schizophrenia, and the heightened vulnerability to cognitive impairment is caused by these mechanisms (Vellucci et al., 2025).

3.2 Role of Microglia in CNS Disorders

3.2.1 Neurodegenerative Diseases

Microglia are crucial immune cells that produce cytokines and chemokines and phagocytose dead cells and foreign materials. They promptly eliminate superfluous materials, including foreign objects and dead cells, and their immune system promotes the repair of the damaged region (Prinz et al., 2021; Rodríguez-Gómez et al., 2020). Many neurodegenerative disorders are associated with microglia activation. In addition to their protective role of phagocytosing and eliminating pathological protein aggregates, microglia can also be harmful when they take up too many protein aggregates, which can impair their ability to phagocytose, cause neuroinflammation, and ultimately cause neurodegeneration. Peripheral immune cell infiltration also causes microglia to adopt a pro-inflammatory phenotype, which speeds up the course of the disease (Gao et al., 2023b). The two varieties of microglia are M1 and M2. When M2 microglia detect transforming growth factor (TGF)-β and IL-4, they produce neurotrophic factors, including brain-derived neurotrophic factor and anti-inflammatory cytokines like IL-10 (Butovsky et al., 2014; Parkhurst et al., 2013; Tang & Le, 2016; S. Yi et al., 2020), which prevent many neurodegenerative diseases from progressing. Proinflammatory cytokines released by activated microglia in diseases like MS and AD worsen oligodendrocyte damage and impede the healing process (Zhang et al., 2024). Amyloid-β (Aβ) plaque formation, neuronal cell death, and microglia activation are the hallmarks of AD, which progressively worsen memory and cognitive function (Fakhoury, 2017; Shi et al., 2019; Tan et al., 2021). Numerous investigations show that AD brains have microglial activation (Dal Bianco et al., 2008; Sastre et al., 2006), and in Aβ plaques, microglial activity was noticeably greater than in cortical regions devoid of plaque (Dal Bianco et al., 2008; Liu & Hong, 2003). The first evidence of reactive microglia was found in the substantia nigra (SN) of PD patients' postmortem brain tissue (McGeer et al., 1988). Furthermore, PD patients' PET scans showed extensive microglial activation. Microglial activation was shown to be present in both newly diagnosed patients and those with chronic diseases (Gerhard et al., 2006; Janda et al., 2018). The process of microglial recruitment and activation, or microgliosis, is accompanied by elevated cytokine levels (Imamura et al., 2003). The precise etiology of MSA is still unknown, even though several factors may be involved in neurodegeneration. The microglial activation and neuroinflammation are key characteristics of MSA, according to pathological investigations and PET imaging of MSA brains (Ishizawa et al., 2004; Jucaite et al., 2022; Kübler et al., 2019; Nykjær et al., 2017; Salvesen et al., 2017). Microglial activation is prevalent in areas of motor-related structures, such as cerebellar input, extrapyramidal, and pyramidal motor structures. This showed that the manner of microglial activation was in line with the pattern of system degradation peculiar to MSA (Ishizawa et al., 2004). The MSA microglial activation encourages neurodegeneration. Microglia have been linked to ALS, a neurodegenerative condition marked by a large loss of motor neurons that causes paralysis and early death (Clarke & Patani, 2020). Microglia promote neuronal health in homeostatic settings by removing protein aggregates, producing anti-inflammatory and neurotrophic substances, and conducting immunological monitoring. However, these cells progressively take on DAM (disease-associated microglia) characteristics in ALS, which can be detrimental as well as beneficial. While DAM may assist in eliminating toxic aggregates and preserving synaptic integrity, it can also produce pro-inflammatory cytokines, including TNF-α and IL-1β, as well as substances that cause astrocytes to become neurotoxically reactive. This promotes the death of motor neurons, causes tissue inflammation, and initiates excessive synaptic pruning. Therefore, ALS progression is largely determined by the dynamic balance between DAM-mediated neuroprotection and neurotoxicity (Bond et al., 2025). Polarization of M1/M2 microglia during motor neuron degeneration brought on by ALS. Activated microglia serve as a bridge between the toxic M1 phenotype, which generates cytokines that increase inflammation, and the neuroprotective M2 phenotype, which releases neuroprotective molecules to sustain neuron survival and promote tissue repair throughout ALS. This causes tissue inflammation, excessive synaptic pruning, and promotes motor neuron death. Encouraging M1 polarisation even further, which leads to neuronal death. Promising methods to reduce local neurodegeneration and enhance the disease's clinical outcome involve therapeutic techniques that target microglia polarisation and induce the M2 phenotype (Geloso et al., 2017).

Figure 3. Neuroprotective and neurotoxic roles of microglia

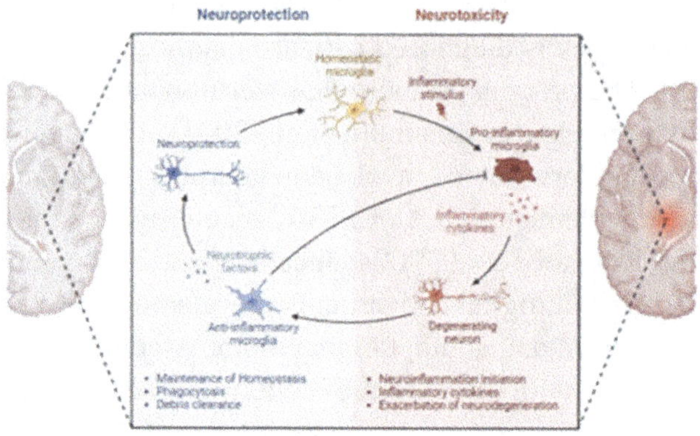

3.2.2 Traumatic CNS Disorders

Microglia, the CNS's main innate immune cells, are essential for coordinating the neuroinflammatory reaction to TBI. Microglia quickly activate and alter phenotypically after injury, which can either worsen secondary damage or encourage tissue structural regeneration and neuroprotection (Zhao et al., 2025). Microglia are an essential part of neuroinflammation after TBI, as in the CNS, they are the first line of defense and the key cellular constituent of the innate immune system whenever damage or illness occurs. Following TBI, microglia in the wounded brain can generate neuroprotective substances, remove cellular debris, and coordinate neuro-restorative processes that aid in neurological recovery. Microglia can, however, also become dysregulated and generate large amounts of cytotoxic and pro-inflammatory mediators, which impede CNS repair and lead to cell death and neuronal dysfunction. Their polarization status and post-injury functional responses may explain the dual role of microglial activation in promoting both positive and negative effects on neurons. (Loane & Kumar, 2016). Microglia are the principal innate immune cells in the CNS; therefore, they play a critical role in the management of neuroinflammation reactions to TBI. One injured, they instantly become activated and undergo significant phenotypic alteration that might worsen secondary damages or facilitate the structure rebuilding of tissues and neuroprotection (Zhao et al., 2025). Microglia are deemed to act as a fundamental element during neuroinflammation that arises after TBI, as they are the most significant cellular agent of the innate immune system in the CNS and act as an initial line of defense in the event of disease and injury. In the context of the injured brain, they can generate neuroprotective factors, coordinate neurorestorative processes, and eliminate cellular debris that are neurotherapeutically significant in the process of neurological recovery after TBI. They may become dysregulated, which leads them to produce increased levels of cytotoxic and pro-inflammatory mediators that hamper CNS repair and cause neuronal dysfunction and ultimately cause cell death. The dual function of microglial activation promotes both useful and harmful effects on neurons, which may be attributed to their polarization and functional responses to injury (Loane & Kumar, 2016). TBI also influenced the critical role of neurogenesis in the hippocampus, such as formation, maturation, and integration of new nerve cells, as a result of microglia activation and subsequent inflammatory responses. They also participate in the formation and elimination of synapses, the determination of the neuron population, and the restoration of damage in the development of the brain. When responding to injury, activated microglia secrete a range of chemokines, pro-inflammatory cytokines, and other neurotoxic agents that worsen the post-TBI injury (Navabi et al., 2024). Microglia experience both positive and negative effects following traumatic spinal cord injury (SCI), including modification, proliferation, and activation, in morphology, protein, and gene expression. Numerous substances, such as nitric oxide, superoxide species, chemokines, growth factors, anti- and pro-inflammatory cytokines, and prostaglandins, are released by microglia and influence secondary damage and post-injury recovery (Loane & Byrnes, 2010). Proteases, reactive oxygen species, and proinflammatory cytokines are produced by activated microglia, which results in secondary neuronal damage. However, by secreting cytokines and anti-inflammatory growth factors, active microglia also aid in neuronal repair. Proinflammatory cytokines worsen brain damage get worse as proinflammatory cytokines increase axonal demyelination, endothelial permeability, decrease neural stem/progenitor cells (NSPCs), and promote A1 astrocyte activation. On the other hand, anti-inflammatory substances aid in locomotor recovery and tissue repair by promoting axonal remyelination, reducing reactive astrocytes, and facilitating angiogenesis and NSPC growth. The CNS needs helpful microglia for ongoing harm prevention because of its limited ability for regeneration (L. Xu et al., 2022).

3.2.3 Psychiatric disorders

One of the main contributing factors and possible contributors to the pathophysiology of schizophrenia has been suggested to be excessive synaptic pruning by microglia (Singhal & Baune, 2017). Microglia contribute to the worsening and development of cognitive impairment and other symptoms associated with schizophrenia (Salter & Stevens, 2017). In schizophrenia, microglia are activated, particularly those in the brain's grey matter. Key free radicals and proinflammatory cytokines are released by activated microglia and are known neurotoxic variables that contribute to cognitive impairment. Resting microglia may become activated in response to stressful situations, sometimes referred to as psychological stress, which is one of the major environmental risk factors for schizophrenia. These activated cells generate free radicals and proinflammatory cytokines, which may be linked to significant structural alterations in the brain, primarily in the GMV, especially in regions of the brain involved in memory processing. Microglial activation may therefore play a significant role in schizophrenia-related cognitive impairment (Zhuo et al., 2023). Microglia may become persistently active in pathological conditions, such as schizophrenia, resulting in an inflammatory state. Given the growing amount of research indicating immune system abnormalities and elevated neuro-inflammation in a significant proportion of schizophrenia patients, this is of tremendous relevance (Murphy et al., 2021; Trépanier et al., 2016; Van Kesteren et al., 2017) and some antipsychotics' anti-inflammatory properties (Patel et al., 2023). In postmortem investigations of schizophrenia, alterations in astrocytic density and/or several astrocytic markers have been documented. The quantity, shape, and function of astrocytes and microglia may alter as schizophrenia develops (Laricchiuta et al., 2024; Notter, 2021).

3.3 Role of Oligodendrocytes in CNS Disorders

3.3.1 Neurodegenerative Diseases

Oligodendrocytes (OLs) are essential to brain function because they are the CNS's myelinating glial cells, which promote the quick passage of electrical impulses along axons and maintain the health of axons. The consequences of OL injury and the molecular processes that modulate its influence are relevant questions. The loss or malfunction of mature OL and oligodendrocyte precursor cells (OPCs) results in poor myelination and consequent neurological impairments because of their involvement in facilitating saltatory conduction. Demyelination is frequently linked to trauma or inflammation, although hereditary factors can also cause it (Duncan & Radcliff, 2016). In addition to their function in myelination, OLs actively contribute to the central nervous system's inflammatory reactions. In many neurodegenerative illnesses, neuroinflammation is a major factor that leads to OL destruction and impaired remyelination (Tamberi et al., 2024). Microglia and astrocytes interact dynamically with OLs and OPCs to shape the inflammatory environment (Tabatabaee, 2024). Additionally, OLs themselves can release inflammatory mediators, which can further impact the course of illness (Wang et al., 2024). To design therapies that reduce neuroinflammation and restore myelin integrity, it is essential to comprehend the reciprocal interaction between OLs and immune signals. In the CNS, OLs also play a crucial role in maintaining iron homeostasis, which is necessary for mitochondrial activity, neurotransmitter production, and general neuronal metabolism (Kulaszyńska et al., 2024). In many neurodegenerative illnesses, dysregulation of iron management, particularly its buildup in the brain, is a common pathological characteristic (S. Lee & Kovacs, 2024). Long investigated classical myelin diseases, OLs are now understood to have a significant role in more general NDs, including AD, PD, and ALS/MND (Han et al., 2022). The intricate relationships between OPCs, OLs, and neurons, as well as adaptive myelination in memory and learning, and myelin plasticity, have been clarified by recent developments in neuroscientific research, providing insight into how disturbances in these relationships contribute to the pathophysiology of NDs (Spaas et al., 2021). Post-mortem brains from AD patients show myelin degradation and OL loss, especially in areas like the cortex and hippocampus that are vulnerable to AD disease (Nasrabady et al., 2018). The association between this cell loss and the advancement of the illness and cognitive impairment raises the possibility that OL dysfunctions play a role in the neurodegenerative process of AD. It has been demonstrated that the buildup of amyloid-beta (Aβ) plaques, a defining feature of AD, damages OLs and compromises the integrity of myelin (Depp et al., 2023). Myelin integrity loss in AD might result from both OLs' inability to properly maintain and repair myelin sheaths and the direct impact of Aβ on these cells (Zou et al., 2023). Moreover, tau pathology, another important aspect of AD, and signal transmission impairment caused by myelin changes may worsen neuronal dysfunction (Depp et al., 2022). White matter hyperintensities and structural abnormalities, which suggest myelin loss and axonal injury, have been repeatedly seen in AD patients' magnetic resonance imaging (MRI) investigations (Ferris et al., 2022). The loss of dopaminergic neurons in the substantia nigra pars compacta and the appearance of Lewy bodies made of α-synuclein (α-syn) are the main characteristics of PD, which is also significantly influenced by OLs (Zhang et al., 2022). The intricacies of OL participation in PD are being unraveled by recent research, which offers fresh perspectives on the disease's pathophysiology and possible treatment options. PD is characterized by α-syn accumulation in OLs and neurons, which results in glial cytoplasmic inclusions (GCIs). α-syn buildup in OLs is harmful and causes myelin breakdown and OL death (Jeon et al., 2020; Mavroeidi & Xilouri, 2021). These inclusions are seen in both conventional PD cases and multiple system atrophy (MSA), a Parkinsonian condition (Kaji et al., 2020). The disease's complex nature is shown by the role of OLs in ALS/MND, a debilitating ND marked by the gradual loss of motor neurons in the spinal cord, brain stem, and motor cortex (Traiffort et al., 2021). In ALS/MND, OLs and myelin malfunction play crucial roles. Recent scientific discoveries have started to clarify these functions, pointing to prospective treatment targets and processes that contribute to motor neuron degeneration (Z. Gong et al., 2022; Traiffort et al., 2021). Beyond the axonal pathology in white matter, the loss of OLs around degenerating motor neurons indicates a supporting role for these glial cells in motor neuron health and function (Lorente Pons et al., 2020; Raffaele et al., 2021). OLs promote axon metabolism by using the monocarboxylate transporter to use lactate as an energy source (Nave et al., 2023). HD and other neurological illnesses may be linked to OLs, according to emerging research (Ferrari Bardile et al., 2021; Goldman, 2024; Hedreen et al., 2024). According to research, people with HD have notable changes in their white matter and myelin integrity, which may imply that myelin degradation contributes to the course of the disease (Casella et al., 2020; Sun et al., 2022).

Figure 4. Role of oligodendrocytes in remyelination after CNS disorder

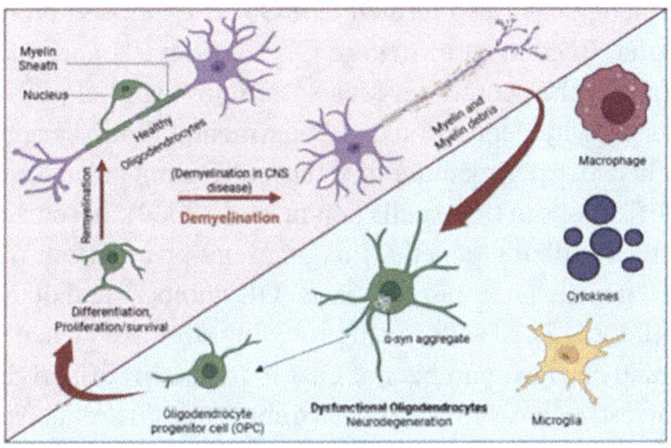

3.3.2 Traumatic CNS Disorders

Myelin is produced and maintained throughout the central nervous system by oligodendrocytes. Axons in white matter tracts gradually demyelinate and degenerate, which is one of the pathological characteristics seen after TBI (Dent et al., 2015). This leads to extended, perceptual (Vaessen et al., 2016), sensorimotor (Corbetta et al., 2015), and cognitive deficits (Croall et al., 2014; Taib et al., 2017). The loss of the myelin sheath, an insulating coating rich in cholesterol that envelops nervous system axons, is a significant effect of traumatic brain and spinal cord injury. Oligodendrocytes are responsible for the production and maintenance of myelin in the CNS. The death of oligodendrocyte cells and consequent myelin loss brought on by CNS damage may have detrimental effects on functional recovery. Numerous neurodegenerative illnesses have been linked to demyelination, which slows down signal transmission along axons and inhibits neuronal function (Huntemer-Silveira et al., 2021). Following spinal cord injury SCI, OLs are more vulnerable to the toxicity of the initial lesion environment. They experience both acute necrosis and apoptosis, with the latter occurring over an extended period of time. Axon survival and function are hampered by demyelination caused by OL loss. Concurrently, there is a prolonged and rapid proliferation of OL progenitor cells, particularly at the edges of the lesions (Almad et al., 2011). In the pathophysiology of SCI, oligodendrocytes and oligodendrocyte precursor cells (ODC/OPC) are essential; the regeneration of myelin sheaths and axons is greatly influenced by their level of development. (J. Yang et al., 2018). In the adult central nervous system, OPCs are the most abundant and widely distributed population of progenitor cells, enabling them to react quickly to injury. Following traumatic SCI, oligodendrocytes and myelin are destroyed, which risks effective conduction and may affect the long-term health of axons. As a result, OPCs multiply and undergo differentiation into Schwann cells and new oligodendrocytes to produce remyelinated axons. OPCs limit the regeneration of damaged axons and aid in the creation of scars following SCI. Additionally, OPCs produce cytokines, change their gene expression after demyelination, and sustain the immune response (G. J. Duncan et al., 2020).

3.3.3 Psychiatric disorders

Mature oligodendrocytes (OLs) and OPCs are two types of OL lineage cells seen in the mammalian CNS. Because NG2-proteoglycan (also termed CSPG4) is expressed on the cell surface, OPCs are often referred to as NG2 glia. In contrast to mature OLs, astrocytes, and microglia, they make up the fourth major glial cell group in the central nervous system (What et al., 2016). Strong circumstantial evidence suggests that persons with depression and other mental illnesses regularly have changed OLs. Furthermore, there is conclusive experimental proof that significant behavioral changes may be brought about only by genetic modifications in these cells (Zhou et al., 2020). In schizophrenia, evidence for OL disruption has been described utilizing genetic, imaging, and postmortem research. In schizophrenia, there is reduced OL-related gene expression, decreased OL numbers and density, and abnormal myelin morphology (Edgar & Sibille, 2012). In addition to providing strong evidence for disrupted myelination, aberrant gene expression, and changed numbers of OLs in the brains of individuals with schizophrenia, imaging and postmortem investigations have also shown abnormal OL-related processes in these patients. The cognitive abnormalities of schizophrenia, which are closely linked to a poor long-term outcome, may be caused by OL deficiencies, which may be the consequence of failed maturation and disrupted regeneration (Raabe et al., 2019). It is also believed that oligodendrocytes can reduce inflammatory damage (Boccazzi et al., 2021; Madsen et al., 2020). Patients with schizophrenia have been found to have aberrant myelination of connecting fibers in the left frontotemporal region, a part of the brain thought to be important in the formation of verbal and auditory hallucinations (Vanes et al., 2019).

3.4 Role of Ependymal Cells in CNS Disorders

3.4.1 Neurodegenerative Diseases

The elimination of metabolic waste and protein aggregates, especially amyloid-beta monomers, from the CSF depends on ependymal cells; hence, their malfunction may hinder the regular detoxification process of the CSF and result in amyloid neurotoxicity (Luca et al., 2018). A biochemical process believed to precede and accelerate the production of neurofibrillary tangles, the oligomerization of amyloid-beta monomers into amyloid plaques, may be facilitated by dysfunctional ependymal cells (Ashton et al., 2022; Serot et al., 2012). In individuals with AD, deposits of amyloid plaques and neurofibrillary tangles have been seen in the periventricular areas, indicating an imbalance between the ependymal cell line's generation and removal of amyloid plaques (Castaño et al., 2013; Nagele et al., 2004). The connection between the pathophysiology of AD and poor lipid metabolism has been seen in the lining of ependymal cells (F. Yin, 2023). PD patients have been shown to have deposits of aggregated alpha-synuclein in subependymal tissues and ependymal cells, which correlate to Braak stages 5 and 6 (Kovacs, 2014). This is a proven method for describing the course of PD (Visanji et al., 2014). The pathophysiology of PD has also been connected to loss-of-function mutations in the parkin co-regulated gene (PACRG), which is a gene involved in the creation of the axoneme microtubule structure in both motile and non-motile cilia (Khan et al., 2021; Wilson et al., 2009). Because PACRG is found in the ependymal cilia that line the ventricular system and is strongly expressed there (Wilson et al., 2009), dysfunction of PACRG mutations could indicate compromised ependymal cilia activity, which could be a factor in the etiology of PD. The relationship between PACRG dysregulation and ependymal cell damage, as well as whether PACRG mutations brought on by environmental or genetic factors might induce ependymal cell malfunction, is yet unknown (Alecu & Bennett, 2019). The pathogenic mechanism of Huntington's disease has recently been linked to ciliogenesis and ependymal ciliary activity (Kaliszewski et al., 2015). Ependymal cells in HD patients and HTT mutant animals showed hydrocephalus, disordered cilia, delocalized pericentriolar material 1 (PCM1), and changed CSF flow. It is believed that the HTT-PCM1-HAP1 (Huntington-associated protein 1) pathway controls ciliogenesis and CSF balance. Mutant mice deficient in HTT showed aberrant CSF synthesis and circulation, altered ciliogenesis, and an uncoordinated cilia beating pattern (Dietrich et al., 2009; Keryer et al., 2011). The ependymal cell malfunction is the cause of the neurotoxic substances and inflammatory cytokines found in the CSF of ALS patients with sporadic ALS (Payne, 2009). When the CSF from ALS patients was introduced to isolated and cultured motor and spinal cord neurons, they showed indications of ALS-specific neurodegeneration (Sen et al., 2005; Vijayalakshmi et al., 2009). There is a connection between the pathophysiology of MS and ependymal cell malfunction (Hatrock et al., 2020). Ependymal cells are, in fact, very susceptible to inflammation (Hatrock et al., 2020; Schubert et al., 2019). They exhibit morphological anomalies, known as the "Dot-Dash" sign, which is thought to be brought on by inflammation (Lisanti et al., 2005). Therefore, inflammation-induced ependymal cell lining destruction might be a possible source of MS etiology. The notion that ependymal cells malfunction is further supported by the changed CSF composition seen in MS patients (Hatrock et al., 2020; Schubert et al., 2019).

3.4.2 Traumatic CNS Disorders

Ependymal cells in SCI have been well investigated. After SCI, ependymal cells in the central canal of the spinal cord (SCECs) proliferate, differentiate, and migrate to the lesion site and act as a potential repair cell source (Wang et al., 2019). Ependymal cells are stimulated, multiply, move out of the central canal, differentiate, and exhibit spontaneous Ca2+ activity in an ex vivo SCI model (Fernandez-Zafra et al., 2017).

Figure 5. Manifestation of ependymal cell dysfunction

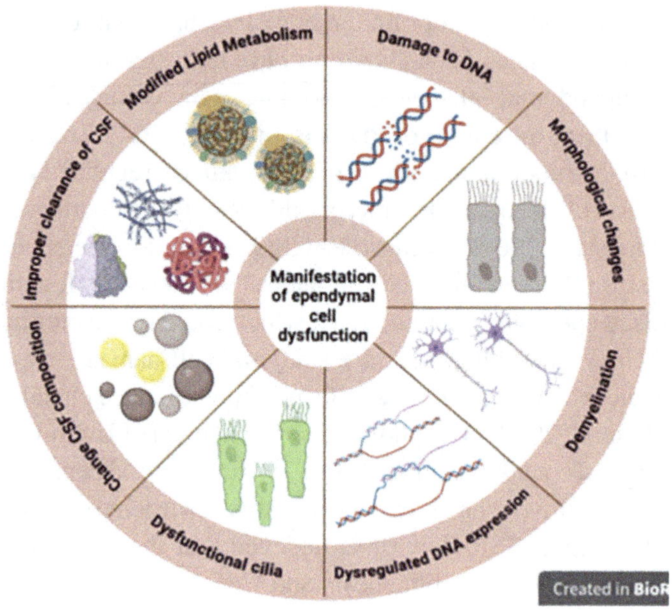

Ependymal cells displayed multilineage differentiation following SCI, mostly developing into astrocyte-like cells and, to a lesser extent, oligodendrocytes that resembled cell types. Ependymal cells go to the site of damage to form the glial scar and even promote local growth (Lacroix et al., 2014; Meletis et al., 2008).

3.4.3 Psychiatric Disorders

Ependymal cells use a range of signaling pathways to alter their shape and function under certain pathological circumstances. The cerebrospinal fluid (CSF) is propelled by ependymal cells, and disruption of ependymal planar cell polarity results in abnormal CSF circulation, which may have an impact on psychiatric and neurological disorders. According to a prior study, p11 levels were lower in individuals with MDD and a chronic stress-induced mouse depression model. This was because ependymal planar cell polarity was disrupted and CSF flow was reduced, which led to depression and anxiety (S. Deng et al., 2023; Seo et al., 2021). One of the most common and crippling mental illnesses is major depression (MD), which is characterized by a chronically low mood, anhedonia (loss of interest or enjoyment in once-enjoyed activities), irregular eating and sleeping patterns, and, in extreme situations, cognitive impairment and suicidal thoughts. The pathophysiology of depression involves glial dysfunction, particularly that involving astrocytes and microglia, as well as inflammatory processes (Verkhratsky et al., 2013). Several instances of depression are associated with hyperactive microglia in the hippocampus and prefrontal cortex, among other parts of the brain (e.g., those induced by chronic stress) (Pittenger & Duman, 2008). In areas crucial for mood regulation, such as the hippocampus, astrocyte failure can result in glutamate buildup in the synaptic cleft, which can cause excitotoxicity and synaptic loss (Bacchi, 2025; Leonard, 2018; Scuderi et al., 2021).

4 POTENTIAL EFFECT OF NATURAL BIOMOLECULES IN CNS DISORDERS

4.1 Role of Polyphenols

The potential role of polyphenols in slowing the progression of neurodegenerative disorders and improving brain function has drawn more attention (Jalouli et al., 2025). Polyphenols have been thoroughly investigated in recent decades for their potential to prevent and cure age-related NDs. Thus, it is recognized that a variety of polyphenols affect how neurons operate (Rojas-García et al., 2023; Sriganesh et al., 2016) can provide neuronal cells with pleiotropic actions (Arias-Sánchez et al., 2023; Figueira et al., 2016). There are no significant negative side effects from consuming polyphenols (Arias-Sánchez et al., 2023). Consuming plant polyphenols in older adults enhances neurocognitive abilities, prevents the onset of neurodegenerative disorders, or slows their progression (Grabska-Kobyłecka et al., 2023). The neuroprotective polyphenols resveratrol, kaempferol, baicalein, naringenin, EGCG, and caffeic acid have been shown to reduce brain inflammation in Parkinson's disease patients (De Andrade Teles et al., 2018; Jalouli et al., 2025; Lou et al., 2014; Tavan et al., 2024; Xue et al., 2024; X. Zhang et al., 2017). Myricetin and piceatannol, two polyphenols, have demonstrated promising therapeutic potential in the treatment of multiple sclerosis (Rudrapal et al., 2024). Because of their prospective qualities, they have been proposed as a promising therapeutic target for the treatment of age-related multiple sclerosis (MS) and amyotrophic lateral sclerosis (ALS) (Chatterjee et al., 2024). Green tea polyphenols (GTPs) are polyhydroxy phenolic chemicals, usually flavonoids or catechins. GTPs can lessen the ischemic region's increased BBB permeability (Zhang et al., 2010). The flavanols catechin, (-)-epicatechin (EC), and (-)-epigallocatechin-3-gallate (EGCG) have pleiotropic actions that impact several pathological features of traumatic brain injury. After TBI, both catechin and EC therapy dramatically decreased the development of cerebral edema and brain tissue damage (Carecho et al., 2023; Jiang et al., 2017). Additionally, polyphenols contain anti-inflammatory qualities that mitigate the elevated inflammation found in the pathophysiology of anxiety and depression. Additional advantages of polyphenols include their capacity to restore cognitive and behavioural impairments and modify cellular signaling pathways to provide neuro-protective effects (Gomez-Pinilla & Nguyen, 2012; Lin et al., 2021). The natural catechin polyphenol EGCG, which is abundant in green tea, is well known for its strong antioxidant properties and capacity to reduce depressive and stress symptoms. When psychological stress impairs cognitive function, green tea polyphenol supplementation improves it (W. Q. Chen et al., 2009; Gomez-Pinilla & Nguyen, 2012). Piperine, resveratrol, eugenol, and curcumin are used as antidepressants (Kulkarni et al., 2008, 2009; Z. Zhang et al., 2019).

Table 1. Bioactive compounds used in clinical trials of CNS disorders

Sr No	Plant-derived compounds	Dose and route	Animal model/cell line	Reference
Neurodegenerative Disorders				
Polyphenols and Flavonoids				
1.	Quercitin	25-75mg/kg Intraperitoneally; 4 days	Rotenone model of Parkinson's disease in rats	(Karuppagounder et al., 2013)

continued on following page

Table 1. Continued

2.	Resveratrol	20 mg/kg/day Orally; 3 weeks	Rotenone model of Parkinson's disease in rats	(Gaballah et al., 2016)
3.	Curcumin	0–1,000 μM	N27 dopaminergic neuronal cell cultures	(Chithra et al., 2023)
4.	Mangiferin	0.25 mM	Murine neuroblastoma cell line N2A	(Jalouli et al., 2025)
5.	Naringenin	70 mg/kg orally;4 days 20, to 80 mM	6-OHDA-induced PD model both in mice & n SH-SY5Y cells	(Lou et al., 2014)
6.	Baicalein	2.5mg/kg,100mg/kg &200mg/kg orally; day 15 to 42 0.1, 1, 10μM	Rotenone model of Parkinson's disease in rats rotenone-treated SH-SY5Y cells	(Zhang et al., 2017).
7.	Rutin	25mg/kg Orally, 3 weeks	6-OHDA-induced PD model in rats	(Islam et al., 2012).
8.	Hesperidin	50 mg/kg Orally; Till 28 days	6-OHDA-induced PD model in rats	(Antunes et al., 2014).
9.	Apeginin	20 mg/kg ip	Rotenone model of Parkinson's disease in rats	(Anusha et al., 2017)
10.	Epigallocatechin-3-gallate (EGCG)	100-400mg/kg; Orally, 36mM	Rat hippocampal injury model, immortalized mouse hippocampal neuronal cell line	(Kang et al., 2010)
11.	Anthocyanin	12 μg/g/ day i.p. 14 days anthocyanin-loaded PEG-AuNPs	Aβ1–42 mouse model HT22 neuronal cells	(Ali et al., 2017)
12.	Hesperidin	100 mg/kg;orally;10 day	Mouse model of cerebral amyloidosis	(C. Li et al., 2015)
13.	Kaempferol	5, 7/5, 10 mg/kg; i.p; 21 days	Intra -cerebroventricular STZ-induced experimental AD model in Wistar rats	(Darbandi et al., 2016)
14.	Resveratrol	5 μ(G. Y. Choi et al., 2023) M	Alzheimer's Mouse Model (Tg2576 line) (AD)	(Lymperopoulos et al., 2024)
15.	Naringenin	100 mg/kg/day; orally;20 days	Rat model of Alzheimer's disease	(Choi et al., 2023)
16.	Apigenin	110 mg/kg per day in the diet	C57BL/6 and GFAP-IL6 heterozygous mice model	(Chesworth et al., 2021)
17.	Apegeninn, baicalein, catechin, epicatechin, epigallocatechin gallate (EGCG), kaempferol, luteolin, myricetin, quercetin, and rutin	30 μM	Cultured Mouse Cortical Neurons for AD	(Choi et al., 2014)
18.	Hesperidine Naringin	100mg/kg each (phase 1) 50mg/kg each (phase 11); i.p 14 days	3-NP-induced HD in the rat model	(Kumar & Kumar, 2010)
19.	Hesperidine	25, 50, and 100 mg/kg;p.o 21 days	Quinolinic acid-induced HD in a rat model	(Kumar et al., 2013)

continued on following page

Table 1. Continued

20.	Quercitin	25mg/kg; orally 21 days	3-NP-induced HD in the rat model	(Sandhir & Mehrotra, 2013)
21.	Rutin	25mg/kg &50mg/kg; Orally 14 days	3-NP-induced HD in the rat model	(Suganya & Sumathi, 2017)
22.	Fisetin	0.05%in chow 10 mM	R6/2 mouse model of HD PC12 cells expressing mutant Httex1	(Maher et al., 2011)
Alkaloids and Terpenoids				
1.	Oxymetrine	(5, 10, or 20 mg/kg) i.p for 7 days 5 mM	(MPTP)-stimulated mice mice primary neuron-microglia cocultures	(P. Gan et al., 2020)
2.	Myrcene	50mg/kg i.p	Rotenone model of Parkinson's disease in rats	(Azimullah et al., 2023)
3.	Berberine	260 mg/kg in the diet for 3 months 5 μM	APP/PS1 transgenic mice model, mouse hippocampal neuron cell line HT22	(Wu et al., 2021)
4.	Caffeine	0.3 g/L through drinking water 1.5 mg/0.2 ml	AD Tg mice model THY-Tau22 Tg mouse model	(Merighi et al., 2023)
5.	Tetrahydropalmatine	20, 40, and 80 mg/kg; orally, 8 weeks	d-galactose-induced rat model of AD	(Qu et al., 2016)
6.	Tetradine	40mg/kg ;orally; for 14 days	Amyloid-β (Aβ) induced rat model of AD	(F. Q. He et al., 2011).
7.	Sinomenine	100μM	C8D1A cell line and hippocampal HT22 cells	(Singh et al., 2020)
8.	Oxymatrine	40 mg/mL;i.p; 8weeks	hAPP transgenic mice model	(Y. Chen et al., 2019)
9.	Linalool	50 &100mg/kg;i.p 3 weeks	Amyloid-β (Aβ) induced rat model of AD	(C. Yuan et al., 2021)
10.	Huperzine-A	5, 10, 20, 50, 100, 200, 500 nM and 1 μM in diet for 24hrs 50 nM	Cadmium chloride-induced HD in Drosophila melanogaster model SH-SY5Y neuroblastoma cell	(Subaraja et al., 2024)
11.	Ginsenoside	30mg/kg to 60mg/kg Orally 5 weeks	R6/2 transgenic mouse model of HD	(Hua et al., 2022)
Tannins				
1.	Condensed tannins	5, 10, 20, and 40mg/kg, p.o.	Parkinson's Disease Model of Rats	(Calou et al., 2014)
Traumatic CNS Disorders				
Polyphenols and Flavonoids				
	Naringenin	50mg/kg&100mg/kg;i.p 7days	Mice model of traumatic brain injury	(C. Deng et al., 2021)
	Quercitin	5, 20 or 50 mg/kg;i.p 3days	Rat model of traumatic brain injury	(Song et al., 2021)

continued on following page

Table 1. Continued

	Luteolin	30mg/kg; i.p Once after TBI	Mice model of traumatic brain injury	(J. Xu et al., 2014)
	Curcumin	10,20,30 and 50 mg/Kg; i.p 28days	Rat model of traumatic brain injury	(G. Sun et al., 2020)
	Resveratrol	100mg/kg;i.p 3 days	Rat model of traumatic brain injury	(Feng et al., 2016)
		100mg/kg;i.p 30 min before TBI	Rat model of traumatic brain injury	(Zou et al., 2018)
	EGCG	0.1% (w/v) EGCG in drinking water; 6-8 weeks of age	Rat model of traumatic brain injury	(Itoh et al., 2011)
	Chrysin	25, 50, or 100 mg/kg;i.p Orally 2 weeks	Rat model of traumatic brain injury	(Rashno et al., 2023)
	Baicalein	30mg/kg; i.p for 4 days (immediately, 25, 49 and 73 h)	Rat model of traumatic brain injury	(Chen et al., 2008)
	GTP,s Green tea polyphenols	400 mg/kg/day; Intragastrically 30days	Rat model of cerebral ischemia	(Zhang et al., 2010)
	Rutin	50mg/kg &100mg/kg; i.p 5 days	Rat model of cortical focal ischemia	(Rodrigues et al., 2013)
	Hesperidin	50mg/kg and 100 mg/kg, po 7 days	Rat model of cerebral ischemic reprofusion injury	(Gaur & Kumar, 2010)
	Epicatechin	30 mg/kg;i.p 1 week	Mice model of spinal cord injury	(Che et al., 2017).
	Epicatechin	5 μM	Organotypic spinal cord cultures	(Pérez et al., 2016)
	Quercitin	100mg/kg; i.p Three days	Rat model of spinal cord injury	(Ocal et al., 2019)
	Apigenin	10mg/kg-20mg/kg;i.p Three days	Rat model of spinal cord injury	(Zhang et al., 2014)
	Apigenin	10mg/kg-20mg/kg;i.p Three days	Rat model of spinal cord injury	(Zhang et al., 2014)
	Luteolin	50mg/kg&100mg/kg;i.p	Ischemia–reperfusion injury of the spinal cord (SCI) in a rat model	(J. Fu et al., 2018).
	Resveratrol	200 mg/kg/day;i.p Three days	Rat model of spinal cord injury	(P. Wang et al., 2018)
Alkaloids and Terpenoids				
	Berberine	10mg/kg 28 days	Rat model of traumatic brain injury	(Chen et al., 2014)
	Tanshinone	20 to 50 mg/kg; i.p 7 days	Rat model of spinal cord injury	(X. Yin et al., 2012)
	Aloperine	80 mg/kg and 40 mg/kg, i.p	Mice model of chronic constriction injury of the spinal cord	(Xu et al., 2014)

continued on following page

Table 1. Continued

	Oxymatrine	120 mg/kg	Rat model of hypoxic-ischemic brain injury	(Ge et al., 2018)
	Tetramethylpyrazine	30mg/kg; through the tail vein 30 mins before ischemia	Rat model of spinal cord ischemic-reperfusion injury	(Fan et al., 2011)
	Tetrandrine	0.1, 1, 10, 20 μM	Spinal cord ischemia in primary rat spinal cord astrocytes	(Bao et al., 2016).
	Asiatic acid	75 mg/kg; i.p after trauma	Rat model of spinal cord injury	(Gurcan et al., 2017)
Psychiatric Disorders				
Polyphenols and Flavonoids				
	Green Tea polyphenols	0.1%, 0.5% and 1% doses in the diet for 3 weeks	Psychological stress model of rats	(Chen et al., 2009)
	Curcumin	10-80 mg/kg, i.p.	Forced depression model of mice	(Kulkarni et al., 2008)
	Honokiol and magnolol	10 and 30 mg/kg gastric gavage for 1 week	Stressed model of rats	(Xu et al., 2008)
	Amentoflavone	100mg/kg for FST, 50mg/kg for TST 25mg/kg for anxiolytic test orally	Psychological stress model of rats	(Ishola et al., 2012
	Apegenin	20 mg/kg, intragastrically 3 weeks	Chronic unpredictable mild stress model of the rat	(Li et al., 2016)
	Baicalein	10, 20, or 40 mg/kg; i.p 14 days	Rat model of repeated restraint stress	(Lee et al., 2013)
	Luteolin	1 to 10 μM	Stress-induced human neuroblastoma cells	(Ishisaka et al., 2011)
	Nobiletin	25, 50, and 100 mg/kg, p.o.	Psychological stress model of mice	(L. T. Yi et al., 2011)
	Fisetin	20, 40, and 80 mg/kg p.o. for 7 days	LPS-induced depression in a mouse model	(X. Yu et al., 2016)
	Icariin	20 or 40 mg/kg; p.o 35 days	Chronic mild stress model of the rat	(Liu et al., 2015)
		60 mg/kg by gastric instillation for 21 days	Corticosterone-induced depression model of the rat	(M. juan Gong et al., 2016)
	Kaempferitrin	1.0, 5.0, 10 and 20 mg/kg for TST & FST 10, and 20 mg/kg for corticosterone levels	Psychological stress model of mice	(Cassani et al., 2014)
	Kaempferol Quercetin kaempferol-3- O-β-D-glucose quercetin-3-O-β-D-glucose	0.35 mM/kg; i.p Each at a volume of 0.1 mL/10 g, 60 min before the test session	Antidepressant model of mice	(Yan et al., 2015)
	Myricitrin	3–30 mg/kg single &repeated administration	Antidepressant model of mice	(Meyer et al., 2017)
	Quercitin	50 mg/kg;i.p for 21 days	Antidepressant effect in diabetic mice model	(Demir et al., 2015)
	Hesperidine	0.1, 0.3 and 1 mg/kg;i.p single administration	Antidepressant model of mice	(Souza et al., 2013)

continued on following page

Table 1. Continued

	Naringenin	1-day (20 mg/kg), 7-day (10, 20 mg/kg) and 14-day (5, 10, 20 mg/kg) for TST	Antidepressant effect in the tail suspension test of mice	(L. T. Yi et al., 2012)
	Dihydromyrecitin	20mg/kg;i.p 7consecutive doses day18-24	Antidepressant model of mice	(Huang et al., 2022)
	Silymarin	100 mg/kg and 200 mg/kg, orally	Experimentally induced depression model of rats	(Thakare et al., 2017)
Alkaloids and Terpenoids				
	Berberine	10-50 mg/kg for 10days; i.p 10 days	Antidepressant effect in morphine-dependent rats	(Lee et al., 2012)
	Diterpene Alkaloids	.025mg/kg orally for 5days	Antidepressant model of mice	(Nesterova et al., 2011)
	Punarnavine	20 and 40 mg/kg; p.o 14 days	Chronic unpredictable mild stress model of mice	(Dhingra & Valecha, 2014)
	mesembrine	20 mg/kg; i.p	Forced swim test in rats	(Loria et al., 2014)
	D-limonene	10mg/kg; orally 21 days	Antidepressant model of rats	(Alkanat & Alkanat, 2024)
	Oleanolic acid derivatives	5, 10, or 20mg/kg; p.o	Antidepressant model of mice	(Fajemiroye et al., 2015)
	Boswellic acids	50mg/kg, 100mg/kg & 200mg/kg; p.o	Acute model of depression in mice	(Adake et al., 2013)
Tannins				
	Drymaria cordata L.	50 and 100 mg/kg	Anxiolytic-like model of mice	(Barua et al., 2009)
	Terminalia chebula	50 and 100 mg/kg	Anxiolytic-like model of mice	(Chandrasekhar et al., 2018),

4.2 Role of Flavonoids

Flavonoids can pass the blood-brain barrier with chronic or acute administration, showing that these compounds might potentially have a direct influence on the brain; hence, these chemical compounds could be utilized as a preventive, to slow down the course of disorders such as AD and PD (De Andrade Teles et al., 2018). As the foundation of the neuroprotection shown in Parkinsonism caused by mitochondrial neurotoxins, quercetin may be able to restore mitochondrial electron transport abnormalities and enhance its function (Biswas et al., 2022; Ferreira et al., 2024; Gugliandolo et al., 2020; Karuppagounder et al., 2013). One potential treatment option for stroke is rutin. Rutin treatment has demonstrated beneficial effects in models of localized and global brain ischemia (Rodrigues et al., 2013). Rutin decreases lipid peroxidation and cytokine production, scavenges superoxide radicals, boosts antioxidant enzymatic activity in vitro, and shields rat models from cognitive impairments such as CNS damage (Javed et al., 2012; P. xin Xu et al., 2014). By decreasing microgliosis, lowering blood and/or brain levels of TNF-α, iNOS, IL-1β, and IL-6 mRNA and protein, and raising IL-10 levels, quercetin reduces the inflammatory response following traumatic brain injury (Schültke et al., 2005; Song et al., 2021; Zhang et al., 2018). In rats, silymarin has an antidepressant effect (Thakare et al., 2017). Fisetin is a strong option for treating neurological and psychological disorders since it has antidepressant properties and has been shown to alter depressed behavior in an animal model of acute neuroinflammation (X. Yu et al., 2016).

4.3 Role of Alkaloids

Alkaloids are well recognized for being a constant supply of medication for the treatment of long-term conditions like diabetes, cancer, and neurological disorders. For instance, alkaloid derivatives galanthamine and huperzine A are now being utilized to treat the symptoms of neurodegenerative diseases (Kong et al., 2021). The bark and roots of several plants, including *Berberis vulgaris, B. aristotelia, B. aquifolium, Hydrastus canadensis, Pellodendron chenins, and Coptis rhizomes,* contain berberine, a quaternary botanical isoquinoline alkaloid (de Lima et al., 2025). By blocking the hyperphosphorylation of Tau protein, preventing the synthesis of Aβ, and blocking four important enzymes (acetylcholinesterase, butylcholinesterase, and two isomers of monoamine oxidase) involved in the pathophysiology of AD, BBR can mitigate the effects of AD (Sunhe et al., 2024; Wu et al., 2021). Because of its many physical and chemical characteristics, such as its quick absorption and organic dispersion, caffeine has a neuroprotective impact (Hu et al., 2024; Merighi et al., 2023). One treatment for AD is tetrandrine, a bisbenzylisoquinoline alkaloid that dramatically reduces Aβ (1–42)-induced memory loss and spatial learning. Because of their potential as a treatment for neurodegenerative diseases like Parkinson's, Alzheimer's, and other conditions involving neuronal degeneration, terpenes have been studied in preclinical research (Mony et al., 2022).

Figure 6. Phytochemical shows protective effects against various CNS disorders

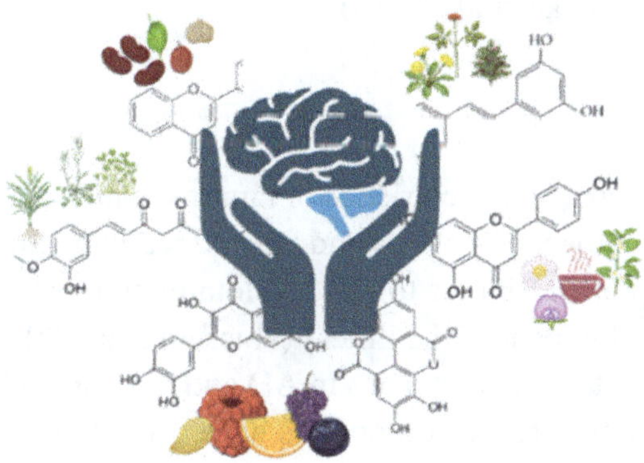

In the Forced Swim Test, berberine treatment dramatically reduced immobility and enhanced climbing behavior, confirming that the antidepressant-like action (Erel, 2005; B. Lee et al., 2012). In several scenarios, the alkaloid punaravine, which comes from Boerhaavia diffusa Linn, demonstrated notable antidepressant effects in both stressed and unstressed mice (Dhingra & Valecha, 2014).

4.4 Role of Terpenoids:

Numerous organisms, including plants, bacteria, fungi, animals, marine life, sedimentary rocks, and oils, contain terpenoids (Roshan Kumar et al., 2022). These are naturally occurring compounds that are produced via the process of biosynthesis; they are commonly referred to as "terpenes" and have a variety of structural types. Citrus peel oils include limonene, a monocyclic monoterpene. In rats with ROT-induced Parkinson's disease (PD), it prevented behavioral deficits, decreased α-synuclein expression, saved dopaminergic neurons from dying, reduced oxidative stress, restored Complex-I activity, decreased lipid peroxidation, and downregulated proinflammatory cytokines. It also favorably modulated apoptotic pathways, such as MAPK, mTOR, and Hippo signaling (Eddin et al., 2023; Ojha, 2021). Terpenes, like ginkgolides that are taken from the Ginkgo biloba tree, also have encouraging anti-inflammatory and antioxidant properties, shielding neurons from synaptic damage and restoring neuronal function in models of Alzheimer's disease (da Silva Carneiro et al., 2025; Gu et al., 2012). It has been demonstrated that linalool lowers ROS generation and shields neuronal cells from oxidative injury. This neuroprotective effect is especially important for conditions like Alzheimer's, where oxidative stress is a major factor in the development of the illness (Galipoğlu et al., 2014; C. Yuan et al., 2021). In a mouse model of Parkinson's disease (PD) produced by rotenone (ROT), myrcene, a monoterpene, prevented the death of dopaminergic neurons and clarified the underlying processes. Myrcene helps decrease the course of Parkinson's disease by protecting dopaminergic neurons and reducing oxidative stress (Azimullah et al., 2023). D-limonene has been related to a decrease in inflammatory cytokines, including TNF-α and IL-1, which are connected to depression and neuroinflammation (Del Prado-Audelo et al., 2021). Because dlimonene lowers nitrite levels in the hippocampus, it has antidepressant effects that may be directly related to neuroinflammation (Alkanat & Alkanat, 2024). It has been discovered that boswellic acids have antidepressant properties (Adake et al., 2013).

4.5 Role of Tannins

One of the many pharmacological characteristics of tannins, which are biological substances, is neuroprotection. By delaying the pathogenesis of AD and PD and reducing the synthesis of several cytotoxic agents, they reduce their symptoms. Similar to AD, gallotannins prevent the buildup of Aβ peptides, protecting the etiology and development of the illness (Hussain et al., 2019). It has been discovered that condensed tannins have neuroprotective properties in AD and have anticholinesterase activity (Duval & Avérous, 2016; Zengin et al., 2016). These compounds have been demonstrated to shield neurons from oxidative stress and the buildup of Tau and amyloid-beta (Aβ) proteins, which are crucial to the etiology of Alzheimer's disease (Dufourc, 2024). The effects of tannins on Parkinson's disease involve several channels, including cellular signaling, neuroprotective, anti-inflammatory, and antioxidant pathways (PD). To eliminate misfolded proteins and prevent neurodegeneration, the proteasome and autophagic systems are crucial; tannins support protein homeostasis by boosting their activity (Satapathy & Kumar, 2024). The primary component of Lewy bodies, which reflect the distinct clinical characteristics of Parkinson's disease, is alpha-synuclein (α-Syn) heaps. Remarkably, the breakdown of α-syn in PD is favorably correlated with gallic acid from the HTs (Dumitriu et al., 2012). In TBI mice, tannic acid decreased the amount of water in the brain and prevented neuronal damage. Increased glutathione (GSH) levels suggest that it may mitigate oxidative stress (Salman et al., 2020). Tannic acid improves antibacterial activity, reduces oxidative damage, and reduces the inflammatory response (X. He et al.,

2020; Y. Li et al., 2021; L. Zhou et al., 2018). Tannic acid improves antidepressant activity, reduces oxidative damage, and reduces the inflammatory response (Sinclair et al., 2007). The administration of T. chebula's (TCHE) tannin-rich extract demonstrated anxiolytic effects, making it a viable option for creating a medication to treat anxiety (Chandrasekhar et al., 2018). Mice that were given an oral extract of the stem barks of Hallea ciliate, which is rich in alkaloids, flavonoids, and tannins, showed anxiolytic effects (Njapdounke et al., 2016). Alkaloids, tannins, glycosides, phenolic acids, flavonoids, terpenes, and essential oils are among the several phytochemical groups that include compounds that are effective against schizophrenia (Saleem et al., 2022).

5 CONCLUSION

The interplay between natural bioactive molecules and neuroglia represents a critical and emerging area in CNS therapeutics. Neuroglial cells, through their diverse functional states, regulate neuronal survival, repair, and homeostasis, but their dysregulation contributes significantly to the onset and progression of neurodegenerative and other CNS disorders. Natural bioactive compounds, including alkaloids, terpenoids, and polyphenols, offer multifaceted neuroprotection by modulating glial activity, reducing neuroinflammation, and mitigating oxidative stress. Agents such as curcumin, resveratrol, and ginkgolides have demonstrated beneficial effects across Alzheimer's disease, Parkinson's disease, depression, epilepsy, and traumatic injuries. Unlike conventional treatments, these compounds often act on multiple pathological pathways, offering broader therapeutic potential with fewer adverse effects. The integration of such molecules into treatment regimens could complement existing therapies, enhance patient responsiveness, and improve quality of life. Continued research is essential to elucidate the precise molecular mechanisms underlying this cross-talk, optimize dosing strategies, and translate these findings into clinically viable interventions. This approach holds promise for reshaping the therapeutic landscape of CNS disorders and advancing toward more effective, personalized, and holistic care strategies.

REFERENCES

Adake, P., Chandrashekar, R., & Rao, S. N. (2013). *THE PHARMA INNOVATION - JOURNAL Pre-clinical evaluation of antidepressant activity of Boswellia serrata by Tail Suspension Test. 2*(5), 13–19.

Aguzzi, A., & O'Connor, T. (2010). Protein aggregation diseases: Pathogenicity and therapeutic perspectives. In *Nature Reviews Drug Discovery* (Vol. 9, Issue 3, pp. 237–248). Nat Rev Drug Discov. DOI: 10.1038/nrd3050

Alecu, I., & Bennett, S. A. L. (2019). Dysregulated lipid metabolism and its role in α-synucleinopathy in Parkinson's disease. In *Frontiers in Neuroscience* (Vol. 13, Issue APR). Frontiers Media S.A. DOI: 10.3389/fnins.2019.00328

Ali, T., Kim, M. J., Rehman, S. U., Ahmad, A., & Kim, M. O. (2017). Anthocyanin-Loaded PEG-Gold Nanoparticles Enhanced the Neuroprotection of Anthocyanins in an Aβ1–42 Mouse Model of Alzheimer's Disease. *Molecular Neurobiology, 54*(8), 6490–6506. DOI: 10.1007/s12035-016-0136-4 PMID: 27730512

Almad, A., Sahinkaya, F. R., & McTigue, D. M. (2011). Oligodendrocyte Fate after Spinal Cord Injury. In *Neurotherapeutics* (Vol. 8, Issue 2, pp. 262–273). Neurotherapeutics. DOI: 10.1007/s13311-011-0033-5

Álvarez-Pérez, B., Homs, J., Bosch-Mola, M., Puig, T., Reina, F., Verdú, E., & Boadas-Vaello, P. (2016). Epigallocatechin-3-gallate treatment reduces thermal hyperalgesia after spinal cord injury by down-regulating RhoA expression in mice. *European Journal of Pain (London, England), 20*(3), 341–352. DOI: 10.1002/ejp.722 PMID: 25913854

Am, J., & a. (2023). Medicinal Properties of Plant-Derived Compounds and its Significance. *Medicinal & Aromatic Plants, 12*(3), 1–2. DOI: 10.35248/2167-0412.23.12.455

Antunes, M. S., Goes, A. T. R., Boeira, S. P., Prigol, M., & Jesse, C. R. (2014). Protective effect of hesperidin in a model of Parkinson's disease induced by 6-hydroxydopamine in aged mice. *Nutrition (Burbank, Los Angeles County, Calif.), 30*(11–12), 1415–1422. DOI: 10.1016/j.nut.2014.03.024 PMID: 25280422

Anusha, C., Sumathi, T., & Joseph, L. D. (2017). Protective role of apigenin on rotenone induced rat model of Parkinson's disease: Suppression of neuroinflammation and oxidative stress mediated apoptosis. *Chemico-Biological Interactions, 269*, 67–79. DOI: 10.1016/j.cbi.2017.03.016 PMID: 28389404

Arias-Sánchez, R. A., Torner, L., & Fenton Navarro, B. (2023). Polyphenols and Neurodegenerative Diseases: Potential Effects and Mechanisms of Neuroprotection. *Molecules (Basel, Switzerland), 28*(14), 1–15. DOI: 10.3390/molecules28145415 PMID: 37513286

Aryal, B., Raut, B. K., Bhattarai, S., Bhandari, S., Tandan, P., Gyawali, K., Sharma, K., Ranabhat, D., Thapa, R., Aryal, D., Ojha, A., Devkota, H. P., & Parajuli, N. (2022). Potential Therapeutic Applications of Plant-Derived Alkaloids against Inflammatory and Neurodegenerative Diseases. *Evidence-Based Complementary and Alternative Medicine : eCAM, 7299778*, 1–18. Advance online publication. DOI: 10.1155/2022/7299778 PMID: 35310033

Ashton, N. J., Benedet, A. L., Pascoal, T. A., Karikari, T. K., Lantero-Rodriguez, J., Brum, W. S., Mathotaarachchi, S., Therriault, J., Savard, M., Chamoun, M., Stoops, E., Francois, C., Vanmechelen, E., Gauthier, S., Zimmer, E. R., Zetterberg, H., Blennow, K., & Rosa-Neto, P. (2022). Cerebrospinal fluid p-tau231 as an early indicator of emerging pathology in Alzheimer's disease. *EBioMedicine*, *76*, 103836. Advance online publication. DOI: 10.1016/j.ebiom.2022.103836 PMID: 35158308

Attwell, D., Buchan, A. M., Charpak, S., Lauritzen, M., MacVicar, B. A., & Newman, E. A. (2010). Glial and neuronal control of brain blood flow. In *Nature* (Vol. 468, Issue 7321, pp. 232–243). Nature Publishing Group. DOI: 10.1038/nature09613

Awuchi, C. G. (2023). Plants, phytochemicals, and natural practices in complementary and alternative system of medicine for treatment of central nervous system disorders. In *International Journal of Food Properties* (Vol. 26, Issue 1, pp. 1190–1213). Taylor and Francis Ltd. DOI: 10.1080/10942912.2023.2205039

Bacchi, A. D. (2025). Beyond the Neuron: The Integrated Role of Glia in Psychiatric Disorders. *Neuroglia (Basel, Switzerland)*, *6*(2), 15. Advance online publication. DOI: 10.3390/neuroglia6020015

Badjakov, I., Kaluza, M., Ksiazek-Winiarek, D., Szpakowski, P., Czpakowska, J., Fijalkowska, J., & Glabinski, A. (2025). Polyphenols in the Central Nervous System: Cellular Effects and Liposomal Delivery Approaches. *International Journal of Molecular Sciences 2025, Vol. 26, Page 6477, 26*(13), 6477. DOI: 10.3390/ijms26136477

Bao, G., Li, C., Qi, L., Wang, N., & He, B. (2016). Tetrandrine protects against oxygen-glucose-serum deprivation/reoxygenation-induced injury via PI3K/AKT/NF-κB signaling pathway in rat spinal cord astrocytes. *Biomedicine and Pharmacotherapy*, *84*(277), 925–930. DOI: 10.1016/j.biopha.2016.10.007 PMID: 27764754

Barua, C. C., Roy, J. D., Buragohain, B., Barua, A. G., Borah, P., & Lahkar, M. (2009). Anxiolytic effect of hydroethanolic extract of Drymaria cordata L Willd. *Indian Journal of Experimental Biology*, *47*(12), 969–973. PMID: 20329700

Bazira, P. J. (2021). Clinically applied anatomy of the vertebral column. [oxford]. *Surgery*, *39*(6), 315–323.

Benarroch, E. E. (2005). Neuron-astrocyte interactions: Partnership for normal function and disease in the central nervous system. *Mayo Clinic Proceedings*, *80*(10), 1326–1338. DOI: 10.4065/80.10.1326 PMID: 16212146

Bernier, L. P., York, E. M., Kamyabi, A., Choi, H. B., Weilinger, N. L., & MacVicar, B. A. (2020). Microglial metabolic flexibility supports immune surveillance of the brain parenchyma. *Nature Communications 2020 11:1, 11*(1), 1–17. DOI: 10.1038/s41467-020-15267-z

Biswas, P., Dey, D., Biswas, P. K., Rahaman, T. I., Saha, S., Parvez, A., Khan, D. A., Lily, N. J., Saha, K., Sohel, M., Hasan, M. M., Al Azad, S., Bibi, S., Hasan, M. N., Rahmatullah, M., Chun, J., Rahman, M. A., & Kim, B. (2022). A Comprehensive Analysis and Anti-Cancer Activities of Quercetin in ROS-Mediated Cancer and Cancer Stem Cells. *International Journal of Molecular Sciences*, *23*(19), 11746. Advance online publication. DOI: 10.3390/ijms231911746 PMID: 36233051

Boccazzi, M., Van Steenwinckel, J., Schang, A. L., Faivre, V., Le Charpentier, T., Bokobza, C., Csaba, Z., Verderio, C., Fumagalli, M., Mani, S., & Gressens, P. (2021). The immune-inflammatory response of oligodendrocytes in a murine model of preterm white matter injury: The role of TLR3 activation. *Cell Death & Disease*, *12*(2), 166. Advance online publication. DOI: 10.1038/s41419-021-03446-9 PMID: 33558485

Bond, S., Saxena, S., & Sierra-Delgado, J. A. (2025). Microglia in ALS: Insights into Mechanisms and Therapeutic Potential. *Cells 2025, Vol. 14, Page 421*, *14*(6), 421. DOI: 10.3390/cells14060421

Butovsky, O., Jedrychowski, M. P., Moore, C. S., Cialic, R., Lanser, A. J., Gabriely, G., Koeglsperger, T., Dake, B., Wu, P. M., Doykan, C. E., Fanek, Z., Liu, L., Chen, Z., Rothstein, J. D., Ransohoff, R. M., Gygi, S. P., Antel, J. P., & Weiner, H. L. (2014). Identification of a unique TGF-β-dependent molecular and functional signature in microglia. *Nature Neuroscience*, *17*(1), 131–143. DOI: 10.1038/nn.3599 PMID: 24316888

Calou, I., Bandeira, M. A., Aguiar-Galvão, W., Cerqueira, G., Siqueira, R., Neves, K. R., Brito, G. A., & Viana, G. (2014). Neuroprotective Properties of a Standardized Extract from Myracrodruon urundeuva Fr. All. (Aroeira-Do-Sertão), as Evaluated by a Parkinson's Disease Model in Rats. *Parkinson's Disease*, *2014*(1), 519615. DOI: 10.1155/2014/519615 PMID: 25061534

Câmara, J. S., Perestrelo, R., Ferreira, R., Berenguer, C. V., Pereira, J. A. M., & Castilho, P. C. (2024). Plant-Derived Terpenoids: A Plethora of Bioactive Compounds with Several Health Functions and Industrial Applications—A Comprehensive Overview. *Molecules (Basel, Switzerland)*, *29*(16), 3861. DOI: 10.3390/molecules29163861 PMID: 39202940

Camberos-Barraza, J., Camacho-Zamora, A., Bátiz-Beltrán, J. C., Osuna-Ramos, J. F., Rábago-Monzón, Á. R., Valdez-Flores, M. A., & De la Herrán-Arita, A. K. (2024). Sleep, glial function, and the endocannabinoid system: Implications for neuroinflammation and sleep disorders. *International Journal of Molecular Sciences*, *25*(6), 3160.

Carecho, R., Carregosa, D., Ratilal, B. O., Angeles, M., & Loncarevic-vasiljkovic, N. (2023). *Dietary (Poly) phenols in Traumatic Brain Injury*.

Casella, C., Lipp, I., Rosser, A., Jones, D. K., & Metzler-Baddeley, C. (2020). A Critical Review of White Matter Changes in Huntington's Disease. In *Movement Disorders* (Vol. 35, Issue 8, pp. 1302–1311). John Wiley and Sons Inc. DOI: 10.1002/mds.28109

Cassani, J., Dorantes-Barrón, A. M., Novales, L. M., Real, G. A., & Estrada-Reyes, R. (2014). Antidepressant-like effect of kaempferitrin isolated from justicia spicigera schltdl (Acanthaceae) in two behavior models in mice: Evidence for the involvement of the serotonergic system. *Molecules (Basel, Switzerland)*, *19*(12), 21442–21461. DOI: 10.3390/molecules191221442 PMID: 25532842

Castaño, E. M., Maarouf, C. L., Wu, T., Leal, M. C., Whiteside, C. M., Lue, L. F., Kokjohn, T. A., Sabbagh, M. N., Beach, T. G., & Roher, A. E. (2013). Alzheimer disease periventricular white matter lesions exhibit specific proteomic profile alterations. *Neurochemistry International*, *62*(2), 145–156. DOI: 10.1016/j.neuint.2012.12.001 PMID: 23231993

Chandrasekhar, Y., Phani Kumar, G., Navya, K., Ramya, E. M., & Anilakumar, K. R. (2018). Tannins from Terminalia chebula fruits attenuates GABA antagonist-induced anxiety-like behaviour via modulation of neurotransmitters. *The Journal of Pharmacy and Pharmacology*, *70*(12), 1662–1674. DOI: 10.1111/jphp.13007 PMID: 30198561

Chang, J., Qian, Z., Wang, B., Cao, J., Zhang, S., Jiang, F., Kong, R., Yu, X., Cao, X., Yang, L., & Chen, H. (2023). Transplantation of A2 type astrocytes promotes neural repair and remyelination after spinal cord injury. *Cell Communication and Signaling : CCS*, *21*(1), 1–20. DOI: 10.1186/s12964-022-01036-6 PMID: 36797790

Chatterjee, A., Kumar, S., Roy Sarkar, S., Halder, R., Kumari, R., Banerjee, S., & Sarkar, B. (2024). Dietary polyphenols represent a phytotherapeutic alternative for gut dysbiosis associated neurodegeneration: A systematic review. *The Journal of Nutritional Biochemistry*, *129*, 109622. DOI: 10.1016/j.jnutbio.2024.109622 PMID: 38490348

Che, F., Wang, G., Yu, J., Wang, X., Lu, Y., Fu, Q., Su, Q., Jiang, J., & Du, Y. (2017). Effects of epigallocatechin-3-gallate on iron metabolismin spinal cord motor neurons. *Molecular Medicine Reports*, *16*(3), 3010–3014. DOI: 10.3892/mmr.2017.6919 PMID: 28677731

Chen, C. G., Hung, T. H., Lee, C. Y., Wang, L. F., Wu, C. H., Ke, C. H., & Chen, S. F. (2014). Berberine protects against neuronal damage via suppression of glia-mediated inflammation in traumatic brain injury. *PLoS One*, *9*(12), 1–30. DOI: 10.1371/journal.pone.0115694 PMID: 25546475

Chen, S. F., Hsu, C. W., Huang, W. H., & Wang, J. Y. (2008). Post-injury baicalein improves histological and functional outcomes and reduces inflammatory cytokines after experimental traumatic brain injury. *British Journal of Pharmacology*, *155*(8), 1279–1296. DOI: 10.1038/bjp.2008.345 PMID: 18776918

Chen, W. Q., Zhao, X. L., Hou, Y., Li, S. T., Hong, Y., Wang, D. L., & Cheng, Y. Y. (2009). Protective effects of green tea polyphenols on cognitive impairments induced by psychological stress in rats. *Behavioural Brain Research*, *202*(1), 71–76. DOI: 10.1016/j.bbr.2009.03.017 PMID: 19447283

Chen, Y., Qi, Z., Qiao, B., Lv, Z., Hao, Y., & Li, H. (2019). Oxymatrine can attenuate pathological deficits of Alzheimer's disease mice through regulation of neuroinflammation. *Journal of Neuroimmunology*, *334*, 576978. DOI: 10.1016/j.jneuroim.2019.576978 PMID: 31177033

Chesworth, R., Gamage, R., Ullah, F., Sonego, S., Millington, C., Fernandez, A., Liang, H., Karl, T., Münch, G., Niedermayer, G., & Gyengesi, E. (2021). Spatial Memory and Microglia Activation in a Mouse Model of Chronic Neuroinflammation and the Anti-inflammatory Effects of Apigenin. *Frontiers in Neuroscience*, *15*(July), 1–14. DOI: 10.3389/fnins.2021.699329 PMID: 34393713

Chiou, B., & Piao, X. (2021). Linking adhesion GPCRs to glial cell development and function. *Factors Affecting Neurodevelopment: Genetics, Neurology, Behavior, and Diet*, 93–103. DOI: 10.1016/B978-0-12-817986-4.00009-2

Chithra, Y., Dey, G., Ghose, V., Chandramohan, V., Gowthami, N., Vasudev, V., & Srinivas Bharath, M. M. (2023). Mitochondrial Complex I Inhibition in Dopaminergic Neurons Causes Altered Protein Profile and Protein Oxidation: Implications for Parkinson's disease. *Neurochemical Research*, *48*(8), 2360–2389. DOI: 10.1007/s11064-023-03907-x PMID: 36964824

Choi, G. Y., Kim, H. B., Hwang, E. S., Park, H. S., Cho, J. M., Ham, Y. K., Kim, J. H., Mun, M. K., Maeng, S., & Park, J. H. (2023). Naringin enhances long-term potentiation and recovers learning and memory deficits of amyloid-beta induced Alzheimer's disease-like behavioral rat model. *Neurotoxicology, 95*, 35–45. DOI: 10.1016/j.neuro.2022.12.007 PMID: 36549596

Choi, S.-M., Kim, B. C., Cho, Y.-H., Choi, K.-H., Chang, J., Park, M.-S., Kim, M.-K., Cho, K.-H., & Kim, J.-K. (2014). Effects of Flavonoid Compounds on β-amyloid-peptide-induced Neuronal Death in Cultured Mouse Cortical Neurons. *Chonnam Medical Journal, 50*(2), 45. DOI: 10.4068/cmj.2014.50.2.45 PMID: 25229015

Cichon, N., Saluk-Bijak, J., Gorniak, L., Przyslo, L., & Bijak, M. (2020). Flavonoids as a Natural Enhancer of Neuroplasticity—An Overview of the Mechanism of Neurorestorative Action. *Antioxidants, 9*(11), 1035. DOI: 10.3390/antiox9111035 PMID: 33114058

Clarke, B. E., & Patani, R. (2020). The microglial component of amyotrophic lateral sclerosis. *Brain, 143*(12), 3526–3539. DOI: 10.1093/brain/awaa309 PMID: 33427296

Clarke, L. E., Liddelow, S. A., Chakraborty, C., Münch, A. E., Heiman, M., & Barres, B. A. (2018). Normal aging induces A1-like astrocyte reactivity. *Proceedings of the National Academy of Sciences of the United States of America, 115*(8), E1896–E1905. DOI: 10.1073/pnas.1800165115 PMID: 29437957

Corbetta, M., Ramsey, L., Callejas, A., Baldassarre, A., Hacker, C. D., Siegel, J. S., Astafiev, S. V., Rengachary, J., Zinn, K., Lang, C. E., Connor, L. T., Fucetola, R., Strube, M., Carter, A. R., & Shulman, G. L. (2015). Common behavioral clusters and subcortical anatomy in stroke. *Neuron, 85*(5), 927–941. DOI: 10.1016/j.neuron.2015.02.027 PMID: 25741721

Cosme, F., Aires, A., Pinto, T., Oliveira, I., Vilela, A., & Gonçalves, B. (2025). A Comprehensive Review of Bioactive Tannins in Foods and Beverages: Functional Properties, Health Benefits, and Sensory Qualities. In *Molecules* (Vol. 30, Issue 4, p. 800). Multidisciplinary Digital Publishing Institute. DOI: 10.3390/molecules30040800

Crawford, A. H., Tripathi, R. B., Foerster, S., McKenzie, I., Kougioumtzidou, E., Grist, M., Richardson, W. D., & Franklin, R. J. M. (2016). Pre-Existing Mature Oligodendrocytes Do Not Contribute to Remyelination following Toxin-Induced Spinal Cord Demyelination. *American Journal of Pathology, 186*(3), 511–516. DOI: 10.1016/j.ajpath.2015.11.005 PMID: 26773350

Croall, I. D., Cowie, C. J. A., He, J., Peel, A., Wood, J., Aribisala, B. S., Mitchell, P., Mendelow, A. D., Smith, F. E., Millar, D., Kelly, T., & Blamire, A. M. (2014). White matter correlates of cognitive dysfunction after mild traumatic brain injury. *Neurology, 83*(6), 494–501. DOI: 10.1212/WNL.0000000000000666 PMID: 25031282

da Silva Carneiro, R., de Almeida Da Costa, M. H., Santos, J. Z. L. V., Uchôa, V. T., da Silva Lopes, L., & Mendes, A. N. (2025). Therapeutic actions of terpenes in neurodegenerative disorders and their correlations with the regulation and remodeling of the extracellular matrix. *Biocell, 49*(1), 109–125. DOI: 10.32604/biocell.2024.058405

Dal Bianco, A., Bradl, M., Frischer, J., Kutzelnigg, A., Jellinger, K., & Lassmann, H. (2008). Multiple sclerosis and Alzheimer's disease. *Annals of Neurology*, *63*(2), 174–183. DOI: 10.1002/ana.21240 PMID: 17924575

Dar, N. J., Bhat, J. A., John, U., & Bhat, S. A. (2024). Neuroglia in Neurodegeneration: Exploring Glial Dynamics in Brain Disorders. *Neuroglia 2024, Vol. 5, Pages 488-504, 5*(4), 488–504. DOI: 10.3390/neuroglia5040031

Darbandi, N., Ramezani, M., Khodagholi, F., & Noori, M. (2016). Kaempferol promotes memory retention and density of hippocampal CA1 neurons in intra-cerebroventricular STZ-induced experimental AD model in Wistar rats. *Biologija (Vilnius, Lithuania)*, *62*(3), 157–168. DOI: 10.6001/biologija.v62i3.3368

de Andrade Teles, R. B., Diniz, T. C., Costa Pinto, T. C., de Oliveira, R. G.Junior, Gama e Silva, M., de Lavor, É. M., & da Silva Almeida, J. R. G. (2018). Flavonoids as therapeutic agents in Alzheimer's and Parkinson's diseases: A systematic review of preclinical evidences. *Oxidative Medicine and Cellular Longevity*, *2018*(1), 7043213.

de Lima, E. P., Laurindo, L. F., Catharin, V. C. S., Direito, R., Tanaka, M., Jasmin Santos German, I., Lamas, C. B., Guiguer, E. L., Araújo, A. C., Fiorini, A. M. R., & Barbalho, S. M. (2025). Polyphenols, Alkaloids, and Terpenoids Against Neurodegeneration: Evaluating the Neuroprotective Effects of Phytocompounds Through a Comprehensive Review of the Current Evidence. *Metabolites*, *15*(2), 124. Advance online publication. DOI: 10.3390/metabo15020124 PMID: 39997749

De Strooper, B., & Karran, E. (2016). The Cellular Phase of Alzheimer's Disease. In *Cell* (Vol. 164, Issue 4, pp. 603–615). Cell Press. DOI: 10.1016/j.cell.2015.12.056

Del Prado-Audelo, M. L., Cortés, H., Caballero-Florán, I. H., González-Torres, M., Escutia-Guadarrama, L., Bernal-Chávez, S. A., Giraldo-Gomez, D. M., Magaña, J. J., & Leyva-Gómez, G. (2021). Therapeutic Applications of Terpenes on Inflammatory Diseases. *Frontiers in Pharmacology*, *12*, 704197. DOI: 10.3389/fphar.2021.704197 PMID: 34483907

Demir, E. A., Gergerlioglu, H. S., & Oz, M. (2015). Antidepressant-like effects of quercetin in diabetic rats are independent of hypothalamic-pituitary-adrenal axis. *Acta Neuropsychiatrica*, *28*(1), 23–30. DOI: 10.1017/neu.2015.45 PMID: 26234153

Deng, C., Yi, R., Fei, M., Li, T., Han, Y., & Wang, H. (2021). Naringenin attenuates endoplasmic reticulum stress, reduces apoptosis, and improves functional recovery in experimental traumatic brain injury. *Brain Research*, *1769*, 147591. DOI: 10.1016/j.brainres.2021.147591 PMID: 34324877

Deng, Q., Wu, C., Parker, E., Liu, T. C. Y., Duan, R., & Yang, L. (2024). Microglia and Astrocytes in Alzheimer's Disease: Significance and Summary of Recent Advances. In *Aging and Disease* (Vol. 15, Issue 4, pp. 1537–1564). International Society on Aging and Disease. DOI: 10.14336/AD.2023.0907

Deng, S., Gan, L., Liu, C., Xu, T., Zhou, S., Guo, Y., Zhang, Z., Yang, G. Y., Tian, H., & Tang, Y. (2023). Roles of Ependymal Cells in the Physiology and Pathology of the Central Nervous System. *Aging and Disease*, *14*(2), 468. DOI: 10.14336/AD.2022.0826-1 PMID: 37008045

Dent, K. A., Christie, K. J., Bye, N., Basrai, H. S., Turbic, A., Habgood, M., Cate, H. S., & Turnley, A. M. (2015). Oligodendrocyte birth and death following traumatic brain injury in adult mice. *PLoS One*, *10*(3), e0121541. DOI: 10.1371/journal.pone.0121541 PMID: 25798924

Depp, C., Sun, T., Sasmita, A. O., Spieth, L., Berghoff, S. A., Nazarenko, T., Overhoff, K., Steixner-Kumar, A. A., Subramanian, S., Arinrad, S., Ruhwedel, T., Möbius, W., Göbbels, S., Saher, G., Werner, H. B., Damkou, A., Zampar, S., Wirths, O., Thalmann, M., & Nave, K. A. (2023). Myelin dysfunction drives amyloid-β deposition in models of Alzheimer's disease. *Nature*, *618*(7964), 349–357. DOI: 10.1038/s41586-023-06120-6 PMID: 37258678

Depp, C. M., Nave, K., & Lab, K. N. (2022). Ageing-associated myelin dysfunction drives amyloid deposition in mouse models of Alzheimer's disease. *Alzheimer's & Dementia*, *18*(S4), e061183. Advance online publication. DOI: 10.1002/alz.061183

Dhingra, D., & Valecha, R. (2014). Punarnavine, an alkaloid isolated from ethanolic extract of Boerhaavia diffusa Linn. Reverses depression-like behaviour in mice subjected to chronic unpredictable mild stress. *Indian Journal of Experimental Biology*, *52*(8), 799–807. PMID: 25141543

Dietrich, P., Shanmugasundaram, R., Shuyu, E., & Dragatsis, I. (2009). Congenital hydrocephalus associated with abnormal subcommissural organ in mice lacking huntingtin in Wnt1 cell lineages. *Human Molecular Genetics*, *18*(1), 142–150. DOI: 10.1093/hmg/ddn324 PMID: 18838463

Duncan, G. J., Manesh, S. B., Hilton, B. J., Assinck, P., Plemel, J. R., & Tetzlaff, W. (2020). The fate and function of oligodendrocyte progenitor cells after traumatic spinal cord injury. In *GLIA* (Vol. 68, Issue 2, pp. 227–245). John Wiley and Sons Inc. DOI: 10.1002/glia.23706

Duncan, I. D., & Radcliff, A. B. (2016). Inherited and acquired disorders of myelin: The underlying myelin pathology. In *Experimental Neurology* (Vol. 283, pp. 452–475). Academic Press., DOI: 10.1016/j.expneurol.2016.04.002

Duval, A., & Avérous, L. (2016). Characterization and Physicochemical Properties of Condensed Tannins from Acacia catechu. *Journal of Agricultural and Food Chemistry*, *64*(8), 1751–1760. DOI: 10.1021/acs.jafc.5b05671 PMID: 26853705

Edgar, N., & Sibille, E. (2012). A putative functional role for oligodendrocytes in mood regulation. *Translational Psychiatry*, *2*(5), e109–e109.

Escartin, C., Galea, E., Lakatos, A., O'Callaghan, J. P., Petzold, G. C., Serrano-Pozo, A., Steinhäuser, C., Volterra, A., Carmignoto, G., Agarwal, A., Allen, N. J., Araque, A., Barbeito, L., Barzilai, A., Bergles, D. E., Bonvento, G., Butt, A. M., Chen, W. T., Cohen-Salmon, M., … Verkhratsky, A. (2021). Reactive astrocyte nomenclature, definitions, and future directions. *Nature Neuroscience 2021 24:3, 24*(3), 312–325. DOI: 10.1038/s41593-020-00783-4

Esmeeta, A., Adhikary, S., Dharshnaa, V., Swarnamughi, P., Maqsummiya, Z. U., Banerjee, A., & Duttaroy, A. K. (2022). Plant-derived bioactive compounds in colon cancer treatment: An updated review. *Biomedicine and Pharmacotherapy*, *153*, 113384.

Fakhoury, M. (2018). Microglia and astrocytes in Alzheimer's disease: Implications for therapy. *Current Neuropharmacology*, *16*(5), 508–518.

Feng, Y., Cui, Y., Gao, J. L., Li, M. H., Li, R., Jiang, X. H., Tian, Y. X., Wang, K. J., Cui, C. M., & Cui, J. Z. (2016). Resveratrol attenuates neuronal autophagy and inflammatoryinjury by inhibiting the TLR4/NF-κB signaling pathwayin experimental traumatic brain injury. *International Journal of Molecular Medicine*, *37*(4), 921–930. DOI: 10.3892/ijmm.2016.2495 PMID: 26936125

Fernandez-Zafra, T., Codeluppi, S., & Uhlén, P. (2017). An ex vivo spinal cord injury model to study ependymal cells in adult mouse tissue. *Experimental Cell Research*, *357*(2), 236–242.

Ferrari Bardile, C., Sidik, H., Quek, R., Yusof, N. A. B. M., Garcia-Miralles, M., & Pouladi, M. A. (2021). Abnormal spinal cord myelination due to oligodendrocyte dysfunction in a model of Huntington's disease. *Journal of Huntington's Disease*, *10*(3), 377–384.

Ferreira, C., Vieira, P., Sá, H., Malva, J., Castelo-Branco, M., Reis, F., & Viana, S. (2024). Polyphenols: Immunonutrients tipping the balance of immunometabolism in chronic diseases. *Frontiers in Immunology*, *15*(March), 1–26. DOI: 10.3389/fimmu.2024.1360065 PMID: 38558823

Ferris, J. K., Greeley, B., Vavasour, I. M., Kraeutner, S. N., Rinat, S., Ramirez, J., & Boyd, L. A. (2022). In vivo myelin imaging and tissue microstructure in white matter hyperintensities and perilesional white matter. *Brain Communications*, *4*(3), fcac142.

Figueira, I., Menezes, R., Macedo, D., Costa, I., & dos Santos, C. N. (2016). Polyphenols Beyond Barriers: A Glimpse into the Brain. *Current Neuropharmacology*, *15*(4), 562–594. DOI: 10.2174/1570159X14666161026151545 PMID: 27784225

Fraga-Corral, M., Otero, P., Echave, J., Garcia-Oliveira, P., Carpena, M., Jarboui, A., & Prieto, M. A. (2021). By-products of agri-food industry as tannin-rich sources: A review of tannins' biological activities and their potential for valorization. *Foods*, *10*(1), 137.

Frost, G. R., & Li, Y. M. (2017). The role of astrocytes in amyloid production and Alzheimer's disease. *Open Biology*, *7*(12), 170228.

Fu, J., Sun, H., Zhang, Y., Xu, W., Wang, C., Fang, Y., & Zhao, J. (2018). Neuroprotective Effects of Luteolin Against Spinal Cord Ischemia-Reperfusion Injury by Attenuation of Oxidative Stress, Inflammation, and Apoptosis. *Journal of Medicinal Food*, *21*(1), 13–20. DOI: 10.1089/jmf.2017.4021 PMID: 28976796

Fu, Y., Yang, J., Wang, X., Yang, P., Zhao, Y., Li, K., Chen, Y., & Xu, Y. (2018). Herbal Compounds Play a Role in Neuroprotection through the Inhibition of Microglial Activation. *Journal of Immunology Research*, *2018*(1), 9348046. DOI: 10.1155/2018/9348046 PMID: 29850641

Gan, H., Ma, Q., Hao, W., Yang, N., Chen, Z. S., Deng, L., & Chen, J. (2024). Targeting autophagy to counteract neuroinflammation: A novel antidepressant strategy. In *Pharmacological Research* (Vol. 202). Academic Press., DOI: 10.1016/j.phrs.2024.107112

Gan, P., Ding, L., Hang, G., Xia, Q., Huang, Z., Ye, X., & Qian, X. (2020). Oxymatrine Attenuates Dopaminergic Neuronal Damage and Microglia-Mediated Neuroinflammation Through Cathepsin D-Dependent HMGB1/TLR4/NF-κB Pathway in Parkinson's Disease. *Frontiers in Pharmacology*, *11*(May), 1–13. DOI: 10.3389/fphar.2020.00776 PMID: 32528295

Gao, C., Jiang, J., Tan, Y., & Chen, S. (2023a). Microglia in neurodegenerative diseases: mechanism and potential therapeutic targets. *Signal Transduction and Targeted Therapy 2023 8:1, 8*(1), 1–37. DOI: 10.1038/s41392-023-01588-0

Ge, X. H., Shao, L., & Zhu, G. J. (2018). Oxymatrine attenuates brain hypoxic-ischemic injury from apoptosis and oxidative stress: Role of p-Akt/GSK3β/HO-1/Nrf-2 signaling pathway. *Metabolic Brain Disease, 33*(6), 1869–1875. DOI: 10.1007/s11011-018-0293-4 PMID: 30032345

Geloso, M. C., Corvino, V., Marchese, E., Serrano, A., Michetti, F., & D'Ambrosi, N. (2017). The dual role of microglia in ALS: Mechanisms and therapeutic approaches. *Frontiers in Aging Neuroscience, 9*, 242.

Gerhard, A., Pavese, N., Hotton, G., Turkheimer, F., Es, M., Hammers, A., & Brooks, D. J. (2006). In vivo imaging of microglial activation with [11C](R)-PK11195 PET in idiopathic Parkinson's disease. *Neurobiology of Disease, 21*(2), 404–412.

Goldman, J. G., Volpe, D., Ellis, T. D., Hirsch, M. A., Johnson, J., Wood, J., & Wallock, K. (2024). Delivering multidisciplinary rehabilitation care in Parkinson's disease: An international consensus statement. *Journal of Parkinson's Disease, 14*(1), 135–166.

Gomez-Pinilla, F., & Nguyen, T. T. (2012). Natural mood foods: The actions of polyphenols against psychiatric and cognitive disorders. *Nutritional Neuroscience, 15*(3), 127–133.

Gong, M., Han, B., Wang, S., Liang, S., & Zou, Z. (2016). Icariin reverses corticosterone-induced depression-like behavior, decrease in hippocampal brain-derived neurotrophic factor (BDNF) and metabolic network disturbances revealed by NMR-based metabonomics in rats. *Journal of Pharmaceutical and Biomedical Analysis, 123*, 63–73. DOI: 10.1016/j.jpba.2016.02.001 PMID: 26874256

Gong, Z., Ba, L., & Zhang, M. (2022). Dysfunction of the oligodendrocytes in amyotrophic lateral sclerosis. *Journal of Biomedical Research, 36*(5), 336–342. DOI: 10.7555/JBR.36.20220009 PMID: 36106501

Grabska-Kobyłecka, I., Szpakowski, P., Król, A., Książek-Winiarek, D., Kobyłecki, A., Głąbiński, A., & Nowak, D. (2023). Polyphenols and Their Impact on the Prevention of Neurodegenerative Diseases and Development. *Nutrients, 15*(15), 1–34. DOI: 10.3390/nu15153454 PMID: 37571391

Gu, J. H., Ge, J., Li, M., Wu, F., Zhang, W., & Qin, Z.-H. (2012). Inhibition of NF-κB activation is associated with anti-inflammatory and anti-apoptotic effects of Ginkgolide B in a mouse model of cerebral ischemia/reperfusion injury. *European Journal of Pharmaceutical Sciences, 47*(4), 652–660. DOI: 10.1016/j.ejps.2012.07.016 PMID: 22850444

Gugliandolo, E., Peritore, A. F., D'amico, R., Licata, P., & Crupi, R. (2020). Evaluation of neuroprotective effects of quercetin against aflatoxin b1-intoxicated mice. *Animals (Basel), 10*(5), 1–13. DOI: 10.3390/ani10050898 PMID: 32455780

Guo, F., Liu, X., Cai, H., & Le, W. (2018). Autophagy in neurodegenerative diseases: pathogenesis and therapy. In *Brain Pathology* (Vol. 28, Issue 1, pp. 3–13). Blackwell Publishing Ltd. DOI: 10.1111/bpa.12545

Gurcan, O., Gurcay, A. G., Kazanci, A., Senturk, S., Bodur, E., Karaca, E. U., Turkoglu, O. F., & Bavbek, M. (2017). Effect of asiatic acid on the treatment of spinal cord injury: An experimental study in rats. *Turkish Neurosurgery*, *27*(2), 259–264. DOI: 10.5137/1019-5149.JTN.15747-15.2 PMID: 27593768

Han, S., Gim, Y., Jang, E. H., & Hur, E. M. (2022). Functions and dysfunctions of oligodendrocytes in neurodegenerative diseases. In *Frontiers in Cellular Neuroscience* (Vol. 16, p. 1083159). Frontiers Media S.A., DOI: 10.3389/fncel.2022.1083159

Hao, T., Du, X., Yang, S., Zhang, Y., & Liang, F. (2020). Astrocytes-induced neuronal inhibition contributes to depressive-like behaviors during chronic stress. *Life Sciences*, *258*, 118099. DOI: 10.1016/j.lfs.2020.118099 PMID: 32682917

Hasan, S., Khatri, N., Rahman, Z. N., Menezes, A. A., Martini, J., Shehjar, F., Mujeeb, N., & Shah, Z. A. (2023). Neuroprotective Potential of Flavonoids in Brain Disorders. *Brain Sciences*, *13*(9), 1258. DOI: 10.3390/brainsci13091258 PMID: 37759859

Hatrock, D., Caporicci-Dinucci, N., & Stratton, J. (2020). Ependymal cells and multiple sclerosis: Proposing a relationship. In *Neural Regeneration Research* (Vol. 15, Issue 2, pp. 263–264). Wolters Kluwer Medknow Publications. DOI: 10.4103/1673-5374.265551

He, F. Q., Qiu, B. Y., Zhang, X. H., Li, T. K., Xie, Q., Cui, D. J., Huang, X. L., & Gan, H. T. (2011). Tetrandrine attenuates spatial memory impairment and hippocampal neuroinflammation via inhibiting NF-κB activation in a rat model of Alzheimer's disease induced by amyloid-β(1-42). *Brain Research*, *1384*, 89–96. DOI: 10.1016/j.brainres.2011.01.103 PMID: 21300035

He, X., Liu, X., Yang, J., Du, H., Chai, N., Sha, Z., Geng, M., Zhou, X., & He, C. (2020). Tannic acid-reinforced methacrylated chitosan/methacrylated silk fibroin hydrogels with multifunctionality for accelerating wound healing. *Carbohydrate Polymers*, *247*, 116689. DOI: 10.1016/j.carbpol.2020.116689 PMID: 32829817

Hedreen, J. C., Berretta, S., & White, C. L.III. (2024). Postmortem neuropathology in early Huntington disease. *Journal of Neuropathology and Experimental Neurology*, *83*(5), 294–306. DOI: 10.1093/jnen/nlae022 PMID: 38553027

Hong, B., Ohtake, Y., Itokazu, T., & Yamashita, T. (2024). Glial senescence enhances α-synuclein pathology owing to its insufficient clearance caused by autophagy dysfunction. *Cell Death Discovery*, *10*(1), 50. Advance online publication. DOI: 10.1038/s41420-024-01816-8 PMID: 38272865

Hossain, M. T., Furhatun-Noor, A., Matin, A., Tabassum, F., & Ar Rashid, H. (2021). A review study on the pharmacological effects and mechanism of action of tannins. *European Journal Pharmaceutical and Medical Research*, (August), 5–10.

Hu, Q., Zhang, R., Dong, X., Yang, D., Du, Q., & Yu, W. (2024). Corrigendum to: Huperzine A ameliorates neurological deficits after spontaneous subarachnoid hemorrhage through endothelial cell pyroptosis inhibition. *Acta Biochimica et Biophysica Sinica*, *56*(5), 824. DOI: 10.3724/abbs.2024067 PMID: 38712432

Hua, K. F., Chao, A. C., Lin, T. Y., Chen, W. T., Lee, Y. C., Hsu, W. H., Lee, S. L., Wang, H. M., Yang, D. I., & Ju, T. C. (2022). Ginsenoside compound K reduces the progression of Huntington's disease via the inhibition of oxidative stress and overactivation of the ATM/AMPK pathway. *Journal of Ginseng Research*, *46*(4), 572–584. DOI: 10.1016/j.jgr.2021.11.003 PMID: 35818427

Huang, J., Chen, B., Wang, H., Hu, S., Yu, X., Reilly, J., He, Z., You, Y., & Shu, X. (2022). *Dihydromyricetin Attenuates Depressive-like Behaviors in Mice by Inhibiting the AGE-RAGE Signaling Pathway.*

Huntemer-Silveira, A., Patil, N., Brickner, M. A., & Parr, A. M. (2021). Strategies for Oligodendrocyte and Myelin Repair in Traumatic CNS Injury. In *Frontiers in Cellular Neuroscience* (Vol. 14). Frontiers Media S.A., DOI: 10.3389/fncel.2020.619707

Hussain, G., Huang, J., Rasul, A., Anwar, H., Imran, A., Maqbool, J., Razzaq, A., Aziz, N., Makhdoom, E., Konuk, M., & Sun, T. (2019). Putative Roles of Plant-Derived Tannins in Neurodegenerative and Neuropsychiatry Disorders: An Updated Review. *Molecules (Basel, Switzerland)*, *24*(12), 2213. Advance online publication. DOI: 10.3390/molecules24122213 PMID: 31200495

Ibrahim, A. M., Pottoo, F. H., Dahiya, E. S., Khan, F. A., & Kumar, J. S. (2020). Neuron-glia interactions: Molecular basis of alzheimer's disease and applications of neuroproteomics. *The European Journal of Neuroscience*, *52*(2), 2931–2943.

Imamura, K., Hishikawa, N., Sawada, M., Nagatsu, T., Yoshida, M., & Hashizume, Y. (2003). Distribution of major histocompatibility complex class II-positive microglia and cytokine profile of Parkinson's disease brains. *Acta Neuropathologica*, *106*(6), 518–526. DOI: 10.1007/s00401-003-0766-2 PMID: 14513261

Ishisaka, M., Kakefuda, K., Yamauchi, M., Tsuruma, K., Shimazawa, M., Tsuruta, A., & Hara, H. (2011). Luteolin shows an antidepressant-like effect via suppressing endoplasmic reticulum stress. *Biological & Pharmaceutical Bulletin*, *34*(9), 1481–1486. DOI: 10.1248/bpb.34.1481 PMID: 21881237

Ishizawa, K., Komori, T., Sasaki, S., Arai, N., Mizutani, T., & Hirose, T. (2004). Microglial Activation Parallels System Degeneration in Multiple System Atrophy. *Journal of Neuropathology and Experimental Neurology*, *63*(1), 43–52. DOI: 10.1093/jnen/63.1.43 PMID: 14748560

Islam, F., Khan, M. M., Raza, S. S., Javed, H., Ahmad, A., Khan, A., Islam, F., & Safhi, M. M. (2012). Rutin protects dopaminergic neurons from oxidative stress in an animal model of parkinson's disease. *Neurotoxicity Research*, *22*(1), 1–15. DOI: 10.1007/s12640-011-9295-2 PMID: 22194158

Itoh, T., Imano, M., Nishida, S., Tsubaki, M., Hashimoto, S., Ito, A., & Satou, T. (2011). (-)-Epigallocatechin-3-gallate protects against neuronal cell death and improves cerebral function after traumatic brain injury in rats. *Neuromolecular Medicine*, *13*(4), 300–309. DOI: 10.1007/s12017-011-8162-x PMID: 22038400

Jaeger, B. N., Parylak, S. L., & Gage, F. H. (2018). Mechanisms of dietary flavonoid action in neuronal function and neuroinflammation. *Molecular Aspects of Medicine*, *61*, 50–62. DOI: 10.1016/j.mam.2017.11.003 PMID: 29117513

Jäkel, S., Agirre, E., Mendanha Falcão, A., van Bruggen, D., Lee, K. W., Knuesel, I., Malhotra, D., ffrench-Constant, C., Williams, A., & Castelo-Branco, G. (2019). Altered human oligodendrocyte heterogeneity in multiple sclerosis. *Nature 2019 566:7745*, *566*(7745), 543–547. DOI: 10.1038/s41586-019-0903-2

Jalouli, M., Rahman, M. A., Biswas, P., Rahman, H., Harrath, A. H., Lee, I. S., Kang, S., Choi, J., Park, M. N., & Kim, B. (2025). Targeting natural antioxidant polyphenols to protect neuroinflammation and neurodegenerative diseases: a comprehensive review. In *Frontiers in Pharmacology* (Vol. 16, p. 1492517). Frontiers Media SA. DOI: 10.3389/fphar.2025.1492517

Janda, E., Boi, L., & Carta, A. R. (2018). Microglial phagocytosis and its regulation: A therapeutic target in parkinson's disease? In *Frontiers in Molecular Neuroscience* (Vol. 11). Frontiers Media S.A., DOI: 10.3389/fnmol.2018.00144

Javed, H., Khan, M. M., Ahmad, A., Vaibhav, K., Ahmad, M. E., Khan, A., Ashafaq, M., Islam, F., Siddiqui, M. S., Safhi, M. M., & Islam, F. (2012). Rutin prevents cognitive impairments by ameliorating oxidative stress and neuroinflammation in rat model of sporadic dementia of Alzheimer type. *Neuroscience, 210*, 340–352. DOI: 10.1016/j.neuroscience.2012.02.046 PMID: 22441036

Jeon, Y. M., Kwon, Y., Jo, M., Lee, S., Kim, S., & Kim, H. J. (2020). The Role of Glial Mitochondria in α-Synuclein Toxicity. In *Frontiers in Cell and Developmental Biology* (Vol. 8). Frontiers Media S.A., DOI: 10.3389/fcell.2020.548283

Jiang, Z., Zhang, J., Cai, Y., Huang, J., & You, L. (2017). Catechin attenuates traumatic brain injury-induced blood–brain barrier damage and improves longer-term neurological outcomes in rats. *Experimental Physiology*, *102*(10), 1269–1277. DOI: 10.1113/EP086520 PMID: 28678393

Jucaite, A., Cselényi, Z., Kreisl, W. C., Rabiner, E. A., Varrone, A., Carson, R. E., Rinne, J. O., Savage, A., Schou, M., Johnström, P., Svenningsson, P., Rascol, O., Meissner, W. G., Barone, P., Seppi, K., Kaufmann, H., Wenning, G. K., Poewe, W., & Farde, L. (2022). Glia Imaging Differentiates Multiple System Atrophy from Parkinson's Disease: A Positron Emission Tomography Study with [11C]PBR28 and Machine Learning Analysis. *Movement Disorders*, *37*(1), 119–129. DOI: 10.1002/mds.28814 PMID: 34609758

Kaji, S., Maki, T., Ishimoto, T., Yamakado, H., & Takahashi, R. (2020). Insights into the pathogenesis of multiple system atrophy: Focus on glial cytoplasmic inclusions. In *Translational Neurodegeneration* (Vol. 9, Issue 1, pp. 1–15). BioMed Central. DOI: 10.1186/s40035-020-0185-5

Kaliszewski, M., Knott, A. B., & Bossy-Wetzel, E. (2015). Primary cilia and autophagic dysfunction in Huntington's disease. *Cell Death and Differentiation*, *22*(9), 1413–1424.

Kang, K. S., Wen, Y., Yamabe, N., Fukui, M., Bishop, S. C., & Zhu, B. T. (2010). Dual beneficial effects of (-)-epigallocatechin-3-gallate on levodopa methylation and hippocampal neurodegeneration: In vitro and in vivo studies. *PLoS One*, *5*(8), e11951. Advance online publication. DOI: 10.1371/journal.pone.0011951 PMID: 20700524

Khan, N., Pelletier, D., McAlear, T. S., Croteau, N., Veyron, S., Bayne, A. N., Black, C., Ichikawa, M., Khalifa, A. A. Z., Chaaban, S., Kurinov, I., Brouhard, G., Bechstedt, S., Bui, K. H., & Trempe, J. F. (2021). Crystal structure of human PACRG in complex with MEIG1 reveals roles in axoneme formation and tubulin binding. *Structure (London, England)*, *29*(6), 572–586.e6. DOI: 10.1016/j.str.2021.01.001 PMID: 33529594

Khelfaoui, H., Ibaceta-Gonzalez, C., & Angulo, M. C. (2024). Functional myelin in cognition and neurodevelopmental disorders. *Cellular and Molecular Life Sciences 2024 81:1, 81*(1), 1–14. DOI: 10.1007/s00018-024-05222-2

Kim, S., Chun, H., Kim, Y., Kim, Y., Park, U., Chu, J., Bhalla, M., Choi, S. H., Yousefian-Jazi, A., Kim, S., Hyeon, S. J., Kim, S., Kim, Y., Ju, Y. H., Lee, S. E., Lee, H., Lee, K., Oh, S. J., Hwang, E. M., & Ryu, H. (2024). Astrocytic autophagy plasticity modulates Aβ clearance and cognitive function in Alzheimer's disease. *Molecular Neurodegeneration, 19*(1), 1–26. DOI: 10.1186/s13024-024-00740-w PMID: 39044253

Kister, A., & Kister, I. (2023). Overview of myelin, major myelin lipids, and myelin-associated proteins. *Frontiers in Chemistry, 10*, 1041961.

Kong, Y. R., Tay, K. C., Su, Y. X., Khaw, K. Y., Wong, C. K., & Tan, W. N. (2021). Potential of Naturally Derived Alkaloids as Multi-Targeted Therapeutic Agents for Neurodegenerative Diseases. *Molecules 2021, Vol. 26, Page 728, 26*(3), 728. DOI: 10.3390/molecules26030728

Kübler, S., Soutschek, A., & Schubert, T. (2019). The causal role of the lateral prefrontal cortex for task-order coordination in dual-task situations: A study with transcranial magnetic stimulation. *Journal of Cognitive Neuroscience, 31*(12), 1840–1856.

Kulaszyńska, M., Kwiatkowski, S., & Skonieczna-Żydecka, K. (2024). The iron metabolism with a specific focus on the functioning of the nervous system. *Biomedicines, 12*(3), 595.

Kulkarni, S. K., Bhutani, M. K., & Bishnoi, M. (2008). Antidepressant activity of curcumin: Involvement of serotonin and dopamine system. *Psychopharmacology, 201*(3), 435–442. DOI: 10.1007/s00213-008-1300-y PMID: 18766332

Kulkarni, S. K., Dhir, A., & Akula, K. K. (2009). Potentials of curcumin as an antidepressant. *TheScientificWorldJournal, 9*, 1233–1241. DOI: 10.1100/tsw.2009.137 PMID: 19882093

Kumar, A., Chaudhary, T., & Mishra, J. (2013). Minocycline modulates neuroprotective effect of hesperidin against quinolinic acid induced Huntington's disease like symptoms in rats: Behavioral, biochemical, cellular and histological evidences. *European Journal of Pharmacology, 720*(1–3), 16–28. DOI: 10.1016/j.ejphar.2013.10.057 PMID: 24211676

Kumar, G. P., Anilakumar, K. R., & Naveen, S. (2015). Phytochemicals having neuroprotective properties from dietary sources and medicinal herbs. In *Pharmacognosy Journal* (Vol. 7, Issue 1, pp. 1–17). DOI: 10.5530/pj.2015.1.1

Kumar, P., & Kumar, A. (2010). Protective effect of hesperidin and naringin against 3-nitropropionic acid induced Huntington's like symptoms in rats: Possible role of nitric oxide. *Behavioural Brain Research, 206*(1), 38–46. DOI: 10.1016/j.bbr.2009.08.028 PMID: 19716383

Lacroix, S., Hamilton, L. K., Vaugeois, A., Beaudoin, S., Breault-Dugas, C., Pineau, I., Lévesque, S. A., Grégoire, C. A., & Fernandes, K. J. L. (2014). Central canal ependymal cells proliferate extensively in response to traumatic spinal cord injury but not demyelinating lesions. *PLoS One, 9*(1), e85916. Advance online publication. DOI: 10.1371/journal.pone.0085916 PMID: 24475059

Laricchiuta, D., Papi, M., Decandia, D., Panuccio, A., Cutuli, D., Peciccia, M., Mazzeschi, C., & Petrosini, L. (2024). The role of glial cells in mental illness: A systematic review on astroglia and microglia as potential players in schizophrenia and its cognitive and emotional aspects. *Frontiers in Cellular Neuroscience*, *18*(February), 1–26. DOI: 10.3389/fncel.2024.1358450 PMID: 38419655

Lee, B., Sur, B., Park, J., Kim, S. H., Kwon, S., Yeom, M., Shim, I., Lee, H., & Hahm, D. H. (2013). Chronic administration of baicalein decreases depression-like behavior induced by repeated restraint stress in rats. *The Korean Journal of Physiology & Pharmacology; Official Journal of the Korean Physiological Society and the Korean Society of Pharmacology*, *17*(5), 393–403. DOI: 10.4196/kjpp.2013.17.5.393 PMID: 24227939

Lee, B., Sur, B., Yeom, M., Shim, I., Lee, H., & Hahm, D. H. (2012). Effect of berberine on depression- and anxiety-like behaviors and activation of the noradrenergic system induced by development of morphine dependence in rats. *The Korean Journal of Physiology & Pharmacology; Official Journal of the Korean Physiological Society and the Korean Society of Pharmacology*, *16*(6), 379–386. DOI: 10.4196/kjpp.2012.16.6.379 PMID: 23269899

Lee, S., & Kovacs, G. G. (2024). The Irony of Iron: The Element with Diverse Influence on Neurodegenerative Diseases. In *International journal of molecular sciences* (Vol. 25, Issue 8, p. 4269). Multidisciplinary Digital Publishing Institute. DOI: 10.3390/ijms25084269

Leonard, B. E. (2018). Inflammation and depression: A causal or coincidental link to the pathophysiology? *Acta Neuropsychiatrica*, *30*(1), 1–16. DOI: 10.1017/neu.2016.69 PMID: 28112061

Li, C., Zug, C., Qu, H., Schluesener, H., & Zhang, Z. (2015). Hesperidin ameliorates behavioral impairments and neuropathology of transgenic APP/PS1 mice. *Behavioural Brain Research*, *281*, 32–42. DOI: 10.1016/j.bbr.2014.12.012 PMID: 25510196

Li, H., Liu, Y., Sun, Y., Guo, H., Lv, S., Guo, W., Ren, J., Wang, Y., Zu, J., Yan, J., & Wang, N. (2024). Targeting astrocytes polarization after spinal cord injury: a promising direction. In *Frontiers in Cellular Neuroscience* (Vol. 18, p. 1478741). Frontiers Media SA. DOI: 10.3389/fncel.2024.1478741

Li, J., Zhang, L., Chu, Y., Namaka, M., Deng, B., Kong, J., & Bi, X. (2016). Astrocytes in oligodendrocyte lineage development and white matter pathology. *Frontiers in Cellular Neuroscience*, *10*(MAY), 1–13. DOI: 10.3389/fncel.2016.00119 PMID: 27242432

Li, R., Wang, X., Qin, T., Qu, R., & Ma, S. (2016). Apigenin ameliorates chronic mild stress-induced depressive behavior by inhibiting interleukin-1β production and NLRP3 inflammasome activation in the rat brain. *Behavioural Brain Research*, *296*, 318–325. DOI: 10.1016/j.bbr.2015.09.031 PMID: 26416673

Li, Y., Chen, M., Yan, J., Zhou, W., Gao, S., Liu, S., Li, Q., Zheng, Y., Cheng, Y., & Guo, Q. (2021). Tannic acid/Sr2+-coated silk/graphene oxide-based meniscus scaffold with anti-inflammatory and anti-ROS functions for cartilage protection and delaying osteoarthritis. *Acta Biomaterialia*, *126*, 119–131. DOI: 10.1016/j.actbio.2021.02.046 PMID: 33684536

Liddelow, S. A., Guttenplan, K. A., Clarke, L. E., Bennett, F. C., Bohlen, C. J., Schirmer, L., Bennett, M. L., Münch, A. E., Chung, W. S., Peterson, T. C., Wilton, D. K., Frouin, A., Napier, B. A., Panicker, N., Kumar, M., Buckwalter, M. S., Rowitch, D. H., Dawson, V. L., Dawson, T. M., & Barres, B. A. (2017). Neurotoxic reactive astrocytes are induced by activated microglia. *Nature*, *541*(7638), 481–487. DOI: 10.1038/nature21029 PMID: 28099414

Lin, K., Li, Y., Du Toit, E., Wendt, L., & Sun, J. (2021). Effects of Polyphenol Supplementations on Improving Depression, Anxiety, and Quality of Life in Patients With Depression. *Frontiers in Psychiatry*, *12*(November), 1–13. DOI: 10.3389/fpsyt.2021.765485 PMID: 34819888

Lisanti, C. J., Asbach, P., & Bradley, W. G. (2005). The ependymal "Dot-Dash" sign: An MR imaging finding of early multiple sclerosis. *AJNR. American Journal of Neuroradiology*, *26*(8), 2033–2036. DOI: 10.1016/S0098-1672(08)70484-3 PMID: 16155155

Liu, B., & Hong, J. S. (2003). Role of microglia in inflammation-mediated neurodegenerative diseases: Mechanisms and strategies for therapeutic intervention. In *Journal of Pharmacology and Experimental Therapeutics* (Vol. 304, Issue 1, pp. 1–7). J Pharmacol Exp Ther. DOI: 10.1124/jpet.102.035048

Liu, B., Xu, C., Wu, X., Liu, F., Du, Y., Sun, J., Tao, J., & Dong, J. (2015). Icariin exerts an antidepressant effect in an unpredictable chronic mild stress model of depression in rats and is associated with the regulation of hippocampal neuroinflammation. *Neuroscience*, *294*(March), 193–205. DOI: 10.1016/j.neuroscience.2015.02.053 PMID: 25791226

Loane, D. J., & Byrnes, K. R. (2010). Role of Microglia in Neurotrauma. *Neurotherapeutics; the Journal of the American Society for Experimental NeuroTherapeutics*, *7*(4), 366–377. DOI: 10.1016/j.nurt.2010.07.002 PMID: 20880501

Loane, D. J., & Kumar, A. (2016). Microglia in the TBI brain: The good, the bad, and the dysregulated. In *Experimental Neurology* (Vol. 275, Issue 0 3, pp. 316–327). Academic Press Inc. DOI: 10.1016/j.expneurol.2015.08.018

Lorente Pons, A., Higginbottom, A., Cooper-Knock, J., Alrafiah, A., Alofi, E., Kirby, J., Shaw, P. J., Wood, J. D., & Highley, J. R. (2020). Oligodendrocyte pathology exceeds axonal pathology in white matter in human amyotrophic lateral sclerosis. *The Journal of Pathology*, *251*(3), 262–271. DOI: 10.1002/path.5455 PMID: 32391572

Loria, M. J., Ali, Z., Abe, N., Sufka, K. J., & Khan, I. A. (2014). Effects of Sceletium tortuosum in rats. *Journal of Ethnopharmacology*, *155*(1), 731–735. DOI: 10.1016/j.jep.2014.06.007 PMID: 24930358

Lou, H., Jing, X., Wei, X., Shi, H., Ren, D., & Zhang, X. (2014). Naringenin protects against 6-OHDA-induced neurotoxicity via activation of the Nrf2/ARE signaling pathway. *Neuropharmacology*, *79*(December), 380–388. DOI: 10.1016/j.neuropharm.2013.11.026 PMID: 24333330

Luca, A., Calandra, C., & Luca, M. (2018). Molecular bases of Alzheimer's disease and neurodegeneration: The role of neuroglia. In *Aging and Disease* (Vol. 9, Issue 6, pp. 1134–1152). International Society on Aging and Disease. DOI: 10.14336/AD.2018.0201

Lymperopoulos, D., Dedemadi, A. G., Voulgari, M. L., Georgiou, E., Dafnis, I., Mountaki, C., Panagopoulou, E. A., Karvelas, M., Chiou, A., Karathanos, V. T., & Chroni, A. (2024). Corinthian Currants Promote the Expression of Paraoxonase-1 and Enhance the Antioxidant Status in Serum and Brain of 5xFAD Mouse Model of Alzheimer's Disease. *Biomolecules*, *14*(4), 426. Advance online publication. DOI: 10.3390/biom14040426 PMID: 38672443

Madsen, P. M., Desu, H. L., Vaccari, J. P. de R., Florimon, Y., Ellman, D. G., Keane, R. W., Clausen, B. H., Lambertsen, K. L., & Brambilla, R. (2020). Oligodendrocytes modulate the immune-inflammatory response in EAE via TNFR2 signaling. *Brain, Behavior, and Immunity*, *84*, 132–146. DOI: 10.1016/j.bbi.2019.11.017 PMID: 31785393

Maher, P., Dargusch, R., Bodai, L., Gerard, P. E., Purcell, J. M., & Lawrence Marsh, J. (2011). Erk activation by the polyphenols fisetin and resveratrol provides neuroprotection in multiple models of Huntington's disease. *Human Molecular Genetics*, *20*(2), 261–270. DOI: 10.1093/hmg/ddq460 PMID: 20952447

Martinez-Vicente, M. (2017). Neuronal mitophagy in neurodegenerative diseases. In *Frontiers in Molecular Neuroscience* (Vol. 10). Frontiers Research Foundation., DOI: 10.3389/fnmol.2017.00064

Matejuk, A., & Ransohoff, R. M. (2020). Crosstalk Between Astrocytes and Microglia: An Overview. In *Frontiers in Immunology* (Vol. 11, p. 507878). Frontiers Media S.A. DOI: 10.3389/fimmu.2020.01416

Mavroeidi, P., & Xilouri, M. (2021). Neurons and glia interplay in α-synucleinopathies. In *International Journal of Molecular Sciences* (Vol. 22, Issue 9, p. 4994). Multidisciplinary Digital Publishing Institute. DOI: 10.3390/ijms22094994

McGeer, P. L., Itagaki, S., Boyes, B. E., & McGeer, E. G. (1988). Reactive microglia are positive for HLA-DR in the: Substantia nigra of Parkinson's and Alzheimer's disease brains. *Neurology*, *38*(8), 1285–1291. DOI: 10.1212/WNL.38.8.1285 PMID: 3399080

Meletis, K., Barnabé-Heider, F., Carlén, M., Evergren, E., Tomilin, N., Shupliakov, O., & Frisén, J. (2008). Spinal Cord Injury Reveals Multilineage Differentiation of Ependymal Cells. *PLoS Biology*, *6*(7), e182. DOI: 10.1371/journal.pbio.0060182 PMID: 18651793

Merighi, S., Travagli, A., Nigro, M., Pasquini, S., Cappello, M., Contri, C., Varani, K., Vincenzi, F., Borea, P. A., & Gessi, S. (2023). Caffeine for Prevention of Alzheimer's Disease: Is the A2A Adenosine Receptor Its Target? *Biomolecules*, *13*(6), 1–15. DOI: 10.3390/biom13060967 PMID: 37371547

Meyer, E., Mori, M. A., Campos, A. C., Andreatini, R., Guimarães, F. S., Milani, H., & de Oliveira, R. M. W. (2017). Myricitrin induces antidepressant-like effects and facilitates adult neurogenesis in mice. *Behavioural Brain Research*, *316*, 59–65. DOI: 10.1016/j.bbr.2016.08.048 PMID: 27569185

Mony, T. J., Elahi, F., Choi, J. W., & Park, S. J. (2022). Neuropharmacological Effects of Terpenoids on Preclinical Animal Models of Psychiatric Disorders: A Review. *Antioxidants*, *11*(9), 1834. DOI: 10.3390/antiox11091834 PMID: 36139909

Morris, G., Maes, M., Berk, M., Carvalho, A. F., & Puri, B. K. (2020). Nutritional ketosis as an intervention to relieve astrogliosis: Possible therapeutic applications in the treatment of neurodegenerative and neuroprogressive disorders. *European Psychiatry : the Journal of the Association of European Psychiatrists, 63*(1), e8. Advance online publication. DOI: 10.1192/j.eurpsy.2019.13 PMID: 32093791

Mulica, P., Grünewald, A., & Pereira, S. L. (2021). Astrocyte-Neuron Metabolic Crosstalk in Neurodegeneration: A Mitochondrial Perspective. *Frontiers in Endocrinology, 12*, 668517. DOI: 10.3389/fendo.2021.668517 PMID: 34025580

Murphy, C. E., Walker, A. K., & Weickert, C. S. (2021). Neuroinflammation in schizophrenia: The role of nuclear factor kappa B. *Translational Psychiatry, 11*(1), 1–13. DOI: 10.1038/s41398-021-01607-0 PMID: 34650030

Nagele, R. G., Wegiel, J., Venkataraman, V., Imaki, H., Wang, K. C., & Wegiel, J. (2004). Contribution of glial cells to the development of amyloid plaques in Alzheimer's disease. *Neurobiology of Aging, 25*(5), 663–674. DOI: 10.1016/j.neurobiolaging.2004.01.007 PMID: 15172746

Nasrabady, S. E., Rizvi, B., Goldman, J. E., & Brickman, A. M. (2018). White matter changes in Alzheimer's disease: a focus on myelin and oligodendrocytes. In *Acta neuropathologica communications* (Vol. 6, Issue 1, p. 22). BioMed Central. DOI: 10.1186/s40478-018-0515-3

Navabi, S. P., Badreh, F., Khombi Shooshtari, M., Hajipour, S., Moradi Vastegani, S., & Khoshnam, S. E. (2024). Microglia-induced neuroinflammation in hippocampal neurogenesis following traumatic brain injury. In *Heliyon* (Vol. 10, Issue 16, p. e35869). Elsevier. DOI: 10.1016/j.heliyon.2024.e35869

Nave, K. A., Asadollahi, E., & Sasmita, A. (2023). Expanding the function of oligodendrocytes to brain energy metabolism. In *Current Opinion in Neurobiology* (Vol. 83). Elsevier Ltd., DOI: 10.1016/j.conb.2023.102782

Nesterova, Y. V., Povetieva, T. N., Suslov, N. I., Semenov, A. A., & Pushkarskiy, S. V. (2011). Antidepressant activity of diterpene alkaloids of aconitum baicalense turcz. *Bulletin of Experimental Biology and Medicine, 151*(4), 425–428. DOI: 10.1007/s10517-011-1347-3 PMID: 22448357

Ng, Y. P., Or, T. C. T., & Ip, N. Y. (2015). Plant alkaloids as drug leads for Alzheimer's disease. In *Neurochemistry International* (Vol. 89, pp. 260–270). Elsevier Ltd., DOI: 10.1016/j.neuint.2015.07.018

Nichols, E., Szoeke, C. E. I., Vollset, S. E., Abbasi, N., Abd-Allah, F., Abdela, J., Aichour, M. T. E., Akinyemi, R. O., Alahdab, F., Asgedom, S. W., Awasthi, A., Barker-Collo, S. L., Baune, B. T., Béjot, Y., Belachew, A. B., Bennett, D. A., Biadgo, B., Bijani, A., Bin Sayeed, M. S., & Murray, C. J. L. (2019). Global, regional, and national burden of Alzheimer's disease and other dementias, 1990–2016: A systematic analysis for the Global Burden of Disease Study 2016. *Lancet Neurology, 18*(1), 88–106. DOI: 10.1016/S1474-4422(18)30403-4 PMID: 30497964

Njapdounke, S. K. J., Nkantchoua, N. G., Moto Okomolo, F. C., Taiwe, S. G., Neteydji, S., Pale, S., Ayissi Mbomo, E. R., & Ngo Bum, E. (2016). Anxiolytic - like properties of Hallea ciliata in mice. *African Journal of Traditional, Complementary, and Alternative Medicines : AJTCAM, 13*(4), 1–7. DOI: 10.21010/ajtcam.v13i4.1 PMID: 28852713

Notter, T. (2021). Astrocytes in schizophrenia. *Brain and Neuroscience Advances, 5*, 23982128211009148. Advance online publication. DOI: 10.1177/23982128211009148 PMID: 33997293

Nykjær, C. H., Brudek, T., Salvesen, L., & Pakkenberg, B. (2017). Changes in the cell population in brain white matter in multiple system atrophy. *Movement Disorders, 32*(7), 1074–1082. DOI: 10.1002/mds.26979 PMID: 28394027

Oberheim, N. A., Goldman, S. A., & Nedergaard, M. (2012). Heterogeneity of astrocytic form and function. *Methods in Molecular Biology (Clifton, N.J.), 814*, 23–45. DOI: 10.1007/978-1-61779-452-0_3 PMID: 22144298

Ocal, O., Borcek, A. O., Pasaoglu, O., Gundogdu, A. C., Kaplanoglu, G. T., & Baykaner, M. K. (2019). Can quercetin be an option for treatment of spinal cord injury? An experimental study. *Turkish Neurosurgery, 29*(2), 247–253. DOI: 10.5137/1019-5149.JTN.23799-18.1 PMID: 30649798

Ojha, S. (2021). *Containing Natural Products*. 1–26.

Ortiz, G. G., Flores Alvarado, L. J., Pacheco Moisés, F. P., Mireles Ramírez, M. A., González Renovato, E. D., Sánchez López, A. L., ... & Sánchez González, V. J. (2016). Cross-talk between glial cells and neurons: Relationship in Multiple Sclerosis.

Ozogul, Y., Ucar, Y., Tadesse, E. E., Rathod, N., Kulawik, P., Trif, M., Esatbeyoglu, T., & Ozogul, F. (2025). Tannins for food preservation and human health: A review of current knowledge. *Applied Food Research, 5*(1), 100738. DOI: 10.1016/j.afres.2025.100738

Panche, A. N., Diwan, A. D., & Chandra, S. R. (2016). Flavonoids: An overview. *Journal of Nutritional Science, 5*, e47. DOI: 10.1017/jns.2016.41 PMID: 28620474

Parkhurst, C. N., Yang, G., Ninan, I., Savas, J. N., Yates, J. R.III, Lafaille, J. J., Hempstead, B. L., Littman, D. R., & Gan, W. B. (2013). Microglia promote learning-dependent synapse formation through brain-derived neurotrophic factor. *Cell, 155*(7), 1596–1609. DOI: 10.1016/j.cell.2013.11.030 PMID: 24360280

Patel, S., Keating, B. A., & Dale, R. C. (2023). Anti-inflammatory properties of commonly used psychiatric drugs. *Frontiers in Neuroscience, 16*, 1039379. Advance online publication. DOI: 10.3389/fnins.2022.1039379 PMID: 36704001

Pathak, D., & Sriram, K. (2023). Neuron-astrocyte omnidirectional signaling in neurological health and disease. *Frontiers in Molecular Neuroscience, 16*, 1169320.

Payne, A. G. (2009). Experimental regimen targeting the ependyma slows disease progression in four patients with amyotrophic lateral sclerosis. *Medical Hypotheses, 72*(5), 548–550. DOI: 10.1016/j.mehy.2008.12.032 PMID: 19200662

Peng, L., Parpura, V., & Verkhratsky, A. (2014). EDITORIAL Neuroglia as a Central Element of Neurological Diseases: An Underappreciated Target for Therapeutic Intervention. *Current Neuropharmacology, 12*(4), 303–307. DOI: 10.2174/1570159X12999140829152550 PMID: 25342938

Peng, L., Verkhratsky, A., Gu, L., & Li, B. (2015). Targeting astrocytes in major depression. *Expert Review of Neurotherapeutics, 15*(11), 1299–1306. DOI: 10.1586/14737175.2015.1095094 PMID: 26471936

Pittenger, C., & Duman, R. S. (2008). Stress, depression, and neuroplasticity: A convergence of mechanisms. *Neuropsychopharmacology : Official Publication of the American College of Neuropsychopharmacology*, *33*(1), 88–109. DOI: 10.1038/sj.npp.1301574 PMID: 17851537

Prinz, M., Masuda, T., Wheeler, M. A., & Quintana, F. J. (2021). Microglia and Central Nervous System-Associated Macrophages mdash from Origin to Disease Modulation. In *Annual Review of Immunology* (Vol. 39, pp. 251–277). Annual Reviews Inc., DOI: 10.1146/annurev-immunol-093019-110159

Puentes-Orozco, M., Albarracin, S. L., & Velásquez, M. M. (2024). Neuroinflammation and major depressive disorder: Astrocytes at the crossroads. *Frontiers in Cellular Neuroscience*, *18*(November), 1–8. DOI: 10.3389/fncel.2024.1504555 PMID: 39650796

Qu, Z., Zhang, J., Yang, H., Huo, L., Gao, J., Chen, H., & Gao, W. (2016). Protective effect of tetrahydropalmatine against d-galactose induced memory impairment in rat. *Physiology & Behavior*, *154*, 114–125. DOI: 10.1016/j.physbeh.2015.11.016 PMID: 26592138

Quan, L., Uyeda, A., & Muramatsu, R. (2022). Central nervous system regeneration: the roles of glial cells in the potential molecular mechanism underlying remyelination. *Inflammation and Regeneration 2022 42:1*, *42*(1), 1–12. DOI: 10.1186/s41232-022-00193-y

Quincozes-Santos, A., Santos, C. L., de Souza Almeida, R. R., da Silva, A., Thomaz, N. K., Costa, N. L. F., Weber, F. B., Schmitz, I., Medeiros, L. S., Medeiros, L., Dotto, B. S., Dias, F. R. P., Sovrani, V., & Bobermin, L. D. (2021). Gliotoxicity and Glioprotection: The Dual Role of Glial Cells. *Molecular Neurobiology*, *58*(12), 6577–6592. DOI: 10.1007/s12035-021-02574-9 PMID: 34581988

Raabe, F. J., Slapakova, L., Rossner, M. J., Cantuti-Castelvetri, L., Simons, M., Falkai, P. G., & Schmitt, A. (2019). Oligodendrocytes as a new therapeutic target in schizophrenia: From histopathological findings to neuron-oligodendrocyte interaction. *Cells*, *8*(12), 1496. Advance online publication. DOI: 10.3390/cells8121496 PMID: 31771166

Raffaele, S., Boccazzi, M., & Fumagalli, M. (2021). Oligodendrocyte dysfunction in amyotrophic lateral sclerosis: Mechanisms and therapeutic perspectives. In *Cells* (Vol. 10, Issue 3, pp. 1–32). Multidisciplinary Digital Publishing Institute. DOI: 10.3390/cells10030565

Ransom, B. R., & Ransom, C. B. (2011). Astrocytes: multitalented stars of the central nervous system. Astrocytes: Methods and Protocols, 3-7.

Rashno, M., Sarkaki, A., Farbood, Y., Rashno, M., Khorsandi, L., Naseri, M. K. G., & Dianat, M. (2023). Possible mechanisms involved in the neuroprotective effects of chrysin against mild traumatic brain injury-induced spatial cognitive decline: An in vivo study in a rat model. *Brain Research Bulletin*, *204*(October), 110779. DOI: 10.1016/j.brainresbull.2023.110779 PMID: 37827266

Rawani, N. S., Chan, A. W., Todd, K. G., Baker, G. B., & Dursun, S. M. (2025). The Role of Neuroglia in the Development and Progression of Schizophrenia. *Biomolecules*, *15*(1), 10. Advance online publication. DOI: 10.3390/biom15010010 PMID: 39858403

Riaz, T., Akram, M., & Laila, U. RidaZainab, Muhammad Talha Khalil, Momina Iftikhar, FethiAhmet Ozdemir, Gaweł Sołowski, Marcos Altable, AdonisSfera, Hamza Khalifa Ibrahim, P. P. (2023). *(PDF) Therapeutic applications of glycosides obtained from medicinal plants*. https://www.researchgate.net/publication/373272302_Therapeutic_applications_of_glycosides_obtained_from_medicinal_plants

Rodrigues, A. M. G., Marcilio, F. D. S., Frazão Muzitano, M., & Giraldi-Guimarães, A. (2013). Therapeutic potential of treatment with the flavonoid rutin after cortical focal ischemia in rats. *Brain Research*, *1503*, 53–61. DOI: 10.1016/j.brainres.2013.01.039 PMID: 23370003

Rodríguez-Gómez, J. A., Kavanagh, E., Engskog-Vlachos, P., Engskog, M. K. R., Herrera, A. J., Espinosa-Oliva, A. M., Joseph, B., Hajji, N., Venero, J. L., & Burguillos, M. A. (2020). Microglia: Agents of the CNS Pro-Inflammatory Response. In *Cells* (Vol. 9, Issue 7). NLM (Medline). DOI: 10.3390/cells9071717

Rojas-García, A., Fernández-Ochoa, Á., Cádiz-Gurrea, M. de la L., Arráez-Román, D., & Segura-Carretero, A. (2023). Neuroprotective Effects of Agri-Food By-Products Rich in Phenolic Compounds. *Nutrients*, *15*(2), 449. Advance online publication. DOI: 10.3390/nu15020449 PMID: 36678322

Rudrapal, M., Rakshit, G., Singh, R. P., Garse, S., Khan, J., & Chakraborty, S. (2024). Dietary Polyphenols: Review on Chemistry/Sources, Bioavailability/Metabolism, Antioxidant Effects, and Their Role in Disease Management. *Antioxidants*, *13*(4), 429. Advance online publication. DOI: 10.3390/antiox13040429 PMID: 38671877

Saleem, A., Qurat-ul-Ain, , & Akhtar, M. F. (2022). Alternative Therapy of Psychosis: Potential Phytochemicals and Drug Targets in the Management of Schizophrenia. *Frontiers in Pharmacology*, *13*, 895668. DOI: 10.3389/fphar.2022.895668 PMID: 35656298

Salter, M. W., & Stevens, B. (2017). Microglia emerge as central players in brain disease. *Nature Medicine*, *23*(9), 1018–1027. DOI: 10.1038/nm.4397 PMID: 28886007

Salvesen, L., Winge, K., Brudek, T., Agander, T. K., Løkkegaard, A., & Pakkenberg, B. (2017). Neocortical Neuronal Loss in Patients with Multiple System Atrophy: A Stereological Study. *Cerebral Cortex (New York, N.Y.)*, *27*(1), 400–410. DOI: 10.1093/cercor/bhv228 PMID: 26464477

Sandhir, R., & Mehrotra, A. (2013). Quercetin supplementation is effective in improving mitochondrial dysfunctions induced by 3-nitropropionic acid: Implications in Huntington's disease. *Biochimica et Biophysica Acta. Molecular Basis of Disease*, *1832*(3), 421–430. DOI: 10.1016/j.bbadis.2012.11.018 PMID: 23220257

Sastre, M., Klockgether, T., & Heneka, M. T. (2006). Contribution of inflammatory processes to Alzheimer's disease: Molecular mechanisms. *International Journal of Developmental Neuroscience*, *24*(2–3), 167–176. DOI: 10.1016/j.ijdevneu.2005.11.014 PMID: 16472958

Satapathy, T., & Kumar, D. (2024). A comprehensive review and recent advancement in the application of tannins for treating Parkinson disease. *Pharmacological Research. Modern Chinese Medicine*, *12*, 100499. DOI: 10.1016/j.prmcm.2024.100499

Schubert, J. J., Veronese, M., Marchitelli, L., Bodini, B., Tonietto, M., Stankoff, B., Brooks, D. J., Bertoldo, A., Edison, P., & Turkheimer, F. E. (2019). Dynamic 11C-PIB PET shows cerebrospinal fluid flow alterations in Alzheimer disease and multiple sclerosis. *Journal of Nuclear Medicine*, *60*(10), 1452–1460. DOI: 10.2967/jnumed.118.223834 PMID: 30850505

Schültke, E., Kamencic, H., Zhao, M., Tian, G. F., Baker, A. J., Griebel, R. W., & Juurlink, B. H. J. (2005). Neuroprotection following fluid percussion brain trauma: A pilot study using quercetin. *Journal of Neurotrauma*, *22*(12), 1475–1484. DOI: 10.1089/neu.2005.22.1475 PMID: 16379584

Scuderi, C., Verkhratsky, A., Parpura, V., & Li, B. (2021). Neuroglia in Psychiatric Disorders. *Advances in Neurobiology*, *26*, 3–19. DOI: 10.1007/978-3-030-77375-5_1 PMID: 34888828

Sen, I., Nalini, A., Joshi, N. B., & Joshi, P. G. (2005). Cerebrospinal fluid from amyotrophic lateral sclerosis patients preferentially elevates intracellular calcium and toxicity in motor neurons via AMPA/kainate receptor. *Journal of the Neurological Sciences*, *235*(1–2), 45–54. DOI: 10.1016/j.jns.2005.03.049 PMID: 15936037

Seo, J. S., Mantas, I., Svenningsson, P., & Greengard, P. (2021). Ependymal cells-CSF flow regulates stress-induced depression. *Molecular Psychiatry*, *26*(12), 7308–7315. DOI: 10.1038/s41380-021-01202-1 PMID: 34234280

Serot, J. M., Zmudka, J., & Jouanny, P. (2012). A possible role for CSF turnover and choroid plexus in the pathogenesis of late onset alzheimer's disease. In *Journal of Alzheimer's Disease* (Vol. 30, Issue 1, pp. 17–26). IOS Press. DOI: 10.3233/JAD-2012-111964

Sharma, S. H., Thulasingam, S., & Nagarajan, S. (2017). Terpenoids as anti-colon cancer agents – A comprehensive review on its mechanistic perspectives. *European Journal of Pharmacology*, *795*, 169–178. DOI: 10.1016/j.ejphar.2016.12.008 PMID: 27940056

Shi, Y., Manis, M., Long, J., Wang, K., Sullivan, P. M., Serrano, J. R., Hoyle, R., & Holtzman, D. M. (2019). Microglia drive APOE-dependent neurodegeneration in a tauopathy mouse model. *The Journal of Experimental Medicine*, *216*(11), 2546–2561. DOI: 10.1084/jem.20190980 PMID: 31601677

Siddiqui, T., Khan, M. U., Sharma, V., & Gupta, K. (2024). Terpenoids in essential oils: Chemistry, classification, and potential impact on human health and industry. *Phytomedicine Plus : International Journal of Phytotherapy and Phytopharmacology*, *4*(2), 100549. DOI: 10.1016/j.phyplu.2024.100549

Sinclair, A., Begg, D., Mathai, M., & Weisinger, R. (2007). *Omega 3 fatty acids and the brain : review of studies in depression*.

Singh, D., Agrawal, A., Singal, C. M. S., Pandey, H. S., Seth, P., & Sharma, S. K. (2020). Sinomenine inhibits amyloid beta-induced astrocyte activation and protects neurons against indirect toxicity. *Molecular Brain*, *13*(1), 1–10. DOI: 10.1186/s13041-020-00569-6 PMID: 32127013

Singhal, G., & Baune, B. T. (2017). Microglia: An interface between the loss of neuroplasticity and depression. *Frontiers in Cellular Neuroscience*, *11*(September), 1–16. DOI: 10.3389/fncel.2017.00270 PMID: 28943841

Smeriglio, A., Barreca, D., Bellocco, E., & Trombetta, D. (2017). Proanthocyanidins and hydrolysable tannins: Occurrence, dietary intake and pharmacological effects. *British Journal of Pharmacology, 174*(11), 1244–1262. DOI: 10.1111/bph.13630 PMID: 27646690

Sofroniew, M. V. (2020). Astrocyte Reactivity: Subtypes, States, and Functions in CNS Innate Immunity. *Trends in Immunology, 41*(9), 758–770. DOI: 10.1016/j.it.2020.07.004 PMID: 32819810

Song, J., Du, G., Wu, H., Gao, X., Yang, Z., Liu, B., & Cui, S. (2021). Protective effects of quercetin on traumatic brain injury induced inflammation and oxidative stress in cortex through activating Nrf2/HO-1 pathway. *Restorative Neurology and Neuroscience, 39*(1), 73–84. DOI: 10.3233/RNN-201119 PMID: 33612499

Souza, L. C., de Gomes, M. G., Goes, A. T. R., Del Fabbro, L., Filho, C. B., Boeira, S. P., & Jesse, C. R. (2013). Evidence for the involvement of the serotonergic 5-HT1A receptors in the antidepressant-like effect caused by hesperidin in mice. *Progress in Neuro-Psychopharmacology & Biological Psychiatry, 40*(1), 103–109. DOI: 10.1016/j.pnpbp.2012.09.003 PMID: 22996046

Spaas, J., van Veggel, L., Schepers, M., Tiane, A., van Horssen, J., Wilson, D. M., Moya, P. R., Piccart, E., Hellings, N., Eijnde, B. O., Derave, W., Schreiber, R., & Vanmierlo, T. (2021). Oxidative stress and impaired oligodendrocyte precursor cell differentiation in neurological disorders. In *Cellular and Molecular Life Sciences* (Vol. 78, Issue 10, pp. 4615–4637). Springer. DOI: 10.1007/s00018-021-03802-0

Sriganesh, K., Shanthanna, H., & Busse, J. W. (2016). A brief overview of systematic reviews and meta-analyses. *Indian Journal of Anaesthesia, 60*(9), 689–694. DOI: 10.4103/0019-5049.190628 PMID: 27729699

Subaraja, M., Arokiyaraj, S., & Mathew, P. (2024). Neuroprotective effect of huperzine-A against cadmium chloride-induced Huntington's disease in Drosophila melanogaster model. *Journal of King Saud University. Science, 36*(8), 103319. DOI: 10.1016/j.jksus.2024.103319

Subedi, L., & Yumnam, S. (2021). Terpenoids from Abies holophylla Attenuate LPS-Induced Neuroinflammation in Microglial Cells by Suppressing the JNK-Related Signaling Pathway. *International Journal of Molecular Sciences 2021, Vol. 22, Page 965, 22*(2), 965. DOI: 10.3390/ijms22020965

Suganya, S. N., & Sumathi, T. (2017). Effect of rutin against a mitochondrial toxin, 3-nitropropionicacid induced biochemical, behavioral and histological alterations-a pilot study on Huntington's disease model in rats. *Metabolic Brain Disease, 32*(2), 471–481. DOI: 10.1007/s11011-016-9929-4 PMID: 27928694

Sun, G., Miao, Z., Ye, Y., Zhao, P., Fan, L., Bao, Z., Tu, Y., Li, C., Chao, H., Xu, X., & Ji, J. (2020). Curcumin alleviates neuroinflammation, enhances hippocampal neurogenesis, and improves spatial memory after traumatic brain injury. *Brain Research Bulletin, 162*, 84–93. DOI: 10.1016/j.brainresbull.2020.05.009 PMID: 32502596

Sun, Y., Tong, H., Yang, T., Liu, L., Li, X. J., & Li, S. (2022). Insights into White Matter Defect in Huntington's Disease. In *Cells* (Vol. 11, Issue 21, p. 3381). Multidisciplinary Digital Publishing Institute. DOI: 10.3390/cells11213381

Sunhe, Y.-X., Zhang, Y.-H., Fu, R.-J., Xu, D.-Q., & Tang, Y.-P. (2024). Neuroprotective effect and preparation methods of berberine. *Frontiers in Pharmacology, 15*(September), 1–23. DOI: 10.3389/fphar.2024.1429050 PMID: 39309003

Tabatabaee, M. (2024). Microbiome-Glia Crosstalk: Bridging the Communication Divide in the Central Nervous System. In *Neuroglia* (Vol. 5, Issue 2, pp. 89–104). Multidisciplinary Digital Publishing Institute. DOI: 10.3390/neuroglia5020007

Taib, T., Leconte, C., Van Steenwinckel, J., Cho, A. H., Palmier, B., Torsello, E., Kuen, R. L., Onyeomah, S., Ecomard, K., Benedetto, C., Coqueran, B., Novak, A. C., Deou, E., Plotkine, M., Gressens, P., Marchand-Leroux, C., & Besson, V. C. (2017). Neuroinflammation, myelin and behavior: Temporal patterns following mild traumatic brain injury in mice. *PLoS One, 12*(9), e0184811. Advance online publication. DOI: 10.1371/journal.pone.0184811 PMID: 28910378

Takano, T., & Soderling, S. H. (2021). Tripartite synaptomics: Cell-surface proximity labeling in vivo. *Neuroscience Research, 173*, 14–21. DOI: 10.1016/j.neures.2021.05.002 PMID: 34019951

Takano, T., Tian, G. F., Peng, W., Lou, N., Libionka, W., Han, X., & Nedergaard, M. (2006). Astrocyte-mediated control of cerebral blood flow. *Nature Neuroscience, 9*(2), 260–267. DOI: 10.1038/nn1623 PMID: 16388306

Tamberi, L., Belloni, A., Pugnaloni, A., Rippo, M. R., Olivieri, F., Procopio, A. D., & Bronte, G. (2024). The Influence of Myeloid-Derived Suppressor Cell Expansion in Neuroinflammation and Neurodegenerative Diseases. In *Cells* (Vol. 13, Issue 7). Multidisciplinary Digital Publishing Institute (MDPI). DOI: 10.3390/cells13070643

Tan, Y., Zheng, Y., Xu, D., Sun, Z., Yang, H., & Yin, Q. (2021). Galectin-3: a key player in microglia-mediated neuroinflammation and Alzheimer's disease. In *Cell and Bioscience* (Vol. 11, Issue 1, pp. 1–13). BioMed Central. DOI: 10.1186/s13578-021-00592-7

Tang, Y., & Le, W. (2016). Differential Roles of M1 and M2 Microglia in Neurodegenerative Diseases. In *Molecular Neurobiology* (Vol. 53, Issue 2, pp. 1181–1194). Humana Press Inc. DOI: 10.1007/s12035-014-9070-5

Tavan, M., Hanachi, P., de la Luz Cádiz-Gurrea, M., Segura Carretero, A., & Mirjalili, M. H. (2024). Natural Phenolic Compounds with Neuroprotective Effects. *Neurochemical Research, 49*(2), 306–326. DOI: 10.1007/s11064-023-04046-z PMID: 37940760

Thakare, V. N., Aswar, M. K., Kulkarni, Y. P., Patil, R. R., & Patel, B. M. (2017). Silymarin ameliorates experimentally induced depressive like behavior in rats: Involvement of hippocampal BDNF signaling, inflammatory cytokines and oxidative stress response. *Physiology & Behavior, 179*, 401–410. DOI: 10.1016/j.physbeh.2017.07.010 PMID: 28711395

Thau, L., Reddy, V., & Singh, P. (2022). Anatomy, Central Nervous System. *BMJ (Clinical Research Ed.), 1*(4293), 478–478. DOI: 10.1136/bmj.1.4293.478 PMID: 31194336

Traiffort, E., Morisset-Lopez, S., Moussaed, M., & Zahaf, A. (2021). Defective oligodendroglial lineage and demyelination in amyotrophic lateral sclerosis. In *International Journal of Molecular Sciences* (Vol. 22, Issue 7, p. 3426). Multidisciplinary Digital Publishing Institute. DOI: 10.3390/ijms22073426

Trépanier, M. O., Hopperton, K. E., Mizrahi, R., Mechawar, N., & Bazinet, R. P. (2016). Postmortem evidence of cerebral inflammation in schizophrenia: A systematic review. *Molecular Psychiatry*, *21*(8), 1009–1026. DOI: 10.1038/mp.2016.90 PMID: 27271499

Tsunemi, T., Ishiguro, Y., Yoroisaka, A., Valdez, C., Miyamoto, K., Ishikawa, K., Saiki, S., Akamatsu, W., Hattori, N., & Krainc, D. (2020). Astrocytes protect human dopaminergic neurons from α-synuclein accumulation and propagation. *The Journal of Neuroscience : The Official Journal of the Society for Neuroscience*, *40*(45), 8618–8628. DOI: 10.1523/JNEUROSCI.0954-20.2020 PMID: 33046546

Vaessen, M. J., Saj, A., Lovblad, K. O., Gschwind, M., & Vuilleumier, P. (2016). Structural white-matter connections mediating distinct behavioral components of spatial neglect in right brain-damaged patients. *Cortex*, *77*, 54–68. DOI: 10.1016/j.cortex.2015.12.008 PMID: 26922504

Vaillant-Beuchot, L., Mary, A., Pardossi-Piquard, R., Bourgeois, A., Lauritzen, I., Eysert, F., Kinoshita, P. F., Cazareth, J., Badot, C., Fragaki, K., Bussiere, R., Martin, C., Mary, R., Bauer, C., Pagnotta, S., Paquis-Flucklinger, V., Buée-Scherrer, V., Buée, L., Lacas-Gervais, S., & Chami, M. (2021). Accumulation of amyloid precursor protein C-terminal fragments triggers mitochondrial structure, function, and mitophagy defects in Alzheimer's disease models and human brains. *Acta Neuropathologica*, *141*(1), 39–65. DOI: 10.1007/s00401-020-02234-7 PMID: 33079262

Van Kesteren, C. F. M. G., Gremmels, H., De Witte, L. D., Hol, E. M., Van Gool, A. R., Falkai, P. G., Kahn, R. S., & Sommer, I. E. C. (2017). Immune involvement in the pathogenesis of schizophrenia: A meta-analysis on postmortem brain studies. *Translational Psychiatry*, *7*(3), e1075–e11. DOI: 10.1038/tp.2017.4 PMID: 28350400

Vanes, L. D., Mouchlianitis, E., Barry, E., Patel, K., Wong, K., & Shergill, S. S. (2019). Cognitive correlates of abnormal myelination in psychosis. *Scientific Reports*, *9*(1), 1–9. DOI: 10.1038/s41598-019-41679-z PMID: 30914748

Vasile, F., Dossi, E., & Rouach, N. (2017). Human astrocytes: Structure and functions in the healthy brain. *Brain Structure & Function*, *222*(5), 2017–2029. DOI: 10.1007/s00429-017-1383-5 PMID: 28280934

Vauzour, D. (2012). Dietary Polyphenols as Modulators of Brain Functions: Biological Actions and Molecular Mechanisms Underpinning Their Beneficial Effects. *Oxidative Medicine and Cellular Longevity*, *914273*, 1–16. Advance online publication. DOI: 10.1155/2012/914273 PMID: 22701758

Vauzour, D., Vafeiadou, K., Rodriguez-Mateos, A., Rendeiro, C., & Spencer, J. P. E. (2008). The neuroprotective potential of flavonoids: A multiplicity of effects. *Genes & Nutrition*, *3*(3–4), 115–126. DOI: 10.1007/s12263-008-0091-4 PMID: 18937002

Vellucci, L., Mazza, B., Barone, A., Nasti, A., De Simone, G., Iasevoli, F., & de Bartolomeis, A. (2025). The Role of Astrocytes in the Molecular Pathophysiology of Schizophrenia: Between Neurodevelopment and Neurodegeneration. *Biomolecules*, *15*(5), 1–36. DOI: 10.3390/biom15050615 PMID: 40427508

Verkhratsky, A., Hol, E. M., de Witte, L. D., & Aronica, E. (2025). Neuroglia in the healthy brain. *Handbook of Clinical Neurology*, *209*, 1–5. DOI: 10.1016/B978-0-443-19104-6.00008-5 PMID: 40122619

Verkhratsky, A., Matteoli, M., Parpura, V., Mothet, J., & Zorec, R. (2016). Astrocytes as secretory cells of the central nervous system: Idiosyncrasies of vesicular secretion. *The EMBO Journal, 35*(3), 239–257. DOI: 10.15252/embj.201592705 PMID: 26758544

Verkhratsky, A., & Nedergaard, M. (2018). Physiology of astroglia. *Physiological Reviews, 98*(1), 239–389. DOI: 10.1152/physrev.00042.2016 PMID: 29351512

Verkhratsky, A., Rodríguez, J. J., & Parpura, V. (2013). Astroglia in neurological diseases. *Future Neurology, 8*(2), 149–158. DOI: 10.2217/fnl.12.90 PMID: 23658503

Verkhratsky, A., & Semyanov, A. (2025). *Physiology of neuroglia of the central nervous system. 209*, 69–91. DOI: 10.1016/B978-0-443-19104-6.00005-X

Vijayalakshmi, K., Alladi, P. A., Sathyaprabha, T. N., Subramaniam, J. R., Nalini, A., & Raju, T. R. (2009). Cerebrospinal Fluid from sporadic Amyotrophic Lateral Sclerosis patients induces degeneration of a cultured motor neuron cell line. *Brain Research, 1263*, 122–133. DOI: 10.1016/j.brainres.2009.01.041 PMID: 19368830

Visanji, N. P., Brooks, P. L., Hazrati, L. N., & Lang, A. E. (2014). The prion hypothesis in Parkinson's disease: Braak to the future. In *Acta Neuropathologica Communications* (Vol. 2, Issue 1). BioMed Central Ltd. DOI: 10.1186/2051-5960-1-2

Wang, J., Zhen, Y., Yang, J., Yang, S., & Zhu, G. (2024). Recognizing Alzheimer's disease from perspective of oligodendrocytes: Phenomena or pathogenesis? In *CNS Neuroscience and Therapeutics* (Vol. 30, Issue 3, p. e14688). John Wiley and Sons Inc. DOI: 10.1111/cns.14688

Wang, P., Jiang, L., Zhou, N., Zhou, H., Liu, H., Zhao, W., Zhang, H., Zhang, X., & Hu, Z. (2018). Resveratrol ameliorates autophagic flux to promote functional recovery in rats after spinal cord injury. *Oncotarget, 9*(9), 8427–8440. DOI: 10.18632/oncotarget.23877 PMID: 29492205

Wang, X., & Jia, J. (2023). Magnolol improves Alzheimer's disease-like pathologies and cognitive decline by promoting autophagy through activation of the AMPK/mTOR/ULK1 pathway. *Biomedicine and Pharmacotherapy, 161*, 114473. Advance online publication. DOI: 10.1016/j.biopha.2023.114473 PMID: 36889111

Wang, Z., Huang, J., Liu, C., Liu, L., Shen, Y., Shen, C., & Liu, C. (2019). BAF45D Downregulation in Spinal Cord Ependymal Cells Following Spinal Cord Injury in Adult Rats and Its Potential Role in the Development of Neuronal Lesions. *Frontiers in Neuroscience, 13*, 471535. DOI: 10.3389/fnins.2019.01151 PMID: 31736692

Wei, D. C., & Morrison, E. H. (2023). *Histology, Astrocytes*. StatPearls.

What, H., Projects, N., Link, B., Plan, E., Management, E., & Identify, S. (2016). Festivals Toolkit. *Health & Safety : Risk Assessment & Management, 63*(8), 1429–1451. DOI: 10.1002/glia.22859.NG2-glia

White, R. E., & Jakeman, L. B. (2008). Don't fence me in: Harnessing the beneficial roles of astrocytes for spinal cord repair. *Restorative Neurology and Neuroscience, 26*(2–3), 197–214. DOI: 10.3233/RNN-2008-00442 PMID: 18820411

Williamson, M. R., Kwon, W., Woo, J., Ko, Y., Maleki, E., Yu, K., Murali, S., Sardar, D., & Deneen, B. (2024). Learning-associated astrocyte ensembles regulate memory recall. *Nature 2024 637:8045, 637*(8045), 478–486. DOI: 10.1038/s41586-024-08170-w

Wilson, G. R., Tan, J. T., Brody, K. M., Taylor, J. M., Delatycki, M. B., & Lockhart, P. J. (2009). Expression and localization of the Parkin Co-Regulated Gene in mouse CNS suggests a role in ependymal cilia function. *Neuroscience Letters*, *460*(1), 97–101. DOI: 10.1016/j.neulet.2009.05.043 PMID: 19463890

Wink, M. (2000). Interference of alkaloids with neuroreceptors and ion channels. *Studies in Natural Products Chemistry, 21*(PART B), 3–122. DOI: 10.1016/S1572-5995(00)80004-6

Wright-Jin, E. C., & Gutmann, D. H. (2019). Microglia as Dynamic Cellular Mediators of Brain Function. *Trends in Molecular Medicine*, *25*(11), 967–979. DOI: 10.1016/j.molmed.2019.08.013 PMID: 31597593

Wu, Y., Chen, Q., Wen, B., Wu, N., He, B., & Chen, J. (2021). Berberine Reduces Aβ42 Deposition and Tau Hyperphosphorylation via Ameliorating Endoplasmic Reticulum Stress. *Frontiers in Pharmacology*, *12*(July), 1–14. DOI: 10.3389/fphar.2021.640758 PMID: 34349640

Wu, Y., Zhong, L., Yu, Z., & Qi, J. (2019). Anti-neuroinflammatory effects of tannic acid against lipopolysaccharide-induced BV2 microglial cells via inhibition of NF-κB activation. *Drug Development Research*, *80*(2), 262–268. DOI: 10.1002/ddr.21490 PMID: 30724376

Xie, S., & Li, F. (2024). Ependymal cells: roles in central nervous system infections and therapeutic application. *Journal of Neuroinflammation 2024 21:1, 21*(1), 1–14. DOI: 10.1186/s12974-024-03240-2

Xu, B., Bai, L., Chen, L., Tong, R., Feng, Y., & Shi, J. (2022). Terpenoid natural products exert neuroprotection via the PI3K/Akt pathway. *Frontiers in Pharmacology*, *13*, 1036506. DOI: 10.3389/fphar.2022.1036506 PMID: 36313360

Xu, J., Wang, H., Lu, X., Ding, K., Zhang, L., He, J., Wei, W., & Wu, Y. (2014). Posttraumatic administration of luteolin protects mice from traumatic brain injury: Implication of autophagy and inflammation. *Brain Research*, *1582*, 237–246. DOI: 10.1016/j.brainres.2014.07.042 PMID: 25093609

Xu, L., Wang, J., Ding, Y., Wang, L., & Zhu, Y. J. (2022). Current Knowledge of Microglia in Traumatic Spinal Cord Injury. In *Frontiers in Neurology* (Vol. 12, p. 796704). Frontiers Media S.A., DOI: 10.3389/fneur.2021.796704

Xu, P., Wang, S., Yu, X., Su, Y., Wang, T., Zhou, W., Zhang, H., Wang, Y., & Liu, R. (2014). Rutin improves spatial memory in Alzheimer's disease transgenic mice by reducing Aβ oligomer level and attenuating oxidative stress and neuroinflammation. *Behavioural Brain Research*, *264*, 173–180. DOI: 10.1016/j.bbr.2014.02.002 PMID: 24512768

Xu, Q., Yi, L. T., Pan, Y., Wang, X., Li, Y. C., Li, J. M., Wang, C. P., & Kong, L. D. (2008). Antidepressant-like effects of the mixture of honokiol and magnolol from the barks of Magnolia officinalis in stressed rodents. *Progress in Neuro-Psychopharmacology & Biological Psychiatry*, *32*(3), 715–725. DOI: 10.1016/j.pnpbp.2007.11.020 PMID: 18093712

Xu, Y. Q., Jin, S. J., Liu, N., Li, Y. X., Zheng, J., Ma, L., Du, J., Zhou, R., Zhao, C. J., Niu, Y., Sun, T., & Yu, J. Q. (2014). Aloperine attenuated neuropathic pain induced by chronic constriction injury via anti-oxidation activity and suppression of the nuclear factor kappa B pathway. *Biochemical and Biophysical Research Communications*, *451*(4), 568–573. DOI: 10.1016/j.bbrc.2014.08.025 PMID: 25128276

Xue, J., Tao, K., Wang, W., & Wang, X. (2024). What Can Inflammation Tell Us about Therapeutic Strategies for Parkinson's Disease? *International Journal of Molecular Sciences*, *25*(3), 1641. Advance online publication. DOI: 10.3390/ijms25031641 PMID: 38338925

Yan, S. X., Lang, J. L., Song, Y. Y., Wu, Y. Z., Lv, M. H., Zhao, X., Liu, Y. H., & Xu, C. Y. (2015). Studies on anti-depressant activity of four flavonoids isolated from Apocynum venetum linn (Apocynaceae) leaf in mice. *Tropical Journal of Pharmaceutical Research*, *14*(12), 2269–2277. DOI: 10.4314/tjpr.v14i12.17

Yang, J., Xiong, L. L., Wang, Y. C., He, X., Jiang, L., Fu, S. J., Han, X. F., Liu, J., & Wang, T. H. (2018). Oligodendrocyte precursor cell transplantation promotes functional recovery following contusive spinal cord injury in rats and is associated with altered MicroRNA expression. *Molecular Medicine Reports*, *17*(1), 771–782. DOI: 10.3892/mmr.2017.7957 PMID: 29115639

Yang, W., Chen, X., Li, Y., Guo, S., Wang, Z., & Yu, X. (2020). Advances in Pharmacological Activities of Terpenoids. *Natural Product Communications*, *15*(3), 1934578X20903555. Advance online publication. DOI: 10.1177/1934578X20903555

Yi, L. T., Li, J., Li, H. C., Su, D. X., Quan, X. B., He, X. C., & Wang, X. H. (2012). Antidepressant-like behavioral, neurochemical and neuroendocrine effects of naringenin in the mouse repeated tail suspension test. *Progress in Neuro-Psychopharmacology & Biological Psychiatry*, *39*(1), 175–181. DOI: 10.1016/j.pnpbp.2012.06.009 PMID: 22709719

Yi, L. T., Xu, H. L., Feng, J., Zhan, X., Zhou, L. P., & Cui, C. C. (2011). Involvement of monoaminergic systems in the antidepressant-like effect of nobiletin. *Physiology & Behavior*, *102*(1), 1–6. DOI: 10.1016/j.physbeh.2010.10.008 PMID: 20951716

Yi, S., Jiang, X., Tang, X., Li, Y., Xiao, C., Zhang, J., & Zhou, T. (2020). IL-4 and IL-10 promotes phagocytic activity of microglia by up-regulation of TREM2. *Cytotechnology*, *72*(4), 589–602. DOI: 10.1007/s10616-020-00409-4 PMID: 32623621

Yin, F. (2023). Lipid metabolism and Alzheimer's disease: clinical evidence, mechanistic link and therapeutic promise. In *FEBS Journal* (Vol. 290, Issue 6, pp. 1420–1453). John Wiley and Sons Inc. DOI: 10.1111/febs.16344

Yin, X., Yin, Y., Le Cao, F., Chen, Y. F., Peng, Y., Hou, W. G., Sun, S. K., & Luo, Z. J. (2012). Tanshinone IIA attenuates the inflammatory response and apoptosis after traumatic injury of the spinal cord in adult rats. *PLoS One*, *7*(6), 1–10. DOI: 10.1371/journal.pone.0038381 PMID: 22675554

Yoon, C. S. (2025). Natural Products in the Treatment of Neuroinflammation at Microglia: Recent Trend and Features. *Cells 2025, Vol. 14, Page 571*, *14*(8), 571. DOI: 10.3390/cells14080571

Yu, X., Jiang, X., Zhang, X., Chen, Z., Xu, L., Chen, L., Wang, G., & Pan, J. (2016). The effects of fisetin on lipopolysaccharide-induced depressive-like behavior in mice. *Metabolic Brain Disease*, *31*(5), 1011–1021. DOI: 10.1007/s11011-016-9839-5 PMID: 27209403

Yu, Y., Liao, X., Xie, X., Li, Q., Chen, X., & Liu, R. (2024). The role of neuroglial cells communication in ischemic stroke. *Brain Research Bulletin, 209*, 110910. DOI: 10.1016/j.brainresbull.2024.110910 PMID: 38423190

Yuan, M., & Wu, H. (2022). Astrocytes in the Traumatic Brain Injury: the Good and the Bad. In *Experimental Neurology* (Vol. 348, p. 113943). Academic Press. DOI: 10.1016/j.expneurol.2021.113943

Zbieta, E., & Ebas, R. (2025). Role of Flavonoids in Protecting Against Neurodegenerative Diseases—Possible Mechanisms of Action. *International Journal of Molecular Sciences 2025, Vol. 26, Page 4763, 26*(10), 4763. DOI: 10.3390/ijms26104763

Zengin, G., Locatelli, M., Carradori, S., Mocan, A. M., & Aktumsek, A. (2016). Total Phenolics, Flavonoids, Condensed Tannins Content of Eight Centaurea Species and Their Broad Inhibitory Activities against Cholinesterase, Tyrosinase, α-Amylase and α-Glucosidase. *Notulae Botanicae Horti Agrobotanici Cluj-Napoca, 44*(1), 195–200. DOI: 10.15835/nbha44110259

Zhang, C., Qiu, M., & Fu, H. (2024). Oligodendrocytes in central nervous system diseases: the effect of cytokine regulation. In *Neural Regeneration Research* (Vol. 19, Issue 10, pp. 2132–2143). Wolters Kluwer Medknow Publications. DOI: 10.4103/1673-5374.392854

Zhang, F., Li, F., & Chen, G. (2014). Neuroprotective effect of apigenin in rats after contusive spinal cord injury. *Neurological Sciences, 35*(4), 583–588. DOI: 10.1007/s10072-013-1566-7 PMID: 24166720

Zhang, H., Zhang, X., Chai, Y., Wang, Y., Zhang, J., & Chen, X. (2025). Astrocyte-mediated inflammatory responses in traumatic brain injury: mechanisms and potential interventions. In *Frontiers in Immunology* (Vol. 16, p. 1584577). Frontiers Media SA. DOI: 10.3389/fimmu.2025.1584577

Zhang, L., Wang, H., Zhou, Y., Zhu, Y., & Fei, M. (2018). Fisetin alleviates oxidative stress after traumatic brain injury via the Nrf2-ARE pathway. *Neurochemistry International, 118*(February), 304–313. DOI: 10.1016/j.neuint.2018.05.011 PMID: 29792955

Zhang, S., Liu, Y., Zhao, Z., & Xue, Y. (2010). Effects of green tea polyphenols on caveolin-1 of microvessel fragments in rats with cerebral ischemia. *Neurological Research, 32*(9), 963–970. DOI: 10.1179/016164110X12700393823570 PMID: 20444327

Zhang, X., Du, L., Zhang, W., Yang, Y., Zhou, Q., & Du, G. (2017). Therapeutic effects of baicalein on rotenone-induced Parkinson's disease through protecting mitochondrial function and biogenesis. *Scientific Reports, 7*(1), 1–14. DOI: 10.1038/s41598-017-07442-y PMID: 28855526

Zhang, X., Zhang, R., Nisa Awan, M. U., & Bai, J. (2022). The Mechanism and Function of Glia in Parkinson's Disease. In *Frontiers in Cellular Neuroscience* (Vol. 16, p. 903469). Frontiers Media S.A., DOI: 10.3389/fncel.2022.903469

Zhang, Z., Hamada, H., & Gerk, P. M. (2019). Selectivity of dietary phenolics for inhibition of human monoamine oxidases A and B. *BioMed Research International, 2019*, 1–12. Advance online publication. DOI: 10.1155/2019/8361858 PMID: 30809547

Zhao, J. Y., Zhou, Y., Zhou, C. W., & Zhan, K. Bin, Yang, M., Wen, M., & Zhu, L. Q. (2025). Revisiting the critical roles of reactive microglia in traumatic brain injury. In *International journal of surgery (London, England)* (Vol. 111, Issue 6, pp. 3942–3978). Int J Surg. DOI: 10.1097/JS9.0000000000002420

Zhou, B., Zhu, Z., Ransom, B. R., & Tong, X. (2020). Oligodendrocyte lineage cells and depression. *Molecular Psychiatry 2020 26:1, 26*(1), 103–117. DOI: 10.1038/s41380-020-00930-0

Zhou, L., Fan, L., Yi, X., Zhou, Z., Liu, C., Fu, R., Dai, C., Wang, Z., Chen, X., Yu, P., Chen, D., Tan, G., Wang, Q., & Ning, C. (2018). Soft Conducting Polymer Hydrogels Cross-Linked and Doped by Tannic Acid for Spinal Cord Injury Repair. *ACS Nano, 12*(11), 10957–10967. DOI: 10.1021/acsnano.8b04609 PMID: 30285411

Zhou, Y., Dhaher, R., Parent, M., Hu, Q. X., Hassel, B., Yee, S. P., Hyder, F., Gruenbaum, S. E., Eid, T., & Danbolt, N. C. (2019). Selective deletion of glutamine synthetase in the mouse cerebral cortex induces glial dysfunction and vascular impairment that precede epilepsy and neurodegeneration. *Neurochemistry International, 123*, 22–33. DOI: 10.1016/j.neuint.2018.07.009 PMID: 30053506

Zhou, Y., Shao, A., Yao, Y., Tu, S., Deng, Y., & Zhang, J. (2020). Dual roles of astrocytes in plasticity and reconstruction after traumatic brain injury. In *Cell Communication and Signaling* (Vol. 18, Issue 1, pp. 1–16). BioMed Central. DOI: 10.1186/s12964-020-00549-2

Zhuo, C., Tian, H., Song, X., Jiang, D., Chen, G., Cai, Z., Ping, J., Cheng, L., Zhou, C., & Chen, C. (2023). Microglia and cognitive impairment in schizophrenia: Translating scientific progress into novel therapeutic fiinterventions. *Schizophrenia, 9*(1), 42. Advance online publication. DOI: 10.1038/s41537-023-00370-z PMID: 37429882

Zou, P., Liu, X., Li, G., & Wang, Y. (2018). Resveratrol pretreatment attenuates traumatic brain injury in rats by suppressing NLRP3 inflammasome activation via SIRT1. *Molecular Medicine Reports, 17*(2), 3212–3217. DOI: 10.3892/mmr.2017.8241 PMID: 29257276

Zou, P., Wu, C., Liu, T. C. Y., Duan, R., & Yang, L. (2023). Oligodendrocyte progenitor cells in Alzheimer's disease: from physiology to pathology. In *Translational Neurodegeneration* (Vol. 12, Issue 1, pp. 1–20). BioMed Central. DOI: 10.1186/s40035-023-00385-7

Chapter 2
Beyond the Surface: Polyphenols as Cellular Reprogrammers in Neurodegenerative Diseases

Laaraib Nawaz
https://orcid.org/0000-0002-9420-632X
The University of Lahore, Pakistan

Shahzad Irfan
https://orcid.org/0000-0003-3260-4215
Government College University, Faisalabad, Pakistan

Qammar Abbas
The University of Lahore, Pakistan

Nimra Aziz
The University of Lahore, Pakistan

Shazia Batool
The University of Lahore, Pakistan

ABSTRACT

Polyphenols have been extensively acknowledged for their anti-inflammatory and antioxidant properties. Nonetheless, the recent evidence is suggestive of a broader role of polyphenols as cellular programming agents, with the ability to influence fundamental biological pathways and core cellular processes. These processes include epigenetic alterations, such as DNA methylation and histone acetylation, synaptic plasticity, mitochondrial biogenesis, metabolic pathways, and cellular senescence. Polyphenols enhance neuronal survival, promote neurogenesis, and support cognitive resilience by targeting transcription factors and signaling cascades. Their cellular reprogramming potential marks a paradigm shift, from merely offering symptomatic relief to modification of the underlying disease process. This emerging perspective positions polyphenols not only as neuroprotective agents but also as promising tools in advancing precision medicine approaches for neurodegenerative disorders.

DOI: 10.4018/979-8-3373-4139-2.ch002

1. BACKGROUND

Neurodegenerative diseases (NDDs) represent one of the most significant challenges to contemporary medicine, currently affecting approximately 50 million people globally, with a disproportionate impact on the aged population. This challenge will further intensify as statistical reports indicate that the global aging population of 60 years and above will increase from 703 million (9%) in 2019 to 1.5 billion (16%) by 2050, suggesting a substantial rise in the prevalence of NDDs in the coming decades (Kumar et al., 2020). Despite significant advancements in medical science that improved the quality of life and enabled the treatment of previously untreatable conditions, neuronal regeneration remains largely underdeveloped. Existing NDD therapies are confined and often show less efficacy, underscoring the pressing need for the clinical and scientific communities to develop proficient therapeutic strategies. Available regenerative treatments, including allopathic medicine, surgery, organ transplantation, and stem cell therapy, often prove insufficient restoration of damaged tissue or in effectively treating NDDs. The NDDs, including Alzheimer's (AD), Parkinson's (PD), and Huntington's, are multifaceted conditions characterized by the progressive deterioration of neuronal structural integrity and functional loss, consequently leading to cognitive impairment, motor dysfunctions, and compromised quality of life (Mittal & Agrawal, 2024). The multifactorial nature of disease onset, along with overlapping triggers and diverse neurodegenerative mechanisms, including abnormal protein dynamics, defective protein clearance, oxidative stress, mitochondrial dysfunction, disrupted bioenergetics, and heavy metal toxicity, poses a significant barrier to effective therapies. Thus, lacking a definite cure and offering only symptomatic relief (Chopade et al., 2023) to these diseases remain a challenge to neuroscience and public health (Rocca, 2018).

In the pursuit of disease-modifying therapies, growing attention is being directed towards natural compounds capable of operating at diverse biological processes (Nawaz et al., 2024). Polyphenols traditionally valued for their antioxidant activities are now gaining recognition for their capacity to modulate cell fate decisions. Their pleiotropic characteristics enable their interaction with signaling pathways and transcriptional regulators involved in neurodevelopment, inflammation, and cellular homeostasis (Kishimoto et al., 2013). What distinguishes polyphenols is their emerging capacity to modulate fundamental mechanisms of neurodegeneration beyond merely offering protection to cellular trajectories. Latest studies suggest that polyphenols may exert epigenetic modulation, reprogram metabolic pathways, and enhance neuronal resilience by targeting histone deacetylases, mitochondrial regulators, and sirtuins (Ayissi et al., 2014). Cellular reprogramming entails the modulation of transcription factors, epigenetic modifications, or signaling cascades to revert differentiated cells to a more regenerative state. This process involves dedifferentiation, transdifferentiation, and partial reprogramming. Given the brain's limited ability to regenerate, cellular programming presents a promising approach to restore lost neurons, reverse senescence-related cellular dysfunction, and stimulate the brain's intrinsic repair process (Yang et al., 2022). In this context, the cellular reprogramming ability of polyphenols, which modulates epigenetic factors or enhances antioxidant potential, opens up a new avenue for non-invasive, natural reprogramming of dietary and plant-derived compounds (Ayissi et al., 2014). This chapter focuses on the emerging role of polyphenols as cellular reprogramming agents in NDDs, highlighting their ability to modulate gene expression, epigenetic mechanisms, and neural regeneration.

1.1. The Pathogenic Landscape: Systems Requiring Programming

The NDDs are among the primary drivers of worldwide disability and morbidity. Their significant impact on the aging population has made them a major focus of research and healthcare. These diseases are generally characterized by a gradual loss of neuronal structural integrity and function, leading to atrophy of the brain (Liu et al., 2018). As these diseases are multifactorial, the following are core contributing factors of NDDs pathogenesis:

1.1.1. Oxidative Stress: A Consequence of Failed Redox Programming

The human brain is extremely sensitive to free radicals and oxidative stress due to its exceptional metabolic activities. To sustain the ionic gradient and facilitate synaptic transmission, the human brain requires a substantial amount of energy (Harris et al., 2012) and primarily depends on mitochondria to generate the required energy. This intense synaptic transmission requires an action potential, which causes an intracellular elevation of calcium (Ca^{2+}; 0.001μm - 100μm), followed by activation of neuronal nitric oxide synthase. It leads to nitric oxide (NO) formation, which binds to cytochrome c oxidase and stops mitochondrial respiration. In compensation, mitochondria try to buffer elevated intracellular Ca^{2+} levels as the overload induces prolonged opening of the mitochondrial permeability transition pore (mPTP), which can disrupt mitochondrial function, inhibit ATP generation, release of apoptotic proteins, and ultimately cell death (Jurcau, 2021). Secondly, the brain has a relatively weak antioxidant defense system, as neuronal cells have about 50 times less catalase than liver cells and their cytosolic glutathione levels are 50% lower in neurons in comparison to other cells (Cobley et al., 2018), potentially impairing peroxiredoxin activity (Fisher, 2017). Apart from a weak antioxidant defense system, free radicals are frequently generated in the brain by microglia (Mander et al., 2006), which produce superoxides by NADPH oxidase isoforms (Colton & Gilbert, 1987), and the metabolism of neurotransmitters like dopamine (E Edmondson, 2014). Additionally, the brain is enriched with cholesterol, which can also undergo an auto-oxidative pathway, further contributing to oxidative stress (Zerbinati & Iuliano, 2017). Lastly, brain cell membranes are abundant in polyunsaturated fatty acids, making them particularly vulnerable to lipid peroxidation triggered through free radical attack (Cobley et al., 2018). This oxidative stress not only damages neurons directly but also disrupts the epigenetic landscape, for instance, by oxidizing non-coding RNAs and microRNAs, ultimately affecting brain development and plasticity. Lack of protective coverage of histones around these molecules makes them susceptible to free radicals (Simms & Zaher, 2016). In addition, oxidation of mRNA leads to the production of truncated and mutated proteins, which are prone to misfolding (Tanaka et al., 2007). This has sparked significant interest in polyphenols, as these are natural plant-derived antioxidant compounds with potent neuroprotective properties. Some of these flavonoids and non-flavonoids can penetrate the blood-brain barrier and accumulate in brain tissues. Researches also indicate that consumption of polyphenols may improve memory, cognition, and mood in aging people (Grabska-Kobyłecka et al., 2023). This intricate link between oxidative stress and epigenetic modifications provides a critical therapeutic window. In the next sections, we will explore how polyphenols are uniquely equipped to exploit this window both as antioxidants and as epigenetic modifications.

1.1.2. Inflammation: The Need for Microglial Phenotype Reprogramming

Microglia are the brain's resident immune cells, constantly monitoring their environment. In a healthy brain, microglia are resting or quiescent. However, in pathological conditions, they are first to be activated as a result of tissue damage or infection. Once activated, microglia change their morphological features and function (Davis & Carson, 2012). Their core function is to recognize and phagocytose pathogens, remodel synapses, and present antigens. Activation of microglial cells is divided into two types, which are classical M1 and alternative M2 activation. M1 activation is characterized by hyper-ramified or ameboid microglial cells, which synthesize proinflammatory cytokines, i.e., interleukin-1β (IL-1β), tumor necrosis factor-α (TNF-α), IL-6, nitric oxide (NO), and superoxide radicals (Guo et al., 2022; Tang & Le, 2016). Whereas activation of M2 microglia is associated with the release of anti-inflammatory molecules, i.e., IL-10, insulin-like growth factor-1 (IGF-1), transforming growth factor-β (TGF-β), and neurotrophic factors, which assist in neuronal healing and arrest neuronal injury. The severity and type of injury can directly influence the activation and transformation of these microglial phenotypes (Hu et al., 2015; Réus et al., 2015). Release of chemokines, free radicals, and cytokines by these activated microglia contributes to neuronal insult, synaptic delay, and aggravation of Aβ pathology. With time, as microglia become dysfunctional, their phagocytic activity is reduced, leading to the accumulation of Aβ, ultimately generating a vicious cycle of inflammation and neurodegeneration. Inflammation weakens the blood-brain barrier, leading to the entry of peripheral immune cells and cytotoxic molecules into the brain. This further amplifies the inflammatory pathways and exacerbates neurodegeneration. Microglial activation can also be triggered by misfolded proteins, including α-synuclein, amyloid-β, and tau (Sengupta & Kayed, 2022), which in turn triggers an inflammatory cascade. Initially, this activation is a protective response of cells to promote the clearance of toxic metabolites and proteins; however, chronic activation impairs microglial activation, leading to neurodegeneration, as microglia are exhausted by phagocytosis and can no longer clear debris. For example, in Alzheimer's disease (AD), plaques of amyloid-β are accumulated in the extracellular spaces of the brain and persistently trigger microglial activation. These plaques also serve as a pathological hallmark of AD (Gandy, 2005). Inflammatory cascade and microglial activation by misfolded proteins alter neuronal epigenetic machinery. This induces aberrant DNA methylation, silences neuroprotective genes, and locks the cell in a degenerative state.

1.1.3. Genetic and Epigenetic Factors: The Core Programming Code

Epigenetic changes encompass a broad range of modifications mainly affecting gene expression without changing the underlying DNA sequences. Factors like age, stress, diet, and disease conditions shape these changes (Giallongo et al., 2019), including alteration in chromatin structure, either chemically modifying DNA nucleobases and histones (Giallongo et al., 2021), or by replacing canonical histones with non-canonical histone isoforms within the nucleosome (Giallongo, Řeháková, et al., 2022), that primarily affect gene expression without altering. While the classical epigenetic profiles are typically heritable and transferred to offspring, neuro-epigenetics refers to epigenetic modifications in the CNS cells, and it is not heritable, as neurons are non-dividing cells (Hwang et al., 2017). For example, DNA methylation plays a significant role in the development of the brain, with DNA methyltransferases (*DNMT1, DNMT3A*, and *DNMT3B*) playing a key role in memory acquisition and storage (Day & Sweatt, 2010). Studies have reported that deletion of *DNMT1* and *DNMT3A* in mice compromises hippocampus-dependent learning and memory, and ultimately disrupts the process of fear conditioning (Feng et al.,

2010). The latter enhances the activity of *DNMT*, leading to repression of the gene coding the catalytic subunit of memory suppressor protein phosphatase 1 (*PP1*), in coordination with Reelin expression, which encodes a protein associated with plasticity (Giallongo et al., 2022). Moreover, DNA methylation in promoter regions typically inhibits gene transcription. In neurons, this mechanism is crucial for silencing genes that are unnecessary in particular brain regions or developmental phases (Vaissière et al., 2008). It also helps to regulate the intrinsic excitability of the neuronal membrane, as well as maintains synaptic function by modulating the expression of synaptic plasticity-related genes, for example, brain-derived neurotrophic factor (*BDNF*) (Feng et al., 2010). During neurogenesis, patterns of DNA methylation are dynamically remodeled to ensure accurate lineage specification, directing neural stem cells to differentiate into neurons or glial cells (Sun et al., 2021).

Epigenetic modifications include histone modification, DNA methylation, and changes in coding and non-coding RNAs, leading to alterations in the genome expression as a part of epigenetic modifications. These modifications play a significant role in gene and protein expression without causing any change in the DNA sequence. However, aberrant epigenetic modifications can lead to severe neuronal cell degeneration (Xylaki et al., 2019). For instance, abnormal DNA methylation is associated with neurodegenerative diseases such as AD, stress, and schizophrenia. Similarly, hypermethylation of genes crucial for survival can cause neuronal loss, whereas hypomethylation of proinflammatory genes can trigger neuroinflammation. Many of the NDDs, including PD, AD, Huntington's disease, and others, are greatly influenced by environmental changes (Hwang et al., 2017), aging, and lifestyle factors through epigenetic mechanisms. In PD, epigenetic alterations like DNA hypomethylation in the α-synuclein (*SNCA*) gene and histone modification triggered by α-synuclein lead to neurodegeneration. In addition, microRNAs, including miR-7, miR-133b, and miR-153, also influence genes related to PD (Klein & Westenberger, 2012). Similarly, environmental exposure to toxins like manganese can change patterns of DNA methylation (Elkin et al., 2022; Vaiserman & Lushchak, 2021). In AD, histone deacetylases (HDACs), especially HDAC2 and HDAC6, are upregulated, leading to impaired expression of memory-related genes (Broide et al., 2007; Ding et al., 2008). Moreover, dysregulated non-coding RNAs and decreased DNA methylation, such as in the promoter regions of the APP gene, are linked with cognitive decline and increased deposition of Aβ. Overall, these findings underscore the significant role of epigenetic modulation in the progression of AD and Parkinson's disease (Ghosh & Saadat, 2023). Nutri-epigenomics studies how various nutrients, such as vitamins, minerals, and bioactive phytoconstituents, and individual dietary patterns influence the epigenome and lead to modification of gene transcription without altering the sequence within DNA. As it unveils the effect of diet on the epigenome, it gradually became evident that bioactive compounds, such as polyphenols, also have the potential to regulate epigenetic pathways with probable therapeutic implications. Over time, this understanding led to the coining of a new term, "epigenetic diet". Many studies reported potential therapeutic effects of natural phytochemicals against NDD owing to their anti-inflammatory (Yahfoufi et al., 2018), free radical scavenging, and neuroprotective properties in cellular and animal studies. In this chapter, we will explore how polyphenols contribute to the regulation of epigenetic mechanisms.

1.2. Polyphenols as Therapeutic Cellular Reprogrammers

Currently, polyphenols are recognized as some of the most prominent secondary metabolites in plants, exhibiting a wide range of bioactive characteristics (Belščak-Cvitanović et al., 2018), making them highly valued for their beneficial effects in plants and humans (Rana et al., 2022). Polyphenols are a structurally diverse group of compounds, contributing to their physiological functions in plants as well as organoleptic and nutritional values of plant-derived foods (Belščak-Cvitanović et al., 2018). In humans, reported therapeutic activities of polyphenols include anti-inflammatory (Yahfoufi et al., 2018), antioxidant (Croft, 2016), antihyperglycemic, cardioprotective (Nawaz et al., 2024), nephroprotective (Guerreiro et al., 2022), anticancer (Cháirez-Ramírez et al., 2021), antimicrobial (Luo et al., 2021), antiaging (Luo et al., 2021), epigenetic regulation (Ullah et al., 2020). However, polyphenols face several challenges in human clinical trials related to bioavailability, metabolism, dosing, and safety: Polyphenols generally have low bioavailability due to poor water solubility, unstable chemical structure, and limited absorption in the gastrointestinal tract (Lu et al., 2025). They undergo extensive and rapid metabolism in the liver, leading to quick systemic clearance, which reduces their effective concentration in target tissues, particularly the brain (D'Archivio et al., 2010). The variability of polyphenol bioavailability also depends on factors such as food matrix, physical and chemical properties, host-related factors (age, health status), and gut microbiota composition (Mignet et al., 2013). Once ingested, polyphenols are extensively metabolized into various conjugates (glucuronides, sulfates), which complicates the evaluation of their pharmacokinetics and true biological activity in human tissues. This metabolic transformation can alter their potency and make it challenging to relate plasma levels to actual therapeutic effects (D'Archivio et al., 2010). Apart from bioavailability and metabolism issues, determining effective dosing in humans is difficult due to the low and variable bioavailability and the lack of standardized formulations. Clinical trials often rely on high doses that are hard to achieve through diet alone, raising questions about practicality and long-term adherence. Moreover, establishing effective doses is difficult due to variable bioavailability and the need for high doses for clinical efficacy, which may be hard to achieve through diet and could lead to compliance issues (Lu et al., 2025). While phase I trials generally confirm safety at high doses, long-term safety and potential drug interactions require further study. Thus, appropriate strategies are needed to overcome these challenges. Innovative delivery technologies are critical to overcome these challenges in human neurodegenerative disease trials (**Table 1**). Improving bioavailability through encapsulation and prodrug design is a key focus to translate polyphenol efficacy from preclinical models to humans. Advanced delivery systems like liposomes, nanoparticles, and nanoemulsions are being developed to enhance stability, absorption, and targeted delivery, particularly to the brain (Mignet et al., 2013).

Table 1. Key challenges in polyphenols therapeutics in neurodegenerative diseases and promising future strategies.

Challenges / Limitations	Description	Proposed Future Strategy/ Solutions
Low Bioavailability & Stability	Poor water solubility, chemical instability, and limited GI absorption reduce systemic concentrations (Lu et al., 2025).	Advanced Delivery Systems: Develop nano-formulations (e.g., liposomes, nanoparticles, nanoemulsions) to enhance solubility, protect from degradation, and improve absorption (Mignet et al., 2013).
Extensive Metabolism & Rapid Clearance	Polyphenols undergo rapid phase I/II metabolism (glucuronidation, sulfation) in the liver, leading to quick clearance and producing metabolites with potentially altered activity (D'Archivio et al., 2010).	Prodrug Design & Targeted Delivery: Create structural analogs or prodrugs resistant to first-pass metabolism. Use nano-carriers for targeted delivery to bypass metabolic pathways and enhance brain uptake.
Complex Pharmacokinetics	The presence of multiple conjugates in the bloodstream makes it difficult to correlate plasma levels with true biological activity and therapeutic effects in the brain (D'Archivio et al., 2010)	Improved Biomarker Development: Focus research on identifying and measuring active metabolites directly in target tissues. Employ advanced imaging and pharmacokinetic modeling to understand distribution.
Variable Bioavailability & Dosing Issues	Bioavailability is influenced by food matrix, gut microbiota, and individual host factors (age, health), making effective and standardized dosing difficult (Mignet et al., 2013). High doses needed for efficacy are hard to achieve through diet.	Personalized Nutrition (Nutrigenomics): Use genetic and microbiome profiling to tailor polyphenol interventions for individual metabolic responses. Develop concentrated, standardized formulations for reliable dosing.
Translation to Human Efficacy	Inconclusive results in human trials due to the challenges above. Long-term safety and potential drug interactions are not fully established.	Robust Clinical Trial Design: Conduct large-scale, long-term trials with standardized, bioavailability-enhanced formulations. Prioritize trials that include pharmacokinetic monitoring and safety pharmacovigilance.

1.2.1. An Overview of Polyphenolic Compounds With Reprogramming Potential in Neurodegenerative Diseases

Polyphenols are thought to be present in the brain at very low levels (1 nmol/g of tissue), yet this concentration is enough to influence neuronal pathways as they can penetrate the blood-brain barrier either by diffusion in the form of aglycones or through their conjugated forms (Schaffer & Halliwell, 2012). The capability of flavonoids to penetrate the blood-brain barrier is not only due to their lipophilic properties but also to the nature of conjugation. For instance, metabolites that undergo conjugation by methylation in the liver and small intestine are more lipophilic, making the permeation of the blood-brain barrier faster than aglycones (Youdim et al., 2003). On the contrary, less polyphilic metabolites must bind to specialized ATP-dependent transporters. Once in the brain, polyphenols can directly modulate neurotransmitter activity and participate in signaling cascades involving kinases like MAPK, PI3K, and PKB (Di Meo et al., 2020). Moreover, polyphenol-rich diets have been associated with preservation of cognitive function by supporting neuronal survival, promoting differentiation, and enhancing neuronal regeneration. Moreover, polyphenols contribute to halting the progression of NDDs by positively influencing memory, learning, and cognitive function. In the context of AD, the neuroprotective effect of flavonoids is primarily attributed to the regulation of enzymes such as glycogen synthase kinase 3β (GSK3β) and cyclin-dependent kinase 5 (CDK5) (Zeng et al., 2021)

1.2.2. General Mechanism of Neuroprotection: Polyphenols as Antioxidants

Polyphenols are characterized by structural elements such as hydroxyl groups, phenol rings, and conjugated bonds, which give them a broad spectrum of bioactive properties, such as metal-chelating, fibril-destabilizing (Gupta & Dasmahapatra, 2020), estrogenic, enzyme-modulating, and antioxidative properties. These properties enable the polyphenols to counteract the primary pathological characteristics of NDDs, mainly abnormal accumulation of misfolded proteins, generation of free radicals, and impaired mitochondrial function. Their ability to conjugate with divalent metals, interfere with amyloid fibril formation, and bind with estrogenic receptors contributes to their versatile biological effects. Moreover, their strong antioxidative action is either because of direct action of free radical scavenging or indirect effects by activation of the Nrf2/Keap1/ARE pathway, which is the primary cellular defense mechanism against oxidative stress. Overall, these mechanisms contribute to the neuroprotection offered by polyphenols (Renaud & Martinoli, 2019). Polyphenols not only intervene in NDDs as antioxidant agents, but also as epigenetic modulators and cellular reprogrammers.

1.2.3. Polyphenols-Mediated Neuroprotection via Cellular Reprogramming

Polyphenols act as cellular reprogrammers by targeting the following aspects of NDDs:

1.2.3.a. Polyphenols Help Reprogram Microglial Phenotype

Polyphenols contribute to the reprogramming of microglial phenotypes by targeting key signaling pathways (**Figure 1**). They suppress the key proinflammatory pathways such as NF-κB and MAPK, which are involved in M1 activation. Simultaneously, they encourage a transition towards an M2-like reparative state by activating anti-inflammatory pathways and metabolic modulators, for instance, Nrf2, PPAR-γ, and AMPK (Feng et al., 2017). This shift towards M2-state is reflected by upregulation of typical M2 markers, including Arginase-1 and IL-10. Polyphenolic compounds such as resveratrol and curcumin have revealed their reprogramming potential by downregulation of M1-associated typical markers and by promoting M2 transition (Yang et al., 2017). Similarly, epigallocatechin gallate (EGCG), a polyphenol found in green tea, enhanced the amyloid-β clearance and helped in suppressing the neuroinflammation by activating M2-type microglia (Chen et al., 2024).

Figure 1. Microglial activation, phenotype switching, and functional transition in neuroinflammation

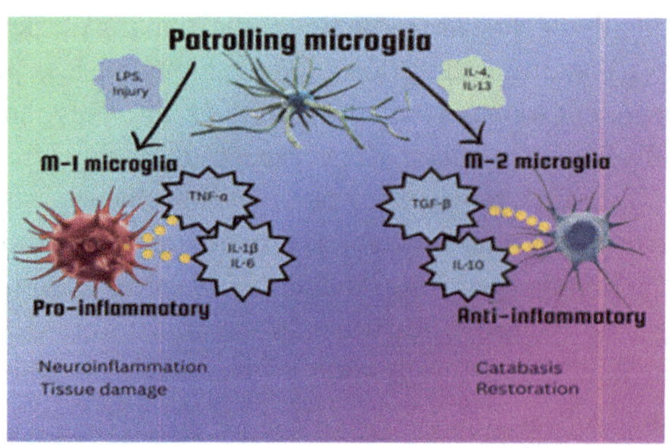

Besides promoting M2-microglial transition, polyphenols also regulate neuroinflammation by suppressing the release of proinflammatory cytokines, as these cytokines are mainly secreted by activated M1-type microglia. These cytokines, such as IL-1β, TNF-α, and IL-6, are the main contributors of neuronal injury and synaptic dysfunction observed in NDDs (Smith et al., 2012). They also inhibit proinflammatory signaling pathways like NF-κB and MAPKs, which regulate the release of these cytokines (Zhang et al., 2019). Ultimately, polyphenols mitigate the inflammation, reduce oxidative stress, and protect the neurons. For example, curcumins are known to inhibit proinflammatory cytokines by modulating the NF-κB pathway (Li et al., 2022). Likewise, another polyphenol, resveratrol, inhibits proinflammatory cytokines via activation of SIRT1 (involved in cellular stress and metabolism) (Zhu et al., 2011). Lastly, EGCG downregulated proinflammatory cytokines by blocking TLR4 signaling (involved in the regulation of immune response) (Wang et al., 2019).

1.2.3.b. Polyphenols Regulate Antioxidant Defense Reprogramming

Polyphenols also modulate the central regulators of the cellular antioxidant defense system by activating the Nrf2 pathway. Upon activation, Nrf2 translocate to the nucleus and upregulate the genes coding for antioxidant enzymes such as hypoxia inducible factior-1 (HO-1), superoxide dismutase (SOD), and glutathione peroxidase. Following activation, these enzymes offer neuronal protection by lowering oxidative stress (Renaud & Martinoli, 2019). Additionally, polyphenols possess the ability to directly neutralize the reactive oxygen and nitrogen species and thereby enhance the effectiveness of the intrinsic cellular antioxidant enzymes and ultimately limit the oxidative and nitrosative stress (Kung et al., 2021). Polyphenols also promote mitochondrial protection and biogenesis by improving mitochondrial performance and activating specific pathways that stimulate new mitochondria formation, such as PGC-1α. This pathway further helps shield against oxidative stress, enhances cellular energy metabolism, and promotes neural repair and survival.

1.2.3.c. Polyphenols Modulate Epigenetic Modifications: In Vitro, In Vivo, and Human Clinical Trial Evidence

Polyphenols, a diverse group of phytochemicals abundant in fruits, vegetables, tea, wine, and spices, are increasingly recognized not only for their antioxidant and anti-inflammatory properties but also as powerful modulators of the epigenome. Rather than functioning merely as radical scavengers, polyphenols interact directly with the molecular machinery that governs gene regulation, thereby reprogramming cellular behavior. Through their ability to alter DNA methylation, histone modifications, and microRNA (miRNA) networks, polyphenols reshape chromatin structure and transcriptional landscapes, restoring protective gene expression programs while suppressing pathogenic ones. This cellular reprogramming is particularly relevant to neurodegenerative diseases (NDDs), where epigenetic dysregulation contributes to neuronal vulnerability, chronic inflammation, and progressive cognitive decline.

Polyphenols have exhibited profound modulatory effects on epigenetic processes. One of the central processes governing the regulation of gene expression is DNA methylation. DNA methylation is a key epigenetic modification involving the transfer of a methyl group to the 5' position of cytosine residues within CpG islands, typically leading to transcriptional repression. Aberrant hypermethylation silences tumor suppressor and neuroprotective genes, contributing to cancer and NDD progression (Minarovits et al., 2015).

Polyphenols contribute to the reactivation of genes essential to neuronal survival, repair, and neural function by facilitating the demethylation of neuroprotective genes. For example, EGCG and other major tea-derived polyphenols, such as catechin and epicatechin, have exhibited the ability to inhibit DNA methyltransferase (DNMT) activity in a concentration-dependent manner, resulting in effective demethylation and reactivation of neuroprotective genes (Fang et al., 2007). Among these, EGCG is well-reputed for direct and indirect modulation of the DNA methylation mechanisms. Mechanistically, it directly inhibits DNMT by forming hydrogen bonds within its active site, which prevents the transfer of methyl groups from the donor S-adenosylmethionine (SAM). This blockade reactivates silenced promoters, restoring normal gene expression (Fang et al., 2007). Moreover, it inhibits histone methyltransferases (HMTs) and affects histone post-translational modifications. Similarly, resveratrol, a polyphenol abundant in mulberries, peanuts, cranberries, and grapes, primarily influences epigenetic mechanisms by histone modifications. It exerts a neuroprotective effect and enhances cognitive functions through activation of the deacetylase enzyme SIRT1 (involved in gene silencing) (Huber & Superti-Furga, 2011), and has also been shown to inhibit DNMT function (Hardy & Tollefsbol, 2011). However, resveratrol exerts its indirect effects on methylation patterns by altering the SAM/SAM/S-adenosylhomocysteine (SAH) ratio, thereby modifying the availability of methyl donors (Royston, 2018). These epigenetic modifications by polyphenols help prevent harmful gene silencing, which is often experienced in NDDs, hence improving cellular resilience and slowing pathogenesis and progression of disease.

In vitro evidence is compelling. Fang et al. (2007) showed that EGCG treatment of human esophageal, colon, and prostate cancer cells reduced DNMT activity, leading to demethylation and reactivation of p16^INK4a, RARβ, and GSTP1. Similar effects were observed with quercetin in bladder and colon cancer cell lines, where it demethylated estrogen receptor and RASSF1A promoters, restoring tumor suppressor function (Mukherjee et al., 2015). In neuronal models, EGCG promoted demethylation of neuroprotective genes linked to synaptic repair and survival. In vivo studies further support this mechanism. In rodent models fed high-fat diets, green tea catechins reversed hypermethylation at metabolic and cognitive loci, improving insulin sensitivity, reducing oxidative stress, and enhancing memory. Resveratrol administration in Wistar rats altered DNA methylation patterns in adipose tissue, reducing fatty acid synthase expression, while also improving hippocampal-dependent learning and memory (Lagouge et al., 2006). Together, these findings illustrate how polyphenols act as epigenetic demethylating agents, reprogramming cells by reactivating protective gene networks silenced by pathological methylation.

Additionally, polyphenols significantly influence other core epigenetic mechanisms involved in the regulation of gene expression and accessibility, such as histone acetylation and chromatin remodeling (Chung, 2010). Chromatin, a compact and dynamic structure, not only condenses the whole genome within the nucleus but also controls access to the genetic information. Whereas histone acetylation modifies this structure by loosening DNA-histone interactions, thereby overcoming chromatin's repressive barrier and enhancing the transcriptional potential of specific genes. (Gregory et al., 2001). Acetylation of histones H3 and H4 typically loosens DNA-histone interactions, increasing transcriptional accessibility, while deacetylation and repressive methylation often silence genes. Dysregulation of these processes contributes to cancer, metabolic disorders, and NDDs (**Table 2**).

Table 2. Summary of polyphenol-mediated epigenetic reprogramming mechanism: a synthesis of in vitro and in vivo evidence.

Compound	DNA Methylation	Histone Acetylation	MicroRNA	Models	Mechanism	References
EGCG	Inhibits DNMT activity; demethylates p16^INK4a, RARβ, GSTP1 promoters; demethylates hTERT; reactivates neuroprotective genes	HDAC inhibitor; increases H3/ H4 acetylation; inhibits HMTs; alters H3K9/ H4K16 acetylation	↓miR-92/93/106b; ↑miR-34a, miR-99a, miR-7-1, miR-124, miR-132	In vitro: esophageal, colon, prostate, colorectal cancer cells, neuronal cells; In vivo: rodent high-fat diet, AD models	Direct DNMT1 binding via hydrogen bonds; inhibition of HAT/ HDAC; reactivation of tumor suppressors and neuroprotective genes; chromatin remodeling	(Fang et al., 2007; Mokhtari et al., 2020; Zhou & Pang, 2025).
Quercetin	Demethylates estrogen receptor and *RASSF1A* promoters in bladder/colon cancer cells	Inhibits p300 HAT; increases H3 acetylation; modulates SIRT1/HDACs	↑miR-125b, ↑miR-122 in hepatocytes; ↑miR-132 in neurons	In vitro: bladder, colon, leukemia, endothelial cells; In vivo: mice, hepatocytes	Inhibition of DNMTs and HATs; activation of apoptotic pathways via histone acetylation; modulation of inflammatory genes	(Benameur et al., 2021; Číž et al., 2020; Mukherjee et al., 2015).
Curcumin	Inhibits DNMT1 in MCF-7; downregulates DNMT3b in A549 cells	Inhibits p300/ CBP HAT; downregulates HDAC4; modulates H3/ H4 acetylation	Alters multiple miRNAs (pancreatic cancer cells, neuronal models)	In vitro: MCF-7, A549, leukemia cells; In vivo: HFD rodent models	DNMT1 inhibition; dual HAT/HDAC modulation; regulates lipid metabolism genes	(Chen et al., 2024; Ding et al., 2016; Yuan et al., 2012)
Resveratrol	Modulates *FAS* promoter methylation; alters SAM/SAH ratio; DNMT inhibition	Activates SIRT1; increases hippocampal histone acetylation at synaptic genes; reduces H2A.Z	↑miR-124, ↑miR-132 in neurons; inhibits miR-520h in cancer; modulates miR-134	In vitro: cancer cells; In vivo: rodents (adipose tissue, hippocampus, maternal HFD models)	SIRT1 activation; AMPK phosphorylation; restoration of neuroprotective miRNAs; anti-inflammatory via STAT3/NF-κB	(Lagouge et al., 2006; Zhao et al., 2013; Zhou & Pang, 2025).
Sulforaphane (SFN)	Inhibits DNMTs; promoter demethylation of *cyclin D2*; reactivates tumor suppressors	Inhibits HDACs; increases acetylation in animal tissues	Alters tumor suppressor miRNA networks (in vivo models)	In vitro: prostate, breast cancer cells; In vivo: mouse models, cruciferous-rich diets	HDAC/DNMT inhibition; reactivation of protective genes; modulation of inflammatory/oxidative pathways	(Dacosta & Bao, 2017; Ho et al., 2011; Tortorella et al., 2015).

Polyphenols increase the histone acetylation by targeting these histone-modifying enzymes and loosening the condensed chromatin, which facilitates the transcriptional factors to conjugate to DNA, initiate gene expression, and restore chromatin states (Ho et al., 2011). Curcumin, for instance, inhibits p300/ CBP histone acetyltransferase (HAT) activity, lowering acetylation of oncogenic promoters (Yuan et al., 2012). EGCG and sulforaphane act as inhibitors of histone deacetylases (HDACs), resulting in hyperacetylation of histones H3K9 and H4K16, which promotes transcription of protective genes (Tortorella et al., 2015). Resveratrol provides a unique example of activating a histone-modifying enzyme: it enhances the activity of SIRT1, a class III HDAC dependent on NAD^+, thereby linking chromatin remodeling to energy metabolism and mitochondrial health (Gomez de Cedron & Ramírez de Molina, 2023).

In vitro evidence highlights the breadth of these effects. In colorectal cancer cells, EGCG increased H3 and H4 acetylation, restoring transcriptional activity of tumor suppressors. In leukemia models, curcumin downregulated HDAC4, while also inhibiting HAT p300/CBP, tipping the histone code toward apoptosis and growth arrest. Quercetin suppressed COX-2 expression in endothelial cells by inhibiting p300 HAT activity and promoted apoptosis in leukemia cells through histone H3 acetylation (Číž et al., 2020). In vivo studies confirm these roles in neuronal and metabolic contexts. In mice, resveratrol supplementation increased hippocampal histone acetylation at genes involved in synaptic plasticity, enhancing learning and memory (Zhao et al., 2013). Curcumin supplementation in high-fat diet models reduced adiposity and improved insulin sensitivity, likely through modulation of histone modifications regulating lipid metabolism (Ding et al., 2016). Sulforaphane consumption in cruciferous vegetable-rich diets was linked to increased histone acetylation in animal tissues, reactivating tumor suppressor genes. These data show that polyphenols function as histone code modifiers, directly remodeling chromatin accessibility to reprogram transcriptional responses central to cellular resilience By regulating these mechanisms, polyphenols influence the activation of anti-inflammatory response and neuroprotective genes. One study reported that EGCG causes the inhibition of DNMTs and histone deacetylases (HDACs), leading to increased acetylation of H3 and H4 histones, both of which are significant in regulating DNA accessibility. These chromatin modifications allow better access of transcriptional factors to the hTERT promoter, ultimately lowering its expression levels in cancerous cells. Similarly, another polyphenol, Sulforaphane (SFN), present in cruciferous vegetables, is also known to cause inhibition of HDACs and DNMTs, leading to reactivation of tumor suppressor genes. This reactivation (Ho et al., 2011).

MicroRNAs (miRNAs) are another important part of epigenetic modification as these have a significant role in modulating neuronal activity during the developmental stages of the brain and also in adulthood (Rajman & Schratt, 2017). They are required for crucial processes such as neuronal differentiation, proper functions, and survival, as they act as post-transcriptional regulators of gene expression. However, with aging, miRNAs encounter various disruptions that increase the neuronal susceptibility to cellular stress, leading to the onset and progression of NDDs. miRNAs are initially transcribed as precursor hairpin structures that are processed in the cytoplasm and the nucleus by Drosha and Dicer ribonucleases. After processing, mature miRNAs are then incorporated into another protein, Argonaute (Ago2), inside the RNA-induced silencing complex (RISC) (Bartel, 2018). Many researchers are increasingly focusing on the role of miRNAs in the development of the central nervous system (CNS), as it encompasses the dynamic processes involved in the formation as well as refinement of the nervous system from embryogenesis to adulthood. One such study compared the expression of miRNAs in embryonic neurons and mature neuronal (astrocyte) cultures. They revealed that miR-124 is preferentially expressed in neurons, and this expression increases in parallel to neuronal maturation, underscoring the significant role in orchestrating neuronal differentiation and maturation during CNS development (Sun et al., 2015). miR-124 in particular is abundant in the brain and is selectively expressed in nearly all areas of the brain except the pituitary gland. It is approximately 100 times less expressed in non-neuronal tissues. They act as fine-tuners of gene expression networks controlling neuronal differentiation, inflammation, synaptic plasticity, and stress responses. Any dysregulation in miRNAs is strongly implicated in NDDs, where they promote neuroinflammation and synaptic dysfunction. Polyphenols reprogram these networks by altering miRNA expression profiles. EGCG downregulates oncogenic miR-92, miR-93, and miR-106b while upregulating tumor suppressors such as miR-34a and miR-99a (Mokhtari et al., 2020). In neuronal contexts, polyphenols rescue downregulated neuroprotective miRNAs, including miR-124 and miR-132 (Zhou & Pang, 2025). miR-124, abundant in neurons, suppresses microglial activation and inflammation (Yang et al.,

2019). Its restoration by resveratrol promotes anti-inflammatory M2 microglial polarization. miR-132, regulated by CREB, orchestrates dendritic branching and synaptic maturation; resveratrol and quercetin have been shown to enhance its expression, rescuing plasticity deficits (Benameur et al., 2021; Zhang et al., 2019). In vitro studies demonstrate that EGCG modulates miRNA profiles in cancer and neuronal cell lines, reducing oncogenic signals and promoting survival pathways (Dharshini & Mandal, 2024). Quercetin increased miR-125b and miR-122 in hepatocytes, regulating inflammation and lipid metabolism. In vivo, resveratrol increased miR-124 and miR-132 expression in rodent brains, enhancing dendritic complexity and synapse formation, while reducing neuroinflammation (Bigagli, 2015). In Alzheimer's disease models, polyphenol administration restored miR-124 levels, dampening inflammatory cascades, and improved memory outcomes. Sulforaphane treatment altered miRNA networks in mouse models, reactivating tumor suppressor pathways (Dacosta & Bao, 2017). Through these actions, polyphenols function as post-transcriptional reprogrammers, reshaping cellular responses by fine-tuning gene regulatory networks. Furthermore, microglia transition and resultant inflammation are key hallmark pathogenic mechanisms in the progression of NDDs, and these mechanisms are also influenced by miRNAs. In this context, miR-124 has emerged as the main regulatory miRNA as its expression levels were remarkably downregulated in response to pro-inflammatory conditions. This downregulation was consistent with sustained microglial activation and neuroinflammation. On the contrary, miR-124 overexpression has been associated with suppressed inflammatory pathways, implying that restoring miR-124 expression levels might offer a promising therapeutic option to mitigate inflammation and arrest the progression of NDDs (Han et al., 2020). Similarly, miR-132, another miRNA abundant in the brain, also plays a crucial role in neuronal plasticity and morphogenesis. It is regulated by CREB, an activity-dependent transcription factor, and it mainly influences the development of axons and dendrites, and spine maturation through multiple signaling cascades. Many researchers have reported that deletion in miR-132 is linked with impaired dendrite branching, reduced length, and spine density, which consequently disrupts the newly formed neurons and reduces the synapse formation in the adult hippocampus (El Fatimy et al., 2018). Although many in vivo and in vitro studies have been conducted to rule out the epigenetic potential of polyphenols, direct evidence from human clinical trials is extremely limited and remains a major gap in the field. The major challenge lies in measuring the brain-specific epigenetic modifications in living patients. However, there is some indirect and emerging evidence from a trial of resveratrol, which showed activation of the SIRT1 pathway, a key deacetylase involved in epigenetic regulation and stress response, in Alzheimer's patients (Moussa et al., 2017). A few human studies also provided indirect evidence of polyphenols improving neurodegenerative conditions (**Table 3**). Currently, the epigenetic activity of polyphenols in humans is largely inferred from their established mechanisms in cellular and animal models. Therefore, a critical future direction is the incorporation of epigenetic biomarkers as exploratory endpoints in clinical trials to validate this reprogramming potential in humans directly.

Table 3. Overview of human clinical trials evidence for selected polyphenols in neurodegenerative diseases.

Polyphenol	Dietary Source	NDD	Reported Effects in Humans	Key Challenges/Limitations	Reference
Resveratrol	Red grapes, wine, berries	AD	Safe at high doses; stabilized Aβ40 levels in CSF; no significant cognitive improvement.	High doses required; rapid metabolism; unclear if target engagement is sufficient for clinical benefit.	(Moussa et al., 2017)
Green Tea Catechins (EGCG)	Green tea	MIC	Improvements in memory and functional brain connectivity.	Larger, long-term AD trials are lacking; bioavailability is very low.	(Köbe et al., 2017)
Grape Seed Extract	Grapes, wine	AD	Phase 2 trial completed targeting tau pathology; results awaited/pending.	Full results have not yet been widely published; bioavailability is a major constraint.	NCT02033941
Curcumin	Turmeric	PD	Pilot study confirmed detectability in plasma, but at very low concentrations.	Extremely poor bioavailability and rapid clearance limit therapeutic potential.	(Ghodsi et al., 2022)
EGCG	Green tea	PD	Early trials focused on safety and dosing; definitive efficacy has not been established.	Bioavailability and blood-brain barrier penetration are significant hurdles.	NCT00461942
Cocoa Flavanols	Cocoa, dark chocolate	VIC	Improved cerebral blood flow, which may support neuronal health.	Effects are indirect; linking improved blood flow to specific disease modification is complex.	(Sorond et al., 2013)

MIC: Mild cognitive impairment, VIC: Vascular cognitive impairment, AD: Alzheimer's disease, PD: Parkinson's disease, EGCG: Epigallocatechin gallate.

Polyphenols confer neuroprotective effects in NDDs by modulating the expression of specific miRNAs, which are core regulators of inflammation, oxidative stress, and neuronal survival. Notably, miR-132 and miR-124 have been consistently reported to be downregulated miRNAs in AD and Parkinson's disease, highlighting their role in neurodegeneration. Resveratrol, a polyphenolic compound, has exhibited anti-inflammatory properties by downregulating the expression levels of inflammatory cytokines and signal transducer and activator of transcription (STAT3), both of which are hallmarks of neuroinflammation. Besides its anti-inflammatory effects in mitigating the pathogenesis of NDDs, it has also exhibited up-regulation of SIRTI expression, a key protein involved in neuronal protection and survival. Moreover, it modulated the miR-134 levels, which are implicated in neurodegeneration. This modulation strengthens neurite outgrowth, improves synapse formation and function (Zhou & Pang, 2025). Additionally, it increases the miR-124 levels, which support anti-inflammatory pathways and promote microglial polarization towards the M2 phenotype, which is also neuroprotective. By converging on DNA methylation, histone modification, and miRNA regulation, polyphenols orchestrate a multi-layered reprogramming of the epigenome. They restore the expression of silenced neuroprotective genes, remodel chromatin to favor adaptive transcription, and fine-tune post-transcriptional regulators of neuronal differentiation and survival. Through these epigenetic modifications (**Figure 2**), polyphenols contribute to cellular reprogramming, attenuate neuroinflammation, and offer protection against the pathogenesis of NDDs.

Figure 2. Detailed mechanism of polyphenols as cellular reprogrammers in neurodegenerative diseases

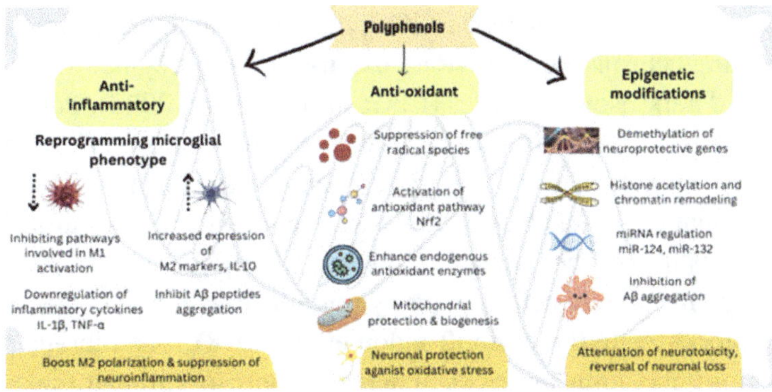

1.2.4. Disease-Specific Reprogramming: Evidence From Alzheimer's Disease

Polyphenols such as EGCG and curcumin have been shown to inhibit amyloid-β aggregation, a central pathological feature of AD. These compounds interact with Aβ monomers and oligomers, disrupting their misfolding and preventing the formation of toxic amyloid plaques. EGCG, a predominant catechin found in green tea, converts Aβ fibrils into non-toxic, off-pathway aggregates (Li et al., 2024). Similarly, curcumin, a bioactive compound derived from turmeric, inhibits the formation of β-sheet, hence decreasing amyloid plaque accumulation (Thapa et al., 2016). Through these actions, polyphenols not only inhibit Aβ aggregation but also promote its clearance, reduce neurotoxicity and microglial activation, and ultimately contribute to slowing the progression of Alzheimer's disease.

1.2.5. Disease-Specific Reprogramming: Evidence From Parkinson's Disease

In Parkinson's disease, polyphenols exert neuroprotection on dopaminergic neurons in the substantia nigra (Zhou et al., 2019). They combat oxidative stress through activation of the Nrf2-mediated antioxidant pathway, and suppress neuroinflammation by promoting shifting of microglia from proinflammatory M1-phenotype to anti-inflammatory M2-phenotype (Renaud & Martinoli, 2019) and diminished the accumulation of α-synuclein (Gautam et al., 2017). Moreover, polyphenols offer support to mitochondria for proper functioning and prevent apoptosis, collectively restoring the integrity of the dopaminergic neurons and slowing the progression of Parkinson's disease. Polyphenols including EGCG, curcumin, and baicalein, have been shown to counter α-synuclein toxicity by inhibiting the misfolding of α-synuclein and accumulation in toxic fibrils (Sharma & Nehru, 2018; Zaidi & Deep, 2020; Zhao et al., 2017). They stabilize α-synuclein into its original form and prevent fibril formation, thereby mitigating its neurotoxicity. Moreover, their anti-inflammatory and antioxidant effects further protect the dopaminergic neurons against the α-synuclein-mediated neurodegeneration. Polyphenols have shown potential as cellular reprogrammers and neuroprotective agents in human clinical and preclinical studies related to neurodegenerative diseases (**Table 4**). They exhibit antioxidant, anti-inflammatory, and signaling regulation properties that could modulate stem cells and neural cells, supporting neurogenesis, reducing oxidative stress, and suppressing neurodegeneration pathways. Polyphenols can activate genes in stem cells and modulate their activity, including key antioxidant pathways (Nrf2) and enzymes like glutathione peroxidase, SOD, and catalase, which reduce oxidative damage in neural cells (Rodríguez-Vera et al., 2022). Studies report polyphenols such as EGCG, curcumin, and quercetin enhancing neural stem cell proliferation, protecting neurons from amyloid toxicity, and limiting neuroinflammation, which are critical in Alzheimer's and Parkinson's diseases (El Gaamouch et al., 2022). Polyphenols modulate cell signaling pathways (Akt, Nrf2, STAT, MAPK) that are crucial for neuronal viability, synaptic plasticity, and immune responses in neurodegenerative disease contexts (Jalouli et al., 2025). Epidemiological and clinical evidence suggest that polyphenol-rich diets may decrease the risk of and alleviate symptoms in neurodegenerative disorders like Alzheimer's, Parkinson's, Multiple Sclerosis, and others (Jalouli et al., 2025). Polyphenols complement stem cell therapy by possibly enhancing endogenous neural stem cell activity and reparative processes in the brain (Tandon et al., 2018). The therapeutic effects are linked to multi-mechanism actions, including antioxidant defenses, anti-inflammatory activity, and prevention of protein misfolding or aggregation typical in neurodegenerative diseases (Gahtani et al., 2024). Polyphenols hold promising roles as cellular reprogrammers in human clinical contexts by supporting neural stem cell function and neurogenesis, reducing oxidative stress and inflammation via multiple molecular pathways, and complementing existing therapies for neurodegenerative diseases through modulation of cell survival and repair mechanisms. However, more clinical trials are needed to fully establish and standardize polyphenol use as advanced therapeutic agents for neurodegenerative diseases (Rodríguez-Vera et al., 2022).

Table 4. Summary of polyphenols' general effects, epigenetic mechanisms, and challenges in NDDs models.

Polyphenols	Target NDDs	Reported Effects	Proposed Mechanism (Incl. Epigenetics)	Limitations	References
Resveratrol, Curcumin, EGCG, Quercetin, Anthocyanins	General (AD, PD, ALS, HD)	Redox balance. Microglial modulation. Enhanced NSC survival, differentiation, & synaptic integration.	Nrf2/ARE pathway activation. BDNF upregulation. Epigenetic: Modulating miRNAs & HDACs/SIRT1 to promote pro-survival gene expression.	The bioavailability of polyphenols is low. Optimal dosing & timing for adjuvant use with stem cells is unknown.	(Rodríguez-Vera et al., 2022)
Resveratrol, Curcumin, EGCG, Oleocanthal, Quercetin	AD	Reduces ROS. Lowers cytokine levels. Inhibits Aβ aggregation & tau hyperphosphorylation, enhances synaptic plasticity.	BACE1 inhibition (reduces Aβ production). Chaperone-like activity prevents protein misfolding. Epigenetic: SIRT1 activation (by Resveratrol) promotes neuroprotection; modulation of DNMTs.	Human trials have shown inconsistent results, largely due to poor bioavailability and variability in polyphenol formulations.	(El Gaamouch et al., 2022)
Curcumin, Resveratrol, EGCG, Quercetin, Ferulic Acid	General (AD, PD)	Potent ROS scavenging, boosts glutathione. Suppresses microglial activation. Reduces neuronal apoptosis.	Primarily through activation of the Nrf2 pathway. Inhibition of NF-κB. Epigenetic: Detailed discussion on how polyphenols can reverse HDAC-mediated gene silencing of antioxidant genes and modulate Nrf2 promoter methylation.	Heavily reliant on pre-clinical data (in vitro & animal studies). Clinical translation remains a significant challenge.	(Jalouli et al., 2025)

continued on following page

Table 4. Continued

Polyphenols	Target NDDs	Reported Effects	Proposed Mechanism (Incl. Epigenetics)	Limitations	References
Resveratrol, Curcumin, EGCG, Luteolin, Genistein	MS & AD	Immunomodulation (esp. for MS). Promotes remyelination (OPC differentiation) in MS & neurogenesis (NSC differentiation) in AD.	Wnt/β-catenin & SHH pathway modulation for stem cell differentiation. Epigenetic: Regulation of miRNAs that control the fate of neural stem cells (NSCs) and oligodendrocyte progenitor cells (OPCs).	The complex interplay between polyphenols, immune cells, and stem cells in the human brain is not fully understood.	(Tandon et al., 2018)
Curcumin, Resveratrol, EGCG, Gingerol, Anthocyanins	PD	Protects dopaminergic neurons. Suppresses nigral inflammation. Inhibits α-synuclein aggregation, mitigates mitochondrial dysfunction.	DJ-1/Nrf2 pathway activation. α-synuclein fibril disruption. Epigenetic: Modulation of SIRT1 and miRNAs (e.g., miR-155) involved in neuroinflammation and protein aggregation.	Most compelling evidence is from cellular and animal models of PD; human data is lacking. The blood-brain barrier remains a major obstacle.	(Gahtani et al., 2024).

AD: Alzheimer's disease, PD: Parkinson's disease, HD: Huntington's disease, ALS: Amyotrophic lateral sclerosis, MS: Multiple sclerosis, EGCG: Epigallocatechin gallate.

CONCLUSION: THE FUTURE OF POLYPHENOLS AS PRECISION PROGRAMMERS IN NEUROLOGY

This chapter establishes the compelling evidence that dietary polyphenols are promising multifunctional agents with the ability to induce cellular reprogramming in neurodegenerative diseases. Through modulation of epigenetic pathways, mitigation of oxidative stress, and suppression of neuroinflammation, they offer a shift from symptomatic relief to true disease-modifying strategies. The synthesis of in vitro and in vivo evidence robustly supports their ability to influence key transcription factors, histone modifications, and microRNA networks, promoting a neuroprotective cellular state. However, the translation of this promising preclinical data into clinical success faces a significant bottleneck, primarily due to profound challenges in bioavailability, extensive metabolism, and variable human responses. Clinical trials, while demonstrating safety, have often yielded mixed or modest efficacy outcomes, underscoring this translational gap. Although limitations such as poor bioavailability persist, their broad-spectrum effects position them as strong candidates for precision medicine. Advancing delivery systems and conducting robust clinical trials will be essential to understand their therapeutic potential in NDDs. Thus, future research must pivot towards innovative delivery strategies, such as nanoencapsulation, to enhance brain-targeted delivery. Furthermore, a personalized medicine approach, informed by nutrigenomics, will be crucial to identify responsive subpopulations. Ultimately, overcoming these pharmacological hurdles is essential to validate the therapeutic potential of polyphenols and unlock their definitive role in a precision nutrition strategy for combating neurodegenerative diseases.

REFERENCES

Ayissi, V. B. O., Ebrahimi, A., & Schluesenner, H. (2014). Epigenetic effects of natural polyphenols: A focus on SIRT1-mediated mechanisms. *Molecular Nutrition & Food Research*, *58*(1), 22–32. DOI: 10.1002/mnfr.201300195 PMID: 23881751

Bartel, D. P. (2018). Metazoan micrornas. *Cell*, *173*(1), 20–51. DOI: 10.1016/j.cell.2018.03.006 PMID: 29570994

Belščak-Cvitanović, A., Durgo, K., Huđek, A., Bačun-Družina, V., & Komes, D. (2018). Overview of polyphenols and their properties. In *Polyphenols: Properties, recovery, and applications* (pp. 3–44). Elsevier.

Benameur, T., Soleti, R., & Porro, C. (2021). The potential neuroprotective role of free and encapsulated quercetin mediated by miRNA against neurological diseases. *Nutrients*, *13*(4), 1318. DOI: 10.3390/nu13041318 PMID: 33923599

Bigagli, E. (2015). Gene and microRNA expression profiles in models of brain aging: effects of bioactive phenolic compounds.

Broide, R. S., Redwine, J. M., Aftahi, N., Young, W., Bloom, F. E., & Winrow, C. J. (2007). Distribution of histone deacetylases 1–11 in the rat brain. *Journal of Molecular Neuroscience*, *31*(1), 47–58. DOI: 10.1007/BF02686117 PMID: 17416969

Cháirez-Ramírez, M. H., de la Cruz-López, K. G., & García-Carrancá, A. (2021). Polyphenols as antitumor agents targeting key players in cancer-driving signaling pathways. *Frontiers in Pharmacology*, *12*, 710304. DOI: 10.3389/fphar.2021.710304 PMID: 34744708

Chen, Y., Liu, Z., & Gong, Y. (2024). Neuron-immunity communication: Mechanism of neuroprotective effects in EGCG. *Critical Reviews in Food Science and Nutrition*, *64*(25), 9333–9352. DOI: 10.1080/10408398.2023.2212069 PMID: 37216484

Chopade, P., Chopade, N., Zhao, Z., Mitragotri, S., Liao, R., & Chandran Suja, V. (2023). Alzheimer's and Parkinson's disease therapies in the clinic. *Bioengineering & Translational Medicine*, *8*(1), e10367. DOI: 10.1002/btm2.10367 PMID: 36684083

Chung, I. R. S. (2010). Dietary polyphenols, deacetylases and chromatin remodeling in inflammation. *World Review of Nutrition and Dietetics*, *101*, 84–94. DOI: 10.1159/000314513 PMID: 20436255

Číž, M., Dvořáková, A., Skočková, V., & Kubala, L. (2020). The role of dietary phenolic compounds in epigenetic modulation involved in inflammatory processes. *Antioxidants*, *9*(8), 691. DOI: 10.3390/antiox9080691 PMID: 32756302

Cobley, J. N., Fiorello, M. L., & Bailey, D. M. (2018). 13 reasons why the brain is susceptible to oxidative stress. *Redox Biology*, *15*, 490–503. DOI: 10.1016/j.redox.2018.01.008 PMID: 29413961

Colton, C. A., & Gilbert, D. L. (1987). Production of superoxide anions by a CNS macrophage, the microglia. *FEBS Letters*, *223*(2), 284–288. DOI: 10.1016/0014-5793(87)80305-8 PMID: 2822487

Croft, K. D. (2016). Dietary polyphenols: Antioxidants or not? *Archives of Biochemistry and Biophysics*, *595*, 120–124. DOI: 10.1016/j.abb.2015.11.014 PMID: 27095227

D'Archivio, M., Filesi, C., Varì, R., Scazzocchio, B., & Masella, R. (2010). Bioavailability of the polyphenols: Status and controversies. *International Journal of Molecular Sciences*, *11*(4), 1321–1342. DOI: 10.3390/ijms11041321 PMID: 20480022

Dacosta, C., & Bao, Y. (2017). The role of MicroRNAs in the chemopreventive activity of sulforaphane from cruciferous vegetables. *Nutrients*, *9*(8), 902. DOI: 10.3390/nu9080902 PMID: 28825609

Davis, D. S., & Carson, M. J. (2012). An introduction to CNS-resident microglia: definitions, assays, and functional roles in health and disease. In *Neural-immune interactions in brain function and alcohol related disorders* (pp. 3–29). Springer.

Day, J. J., & Sweatt, J. D. (2010). DNA methylation and memory formation. *Nature Neuroscience*, *13*(11), 1319–1323. DOI: 10.1038/nn.2666 PMID: 20975755

Dharshini, L. C. P., & Mandal, A. K. A. (2024). Regulation of gene expression by modulating microRNAs through Epigallocatechin-3-gallate in cancer. *Molecular Biology Reports*, *51*(1), 230. DOI: 10.1007/s11033-023-09145-2 PMID: 38281210

Di Meo, F., Valentino, A., Petillo, O., Peluso, G., Filosa, S., & Crispi, S. (2020). Bioactive polyphenols and neuromodulation: Molecular mechanisms in neurodegeneration. *International Journal of Molecular Sciences*, *21*(7), 2564. DOI: 10.3390/ijms21072564 PMID: 32272735

Ding, H., Dolan, P. J., & Johnson, G. V. (2008). Histone deacetylase 6 interacts with the microtubule-associated protein tau. *Journal of Neurochemistry*, *106*(5), 2119–2130. DOI: 10.1111/j.1471-4159.2008.05564.x PMID: 18636984

Ding, L., Li, J., Song, B., Xiao, X., Zhang, B., Qi, M., Huang, W., Yang, L., & Wang, Z. (2016). Curcumin rescues high fat diet-induced obesity and insulin sensitivity in mice through regulating SREBP pathway. *Toxicology and Applied Pharmacology*, *304*, 99–109. DOI: 10.1016/j.taap.2016.05.011 PMID: 27208389

Edmondson, E. (2014). Hydrogen peroxide produced by mitochondrial monoamine oxidase catalysis: Biological implications. *Current Pharmaceutical Design*, *20*(2), 155–160. DOI: 10.2174/13816128113190990406 PMID: 23701542

El Fatimy, R., Li, S., Chen, Z., Mushannen, T., Gongala, S., Wei, Z., Balu, D. T., Rabinovsky, R., Cantlon, A., Elkhal, A., Selkoe, D. J., Sonntag, K. C., Walsh, D. M., & Krichevsky, A. M. (2018). MicroRNA-132 provides neuroprotection for tauopathies via multiple signaling pathways. *Acta Neuropathologica*, *136*(4), 537–555. DOI: 10.1007/s00401-018-1880-5 PMID: 29982852

El Gaamouch, F., Chen, F., Ho, L., Lin, H.-Y., Yuan, C., Wong, J., & Wang, J. (2022). Benefits of dietary polyphenols in Alzheimer's disease. *Frontiers in Aging Neuroscience*, *14*, 1019942. DOI: 10.3389/fnagi.2022.1019942 PMID: 36583187

Elkin, E. R., Higgins, C., Aung, M. T., & Bakulski, K. M. (2022). Metals exposures and DNA methylation: Current evidence and future directions. *Current Environmental Health Reports*, *9*(4), 673–696. DOI: 10.1007/s40572-022-00382-4 PMID: 36282474

Fang, M., Chen, D., & Yang, C. S. (2007). Dietary polyphenols may affect DNA methylation123. *The Journal of Nutrition, 137*(1), 223S–228S. DOI: 10.1093/jn/137.1.223S PMID: 17182830

Feng, J., Zhou, Y., Campbell, S. L., Le, T., Li, E., Sweatt, J. D., Silva, A. J., & Fan, G. (2010). Dnmt1 and Dnmt3a maintain DNA methylation and regulate synaptic function in adult forebrain neurons. *Nature Neuroscience, 13*(4), 423–430. DOI: 10.1038/nn.2514 PMID: 20228804

Feng, X., Yu, W., Li, X., Zhou, F., Zhang, W., Shen, Q., Li, J., Zhang, C., & Shen, P. (2017). Apigenin, a modulator of PPARγ, attenuates HFD-induced NAFLD by regulating hepatocyte lipid metabolism and oxidative stress via Nrf2 activation. *Biochemical Pharmacology, 136*, 136–149. DOI: 10.1016/j.bcp.2017.04.014 PMID: 28414138

Fisher, A. B. (2017). Peroxiredoxin 6 in the repair of peroxidized cell membranes and cell signaling. *Archives of Biochemistry and Biophysics, 617*, 68–83. DOI: 10.1016/j.abb.2016.12.003 PMID: 27932289

Gahtani, R. M., Shoaib, S., Hani, U., Jayachithra, R., Alomary, M. N., Chauhan, W., Jahan, R., Tufail, S., & Ansari, M. A. (2024). Combating Parkinson's disease with plant-derived polyphenols: Targeting oxidative stress and neuroinflammation. *Neurochemistry International, 178*, 105798. DOI: 10.1016/j.neuint.2024.105798 PMID: 38950626

Gandy, S. (2005). The role of cerebral amyloid β accumulation in common forms of Alzheimer disease. *The Journal of Clinical Investigation, 115*(5), 1121–1129. PMID: 15864339

Gautam, S., Karmakar, S., Batra, R., Sharma, P., Pradhan, P., Singh, J., Kundu, B., & Chowdhury, P. K. (2017). Polyphenols in combination with β-cyclodextrin can inhibit and disaggregate α-synuclein amyloids under cell mimicking conditions: A promising therapeutic alternative. *Biochimica et Biophysica Acta (BBA)-. Biochimica et Biophysica Acta. Proteins and Proteomics, 1865*(5), 589–603. DOI: 10.1016/j.bbapap.2017.02.014 PMID: 28238838

Ghodsi, H., Rahimi, H. R., Aghili, S. M., Saberi, A., & Shoeibi, A. (2022). Evaluation of curcumin as add-on therapy in patients with Parkinson's disease: A pilot randomized, triple-blind, placebo-controlled trial. *Clinical Neurology and Neurosurgery, 218*, 107300. DOI: 10.1016/j.clineuro.2022.107300 PMID: 35636380

Ghosh, P., & Saadat, A. (2023). Neurodegeneration and epigenetics: A review. [English Edition]. *Neurología : Publicación Oficial de la Sociedad Española de Neurología, 38*(6), e62–e68. DOI: 10.1016/j.nrl.2021.01.016 PMID: 37344098

Giallongo, S., Lo Re, O., Lochmanová, G., Parca, L., Petrizzelli, F., Zdráhal, Z., Mazza, T., & Vinciguerra, M. (2021). Phosphorylation within intrinsic disordered region discriminates histone variant macroH2A1 splicing isoforms—macroH2A1. 1 and macroH2A1. 2. *Biology (Basel), 10*(7), 659. DOI: 10.3390/biology10070659 PMID: 34356514

Giallongo, S., Lo Re, O., & Vinciguerra, M. (2019). Macro histone variants: Emerging rheostats of gastrointestinal cancers. *Cancers (Basel), 11*(5), 676. DOI: 10.3390/cancers11050676 PMID: 31096699

Giallongo, S., Longhitano, L., Denaro, S., D'Aprile, S., Torrisi, F., La Spina, E., Giallongo, C., Mannino, G., Lo Furno, D., Zappalà, A., Giuffrida, R., Parenti, R., Li Volti, G., Tibullo, D., & Vicario, N. (2022). The role of epigenetics in neuroinflammatory-driven diseases. *International Journal of Molecular Sciences, 23*(23), 15218. DOI: 10.3390/ijms232315218 PMID: 36499544

Giallongo, S., Řeháková, D., Biagini, T., Lo Re, O., Raina, P., Lochmanová, G., Zdráhal, Z., Resnick, I., Pata, P., Pata, I., Mistrík, M., de Magalhães, J. P., Mazza, T., Koutná, I., & Vinciguerra, M. (2022). Histone variant macroH2A1. 1 enhances nonhomologous end joining-dependent DNA double-strand-break repair and reprogramming efficiency of human iPSCs. *Stem Cells (Dayton, Ohio), 40*(1), 35–48. DOI: 10.1093/stmcls/sxab004 PMID: 35511867

Gomez de Cedron, M., & Ramírez de Molina, A. (2023). Metabolo-epigenetic interplay provides targeted nutritional interventions in chronic diseases and ageing. *Frontiers in Oncology, 13*, 1169168–1169168. DOI: 10.3389/fonc.2023.1169168 PMID: 37404756

Grabska-Kobyłecka, I., Szpakowski, P., Król, A., Książek-Winiarek, D., Kobyłecki, A., Głąbiński, A., & Nowak, D. (2023). Polyphenols and their impact on the prevention of neurodegenerative diseases and development. *Nutrients, 15*(15), 3454. DOI: 10.3390/nu15153454 PMID: 37571391

Gregory, P. D., Wagner, K., & Hörz, W. (2001). Histone acetylation and chromatin remodeling. *Experimental Cell Research, 265*(2), 195–202. DOI: 10.1006/excr.2001.5187 PMID: 11302684

Guerreiro, Í., Ferreira-Pêgo, C., Carregosa, D., Santos, C. N., Menezes, R., Fernandes, A. S., & Costa, J. G. (2022). Polyphenols and their metabolites in renal diseases: An overview. *Foods, 11*(7), 1060. DOI: 10.3390/foods11071060 PMID: 35407148

Guo, S., Wang, H., & Yin, Y. (2022). Microglia polarization from M1 to M2 in neurodegenerative diseases. *Frontiers in Aging Neuroscience, 14*, 815347. DOI: 10.3389/fnagi.2022.815347 PMID: 35250543

Gupta, S., & Dasmahapatra, A. K. (2020). Destabilization potential of phenolics on Aβ fibrils: Mechanistic insights from molecular dynamics simulation. *Physical Chemistry Chemical Physics, 22*(35), 19643–19658. DOI: 10.1039/D0CP02459G PMID: 32830209

Han, D., Dong, X., Zheng, D., & Nao, J. (2020). MiR-124 and the underlying therapeutic promise of neurodegenerative disorders. *Frontiers in Pharmacology, 10*, 1555. DOI: 10.3389/fphar.2019.01555 PMID: 32009959

Hardy, T. M., & Tollefsbol, T. O. (2011). Epigenetic diet: Impact on the epigenome and cancer. *Epigenomics, 3*(4), 503–518. DOI: 10.2217/epi.11.71 PMID: 22022340

Harris, J. J., Jolivet, R., & Attwell, D. (2012). Synaptic energy use and supply. *Neuron, 75*(5), 762–777. DOI: 10.1016/j.neuron.2012.08.019 PMID: 22958818

Ho, E., Beaver, L. M., Williams, D. E., & Dashwood, R. H. (2011). Dietary factors and epigenetic regulation for prostate cancer prevention. *Advances in Nutrition, 2*(6), 497–510. DOI: 10.3945/an.111.001032 PMID: 22332092

Hu, X., Leak, R. K., Shi, Y., Suenaga, J., Gao, Y., Zheng, P., & Chen, J. (2015). Microglial and macrophage polarization—New prospects for brain repair. *Nature Reviews. Neurology*, *11*(1), 56–64. DOI: 10.1038/nrneurol.2014.207 PMID: 25385337

Huber, K., & Superti-Furga, G. (2011). After the grape rush: Sirtuins as epigenetic drug targets in neurodegenerative disorders. *Bioorganic & Medicinal Chemistry*, *19*(12), 3616–3624. DOI: 10.1016/j.bmc.2011.01.018 PMID: 21306906

Hwang, J.-Y., Aromolaran, K. A., & Zukin, R. S. (2017). The emerging field of epigenetics in neurodegeneration and neuroprotection. *Nature Reviews. Neuroscience*, *18*(6), 347–361. DOI: 10.1038/nrn.2017.46 PMID: 28515491

Jalouli, M., Rahman, M. A., Biswas, P., Rahman, H., Harrath, A. H., Lee, I.-S., Kang, S., Choi, J., Park, M. N., & Kim, B. (2025). Targeting natural antioxidant polyphenols to protect neuroinflammation and neurodegenerative diseases: A comprehensive review. *Frontiers in Pharmacology*, *16*, 1492517. DOI: 10.3389/fphar.2025.1492517 PMID: 39981183

Jurcau, A. (2021). Insights into the pathogenesis of neurodegenerative diseases: Focus on mitochondrial dysfunction and oxidative stress. *International Journal of Molecular Sciences*, *22*(21), 11847. DOI: 10.3390/ijms222111847 PMID: 34769277

Kishimoto, Y., Tani, M., & Kondo, K. (2013). Pleiotropic preventive effects of dietary polyphenols in cardiovascular diseases. *European Journal of Clinical Nutrition*, *67*(5), 532–535. DOI: 10.1038/ejcn.2013.29 PMID: 23403879

Klein, C., & Westenberger, A. (2012). Genetics of Parkinson's disease. *Cold Spring Harbor Perspectives in Medicine*, *2*(1), a008888. DOI: 10.1101/cshperspect.a008888 PMID: 22315721

Köbe, T., Witte, A. V., Schnelle, A., Tesky, V. A., Pantel, J., Schuchardt, J.-P., Hahn, A., Bohlken, J., Grittner, U., & Flöel, A. (2017). Impact of resveratrol on glucose control, hippocampal structure and connectivity, and memory performance in patients with mild cognitive impairment. *Frontiers in Neuroscience*, *11*, 105. DOI: 10.3389/fnins.2017.00105 PMID: 28326010

Kumar, D., Ambasta, R. K., & Kumar, P. (2020). Ubiquitin biology in neurodegenerative disorders: From impairment to therapeutic strategies. *Ageing Research Reviews*, *61*, 101078. DOI: 10.1016/j.arr.2020.101078 PMID: 32407951

Kung, H.-C., Lin, K.-J., Kung, C.-T., & Lin, T.-K. (2021). Oxidative stress, mitochondrial dysfunction, and neuroprotection of polyphenols with respect to resveratrol in Parkinson's disease. *Biomedicines*, *9*(8), 918. DOI: 10.3390/biomedicines9080918 PMID: 34440122

Lagouge, M., Argmann, C., Gerhart-Hines, Z., Meziane, H., Lerin, C., Daussin, F., Messadeq, N., Milne, J., Lambert, P., Elliott, P., Geny, B., Laakso, M., Puigserver, P., & Auwerx, J. (2006). Resveratrol improves mitochondrial function and protects against metabolic disease by activating SIRT1 and PGC-1α. *Cell*, *127*(6), 1109–1122. DOI: 10.1016/j.cell.2006.11.013 PMID: 17112576

Li, R., Zhou, Y., Zhang, S., Li, J., Zheng, Y., & Fan, X. (2022). The natural (poly) phenols as modulators of microglia polarization via TLR4/NF-κB pathway exert anti-inflammatory activity in ischemic stroke. *European Journal of Pharmacology*, *914*, 174660. DOI: 10.1016/j.ejphar.2021.174660 PMID: 34863710

Li, X., Zhang, Y., Yang, Z., Zhang, S., & Zhang, L. (2024). The Inhibition Effect of Epigallocatechin-3-Gallate on the Co-Aggregation of Amyloid-β and Human Islet Amyloid Polypeptide Revealed by Replica Exchange Molecular Dynamics Simulations. *International Journal of Molecular Sciences*, *25*(3), 1636. DOI: 10.3390/ijms25031636 PMID: 38338914

Liu, Z., Ren, Z., Zhang, J., Chuang, C.-C., Kandaswamy, E., Zhou, T., & Zuo, L. (2018). Role of ROS and nutritional antioxidants in human diseases. *Frontiers in Physiology*, *9*, 360203. DOI: 10.3389/fphys.2018.00477 PMID: 29867535

Lu, Y., Wang, K., & Hu, L. (2025). Advancements in delivery systems for dietary polyphenols in enhancing radioprotection effects: challenges and opportunities. *npj Science of Food*, *9*(1), 51.

Luo, J., Si, H., Jia, Z., & Liu, D. (2021). Dietary anti-aging polyphenols and potential mechanisms. *Antioxidants*, *10*(2), 283. DOI: 10.3390/antiox10020283 PMID: 33668479

Mander, P. K., Jekabsone, A., & Brown, G. C. (2006). Microglia proliferation is regulated by hydrogen peroxide from NADPH oxidase. *The Journal of Immunology : Official Journal of the American Association of Immunologists*, *176*(2), 1046–1052. DOI: 10.4049/jimmunol.176.2.1046 PMID: 16393992

Mignet, N., Seguin, J., & Chabot, G. G. (2013). Bioavailability of polyphenol liposomes: A challenge ahead. *Pharmaceutics*, *5*(3), 457–471. DOI: 10.3390/pharmaceutics5030457 PMID: 24300518

Minarovits, J., Banati, F., Szenthe, K., & Niller, H. H. (2015). Epigenetic regulation. *Patho-Epigenetics of Infectious Disease*, 1-25.

Mittal, P., & Agrawal, N. (2024). An Overview of Neurodegenerative Disorders. *Altered Metabolism: A Major Contributor of Comorbidities in Neurodegenerative Diseases*, 1-27.

Mokhtari, H., Yaghmaei, B., Sirati-Sabet, M., Jafari, N., Mardomi, A., Abediankenari, S., & Mahrooz, A. (2020). Epigallocatechin-3 gallate enhances the efficacy of MicroRNA-34a mimic and MiR-93 inhibitor co-transfection in prostate cancer cell line. *Iranian Journal of Allergy, Asthma, and Immunology*. Advance online publication. DOI: 10.18502/ijaai.v19i6.4930 PMID: 33463130

Moussa, C., Hebron, M., Huang, X., Ahn, J., Rissman, R. A., Aisen, P. S., & Turner, R. S. (2017). Resveratrol regulates neuro-inflammation and induces adaptive immunity in Alzheimer's disease. *Journal of Neuroinflammation*, *14*(1), 1. DOI: 10.1186/s12974-016-0779-0 PMID: 28086917

Mukherjee, N., Kumar, A. P., & Ghosh, R. (2015). DNA methylation and flavonoids in genitourinary cancers. *Current Pharmacology Reports*, *1*(2), 112–120. DOI: 10.1007/s40495-014-0004-8 PMID: 26005633

National Library of Medicine (US). (2007, April 19). *The Effects of Green Tea on Parkinson's Disease* (Clinical Trial Registration No. NCT00461942). ClinicalTrials.gov. https://clinicaltrials.gov/ct2/show/NCT00461942

National Library of Medicine (US). (2014, February 12). *Grape Seed Extract in People With Memory Decline* (Clinical Trial Registration No. NCT02033941). ClinicalTrials.gov. https://clinicaltrials.gov/ct2/show/NCT02033941

Nawaz, L., Grieve, D. J., Muzaffar, H., Iftikhar, A., & Anwar, H. (2024). Methanolic Extract of Phoenix Dactylifera Confers Protection against Experimental Diabetic Cardiomyopathy through Modulation of Glucolipid Metabolism and Cardiac Remodeling. *Cells, 13*(14), 1196. DOI: 10.3390/cells13141196 PMID: 39056777

Rajman, M., & Schratt, G. (2017). MicroRNAs in neural development: From master regulators to fine-tuners. *Development (Cambridge, England), 144*(13), 2310–2322. DOI: 10.1242/dev.144337 PMID: 28676566

Rana, A., Samtiya, M., Dhewa, T., Mishra, V., & Aluko, R. E. (2022). Health benefits of polyphenols: A concise review. *Journal of Food Biochemistry, 46*(10), e14264. DOI: 10.1111/jfbc.14264 PMID: 35694805

Renaud, J., & Martinoli, M.-G. (2019). Considerations for the use of polyphenols as therapies in neurodegenerative diseases. *International Journal of Molecular Sciences, 20*(8), 1883. DOI: 10.3390/ijms20081883 PMID: 30995776

Réus, G. Z., Fries, G. R., Stertz, L., Badawy, M., Passos, I., Barichello, T., Kapczinski, F., & Quevedo, J. (2015). The role of inflammation and microglial activation in the pathophysiology of psychiatric disorders. *Neuroscience, 300*, 141–154. DOI: 10.1016/j.neuroscience.2015.05.018 PMID: 25981208

Rocca, W. A. (2018). The burden of Parkinson's disease: A worldwide perspective. *Lancet Neurology, 17*(11), 928–929. DOI: 10.1016/S1474-4422(18)30355-7 PMID: 30287052

Rodríguez-Vera, D., Abad-García, A., Vargas-Mendoza, N., Pinto-Almazán, R., Farfán-García, E. D., Morales-González, J. A., & Soriano-Ursúa, M. A. (2022). Polyphenols as potential enhancers of stem cell therapy against neurodegeneration. *Neural Regeneration Research, 17*(10), 2093–2101. DOI: 10.4103/1673-5374.335826 PMID: 35259814

Royston, K. J. (2018). *Combinatorial Withaferin A and Sulforaphane and Their Potential Roles in Breast Cancer Prevention and Therapy*. The University of Alabama at Birmingham.

Schaffer, S., & Halliwell, B. (2012). Do polyphenols enter the brain and does it matter? Some theoretical and practical considerations. *Genes & Nutrition, 7*(2), 99–109. DOI: 10.1007/s12263-011-0255-5 PMID: 22012276

Sengupta, U., & Kayed, R. (2022). Amyloid β, Tau, and α-Synuclein aggregates in the pathogenesis, prognosis, and therapeutics for neurodegenerative diseases. *Progress in Neurobiology, 214*, 102270. DOI: 10.1016/j.pneurobio.2022.102270 PMID: 35447272

Sharma, N., & Nehru, B. (2018). Curcumin affords neuroprotection and inhibits α-synuclein aggregation in lipopolysaccharide-induced Parkinson's disease model. *Inflammopharmacology, 26*(2), 349–360. DOI: 10.1007/s10787-017-0402-8 PMID: 29027056

Simms, C. L., & Zaher, H. S. (2016). Quality control of chemically damaged RNA. *Cellular and Molecular Life Sciences : CMLS, 73*(19), 3639–3653. DOI: 10.1007/s00018-016-2261-7 PMID: 27155660

Smith, J. A., Das, A., Ray, S. K., & Banik, N. L. (2012). Role of pro-inflammatory cytokines released from microglia in neurodegenerative diseases. *Brain Research Bulletin, 87*(1), 10–20. DOI: 10.1016/j.brainresbull.2011.10.004 PMID: 22024597

Sorond, F. A., Hurwitz, S., Salat, D. H., Greve, D. N., & Fisher, N. D. (2013). Neurovascular coupling, cerebral white matter integrity, and response to cocoa in older people. *Neurology, 81*(10), 904–909. DOI: 10.1212/WNL.0b013e3182a351aa PMID: 23925758

Sun, J., Yang, J., Miao, X., Loh, H. H., Pei, D., & Zheng, H. (2021). Proteins in DNA methylation and their role in neural stem cell proliferation and differentiation. *Cell Regeneration (London, England), 10*(1), 1–12. DOI: 10.1186/s13619-020-00070-4 PMID: 33649938

Sun, Y., Luo, Z.-M., Guo, X.-M., Su, D.-F., & Liu, X. (2015). An updated role of microRNA-124 in central nervous system disorders: A review. *Frontiers in Cellular Neuroscience, 9*, 193. DOI: 10.3389/fncel.2015.00193 PMID: 26041995

Tanaka, M., Chock, P. B., & Stadtman, E. R. (2007). Oxidized messenger RNA induces translation errors. *Proceedings of the National Academy of Sciences of the United States of America, 104*(1), 66–71. DOI: 10.1073/pnas.0609737104 PMID: 17190801

Tandon, A., Singh, S. J., & Chaturvedi, R. K. (2018). Stem cells as potential targets of polyphenols in multiple sclerosis and Alzheimer's disease. *BioMed Research International, 2018*(1), 1483791. DOI: 10.1155/2018/1483791 PMID: 30112360

Tang, Y., & Le, W. (2016). Differential roles of M1 and M2 microglia in neurodegenerative diseases. *Molecular Neurobiology, 53*(2), 1181–1194. DOI: 10.1007/s12035-014-9070-5 PMID: 25598354

Thapa, A., Jett, S. D., & Chi, E. Y. (2016). Curcumin attenuates amyloid-β aggregate toxicity and modulates amyloid-β aggregation pathway. *ACS Chemical Neuroscience, 7*(1), 56–68. DOI: 10.1021/acschemneuro.5b00214 PMID: 26529184

Tortorella, S. M., Royce, S. G., Licciardi, P. V., & Karagiannis, T. C. (2015). Dietary sulforaphane in cancer chemoprevention: The role of epigenetic regulation and HDAC inhibition. *Antioxidants & Redox Signaling, 22*(16), 1382–1424. DOI: 10.1089/ars.2014.6097 PMID: 25364882

Ullah, H., De Filippis, A., Santarcangelo, C., & Daglia, M. (2020). Epigenetic regulation by polyphenols in diabetes and related complications. *Mediterranean Journal of Nutrition and Metabolism, 13*(4), 289–310. DOI: 10.3233/MNM-200489

Vaiserman, A., & Lushchak, O. (2021). DNA methylation changes induced by prenatal toxic metal exposure: An overview of epidemiological evidence. *Environmental Epigenetics, 7*(1), dvab007. PMID: 34631153

Vaissière, T., Sawan, C., & Herceg, Z. (2008). Epigenetic interplay between histone modifications and DNA methylation in gene silencing. *Mutation Research/Reviews in Mutation Research, 659*(1-2), 40–48. DOI: 10.1016/j.mrrev.2008.02.004 PMID: 18407786

Wang, J., Fan, S. M., & Zhang, J. (2019). Epigallocatechin-3-gallate ameliorates lipopolysaccharide-induced acute lung injury by suppression of TLR4/NF-κB signaling activation. *Brazilian Journal of Medical and Biological Research = Revista Brasileira de Pesquisas Médicas e Biológicas, 52*(7), e8092. DOI: 10.1590/1414-431x20198092 PMID: 31241712

Xylaki, M., Atzler, B., & Outeiro, T. F. (2019). Epigenetics of the Synapse in Neurodegeneration. *Current Neurology and Neuroscience Reports*, *19*(10), 1–10. DOI: 10.1007/s11910-019-0995-y PMID: 31440934

Yahfoufi, N., Alsadi, N., Jambi, M., & Matar, C. (2018). The immunomodulatory and anti-inflammatory role of polyphenols. *Nutrients*, *10*(11), 1618. DOI: 10.3390/nu10111618 PMID: 30400131

Yang, S.-G., Wang, X.-W., Qian, C., & Zhou, F.-Q. (2022). Reprogramming neurons for regeneration: The fountain of youth. *Progress in Neurobiology*, *214*, 102284. DOI: 10.1016/j.pneurobio.2022.102284 PMID: 35533809

Yang, X., Xu, S., Qian, Y., & Xiao, Q. (2017). Resveratrol regulates microglia M1/M2 polarization via PGC-1α in conditions of neuroinflammatory injury. *Brain, Behavior, and Immunity*, *64*, 162–172. DOI: 10.1016/j.bbi.2017.03.003 PMID: 28268115

Yang, Y., Ye, Y., Kong, C., Su, X., Zhang, X., Bai, W., & He, X. (2019). MiR-124 enriched exosomes promoted the M2 polarization of microglia and enhanced hippocampus neurogenesis after traumatic brain injury by inhibiting TLR4 pathway. *Neurochemical Research*, *44*(4), 811–828. DOI: 10.1007/s11064-018-02714-z PMID: 30628018

Youdim, K. A., Dobbie, M. S., Kuhnle, G., Proteggente, A. R., Abbott, N. J., & Rice-Evans, C. (2003). Interaction between flavonoids and the blood–brain barrier: In vitro studies. *Journal of Neurochemistry*, *85*(1), 180–192. DOI: 10.1046/j.1471-4159.2003.01652.x PMID: 12641740

Yuan, Z., Syed, M. A., Panchal, D., Rogers, D., Joo, M., & Sadikot, R. T. (2012). Curcumin mediated epigenetic modulation inhibits TREM-1 expression in response to lipopolysaccharide. *The International Journal of Biochemistry & Cell Biology*, *44*(11), 2032–2043. DOI: 10.1016/j.biocel.2012.08.001 PMID: 22890222

Zaidi, F. K., & Deep, S. (2020). Scutellarin inhibits the uninduced and metal-induced aggregation of α-Synuclein and disaggregates preformed fibrils: Implications for Parkinson's disease. *The Biochemical Journal*, *477*(3), 645–670. DOI: 10.1042/BCJ20190705 PMID: 31939603

Zeng, P., Fang, M., Zhao, H., & Guo, J. (2021). A network pharmacology approach to uncover the key ingredients in Ginkgo Folium and their anti-Alzheimer's disease mechanisms. *Aging (Albany NY)*, *13*(14), 18993–19012. DOI: 10.18632/aging.203348 PMID: 34315132

Zerbinati, C., & Iuliano, L. (2017). Cholesterol and related sterols autoxidation. *Free Radical Biology & Medicine*, *111*, 151–155. DOI: 10.1016/j.freeradbiomed.2017.04.013 PMID: 28428001

Zhang, B., Wei, Y.-Z., Wang, G.-Q., Li, D.-D., Shi, J.-S., & Zhang, F. (2019). Targeting MAPK pathways by naringenin modulates microglia M1/M2 polarization in lipopolysaccharide-stimulated cultures. *Frontiers in Cellular Neuroscience*, *12*, 531. DOI: 10.3389/fncel.2018.00531 PMID: 30687017

Zhang, G., Liu, Y., Xu, L., Sha, C., Zhang, H., & Xu, W. (2019). Resveratrol alleviates lipopolysaccharide-induced inflammation in PC-12 cells and in rat model. *BMC Biotechnology*, *19*(1), 10. DOI: 10.1186/s12896-019-0502-1 PMID: 30691440

Zhao, J., Liang, Q., Sun, Q., Chen, C., Xu, L., Ding, Y., & Zhou, P. (2017). (−)-Epigallocatechin-3-gallate (EGCG) inhibits fibrillation, disaggregates amyloid fibrils of α-synuclein, and protects PC12 cells against α-synuclein-induced toxicity. *RSC Advances*, *7*(52), 32508–32517. DOI: 10.1039/C7RA03752J

Zhao, Y.-N., Li, W.-F., Li, F., Zhang, Z., Dai, Y.-D., Xu, A.-L., Qi, C., Gao, J.-M., & Gao, J. (2013). Resveratrol improves learning and memory in normally aged mice through microRNA-CREB pathway. *Biochemical and Biophysical Research Communications*, *435*(4), 597–602. DOI: 10.1016/j.bbrc.2013.05.025 PMID: 23685142

Zhou, M., & Pang, X. e. (2025). Polyphenols and miRNA interplay: A novel approach to combat apoptosis and inflammation in Alzheimer's disease. *Frontiers in Aging Neuroscience*, *17*, 1571563. DOI: 10.3389/fnagi.2025.1571563 PMID: 40400913

Zhou, Z. D., Xie, S. P., Saw, W. T., Ho, P. G. H., Wang, H. Y., Zhou, L., Zhao, Y., & Tan, E. K. (2019). The therapeutic implications of tea polyphenols against dopamine (DA) neuron degeneration in Parkinson's disease (PD). *Cells*, *8*(8), 911. DOI: 10.3390/cells8080911 PMID: 31426448

Zhu, X., Liu, Q., Wang, M., Liang, M., Yang, X., Xu, X., Zou, H., & Qiu, J. (2011). Activation of Sirt1 by resveratrol inhibits TNF-α induced inflammation in fibroblasts. *PLoS One*, *6*(11), e27081. DOI: 10.1371/journal.pone.0027081 PMID: 22069489

Chapter 3
Bioactive Peptides Produced From Microbial Sources and Their Neuroprotective Effects:
Microbial Peptides

Muhsin Konuk
https://orcid.org/0000-0002-6651-718X
Üsküdar University, Turkey

Safiye Elif Korcan
Uşak University, Turkey

ABSTRACT

Microbial bioactive peptides represent a promising class of natural compounds with significant potential as neuroprotective agents. Produced through fermentation, these peptides show antioxidant, anti-inflammatory, antimicrobial, and immunomodulatory activities. Their ability to modulate the gut-brain axis further enhances their role in supporting neurological health by influencing immune responses, oxidative stress balance, and synaptic function. Key mechanisms of these peptides exert neuroprotection, neuroinflammation, inhibition of oxidative stress, prevention of apoptosis, and enhancement of mitochondrial function and neurotrophic signaling. Additionally, some peptides protect apoptosis in Alzheimer's, Huntington's, and etc. This review comprehensively outlines the sources, production, mechanisms, and future directions of microbial peptides, underscoring their growing importance in the prevention and treatment of neurodegenerative diseases.

INTRODUCTION

The use of natural products is as old as human history and has played an important role in medicine, food, and cosmetics. Since ancient times, people have used plants, fungi, animal products, and minerals in the treatment of diseases. For example, in Ancient Egypt, garlic, honey, and various plant extracts were used for wound healing and infection prevention. In Ancient Greece, Hippocrates and Galen systematically classified plant-derived medicines, laying the foundations of medicine. In China and India,

DOI: 10.4018/979-8-3373-4139-2.ch003

within the frameworks of Ayurveda and Traditional Chinese Medicine, natural products were applied therapeutically for hundreds of years. During the Middle Ages, Avicenna's "The Canon of Medicine" in the Islamic world described the pharmacological effects of natural products in detail. In the 19th century, with the development of chemistry, pure compounds were isolated from natural products (such as morphine, quinine, and salicin), forming the basis of modern pharmacology. In the 20th century, Alexander Fleming's discovery of penicillin from mold in 1928 became one of the most revolutionary examples of natural products in the development of antibiotics and modern medicine (Gaynes, 2017). Today, natural products continue to be significant both in maintaining traditional therapies and as a source of biologically active compounds in modern drug discovery. This shift has led to a growing interest in the bioactive compounds offered by microorganisms, both in agricultural and food technologies and in pharmaceutical research (Choi& Oh, 2015).

Today, both chemical synthesis and biotechnological approaches enable the elucidation of complex interactions of natural components, facilitating the development of synergistic therapeutic platforms. In particular, in neurodegenerative diseases, the loss of nerve cells and chronic inflammation are two major issues that limit the effectiveness of current drugs. The immunomodulatory effects of natural products, which reduce neuroinflammation and protect nerve cells, offer promising potential in the treatment of disorders such as Alzheimer's, Parkinson's, and motor neuron disease (Sinha and Mukhopadhyay, 2021).

Peptides, polysaccharides, and other small molecules derived from microorganisms can contribute to improved neuronal function by regulating intracellular protein homeostasis in the brain. Moreover, the selective activity of these bioactive agents—with minimal harm to healthy cells makes it possible to reduce side effect profiles, as seen in cancer treatment (Cragg et al., 2009), and applies similarly to therapeutic approaches for neurodegenerative disorders (Mohd Sairazi & Sirajudeen, 2020).

This chapter covers the definition, classification, and production methods of bioactive peptides derived from microbial sources. It also examines the neuroprotective mechanisms of these peptides, including their antioxidant, anti-inflammatory, apoptosis-regulating, and synaptic plasticity-enhancing effects. The role of microbial peptides in the gut-brain axis and challenges related to their bioavailability are also discussed. The final section highlights innovative applications such as personalized therapies, targeted nanoformulations, microbial engineering, AI-assisted peptide design, and the bioelectronic potential of electrically conductive microbial peptides. Overall, the chapter emphasizes the promising future role of microbial peptides in the treatment of neurological disorders.

1. BIOACTIVE PEPTIDES: DEFINITION, CLASSIFICATION AND SOURCES

Bioactive peptides are short chains of amino acids released through enzymatic hydrolysis of proteins or microbial fermentation, and they have the ability to regulate physiological functions (Sánchez & Vázquez, 2017; Korhonen & Pihlanto, 2006). These molecules play an important role in food and pharmaceutical industries due to their non-nutritive effects. Typically consisting of 2–20 amino acids, bioactive peptides are protein fragments that can exhibit antimicrobial, antihypertensive, antioxidant, or immunomodulatory effects in the host organism (Udenigwe & Aluko, 2012). Their biological activities are directly related to their amino acid sequences and three-dimensional structures. Bioactive peptides can be classified based on their source, function, and structural characteristics (Table 1).

Natural peptides can be broadly classified into three main categories based on their origin: microbial, plant-derived, and animal-derived peptides. Microbial peptides include compounds such as bacteriocins and non-ribosomal peptides, which are produced by bacteria and fungi and often exhibit antimicrobial or immunomodulatory activities. Plant-derived peptides encompass molecules like cyclotides, defensins, and storage proteins, which contribute to plant defense mechanisms and possess antioxidants, antimicrobial, and therapeutic properties. Animal derived peptides, on the other hand, include bioactive molecules such as antimicrobial peptides (AMPs), hormone-like peptides, and neuropeptides, which play critical roles in innate immunity, cell signaling, and tissue repair. Together, these diverse classes of natural peptides represent a valuable reservoir of bioactive compounds with potential applications in medicine, agriculture, and biotechnology. Comparison of Microbial, Plant derived, and Animal derived peptides are given table 1.

Table 1. Comparison of Microbial, Plant-Derived, and Animal-Derived Peptides

Feature	Microbial Peptides	Plant-Derived Peptides	Animal-Derived Peptides
Source	Produced by bacteria, fungi, and other microorganisms (Godyń et al., 2016; Özben & Özben, 2019)	Extracted from seeds, legumes, cereals, or other plant tissues (Hartmann & Meisel, 2007; Nongonierma & FitzGerald, 2015)	Extracted from milk, eggs, meat, or other animal tissues (Korhonen & Pihlanto, 2006; Meisel, 1997)
Bioactivity	Broad-spectrum antimicrobial, immunomodulatory, neuroprotective, antioxidant (Ma et al., 2022; Liu et al., 2024)	Mostly antioxidant, antihypertensive, anti-inflammatory; some antimicrobial (Udenigwe & Aluko, 2012)	Strong antioxidant, antimicrobial, immunomodulatory, and neuroprotective activities (Korhonen & Pihlanto, 2006)
Production & Scalability	Produced via fermentation; scalable with bioreactors; easier genetic manipulation (Godyń et al., 2016)	Requires plant extraction; yield varies; seasonal dependency (Hartmann & Meisel, 2007)	Extraction may be expensive; limited yield; ethical concerns (Meisel, 1997)
Stability	Often small and stable; can be engineered for enhanced stability (Özben & Özben, 2019)	Generally stable; may require processing (Udenigwe & Aluko, 2012)	Can be less stable; sensitive to enzymatic degradation (Korhonen & Pihlanto, 2006)
Bioavailability & Delivery	Low oral bioavailability; can be enhanced with nanoformulations (Journal of Controlled Release, 2014)	Often stable orally; absorption may vary (Nongonierma & FitzGerald, 2015)	May require enzymatic protection or encapsulation for oral delivery (Meisel, 1997)
Safety & Immunogenicity	Usually, low immunogenicity; engineered peptides may trigger immune response (Godyń et al., 2016)	Generally safe; low immunogenicity (Hartmann & Meisel, 2007)	Potential for allergenicity or immune response (Korhonen & Pihlanto, 2006)
Regulatory Status	Novel: regulatory approval still evolving (Godyń et al., 2016)	Generally recognized as safe (GRAS) (Udenigwe & Aluko, 2012)	Subject to stricter regulatory scrutiny; food vs therapeutic use (Meisel, 1997)
Therapeutic Potential	Strong potential for neurodegeneration, infections, metabolic modulation (Ma et al., 2022; Liu et al., 2024)	Primarily for metabolic, cardiovascular, antioxidant benefits (Udenigwe & Aluko, 2012)	Strong therapeutic potential; already used in nutraceuticals and drugs (Korhonen & Pihlanto, 2006)

Bioactive peptides hold critical potential for future therapeutic and nutraceutical developments through multidisciplinary research. Reducing production costs and designing targeted peptides are key areas of focus for ongoing and future research. However, factors such as bioavailability, blood–brain barrier (BBB) passage, and immunogenicity, lack of clinical validation, cost, and manufacturing scalability limit their therapeutic and diagnostic applications (Table 2). Peptide Therapy also presents some risks and limitations (Table 3). Therefore, nanocarrier systems, stabilization techniques, and comprehensive clinical studies are necessary for peptide-based products to achieve industrial and clinical success.

Table 2. Limitations, Translational Challenges, and Industrial Applicability of Peptides in Therapy and Diagnostics

Category	Limitations / Challenges	Description / Implications
Limitations	Bioavailability & Stability	Peptides are often rapidly degraded by gastrointestinal enzymes, resulting in low oral bioavailability and short half-life.
	Blood–Brain Barrier (BBB) Penetration	CNS-targeted peptides may not effectively cross the BBB, limiting efficacy in neurodegenerative diseases.
	Immunogenicity & Toxicity	Some peptides can trigger immune responses or allergic reactions, potentially causing adverse effects.
	Target Specificity	Limited specificity can lead to off-target effects or reduced therapeutic efficacy.
Translational Challenges	Preclinical-to-Clinical Translation	Efficacy in vitro or in animal models does not always translate to humans, creating uncertainty in clinical outcomes.
	Clinical Trial Limitations	Small cohort sizes, patient heterogeneity, and lack of appropriate biomarkers complicate clinical validation.
	Regulatory Approval	Obtaining approval from agencies like FDA/EMA requires comprehensive safety, pharmacokinetic, and toxicology data, which can be time-consuming.
Industrial Applicability	Production Costs	Peptide synthesis, especially for long or modified peptides, is expensive, limiting large-scale feasibility.
	Scalability	Optimization is needed to maintain purity, stability, and yield during scale-up from lab to industrial production.
	Formulation & Storage	Peptides require advanced formulation (e.g., nanoparticles, liposomes) for stability; storage conditions critically affect shelf-life.

Table 3. Peptide Therapy Risks and Limitations

Category	Microbial Peptides	Plant-Derived Peptides	Animal-Derived Peptides	Common Challenges / Potential Risks	References
Rapid Metabolism / Short Half-Life	Usually short half-life; susceptible to enzymatic degradation	Most peptides are rapidly metabolized; low oral bioavailability	Rapid renal and hepatic elimination	Reduced therapeutic efficacy; frequent dosing required	Lau & Dunn, 2018; Wang et al., 2022
Bioavailability & Delivery	Limited ability to cross blood-brain barrier	Low gastrointestinal stability	Limited tissue penetration	Injection or carrier systems required	Aileen Funke & Willbold, 2012; Xie et al., 2020
Immune Response / Allergic Reactions	Can be recognized as foreign proteins; anti-peptide antibodies possible	May trigger immune response; risk of allergy	Risk of immune response in long-term use	Autoimmune reaction, inflammation	Ciobanasu, 2021; Lau & Dunn, 2018

continued on following page

Table 3. Continued

Category	Microbial Peptides	Plant-Derived Peptides	Animal-Derived Peptides	Common Challenges / Potential Risks	References
Non-Specific Effects / Toxicity	Risk of cell membrane damage; kidney/ liver toxicity	Organ toxicity at high doses	Cytotoxicity; off-target effects	Adverse effects; limits treatment safety	Marqus et al., 2017; Fetse et al., 2023
Translational / Clinical Challenges	Effective in animal models, limited in humans	Pharmacodynamic differences	Challenges in clinical translation	Efficacy in humans uncertain	Wang et al., 2022; Javid et al., 2024
Production / Stability / Cost	Enzymatic degradation; high cost	Low chemical stability; costly production	Storage and transport challenges	Advanced formulation and carrier systems needed	Lau & Dunn, 2018; Apostolopoulos et al., 2021

2. PRODUCTION METHODS OF MICROBIAL BIOACTIVE PEPTIDES

In recent years, microbial bioactive peptides have attracted great attention due to their therapeutic potential and applications in food, pharmaceuticals, and nutraceuticals. Microbial bioactive peptides are produced by various microorganisms such as *Lactobacillus*, *Bacillus*, *Streptomyces*, and *Aspergillus* species. *Bacillus subtilis* is known to produce antimicrobial lipopeptides like surfactin, fengycin, and iturin (Jacques, 2011).

These microbes can synthesize peptides through ribosomal (RiPPs) or non-ribosomal peptide synthesis (NRPS) pathways (Tajbakhsh et al., 2017).

Ribosomally synthesized peptides are encoded directly by genes and undergo post-translational modifications, whereas non-ribosomal peptides are synthesized by large enzyme complexes called non-ribosomal peptide synthetases (NRPS), often incorporating non-standard amino acids (Strieker et al., 2010).

In RiPP biosynthesis, genetic material is transcribed into mRNA, which is then translated by the ribosome into a peptide chain. The resulting precursor peptide is subsequently processed by various modification enzymes to become a mature natural product (Arnison et al., 2013). This pathway resembles classical protein synthesis as it is ribosome-dependent, and the final product typically has antimicrobial or signaling functions.

In contrast, peptide synthesis in the NRPS system occurs independently of the ribosome. In this system, NRPS enzyme complexes directly assemble amino acids. Each module activates a specific amino acid, attaches it to a carrier protein, and incorporates it into the growing peptide chain (Finking & Marahiel, 2004; Fischbach& Walsh, 2006). Peptides produced by this system generally have more complex structures, including D-amino acids, cyclic formations, or branched chains.

The synthesized peptide products can function within the cell or be secreted outside. Notably, NRPS products often include molecules that act in the extracellular environment, such as antibiotics, siderophores, and toxins (Fischbach & Walsh, 2006). The RiPP products, on the other hand, may function as antimicrobial peptides, quorum sensing signals, or defense molecules (Arnison et al., 2013).

The methods used for the production of microbial bioactive peptides range from traditional fermentation processes to advanced biotechnological strategies. The choice of these methods depends on the structural characteristics of the target peptide, production efficiency, cost-effectiveness, and the intended final application. Below, three key approaches in the production process are discussed in detail.

2.1. Traditional Methods: Fermentation and Enzymatic Hydrolysis

Fermentation is a traditional method that allows microorganisms to produce bioactive peptides naturally under controlled conditions. Microorganisms such as *Lactobacillus*, *Bacillus*, *Streptomyces*, and *Pediococcus* are commonly used in this process. During fermentation, microorganisms break down proteins in the growth medium and secrete peptides with antimicrobial, antioxidant, and antihypertensive properties. The success of fermentation depends on optimizing environmental factors such as pH, temperature, oxygen levels, and nutrient composition. For example, lactic acid bacteria generally yield the highest production at low pH and anaerobic conditions, while *Bacillus* species may be active at higher temperatures (Korhonen & Pihlanto, 2006).

Enzymatic hydrolysis is a controlled method in which proteins are broken down into bioactive peptides with the help of specific proteases. The most commonly used enzymes in this process include pepsin, trypsin, papain, alcalase, and flavourzyme. The selection of enzymes plays a crucial role in determining the peptide profile and bioactivity (Agyei &Danquah, 2011). This method is frequently used for evaluating food-derived proteins, such as whey protein or fish protein. Enzymatic hydrolysis is an environmentally friendly, non-toxic method with high selectivity, making it widely preferred in the food and pharmaceutical industries.

2.2. Biotechnological Approaches: Recombinant DNA Technology and Synthetic Biology

Recombinant DNA technology is based on transferring the DNA sequence that carries the genetic code for a specific bioactive peptide into suitable host cells (e.g., *E. coli*, *Saccharomyces cerevisiae*) via vectors, enabling these cells to synthesize the peptide. This method allows for high-yield and standardized peptide production.

Additionally, this technology facilitates peptide modification, secretion through transport systems, and even extracellular production, simplifying purification processes. Moreover, expression levels can be finely controlled using different promoter systems (Piers et al, 1993; Miller, 1979; Khan et al, 2016). Synthetic biology goes beyond genetic engineering by establishing artificial biosynthetic pathways and designing new peptide structures. In this field, modular genetic circuits enable microorganisms to synthesize peptides they do not naturally produce. Synthetic biology studies conducted in intracellular or extracellular systems are pioneering the production of peptides that may not exist in nature but could have significant therapeutic potential. CRISPR-based editing, optimized ribosomal translators, and engineered enzymes are commonly used in these systems (Rallapalli & Komor, 2023; Amen Rallapalli et al, 2025).

2.3. Industrial Scale Production: Optimization and Purification Strategies

Industrial-scale production requires replicating laboratory-scale successes in large volumes. Therefore, it is essential to develop microbial strains with enhanced productivity through mutagenesis or adaptive evolution, as well as optimize growth media, mixing, and aeration strategies. Bioreactor designs also play a critical role in industrial production. Batch, fed-batch, and continuous culture systems are selected based on production time, peptide stability, and cost-effectiveness (Gnasegaran et al., 2017).

The purification of microbial bioactive peptides constitutes a significant portion of production costs. Therefore, developing effective and cost-efficient purification techniques is crucial. Commonly used methods include membrane filtration, ultrafiltration, ion-exchange chromatography, reverse-phase chromatography, and HPLC systems. High-purity peptides are required for pharmaceutical and functional food applications. Additionally, drying techniques such as spray drying, lyophilization, and encapsulation can be applied to improve stability (ThyabGddoa Al-sahlany et al, 2020; Jiang et al, 2024).

3. MICROBIAL BIOACTIVE PEPTIDES

Microbial-derived peptides, including bacteriocins, are either directly synthesized by microorganisms or released from fermented dairy products through enzymatic hydrolysis. Their biological activity is largely determined by their amino acid sequence, hydrophobicity, net charge, and presence of functional chemical groups (Sneath,1966; Tian et al., 2023). Certain antibiotics and siderophores are produced via non-ribosomal peptide synthetase (NRPS) systems (Carroll & Moore, 2018). Due to their low natural abundance in food and microbial sources, these peptides are often isolated using chromatographic or ion-exchange methods or produced in larger quantities through chemical synthesis or recombinant DNA technology. For instance, Roca Pinilla (2020) successfully produced a multifunctional polypeptide containing three antibacterial domains in *Escherichia coli*, demonstrating potent antimicrobial activity. Among these peptides, nisin, a lantibiotic from *Lactococcus lactis*, consists of 34 amino acids and five lanthionine rings, providing a stable 3D structure that interferes with cell wall biosynthesis and modulates inflammation via NF-κB and COX-2 inhibition (Jenab et al, 2020; Luo & Song, 2021).

Similarly, MccJ25, a 21-amino-acid peptide from *E. coli*, adopts a unique "threaded lasso" conformation, making it highly stable and a strong inhibitor of RNase activity; its anti-inflammatory effects include inhibition of NF-κB and p38 MAPK pathways (Yu et al., 2018; Ding et al., 2020). Some bacteriocins, such as BacSp222 from *Staphylococcus pseudintermedius*, display immunomodulatory effects and can even enhance pro-inflammatory cytokines via NF-κB activation (Śmiałek et al., 2021; Śmiałek et al., 2022). In contrast, synthetic hybrid peptides like CC34derived from Cecropin B and a frog peptideexhibit potent antimicrobial activity with favorable safety and stability profiles, and act by suppressing COX-2 and iNOS expression (Dong et al., 2021). Overall, bacteriocins vary widely in structure and function, ranging from lantibiotics like nisin and mersacidin to heat-stable microcins like MccJ25 and MccJ27, produced by various species including *Lactobacillus*, *Bifidobacterium*, and *Staphylococcus*. The neuroprotective and anti-inflammatory properties of these microbial peptides are increasingly being explored for therapeutic applications due to their structural diversity and multiple modes of action. Structural and functional characteristics of selected microbial peptides with potential neuroprotective or anti-inflammatory activity at table 4

Table 4. Structural and Functional Characteristics of Selected Microbial Peptides with Potential Neuroprotective or Anti-inflammatory Activity

Peptide Name	Source Organism	Structural Features	Biological Activity	Effect Mechanisms	References
Nisin A/Z	*Lactococcus lactis*	34 aa, 5 lanthionine rings	Inhibits cell wall synthesis	NF-κB and COX-2 inhibition; IL-10 upregulation	Thoda&Touraki, 2023; Jaye et al, 2022
Microcin J25 (MccJ25)	*Escherichia coli*	21 aa, threaded lasso structure	RNase inhibition; high stability	NF-κB and p38 MAPK inhibition; ↓TNF-α	Parker, & Davies, 2022; Doebber et al., 2019;
BacSp222	*Staphylococcus pseudintermedius*	~50 aa, lantibiotic modifications	Immunomodulatory activity	Activates NF-κB; ↑proinflammatory cytokines	Śmiałek et al., 2022
CC34	Synthetic (Cecropin B + frog peptide)	~30 aa, cationic, +5.5 net charge	Antimicrobial	COX-2 and iNOS inhibition	Dong et al., 2021
Brevibacillin	*Brevibacillus brevis*	31–37 aa, cyclic and non-cyclic forms	Antimicrobial	COX-2 inhibition; ↓inflammation	Mongalo et al, 2016
Plectasin	*Pseudoplectanianigrella* (fungus)	40 aa, disulfide bonds, β-sheet structure	Binds lipid II; antibacterial	Mild NF-κB inhibition	Wan, et al, 2016; Tenland et al., 2018 Sahl et al., 2006
Alamethicin	*Trichoderma arundinaceum* (fungus)	~20 aa, α-helical, contains Aib residues	Forms membrane pores	May modulate cellular signaling	Malmiercaet al, 2016
Microginin	*Microcystis aeruginosa* (cyanobacterium)	10–12 aa, linear, contains Ahp	Protease inhibitor; antimicrobial	Not yet studied in inflammation	da Silva Pinto Neto et al, 2025; Ishida et al., 1998
Enterocin	*Enterococcus faecalis*	Structure not fully characterized	Anti-*Listeria* activity	No data on inflammation	Giraffa, 1995
Pediocin	*Pediococcusacidilactici*	Structure not fully characterized	Anti-*Listeria*, inhibits Gram-positive bacteria	No data on inflammation	Khorshidian et al, 2021

4. NEUROPROTECTIVE EFFECT MECHANISMS OF MICROBIALBIOACTIVE PEPTIDES

4.1. Antioxidant Activity: Free radical scavenging and metal chelation.

The neuroprotective effects of bioactive peptides are largely mediated through free radical scavenging (antioxidant activity) and metal ion chelation mechanisms. These pathways play a crucial role in reducing neuronal damage caused by oxidative stress, thereby contributing to the prevention of neurodegenerative diseases such as Alzheimer's and Parkinson's. In such conditions, reactive oxygen species (ROS) and hydroxyl radicals oxidize neuronal membrane lipids, proteins, and DNA, ultimately leading to cell death.

Free radicals are highly reactive molecules due to their lack of electrons. They can damage lipids, proteins, and DNA in cells, leading to oxidative stress. Microbial peptides counteract these harmful molecules through two main pathways.

I) Direct Free Radical Neutralization: Microbial peptides react directly with free radicals thanks to their amino acid composition. Amino acids such as cysteine, methionine, tryptophan, and tyrosine can donate electrons or hydrogen atoms to stabilize radicals. This reduces the radical's reactivity and renders them harmless.
II) Supporting Antioxidant Enzyme Activation: Peptides can activate cellular signaling pathways that regulate antioxidant defense systems. For example, they may stimulate the Nrf2 (nuclear factor erythroid 2-related factor 2) pathway, increasing the production of enzymes such as superoxide dismutase (SOD), catalase, and glutathione peroxidase. These enzymes remove free radicals more effectively.

Certain metals (especially Fe^{2+} and Cu^{2+}) accelerate oxidative reactions inside cells, leading to free radical formation. Microbial peptides prevent this process by binding to these metals. Amino acid side chains in peptides (e.g., histidine, cysteine, aspartic acid, and glutamic acid) can form coordination bonds with metal ions. These bonds sequester the ions in inactive forms, preventing them from participating in oxidative reactions. By inhibiting the release of metal ions, metal-catalyzed processes such as the Fenton reaction are stopped. This prevents the formation of highly harmful ROS like hydroxyl radicals (OH).

Aromatic amino acids (tyrosine, tryptophan, and phenylalanine) and sulfhydryl-containing cysteine residues in bioactive peptides can neutralize reactive species by donating electrons or hydrogen atoms to radicals such as DPPH and ABTS (Monteiro & Paiva-Martins, 2022). Moreover, transition metals like Fe^{2+} and Cu^{2+} can catalyze ROS formation via Fenton reactions. By binding to these pro-oxidant metal ions, bioactive peptides can inhibit Fenton reactions and lipid peroxidation, thereby exerting neuroprotective effects.

For example, Microcin J25 (MccJ25), a lasso-peptide produced by *Escherichia coli* strains harboring plasmid-encoded antibiotic synthesis and export cassettes, demonstrated radical scavenging assays and showed high binding capacity in ferrozine-Fe^{2+} chelation (López-García et al., 2022). Similarly, an extract from *Streptomyces sp.* MUM212 reduced DPPH and ABTS radicals. It also exhibited notable superoxide ($O_2\bullet^-$) scavenging and metal chelation capacities. Furthermore, the extract inhibited lipid peroxidation by 16.7% and maintained cell viability in Vero cells exposed to H_2O_2-induced oxidative damage (Tan et al, 2017).

Lactic acid bacteria also show comparable protective effects. For example, *Lactobacillus* species can bind copper ions and inhibit amyloid-beta aggregation, which is associated with Alzheimer's disease (Tianet al., 2015; Sharma & Rajawat, 2025). Nisin, derived from *Lactococcus lactis*, significantly reduced lipid peroxidation in food systems and provided protection through its Fe^{2+} and Cu^{2+} chelating capacity (Yuan et al., 2023).

Cordymin, a peptide isolated from the fungus *Cordyceps sinensis*, increased glutathione levels and related enzyme activities but decreasing lipid peroxidation (Wanget al., 2012). Similarly, the FE7 peptide derived from fermented milk significantly reduced ROS production in LPS-stimulated microglial cells (Li et al., 2015). Antioxidant effects are thought to be mediated through activation of antioxidant defense pathways (Table 5) such as Nrf2/ARE.

Table 5. Antioxidant Defense Pathways

Pathway	Key Protein/TF	Activation Mechanism	Target Genes/ Proteins	Main Function	Reference
Nrf2/ARE	Nrf2, Keap1	Oxidative stress modifies cysteine residues in Keap1 → Nrf2 released and translocates to nucleus	HO-1, NQO1, GCL, GPx, SOD	Maintains redox homeostasis, detoxification	(Kaspar et al., 2009)
NF-κB	NF-κB, IκB	ROS/cytokines trigger IκB degradation → NF-κB translocates to nucleus	COX-2, iNOS, some antioxidant genes	Immune response, inflammation, stress adaptation	(Morgan & Liu, 2011)
MAPK (ERK, JNK, p38)	MAP kinases	ROS activate MAPK cascade	Nrf2, c-Jun, ATF2	Cell fate decisions (apoptosis/ proliferation), antioxidant gene expression	(Son et al., 2011)
FoxO	FoxO transcription factors	PI3K/Akt inhibition allows FoxO nuclear translocation	MnSOD, Catalase	Long-term oxidative stress defense, DNA repair, autophagy	(Eijkelenboom & Burgering, 2013)
Mitochondrial Defense	MnSOD (SOD2), GPx, Peroxiredoxins	Triggered by mitochondrial ROS production	MnSOD, GPx, Prx, CoQ10, Glutathione	Mitochondrial ROS detoxification, energy balance	(Sena & Chandel, 2012)

4.2. Anti-inflammatory Effects: Inhibition of NF-κB and COX-2

Microbial peptides can modulate neuroinflammation by reducing the secretion of pro-inflammatory cytokines and chemokines. For example, the FE7 peptide significantly suppressed LPS-induced production of IL-1β and nitric oxide (NO) in BV2 microglial cells (Li et al., 2015). Cordymin, derived from *Cordyceps sinensis*, inhibited the post-stroke increase of IL-1β and TNF-α in brain tissue and reduced neutrophil infiltration (Wang et al., 2012). These actions attenuate neuronal damage through inhibition of the NF-κB pathway and downstream inflammatory cascades.

Bacterial peptides exert their anti-inflammatory effects primarily by targeting the NF-κB signaling pathway. By inhibiting NF-κB nuclear translocation, they reduce the expression of inflammatory mediators such as TNF-α, IL-1β, and IL-6, as well as enzymes like COX-2 and iNOS. For instance, the cationic peptide CC34 significantly suppressed NF-κB activation in LPS-stimulated macrophages by inhibiting the phosphorylation of IKKβ, IκBα, and NF-κB p65, leading to reduced mRNA levels of iNOS and COX-2 (Dong et al., 2021). Similarly, Microcin J25 (MccJ25) reduced the phosphorylation of NF-κB and p38 MAPK in LPS-treated mice, which decreased the production of TNF-α and IL-6 and increased the anti-inflammatory cytokine IL-10 (Yu et al., 2022).

In combination therapies, nisin with lycoricidine or glabridin showed synergistic anti-inflammatory effects by suppressing NF-κB activation in human monocytes stimulated with *Enterococcus faecalis*

(Grenier et al., 2020). Nisin Z also reduced the release of TNF-α, IL-1β, and IL-6 in LPS-stimulated human epithelial cells by inhibiting ERK/p38 MAPK pathways while simultaneously increasing IL-10 production (Huang et al., 2022).

Most of these anti-inflammatory activities have been demonstrated through in vitro experiments. CC34 and MccJ25 were tested on RAW 264.7 murine macrophages, showing reduced pro-inflammatory cytokines and COX-2/iNOS expression (Dong et al., 2021). NF-κB luminescence reporter assays in THP-1 and U937 monocyte lines confirmed the inhibitory effects of nisin-lycoricidine combinations (Grenier et al., 2020). In epithelial cell models, nisin Z demonstrated anti-inflammatory activity in human breast epithelial MCF10A cells (Huang et al., 2022).

In vivo studies further support these findings. In mouse models of LPS-induced mastitis or intestinal inflammation, CC34 treatment significantly reduced TNF-α, IL-1β, and IL-6 levels in jejunal tissues, minimized histological damage, and decreased MPO activity (Dong et al., 2021). In LPS-shocked mice, MccJ25 reduced circulating LPS levels, decreased mortality rates, lowered TNF-α/IL-6 levels, and alleviated intestinal inflammation (Yu et al., 2022).

Additionally, in a piglet model of *Enterotoxigenic Escherichia coli* (ETEC) infection, MccJ25 was shown to reduce diarrhea and intestinal inflammation, potentially through modulation of the NF-κB pathway via dopamine receptor interaction (Shang et al., 2024). In a periodontitis model, treatment with nisin or nisin-producing *Lactococcus* strains significantly reduced bone loss and inflammation in periodontal tissues (Gao et al., 2022). Table 6 summarizes key findings from selected in vitro and in vivo studies on the anti-inflammatory effects of microbial peptides.

Table 6. Examples of In Vitro and In Vivo Studies on the Anti-inflammatory Effects of Microbial Peptides

Peptide	Model Type	System / Cell Line or Animal Model	Key Findings	Reference
CC34	In Vitro	RAW 264.7 (murine macrophages)	LPS-induced secretion of TNF-α, IL-1β, and IL-6 decreased by 60 75%; downregulation of COX-2 and iNOS mRNA.	Dong et al.,2021
CC34	In Vivo	LPS-injected mice (jejunal tissue)	Significant reduction in inflammatory cytokines and MPO activity in jejunum; improved histological integrity.	Dong et al.,20210
MccJ25	In Vitro	RAW 264.7	Phosphorylation of NF-κB p65 and p38 MAPK inhibited by 50–65%; NO production decreased.	Yu et al., 2018
MccJ25	In Vivo	LPS-challenged mice	Plasma TNF-α and IL-6 levels reduced by 40–55%; improved survival rate.	Yu et al., 2022
Nisin Z	In Vitro	MCF10A (human epithelial cells)	Inhibited ERK/p38 MAPK pathway; reduced TNF-α, IL-1β, IL-6 secretion; increased IL-10 levels.	Almeida et al., 2020
Nisin Z	In Vivo	Mouse mastitis model	Significant decrease in mammary gland inflammation and serum TNF-α/IL-6 levels.	Almeida et al., 2020
Nisin + Lycoricidine	In Vitro	THP-1 (human monocytes)	NF-κB activation inhibited by ~70% in *E. faecalis*-stimulated monocytes.	Çetin et al., 2019

The anti-inflammatory effects of bacterial peptides are mediated through various molecular mechanisms. One such mechanism involves receptor interaction; for example, the peptide CC34 neutralizes lipopolysaccharide (LPS), thereby blocking Toll-like receptor 4 (TLR4) activation (Dong et al., 2021). Kinase inhibition is another critical mechanism: CC34 prevents the phosphorylation of IKKβ and IκBα, effectively halting the nuclear translocation of NF-κB (Dong et al., 2021). Similarly, microcin J25 (MccJ25) suppresses the TLR4-MyD88 axis, resulting in the downregulation of both NF-κB and p38 MAPK signaling pathways (Yu et al., 2022).

At the transcriptional level, certain peptides inhibit NF-κB subunit p65 from binding to DNA or interacting with transcriptional coactivators. For instance, the synthetic cationic peptide IDR-1002 blocks the interaction between NF-κB and the coactivator CBP in macrophages, thereby suppressing NF-κB-dependent gene expression, including TNF-α and COX-2 (Huante-Mendoza et al., 2016). In another study, MccJ25 was shown to inhibit NF-κB indirectly via activation of the DRD5–cAMP–PKA signaling axis in a drug-like application model (Shang et al., 2024).

Additionally, epigenetic modulation contributes to anti-inflammatory activity; for example, the inhibition of histone acetyltransferases involved in NF-κB transcriptional activation reduces inflammatory responses, as seen with IDR-1002. These peptides can also attenuate intracellular oxidative stress—CC34 has been reported to decrease reactive oxygen species (ROS) production—and may enhance the release of anti-inflammatory cytokines such as IL-10, as observed with nisin Z.

Experimental data further support these findings. CC34 significantly reduced TNF-α, IL-1β, and IL-6 secretion in LPS-stimulated RAW 264.7 macrophages and downregulated iNOS and COX-2 mRNA levels. In vivo, CC34 treatment in LPS-injected mice led to decreased cytokine levels in both jejunal tissue and serum, reduced MPO activity, and alleviated mucosal damage (Dong et al., 2021).

MccJ25 demonstrated potent anti-inflammatory activity in both in vitro and in vivo models. In LPS-challenged mice, it reduced circulating LPS concentrations and significantly increased survival rates while decreasing TNF-α and IL-6 levels. Furthermore, NF-κB and p38 MAPK phosphorylation were markedly suppressed (Yuet al., 2022). In a porcine model of enterotoxigenic *E. coli* (ETEC) infection, MccJ25 alleviated diarrhea and intestinal inflammation. The underlying mechanism involved downregulation of dopamine receptor D1 and upregulation of D5, ultimately leading to NF-κB inhibition (Shang et al., 2024).

Nisin Z also exhibited strong anti-inflammatory effects. In LPS-stimulated human mammary epithelial MCF10A cells, it reduced the secretion of proinflammatory cytokines while enhancing IL-10 production, mediated through inhibition of the ERK/p38 MAPK pathway. In a mouse mastitis model, injected nisin Z significantly decreased inflammatory cell infiltration in mammary tissue and lowered TNF-α/IL-6 levels in both serum and tissue (Hu et al, 2018). Moreover, nisin combined with lycoricidine inhibited NF-κB activation in human monocytes stimulated by *Enterococcus faecalis* (Grenieret al., 2020).

4.3. Apoptosis Prevention Role: Regulation of Bcl-2 family proteins and caspase activity.

Antimicrobial peptides (AMPs) are increasingly recognized not only for their antibacterial properties but also for their ability to modulate host cell death pathways. These molecules can either induce or inhibit apoptosis (programmed cell death), primarily through mitochondrial-dependent mechanisms (Oyinloye et al., 2015). In figure 1 representation illustrates how antimicrobial peptides (AMPs) induce apoptosis through multiple interconnected pathways. One of the most prominent mechanisms by which AMPs regulate apoptosis is by altering mitochondrial membrane permeability, which triggers the release of cytochrome c into the cytoplasm. This activates the apoptosome complex, leading to caspase-9 and subsequently caspase-3 activation (Wang et al., 2022). This pathway has been consistently observed in various AMP-mediated apoptotic responses. For example, PR-39, a porcine-derived AMP produced by

Actinobacillus pleuropneumoniae, has been shown to inhibit caspase-3 activation and suppress apoptosis in RAW 264.7 macrophages (Ramanathan et al., 2004), suggesting a cytoprotective role against infection-induced cell death. Conversely, several AMPs have demonstrated pro-apoptotic activity, especially in cancer cells. TP4, an AMP derived from the marine fish *Tilapia*, selectively activated mitochondrial apoptosis in MCF-7 breast cancer cells (Ardeshir &Moarefvand, 2025). Similarly, moricin, an AMP from the silkworm *Bombyx mori*, increased ROS levels, inhibited the Notch signaling pathway, and induced caspase-dependent apoptosis in triple-negative breast cancer cells (Ahmad et al., 2023).

CC34, a synthetic cationic AMP structurally similar to bacterial peptides, has been shown to modulate the expression of Bcl-2 family proteins and activate caspase cascades, leading to apoptotic cell death in cancer models (Dong et al., 2025). Likewise, nisin, a lantibiotic AMP produced by *Lactococcus lactis*, was reported to activate p53 expression and trigger apoptosis in human colon cancer cells (Mohan Latha Kumari et al., 2025). Periplanetasin-5, an AMP isolated from the cockroach *Periplaneta americana*, induced mitochondrial dysfunction and ROS-mediated apoptosis in various human cancer cell lines (Kim et al., 2021). In addition, anticancer peptides combined with compounds such as carvacrol have also been shown to induce apoptosis in cancer cells (Elahi et al., 2022). Beyond cancer, AMPs can also affect immune cells. For instance, LL-37, a human cathelicidin, was shown to increase apoptosis and ATP release in urothelial cells at high concentrations (Lee et al., 2013). *Staphylococcus aureus*-derived pro-apoptotic peptide exacerbated pulmonary fibrosis by inducing apoptosis in murine models (D'Alessandro-Gabazza et al., 2020). AMPs also play a role in shaping the immune response by modulating neutrophil apoptosis. Nagaoka et al. (2012) reported that certain AMPs could influence the lifespan of neutrophils and thus affect the resolution of inflammation.

In conclusion, AMPs exhibit diverse effects on apoptotic pathways depending on their structure, cellular context, and biological target. Their ability to either trigger or inhibit apoptosis makes them promising candidates as therapeutic agents in cancer, infectious diseases, and immune regulation. Further studies are needed to fully elucidate the molecular details and clinical potential of AMP-mediated apoptosis.

Figure 1. AMP-mediated apoptosis mechanism

<table>
<tr><td colspan="3">AMPs
↓</td></tr>
<tr><td>Membrane Disruption
(Pore formation, ion imbalance)</td><td>Mitochondrial Pathway
(Cytochrome c → Caspase-9 → Caspase-3)</td><td>Death Receptor Pathway
(Fas/TRAIL-R → Caspase-8)</td></tr>
<tr><td>↓</td><td>↓</td><td>↓</td></tr>
<tr><td>↑ ROS Production
(Oxidative stress, DNA damage)</td><td>Mitochondrial dysfunction
& Cytochrome c release</td><td>ER Stress
(CHOP, Caspase-12)</td></tr>
<tr><td></td><td>↓</td><td></td></tr>
<tr><td colspan="3">Apoptosis
(Cell shrinkage, DNA fragmentation, programmed cell death)</td></tr>
</table>

Bioactive peptides can prevent cell death by suppressing pro-apoptotic pathways (e.g., caspase activation) and stabilizing anti-apoptotic proteins such as Bcl-2. Additionally, some peptides enhance neurotrophic signaling that supports neuronal development. For instance, peptide analogs of neurotrophic factors such as BDNF and NGF have been shown to improve neuronal survival in Alzheimer's disease animal models (Lee et al., 2010). Overall, these mechanisms work synergistically to protect neurons from death induced by oxidative stress and inflammatory signals.

Programmed cell death (apoptosis) is a critical component of tissue homeostasis and organismal development. The intrinsic (mitochondrial) apoptosis pathway is initiated by cellular stress signals that promote the activity of pro-apoptotic Bcl-2 family proteins (e.g., Bax, Bak, BH3-only proteins), which increase mitochondrial outer membrane permeability, leading to cytochrome c release and a cascade of caspase-9 and caspase-3 activation. In contrast, anti-apoptotic Bcl-2 family members (Bcl-2, Bcl-xL, Mcl-1) preserve mitochondrial membrane integrity and thereby block the activation of this apoptotic cascade (Nagahara et al., 2009; Kazimet al., 2014; Cardenas-Aguayo et al., 2013). Microbial peptides may enhance the expression or stability of anti-apoptotic Bcl-2 proteins, protecting cells from death. For example, in streptozotocin-induced diabetic mice, treatment with a composite probiotic formulation containing *Lactobacillus* and *Bifidobacterium* supernatants significantly increased Bcl-2 protein levels in kidney and liver tissues and reduced the Bax/Bcl-2 ratio, thereby suppressing caspase-3 activation and preventing apoptosis (Sharma et al., 2011). In addition, peptide fractions obtained from Gram-positive *Bacillus sp.* SC06 increased Bcl-2 expression in oxidatively stressed intestinal epithelial Caco-2 cells, while preserving mitochondrial membrane potential and decreasing the expression of pro-apoptotic proteins Bax and Bad, which significantly inhibited the activation of caspase-9 and caspase-3 (Tang et al., 2021.).

Through preserving mitochondrial integrity, anti-apoptotic Bcl-2 proteins suppress cytochrome c release, thereby preventing apoptosome formation and subsequent caspase-9 activation (Kale et al.,2018). Beyond this, some microbial peptides directly modulate caspase activity under inflammatory stress conditions. In the study published by Wang et al. in 2019 in the journal *Molecules*, the effects of the antimicrobial peptide JH-3 on inflammatory response and apoptosis in RAW264.7 macrophages infected with *Salmonella enterica* serovar Typhimurium CVCC541 were investigated. The results showed that JH-3 significantly reduced the secretion of pro-inflammatory cytokines such as TNF-α and IL-6 in infected cells. Moreover, JH-3 suppressed both the transcription and protein activity of caspase-8 and caspase-9, leading to approximately a 35% reduction in the activation of caspase-3 and a marked decrease in apoptosis rates. These findings indicate that JH-3 effectively inhibits infection-induced inflammation and cell death. Therefore, JH-3 may be considered a potential antimicrobial and anti-inflammatory agent (Wang et al., 2019).Several core mechanisms underlie the anti-apoptotic effects of microbial peptides: upregulation of Bcl-2 gene expression through enhanced activation of nuclear transcription factors; preservation of mitochondrial membrane integrity via inhibition of Bax/Bak oligomerization; suppression of the caspase cascade by preventing apoptosome formation due to lack of cytochrome c release, as well as by directly repressing caspase-8 and -9 promoter activity; and modulation of inflammatory signaling through inhibition of NF-κB and MAPK pathways, leading to decreased TNF-α and IL-1β production, and enhanced Bcl-2-dependent cytoprotective signaling (Zhaiet al, 2008: Orning & Lien., 2021; Pellegrini & Strasser, 2013).Experimental data supports these mechanisms. In streptozotocin-induced diabetic mice, Yuet al. (2014) reported that the composite probiotic formula reduced malondialdehyde (MDA) levels and increased superoxide dismutase (SOD) activity in kidney tissues, thereby reducing cellular stress and enhancing Bcl-2-dependent protective mechanisms. In a macrophage infection model, Wang et al.

(2019) demonstrated that JH-3 reduced inflammatory cytokine release and directly inhibited caspase-8, -9, and -3 activation, preventing apoptosis by approximately 45%. In oxidative stress–challenged intestinal epithelial cells, *Bacillus subtilis* (or closely related *Bacillus amyloliquefaciens* SC06)–derived peptides have been shown to increase Bcl-2 expression and inhibit caspase-9 activation, thus preserving mitochondrial integrity and reducing apoptosis (Tang et al., 2021). Given their apoptosis-inhibiting capabilities, microbial peptides hold significant promise for protecting against burn injuries, side effects of cytotoxic therapies, and inflammation-related tissue damage. In the future, structure-activity relationship (SAR) studies and nano-carrier delivery technologies may improve their intracellular targeting and biostability. Moreover, phase I/II clinical trials will be critical for evaluating the safety and efficacy of Bcl-2-modulating peptides in human models.

4.4. Synaptic Plasticity and BDNF Increase: Modulation of neurotrophic factors

Neurotrophic factors, particularly Brain-Derived Neurotrophic Factor (BDNF), play a critical role in synaptic plasticity and neuronal survival. BDNF binds to TrkB receptors and activates intracellular signaling pathways such as MAPK/ERK, PI3K/Akt, and PLC-γ. These pathways support processes like long-term potentiation (LTP), which is the fundamental molecular mechanism of learning and memory. In addition, BDNF enhances neuronal survival by promoting the activation of anti-apoptotic genes and maintaining mitochondrial functions. Through these mechanisms, BDNF contributes both to the strengthening of synaptic connections and to the preservation of neuronal integrity(Binder & Scharfman, 2004; Park and Poo, 2013; Lu et al., 2013).

The following Figure 2 representation illustrates how modulation of neurotrophic factors, particularly BDNF, enhances synaptic plasticity and neuronal survival

Figure 2. Synaptic plasticity and BDNF ıncrease: Modulation of neurotrophic factors

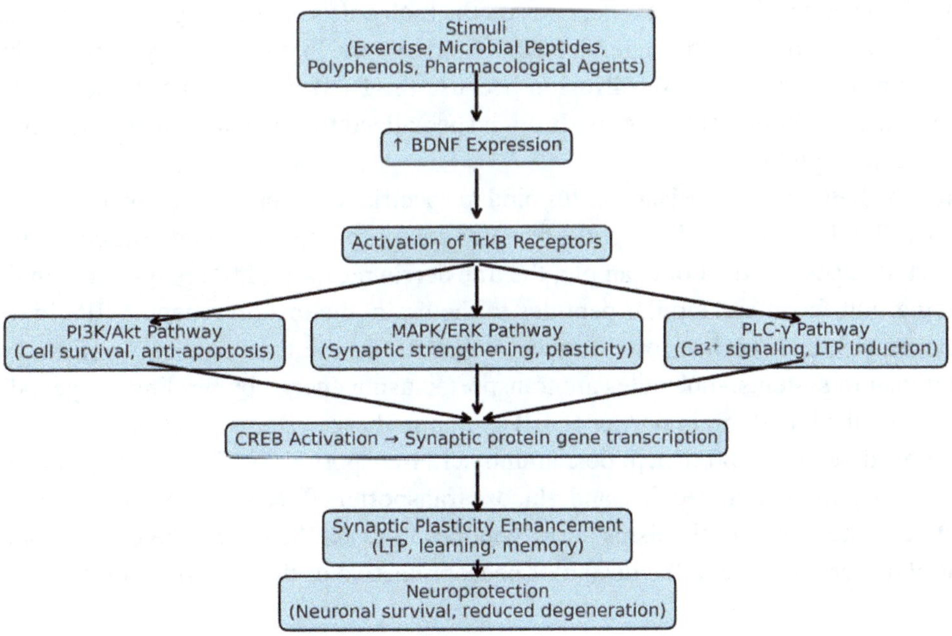

5. MICROBIAL PEPTIDES AND THE GUT-BRAIN AXIS

The Gut–Brain Axis is a complex system that enables bidirectional communication between the gut and the central nervous system (brain and spinal cord). This communication occurs through neural, hormonal, and immune pathways. The vagus nerve, in particular, provides rapid information transfer between the gut and the brain. Trillions of microorganisms residing in the gut (bacteria, fungi, viruses) produce metabolites and neural signals that play a role in this axis; for example, serotonin produced in the gut can directly affect mood. In addition, gut cells secrete hormones, and immune cells transmit inflammatory signals that reach the brain, influencing behavior, mood, immune responses, and digestive functions. Conversely, the brain regulates gut functions through the vagus nerve and hormonal changes influenced by stress, emotions, and hormonal fluctuations. As a result, the gut–brain axis plays a critical role in mood, behavior, digestive health, and immune function, and it is known to be associated with conditions such as depression, anxiety, and neurodegenerative diseases.

The passage of peptides through the Gut–Brain Axis to the brain occurs via both direct and indirect mechanisms, but this process is quite limited because peptides are generally large molecules that tend to be broken down during digestion. Some biologically active peptides can be absorbed intact by gut epithelial cells, a process that usually involves carrier proteins or particle endocytosis. After absorption, peptides enter the bloodstream, and some are protected from protease enzymes, allowing them to reach target tissues. The blood–brain barrier (BBB) plays a critical role in brain entry; some small and hydrophobic peptides can cross the BBB via passive diffusion, while larger or hydrophilic peptides pass through receptor-mediated transcytosis or carrier protein systems. In addition to direct passage, peptides can indirectly affect brain function through metabolites they produce in the gut and their effects on the immune system. For example, certain peptides produced by the gut microbiota can transmit signals to the brain via the vagus nerve. These mechanisms enhance the potential use of peptides as neuroprotective and neuromodulatory agents.

Passive diffusion means that molecules pass through the cell membrane along a concentration gradient without expending energy. This process depends on the molecule's size and lipid solubility; small, lipid-soluble peptides can easily pass through the BBB's lipid bilayer. In contrast, larger or non-hydrophobic peptides cannot pass through passive diffusion because the BBB is a very tight and selective barrier. Such peptides reach the brain using energy through specialized mechanisms such as receptor-mediated transcytosis or carrier protein systems.

In receptor-mediated transcytosis, peptides bind to specific receptors located on the endothelial cells of the BBB. This binding leads to the peptide-receptor complex being taken into the cell via vesicles and transported to the opposite side. For example, the transferrin receptor (TfR) plays a role in the transport of iron-carrying transferrin and similar peptides to the brain, the insulin receptor (IR) in the transport of peptide hormones, and the lipoportein receptor (LRP1) in the transport of ApoE and some other peptides. In transport systems, molecules are transported using energy by binding to specialized carrier proteins located at the blood–brain barrier (BBB). Among these systems, peptide transporters (PepT1, PepT2) transport dipeptides and tripeptides, amino acid transporters (LAT1, CAT1) transport specific amino acids and peptide-like molecules, and glucose transporters (GLUT1) transport glucose and some of its derivatives. These two mechanisms allow peptides to reach the brain when direct transport is not possible and play a critical role in the design of neuroprotective or therapeutic peptides.

6. THE ROLE OF INTESTINAL MICROBIOTA IN PEPTIDE PRODUCTION

The gut microbiota constitutes a complex ecosystem that plays a central role in human physiology and pathology. This microbial community serves as a significant source of peptide production and regulates the host's neurological and immune functions via the gut–brain axis. Microbial peptides are small, often bioactive amino acid chains synthesized through the metabolic activity of bacteria, archaea, and certain fungi. These peptides interact with intestinal epithelial cells in the local gut environment and perform critical functions, including maintaining barrier integrity, modulating inflammatory responses, and controlling pathogenic microbes.

In particular, peptides produced by probiotic bacteria such as *Lactobacillus* and *Bifidobacterium* have been shown to enhance the expression of tight junction proteins (e.g., occludin, claudin, ZO-1) in the intestinal mucosa, thereby supporting epithelial integrity (Roseet al., 2021). This contributes to reduced intestinal permeability and prevents the translocation of toxins or pathogens into systemic circulation. Strengthening the intestinal barrier in this way is crucial for reducing inflammation and systemic effects in diseases such as inflammatory bowel disease (IBD) and irritable bowel syndrome (IBS) (Karczewski et al., 2010; Anderson et al., 2012;). Furthermore, experimental models have demonstrated that microbial peptides suppress inflammatory cytokine production, thereby decreasing both local and systemic inflammation (Wang et al., 2019; Oyinloye et al.,2015). The effects of microbial peptides emerge in synergy with other important microbial metabolites, including short-chain fatty acids (SCFAs) and tryptophan derivatives. SCFAs—particularly acetate, propionate, and butyrate—serve as an energy source for intestinal epithelial cells while also modulating immune responses. Butyrate, acting as a histone deacetylase inhibitor, influences epigenetic regulation by suppressing the expression of inflammatory genes (Chang et al., 2014; Vinolo et al., 2011; Dalile et al., 2019) The concurrent presence of microbial peptides and SCFAs produces synergistic effects, such as strengthening the gut barrier, maintaining immune homeostasis, and reducing neuroinflammation. For instance, animal model studies show that combinations of SCFAs and peptides diminished inflammatory responses and alleviated behavioral stress indicators.

Tryptophan derivatives produced by the intestinal microbiota contribute to the synthesis of neuroactive molecules such as serotonin, melatonin, and kynurenine, playing a key role in gut–brain axis communication and the regulation of mood, cognition, and immune responses (Clarke et al., 2014; Agus et al., 2018).These molecules directly influence neurotransmission, mood regulation, and sleep cycles in the central nervous system (CNS), thereby exerting significant effects on psychiatric and neurodegenerative disorders. Certain metabolites along the kynurenine pathway have either neurotoxic or neuroprotective properties, and microbial peptides may modulate this pathway, offering therapeutic potential for nervous system diseases (Agus, et al., 2018; Schwarcz et al., 2012).

Another essential function of microbial peptides is the preservation of gut barrier integrity. A healthy barrier prevents the systemic translocation of pathogens and toxic substances and closely interacts with immune cells within the intestinal mucosa. A review by Cryan et al. (2019) indicated that microbial peptides upregulate anti-apoptotic genes in epithelial cells, enhancing cell viability and reducing the release of inflammatory cytokines such as IL-1β, TNF-α, and IL-6. In the study by Yan et al. (2011), the soluble protein p40 produced by *Lactobacillus rhamnosus* GG was shown to accelerate mucosal repair and maintain intestinal barrier integrity in mouse models of inflammatory bowel disease. The p40 protein activates the epidermal growth factor receptor (EGFR), which increases the expression of tight junction proteins, thereby contributing to the reduction of inflammation. These findings demonstrate that probiotics can exert significant therapeutic effects on gut health not only as live bacteria but also through their bacterial products. For example, a combination of *Lactobacillus plantarum* and *Bifidobacterium animalis* has been shown to increase IL-10 production in Th2 cells, alleviating disease severity in an experimental autoimmune encephalomyelitis model (Wang et al., 2023). Similarly, *Bacteroides fragilis* has been reported to suppress inflammation and maintain intestinal barrier integrity by inducing IL-10 secretion from Foxp3+ regulatory T cells via TLR2 activation through its polysaccharide A (PSA) (Round & Mazmanian, 2010). Furthermore, dysbiosis is characterized by a decrease in short-chain fatty acid-producing bacteria and an increase in opportunistic pathogens has been associated with reduced IL-10 production in autoimmune neurological diseases, indicating that microbial imbalance influences disease progression (Li, Wu, & Zhou, 2023). These findings suggest that microbial peptides have therapeutic potential by enhancing IL-10 and other anti-inflammatory cytokines, thereby reducing disease severity.

The immunomodulatory effects of microbial peptides are also crucial for regulating the gut–brain axis. Peptides produced in the gut influence immune elements such as dendritic cells, macrophages, and T cells, thereby balancing the immune response. In an experimental autoimmune encephalomyelitis (EAE) model.

Microbial peptides derived from the gut microbiota enhanced the production of IL-10 and other anti-inflammatory cytokines, significantly reducing disease severity. These findings suggest that microbial peptides may regulate immune responses in autoimmune neurological disorders. Additionally, recent research has increasingly emphasized the pivotal role of the gut microbiota and its bioactive peptides in maintaining central nervous system (CNS) health. The bidirectional communication between the gut and brain, known as the gut–brain axis, involves neural (e.g., vagus nerve), immune, and endocrine pathways that collectively regulate mood, cognition, stress responses, and neuroinflammation (Drljača et al., 2023). Microbial metabolites such as short-chain fatty acids (SCFAs), particularly butyrate, have been shown to exert anti-inflammatory effects and reinforce the integrity of the blood–brain barrier. Additionally, tryptophan-derived compoundsincluding serotonin, melatonin, and kynureninemodulate neurotransmission and immune responses. Particular interest are microbial peptides and peptide-like molecules that influence microglial activity by upregulating anti-inflammatory cytokines such as IL-10. In conditions such as depression, anxiety, Alzheimer's disease, and Parkinson’s disease, alterations in gut microbial composition have been associated with increased inflammation, disrupted neurochemical balance, and disease progression. For instance, reduced microbial diversity and increased abundance of opportunistic pathogens have been observed in depression and Parkinson’s disease, while dysbiosis has been linked to amyloid-β aggregation in Alzheimer’s disease. These findings underscore the therapeutic potential of probiotics and postbiotics in restoring microbial balance, reducing neuroinflammation, and supporting neuroprotection. Thus, gut microbiota-derived peptides offer promising avenues for novel interventions in the prevention and treatment of neurodegenerative and psychiatric disorders (Drljača et al., 2023). In this context, microbial peptides hold promise for innovative therapeutic strategies due to their capacity to reduce neuroinflammation, enhance synaptic plasticity, and regulate neurotransmitter metabolism. Clinical studies have yielded promising outcomes regarding the potential of microbiota modulation and peptide-based therapies to improve cognitive functions.

In summary, microbial peptides, as complex metabolic products of the gut microbiota, play multi-faceted roles in maintaining intestinal barrier integrity, reducing systemic inflammation, and regulating the gut–brain axis. Acting synergistically with microbial metabolites such as SCFAs and tryptophan derivatives, these peptides support both local and systemic homeostasis. Current literature provides strong evidence for their therapeutic potential in neurological and immunological diseases, and future research in this area is expected to expand significantly.

6.1. Gut Microbiota, Microbial Peptides, and Their Effects on the Central Nervous System

The gut–microbiota–brain axis is a complex bidirectional communication system through which microbial communities in the gut interact with the central nervous system (CNS). This communication occurs through neural, hormonal, immune-mediated, and microbial metabolite pathways (Cryan et al., 2019). In particular, microbial peptides and other metabolites such as short-chain fatty acids (SCFAs) produced by the gut microbiota have been shown to exert modulatory effects on the CNS.

Several studies have reported that microbial peptides exhibit neuroprotective effects by suppressing neuroinflammation (Table 7). These peptides can support intestinal epithelial integrity, reduce systemic inflammation, limit neuroinflammatory processes, and enhance levels of neurotrophic factors such as brain-derived neurotrophic factor (BDNF) (Dinan&Cryan, 2017). The gut microbiota also influences the serotonergic system by regulating tryptophan metabolism, thereby affecting serotonin synthesis. Moreover, certain microbial species produce gamma-aminobutyric acid (GABA) and other neurotransmitter-like molecules that can modulate behavior (Strandwitz, 2018). Through these mechanisms, probiotic-derived peptides have been shown to alleviate symptoms of depression and anxiety.

In the context of Alzheimer's disease, the aggregation of amyloid-beta plaques and tau proteins-pathological hallmarks has been associated with inflammation. This has led to growing interest in anti-inflammatory peptides derived from microbiota as potential therapeutic agents (Cattaneoet al., 2017). Certain *Lactobacillus* and *Bifidobacterium* species have been reported to reduce such protein accumulations. In Parkinson's disease, increased intestinal permeability and microbial dysbiosis are common. It has been suggested that microbial metabolites can transmit signals to the brain via the vagus nerve, and that α-synuclein protein aggregation may originate in the gut before progressing to the central nervous system. Microbial peptides may modulate this process by suppressing immune responses or inhibiting pathogenic protein propagation.

Table 7. Effects of Gut Microbiota and Microbial Peptides on the CNS and Their Neuroprotective Roles

Disease / Condition	Microbial Mechanism of Action	Potential Role of Microbial Peptides	Reference
Depression	Affects serotonin levels via tryptophan metabolism, mood modulation via GABA production	Anti-inflammatory effects, increases BDNF levels, regulates the HPA axis	Cryan et al., 2019; Strandwitz, 2018
Anxiety	Production of GABA, dopamine, and serotonin; signaling via the vagus nerve	Anxiolytic effects, reduce cortisol levels, modulate stress response	Cryan et al., 2019; Strandwitz, 2018Dinan&Cryan, 2017
Alzheimer's	Dysbiosis may promote amyloid-beta accumulation	Reduces inflammation, inhibits aggregation of amyloid-beta and tau proteins	Dinan &; Cryan, 2017
Parkinson's	α-Synuclein aggregation may originate in the gut; inflammation and gut permeability contribute	Slows α-synuclein propagation, suppresses microglial activation, supports motor function	Sampson et al., 2016
General CNS(MSS) Protection	Neurotransmitter production and immune modulation via the gut–brain axis	Suppresses neuroinflammation, enhances synaptic plasticity, reduces oxidative stress	Cryan et al., 2019

6.2. Synergistic Effects of Peptides with Microbial Metabolites (SCFAs, tryptophan derivatives).

The gut microbiota influences central nervous system (CNS) functions through a variety of bioactive molecules. Microbial peptides act synergistically with other microbial metabolites such as short-chain fatty acids (SCFAs: acetate, propionate, butyrate) and tryptophan derivatives (e.g., indole compounds, kynurenine) to exert significant effects on the immune system, neuronal functions, and behavior. Together, these molecules can induce neuroprotective, anti-inflammatory, and neuromodulatory effects within the CNS.

Table 8. Synergistic Effects of Microbial Peptides with SCFAs and Tryptophan Derivatives

Microbial Product	Mechanism of Action	Synergistic Effect (Peptide + Metabolite)	Potential Neuroprotective Outcomes	Reference
SCFAs (acetate, butyrate, propionate)	Inhibition of histone deacetylases (HDACs); immune regulation; maintenance of intestinal epithelial integrity	Peptides regulate epithelial permeability, while SCFAs enhance anti-inflammatory gene expression	Reduced microglial activity, suppression of neuroinflammation	Dalile et al., 2019
Tryptophan Derivatives (indole, kynurenine, 5-HT)	Involved in serotonin synthesis, activation of aryl hydrocarbon receptor (AhR)	Peptides modulate gut–brain communication, tryptophan derivatives enhance neuronal plasticity	Reduction in depression and anxiety symptoms, increase in BDNF	Agus et al., 2018
GABA and other neurotransmitter analogs	Bind to GABA receptors to regulate CNS activity	Peptides may be synthesized by GABA-producing bacteria; synergy helps balance stress responses	Anxiolytic and antidepressant effects	Strandwitz, 2018
Competition with proinflammatory metabolites (e.g., lipopolysaccharides – LPS)	Activates TLR4 signaling (inflammation)	Microbial peptides may suppress LPS-induced inflammation	Prevention of neuroinflammation and neuronal cell death	Dinan&Cryan, 2017

SCFAs are produced by microbial fermentation of dietary fibers and affect the CNS via immune and epigenetic pathways. Tryptophan derivatives, including serotonin and indole compounds, play a critical role in neuronal communication. The aryl hydrocarbon receptor (AhR) is activated by these derivatives and contributes to the regulation of immune and nervous system homeostasis (table 8).

6.3. Protection of Intestinal Barrier Integrity and Reduction of Systemic Inflammation

The intestinal mucosa serves as a crucial defense line protecting the host from pathogens and toxins originating from the external environment. The integrity of this barrier is maintained by tight junction (TJ) proteins between epithelial cells, such as claudin, occludin, and zonulin. Microbial peptides produced by the gut microbiota contribute to the maintenance of intestinal barrier integrity, modulate immune system responses, and play a crucial role in limiting systemic inflammation by directly or indirectly interacting with host immune and epithelial cells (Kamada et al., 2013).

Microbial peptides produced by the gut microbiota support gut barrier maintenance, regulate immune responses, and help reduce systemic inflammation by directly or indirectly modulating these structures (Cani et al., 2012; Certain microbial peptides can enhance the expression of tight junction proteins in epithelial cells, thereby preventing the "leaky gut" phenomenon (increased intestinal permeability). This helps restrict the translocation of inflammatory molecules such as lipopolysaccharides (LPS) into systemic circulation. Some peptides also stimulate immune cells to produce anti-inflammatory cytokines like interleukin-10 (IL-10), contributing to immune homeostasis. Bacteriocins and similar microbial peptides inhibit the overgrowth of pathogenic bacteria, preventing dysbiosis and limiting the accumulation of pro-inflammatory microbial products. In addition, certain peptides regulate Toll-like receptor (TLR) signaling pathways to prevent excessive inflammatory responses (table 9).

Table 9. Effects of Microbial Peptides on Gut Barrier Integrity and Systemic Inflammation

Mechanism of Action	Related Peptide Type / Effect	Physiological Outcome	Reference
Enhancement of tight junction protein expression	Defensin-like peptides derived from *Lactobacillus* spp.	Decreased intestinal permeability	Tlaskalová-Hogenová et al., 2011
Induction of anti-inflammatory cytokines	Bacteriocins, short peptide immunomodulators	↑ IL-10, ↓ TNF-α	Bermudez-Brito et al., 2012
Inhibition of pathogenic bacteria	Microsidins from lactic acid bacteria	Prevention of dysbiosis	Corr et al., 2007
Prevention of LPS translocation	Peptides with synergistic effects with SCFAs	Reduced endotoxin-induced inflammation	Cani et al., 2012

7. BIOAVAILABILITY AND THERAPEUTIC APPLICATION STRATEGIES

7.1. Gastrointestinal Stability and Absorption Dynamics of Peptides

The gastrointestinal system, as the largest endocrine organ in the body, produces numerous peptide hormones. These hormones not only regulate digestive processes but also play important regulatory roles in energy metabolism, the immune system, and the central nervous system. In recent years, many gastrointestinal peptides have been shown to exhibit neuroprotective effects (Table 10), and these peptides have attracted interest as potential therapeutic targets against neurodegenerative diseases (Holzer & Farzi, 2014). Gastrointestinal peptides exert versatile effects not only within the digestive system but also systemically. Understanding the molecular structures of these peptides and their receptor interactions facilitates the development of new therapeutic approaches for both metabolic diseases and neurodegenerative disorders. In particular, peptides such as ghrelin, GLP-1, and somatostatin have protective effects against oxidative stress and apoptotic pathways in brain tissue, supporting their potential as neuroprotective agents.

The gastrointestinal system is a complex ecosystem that not only performs digestive functions but also regulates immunity, metabolism, and neurological processes through human-microbiota interactions. Microorganisms residing in this system produce bioactive microbial peptides (such as bacteriocins, short-chain peptides, and neuropeptide-like molecules) that influence both local and systemic physiology. In recent years, these microbial peptides have been shown to possess neuroprotective effects, particularly via the brain-gut axis (Cryan et al., 2019). Prominent Gastrointestinal Microbial Peptides are presented in the table below. These microbial peptides communicate signals to the central nervous system directly or indirectly through the vagus nerve, immune responses, and circulating neuropeptide analogs, thereby affecting brain functions.

Notably, some microbial peptides have been demonstrated to enhance serotonergic activity in the gut and influence dopamine metabolism. This may play an important role in diseases such as Parkinson's and depression (El Aidy et al., 2016). Microbial peptides can also interact with human gastrointestinal peptide hormones (e.g., GLP-1, PYY, ghrelin) via cross-signaling. For instance, certain *Lactobacillus* species have been shown to increase GLP-1 secretion. These interactions carry therapeutic potential for both metabolic and neurological disorders (Yadav et al., 2013).

Gastrointestinal peptides derived from microbiota are multifunctional molecules that contribute not only to local gut health but also to central nervous system health. Their neuroprotective effects manifest through mechanisms such as reducing inflammation, suppressing oxidative stress, and regulating neurotransmitter metabolism. These features make them promising biotherapeutics for neurodegenerative diseases.

Table 10. Major Gastrointestinal Peptides: Structure, Source, Function, and Neuroprotective Properties

Peptide Name	Molecular Structure	Site of Synthesis	Physiological Function	Neuroprotective Properties	Reference
Gastrin	17–34 aa; C-terminal tetrapeptide: Trp-Met-Asp-Phe-NH_2	G cells (gastric antrum)	Stimulates gastric acid secretion, supports mucosal growth	Neuromodulatory effect via brain-gut axis; indirect neuroprotective potential	Taché et al., 2018
Cholecystokinin (CCK)	Forms like CCK-8, CCK-33; tyrosine sulfation critical for activity	I cell (duodenum and jejunum)	Stimulates gallbladder contraction and pancreatic enzyme secretion; suppresses appetite	Modulates neurotransmission in hippocampus; may protect against neurodegenerative diseases	Rehfeld, 2017
Secretin	27 aa; similar to glucagon family	S cells (duodenum)	Increases pancreatic bicarbonate secretion; inhibits gastric acid	Suppresses neuroinflammation; potential therapeutic effect in autism spectrum disorders	Feng et al., 2019
Motilin	22 aa linear peptides	M cells (duodenum and jejunum)	Regulates intestinal motility	Indirect interactions with dopamine and serotonin metabolism; limited direct neuroprotective data	Depoortere, 2013
Ghrelin	28 aa; includes n-octanoylation modification	X/A cells (gastric fundus)	Stimulates appetite; triggers growth hormone release	Inhibits oxidative stress and apoptosis; shows neuroprotective effects in Parkinson's and Alzheimer's disease	Moon et al., 2009

continued on following page

Table 10. Continued

Peptide Name	Molecular Structure	Site of Synthesis	Physiological Function	Neuroprotective Properties	Reference
GLP-1	30–31 aa; posttranslational product of proglucagon	L cells (ileum, colon)	Increases insulin secretion; suppresses glucagon release	Protects against beta-amyloid toxicity, promising for Alzheimer's treatment	Hölscher, 2014
Peptide YY (PYY)	36 aa; most active form is PYY_{3-36}	L cells (ileum and colon)	Suppresses appetite; slows intestinal motility	Anti-apoptotic effects in hypothalamic neurons; supports energy balance and neuronal health	Karhunen et al., 2019
Somatostatin	Somatostatin-14 and -28 forms	D cells (pancreas, stomach, duodenum); hypothalamus	Inhibits hormones like GH, insulin, gastrin	GABAergic effect; reduces neuronal damage in epilepsy and Alzheimer's	Patel, 1999
Enteroglucagon	Derived from proglucagon chain	L cells (ileum and colon)	Slows gastric emptying; regulates intestinal absorption	May have neuroprotective effects due to similarity to GLP-1	Sandoval & Seeley, 2004

Table 11: Gastrointestinal Microbial Peptides and Their Neuroprotective Properties

Peptide Name	Source Microorganism	Molecular Structure	Physiological Function	Neuroprotective Properties	Mechanism of Action	Targeted Neurological Processes	Clinical Potential / Application	References
Nisin	*Lactococcus lactis*	34 aa lantibiotic with lanthionine rings	Antimicrobial against Gram-positive bacteria	Suppresses microglia, reduces inflammation in Alzheimer's model	Reduces inflammatory cytokine production via TLR4-NFκB pathway	Neuroinflammation, microglial activation	Supportive treatment in Alzheimer's, inflammation-based diseases	Sunet al., 2020;
Reuterin	*Lactobacillus reuteri*	Low MW aldehyde structure (peptidomimetic)	Inhibits pathobiont bacteria	Suppresses LPS-mediated inflammation, protects intestinal barrier	Reduces ROS and inflammatory mediator production	Gut permeability, neuronal inflammation	Use in microbiota-related diseases like IBS, depression, anxiety	Talarico et al., 1988 ;Xu et al., 2022
Microcin J25 (MccJ25)	*Escherichia coli*	21 aa lasso peptides	Antimicrobial via RNA polymerase inhibition	Indirectly modulates dopaminergic pathways to support behavioral balance	RNA polymerase inhibition + preservation of serotonin/ dopamine-regulating microbiota	Neurotransmitter regulation	Supportive in Parkinson's and dopamine imbalance disorders	Bellomio et al., 2007 ; Shang, et al., 2024;
Bifidocin	*Bifidobacteriumsp*, etc.	39–60 aa linear bacteriocin	Supports mucosal immunity, inhibits pathogens	Protects gut-brain barrier, prevents systemic LPS translocation	Strengthens mucosal barrier, inhibits TLR4 pathway	Systemic inflammation prevention	Autism spectrum disorders, immune-mediated neurological diseases	Liu et al, 2017,Coretti et al., 2018
Clostridial Peptides	*Clostridium butyricum*	Short-chain peptides and SCFAs (especially butyrate)	Epithelial regeneration, immune regulation	Supports synaptic plasticity via HDAC inhibition	HDAC inhibition, epigenetic control of neuronal gene expression	Neurogenesis, synaptic plasticity	improves cognitive dysfunction in **ICV-STZ** Depression, anxiety, Alzheimer's, autism	Sunet al., 2021; Su et al., 2023

continued on following page

Table 11. Continued

Peptide Name	Source Microorganism	Molecular Structure	Physiological Function	Neuroprotective Properties	Mechanism of Action	Targeted Neurological Processes	Clinical Potential / Application	References
Subtilin	*Bacillus subtilis*	Lantibiotic class, 32 aa, modified ring structure	Antimicrobial; symbiotic with probiotics	Supports gut homeostasis and suppresses inflammation	Inhibits pro-inflammatory cytokines, supports GABA production	Inflammation, GABAergic system	Anxiety disorders, brain-gut axis dysfunction in IBS	Raheem et al., 2021; Francisco and Grau, 2025
Enterocin	*Enterococcus faecium*	Linear bacteriocin, approx. 70 aa	Antimicrobial against Gram-positive pathogens	Balances gutmicrobiota, reduces neuroinflammation	Supports commensal flora and modulates immune response	Mood regulation, neuronal inflammation	Depression, microbiota-relatedneuroimmune modulation	Chauhan et al., 2014; Dinan & Cryan, 2017

Bioactive peptides must remain stable in the gastrointestinal (GI) environment and be biologically absorbable in order to exert their physiological effects in the digestive system. However, most peptides have limited bioavailability due to the acidic environment of the GI system, enzymatic degradation, and challenges in epithelial transport (Dahanet al., 2009; Chakrabartiet al., 2018). In this context, various biotechnological and pharmaceutical strategies have been developed both to enhance stability and to provide targeted bioavailability.

Gastrointestinal stability is defined as the resistance of peptide to proteolytic enzymes such as pepsin, trypsin, and chymotrypsin found in the stomach and intestines. Many natural peptides are rapidly degraded by these enzymes, losing their biological activity (Udenigwe& Aluko, 2012). Peptides with cyclic structures are more resistant to proteases and have longer half-lives. Additionally, they show resistance to gastric acidity due to intramolecular hydrogen bonds and hydrophobic cores (Hamman et al., 2005). To exhibit biological effects, peptides must cross the intestinal epithelial layer and enter systemic circulation. Absorption routes generally occur via transcellular transport (active transport and diffusion), paracellular transport (between epithelial cells), or endocytosis/mediated transport mechanisms. Small and lipophilic peptides typically cross by passive diffusion, whereas larger and more hydrophilic peptides require carrier proteins or vesicular transport systems. However, the mucus barrier and peptidases produced by the intestinal epithelium can significantly limit this process (Lemes et al., 2016).

In recent years, carrier systems such as nanocapsules, liposomes, and microparticles have been developed to facilitate peptide transport. These systems protect peptides and can enhance both stability and bioavailability (Agyei et al., 2016). Improving the GI stability of peptides is a fundamental factor determining their suitability for oral administration. Stable and absorbable peptides are attracting significant interest for clinical use in neurodegenerative diseases, metabolic disorders, and immune regulation. In the future, with advances in targeted peptide design and carrier systems, the therapeutic use of orally administered bioactive peptides will become more widespread.

7.2. Ability to Cross the Blood-Brain Barrier and Targeted Carrier Systems (Nanoparticles)

The blood-brain barrier (BBB), one of the most important physiological barriers protecting the central nervous system (CNS), maintains brain homeostasis by providing selective permeability against neurotoxic substances and pathogens. However, this characteristic also significantly limits the access of drugs and bioactive molecules to the CNS (Abbott et al., 2009). For bioactive peptides to exert their therapeutic effects, crossing this barrier is critically important. Traditional peptides struggle to cross the BBB due to low lipophilicity and susceptibility to enzymatic degradation. Carrier systems (Table 12) facilitate BBB penetration and enable targeted therapy. These systems are generally nanotechnology-based. Therefore, nanoparticle (NP)-based carriers have been developed to increase peptide stability, provide targeted distribution, and overcome the BBB.

The BBB consists of brain capillary endothelial cells, tight junctions, astrocyte end feet, and pericytes. This structure allows the passage of small hydrophobic molecules but is highly impermeable to hydrophilic and large molecules (Pardridge, 2012). Carrier proteins such as GLUT1 and LAT1, receptor-mediated endocytosis, adsorptive endocytosis mechanisms, and transcytosis pathways offer alternative routes for drug passage.

Nanoparticles are among the most effective systems used to enhance bioavailability and targeted delivery of peptides. Common types include polymeric nanoparticles, lipid-based nanoparticles (e.g., solid lipid nanoparticles - SLNs, nanostructured lipid carriers - NLCs), dendrimers, nanoemulsions, and inorganic nanoparticles (e.g., gold, silica) (Saraiva et al., 2016). Transporting peptides with nanoparticles provides key advantages such as protection against enzymatic degradation, reduction of systemic side effects via cell and tissue targeting, and surface functionalization with ligands (e.g., transferrin, apolipoprotein E) to cross the BBB. For example, polymeric nanoparticles surface-modified to target transferrin receptors can successfully cross the BBB through receptor-mediated transcytosis Lu et al., 2018). Nanoparticles stabilize peptides against environmental stress while enabling slow and controlled release. This improves both pharmacokinetic profiles and therapeutic efficacy. In particular, peptide-based delivery systems targeted to inflammatory areas in the brain hold promise for slowing disease progression in conditions such as Alzheimer's and Parkinson's disease (Agyei et al., 2016). Additionally, stimuli-responsive nanoparticles (sensitive to pH, enzymes, temperature) provide controlled release by responding to microenvironmental conditions. This feature is important for precise targeting therapeutic peptides. Effective transport of peptides across the BBB opens new horizons for the treatment of neurodegenerative diseases. Nanoparticle-based systems offer a powerful tool to achieve this goal; however, challenges such as biocompatibility, toxicity, and scalability in clinical applications remain critical areas for ongoing research.

Table 12. Traditional Peptide Carrier Systems

Carrier System	Structure/Feature	BBB Crossing Mechanism	Advantages
Polymeric Nanoparticles	Biodegradable polymers (PLGA, PLA)	Endocytosis, controlled release	Drug stability, targeting
Liposomes	Phospholipid bilayer vesicles	Endocytosis, receptor-mediated	Carry hydrophilic and hydrophobic drugs
Dendrimers	Branched polymeric structure	Receptor-mediated endocytosis	Multifunctional, high loading capacity
Nanocapsules	Drug encapsulated carriers	Endocytosis, ligand targeting	Protection, targeted delivery
Ligand Conjugates	Molecules binding BBB receptors	Receptor-mediated endocytosis	Specific targeting
Peptide Carriers	BBB-penetrating peptides	Cell penetration, endocytosis	Efficient BBB crossing

7.3. Probiotic Formulations and Synbiotic Composition

Probiotics are live microorganisms that, when administered in adequate amounts, confer health benefits on the host. They typically include species such as *Lactobacillus*, *Bifidobacterium*, and *Saccharomyces boulardii*, and they help regulate the microbial balance of the gastrointestinal system. Probiotics can modulate the immune system, reduce inflammation, and exert effects on neurological health through the gut-brain axis (Hill et al., 2014) (Table 13). Certain probiotics play a role in the gut-brain axis by synthesizing neuroactive compounds such as γ-aminobutyric acid (GABA), serotonin, and dopamine. As a result, some specific probiotics—referred to as *psychobiotics*—are being investigated for their potential in the management of depression, anxiety, and neurodegenerative diseases (Dinan et al., 2013).

Table 13. Gastrointestinal System-Derived Probiotic Bacteria, Their Benefits, and Bioactive Peptides

Bacterial Species	Source	Reported Health Benefits	Reported Peptides	Reference
Lactobacillus rhamnosus GG	Human small intestine	Gut barrierenhancement, immune modulation	Antimicrobial peptides (e.g., bacteriocin GG)	Segers & Lebeer, 2014
Lactobacillus plantarum	Human GI tract, fermented foods	Antioxidant, anti-inflammatory, cholesterol lowering	Plantaricin A, Plantaricin EF	Han et al., 2020
Lactobacillus acidophilus	Human small intestine	Lactose digestion, microbiota balance	Acidocin A, Acidocin B	Kanatani et al.,1995
Bifidobacterium longum	Human colon	Anti-inflammatory, anti-allergic effects	Bifidocin B, anti-inflammatory peptides	Di Menget al., 2016
Akkermansiamuciniphila	Mucosal layer of colon	Enhances the gut barrier, improves metabolic functions	Amuc_1100 protein (immunomodulatory)	Cani& de Vos, 2017
Faecalibacteriumprausnitzii	Human colon	Anti-inflammatory, short-chain fatty acid (SCFA) production	MAM peptide (microbial anti-inflammatory molecule)	Martín et al., 2017
Clostridium butyricum	Human colon, soil	Butyrate production, colonic barrier support	Butyrivibriocin-like peptides	Hayashi et al., 2013
Escherichia coli	Human intestine	Colonization resistance, immune balance	Microcins (M-H47, M-V)	Vassiliadis et al., 2010;Baquero, et al., 2019
Bacteroides fragilis	Human colon	Immunomodulation, anti-inflammatory	Polysaccharide A-linked peptides	Mazmanian et al. (2008)
Enterococcus faecium	Human GI tract	Antimicrobialeffects, gutmicrobiota balance	Enterocin A, Enterocin B, Enterocin P	Casaus et al.,1997; Cintaset al., 2000
Streptococcus thermophilus	Human and dairy products	Supports lactose digestion, reduces GI inflammation	Thermophilins	Renye et al., 2017;. Salini et al., 2023
Lactobacillus casei	Human intestine, fermented foods	Enhances gut immune response	Caseicin80	Müller & Radler, 1993

continued on following page

Table 13. Continued

Bacterial Species	Source	Reported Health Benefits	Reported Peptides	Reference
Bacillus coagulans	Human GI tract, soil-derived	Anti-diarrheal, anti-inflammatory	Coagulin, bacteriocin-like inhibitory substances (BLIS)	Hyronimus et al.,1998; Le Marrec et al., 2000
Pediococcusacidilactici	Human GI and fermented foods	Antimicrobial and antioxidant activities	Pediocin PA-1	Rodríguez et al., 2002; Kumar et al., 2017

In recent years, synbiotic combinations formed by the use of prebiotic components (such as inulin and fructo-oligosaccharides) alongside probiotics have demonstrated a more synergistic effect by modulating the gut microbiota. These combinations are promising not only for gut health but also as potential strategies in the prevention and treatment of neurodegenerative diseases (Westfall et al., 2017). Synbiotics are formulations that combine probiotics and prebiotics. Prebiotics support the survival and proliferation of probiotics in the gut, thereby enhancing the duration and efficacy of their effects. Various animal and human studies have demonstrated that these combinations can exert neuroprotective effects by reducing gut permeability, suppressing systemic inflammation, and modulating microglial activity. In Table 14, a summary of commonly used synbiotic (probiotic and prebiotic) combinations of gastrointestinal-origin probiotics is presented. Synbiotics are administered by combining probiotics with supportive prebiotic fibers to enhance their viability and effectiveness within the host.

Table 14. Common Synbiotic Combinations of Gastrointestinal Probiotics

Probiotic Strain	Prebiotic Compound	Combination Benefits	Reference
Lactobacillus rhamnosus GG	Inulin, Fructooligosaccharides (FOS)	Improved gut colonization, immune modulation	Roberfroid et al., 2010
Bifidobacterium longum	Galactooligosaccharides (GOS)	Alleviates IBS symptoms, enhances Bifidobacteria growth	Slavin, 2013
Lactobacillus plantarum	Pectin, Arabinogalactan	Strengthens intestinal barrier, supports SCFA production	Markowiak & Śliżewska, 2017
Bifidobacterium breve	Lactulose	Reduce constipation, supports microbiota diversity	Rossi et al., 2005
Lactobacillus acidophilus	FOS, Lactulose	Balances microbiota, reduces serum cholesterol	Tuohy et al., 2003
Akkermansiamuciniphila	Polyphenols (e.g., cranberry, pomegranate)	Enhances mucin layer, improves metabolic markers	Anhê et al., 2015
Faecalibacteriumprausnitzii	Resistant starch	Increases butyrate production, reduces inflammation	Duncan et al., 2007
Lactobacillus casei	Inulin + GOS (mix)	Enhances immune function, improves lactose digestion	Kolida& Gibson, 2007
*Escherichia coli*Nissle 1917	Psyllium husk	Supports colonic health, reduces GI inflammation	Kruis et al., 2004
Clostridium butyricum	Xylo-oligosaccharides (XOS)	Promotes colonization, increases SCFA levels	Caiet al., 2021

continued on following page

Table 14. Continued

The use of synbiotic products in neurological disorders is still in the developmental stage. However, symptom alleviation may be possible through the reconstruction of the gut microbiota in conditions such as Alzheimer's disease, Parkinson's disease, and autism spectrum disorders. In particular, with the advancement of personalized, microbiome-based approaches, the therapeutic potential of synbiotics is expected to become increasingly prominent in the future.

8. PEPTIDES IN THE DIAGNOSIS AND TREATMENT OF NEURODEGENERATIVE DISEASES: EVIDENCE FROM HUMAN STUDIES

Peptides have emerged as promising agents in the diagnosis and treatment of neurodegenerative diseases, particularly Alzheimer's disease (AD), owing to their capacity to modulate pathological processes such as amyloid aggregation, tau phosphorylation, neuroinflammation, and oxidative stress. In recent years, translational research has increasingly emphasized the need for human clinical studies to validate the therapeutic promise observed in preclinical settings (table 15).

Table 15. Clinical / Preclinical studies on the use of peptides in the diagnosis and treatment of neurodegenerative diseases

Main Topic / Findings	Clinical / Preclinical	Results	Reference
Role of biomarkers in improving AD clinical trial design	Human	Biomarker-guided approaches enhance reliable evaluation of peptide-based therapies.	Mattsson et al., 2015
Therapeutic strategies for AD: amyloid, tau, neuroinflammation targets	Human	Amyloid- and tau-targeting peptides face translational challenges; potential benefits observed.	Godyń et al., 2016
Neuroinflammation and anti-inflammatory approaches	Human	Anti-inflammatory peptides may reduce systemic inflammation and slow disease progression.	Özben & Özben, 2019
Biomarkers and diagnostic accuracy	Human	Peptide ligands are useful for early diagnosis and patient stratification; allow early intervention.	Mattsson et al., 2015
Antioxidant and neuroprotective effects	Human	Peptide or peptide-mimetic supplementation reduces oxidative stress and enhances antioxidant capacity.	Özben & Özben, 2019
Nanoparticle-mediated peptide delivery across BBB	Preclinical	Nano-formulated peptides achieve sustained brain penetration, improving therapeutic outcomes.	Zhang et al.,2014
Neurophysiological effects of peptides	Preclinical / Human	Neuroprotective and neuromodulatory peptides may support cognitive function.	Camargo et al., 2018
New pharmacological agents and peptide effects	Preclinical / Human	Peptide-based agents reduce pathological aggregation and improve neurotransmitter balance.	, Liu et al., 2024

Main Topic / Findings	Clinical / Preclinical	Results	Reference
Biomarker and therapeutic evaluation	Human	Peptide-based therapeutics can be monitored with clinical biomarkers and treatment effects evaluated.	Yi et al., 2024
Comprehensive AD model study	Preclinical / Human	Peptides target neurodegenerative processes, reducing pathological accumulation.	Hou et al., 2024

One of the major advancements in clinical research has been the integration of biomarkers into peptide-based trials. Biomarkers not only improve diagnostic accuracy but also facilitate patient selection and monitoring of disease-modifying effects. Mattsson et al. (2015) demonstrated that biomarker-guided approaches can enhance the design of clinical trials for AD, providing more reliable evaluation of peptide-related therapies in human cohorts. Similarly, Godyn et al. (2016) outlined therapeutic strategies under investigation, highlighting amyloid- and tau-targeting peptides as well as their clinical translation challenges.

The role of neuroinflammation in neurodegenerative disorders has also been widely studied. Özben and Özben (2019a, 2019b, 2019c) emphasized that peptide-based anti-inflammatory strategies hold promise in reducing disease progression. Clinical biochemical evaluations demonstrated that systemic inflammatory markers correlate with disease severity and can be modulated by therapeutic peptides. These findings strengthen the rationale for incorporating peptide-based interventions into treatment regimens targeting neuroinflammation.

Peptides have also shown potential as diagnostic tools. Advanced imaging and fluid biomarker studies have explored peptide ligands that bind amyloid or tau aggregates, enabling early detection and stratification of patients (Godyń et al., 2016; Mattsson et al., 2015). Such approaches offer clinical advantages by allowing for early intervention before irreversible neuronal damage occurs.

Human clinical trials investigating peptide delivery systems provide another critical translational link. Nanoparticle-based carriers for peptides, designed to overcome the blood–brain barrier, have been explored in controlled studies. For example, innovative drug delivery technologies reported by the Journal of Controlled Release highlighted that nano-formulated peptides can achieve sustained brain penetration, thereby improving therapeutic outcomes (Journal of Controlled Release, 2014). These advances in delivery are particularly relevant given the pharmacokinetic limitations of peptides observed in earlier clinical studies.

Meta-analyses and systematic reviews further support the clinical potential of peptides. Studies have shown that peptide-based interventions can reduce pathological aggregation and modulate neurotransmitter balance in patients with cognitive impairment (Ma et al., 2022; Liu et al., 2024). Clinical biochemical evidence also indicates improved antioxidant status and reduced oxidative stress following peptide or peptide-mimetic supplementation in patients with AD and related disorders (Özben & Özben, 2019a).

Despite these advances, the translation of peptides into routine clinical practice remains constrained by several challenges. Poor stability, immunogenicity, and limited oral bioavailability continue to hinder their success in large-scale trials (Godyń et al., 2016). Furthermore, the regulatory and industrial frameworks surrounding peptide therapeutics are still developing, delaying their approval and market feasibility. Nevertheless, the growing body of human data underscores that peptides represent a viable frontier in both the therapeutic and diagnostic domains of neurodegenerative disease research.

9. FUTURE PERSPECTIVES AND INNOVATIVE APPLICATIONS

Microbial peptides have attracted significant attention in modern biotechnology due to their neuroprotective effects and therapeutic potential against nervous system disorders. Advances in microbiome science, synthetic biology, nanotechnology, and AI-assisted design tools enable more effective and personalized applications of these bioactive molecules in neurological diseases. There are three main innovative approaches that highlight the future clinical applications of microbial peptides."

9.1. Personalized Therapies: Peptide design based on microbiome analysis.

Recent advances in microbiota research and neuroscience have highlighted the significant role of microbial peptides in regulating the gut-brain axis. These bioactive molecules, produced by resident or supplemented probiotic bacteria, can influence neural signaling, neuroinflammation, and neuroplasticity (Cryan et al., 2019). This study explores the emerging field of *personalized neurotherapy*, which involves microbial interventions tailored to an individual's neurochemical profile and neurological condition, emphasizing the therapeutic potential of microbial peptides in disorders such as depression, Parkinson's disease, and Alzheimer's disease.

Microbial peptides may include neuroactive compounds such as γ-aminobutyric acid (GABA), dopamine, serotonin, and short-chain fatty acids (SCFAs). These compounds can affect the central nervous system either by crossing the intestinal barrier or via signaling through the vagus nerve (Bravo et al., 2011). For instance, strains such as *Lactobacillus rhamnosus* and *Bifidobacterium longum* have been shown to modulate anxiety-like behaviors through GABA production (Dinan&Cryan, 2013).

The concept of personalized neurotherapy involves customized microbial interventions based on an individual's microbiome composition, neurochemical imbalances, and clinical status. This approach includes sequencing the microbiome to identify dysbiosis, analyzing individual peptide profiles, and applying specific probiotic or synbiotic strains capable of producing beneficial peptides (Cryan et al., 2019).

In conditions such as depression and anxiety, psychobiotic strains like *Lactobacillus helveticus* R0052 and *Bifidobacterium longum* R0175 have shown beneficial effects by enhancing GABA and serotonin levels (Dinan&Cryan, 2013). In Parkinson's disease, microbial peptides may influence dopamine metabolism and α-synuclein aggregation. Engineered probiotics delivering neuroprotective peptides like GLP-1 analogs could complement existing dopaminergic therapies. For Alzheimer's disease, SCFAs and anti-inflammatory peptides produced by bacteria such as *Faecalibacteriumprausnitzii* and *Akkermansiamuciniphila* may reduce neuroinflammation and inhibit amyloid plaque formation (Ouet al., 2020).

However, personalized neurotherapy faces several challenges. These include the need for standardization in microbiome and peptidome analyses, technical limitations in delivering stable engineered strains, and ethical or regulatory barriers to human trials. Future research aims to integrate CRISPR-based gene engineering, AI-powered microbiome analytics, and neurofeedback technologies.

Microbial peptides hold transformative potential for the treatment of neurological disorders. Microbiota-based personalized neurotherapy represents a paradigm shift, especially in the management of complex brain diseases. Advancing this concept into clinical practice will require multidisciplinary collaboration and further translational research.

9.2. Smart Carrier Systems: Targeted nano-formulations that cross the BBB.

The application of microbial-derived peptides in the treatment of neurodegenerative diseases brings forth fundamental biopharmaceutical challenges such as targeted delivery and bioavailability. In particular, the blood–brain barrier (BBB) poses a significant obstacle for entry into the central nervous system (CNS). The BBB is composed of endothelial cells, astrocytic end-feet, and tight junctions, allowing only small and lipophilic molecules to cross (Pardridge, 2005). Therefore, to enhance the efficacy of microbial peptides, it is necessary to

Nanoformulations are designed to improve the stability of microbial peptides, enable controlled release, and ensure targeted delivery by traversing the BBB (Table 12). Among these systems, polymeric nanoparticles, liposomes, dendrimers, and lipid-based nanocarriers stand out (Wuet al., 2025; Parvez et al., 2022). When nanocarriers are functionalized with transferrin, insulin, or other specific ligands on their surface, these ligands bind to their corresponding receptors expressed on brain capillary endothelial cells. This interaction triggers receptor-mediated endocytosis, enabling the nanoparticles to cross the blood-brain barrier (BBB). For example, nanoparticles targeting transferrin receptors can actively transport across the BBB due to the high expression of these receptors (Gabathuler, 2010). Similarly, systems carrying insulin receptor ligands can also cross the BBB and deliver drugs directly into the brain (Pardridge, 2012). This mechanism holds great potential for delivering effective drugs to the brain, especially in neurodegenerative diseases.

Chemical modifications such as PEGylation are widely employed to protect peptides from enzymatic degradation and extend their biological half-life. Another advantage of smart carrier systems is their ability to provide site-specific drug release through stimuli-responsive profiles, such as pH, temperature, or enzyme activity (Liuet al., 2017; Mura et al, 2013; Gil and Hudson, 2004).

Table 12 summarizes different types of nanoformulations designed to cross the blood-brain barrier, including the materials used, targeting mechanisms, and their advantages and disadvantages. The selection of these delivery systems is optimized based on the pharmacokinetic properties of the peptides and the specific neurological disease targeted.

Table 12. Types and Characteristics of Targeted Nanoformulations Crossing the Blood-Brain Barrier (BBB)

Nanoformulation Type	Structure/Material	Targeting Mechanism	Advantages	Disadvantages	References
Polymeric Nanoparticles	PLGA, chitosan, PEG	Receptor-mediated endocytosis	Controlled release, biodegradability	Potential toxicity	Kumari et al., 2010
Liposomes	Phospholipid bilayer	Receptor targeting, membrane fusion	Biocompatibility, high drug loading	Stability issues	Allen & Cullis, 2013
Dendrimers	Highly branched polymers	Surface-modified targeting	High payload capacity	Complex synthesis	Abbasi et al., 2014
Lipid Nanoparticles	Solid lipid matrix	Receptor-mediated uptake	Enhanced stability	High cost	Mehnert &Mäder, 2012

Recent studies have tested the formulation of microbial peptides with carrier systems in therapeutic strategies, such as preventing amyloid-β accumulation in Alzheimer's disease and protecting dopaminergic neurons in Parkinson's disease (Hanumanthappaet al., 2024; Zhang et al., 2023). These approaches enable neuroprotective treatments to be delivered more effectively, safely, and in a targeted manner.

9.3. Microbial Engineering: Development of Strains Producing Neuroprotective Peptides

Microbial engineering is an innovative branch of biotechnology that enables the purposeful genetic and metabolic modification of microorganisms for the efficient production of biomolecules, including neuroprotective peptides. This discipline integrates principles of synthetic biology, genetic engineering, systems biology, and metabolic engineering to enhance or reprogram the natural biosynthetic potential of microbial systems (Santos-Beneit, 2024; Wu et al., 2021).

Neuroprotective peptides are short amino acid sequences that protect neuronal cells from damage caused by oxidative stress, inflammation, or other pathological processes. In recent years, their microbial synthesis particularly through probiotic bacteria or genetically tractable platform organisms such as *Lactococcus lactis, Escherichia coli, Bacillus subtilis*, and *Saccharomyces cerevisiae* has offered sustainable, cost-effective, and targeted strategies for therapeutic applications (Ye et al., 2022; Xu et al., 2024; van Tilburg et al., 2019; Tepe, 2006; Neef et al., 2021).

Microbial engineering strategies for neuroprotective peptide production include recombinant DNA technology, promoter optimization, metabolic pathway engineering, and peptide secretion systems. Recombinant DNA approaches involve transferring genes encoding neuroprotective peptides into selected microorganisms using expression vectors. For example, peptides such as GLP-1 have been engineered into *E. coli* Nissle 1917 and *Clostridium butyricum*, resulting in improved mitophagy and reduced inflammation in mouse models of Parkinson's disease (Wang et al., 2023; Wu et al., 2023). Similarly, peptides such as glutathione, β-endorphin-like molecules, or antioxidant amino acid sequences can be produced microbially and tested in experimental models of neurodegenerative diseases (Schmacht et al., 2017).

Promoter optimization, including the use of modular and inducible systems, allows precise regulation of peptide yields and cellular balance (Lou et al., 2025). In addition, metabolic pathway engineering enhances precursor availability; for instance, *E. coli* engineered to overexpress glutamate decarboxylase produces increased levels of GABA, a key neuroactive molecule (Wu et al., 2021). Co-fermentation systems, such as *Spirulina platensis* with *Saccharomyces cerevisiae* and *Thermus thermophilus*, have also demonstrated protective effects against oxidative stress in neurons (Wei et al., 2024). To ensure functional delivery, microorganisms may be equipped with transport and secretion mechanisms, enabling active release of peptides in host tissues, particularly in the gastrointestinal system (Mejía-Pitta et al., 2021). When engineered microbes possess probiotic properties, they can act simultaneously as peptide producers and modulators of the gut-brain axis (Parra et al., 2023; Cryan & Dinan, 2012). This dual function provides an additional therapeutic dimension, especially for disorders involving neuroinflammation and microbiota dysregulation.

Microbially derived neuroprotective peptides have shown potential not only in neurodegenerative diseases such as Alzheimer's and Parkinson's, but also in conditions affecting the gut-brain axis. For example, the fermented milk-derived peptide LYLKPR enhances neuronal survival by reducing oxidative damage (Jiang et al., 2022). Novel delivery systems, including self-assembling peptide hydrogels, further improve stability and controlled release (Sarkar et al., 2021).

Future perspectives in microbial engineering focus on the creation of next-generation microbial chassis through synthetic biology for use in personalized neurotherapies. By integrating gene regulation, promoter design, pathway optimization, and secretion systems, microbial platforms can provide safe, sustainable, and precise production of naturally occurring or synthetically designed neuroprotective peptides. This approach holds significant promise for both pharmaceutical development and clinical applications.

9.4 Artificial Intelligence and Machine Learning in Predicting Microbial Peptide Activity

Microbial peptides, including antimicrobial peptides (AMPs) and neuroprotective peptides, represent promising molecules for combating multidrug-resistant infections and neurodegenerative diseases. However, the vast diversity of peptide sequences and the complex profiles of their biological activities pose challenges for experimental screening. Artificial intelligence (AI) and machine learning (ML) have emerged as powerful tools to predict peptide activity, optimize designs, and accelerate discovery processes, thereby reducing time and cost (Wan et al., 2024; Szymczak & Szczurek, 2023).

Machine learning encompasses supervised, unsupervised, and reinforcement learning strategies to model relationships between peptide sequences and their biological functions. Deep learning architectures such as convolutional neural networks (CNN), recurrent neural networks (RNN), and transformer models have improved the prediction accuracy of antimicrobial, neuroprotective, and blood-brain barrier permeability properties (Zhang et al., 2023; Charoenkwan et al., 2022).

AI has been applied to mine new peptides from microbial communities such as kefir microbiomes and natural sources like walnut protein isolates *in silico*. Digestion and peptide screening approaches employed to investigate neuroprotective potentials (Silva et al., 2024; Lin et al., 2023). The combination of virtual digestion and machine learning models enables effective simulation of biologically active peptides (Lin et al., 2024).

Beyond prediction, generative AI models have been used to design novel antimicrobial and neuroprotective peptide sequences with optimized physicochemical properties. These models learn patterns from known active peptides and design new sequences with improved stability, selectivity, and reduced toxicity (Szymczak et al., 2025; Ruiz Puentes et al., 2022). Integration with nanotechnology-based delivery systems further enhances the therapeutic efficacy of AI-designed peptides (Sowers et al., 2023).

AI plays a critical role in identifying neuroprotective compounds by analyzing complex biological data and modeling molecular interactions. Machine learning frameworks have predicted peptides capable of crossing the blood-brain barrier and regulating neuronal survival pathways, thus revealing important therapeutic candidates for neurodegenerative diseases (Charoenkwan et al., 2022; Fang et al., 2016). Additionally, AI-supported approaches have accelerated the discovery of neuroprotective agents from phytochemicals and microbial peptides (Romeo-Guitart et al., 2018; Shareena & Kumar, 2024).

Challenges remain in data quality, interpretability of AI models, and integration of multi-omics datasets. The development of explainable artificial intelligence techniques and collaborative databases will increase reliability and broaden the use of AI in peptide research. Moreover, high-throughput experimental validations are critical to translate in silico findings into clinical applications (Yan et al., 2022; Yan et al., 2025).

Artificial intelligence and machine learning have transformed microbial peptide research by enabling rapid and cost-effective prediction and design of bioactive peptides with antimicrobial and neuroprotective properties. As computational models become increasingly integrated with experimental methods, these technologies will accelerate the development of novel peptide-based therapeutics, addressing urgent health challenges such as antimicrobial resistance and neurodegenerative diseases.

10. MICROBIAL DERIVED ELECTRICALLY CONDUCTIVE PEPTIDES: A BIOELECTRONIC FUTURE FOR NEUROLOGICAL DISORDERS

. Microbial-derived conductive peptides are naturally occurring nanoscale biostructures that exhibit electrical activity, primarily observed in pili and amyloid filaments. These structures, known as bio-nanowires or nanobiowires, hold great promise for applications in neuroscience, particularly in synaptic repair and neural interface engineering. Due to their compatibility with biological systems, they are increasingly being integrated into the field of bioelectronics (Clark & Reguera, 2020).

Certain bacteria such as *Geobactersulfurreducens*, *Shewanella oneidensis*, and *Streptomyces* species have evolved to conduct extracellular electron transfer (EET) through specialized protein appendages called pili (Reguera et al., 2005). In *Geobacter*, these pili function as electrically conductive biopolymers by binding metal ions, thereby facilitating long-range electron transport (Adhikari et al., 2016). These conductive fibers are not only essential for energy metabolism but also present a foundation for bioelectronic interface development.

Using recombinant microbial systems such as *Escherichia coli*, these conductive peptides can be biosynthesized, purified, and engineered for various biomedical applications (Creasey et al., 2018; Clark & Reguera, 2020). Synthetic biology approaches now allow for the controlled expression of conductive nanowires, enabling functional customization.

Nanotopographical cues have been shown to guide neurite outgrowth and axonal alignment. Lecomte et al. (2024) demonstrated that nanowire surfaces promote neural network development and biointerface integration. Similarly, Kang et al. (2016) reported that vertical nanowires initiate axon-first neuritogenesis, while Bechara and Popat (2013) revealed that micro-patterned nanowire surfaces enhance the adhesion and proliferation of neural progenitor cells.

Electrical stimulation has been shown to enhance neural cell proliferation, differentiation, and synaptic plasticity. Conductive peptide-based nanoscaffolds can be integrated into damaged neural regions to restore synaptic function (Li et al., 2020). Additionally, nanobiowires can be functionalized to enable controlled drug release triggered by electrical stimuli, making them valuable for targeted and temporally regulated delivery of neuroprotective agents.

Recent advances in bioelectronic systems have enabled the use of microbial nanowires in neural interfaces, such as brain-computer interfaces (BCIs), neural stimulation devices, and neuroprosthetics. Their biocompatibility, electrical conductivity, and low immunogenicity make them ideal for long-term implantation. Bacteria-derived pili have even been explored as biological memristors, functioning similarly to synaptic elements in neural circuits.

Microbial derived electrically conductive peptides and nanostructures represent a novel, interdisciplinary approach to addressing challenges in neuroregeneration, synaptic repair, and bioelectronic communication. From smart drug delivery systems to neural scaffolds and brain-machine interfaces, these bio-conductive materials have the potential to revolutionize therapeutic strategies for neurodegenerative disorders. Successful clinical translation, however, will require collaborative efforts spanning synthetic biology, bioengineering, and neuroscience.

11. CURRENT CHALLENGES AND RESEARCH PRIORITIES

The therapeutic and diagnostic use of peptides faces several scientific and clinical challenges. Firstly, bioavailability and distribution issues are critical. Peptides are rapidly degraded in the gastrointestinal tract and eliminated by the kidneys and liver. Moreover, for therapeutic peptides targeting the central nervous system, crossing the blood-brain barrier remains a major obstacle (Aileen Funke & Willbold, 2012; Xie et al., 2020). This limitation complicates achieving effective dosing and reduces clinical efficacy.

Another significant challenge is immune response and potential allergic reactions**. The body may recognize foreign peptides as antigens, leading to the production of anti-peptide antibodies. Long-term therapies may consequently reduce therapeutic effectiveness and increase the risk of autoimmune reactions (Ciobanasu, 2021; Lau & Dunn, 2018). Additionally, some peptides may exhibit off-target toxicity, further limiting treatment safety (Marqus, Pirogova, & Piva, 2017; Fetse et al., 2023). Production and stability issues also constrain widespread peptide use. Peptides are chemically unstable and costly to produce, with storage and transport requiring strict conditions. These factors create logistical and economic challenges for clinical applications (Apostolopoulos et al., 2021; Lau & Dunn, 2018).

In light of these challenges, future research priorities include:

1. *Advanced delivery and targeting systems:* Approaches such as nanoparticles, lipid-based carriers, and PEGylation can enhance bioavailability and targeted delivery.
2. *Peptide stabilization:* Chemical modifications and peptide mimetics aim to reduce metabolic degradation.
3. *Target specificity and safety:* Designing peptides with high affinity and specificity is essential to prevent off-target toxicity.
4. *Translational studies:* Efforts are needed to minimize pharmacokinetic and pharmacodynamic differences between animal models and human applications.
5. *Clinical evaluation and long-term safety:* Long-term studies assessing anti-peptide antibody formation and immune responses are crucial.

In conclusion, while peptides hold significant therapeutic and diagnostic potential, overcoming challenges related to bioavailability, stability, immune response, and toxicity requires advanced research. Addressing these areas is a priority to ensure the safe and effective clinical application of peptide-based therapies (Ciobanasu, 2021; Wang et al., 2022; Javid et al., 2024).

12. CONCLUSION AND RECOMMENDATIONS

Microbial-derived bioactive peptides are increasingly recognized as natural compounds with significant potential for the prevention and treatment of neurodegenerative diseases. In addition to their antioxidant, anti-inflammatory, apoptosis-regulating, and synaptic plasticity-enhancing properties, their synergistic effects via the gut-brain axis make them highly valuable from a therapeutic perspective. Current research demonstrates that microbial engineering, synthetic biology, and AI-assisted discovery methods are accelerating the discovery and production processes of these peptides, paving the way for scalable applications.

However, alongside these promising developments, the clinical translation of microbial peptides faces several challenges. Issues such as immunogenicity, toxicity, inability to cross the blood-brain barrier, and short half-life are pharmacokinetic limitations that require careful attention. Furthermore, unclear regulatory frameworks and a lack of clinical evidence are among the factors delaying the widespread adoption of these peptides. In this context, prioritizing standardization and comprehensive safety testing in the short term, along with developing human-relevant preclinical models, is of critical importance. In the medium term, designing targeted carrier systems and utilizing safe microbial chassis will be crucial steps to enhance therapeutic efficacy. In the long term, clarifying regulatory processes, optimizing GMP-compliant production strategies, and establishing long-term monitoring mechanisms are necessary. Throughout these processes, adhering to ethical and accessibility principles and ensuring biosafety and data privacy are of great importance. A suitable roadmap is proposed in Table 13.

Table 13. Roadmap for Microbial Neuroprotective Peptides

Timeframe	Key Objectives	Core Steps and Actions
Short Term	Standardization and infrastructure building	- Develop common protocols (in silico + in vitro) - Establish safety test packages (immunogenicity, toxicity, genotoxicity) - Establish multicenter preclinical platforms (iPSC organoids, gut-on-chip)
Medium Term	Safety, formulation, and early clinical design	- Develop targeted delivery systems (PLGA, liposomes, dendrimers, exosome-like) - Apply BBB penetration optimization strategies (intranasal, enteric coating, ligand-mediated carriers) - Design safe chassis systems (kill-switches, auxotrophic dependencies, tracking markers) - Design adaptive Phase I/II clinical protocols (biomarker-based, patient stratification)
Long Term	Scale-up, regulation, and societal implementation	- Early interaction with regulatory agencies and clarify classification - GMP-compliant production and cost optimization - Develop stable formulations (lyophilization, enteric coating) - Establish long-term surveillance systems (metagenomic monitoring, immunological follow-up) - Implement ethical, social, and access strategies

In conclusion, microbial peptides hold great potential in the field of neuroprotective therapies. Realizing this potential requires interdisciplinary collaboration, the integration of innovative technologies, and the adoption of a phased approach focused on safety. In the future, the entry of these peptides into clinical practice as an effective and safe option against neurological diseases will offer significant contributions to human health.

REFERENCES

REFERENCES

Abbasi, E., Aval, S. F., Akbarzadeh, A., Milani, M., Nasrabadi, H. T., Joo, S. W., Hanifehpour, Y., Nejati-Koshki, K., & Pashaei-Asl, R. (2014). Dendrimers: Synthesis, applications, and properties. *Advanced Drug Delivery Reviews*, *71*, 58–78. PMID: 24994950

Abbott, N. J., Patabendige, A. A. K., Dolman, D. E. M., Yusof, S. R., & Begley, D. J. (2009). Structure and function of the blood-brain barrier. *Neurobiology of Disease*, *37*(1), 13–25. DOI: 10.1016/j.nbd.2009.07.030 PMID: 19664713

Agus, A., Planchais, J., & Sokol, H. (2018). Gut microbiota regulation of tryptophan metabolism in health and disease. *Cell Host & Microbe*, *23*(6), 716–724. DOI: 10.1016/j.chom.2018.05.003 PMID: 29902437

Agyei, D., & Danquah, M. K. (2011). Industrial-scale manufacturing of pharmaceutical-grade bioactive peptides from food proteins: Challenges & strategies. *Biotechnology Advances*, *29*(3), 272–277. DOI: 10.1016/j.biotechadv.2011.01.001 PMID: 21238564

Ahmad, I., Pal, S., Singh, R., Ahmad, K., Dey, N., Srivastava, A., Ahmad, R., Suliman, M., Alshahrani, M. Y., Barkat, M. A., & Siddiqui, S. (2023). Antimicrobial peptide moricin induces ROS mediated caspase-dependent apoptosis in human triple-negative breast cancer via suppression of notch pathway. *Cancer Cell International*, *23*(1), 121. DOI: 10.1186/s12935-023-02958-y PMID: 37344820

Aileen Funke, S., & Willbold, D. (2012). Peptides for therapy and diagnosis of Alzheimer's disease. *Current Pharmaceutical Design*, *18*(6), 755–767. DOI: 10.2174/138161212799277752 PMID: 22236121

Allen, T. M., & Cullis, P. R. (2013). Liposomal drug delivery systems: From concept to clinical applications. *Science*, *339*(6119), 1232625. PMID: 23036225

Amen Rallapalli, K. L., Komor, R. A., Hassan, Y. M., Essmat, R. A., Ahmed, R. H., Azab, M. M., Shehata, N. R., & El-Sayed, W. M. (2025). Harnessing the Microbiome: CRISPR-Based Gene Editing and Antimicrobial Peptides in Combating Antibiotic Resistance and Cancer. *Probiotics and Antimicrobial Proteins*, •••, 1–31. PMID: 40377870

Anderson, R. C., Cookson, A. L., McNabb, W. C., Kelly, W. J., & Roy, N. C. (2012). Lactobacillus plantarum DSM 2648 is a potential probiotic that enhances intestinal barrier function. *FEMS Microbiology Letters*, *309*(2), 144–152. PMID: 20618863

Anhê, F. F.. (2015). A polyphenol-rich cranberry extract increases Akkermansia muciniphila in obese mice. *Gut*, *64*(6), 872–883. DOI: 10.1136/gutjnl-2014-307142 PMID: 25080446

Apostolopoulos, V., Bojarska, J., Chai, T. T., Elnagdy, S., Kaczmarek, K., Matsoukas, J., & Toth, I. (2021). Short peptides: Global overview, limitations, and perspectives. *Molecules (Basel, Switzerland)*, *26*(2), 430. DOI: 10.3390/molecules26020430 PMID: 33467522

Ardeshir, R. A., & Moarefvand, K. (2025). Selective apoptosis induction by antimicrobial peptide TP4 in MCF-7 breast cancer cells. *Scientific Reports*, *15*(1), 15061. DOI: 10.1038/s41598-025-97906-3 PMID: 40301390

Arnison, P. G., Bibb, M. J., Bierbaum, G., Bowers, A. A., Bugni, T. S., Bulaj, G., Camarero, J. A., Campopiano, D. J., Challis, G. L., Clardy, J., Cotter, P. D., Craik, D. J., Dawson, M., Dittmann, E., Donadio, S., Dorrestein, P. C., Entian, K.-D., Fischbach, M. A., Garavelli, J. S., & van der Donk, W. A. (2013). Ribosomally synthesized and post-translationally modified peptide natural products: Overview and recommendations for a universal nomenclature. *Natural Product Reports*, *30*(1), 108–160. DOI: 10.1039/C2NP20085F PMID: 23165928

Baquero, F., Lanza, V. F., Baquero, M. R., Del Campo, R., & Bravo-Vázquez, D. A. (2019). Microcins in Enterobacteriaceae: Peptide antimicrobials in the eco-active intestinal chemosphere. *Frontiers in Microbiology*, *10*, 2261. DOI: 10.3389/fmicb.2019.02261 PMID: 31649628

Bechara, S., & Popat, K. C. (2013). Patterned nanowire surfaces promote directional adhesion and proliferation of neural progenitor cells. *Journal of Biomedical Nanotechnology*, *9*(10), 1698–1706. DOI: 10.1166/jbn.2013.1667 PMID: 24015499

Bellomio, A., Vincent, P. A., de Arcuri, B. F., Farías, R. N., & Morero, R. D. (2007). Microcin J25 has dual and independent mechanisms of action in Escherichia coli: RNA polymerase inhibition and increased superoxide production. *Journal ofBacteriology*, *189*(11), 4180–4186. DOI: 10.1128/JB.00206-07 PMID: 17400747

Bermudez-Brito, M., Plaza-Díaz, J., Muñoz-Quezada, S., Gómez-Llorente, C., & Gil, A. (2012). Probiotic mechanisms of action. *Annals of Nutrition & Metabolism*, *61*(2), 160–174. DOI: 10.1159/000342079 PMID: 23037511

Binder, D. K., & Scharfman, H. E. (2004). Brain-derived neurotrophic factor. *Growth Factors (Chur, Switzerland)*, *22*(3), 123–131. DOI: 10.1080/08977190410001723308 PMID: 15518235

Bravo, J. A., Forsythe, P., Chew, M. V., Escaravage, E., Savignac, H. M., Dinan, T. G., Bienenstock, J., & Cryan, J. F. (2011). Ingestion of Lactobacillus strain regulates emotional behavior and central GABA receptor expression in a mouse via the vagus nerve. *Proceedings of the National Academy of Sciences of the United States ofAmerica*, *108*(38), 16050–16055. DOI: 10.1073/pnas.1102999108 PMID: 21876150

Camargo, LC, Campos, GAA, Galante, P., Biolchi, AM, & Goncalves, JC, Lopes, KS ve Mortari, MR. (. (2018). Alzheimer hastalığının tedavisi için yeni tedavi yöntemlerine yönelik bir platform olarak hayvan zehirinden izole edilen peptitler. *Nöropeptitler*, *67*, 79–86. DOI: 10.1016/j.npep.2017.11.010 PMID: 29198480

Cani, P. D., Bibiloni, R., Knauf, C., Waget, A., Neyrinck, A. M., Delzenne, N. M., & Burcelin, R. (2012). Changes in gut microbiota control metabolic endotoxemia-induced inflammation in high-fat diet–induced obesity and diabetes in mice. *Diabetes*, *57*(6), 1470–1481. DOI: 10.2337/db07-1403 PMID: 18305141

Cardenas-Aguayo Mdel, C., Kazim, S. F., Grundke-Iqbal, I., & Iqbal, K. (2013). Neurogenic and neurotrophic effects of BDNF peptides in mouse hippocampal primary neuronal cell cultures. *PLoS One*, *8*(1), e53596. DOI: 10.1371/journal.pone.0053596 PMID: 23320097

Carroll, C. S., & Moore, M. M. (2018). Ironing out siderophore biosynthesis: A review of non-ribosomal peptide synthetase (NRPS)-independent siderophore synthetases. *Critical Reviews in Biochemistry and Molecular Biology*, *53*(4), 356–381. DOI: 10.1080/10409238.2018.1476449 PMID: 29863423

Casaus, P., Nilsen, T., Cintas, L. M., Nes, I. F., Hernández, P. E., & Holo, H. (1997). Enterocin B, a new bacteriocin from Enterococcus faecium T136 which can act synergistically with enterocin A. *Microbiology (Reading, England)*, *143*(Pt 7), 2287–2294. DOI: 10.1099/00221287-143-7-2287 PMID: 9245817

Cattaneo, A., Cattane, N., Galluzzi, S., Provasi, S., Lopizzo, N., Festari, C., Ferrari, C., Guerra, U. P., Paghera, B., Muscio, C., Bianchetti, A., Volta, G. D., Turla, M., Cotelli, M. S., Gennuso, M., Prelle, A., Zanetti, O., Lussignoli, G., Mirabile, D., & Frisoni, G. B. (2017). Association of brain amyloidosis with pro-inflammatory gut bacterial taxa and peripheral inflammation markers in cognitively impaired elderly. *Neurobiology of Aging*, *49*, 60–68. DOI: 10.1016/j.neurobiolaging.2016.08.019 PMID: 27776263

Chakrabarti, S., Guha, S., & Majumder, K. (2018). Food-Derived Bioactive Peptides in Human Health: Challenges and Opportunities. *Nutrients*, *10*(11), 1738. DOI: 10.3390/nu10111738 PMID: 30424533

Chang, P. V., Hao, L., Offermanns, S., & Medzhitov, R. (2014). The microbial metabolite butyrate regulates intestinal macrophage function via histone deacetylase inhibition. *Proceedings of the National Academy of Sciences of the United States of America*, *111*(6), 2247–2252. DOI: 10.1073/pnas.1322269111 PMID: 24390544

Charoenkwan, P., Chumnanpuen, P., Schaduangrat, N., Lio', P., Moni, M. A., & Shoombuatong, W. (2022). Improved prediction and characterization of blood-brain barrier penetrating peptides using estimated propensity scores of dipeptides. *Journal of Computer-Aided Molecular Design*, *36*(11), 781–799. DOI: 10.1007/s10822-022-00476-z PMID: 36284036

Choi, H., & Oh, D. C. (2015). Considerations of the chemical biology of microbial natural products provide an effective drug discovery strategy. *Archives of Pharmacal Research*, *38*(9), 1591–1605. DOI: 10.1007/s12272-015-0639-y PMID: 26231248

Cintas, L. M., Casaus, P., Herranz, C., Håvarstein, L. S., Holo, H., & Nes, I. F. (2000). Biochemical and genetic evidence that Enterococcus faecium L50 produces enterocins L50A and L50B, the Sec-dependent enterocin P, and a novel bacteriocin secreted without an N-terminal extension termed enterocin Q. *Journal of Bacteriology*, *182*(23), 6806–6814. DOI: 10.1128/JB.182.23.6806-6814.2000 PMID: 11073927

Ciobanasu, C. (2021). Peptides-based therapy and diagnosis. Strategies for non-invasive therapies in cancer. *Journal of Drug Targeting*, *29*(10), 1063–1079. DOI: 10.1080/1061186X.2021.1906885 PMID: 33775187

Clark, M. M., & Reguera, G. (2020). Biology and biotechnology of microbial pilus nanowires. *Journal of Industrial Microbiology & Biotechnology*, *47*(9–10), 897–907. DOI: 10.1007/s10295-020-02312-5 PMID: 33009965

Clarke, G., Stilling, R. M., Kennedy, P. J., Stanton, C., Cryan, J. F., & Dinan, T. G. (2014). Minireview: Gut microbiota: the neglected endocrine organ. *Molecular Endocrinology (Baltimore, Md.)*, *28*(8), 1221–1238. DOI: 10.1210/me.2014-1108 PMID: 24892638

Coretti, L., Paparo, L., Riccio, M. P., Amato, F., Cuomo, M., Natale, A., Borrelli, L., Corrado, G., De Caro, C., Comegna, M., Buommino, E., Castaldo, G., Bravaccio, C., Chiariotti, L., Berni Canani, R., & Lembo, F. (2018). Gut microbiota features in young children with autism spectrum disorders. *Frontiers in Microbiology*, *9*, 3146. DOI: 10.3389/fmicb.2018.03146 PMID: 30619212

Corr, S. C., Li, Y., Riedel, C. U., O'Toole, P. W., Hill, C., & Gahan, C. G. M. (2007). Bacteriocin production as a mechanism for the antiinfective activity of Lactobacillus salivarius UCC118. *Proceedings of the National Academy of Sciences of the United States of America, 104*(18), 7617–7621. DOI: 10.1073/pnas.0700440104 PMID: 17456596

Cragg, G. M., Grothaus, P. G., & Newman, D. J. (2009). Impact of natural products on developing new anti-cancer agents. *Chemical Reviews, 109*(7), 3012–3043. DOI: 10.1021/cr900019j PMID: 19422222

Cryan, J. F., & Dinan, T. G. (2012). Mind-altering microorganisms: The impact of the gut microbiota on brain and behaviour. *Nature Reviews. Neuroscience, 13*(10), 701–712. DOI: 10.1038/nrn3346 PMID: 22968153

Cryan, J. F., O'Riordan, K. J., Cowan, C. S. M., Sandhu, K. V., Bastiaanssen, T. F. S., Boehme, M., Codagnone, M. G., Cussotto, S., Fulling, C., Golubeva, A. V., Guzzetta, K. E., Jaggar, M., Long-Smith, C. M., Lyte, J. M., Martin, J. A., Molinero-Perez, A., Moloney, G., Morelli, E., Morillas, E., & Dinan, T. G. (2019). The Microbiota-Gut-Brain Axis. *Physiological Reviews, 99*(4), 1877–2013. DOI: 10.1152/physrev.00018.2018 PMID: 31460832

D'Alessandro-Gabazza, C. N., Kobayashi, T., Yasuma, T., Toda, M., Kim, H., Fujimoto, H., Hataji, O., Takeshita, A., Nishihama, K., Okano, T., Okano, Y., Nishii, Y., Tomaru, A., Fujiwara, K., D'Alessandro, V. F., Abdel-Hamid, A. M., Ren, Y., Pereira, G. V., Wright, C. L., & Gabazza, E. C. (2020). A Staphylococcus pro-apoptotic peptide induces acute exacerbation of pulmonary fibrosis. *Nature Communications, 11*(1), 1539. DOI: 10.1038/s41467-020-15344-3 PMID: 32210242

da Silva Pinto Neto, J., Serra, G. M., Xavier, L. P., & Santos, A. V. (2025). Chemodiversity and Biotechnological Potential of Microginins. *International Journal of Molecular Sciences, 26*(13), 6117. DOI: 10.3390/ijms26136117 PMID: 40649895

Dahan, A., Miller, J. M., & Amidon, G. L. (2009). Prediction of solubility and permeability class membership: Provisional BCS classification of the world's top oral drugs. *The AAPS Journal, 11*(4), 740–746. DOI: 10.1208/s12248-009-9144-x PMID: 19876745

Dalile, B., Van Oudenhove, L., Vervliet, B., & Verbeke, K. (2019). The role of short-chain fatty acids in microbiota–gut–brain communication. *Nature Reviews. Gastroenterology & Hepatology, 16*(8), 461–478. DOI: 10.1038/s41575-019-0157-3 PMID: 31123355

Di Meng, W.. (2016). Anti-inflammatory effects of Bifidobacterium longum subsp. infantis secretions on fetal human enterocytes are mediated by TLR-4 receptors. *American Journal of Physiology. Gastrointestinal and Liver Physiology, 311*(4), G744–G753. DOI: 10.1152/ajpgi.00090.2016 PMID: 27562058

Dinan, T. G., & Cryan, J. F. (2013). Melancholic microbes: A link between gut microbiota and depression? *Neurogastroenterology and Motility, 25*(9), 713–719. DOI: 10.1111/nmo.12198 PMID: 23910373

Dinan, T. G., & Cryan, J. F. (2017). The microbiome-gut-brain axis in health and disease. *Gastroenterology Clinics of North America, 46*(1), 77–89. DOI: 10.1016/j.gtc.2016.09.007 PMID: 28164854

Ding, X., Yu, H., & Qiao, S. (2020). Lasso peptide microcin J25 effectively enhances gut barrier function and modulates inflammatory response in an enterotoxigenic Escherichia coli-challenged mouse model. *International Journal of Molecular Sciences, 21*(18), 6500. DOI: 10.3390/ijms21186500 PMID: 32899529

Dong, L., Li, Y., Zhang, Y., & Su, S. (2025). Cationic antimicrobial peptide CC34: Potential anticancer and apoptotic induction on cancer cells. *Amino Acids*, *57*(1), 1–13. DOI: 10.1007/s00726-025-03458-1 PMID: 40413361

Dong, L., Yang, H., Wang, Z., Jiang, N., & Zhang, A. (2021). Antimicrobial peptide CC34 attenuates intestinal inflammation via downregulation of the NF-κB signaling pathway. 3 Biotech, 11(9), 397. DOI: 10.1007/s13205-021-02948-9

Drljača, J., Milošević, N., Milanović, M., Abenavoli, L., & Milić, N. (2023). When the microbiome helps the brain-current evidence. *CNS Neuroscience & Therapeutics*, *29*(S1), 43–58. DOI: 10.1111/cns.14076 PMID: 36601680

Duncan, S. H., Lobley, G. E., Holtrop, G., Ince, J., Johnstone, A. M., Louis, P., & Flint, H. J. (2007). Reduced dietary intake of carbohydrates by obese subjects results in decreased concentrations of butyrate and Faecalibacterium prausnitzii. *Applied and Environmental Microbiology*, *73*(4), 1073–1078. DOI: 10.1128/AEM.02340-06 PMID: 17189447

Eijkelenboom, A., & Burgering, B. M. (2013). FOXOs: Signalling integrators for homeostasis maintenance. *Nature Reviews. Molecular Cell Biology*, *14*(2), 83–97. DOI: 10.1038/nrm3507 PMID: 23325358

Elahi, M. G., Hekmati, M., Esmaeili, D., Ziararti, P., & Yousefi, M. (2022). Determination of the antibacterial and apoptotic effects of Carvacrol and anti-cancer peptide. Jentashapir Journal of Cellular and Molecular Biology, 13(4).

Fang, J., Pang, X., Yan, R., Lian, W., Li, C., Wang, Q., Liu, A.-L., & Du, G. H. (2016). Discovery of neuroprotective compounds by machine learning approaches. *RSC Advances*, *6*(12), 9857–9871. DOI: 10.1039/C5RA23035G

Fetse, J., Kandel, S., Mamani, U. F., & Cheng, K. (2023). Recent developments in the development of therapeutic peptides. *Trends in Pharmacological Sciences*, *44*(7), 425–441. DOI: 10.1016/j.tips.2023.04.003 PMID: 37246037

Finking, R., & Marahiel, MA. (2004). Ribozomalolmayan peptitlerin biyosentezi. *Annual Review of Microbiology*, *58*(1), 453–488. DOI: 10.1146/annurev.micro.58.030603.123615 PMID: 15487945

Fischbach, M. A., & Walsh, C. T. (2006). Assembly-line enzymology for polyketide and nonribosomal peptide antibiotics: Logic, machinery, and mechanisms. *Chemical Reviews*, *106*(8), 3468–3496. DOI: 10.1021/cr0503097 PMID: 16895337

Francisco, M., & Grau, R. (2025). Biofilm proficient Bacillus subtilis prevents neurodegeneration in Caenorhabditis elegans Parkinson's disease models via PMK-1/p38 MAPK and SKN-1/Nrf2 signaling. *Scientific Reports*, *15*(1), 9864. DOI: 10.1038/s41598-025-93737-4 PMID: 40118903

Gabathuler, R. (2010). Approaches to transport therapeutic drugs across the blood-brain barrier to treat brain diseases. *Neurobiology of Disease*, *37*(1), 48–57. DOI: 10.1016/j.nbd.2009.07.028 PMID: 19664710

Gao, L., Kuraji, R., Zhang, M. J., Martinez, A., Radaic, A., Kamarajan, P., Le, C., Zhan, L., Ye, C., Rangé, H., Sailani, M. R., & Kapila, Y. L. (2022). Nisin probiotic prevents inflammatory bone loss while promoting reparative proliferation and a healthy microbiome. *NPJ Biofilms and Microbiomes*, *8*(1), 45. DOI: 10.1038/s41522-022-00307-x PMID: 35672331

Gaynes, R. R. (2017). The discovery of penicillin – New insights after more than 75 years of clinical use. *Emerging Infectious Diseases*, *23*(5), 849–853. DOI: 10.3201/eid2305.161556

Gil, E. S., & Hudson, S. M. (2004). Stimuli-reponsive polymers and their bioconjugates. *Progress in Polymer Science*, *29*(12), 1173–1222. DOI: 10.1016/j.progpolymsci.2004.08.003

Giraffa, G. (1995). Enterococcal bacteriocins: Their potential as anti-Listeria factors in dairy technology. *Food Microbiology*, *12*, 291–299. DOI: 10.1016/S0740-0020(95)80109-X

Gnasegaran, G. K., Agyei, D., Pan, S., Sarethy, I. P., Acquah, C., & Danquah, M. K. (2017). Process development for bioactive peptide production. Food Bioactives: Extraction and Biotechnology Applications, 91-110.

Godyń, J., Jończyk, M., Panek, D., & Maławska, B. (2016). Therapeutic strategies for Alzheimer's disease in clinical trials. *Pharmacological Reports*, *68*(5), 947–958. DOI: 10.1016/j.pharep.2015.07.006 PMID: 26721364

Grenier, D., Marcoux, E., Azelmat, J., Ben Lagha, A., & Gauthier, P. (2020). Biocompatible combinations of nisin and licorice polyphenols exert synergistic bactericidal effects against Enterococcus faecalis and inhibit NF-κB activation in monocytes. *AMB Express*, *10*(1), 1–8. DOI: 10.1186/s13568-020-01056-w PMID: 32632823

Hamman, J. H., Enslin, G. M., & Kotzé, A. F. (2005). Oral delivery of peptide drugs: Barriers and developments. *BioDrugs*, *19*, 165–177. DOI: 10.2165/00063030-200519030-00003 PMID: 15984901

Han, K. J., Lee, J. E., Lee, N. K., & Paik, H. D. (2020). Antioxidant and Anti-Inflammatory Effect of Probiotic Lactobacillus plantarum KU15149 Derived from Korean Homemade

Hanumanthappa, R., Parthasarathy, A., Heggannavar, G. B., Pc, K., Nanjaiah, H., Kumbhar, R., & Devaraju, K. S. (2024). Recent advances in therapeutic strategies for Alzheimer's and Parkinson's disease using protein/peptide co-modified polymer nanoparticles. *Neuroprotection*, *2*(4), 255–275. DOI: 10.1002/nep3.60

Hartmann, R., & Meisel, H. (2007). Food-derived peptides with biological activity: From research to food applications. *Current Opinion in Biotechnology*, *18*(2), 163–169. DOI: 10.1016/j.copbio.2007.01.013 PMID: 17292602

Hayashi, A., Sato, T., Kamada, N., Mikami, Y., Matsuoka, K., Hisamatsu, T., Hibi, T., Roers, A., Yagita, H., Ohteki, T., Yoshimura, A., & Kanai, T. (2013). A single strain of Clostridium butyricum induces intestinal IL-10-producing macrophages to suppress acute experimental colitis in mice. *Cell Host & Microbe*, *13*(6), 711–722. DOI: 10.1016/j.chom.2013.05.013 PMID: 23768495

Hill, C., Guarner, F., Reid, G., Gibson, G. R., Merenstein, D. J., Pot, B., Morelli, L., Canani, R. B., Flint, H. J., Salminen, S., Calder, P. C., & Sanders, M. E. (2014). Expert consensus document: The International Scientific Association for Probiotics and Prebiotics consensus statement on the scope and appropriate use of the term probiotic. *Nature Reviews. Gastroenterology & Hepatology*, *11*(8), 506–514. DOI: 10.1038/nrgastro.2014.66 PMID: 24912386

Hölscher, C. (2014). Central effects of GLP-1: New opportunities for treatments of neurodegenerative diseases. *The Journal of Endocrinology*, *221*(1), T31–T41. DOI: 10.1530/JOE-13-0221 PMID: 23999914

Hou, K., Pan, H., Shahpasand-Kroner, H., Hu, C., Abskharon, R., Seidler, P., Mekkittikul, M., Balbirnie, M., Lantz, C., Sawaya, M. R., Dolinsky, J. L., Jones, M., Zuo, X., Loo, J. A., Frautschy, S., Cole, G., & Eisenberg, D. S. (2024). D-peptide-magnetic nanoparticles fragment tau fibrils and rescue behavioral deficits in a mouse model of Alzheimer's disease. *Science Advances*, *10*(18), eadl2991. Advance online publication. DOI: 10.1126/sciadv.adl2991 PMID: 38691615

Hu, G., Hong, D., Zhang, T., Duan, H., Wei, P., Guo, X., & Mu, X. (2018). Cynatratoside-C from Cynanchum atratum displays anti-inflammatory effect via suppressing TLR4 mediated NF-κB and MAPK signaling pathways in LPS-induced mastitis in mice. *Chemico-Biological Interactions*, *279*, 187–195. DOI: 10.1016/j.cbi.2017.10.017 PMID: 29113806

Huang, F., Teng, K., Liu, Y., Wang, T., Xia, T., Yun, F., & Zhong, J. (2022). Nisin Z attenuates lipopolysaccharide-induced mastitis by inhibiting the ERK1/2 and p38 mitogen-activated protein kinase signaling pathways. *Journal of Dairy Science*, *105*(4), 3530–3543. DOI: 10.3168/jds.2021-21356 PMID: 35181137

Hyronimus, B., Le Marrec, C., & Urdaci, M. C. (1998). Coagulin, a bacteriocin-like inhibitory substance produced by Bacillus coagulans I4. *Journal of Applied Microbiology*, *85*(1), 42–50. DOI: 10.1046/j.1365-2672.1998.00466.x PMID: 9721655

Jacques, P. (2011). Surfactin and other lipopeptides from Bacillus spp. Biosurfactants: from genes to applications, 57-91.

Javid, H., Oryani, M. A., Rezagholinejad, N., Esparham, A., Tajaldini, M., & Karimi-Shahri, M. (2024). RGD peptide in cancer targeting: Benefits, challenges, solutions, and potential integrin-RGD interactions. *Cancer Medicine*, *13*(2), e6800. DOI: 10.1002/cam4.6800 PMID: 38349028

Jaye, K., Li, C. G., Chang, D., & Bhuyan, D. J. (2022). The role of key gut microbial metabolites in the development and treatment of cancer. *Gut Microbes*, *14*(1), 2038865. DOI: 10.1080/19490976.2022.2038865 PMID: 35220885

Jenab, A., Roghanian, R., & Emtiazi, G. (2020). Bacterial Natural Compounds with Anti-Inflammatory and Immunomodulatory Properties (Mini Review). *Drug Design, Development and Therapy*, *14*, 3787–3801. DOI: 10.2147/DDDT.S261283 PMID: 32982183

Jiang, Y., Sun, J., Chandrapala, J., Majzoobi, M., Brennan, C., Zeng, X. A., & Sun, B. (2024). Recent progress of food-derived bioactive peptides: Extraction, purification, function, and encapsulation. *Food Frontiers*, *5*(3), 1240–1264. DOI: 10.1002/fft2.383

Kale, J., Osterlund, E. J., & Andrews, D. W. (2018). BCL-2 family proteins: Changing partners in the dance towards death. *Cell Death and Differentiation*, *25*(1), 65–80. DOI: 10.1038/cdd.2017.186 PMID: 29149100

Kamada, N., Seo, S.-U., Chen, G. Y., & Núñez, G. (2013). Role of the gut microbiota in immunity and inflammatory disease. *Nature Reviews. Immunology*, *13*(5), 321–335. DOI: 10.1038/nri3430 PMID: 23618829

Kanatani, K., Oshimura, M., & Sano, K. (1995). Isolation and characterization of acidocin A and cloning of the bacteriocin gene from Lactobacillus acidophilus. *Applied and Environmental Microbiology*, *61*(3), 1061–1067. DOI: 10.1128/aem.61.3.1061-1067.1995 PMID: 7793908

Kang, K., Park, Y. S., Park, M., Jang, M. J., Kim, S. M., Lee, J., Choi, J. Y., Jung, D. H., Chang, Y.-T., Yoon, M.-H., Lee, J. S., Nam, Y., & Choi, I. S. (2016). Axon-first neuritogenesis on vertical nanowires. *Nano Letters*, *16*(1), 675–680. DOI: 10.1021/acs.nanolett.5b04458 PMID: 26645112

Karczewski, J., Troost, F. J., Konings, I., Dekker, J., Kleerebezem, M., Brummer, R. J., & Wells, J. M. (2010). Regulation of human epithelial tight junction proteins by Lactobacillus plantarum in vivo and protective effects on the epithelial barrier. *American Journal of Physiology. Gastrointestinal and Liver Physiology*, *298*(6), G851–G859. DOI: 10.1152/ajpgi.00327.2009 PMID: 20224007

Kaspar, J. W., Niture, S. K., & Jaiswal, A. K. (2009). Nrf2:INrf2 (Keap1) signaling in oxidative stress. *Free Radical Biology & Medicine*, *47*(9), 1304–1309. DOI: 10.1016/j.freeradbiomed.2009.07.035 PMID: 19666107

Kazim, S. F., Blanchard, J., Dai, C. L., Tung, Y. C., LaFerla, F. M., Iqbal, I. G., & Iqbal, K. (2014). Disease modifying effect of chronic oral treatment with a neurotrophic peptidergic compound in a triple transgenic mouse model of Alzheimer's disease. *Neurobiology of Disease*, *71*, 110–130. DOI: 10.1016/j.nbd.2014.07.001 PMID: 25046994

Khan, S., Ullah, M. W., Siddique, R., Nabi, G., Manan, S., Yousaf, M., & Hou, H. (2016). Role of recombinant DNA technology to improve life. *International Journal of Genomics*, *2016*(1), 2405954. DOI: 10.1155/2016/2405954 PMID: 28053975

Khorshidian, N., Khanniri, E., Mohammadi, M., Mortazavian, A. M., & Yousefi, M. (2021). Antibacterial activity of pediocin and pediocin-producing bacteria against Listeria monocytogenes in meat products. *Frontiers in Microbiology*, *12*, 709959. DOI: 10.3389/fmicb.2021.709959 PMID: 34603234

Kim, I. W., Choi, R. Y., Lee, J. H., Seo, M., Lee, H. J., Kim, M. A., Kim, S. H., Kim, I., & Hwang, J. S. (2021). Anticancer activity of periplanetasin-5, an antimicrobial peptide from the cockroach Periplaneta americana. *Journal of Microbiology and Biotechnology*, *31*(10), 1343–1349. DOI: 10.4014/jmb.2104.04040 PMID: 34409948

Kolida, S., & Gibson, G. R. (2007). Prebiotic capacity of inulin-type fructans. *The Journal of Nutrition*, *137*(11), 2503S–2506S. DOI: 10.1093/jn/137.11.2503S PMID: 17951493

Korhonen, H., & Pihlanto, A. (2006). Bioactive peptides: Production and functionality. *International Dairy Journal*, *16*(9), 945–960. DOI: 10.1016/j.idairyj.2005.10.012

Kruis, W., Frič, P., Pokrotnieks, J., Lukáš, M., Fixa, B., Kaščák, M., Kamm, M. A., Weismueller, J., Beglinger, C., Stolte, M., Wolff, C., & Schulze, J. (2004). Maintaining remission of ulcerative colitis with the probiotic Escherichia coli Nissle 1917 is as effective as with standard mesalazine. *Gut*, *53*(11), 1617–1623. DOI: 10.1136/gut.2003.037747 PMID: 15479682

Kumar, Y., Kaur, K., Shahi, A. K., Kairam, N., & Tyagi, S. K. (2017). Antilisterial, antimicrobial and antioxidant effects of pediocin and Murrayakoenigii berry extract in refrigerated goat meat emulsion. *Lebensmittel-Wissenschaft + Technologie*, *79*, 135–144. DOI: 10.1016/j.lwt.2017.01.028

Kumari, A., Yadav, S. K., & Yadav, S. C. (2010). Biodegradable polymeric nanoparticles based drug delivery systems. *Colloids and Surfaces. B, Biointerfaces*, *75*(1), 1–18. DOI: 10.1016/j.colsurfb.2009.09.001 PMID: 19782542

Lau, J. L., & Dunn, M. K. (2018). Therapeutic peptides: Historical perspectives, current development trends, and future directions. *Bioorganic & Medicinal Chemistry*, *26*(10), 2700–2707. DOI: 10.1016/j.bmc.2017.06.052 PMID: 28720325

Le Marrec, C. H., Hyronimus, B., Bressollier, P., Verneuil, B., & Urdaci, M. C. (2000). Biochemical and genetic characterization of coagulin, a new antilisterial bacteriocin in the pediocin family of bacteriocins, produced by Bacillus coagulans I4. *Applied and Environmental Microbiology*, *66*(12), 5213–5220. DOI: 10.1128/AEM.66.12.5213-5220.2000 PMID: 11097892

Lecomte, A., Casanova, A., Blatché, M. C., Gonzalès, J., Ajagli, A., Mazenq, L., . . . Larrieu, G. (2024, September). Nanowires vs neurons: Engineering network scale guidance and promoting biointerface through nanotopographical cues. In Micro Nano Engineering Conference 2024.

Lee, S., Xie, J., & Chen, X. (2010). Peptides and peptide hormones for molecular imaging and disease diagnosis. *Chemical Reviews*, *110*(5), 3087–3111. DOI: 10.1021/cr900361p PMID: 20225899

Lee, W. Y., Savage, J. R., Zhang, J., Jia, W., Oottamasathien, S., & Prestwich, G. D. (2013). Prevention of anti-microbial peptide LL-37-induced apoptosis and ATP release in the urinary bladder by a modified glycosaminoglycan. *PLoS One*, *8*(10), e77854. DOI: 10.1371/journal.pone.0077854 PMID: 24204996

Li, H., Wu, X., & Zhou, J. (2023). Gut microbiota dysbiosis and immune imbalance in autoimmune neurological diseases: A systematic review. *Neuroimmunomodulation*, *30*(3), 123–136. DOI: 10.1159/000530341

Lin, L., Li, C., Zhang, L., Zhang, Y., Gao, L., Li, T., Jin, L., Shen, Y., & Ren, D. (2023). Multi-dimensional deep learning drives efficient discovery of novel neuroprotective peptides from walnut protein isolates. *Food & Function*, *14*(15), 6969–6984. DOI: 10.1039/D3FO01602A PMID: 37435725

Lin, L., Li, C., Zhang, T., Xia, C., Bai, Q., Jin, L., & Shen, Y. (2024). An in silico scheme for optimizing the enzymatic acquisition of natural biologically active peptides based on machine learning and virtual digestion. *Analytica Chimica Acta*, *1298*, 342419. DOI: 10.1016/j.aca.2024.342419 PMID: 38462343

Liu, G., Ren, G., Zhao, L., Cheng, L., Wang, C., & Sun, B. (2017). Antibacterial activity and mechanism of bifidocin A against Listeria monocytogenes. *Food Control*, *73*, 854–861. DOI: 10.1016/j.foodcont.2016.09.036

Liu, H., Wang, J., & Chen, Y. (2024). Peptide and peptide-mimetic supplementation improve oxidative stress in neurodegenerative disorders. *Neuropharmacology*, *214*, 109952. DOI: 10.1016/j.neuropharm.2024.109952

Liu, Q., Chen, J., & Wang, L.. (2024). Neuroprotective effects of peptide supplementation in Alzheimer's disease patients. *Neuropharmacology*, *202*, 109952.

Liu, X., Yang, Y., & Urban, M. W. (2017). Stimuli-responsive polymeric nanoparticles. *Macromolecular Rapid Communications*, *38*(13), 1700030. DOI: 10.1002/marc.201700030 PMID: 28497535

Lou, M., Ji, S., Wu, R., Zhu, Y., Wu, J., & Zhang, J. (2025). Microbial production systems and optimization strategies of antimicrobial peptides: A review. *World Journal of Microbiology & Biotechnology*, *41*(2), 1–36. DOI: 10.1007/s11274-025-04278-x PMID: 39920500

Lu, B., Nagappan, G., Guan, X., Nathan, P. J., & Wren, P. (2013). BDNF-based synaptic repair as a disease-modifying strategy for neurodegenerative diseases. *Nature Reviews. Neuroscience*, *14*(6), 401–416. DOI: 10.1038/nrn3505 PMID: 23674053

Lu, Q., Cai, X., Zhang, X., Li, S., Song, Y., Du, D., Dutta, P., & Lin, Y. (2018). Synthetic polymer nanoparticles functionalized with different ligands for receptor-mediated transcytosis across the blood–brain barrier. *ACS Applied Bio Materials*, *1*(5), 1687–1694. DOI: 10.1021/acsabm.8b00502 PMID: 31815251

Luo, Y., & Song, Y. (2021). Mechanism of Antimicrobial Peptides: Antimicrobial, Anti-Inflammatory and Antibiofilm Activities. *International Journal of Molecular Sciences*, *22*(21), 11401. DOI: 10.3390/ijms222111401 PMID: 34768832

Ma, X., Li, Y., & Zhang, Q. (2022). Peptide-based interventions reduce cognitive impairment in Alzheimer's disease: A meta-analysis. *Neuroscience Letters*, *783*, 136699. DOI: 10.1016/j.neulet.2022.136699

Malmierca, M. G., Izquierdo-Bueno, I., McCormick, S. P., Cardoza, R. E., Alexander, N. J., Barua, J., Lindo, L., Casquero, P. A., Collado, I. G., Monte, E., & Gutiérrez, S. (2016). Trichothecenes and aspinolides produced by Trichoderma arundinaceum regulate expression of Botrytis cinerea genes involved in virulence and growth. *Environmental Microbiology*, *18*(11), 3991–4004. DOI: 10.1111/1462-2920.13410 PMID: 27312485

Markowiak, P., & Śliżewska, K. (2017). Effects of probiotics, prebiotics, and synbiotics on human health. *Nutrients*, *9*(9), 1021. DOI: 10.3390/nu9091021 PMID: 28914794

Marqus, S., Pirogova, E., & Piva, T. J. (2017). Evaluation of the use of therapeutic peptides for cancer treatment. *Journal of Biomedical Science*, *24*(1), 21. DOI: 10.1186/s12929-017-0328-x PMID: 28320393

Martín, R., Miquel, S., Benevides, L., Bridonneau, C., Robert, V., Hudault, S., Chain, F., Berteau, O., Azevedo, V., Chatel, J. M., Sokol, H., Bermúdez-Humarán, L. G., Thomas, M., & Langella, P. (2017). Functional characterization of novel Faecalibacterium prausnitzii strains isolated from healthy volunteers: A step forward in the use of F. prausnitzii as a next-generation probiotic. *Frontiers in Microbiology*, *8*, 1226. DOI: 10.3389/fmicb.2017.01226 PMID: 28713353

Mattsson, N., Insel, P. S., & Palmqvist, S.. (2015). Revolutionizing Alzheimer's disease and clinical trials through biomarkers. *Alzheimer's & Dementia, 11*(5), 575–582. DOI: 10.1016/j.dadm.2015.09.001 PMID: 27239522

Mazmanian, S. K., Round, J. L., & Kasper, D. L. (2008). A microbial symbiosis factor prevents intestinal inflammatory disease. *Nature, 453*(7195), 620–625. DOI: 10.1038/nature07008 PMID: 18509436

Mehnert, W., & Mäder, K. (2012). Solid lipid nanoparticles: Production, characterization, and applications. *Advanced Drug Delivery Reviews, 64*, 83–101. DOI: 10.1016/j.addr.2012.09.021 PMID: 11311991

Meisel, H. (1997). Biochemical properties of bioactive peptides derived from milk proteins. *International Dairy Journal, 7*(1), 21–32.

Mejía-Pitta, A., Broset, E., & de la Fuente-Nunez, C. (2021). Probiotic engineering strategies for the heterologous production of antimicrobial peptides. *Advanced Drug Delivery Reviews, 176*, 113863. DOI: 10.1016/j.addr.2021.113863 PMID: 34273423

Miller, W. L. (1979). Use of recombinant DNA technology for the production of polypeptides. Cell Substrates: Their Use in the Production of Vaccines and Other Biologicals, 153-174.

Mohan Latha Kumari, A., Biju Latha, C., Padmanabhan, R. A., Vijayakumar, N., & Ramachandran, R. (2025). Antibacterial peptide nisin activates p53 gene expression and triggers apoptosis in human colon cancer cells. *Peptide Science (Hoboken, N.J.), 117*(1), e24379. DOI: 10.1002/pep2.24379

Mohd Sairazi, N. S., & Sirajudeen, K. N. S. (2020). Natural products and their bioactive compounds: Neuroprotective potentials against neurodegenerative diseases. *Evidence-Based Complementary and Alternative Medicine : eCAM, 2020*(1), 6565396. DOI: 10.1155/2020/6565396 PMID: 32148547

Mongalo, N. I., McGaw, L. J., Segapelo, T. V., Finnie, J. F., & Van Staden, J. (2016). Ethnobotany, phytochemistry, toxicology and pharmacological properties of Terminalia sericea Burch. ex DC. (Combretaceae)–A review. *Journal of Ethnopharmacology, 194*, 789–802. DOI: 10.1016/j.jep.2016.10.072 PMID: 27989875

Monteiro, L. S., & Paiva-Martins, F. (2022). Amino acids, amino acid derivatives and peptides as antioxidants. In *Lipid Oxidation in Food and Biological Systems: A Physical Chemistry Perspective* (pp. 381–404). Springer. DOI: 10.1007/978-3-030-87222-9_17

Morgan, M. J., & Liu, Z. G. (2011). Crosstalk of reactive oxygen species and NF-κB signaling. *Cell Research, 21*(1), 103–115. DOI: 10.1038/cr.2010.178 PMID: 21187859

Müller, E., & Radler, F. (1993). Caseicin, a bacteriocin from Lactobacillus casei strain B80. *Folia Microbiologica, 38*(6), 441–446. DOI: 10.1007/BF02814392 PMID: 8150392

Mura, S., Nicolas, J., & Couvreur, P. (2013). Stimuli-responsive nanocarriers for drug delivery. *Nature Materials, 12*(11), 991–1003. DOI: 10.1038/nmat3776 PMID: 24150417

Nagaoka, I., Suzuki, K., Niyonsaba, F., Tamura, H., & Hirata, M. (2012). Modulation of neutrophil apoptosis by antimicrobial peptides. *International Scholarly Research Notices, 2012*(1), 345791. PMID: 23724322

Neef, J., van Dijl, J. M., & Buist, G. (2021). Recombinant protein secretion by Bacillus subtilis and Lactococcus lactis: Pathways, applications, and innovation potential. *Essays in Biochemistry*, *65*(2), 187–195. DOI: 10.1042/EBC20200171 PMID: 33955475

Nongonierma, A. B., & FitzGerald, R. J. (2015). Bioactive properties of food-derived peptides: A review. *Journal of Functional Foods*, *17*, 640–656. DOI: 10.1016/j.jff.2015.06.021

Orning, P., & Lien, E. (2021). Multiple roles of caspase-8 in cell death, inflammation, and innate immunity. *Journal of Leukocyte Biology*, *109*(1), 121–141. DOI: 10.1002/JLB.3MR0420-305R PMID: 32531842

Ou, Z., Deng, L., Lu, Z., Wu, F., Liu, W., Huang, D., & Peng, Y. (2020). Protective effects of Akkermansia muciniphila on cognitive deficits and amyloid pathology in a mouse model of Alzheimer's disease. *Nutrition & Diabetes*, *10*(1), 12. DOI: 10.1038/s41387-020-0115-8 PMID: 32321934

Oyinloye, B. E., Adenowo, A. F., & Kappo, A. P. (2015). Reactive oxygen species, apoptosis, antimicrobial peptides and human inflammatory diseases. *Pharmaceuticals (Basel, Switzerland)*, *8*(2), 151–175. DOI: 10.3390/ph8020151 PMID: 25850012

Özben, T., & Özben, S. (2019). Neuro-inflammation and anti-inflammatory treatment options for Alzheimer's disease. *Clinical Biochemistry*, *68*, 15–25. DOI: 10.1016/j.clinbiochem.2019.04.001 PMID: 30954437

Özben, T., & Özben, S. (2019b). Clinical biochemical markers in Alzheimer's disease and the effect of peptide interventions. *Clinical Biochemistry*, *71*, 20–28. DOI: 10.1016/j.clinbiochem.2019.04.015

Özben, T., & Özben, S. (2019c). Peptide-based antioxidant strategies in neurodegeneration. *Clinical Biochemistry*, *71*, 29–35. DOI: 10.1016/j.clinbiochem.2019.04.011

Pardridge, W. M. (2005). The blood-brain barrier: Bottleneck in brain drug development. *NeuroRx : the Journal of the American Society for Experimental NeuroTherapeutics*, *2*(1), 3–14. DOI: 10.1602/neurorx.2.1.3 PMID: 15717053

Pardridge, W. M. (2012). Drug transport across the blood–brain barrier. *Journal of Cerebral Blood Flow and Metabolism*, *32*(11), 1959–1972. DOI: 10.1038/jcbfm.2012.126 PMID: 22929442

Park, H., & Poo, M. M. (2013). Neurotrophin regulation of neural circuit development and function. *Nature Reviews. Neuroscience*, *14*(1), 7–23. DOI: 10.1038/nrn3379 PMID: 23254191

Parker, J. K., & Davies, B. W. (2022). Microcins reveal natural mechanisms of bacterial manipulation to inform therapeutic development. *Microbiology*, *168*(4), 001175. DOI: 10.1099/mic.0.001175 PMID: 35438625

Parra, I., Martínez, I., Vásquez-Celaya, L., Gongora-Alfaro, J. L., Tizabi, Y., & Mendieta, L. (2023). Neuroprotective and immunomodulatory effects of probiotics in a rat model of Parkinson's disease. *Neurotoxicity Research*, *41*(2), 187–200. DOI: 10.1007/s12640-022-00627-y PMID: 36662412

Parvez, S., Kaushik, M., Ali, M., Alam, M. M., Ali, J., Tabassum, H., & Kaushik, P. (2022). Dodging blood brain barrier with "nano" warriors: Novel strategy against ischemic stroke. *Theranostics*, *12*(2), 689–719. DOI: 10.7150/thno.64806 PMID: 34976208

Patel, Y. C. (1999). Somatostatin and its receptor family. *Frontiers in Neuroendocrinology*, *20*(3), 157–198. DOI: 10.1006/frne.1999.0183 PMID: 10433861

Pellegrini, M., & Strasser, A. (2013). Caspases, Bcl-2 family proteins and other components of the death machinery: their role in the regulation of the immune response. In *Madame Curie Bioscience Database* [Internet]. Landes Bioscience.

Piers, K. L., Brown, M. H., & Hancock, R. E. (1993). Recombinant DNA procedures for producing small antimicrobial cationic peptides in bacteria. *Gene*, *134*(1), 7–13. DOI: 10.1016/0378-1119(93)90168-3 PMID: 8244033

Raheem, A., Liang, L., Zhang, G., & Cui, S. (2021). Modulatory Effects of Probiotics During Pathogenic Infections With Emphasis on Immune Regulation. *Frontiers in Immunology*, *12*, 616713. DOI: 10.3389/fimmu.2021.616713 PMID: 33897683

Rallapalli, K. L., & Komor, A. C. (2023). The design and application of DNA-editing enzymes as base editors. *Annual Review of Biochemistry*, *92*(1), 43–79. DOI: 10.1146/annurev-biochem-052521-013938 PMID: 37018843

Ramanathan, B., Wu, H., Ross, C. R., & Blecha, F. (2004). PR-39, a porcine antimicrobial peptide, inhibits apoptosis: Involvement of caspase-3. *Developmental and Comparative Immunology*, *28*(2), 163–169. DOI: 10.1016/S0145-305X(03)00135-6 PMID: 12969801

Reguera, G., McCarthy, K. D., Mehta, T., Nicoll, J. S., Tuominen, M. T., & Lovley, D. R. (2005). Extracellular electron transfer via microbial nanowires. *Nature*, *435*(7045), 1098–1101. DOI: 10.1038/nature03661 PMID: 15973408

Rehfeld, J. F. (2017). Cholecystokinin—From local gut hormone to ubiquitous messenger. *Frontiers in Endocrinology*, *8*, 47. DOI: 10.3389/fendo.2017.00047 PMID: 28450850

Renye, J. A.Jr, Needleman, D. S., Somkuti, G. A., & Steinberg, D. H. (2017). Complete genome sequence of Streptococcus thermophilus strain B59671, which naturally produces the broad-spectrum bacteriocin thermophilin 110. *Genome Announcements*, *5*(45), e01213–e01217. DOI: 10.1128/genomeA.01213-17 PMID: 29122869

Roberfroid, M., Gibson, G. R., Hoyles, L., McCartney, A. L., Rastall, R., Rowland, I., Wolvers, D., Watzl, B., Szajewska, H., Stahl, B., Guarner, F., Respondek, F., Whelan, K., Coxam, V., Davicco, M.-J., Léotoing, L., Wittrant, Y., Delzenne, N. M., Cani, P. D., & Meheust, A. (2010). Prebiotic effects: Metabolic and health benefits. *British Journal of Nutrition*, *104*(S2), S1–S63. DOI: 10.1017/S0007114510003363 PMID: 20920376

Roca Pinilla, R. (2020). Development of a new generation of antimicrobial proteins based on a versatile nanoparticulated format and multidomain structure.

Rodríguez, J. M., Martínez, M. I., & Kok, J. (2002). Pediocin PA-1, a wide-spectrum bacteriocin from lactic acid bacteria. *Critical Reviews in Food Science and Nutrition*, *42*(2), 91–121. DOI: 10.1080/10408690290825475 PMID: 11934133

Rose, E. C., Odle, J., Blikslager, A. T., & Ziegler, A. L. (2021). Probiotics, prebiotics and epithelial tight junctions: A promising approach to modulate intestinal barrier function. *International Journal of Molecular Sciences*, *22*(13), 6729. DOI: 10.3390/ijms22136729 PMID: 34201613

Rossi, M., Corradini, C., Amaretti, A., Nicolini, M., Pompei, A., Zanoni, S., & Matteuzzi, D. (2005). Fermentation of fructooligosaccharides and inulin by bifidobacteria: A comparative study of pure and fecal cultures. *Applied and Environmental Microbiology*, *71*(10), 6150–6158. DOI: 10.1128/AEM.71.10.6150-6158.2005 PMID: 16204533

Round, J. L., & Mazmanian, S. K. (2010). Inducible Foxp3+ regulatory T-cell development by a commensal bacterium of the intestinal microbiota. *Cell Host & Microbe*, *8*(5), 377–387. DOI: 10.1016/j.chom.2010.10.006 PMID: 20566854

Ruiz Puentes, P., Henao, M. C., Cifuentes, J., Muñoz-Camargo, C., Reyes, L. H., Cruz, J. C., & Arbeláez, P. (2022). Rational discovery of antimicrobial peptides by means of artificial intelligence. *Membranes*, *12*(7), 708. DOI: 10.3390/membranes12070708 PMID: 35877911

Salini, F., Iacumin, L., Comi, G., & Dicks, L. M. T. (2023). Thermophilin 13: In silico analysis provides new insight in genes involved in bacteriocin production. *Microorganisms*, *11*(3), 611. DOI: 10.3390/microorganisms11030611 PMID: 36985185

Sampson, T. R., Debelius, J. W., Thron, T., Janssen, S., Shastri, G. G., Ilhan, Z. E., Challis, C., Schretter, C. E., Rocha, S., Gradinaru, V., Chesselet, M.-F., Keshavarzian, A., Shannon, K. M., Krajmalnik-Brown, R., Wittung-Stafshede, P., Knight, R., & Mazmanian, S. K. (2016). Gut microbiota regulate motor deficits and neuroinflammation in a model of Parkinson's disease. *Cell*, *167*(6), 1469–1480. DOI: 10.1016/j.cell.2016.11.018 PMID: 27912057

Sánchez, A., & Vázquez, A. (2017). Bioactive peptides: A review. *Food Quality and Safety*, *1*(1), 29–46. DOI: 10.1093/fqs/fyx006

Santos-Beneit, F. (2024). What is the role of microbial biotechnology and genetic engineering in medicine? *MicrobiologyOpen*, *13*(2), e1406. DOI: 10.1002/mbo3.1406 PMID: 38556942

Saraiva, C., Praça, C., Ferreira, R., Santos, T., Ferreira, L., & Bernardino, L. (2016). Nanoparticle-mediated brain drug delivery: Overcoming blood–brain barrier to treat neurodegenerative diseases. *Journal of Controlled Release*, *235*, 34–47. DOI: 10.1016/j.jconrel.2016.05.044 PMID: 27208862

Sarkar, B., Ma, X., Agas, A., Siddiqui, Z., Iglesias-Montoro, P., Nguyen, P. K., Kim, K. K., Haorah, J., & Kumar, V. A. (2021). In vivo neuroprotective effect of a self-assembled peptide hydrogel. *Chemical Engineering Journal*, *408*, 127295. DOI: 10.1016/j.cej.2020.127295 PMID: 37842134

Schmacht, M., Lorenz, E., & Senz, M. (2017). Microbial production of glutathione. *World Journal of Microbiology & Biotechnology*, *33*(6), 106. DOI: 10.1007/s11274-017-2277-7 PMID: 28466304

Schwarcz, R., Bruno, J. P., Muchowski, P. J., & Wu, H. Q. (2012). Kynurenines in the mammalian brain: When physiology meets pathology. *Nature Reviews. Neuroscience*, *13*(7), 465–477. DOI: 10.1038/nrn3257 PMID: 22678511

Segers ME, Lebeer S. Towards a better understanding of Lactobacillus rhamnosus GG--host interactions. Microb Cell Fact. 2014;13 Suppl 1(Suppl 1):S7. .DOI: 10.1186/1475-2859-13-S1-S7

Sena, L. A., & Chandel, N. S. (2012). Physiological roles of mitochondrial reactive oxygen species. *Molecular Cell*, *48*(2), 158–167. DOI: 10.1016/j.molcel.2012.09.025 PMID: 23102266

Shang, L., Yang, F., Chen, Q., Dai, Z., Yang, G., Zeng, X., Qiao, S., & Yu, H. (2024). Bacteriocin Microcin J25's antibacterial infection effects and novel non-microbial regulatory mechanisms: Differential regulation of dopaminergic receptors. *Journal of Animal Science and Biotechnology*, *15*(1), 156. DOI: 10.1186/s40104-024-01115-3 PMID: 39533384

Shareena, G., & Kumar, D. (2024). Deciphering the deep learning and machine learning tactics in advancement of neuroprotection by phytochemicals. In *NeuroPhytomedicine* (pp. 205–220). CRC Press.

Sharma, M., & Rajawat, N. K. (2025). Neurotoxicity study of copper oxide nanoparticles and the protective role of a probiotic (Lactobacillus acidophilus) in Swiss albino mice. *Toxicology and Industrial Health*, *41*(7), 07482337251350165. DOI: 10.1177/07482337251350165 PMID: 40492926

Sharma, S., Singh, R. L., & Kakkar, P. (2011). Modulation of Bax/Bcl-2 and caspases by probiotics during acetaminophen induced apoptosis in primary hepatocytes. *Food and Chemical Toxicology*, *49*(4), 770–779. DOI: 10.1016/j.fct.2010.11.041 PMID: 21130831

Silva, M. H., Batista, L. L., Malta, S. M., Santos, A. C., Mendes-Silva, A. P., Bonetti, A. M., Ueira-Vieira, C., & Dos Santos, A. R. (2024). Unveiling the Brazilian kefir microbiome: Discovery of a novel Lactobacillus kefiranofaciens (LkefirU) genome and in silico prospection of bioactive peptides with potential anti-Alzheimer properties. *BMC Genomics*, *25*(1), 884. DOI: 10.1186/s12864-024-10695-3 PMID: 39304820

Sinha, K., & Mukhopadhyay, C. D. (2021). Immunomodulatory Effect of Plant-Based Extracts on Neurodegeneration. In Advanced Studies in Experimental and Clinical Medicine (pp. 31-52). Apple Academic Press. DOI: 10.1201/9781003057451-3

Slavin, J. (2013). Fiber and prebiotics: Mechanisms and health benefits. *Nutrients*, *5*(4), 1417–1435. DOI: 10.3390/nu5041417 PMID: 23609775

Śmiałek, J., Bzowska, M., Hinz, A., Mężyk-Kopeć, R., Sołtys, K., & Mak, P. (2022). Bacteriocin BacSp222 and its succinylated forms exhibit proinflammatory activities toward innate immune cells. *Journal of Inflammation Research*, *15*, 4601–4621. DOI: 10.2147/JIR.S362066 PMID: 35982757

Śmiałek, J., Nowakowski, M., Bzowska, M., Bocheńska, O., Wlizło, A., Kozik, A., Dubin, G., & Mak, P. (2021). Structure, biosynthesis, and biological activity of succinylated forms of bacteriocin BacSp222. *International Journal of Molecular Sciences*, *22*(12), 6256. DOI: 10.3390/ijms22126256 PMID: 34200765

Sneath, P. H. A. (1966). Relations between chemical structure and biological activity in peptides. *Journal of Theoretical Biology*, *12*(2), 157–195. DOI: 10.1016/0022-5193(66)90112-3 PMID: 4291386

Son, Y., Cheong, Y. K., Kim, N. H., Chung, H. T., Kang, D. G., & Pae, H. O. (2011). Mitogen-activated protein kinases and reactive oxygen species: How can ROS activate MAPK pathways? *Journal of Signal Transduction*, *792639*, 1–6. Advance online publication. DOI: 10.1155/2011/792639 PMID: 21637379

Sowers, A., Wang, G., Xing, M., & Li, B. (2023). Advances in antimicrobial peptide discovery via machine learning and delivery via nanotechnology. *Microorganisms, 11*(5), 1129. DOI: 10.3390/microorganisms11051129 PMID: 37317103

Strandwitz, P. (2018). Neurotransmitter modulation by the gut microbiota. *Brain Research, 1693*, 128–133. DOI: 10.1016/j.brainres.2018.03.015 PMID: 29903615

Stricker, M., Tanović, A., & Marahiel, M. A. (2010). Nonribosomal peptide synthetases: Structures and dynamics. *Current Opinion in Structural Biology, 20*(2), 234–240. DOI: 10.1016/j.sbi.2010.01.009 PMID: 20153164

Su, Y., Wang, D., Liu, N., Yang, J., Sun, R., & Zhang, Z. (2023). Clostridium butyricum improves cognitive dysfunction in ICV-STZ-induced Alzheimer's disease mice via suppressing TLR4 signaling pathway through the gut-brain axis. *PLoS One, 18*(6), e0286086. DOI: 10.1371/journal.pone.0286086 PMID: 37267300

Sun, J., Xu, J., Yang, B., Chen, K., Kong, Y., Fang, N., Gong, T., Wang, F., Ling, Z., & Liu, J. (2020). Effect of Clostridium butyricum against Microglia-Mediated Neuroinflammation in Alzheimer's Disease via Regulating Gut Microbiota and Metabolites Butyrate. *Molecular Nutrition & Food Research, 64*(2), e1900636. DOI: 10.1002/mnfr.201900636 PMID: 31835282

Szymczak, P., & Szczurek, E. (2023). Artificial intelligence-driven antimicrobial peptide discovery. *Current Opinion in Structural Biology, 83*, 102733. DOI: 10.1016/j.sbi.2023.102733 PMID: 37992451

Szymczak, P., Zarzecki, W., Wang, J., Duan, Y., Wang, J., Coelho, L. P., de la Fuente-Nunez, C., & Szczurek, E. (2025). AI-driven antimicrobial peptide discovery: Mining and generation. *Accounts of Chemical Research, 58*(12), 1831–1846. Advance online publication. DOI: 10.1021/acs.accounts.0c00594 PMID: 40459283

Taché, Y., Bonaz, B., & Million, M. (2018). Brain-gut axis and its role in gastrointestinal motility and functional bowel disorders. *Gastroenterology Clinics of North America, 47*(2), 321–339. DOI: 10.1016/j.gtc.2018.01.007

Tajbakhsh, M., Karimi, A., Fallah, F., & Akhavan, M. M. (2017). Overview of ribosomal and non-ribosomal antimicrobial peptides produced by Gram positive bacteria. *Cellular and Molecular Biology, 63*(10), 20–32. DOI: 10.14715/cmb/2017.63.10.4 PMID: 29096754

Talarico, T. L., Casas, I. A., Chung, T. C., & Dobrogosz, W. J. (1988). Production and isolation of reuterin, a growth inhibitor produced by Lactobacillus reuteri. *Antimicrobial Agents and Chemotherapy, 32*(12), 1854–1858. DOI: 10.1128/AAC.32.12.1854 PMID: 3245697

Tang, L., Zeng, Z., Zhou, Y., Wang, B., Zou, P., Wang, Q., Ying, J., Wang, F., Li, X., Xu, S., Zhao, P., & Li, W. (2021). Bacillus amyloliquefaciens SC06 Induced AKT–FOXO signaling pathway-mediated autophagy to alleviate oxidative stress in IPEC-J2 Cells. *Antioxidants, 10*(10), 1545. DOI: 10.3390/antiox10101545 PMID: 34679680

Tenland, E., Krishnan, N., Rönnholm, A., Kalsum, S., Puthia, M., Mörgelin, M., Davoudi, M., Otrocka, M., Alaridah, N., Glegola-Madejska, I., Sturegård, E., Schmidtchen, A., Lerm, M., Robertson, B. D., & Godaly, G. (2018). A novel derivative of the fungal antimicrobial peptide plectasin is active against Mycobacterium tuberculosis. *Tuberculosis (Edinburgh, Scotland)*, *113*, 231–238. DOI: 10.1016/j.tube.2018.10.008 PMID: 30514507

Thoda, C., & Touraki, M. (2023). Probiotic-derived bioactive compounds in colorectal cancer treatment. *Microorganisms*, *11*(8), 1898. DOI: 10.3390/microorganisms11081898 PMID: 37630458

Thyab Gddoa Al-sahlany, S., Altemimi, A., Al-Manhel, A., Niamah, A., Lakhssassi, N., & Ibrahim, S.ThyabGddoa Al-sahlany. (2020). Purification of bioactive peptide with antimicrobial properties produced by Saccharomyces cerevisiae. *Foods*, *9*(3), 324. DOI: 10.3390/foods9030324 PMID: 32168785

Tian, F., Xiao, Y., Li, X., Zhai, Q., Wang, G., Zhang, Q., Zhang, H., & Chen, W. (2015). Protective effects of Lactobacillus plantarum CCFM8246 against copper toxicity in mice. *PLoS One*, *10*(11), e0143318. DOI: 10.1371/journal.pone.0143318 PMID: 26605944

Tian, T., Xie, W., Liu, L., Fan, S., Zhang, H., Qin, Z., & Yang, C. (2023). Industrial application of antimicrobial peptides based on their biological activity and structure-activity relationship. *Critical Reviews in Food Science and Nutrition*, *63*(21), 5430–5445. DOI: 10.1080/10408398.2021.2019673 PMID: 34955061

Tlaskalová-Hogenová, H., Štěpánková, R., Kozáková, H., Hudcovic, T., Vannucci, L., Tučková, L., Rossmann, P., Bártová, J., Sokol, D., Funda, D. P., Borovská, D., Zákostelská, Z., & Hrncir, T. (2011). The role of gut microbiota (commensal bacteria) and the mucosal barrier in the pathogenesis of inflammatory and autoimmune diseases and cancer: Contribution of germ-free and gnotobiotic animal models of human diseases. *Cellular & Molecular Immunology*, *8*(2), 110–120. DOI: 10.1038/cmi.2010.67 PMID: 21278760

Tuohy, K. M., Probert, H. M., Smejkal, C. W., & Gibson, G. R. (2003). Using probiotics and prebiotics to improve gut health. *Digestive and Liver Disease*, *35*(5), 282–289. DOI: 10.1016/S1590-8658(03)00088-0 PMID: 12927512

Udenigwe, C. C., & Aluko, R. E. (2012). Food protein-derived bioactive peptides: Production, processing, and potential health benefits. *Journal of Food Science*, *77*(1), R11–R24. DOI: 10.1111/j.1750-3841.2011.02455.x PMID: 22260122

van Tilburg, A. Y., Cao, H., van der Meulen, S. B., Solopova, A., & Kuipers, O. P. (2019). Metabolic engineering and synthetic biology employing Lactococcus lactis and Bacillus subtilis cell factories. *Current Opinion in Biotechnology*, *59*, 1–7. DOI: 10.1016/j.copbio.2019.01.007 PMID: 30784872

Vassiliadis, G., Destoumieux-Garzón, D., Lombard, C., Rebuffat, S., & Peduzzi, J. (2010). Isolation and characterization of two members of the siderophore-microcin family, microcins M and H47. *Antimicrobial Agents and Chemotherapy*, *54*(1), 288–297. DOI: 10.1128/AAC.00744-09 PMID: 19884380

Vinolo, M. A. R., Rodrigues, H. G., Nachbar, R. T., & Curi, R. (2011). Regulation of inflammation by short chain fatty acids. *Nutrients*, *3*(10), 858–876. DOI: 10.3390/nu3100858 PMID: 22254083

Wan, F., Wong, F., Collins, J. J., & de la Fuente-Nunez, C. (2024). Machine learning for antimicrobial peptide identification and design. *Nature Reviews Bioengineering*, *2*(5), 392–407. DOI: 10.1038/s44222-024-00152-x PMID: 39850516

Wan, J., Li, Y., Chen, D., Yu, B., Zheng, P., Mao, X., Yu, J., & He, J. (2016). Expression of a tandemly arrayed plectasin gene from Pseudoplectania nigrella in Pichia pastoris and its antimicrobial activity. *Journal of Microbiology and Biotechnology*, *26*(3), 461–468. DOI: 10.4014/jmb.1508.08091 PMID: 26643963

Wang, H., Zhang, C., Li, M., Liu, C., Wang, J., Ou, X., & Han, Y. (2022). Antimicrobial peptides mediate apoptosis by changing mitochondrial membrane permeability. *International Journal of Molecular Sciences*, *23*(21), 12732. DOI: 10.3390/ijms232112732 PMID: 36361521

Wang, J., Liu, Y. M., Cao, W., Yao, K. W., Liu, Z. Q., & Guo, J. Y. (2012). Anti-inflammation and antioxidant effect of Cordymin, a peptide purified from the medicinal mushroom Cordyceps sinensis, in middle cerebral artery occlusion-induced focal cerebral ischemia in rats. *Metabolic Brain Disease*, *27*(2), 159–165. DOI: 10.1007/s11011-012-9282-1 PMID: 22327557

Wang, L., Wang, N., Zhang, W., Cheng, X., Yan, Z., Shao, G., Wang, X., Wang, R., & Fu, C. (2022). Therapeutic peptides: Current applications and future directions. *Signal Transduction and Targeted Therapy*, *7*(1), 48. DOI: 10.1038/s41392-022-00904-4 PMID: 35165272

Wang, Y., Chen, W. J., Han, Y. Y., Xu, X., Yang, A. X., Wei, J., Hong, D., Fang, X., & Chen, T. T. (2023). Neuroprotective effect of engineered Clostridium butyricum-pMTL007-GLP-1 on Parkinson's disease mice models via promoting mitophagy. *Bioengineering & Translational Medicine*, *8*(3), e10505. DOI: 10.1002/btm2.10505 PMID: 37206220

Wang, Y., Zhang, X., Liu, Z., & Chen, L. (2023). Probiotic-induced modulation of Th2 cells and IL-10 production alleviates experimental autoimmune encephalomyelitis

Xie, J., Bi, Y., Zhang, H., Dong, S., Teng, L., Lee, R. J., & Yang, Z. (2020). Cell-penetrating peptides in diagnosis and treatment of human diseases: From preclinical research to clinical application. *Frontiers in Pharmacology*, *11*, 697. DOI: 10.3389/fphar.2020.00697 PMID: 32508641

Zhang, C., Zheng, X., Wan, X., Shao, X., Liu, Q., Zhang, Z., & Zhang, Q. (2014). The potential use of H102 peptide-loaded dual-functional nanoparticles in the treatment of Alzheimer's disease. *Journal of Controlled Release*, *192*, 317–324. DOI: 10.1016/j.jconrel.2014.07.050 PMID: 25102404

Chapter 4
Blueberries and Brainpower: Exploring the Role of Anthocyanins in Cognitive Health

Richa Sharma
https://orcid.org/0009-0006-1174-2888
AAFT University of Media and Arts, Raipur, India

Khushi Kashyap
https://orcid.org/0000-0003-0663-2132
AAFT University of Media and Arts, Raipur, India

Ankita Shrivastava
https://orcid.org/0009-0005-1472-9286
AAFT University of Media and Arts, Raipur, India

Pankaj Singh Chandel
https://orcid.org/0000-0002-4135-8194
AAFT University of Media and Arts, Raipur, India

ABSTRACT

This chapter looks at how blueberries, a fruit rich in natural plant compounds called anthocyanins,can help support brain health as people age. It explains how these compounds work in the body to reduce inflammation, protect brain cells, and improve memory and focus. The chapter also discusses how blueberries fit into popular brain-friendly diets like the MIND and DASH diets. Readers will learn about different ways to include blueberries in everyday meals whether fresh, frozen, dried, powdered, or as juice and how they're being used in different cultures and cuisines. The chapter also touches on why some people may benefit more than others based on their gut health or genetics. With a mix of science and real-life applications, this chapter highlights how a simple fruit like the blueberry can play a powerful role in keeping the brain healthy over time.

DOI: 10.4018/979-8-3373-4139-2.ch004

1. INTRODUCTION TO ANTHOCYANINS AND BRAIN HEALTH

The beginning of the 21st century has marked an era of remarkable improvements in human longevity across the world. However, this rise in life expectancy has also brought with it a steady increase in age-related disorders such as Alzheimer's disease (AD), Parkinson's disease (PD), and other forms of dementia (World Health Organization, 2022; Prince et al., 2016). According to the World Health Organization (2022), an estimated 55 million people are currently living with dementia worldwide, and this number is expected to double every two decades. The consequences of these disorders extend far beyond the affected individuals, placing a significant emotional, physical, and financial burden on caregivers and healthcare systems alike (Gauthier et al., 2018).

Amid this growing concern, researchers have turned their attention toward natural compounds that may help protect against neurodegeneration. Anthocyanins, the pigments responsible for the vivid red, purple, and blue hues of many fruits and vegetables, are one such group of compounds (Tsuda, 2012). Structurally, anthocyanins belong to the polyphenol family and are abundant in foods such as berries, red cabbage, purple corn, eggplant skin, and black rice. Among these, blueberries have emerged as one of the richest natural sources, containing over 15 distinct anthocyanin compounds identified to date (Joseph et al., 1999).

Beyond their vibrant color, anthocyanins are known for their powerful antioxidant and anti-inflammatory properties. In terms of brain health, these compounds can cross the blood–brain barrier and influence molecular signaling pathways associated with memory, learning, and cognitive function (Williams et al., 2008). This ability makes them promising candidates for supporting neurological resilience and potentially slowing the progression of age-related cognitive decline. Blueberries are not only tasty and accessible fruits but also one of the highest in antioxidant capacity. Numerous preclinical and clinical studies have shown that blueberry anthocyanins exert neuroprotective effects. For example, Krikorian et al. (2010) showed in a placebo-controlled trial that elderly adults drinking blueberry juice for 12 weeks exhibited improved memory and delayed cognitive aging.

Table 1. The anthocyanin content in various berries

Berry Type	Total Anthocyanin Content (mg/100g FW)
Blueberries	150–300
Blackberries	70–235
Raspberries	10–60
Strawberries	30–45
Elderberries	450–600

Anthocyanins occur in noticeably varied forms in blueberries, namely malvidin, delphinidin, cyanidin, and petunidin glycosides. These substances work synergistically to ameliorate neuroinflammation, boost neuronal signaling, and foster synaptic plasticity.

The term "nutritional neuroscience" is relatively new, having evolved in the past 10 years. It is now well accepted that diet influences brain function, and therefore mental well-being. Among other dietary patterns, the Mediterranean diet, DASH diet, and MIND diet have been established as correlates of cognitive outcomes.

In this regard, there are functional foods, such as blueberries, that can be examined for their nutritional worth and pharmacological action. "Brain foods" are being used in an emerging trend in preventive modes of health and personal nutrition plans.

The mechanisms by which anthocyanins provide those cognitive benefits are numerous and will be discussed in more detail in later sections. From modulating oxidative stress and reducing neuro-inflammation to supporting mitochondrial efficiency and enhancing synaptic communication, anthocyanins make a good case for loophole inclusion in daily dietary practices that preserve mental acuity.

2. CHEMISTRY AND CLASSIFICATION OF ANTHOCYANINS

Anthocyanins represent a subclass of the flavonoids water-soluble pigments responsible for the red, purple, and blue colors in numerous fruits and vegetables (He & Giusti, 2010; Khoo, Azlan, Tang, & Lim, 2017). Their chemistry is quite complex, yet it is very fascinating, particularly when examined in terms of their biological roles, including antioxidant, anti-inflammatory, and neuroprotective activities (Tsuda, 2012; Wallace, 2011). This section will cover the structural diversity, classification, and biochemical behavior of anthocyanins, with special emphasis on blueberries, one of their most potent natural sources (Giusti & Wrolstad, 2003).

There are **six major anthocyanidins** commonly found in nature:

Table 2. Structural variants and color characteristics of common anthocyanidins

Anthocyanidin	R1 Substituent	R2 Substituent	Color
Delphinidin	OH	OH	Bluish-purple
Cyanidin	OH	H	Reddish-purple
Petunidin	OCH_3	OH	Purple
Peonidin	OCH_3	H	Red-purple
Malvidin	OCH_3	OCH_3	Deep purple
Pelargonidin	H	H	Orange-red

Usually, plants contain anthocyanins in glycosylated forms, linked to sugars such as glucose, galactose, rhamnose, and arabinose. Acylation by organic acids (like caffeic and ferulic acid) is common on these glycosylated systems and often increases stability and solubility in water.

2.1 Anthocyanin content in blueberries vs. other berries

Blueberries (*Vaccinium* spp.), particularly the wild varieties, are recognized as being among the richest sources of anthocyanins in nature (Prior, Wu, & Schaich, 2007; Kalt et al., 2008). A single serving (100 g) of fresh blueberries varies in anthocyanin content depending on the variety, growing conditions, and ripeness, with reported values ranging from 25 to 495 mg per 100 g of fresh fruit (Brouillard & Delaporte, 1977; Wu et al., 2006).

Table 3. Comparison of total anthocyanins and key anthocyanidins in selected berries

Berry Type	Total Anthocyanins (mg/100g fresh weight)	Dominant Anthocyanidins
Wild Blueberry	400–500	Malvidin, Delphinidin, Cyanidin
Cultivated Blueberry	200–300	Cyanidin, Peonidin, Petunidin
Blackcurrant	250–400	Delphinidin, Cyanidin
Blackberry	150–300	Cyanidin, Pelargonidin
Raspberry	50–100	Cyanidin
Strawberry	30–60	Pelargonidin

Blueberries score on all six points with all the dominant anthocyanidins, which appears to be the reason behind the spectrum neuroprotective effects they provide. These compounds thus appear to explain the enhanced cognition exhibited through research conducted with animals and strong human studies.

2.2 Stability, solubility, and factors affecting anthocyanin degradation

Anthocyanins are highly reactive molecules sensitive to external conditions such as pH, light, temperature, and enzymatic action. Stability under these conditions is important for the development of functional foods and for ensuring bioactivity in the body after ingestion.

pH-Sensitivity: Anthocyanins undergo structural changes as a function of pH:

- pH < 3: Red flavylium cation (most stable)
- pH 4–5: Colorless carbinol pseudobase
- pH 6-7: Blue quinonoidal base
- pH > 7: Degradation into chalcones or polymerization

Temperature: High temperatures lead to hydrolysis of glycoside bonds and ring-opening, which causes the loss of color and less antioxidant activity.

Light and Oxygen: Photo-oxidation and UV irradiation greatly accelerate anthocyanin degradation, especially in unprotected beverages or products.

Enzymatic Degradation: Fresh commodities-polyphenol oxidases and peroxidases can catalyze degradation unless inactivated by heat during processing.

Table 4. Key factors affecting anthocyanin stability

Factor	Effect on Anthocyanins
pH	Alters color and structural form
Temperature	Causes thermal degradation and color fading
Light	Promote photo-oxidation
Oxygen	Accelerates oxidation reactions
Enzymes	Degrade anthocyanin structure in raw produce

To mitigate these challenges, innovations such as microencapsulation, pH-controlled formulations, and low-temperature processing are employed to preserve anthocyanin content in commercial products.

2.3 Analytical methods for quantification

Anthocyanin content is an important factor in both research and industry for determining quality, shelf-life, and potential therapeutic effect. Methods mostly used are:

pH-Differential Method: A simple spectrophotometric measure based on absorbance at 520 and 700 nm across two pH buffers (pH 1.0 and 4.5). Specific for total monomeric anthocyanin estimation.
High-Performance Liquid Chromatography (HPLC): The gold standard method for profiling individual anthocyanin compounds. Coupling with diode-array detectors (DAD) or mass spectrometry (MS) enhances resolution.
UV-Vis Spectrophotometry: Quick screening tool for bulk anthocyanin quantification but not as specific as HPLC.
Capillary Electrophoresis (CE): Used for complex mixture characterization, particularly in cases of acylated anthocyanins (Giusti & Wrolstad, 2001; He & Giusti, 2010).

Table 5. Comparison of analytical techniques

Method	Specificity	Time Efficiency	Application
pH Differential	Low	High	Total anthocyanin estimation
HPLC-DAD/MS	High	Moderate	Individual compound analysis
UV-Vis Spectroscopy	Moderate	High	Rapid estimation of food matrices
Capillary Electrophoresis	High	Low	Advanced anthocyanin profile analysis

3. MECHANISMS OF ACTION IN BRAIN FUNCTION

The distinctive blue-purple pigments of blueberries belong to a group of flavonoids known as anthocyanins, which have been the focus of extensive scientific investigations regarding their neuroprotective properties. In the brain, these compounds interact with a variety of cellular pathways influencing oxidative stress and inflammatory responses, mitochondrial efficiency, and neuroplasticity. Interventions in nutritional neuroscience are now beginning to shed light on the fact that anthocyanins do not act in isolation but rather through a network of overlapping mechanisms that together lend support to brain function and resilience.

3.1 Oxidative stress modulation

Due to its unusually high metabolic rate, the human brain weighs around 2% of body weight yet consumes more than 20% of the body's oxygen; thus, it is extremely vulnerable to oxidative stress (Friedman,

2017). During normal cellular respiration, reactive oxygen species (ROS) are formed; however, under stress conditions, the accumulation of ROS may pose a serious risk for oxidative damage of neurons (Halliwell & Gutteridge, 2015). Anthocyanins function as excellent antioxidants by directly scavenging ROS and stimulating the expression of endogenous antioxidant enzymes, such as superoxide dismutase (SOD), catalase, and glutathione peroxidasc. Anthocyanins protect free radicals from being attacked by donating electrons, thus preventing lipid peroxidation that damages the phospholipid-rich membranes of neurons themselves. Guarding against lipid peroxides preserves not only structural integrity but also mitigates neurotoxicity relating to cognitive impairment (Basu & Lyons, 2012).

3.2 Anti-inflammatory effects

Chronic inflammation of the CNS, commonly called neuroinflammation, constitutes a major risk factor for neurodegenerative diseases like Alzheimer's and Parkinson's disease. Anthocyanins exert profound anti-inflammatory effects through modulation of the inflammatory signaling pathways. In particular, they inhibit the NF-kappa B pathway, which is a crucial regulator of pro-inflammatory gene expression. Such inhibition results in decreased expression of cytokines, including TNF-alpha, IL-6, and IL-1beta. Furthermore, anthocyanins also inhibit COX-2, an enzyme involved in generating prostaglandins, thereby reducing the inflammatory-associated pain and tissue damage resulting from these effects.

3.3 Mitochondrial protection

Neurons predominantly rely on mitochondrial functioning for energy production. Mitochondrial dysfunctions have been known hallmarks of cognitive decline and neurodegeneration in older populations. Anthocyanins contribute to improved mitochondrial performance through increased ATP synthesis, mitochondrial membrane potential preservation, and less damage to mitochondrial DNA. Experimental models show an increase in anthocyanin treatment activity concerning mitochondrial complexes I and IV, which are essential in oxidative phosphorylation; this thus puts energy efficiently to the brain cells harnessing cognitive processes such as memory formation, attention, and learning. Further, anthocyanins reduce mitochondria ROS production for the maintenance of integrity in mitochondria.

3.4 Neuroplasticity & synaptic health

One of the most compelling areas of anthocyanin research is its role in promoting neuroplasticity the brain's ability to reorganize and form new neural connections. Anthocyanins stimulate the expression of brain-derived neurotrophic factor (BDNF), a protein essential for the growth, survival, and differentiation of neurons.

BDNF activation triggers downstream signaling pathways like the cAMP-response element-binding protein (CREB) and the PI3K/Akt pathway, which are vital for synaptic plasticity. Enhanced BDNF signaling improves long-term potentiation (LTP), the molecular foundation of learning and memory. In addition to BDNF, anthocyanins upregulate nerve growth factor (NGF) and synapsin I, reinforcing synaptic strength and connectivity.

4. GUT-BRAIN AXIS AND MICROBIOME MODULATION

The gut–brain axis represents a dynamic and bidirectional communication pathway linking the gastrointestinal tract with the CNS. It encompasses complex interactions among neural, hormonal, and immune pathways. Increasing evidence pinpoints gut microbiota as a critical mediator in cognitive function, emotional behavior, and brain health (Cryan et al., 2019). This microbial consortium influences the brain via metabolites, immune modulation, and especially through signaling pathways mediated indirectly by the vagus nerve (Fülling et al., 2019). Within this context, anthocyanins, particularly the blueberry varieties in which they are abundant, have gained ground as significant dietary modulators of gut–brain communication (Kalt et al., 2020).

4.1 Anthocyanin influence on beneficial gut microbiota

Most anthocyanins are excreted into the large intestine because their absorption in the small intestine is limited (Kay et al., 2017). In the colon, they encounter intestinal microflora that transform anthocyanins into smaller phenolic metabolites (Cardona et al., 2013). This interaction is mutual; microorganisms degrade anthocyanins, while anthocyanins modify the composition and functionality of the gut flora. Studies show selective enhancement of beneficial bacteria such as *Bifidobacterium*, *Lactobacillus*, and *Akkermansia muciniphila*, while suppressing potential pathogenic strains like *Clostridium perfringens* and *Escherichia coli* (Cassidy & Minihane, 2017). This shift promotes an anti-inflammatory and neuroprotective gut environment. A healthier microbial profile improves intestinal barrier integrity, reduces endotoxin production, and lowers systemic inflammation, thereby supporting brain homeostasis (Anhê et al., 2015). These effects are strain-specific and dose-dependent, highlighting the importance of dietary patterns rich in anthocyanins. Longitudinal findings further show that consistent blueberry anthocyanin intake can beneficially reshape gut microbiota composition over time, contributing to enhanced cognitive resilience (Vendrame et al., 2011).

4.2 Role of short-chain fatty acids (SCFAs) and metabolites

One of the most crucial outcomes of anthocyanin-mediated microbial modulation is the consequent increase in short-chain fatty acids (SCFAs)—namely acetate, propionate, and butyrate. These SCFAs are produced by gut microbial fermentation of dietary fibers and polyphenols and play an essential role in both gut and brain health (Dalile et al., 2019). Butyrate, in particular, exhibits strong neuroprotective action; it strengthens the blood–brain barrier (BBB), reduces neuroinflammation, and functions as a histone deacetylase (HDAC) inhibitor, thereby promoting gene expression responsible for neuroplasticity (Bourassa et al., 2022). Acetate and propionate relay metabolic and neural signals from the gut to the hypothalamus, influencing appetite regulation, energy balance, and mood-related pathways (Frost et al., 2014).

SCFAs also modulate the expression of brain-derived neurotrophic factor (BDNF) and regulate neurotransmitter synthesis, including serotonin, which enhances synaptic plasticity and emotional stability (Stilling et al., 2016). Additionally, gut microbiota degrade anthocyanins into smaller phenolic metabolites such as protocatechuic acid, which can cross the BBB and exert direct antioxidant and anti-inflammatory effects on neural tissue (Zhang et al., 2021). These secondary metabolites often demonstrate higher bioavailability and longer retention than parent anthocyanin molecules, further contributing to neuroprotection

4.3 Vagus nerve and microbial communication with the brain

The vagus nerve is the primary bidirectional communication pathway linking the gut and the brain, transmitting information about intestinal status to the central nervous system and modulating motility, secretion, and immune responses through efferent signaling back to the gut (Breit et al., 2018). Anthocyanins indirectly modulate vagal activity by altering gut microbiota composition and enhancing metabolite production. For instance, short-chain fatty acids (SCFAs) and microbial metabolites stimulate enteroendocrine cells and immune cells within the intestinal lining, triggering the release of serotonin, dopamine, and cytokines, all of which can activate or influence vagal afferent terminals (Fülling et al., 2019). These neuroactive compounds subsequently influence the hypothalamic–pituitary–adrenal (HPA) axis, contributing to reductions in cortisol levels and improvements in mood and stress regulation.

Preclinical studies further demonstrate that anthocyanin-rich diets enhance vagal tone, which contributes to improved cognitive flexibility, neuroplasticity, and resilience to stress (Huang et al., 2016). Animal models also show that blueberry supplementation reduces anxiety-like behaviors and enhances memory performance, effects attributed at least partly to gut–brain axis modulation mediated through vagal signaling (Del Rosario et al., 2021).

4.4 Effects on anxiety, mood, and behavior (preclinical and clinical)

Although human studies are still emerging, current evidence supports the beneficial effects of anthocyanin-rich diets on mood, cognition, and behavior. Several interesting findings have been reported. For example, consumption of blueberry-based beverages has been shown to improve verbal memory and executive function in school-aged children (Whyte et al., 2016). Among older adults, blueberry supplementation has been associated with enhanced working memory and reductions in depressive symptoms (Krikorian et al., 2010; Travica et al., 2020).

Pilot trials also suggest that anthocyanin intake may reduce systemic inflammation, improve sleep quality, and support overall psychological well-being (Kent et al., 2017). These improvements are believed to arise not only from the direct neuroprotective actions of anthocyanins but also from gut microbiota–mediated modulation of the hypothalamic–pituitary–adrenal (HPA) axis, enhanced vagal nerve signaling, and reduced systemic inflammation.

Table 6. Anthocyanin effects on the gut–brain axis

Component	Anthocyanin Influence	Cognitive/Emotional Outcome
Gut Microbiota	↑ *Lactobacillus*, *Bifidobacterium*, ↓ *Clostridium* spp.	Improved mood, reduced gut inflammation
SCFA Production	↑ Butyrate, propionate, acetate	Neuroprotection, BBB integrity, anti-inflammation
Neurotransmitter Modulation	↑ Serotonin, dopamine, GABA via microbial stimulation	Enhanced mood, reduced anxiety
Vagal Nerve Stimulation	SCFA-triggered vagal afferent activation	HPA axis regulation, better stress response
Secondary Metabolite Formation	↑ Phenolic acids like protocatechuic acid	Direct antioxidant effects in brain tissue

5. EVIDENCE FROM ANIMAL STUDIES AND IN-VITRO MODELS

Anthocyanins, the flavonoid compounds responsible for the deep blue and purple hues of blueberries, have gained substantial attention for their neuroprotective effects. Emerging data from preclinical studies, including animal models and cell culture experiments, provide compelling evidence that anthocyanins contribute to enhanced cognitive function and a reduced risk of neurodegenerative disorders. These findings lay the groundwork for understanding how dietary anthocyanins may help mitigate age-related cognitive decline, memory impairments, and pathological processes associated with conditions such as Alzheimer's disease (AD) (Shukitt-Hale et al., 2015).

Rodent Studies: Cognitive and Memory Enhancements

Multiple rodent studies show that anthocyanin-rich blueberry supplementation improves learning, memory, and cognitive flexibility. In aged rats and transgenic AD models, dietary blueberry interventions produce measurable improvements in behavior, synaptic signaling, and neuroinflammatory markers (Malin, 2016).

A foundational study by Joseph et al. (1999) demonstrated that aged Fischer 344 rats fed a blueberry-supplemented diet for eight weeks exhibited enhanced motor performance, working memory, and spatial learning compared to controls. These results suggested that blueberry polyphenols, particularly anthocyanins, may reverse or attenuate age-related functional decline.

In Alzheimer's disease transgenic mouse models such as APP/PS1 mice, chronic blueberry supplementation has been shown to reduce amyloid-beta (Aβ) plaque deposition and hippocampal oxidative stress, while also decreasing tau hyperphosphorylation—key pathological hallmarks of AD (Vauzour et al., 2010; Shukitt-Hale et al., 2008). These findings strongly support the protective role of anthocyanins against AD-like neuropathology.

Behavioral Test Outcomes

Morris Water Maze (MWM)-The MWM evaluates spatial learning and memory. Rodents receiving blueberry extracts consistently show superior navigation performance, including reduced latency to locate the hidden platform and increased time spent in the target quadrant during probe trials, indicating enhanced spatial memory consolidation (Williams et al., 2008).

Y-Maze Test-This test measures working memory through spontaneous alternation behavior. Blueberry-fed rodents demonstrate significantly higher alternation rates, reflecting stronger hippocampal-dependent working memory and cognitive flexibility (Galli et al., 2018).

Novel Object Recognition (NOR)-In the NOR paradigm, blueberry-supplemented animals show greater exploration of novel objects, indicating enhanced recognition memory. This improvement aligns with anthocyanin-mediated enhancement of synaptic plasticity and antioxidant defense (Shukitt-Hale et al., 2015).

T-Maze and Passive Avoidance Tests-Blueberry supplementation has also been associated with better performance in T-maze alternation and passive avoidance learning, revealing improvements in associative learning and long-term memory recall (Malin, 2016). These behavioral outcomes parallel anthocyanin-induced modulation of neuronal signaling pathways and reduced neuroinflammation.

In-Vitro Models: Insights into Mechanisms

While animal models provide behavioral evidence, in-vitro studies using neuronal cell lines and primary neurons offer deeper insight into the cellular and molecular mechanisms through which anthocyanins exert neuroprotective effects (Kropat et al., 2013).

Neuroinflammation

Anthocyanins have demonstrated strong anti-inflammatory action in vitro. In microglial and astrocyte cultures, blueberry-derived anthocyanins suppress the production of pro-inflammatory cytokines such as TNF-α, IL-6, and IL-1β, largely through inhibition of the NF-κB signaling pathway—a key regulator of inflammatory responses (Youdim et al., 2000; Tsuda, 2012).

In lipopolysaccharide (LPS)-stimulated microglia, anthocyanin treatment reduces nitric oxide (NO) production and downregulates inducible nitric oxide synthase (iNOS), thereby preventing neurotoxic inflammatory cascades (Speciale et al., 2014).

Apoptosis and Oxidative Stress

- Neuronal apoptosis, a hallmark of cognitive decline and Alzheimer's disease, is a major focus of in-vitro anthocyanin research. In SH-SY5Y neuroblastoma cells, blueberry anthocyanins significantly reduce caspase-3 activation, preserve mitochondrial membrane potential, and increase the expression of anti-apoptotic proteins such as Bcl-2 (Wang et al., 2015).
- Anthocyanins also enhance antioxidant defenses by upregulating enzymes such as superoxide dismutase (SOD) and glutathione peroxidase (GPx). These enhancements are linked to activation of the Nrf2 signaling pathway, a master regulator of cellular antioxidant responses (Kelsey et al., 2010; Tsuda, 2016).

6. HUMAN CLINICAL TRIALS AND OBSERVATIONAL DATA

The neuroprotective potential of anthocyanins, particularly those derived from blueberries, has garnered increasing scientific interest in the realm of human health. Following promising preclinical studies, a growing number of human clinical trials and observational studies have explored the impact of dietary anthocyanin intake on cognitive performance, memory recall, and age-related cognitive decline. This section reviews the existing evidence from human studies, highlights findings in both healthy older adults and individuals with mild cognitive impairment (MCI), and outlines the limitations and strengths of current data, including recent systematic reviews and meta-analyses.

6.1 Cognitive Performance and Memory Recall Studies

Several controlled human trials have evaluated whether blueberry supplementation or anthocyanin intake can enhance episodic memory, attention, executive function, and other cognitive domains. One of the most cited studies was conducted by Krikorian et al. (2010), where older adults with early memory decline consumed blueberry juice daily for 12 weeks. The study reported significant improvements in verbal learning and recall, along with trends toward improved depressive symptoms, supported by fMRI findings showing increased activation in memory-related regions such as the hippocampus and prefrontal cortex.

In a randomized, double-blind, placebo-controlled trial, Whyte et al. (2016) demonstrated that young adults who consumed a single dose of wild blueberry drink showed significantly improved word recall and task-switching abilities within two hours of ingestion. This suggests that anthocyanins can exert acute cognitive-enhancing effects, possibly through enhanced neuronal signaling or improved cerebral blood flow.

Similarly, Miller et al. (2018) investigated the effects of daily blueberry powder supplementation over 90 days in older adults. The intervention group exhibited significant improvements in processing speed and executive function, particularly in tasks requiring complex attention and working memory. Participants with the lowest baseline cognitive scores benefited the most, suggesting a potentially greater therapeutic effect in individuals with existing cognitive impairment.

6.2 Trials in Older Adults and Mild Cognitive Impairment (MCI)

The majority of human anthocyanin trials have focused on older populations both healthy individuals and those with mild cognitive impairment (MCI), a prodromal stage of Alzheimer's disease. In a 12-week randomized trial involving individuals with MCI, Krikorian et al. (2020) administered blueberry powder equivalent to one cup of fresh blueberries per day. Results demonstrated significant improvements in executive function particularly task switching and inhibition—as well as enhanced memory encoding. These cognitive benefits were accompanied by reductions in fasting insulin levels, suggesting a possible relationship between improved glucose regulation and cognitive performance.

Supporting these findings, Boespflug et al. (2018) used functional MRI to evaluate the effects of daily blueberry supplementation in older adults with MCI. After 16 weeks, participants showed increased activation in prefrontal regions associated with working memory and decision-making, which corresponded with improvements in cognitive test scores.

While improvements are not consistently observed across all human trials, the recurring pattern of cognitive enhancement in aging and MCI populations strengthens the hypothesis that anthocyanins may help delay or mitigate cognitive decline through mechanisms such as reduced oxidative stress, enhanced cerebrovascular function, and improved neuroplasticity.

6.3 Limitations of Current Data

Despite these encouraging findings, human research on anthocyanins and cognitive function faces several limitations that should be critically examined:

- Sample Size

Many studies in this domain involve small participant groups, often ranging between 20–60 individuals. These limited sample sizes reduce the statistical power of the findings and hinder the ability to generalize results to broader populations. Larger, more diverse cohorts are necessary to confirm efficacy.

- Dosage and Standardization

There is considerable variability in the form, dose, and anthocyanin content used across trials. Some studies use whole blueberries; others use juice or freeze-dried powder. The anthocyanin content can vary from 50 mg to over 500 mg per day, with little consensus on an optimal dose for cognitive benefits. This makes it challenging to compare results and establish dosing guidelines.

- Duration of Intervention

The majority of trials are short-term, typically ranging from 4 to 24 weeks. While some improvements are observed during this period, cognitive decline is a long-term process, and short studies may not capture the full extent of anthocyanins' preventive or protective effects. Longitudinal studies spanning several years are needed to assess sustained cognitive outcomes.

- Blinding and Placebo Controls

Given the distinct color and taste of blueberry products, ensuring effective blinding can be difficult. Participants might recognize the treatment, potentially introducing bias. Additionally, not all studies use matched placebo controls in terms of calorie content or flavor, further complicating interpretation.

- Biomarker Assessment

Many studies rely solely on cognitive test outcomes, without incorporating biomarkers of inflammation, oxidative stress, or neurodegeneration. Inclusion of biochemical and neuroimaging markers would enhance mechanistic understanding and strengthen causal inferences.

6.4 Meta-Analyses and Systematic Reviews

To address individual study variability, recent systematic reviews and meta-analyses have aggregated findings to identify broader patterns of cognitive benefit. A meta-analysis by Travica et al. (2020) reviewed 14 randomized controlled trials (RCTs) investigating berry polyphenols and cognitive function, reporting significant improvements in episodic memory particularly among older adults with anthocyanin-rich interventions such as blueberries producing the most consistent effects. Similarly, Godos et al. (2020) conducted a systematic review examining dietary polyphenols, including anthocyanins, and found that higher flavonoid intake was associated with a reduced risk of cognitive impairment, despite substantial heterogeneity among study methodologies and outcome measures. Adding to this evidence, Bell et al. (2021) reported that acute anthocyanin consumption enhances attention and psychomotor speed, whereas chronic intake improves memory and executive functioning, especially in individuals at elevated risk for cognitive decline.

While pooled effect sizes in these meta-analyses are generally moderate, they collectively indicate a reliable trend of cognitive improvement associated with anthocyanin supplementation. These findings support the need for more rigorously designed, long-term trials to fully elucidate dosage, duration, and population-specific effects.

7. BIOAVAILABILITY AND METABOLIC CHALLENGES

Despite the growing body of evidence supporting the neuroprotective benefits of anthocyanins, their clinical efficacy is significantly influenced by bioavailability a measure of how much of a compound is absorbed and reaches systemic circulation in an active form. Anthocyanins, while abundant in foods like blueberries, face several metabolic challenges that limit their presence in the plasma and target organs like the brain. This section explores the key metabolic processes affecting anthocyanins, including first-pass degradation, Phase II metabolism, and the critical role of the gut microbiota in generating bioactive metabolites.

7.1 First-Pass Metabolism and Degradation in the Gastrointestinal Tract

Upon ingestion, anthocyanins must survive the harsh conditions of the gastrointestinal (GI) tract, including acidic pH in the stomach and enzymatic digestion in the small intestine. Unfortunately, anthocyanins are chemically unstable in these environments, often undergoing rapid degradation and transformation before absorption can occur (Del Rio et al., 2010; Manach et al., 2005). Studies have shown that only a small fraction—less than 1%—of ingested anthocyanins enter the bloodstream in their original glycosylated forms (Kay et al., 2009). Enzymatic hydrolysis by β-glucosidases and lactase-phlorizin hydrolase (LPH) in the small intestine can cleave sugar moieties, releasing unstable aglycones that are prone to degradation (McGhie & Walton, 2007).

Furthermore, anthocyanins undergo first-pass metabolism in enterocytes and the liver, reducing their systemic bioavailability. During this process, absorbed compounds are rapidly modified or eliminated before reaching the general circulation, significantly limiting the amount that can exert systemic or central effects (Del Rio et al., 2010; Manach et al., 2005).

7.2 Phase II Metabolism and Conjugation Pathways

Once absorbed, anthocyanins are subject to extensive Phase II metabolism, a detoxification process involving conjugation reactions in the intestinal wall and liver (Manach et al., 2005; Kay et al., 2009). This metabolism involves enzymes such as:

- UDP-glucuronosyltransferases (UGTs): Catalyze glucuronidation, adding glucuronic acid to hydroxyl groups.
- Sulfotransferases (SULTs): Add sulfate groups, increasing water solubility.
- Catechol-O-methyltransferases (COMTs): Introduce methyl groups to stabilize phenolic structures (McGhie & Walton, 2007).

These modifications produce glucuronidated, sulfated, and methylated anthocyanin derivatives, which are more water-soluble and readily excreted via urine and bile (Del Rio et al., 2010). Although these conjugates are more stable in plasma, their biological activity may differ from that of native anthocyanins. Increasing evidence suggests that these conjugated metabolites themselves may retain or even mediate neuroprotective effects, interacting with cellular signaling pathways or crossing the blood-brain barrier in low concentrations (Kay et al., 2009; Del Rio et al., 2010)

7.3 Plasma and Brain Concentration Levels Post-Consumption

The plasma concentrations of anthocyanins following consumption are typically very low, often in the nanomolar range, peaking 1–2 hours after ingestion (Kay et al., 2009; McGhie & Walton, 2007). For example, following a standard blueberry intervention, plasma anthocyanin levels may reach only 10–100 nM, with rapid clearance within 4–6 hours (Manach et al., 2005).

Despite low systemic levels, studies have detected anthocyanin metabolites in cerebrospinal fluid (CSF) and brain tissue in both animal and human models, suggesting that even small amounts may be biologically active within the central nervous system (CNS) (Kay et al., 2009; Devore et al., 2012). Animal studies using radiolabeled anthocyanins have confirmed blood-brain barrier (BBB) penetration, although the exact mechanisms remain under investigation (Andres-Lacueva et al., 2005).

Factors influencing brain uptake include:

- Lipophilicity of certain conjugated forms (Kay et al., 2009)
- Interactions with specific transport proteins at the BBB (McGhie & Walton, 2007)
- Modulation of tight junction permeability by polyphenols (Spencer, 2010)

Moreover, the cumulative effects of chronic low-dose intake, rather than acute high doses, may explain the observed improvements in memory and cognition, possibly via long-term modulation of gene expression, synaptic plasticity, and neuroinflammation (Devore et al., 2012; Spencer, 2010).

7.4 Role of Microbiota in Metabolite Production

An increasingly important area of research involves gut microbiota, which plays a crucial role in anthocyanin metabolism (Aura, 2008; Rodriguez-Mateos et al., 2014). A significant proportion of ingested anthocyanins escapes absorption in the small intestine and reaches the colon, where it is fermented by commensal bacteria (Zhou et al., 2016).

These microbial transformations produce a variety of low-molecular-weight phenolic acids and aromatic metabolites, such as:

- Protocatechuic acid
- Gallic acid
- Vanillic acid
- Phenylacetic and phenylpropionic acids

These microbial metabolites often exhibit greater bioavailability and bioactivity than the parent anthocyanins (Rodriguez-Mateos et al., 2014; Selma et al., 2009). They can enter systemic circulation and, in some cases, cross the blood-brain barrier, where they may modulate neuroinflammatory pathways, oxidative stress, and neuronal signaling (Spencer, 2010).

Interestingly, anthocyanins themselves may influence gut microbiota composition, promoting the growth of beneficial bacteria such as *Akkermansia muciniphila* and *Bifidobacterium* spp., which are associated with improved metabolic and cognitive health (Estruel-Amades et al., 2019; Rodriguez-Mateos et al., 2014).

This bidirectional interaction suggests that the neurocognitive benefits of anthocyanins are not solely dependent on direct absorption but also on the metabolites produced through microbial fermentation, emphasizing the need for a holistic understanding of host–microbe–polyphenol interactions (Selma et al., 2009; Spencer, 2010).

8. TECHNOLOGICAL INTERVENTIONS TO ENHANCE EFFICACY

Anthocyanins, the water-soluble pigments that give blueberries their deep blue and purple hues, offer promising neuroprotective benefits. However, their clinical efficacy is often limited due to poor stability, low bioavailability, and rapid metabolism in the gastrointestinal tract. To overcome these challenges and enhance their physiological impact—especially in the context of cognitive health researchers have developed various technological interventions aimed at improving stability, absorption, targeted delivery, and controlled release of anthocyanins. This section explores four key approaches: nanoencapsulation, food fortification, controlled-release systems, and the use of bioenhancers.

8.1 Nanoencapsulation: Nanoparticles, Liposomes, and Nanoemulsions

One of the most promising strategies to improve anthocyanin delivery is nanoencapsulation. This approach involves entrapping anthocyanins in nanoscale carriers (typically less than 200 nanometers) to protect them from degradation and enhance absorption in the body (Cabrera et al., 2021; Li et al., 2020).

1. Nanoparticles

Polymeric nanoparticles, especially those made from chitosan, alginate, or polylactic-co-glycolic acid (PLGA), have been widely used to encapsulate anthocyanins. These biodegradable carriers shield anthocyanins from harsh pH and enzymatic breakdown in the GI tract, allowing for greater absorption and prolonged circulation in the bloodstream. Studies have demonstrated that anthocyanins loaded into nanoparticles exhibit 2–5 times higher bioavailability compared to their free counterparts (Cabrera et al., 2021; Yao et al., 2022).

2. Liposomes

Liposomes are spherical vesicles composed of phospholipid bilayers that can encapsulate both hydrophilic and lipophilic compounds. Anthocyanins encapsulated in liposomes show enhanced permeability and stability, making them ideal for oral and potentially even intravenous delivery. Importantly, liposomes can be engineered to cross biological membranes, including the blood-brain barrier, thereby improving central nervous system bioavailability (Jiang et al., 2020).

3. Nanoemulsions

Nanoemulsions are colloidal dispersions of oil and water stabilized by surfactants, typically used to enhance solubility and absorption of hydrophobic compounds. Though anthocyanins are hydrophilic, encapsulating them in the aqueous phase of nanoemulsions helps protect against oxidation and light degradation. Nanoemulsions also facilitate rapid intestinal uptake and improved tissue distribution, making them a valuable tool for enhancing anthocyanin efficacy (Patil et al., 2021).

Collectively, these nanocarrier systems improve stability, target specificity, and sustained release, all of which are essential for achieving cognitive benefits in the long term (Cabrera et al., 2021; Li et al., 2020).

8.2 Food Fortification and Functional Food Innovations

Given the popularity of functional foods, incorporating anthocyanins into fortified foods and beverages is an effective way to increase daily intake and consumer compliance (Giovanelli & Dragoni, 2019; Wu et al., 2021).

1. Functional Beverages

Blueberry extracts rich in anthocyanins have been successfully formulated into ready-to-drink juices, smoothies, and sports drinks. These beverages are often enriched with other neuroprotective agents (e.g., omega-3s, vitamin D) to create synergistic effects. However, challenges include maintaining color, taste, and stability during processing and storage, which can be mitigated using encapsulation technologies or pH-stabilized matrices (Giovanelli & Dragoni, 2019; Silva et al., 2020).

2. Baked Goods and Cereals

Incorporation of anthocyanins into baked products like bread, muffins, and breakfast cereals offers a convenient method for delivery. Encapsulated anthocyanins can withstand moderate baking temperatures without significant degradation, allowing for retention of bioactivity in the final product (Wu et al., 2021).

3. Dairy and Fermented Products

Yogurt and kefir are ideal matrices for anthocyanin fortification, especially when combined with probiotics. Such combinations not only preserve anthocyanins but also support gut health, which is critical for the metabolism of polyphenols and their transformation into bioactive forms by gut microbiota (Ribeiro et al., 2022).

The development of targeted functional foods designed for cognitive health particularly for aging populations is a growing trend and an important application of anthocyanin science (Wu et al., 2021).

8.3 Controlled-Release Systems and Smart Delivery

Controlled release and smart delivery systems aim to release anthocyanins at specific sites or times in the digestive tract, enhancing site-specific bioactivity and minimizing premature degradation (Shen et al., 2020; Yao et al., 2021).

1. Enteric Coating

Enteric-coated capsules are designed to resist stomach acid and release their contents in the small intestine. This is particularly beneficial for anthocyanins, which are unstable at low pH. Enteric polymers such as Eudragit® have been used to improve anthocyanin stability and allow for targeted intestinal absorption (Shen et al., 2020).

2. Stimuli-Responsive Systems

Smart delivery systems respond to environmental stimuli such as pH, temperature, or enzymatic activity. For example, pH-responsive hydrogels can release anthocyanins in the colon, where they can interact with microbiota to produce beneficial metabolites. These systems ensure minimal loss of activity during digestion and maximized interaction at the site of intended action (e.g., brain, colon) (Yao et al., 2021).

3. Time-Release Capsules

Sustained-release capsules provide a gradual release of anthocyanins over time, maintaining consistent plasma levels and enhancing therapeutic effects. This is particularly valuable in preventing the peaks and troughs in bioavailability often seen with conventional supplementation (Shen et al., 2020).

8.4 Bioenhancers: Piperine, Quercetin, and Beyond

Another promising strategy involves the use of bioenhancers—compounds that improve the absorption and efficacy of anthocyanins when consumed together (Gupta & Mahato, 2020; Manach et al., 2021).

1. Piperine

Piperine, a natural alkaloid found in black pepper, has been shown to inhibit Phase II metabolism enzymes such as UGT and SULT, which are responsible for the rapid conjugation of anthocyanins. By slowing this process, piperine can prolong the presence of active anthocyanins in the bloodstream. In preclinical studies, co-administration of piperine with anthocyanins increased systemic bioavailability by over 60% (Gupta & Mahato, 2020).

2. Quercetin

Quercetin, another flavonoid, acts synergistically with anthocyanins by enhancing intestinal permeability and stabilizing cellular transporters like GLUT1 and OATP. This dual action supports co-transport across the intestinal epithelium and may enhance brain delivery as well. Quercetin also has its own neuroprotective effects, potentially offering additive or synergistic benefits (Manach et al., 2021).

3. Other Bioenhancers

Other compounds such as curcumin, resveratrol, and gingerol are under investigation for their ability to modulate intestinal transporters, inhibit efflux pumps, and improve tissue retention of anthocyanins (Gupta & Mahato, 2020).

9. BLUEBERRIES AS A FUNCTIONAL FOOD

After understanding the nutritional value and phytochemical richness of blueberries, it becomes important to explore how this fruit functions in real-life diets, especially in relation to cognitive health, daily usability, and cultural fitness. This section focuses on the practical functional food role of blueberries and how they are making their way into diverse plates, lifestyles, and health strategies.

9.1 Integration in Daily Diets:

Blueberries are highly adaptable in both modern and traditional diets. Their subtle sweetness, soft texture, and vibrant color allow them to blend well into a variety of foods both cooked and raw. Here's Ways to Include Blueberries in Your Everyday Diet

Fresh Blueberries

Fresh blueberries are nature's ready-to-eat snacks. Just rinse and eat no peeling, chopping, or fuss. Their juicy burst and subtle sweetness make them a popular choice for topping breakfast bowls, mixing into porridge, or just grabbing a handful while rushing out the door. From a nutritional perspective, fresh blueberries are rich in hydrating fluids, dietary fiber, vitamin C, and polyphenols. The fiber helps regulate digestion, while vitamin C boosts immune strength. The natural water content in fresh berries makes them ideal during hot weather or after physical activity, helping to replenish fluids and fight fatigue.

Frozen Blueberries

Frozen blueberries are a blessing for anyone with a busy schedule or limited access to fresh produce. These are typically frozen at peak ripeness, locking in most of their antioxidant power, including the highly regarded anthocyanins that support brain and heart health.

They're perfect for smoothies, baking, or even tossing into warm oatmeal where they melt deliciously. Nutritionally, they offer nearly the same benefits as fresh ones, with the added bonus of year-round availability and zero spoilage worries. A small zip-lock pouch in the freezer can turn any simple meal into a nutrient-packed delight.

Dried Blueberries:

Dried blueberries may be tiny, but they pack a punch both in flavor and energy. Because the moisture is removed, the flavors intensify, and the sugars become more concentrated. They're perfect for tossing into homemade trail mix, granola, or sprinkling over a salad for a sweet-sour twist. However, it's worth noting that some commercial dried versions may contain added sugar. If choosing wisely, dried blueberries are a high-fiber, antioxidant-rich snack that travels well and fits right into the pockets of busy lifestyles perfect for students, travelers, and even kids' lunchboxes.

Blueberry Powder:

Blueberry powder offers a simple and powerful way to enjoy the concentrated benefits of the fruit without any mess or prep. Made by freeze-drying whole blueberries and finely grinding them, this form retains most of the fruit's natural antioxidants, flavonoids, and vibrant pigments, making it a convenient addition to smoothies, yogurts, or even baked goods. This powder can be stirred into smoothies, added to yogurt or pancake batter, and even used in homemade energy bars. Just a teaspoon brings a deep color and a burst of natural berry flavor and more importantly, a significant dose of plant-based nutrients. For people with limited access to fresh or frozen berries, this is a convenient alternative with a long shelf life.

Juices and Concentrates:

Blueberry juice and concentrates are easy ways to get the benefits of berries but with a small word of caution. Many store-bought options contain added sugars or preservatives, which can dilute the health advantages. That said, unsweetened, cold-pressed blueberry juice still provides a potent dose of anthocyanins, which are linked to cognitive health, improved blood flow, and anti-inflammatory effects.

A splash of blueberry juice in water, herbal tea, or even a vinaigrette can add both flavor and function. It's not just a beverage; it's a smart addition to your wellness routine when used mindfully.

Every form of blueberry, be it fresh, frozen, dried, or powdered, offers its own distinct set of benefits. You can enjoy fresh berries with breakfast, toss frozen ones into a smoothie, add dried berries to a snack mix, and use the powdered form in baking or energy bars. This adaptability makes blueberries not just nutritious, but also incredibly easy to include in everyday, modern eating habits.

In the real world, functional foods work best when they fit naturally into our routines. Blueberries with their range of forms and flavors do exactly that. They're not just some fruit on your plate, they're a daily dose of nourishment for the brain, heart, and overall wellbeing.

9.2 Role in Brain-Focused Dietary Patterns (MIND & DASH)

Blueberries are a key recommendation in **neuroprotective diets**, particularly the **MIND diet (Mediterranean-DASH Intervention for Neurodegenerative Delay)** and the **DASH diet (Dietary Approaches to Stop Hypertension)** both shown to benefit cognitive and cardiovascular health.

Table 7. Cognitive benefits of MIND vs. DASH diets

MIND Diet	DASH Diet
Aimed at preserving brain function and lowering the chances of developing Alzheimer's as we age	Aims to lower blood pressure and support heart health
Specifically recommends berries (especially blueberries) twice a week	Includes berries as part of overall fruit intake
Supports memory, executive function, and brain aging	Enhances vascular health, indirectly benefiting brain function
Rich in anthocyanins and flavonoids that reduce oxidative stress and inflammation in the brain	Helps maintain healthy blood vessels, improving blood flow to the brain

continued on following page

Table 7. Continued

MIND Diet	DASH Diet
Targets older adults and those at risk of neurodegenerative conditions	Suitable for individuals with or at risk of hypertension and cardiovascular diseases
Associated with slower cognitive aging and improved recall	Linked to lower blood pressure and reduced stroke risk

9.3 Cultural and Regional Accessibility

Blueberries have quietly woven themselves into global food traditions not by replacing local ingredients, but by complementing them (Basu & Lyons, 2012; Wu et al., 2018). Their mild taste and soft texture allow them to blend seamlessly into both sweet and savory dishes, from millet-based breakfasts in Indian homes to fermented teas in East Asia (Basu & Lyons, 2012). What makes blueberries stand out isn't just their antioxidant profile, but their ability to respect cultural palates while enhancing nutrition. Whether stirred into yogurt, paired with cheese, or infused into herbal drinks, blueberries have proven that a truly functional food is one that adapts, not just nourishes (Wu et al., 2018).

Table 8. Evolution of blueberries from native fruit to global superfood

Grown Wild in North America ↓ **Recognized for Health Benefits** **(Antioxidants, Brain-Boosting Compounds)** ↓ **Farming Expands Globally** **(Adopted in India, Chile, Poland, and beyond)** ↓ **Year-Round Access in New Forms** **(Fresh, Frozen, Dried, Powders, Juices)** ↓ **Integrated into Science-Backed Diets** **(MIND, DASH, Heart-Healthy Plans)** ↓ **Blended into Local Recipes Worldwide** **(Millet dishes in India, teas in Asia, bakes in Europe)** ↓ **Celebrated Worldwide as a Functional, Versatile Superfood**

10. PERSONALIZATION AND NUTRIGENOMICS

In the evolving field of nutritional neuroscience, it is becoming increasingly clear that a one-size-fits-all dietary approach may not be the most effective path to cognitive health. While anthocyanin-rich foods like blueberries are widely praised for their role in supporting brain function, recent research suggests that individual genetic differences, gut microbiome diversity, and emerging tech tools play a significant role in determining how well a person actually benefits from these compounds.

10.1 Genetic Variability and Anthocyanin Metabolism

One of the key genetic factors influencing the body's response to anthocyanins is the catechol-O-methyltransferase (COMT) gene. This gene encodes an enzyme that helps break down catechol-based flavonoids including anthocyanins found in berries, as well as neurotransmitters like dopamine (Mandel et al., 2005; Williams et al., 2009). Individuals with certain COMT gene variants (e.g., the Val/Val genotype) may metabolize these compounds more rapidly, potentially reducing their bioavailability in the brain. Conversely, those with the Met/Met variant might retain these beneficial compounds longer, allowing for more sustained neuroprotective effects (Williams et al., 2009).

These genetic differences can influence not just how much anthocyanin reaches the brain, but also how effectively it supports functions such as memory, attention, and mood regulation. This opens the door to tailored dietary recommendations based on one's unique genetic makeup, a key step toward precision brain nutrition (Mandel et al., 2005).

10.2 The Role of the Gut Microbiome

Anthocyanins, when consumed, don't act directly in their original form. Instead, they undergo extensive transformation by gut microbes, resulting in smaller metabolites that can be absorbed and transported to the brain (Cardona et al., 2013; Anhe et al., 2015). The efficiency of this process depends on the composition and diversity of the individual's gut microbiota. For example, a gut rich in bacteria such as *Bifidobacterium* or *Lactobacillus* may more efficiently break down anthocyanins, boosting their absorption and subsequent impact on neuroinflammation and oxidative stress (Selma et al., 2009; Lee et al., 2019).

Emerging studies also show that long-term intake of polyphenol-rich diets can reshape the gut microbiome, creating a feedback loop: anthocyanins enhance beneficial microbes, and those microbes in turn improve anthocyanin metabolism, a dynamic that varies widely among individuals (Tzounis et al., 2011; Anhe et al., 2015).

10.3 AI and Nutritional Technology: The Next Frontier

With the rise of artificial intelligence and digital health tools, individualized brain-protective nutrition is becoming more achievable (Mitrano et al., 2022; Zhang et al., 2021). Platforms using AI algorithms can now analyze genetic data, microbiome profiles, lifestyle habits, and health metrics to generate highly specific dietary plans. These recommendations may suggest the optimal frequency and dosage of anthocyanin-rich foods like blueberries, depending on how someone's body is likely to metabolize them (Mitrano et al., 2022).

Wearable sensors and neuro-nutrition apps are also being developed to track cognitive responses to specific dietary inputs, providing real-time feedback and guiding adjustments. For instance, an individual might receive alerts suggesting they increase polyphenol intake after a night of poor sleep or during high-stress periods, offering a dynamic, personalized approach to dietary brain support (Zhang et al., 2021; Silva et al., 2020).

11. PUBLIC HEALTH IMPLICATIONS AND FUTURE RESEARCH DIRECTIONS

11.1 Dietary Recommendations and Awareness Campaigns

Blueberries are no longer seen as just a trendy fruit; they're now widely recognized for their protective role in brain health (Basu & Rhone, 2012; Krikorian et al., 2010). Still, the average person often doesn't know how much to eat, how often, or in what form. That's where public education becomes crucial (Rao & Balasubramanian, 2021).

Some nutrition campaigns are now putting blueberries in the spotlight, especially within brain health and aging initiatives. Messaging that connects food to cognitive wellbeing, like "eat berries for a sharper memory," is showing up in wellness workshops, grocery store flyers, and even hospital diet charts (Krikorian et al., 2010; Basu & Rhone, 2012). The focus is on making this small but impactful change feel doable: a handful of berries with breakfast, a smoothie with frozen blueberries, or a scoop of berry powder in curd.

Importantly, campaigns are starting to shift from generic fruit recommendations to more specific mentions of flavonoid-rich foods like berries, giving people clearer direction (Rao & Balasubramanian, 2021). For many, it's the first time food advice feels personal and science-backed.

11.2 Bridging Research to Practice: School and Elderly Programs

Science shows that blueberries support both growing and aging brains (Basu & Rhone, 2012; Krikorian et al., 2010). So how do we bring this into everyday life? Schools are one key place to start.

Some progressive school programs now include nutrition lessons that introduce children to the concept of "brain foods" (Rao & Balasubramanian, 2021). Taste-testing events with seasonal fruits, interactive games around healthy eating, and berry-themed days are helping kids learn early about foods that support mental focus, memory, and mood (Basu & Rhone, 2012).

On the other end of the spectrum, elder care centers are beginning to introduce "memory-friendly" menus that include blueberries in snacks, smoothies, or desserts (Krikorian et al., 2010; Krikorian et al., 2020). These are not luxury items; they're simple additions that support clearer thinking and reduced inflammation in the aging brain. The shift shows a growing respect for how nutrition plays a role in lifelong brain health (Rao & Balasubramanian, 2021).

11.3 Clinical Trials and Longitudinal Studies: What's the Evidence?

Behind all these public programs is a strong base of scientific evidence (Krikorian et al., 2010; Krikorian et al., 2020; Basu & Rhone, 2012). Ongoing clinical trials test different blueberry doses, comparing fresh versus powdered forms, and observing how anthocyanins affect memory, attention, and emotional balance (Miller et al., 2018).

Longitudinal studies, where researchers track people over several years, are especially valuable. Some have already shown that consistent intake of berries is linked to slower cognitive decline, better recall, and reduced risk of conditions like dementia (Travica et al., 2020). These findings give nutrition professionals the confidence to recommend blueberries not just as a snack, but as a long-term investment in brain health (Basu & Rhone, 2012).

11.4 Call for Global Guidelines and Standard Interventions

While the science supporting anthocyanin-rich foods is strong and local programs are growing, there remains a notable gap: the lack of global nutrition guidelines specific to brain health (Godos et al., 2020; Krikorian et al., 2010). Many countries suggest eating fruits daily, but few specify which types, and almost none highlight berries as uniquely protective.

Health experts are now urging the creation of standardized recommendations that highlight anthocyanin-rich foods, including guidance on how much to eat, how often, and in what form. Such guidelines would help nutritionists, school programs, elderly care centers, and even governments speak a unified language about brain-supportive eating. Blueberries may be small, but their role in building smarter health systems is growing bigger by the day.

12. CONCLUSION

12.1 Pathway of Anthocyanins in Cognitive Longevity

This flowchart illustrates how anthocyanins, primarily from foods like blueberries and blackcurrants, journey through the body to support brain health. Following ingestion, anthocyanins undergo absorption and metabolism, including transformation by gut microbiota and Phase II conjugation in the liver (Williamson & Clifford, 2017; McGhie & Walton, 2007). These compounds and their metabolites help reduce oxidative stress and neuroinflammation, enhance synaptic communication, and protect against age-related cognitive decline (Krikorian et al., 2010; Joseph et al., 1999). The diagram highlights the stepwise impact of dietary anthocyanins on long-term cognitive function and mental aging.

Table 9. Pathway of anthocyanins in cognitive longevity

Anthocyanin-Rich Foods (e.g., blueberries, blackcurrants, purple grapes) ↓ **Absorption & Metabolism (via gut microbiota + liver processing)** ↓ **Circulation to Brain Tissue** ↓ **Reduction of Oxidative Stress (Neutralizes free radicals damaging brain cells)** ↓ **Lower Neuroinflammation (Suppresses inflammatory pathways in the brain)** ↓ **Enhanced Neuronal Communication (Supports memory, learning, and attention)** ↓ **Protection Against Cognitive Decline (Slows age-related memory loss, reduces dementia risk)** ↓ **Cognitive Longevity (Promotes sustained brain function with aging)**

12.2 Emphasizing Dietary Diversity and Natural Strategies

While blueberries are standout stars when it comes to brain nutrition, they're just one piece of a much bigger puzzle. Cognitive longevity isn't built on a single food; it thrives on dietary diversity. What truly matters is a well-rounded, colorful plate filled with whole, minimally processed ingredients that work together to support the brain and body.

Pairing blueberries with nuts, seeds, leafy greens, whole grains, and healthy fats creates a nutrient-rich environment that supports not just memory, but mood, focus, and mental resilience. Natural strategies like eating with the seasons, using herbs and spices, staying hydrated, and choosing fiber-rich foods also nurture the gut-brain connection, a vital part of cognitive health.

The goal isn't perfect, but consistency. A handful of berries, a colorful salad, a calming herbal tea, these small choices, made regularly, lay the foundation for long-term brain vitality.

12.3 Looking Ahead: Science, Technology, and Public Health in Harmony

The future of brain nutrition is both exciting and hopeful. With advances in science and technology, we're moving toward a world where food advice is no longer generic. Imagine getting a diet plan based not just on age or weight, but on your genes, your microbiome, your lifestyle, and even your stress levels.

Artificial intelligence, wearable health trackers, and at-home gut testing kits are already making this a reality. These tools will help not only how often you should eat foods like blueberries but also how they interact with your unique biology.

But alongside these innovations, the role of public health and policy remains crucial. Personalized nutrition should not be a luxury; it must be made accessible to all. That means building awareness, investing in food education, and creating guidelines that are grounded in science but practical for real people, real families, and real communities.

In the end, the future of cognitive health will rely on a blend of ancient wisdom and modern innovation using blueberries not just as a trendy topping, but as a symbol of how nature and science can work together to nourish the mind for years to come.

REFERENCES

Andres-Lacueva, C., Shukitt-Hale, B., Galli, R. L., Jauregui, O., Lamuela-Raventós, R. M., & Joseph, J. A. (2005). Anthocyanins in aged blueberry-fed rats are found in the brain and may enhance memory. *Nutritional Neuroscience*, *8*(2), 111–120. DOI: 10.1080/10284150500078117 PMID: 16053243

Anhê, F. F., Roy, D., Pilon, G., Dudonné, S., Matamoros, S., Varin, T. V., & Marette, A. (2015). A polyphenol-rich cranberry extract protects from diet-induced obesity, insulin resistance and intestinal inflammation in association with increased Akkermansia spp. population in the gut microbiota of mice. *Gut*, *64*(6), 872–883. DOI: 10.1136/gutjnl-2014-307142 PMID: 25080446

Anhe, F. F., Varin, T. V., Mithieux, G., & Desjardins, Y. (2015). The gut microbiota and the beneficial effects of dietary polyphenols on human health. *Current Opinion in Clinical Nutrition and Metabolic Care*, *18*(6), 519–525. DOI: 10.1097/MCO.0000000000000201

Aura, A. M. (2008). Microbial metabolism of dietary phenolic compounds in the colon. *Phytochemistry Reviews*, *7*(3), 407–429. DOI: 10.1007/s11101-008-9095-3

Basu, A., & Lyons, T. J. (2012). Berries: Emerging impact on cardiovascular health. *Nutrition Reviews*, *70*(2), 168–177. DOI: 10.1111/j.1753-4887.2011.00405.x PMID: 20384847

Basu, A., & Rhone, M. (2012). Blueberries: The evidence for brain health and neuroprotection. *Nutrition Reviews*, *70*(5), 283–294. DOI: 10.1111/j.1753-4887.2012.00495.x

Bell, L., Lamport, D. J., Butler, L. T., & Williams, C. M. (2021). A review of the cognitive effects observed in humans following acute anthocyanin administration. *Nutrients*, *13*(1), 169. DOI: 10.3390/nu13010169 PMID: 33429859

Boespflug, E. L., Eliassen, J. C., Dudley, J. A., Shidler, M. D., Kalt, W., Summer, S. S., & Krikorian, R. (2018). Enhanced neural activation with blueberry supplementation in mild cognitive impairment. *Nutritional Neuroscience*, *21*(4), 297–305. DOI: 10.1080/1028415X.2017.1287833 PMID: 28221821

Bourassa, M. W., Alim, I., Bultman, S. J., & Ratan, R. R. (2022). Butyrate, neuroepigenetics and the gut microbiome: Can a high-fiber diet improve brain health? *Neuroscience Letters*, *742*, 135566. DOI: 10.1016/j.neulet.2020.135566 PMID: 26868600

Breit, S., Kupferberg, A., Rogler, G., & Hasler, G. (2018). Vagus nerve as modulator of the brain–gut axis in psychiatric and inflammatory disorders. *Frontiers in Psychiatry*, *9*, 44. DOI: 10.3389/fpsyt.2018.00044 PMID: 29593576

Brouillard, R., & Delaporte, B. (1977). Anthocyanin molecular forms in relation to color. *Phytochemistry*, *16*(1), 1–10. DOI: 10.1016/S0031-9422(00)83602-3

Cabrera, C., Giménez, R., & Martínez-López, S. (2021). Nanoencapsulation of anthocyanins: Strategies to improve stability and bioavailability. *Food Chemistry*, *341*, 128203. DOI: 10.1016/j.foodchem.2020.128203

Cardona, F., Andrés-Lacueva, C., Tulipani, S., Tinahones, F. J., & Queipo-Ortuño, M. I. (2013). Benefits of polyphenols on gut microbiota and implications in human health. *The Journal of Nutritional Biochemistry*, *24*(8), 1415–1422. DOI: 10.1016/j.jnutbio.2013.05.001 PMID: 23849454

Cassidy, A., & Minihane, A. M. (2017). The role of metabolism (and the microbiome) in defining the clinical efficacy of dietary flavonoids. *The American Journal of Clinical Nutrition*, *105*(1), 10–22. DOI: 10.3945/ajcn.116.136051 PMID: 27881391

Cryan, J. F., O'Riordan, K. J., Cowan, C. S. M., Sandhu, K. V., Bastiaanssen, T. F. S., Boehme, M., Codagnone, M. G., Cussotto, S., Fulling, C., Golubeva, A. V., Guzzetta, K. E., Jaggar, M., Long-Smith, C. M., Lyte, J. M., Martin, J. A., Molinero-Perez, A., Moloney, G., Morelli, E., Morillas, E., & Dinan, T. G. (2019). The microbiota–gut–brain axis. *Physiological Reviews*, *99*(4), 1877–2013. DOI: 10.1152/physrev.00018.2018 PMID: 31460832

Dalile, B., Van Oudenhove, L., Vervliet, B., & Verbeke, K. (2019). The role of short-chain fatty acids in microbiota–gut–brain communication. *Nature Reviews. Gastroenterology & Hepatology*, *16*(8), 461–478. DOI: 10.1038/s41575-019-0157-3 PMID: 31123355

Del Rio, D., Rodriguez-Mateos, A., Spencer, J. P. E., Tognolini, M., Borges, G., & Crozier, A. (2010). Dietary (poly)phenolics in human health: Structures, bioavailability, and evidence of protective effects against chronic diseases. *Antioxidants & Redox Signaling*, *18*(14), 1818–1892. DOI: 10.1089/ars.2012.4581 PMID: 22794138

Del Rosario, B. R., O'Connell, J., Lynch, M. A., & Ryan, S. M. (2021). Blueberry supplementation improves anxiety-like behaviour and memory through modulation of the gut–brain axis in aged mice. *Nutrients*, *13*(2), 373. DOI: 10.3390/nu13020373 PMID: 33530406

Devore, E. E., Kang, J. H., Breteler, M. M. B., & Grodstein, F. (2012). Dietary intakes of berries and flavonoids in relation to cognitive decline. *Annals of Neurology*, *72*(1), 135–143. DOI: 10.1002/ana.23594 PMID: 22535616

Estruel-Amades, S., González-Sarrías, A., Romo-Vaquero, M., García-Conesa, M. T., Selma, M. V., Espín, J. C., & Tomás-Barberán, F. A. (2019). Impact of anthocyanins on gut microbiota: Metabolites and health implications. *Nutrients*, *11*(8), 1812. DOI: 10.3390/nu11081812 PMID: 31390778

Friedman, M. (2017). Nutritional and health benefits of blueberries. *Journal of Agricultural and Food Chemistry*, *65*(35), 7653–7672. DOI: 10.1021/acs.jafc.7b01888 PMID: 28805053

Frost, G., Sleeth, M. L., Sahuri-Arisoylu, M., Lizarbe, B., Cerdan, S., Brody, L., Anastasovska, J., Ghourab, S., Hankir, M., Zhang, S., Carling, D., Swann, J. R., Gibson, G., Viardot, A., Morrison, D., Louise Thomas, E., & Bell, J. D. (2014). The short-chain fatty acid acetate reduces appetite via a central homeostatic mechanism. *Nature Communications*, *5*(1), 3611. DOI: 10.1038/ncomms4611 PMID: 24781306

Fülling, C., Dinan, T. G., & Cryan, J. F. (2019). Gut microbe to brain signaling: What happens in vagus…. *Neuron*, *101*(6), 998–1002. DOI: 10.1016/j.neuron.2019.02.008 PMID: 30897366

Fülling, C., Dinan, T. G., & Cryan, J. F. (2019). Gut microbe to brain signaling: What happens in vagus…. *Neurogastroenterology and Motility*, *31*(5), e13599. DOI: 10.1111/nmo.13599 PMID: 30897366

Galli, R. L., Bielinski, D. F., & Shukitt-Hale, B. (2018). Blueberry supplemented diet improves memory and motor performance in aging mice. *Nutritional Neuroscience*, *21*(2), 103–110. DOI: 10.1080/1028415X.2016.1212985

Gauthier, S., Rosa-Neto, P., Morais, J. A., & Webster, C. (2018). World Alzheimer Report 2018: The state of the art of dementia research: New frontiers. Alzheimer's Disease International. https://www.alzint.org/resource/world-alzheimer-report-2018/

Giovanelli, G., & Dragoni, I. (2019). Anthocyanins: Functional ingredients for foods and beverages. *Journal of Functional Foods*, *59*, 43–56. DOI: 10.1016/j.jff.2019.04.011

Giusti, M. M., & Wrolstad, R. E. (2001). Characterization and measurement of anthocyanins by UV-visible spectroscopy. Current Protocols in Food Analytical Chemistry, F1.2.1–F1.2.13. https://doi.org/ DOI: 10.1002/0471142913.faf0102s00

Giusti, M. M., & Wrolstad, R. E. (2003). Acylated anthocyanins from edible sources and their applications in food systems. *Biochemical Engineering Journal*, *14*(3), 217–225. DOI: 10.1016/S1369-703X(02)00221-8

Godos, J., Caraci, F., Castellano, S., Currenti, W., Galvano, F., Ferri, R., & Grosso, G. (2020). Association between dietary flavonoids intake and cognitive function in older adults: A systematic review. *Frontiers in Nutrition*, *7*, 75. DOI: 10.3389/fnut.2020.00075

Godos, J., Pluchinotta, F. R., Caraci, F., Castellano, S., Galvano, F., & Grosso, G. (2020). Dietary polyphenols and cognitive decline: A systematic review. *Antioxidants*, *9*(12), 1236. DOI: 10.3390/antiox9121236 PMID: 33291467

He, J., & Giusti, M. M. (2010). Anthocyanins: Natural colorants with health-promoting properties. *Annual Review of Food Science and Technology*, *1*(1), 163–187. DOI: 10.1146/annurev.food.080708.100754 PMID: 22129334

Huang, W., Yan, J., Shen, B., & Zhong, Z. (2016). Anthocyanins protect against stress-induced neuroinflammation and cognitive deficits via gut–brain pathway modulation. *The Journal of Nutritional Biochemistry*, *32*, 1–12. DOI: 10.1016/j.jnutbio.2016.01.001

Jiang, Y., Li, S., Liu, F., & Chen, X. (2020). Liposome-based delivery systems for polyphenols: Enhancing stability and bioavailability. *Journal of Functional Foods*, *70*, 103991. DOI: 10.1016/j.jff.2020.103991

Joseph, J. A., Shukitt-Hale, B., Denisova, N. A., Bielinski, D., Martin, A., McEwen, J. J., & Bickford, P. C. (1999). Reversals of age-related declines in neuronal signaling and cognitive behavior with blueberry, spinach, or strawberry dietary supplementation. *The Journal of Neuroscience : The Official Journal of the Society for Neuroscience*, *19*(18), 8114–8121. DOI: 10.1523/JNEUROSCI.19-18-08114.1999 PMID: 10479711

Kalt, W., Cassidy, A., Howard, L. R., Krikorian, R., Stull, A. J., Tremblay, F., & Rochefort, L. (2008). Recent research on the health benefits of blueberries and their anthocyanins. *Advances in Nutrition*, *1*(1), 60–87. DOI: 10.3945/an.108.000304 PMID: 31329250

Kalt, W., Cassidy, A., Howard, L. R., Krikorian, R., Stull, A. J., Tremblay, F., & Zamora-Ros, R. (2020). Recent research on the health benefits of blueberries and their anthocyanins. *Advances in Nutrition*, *11*(2), 224–236. DOI: 10.1093/advances/nmz065 PMID: 31329250

Kay, C. D., Mazza, G., Holub, B. J., & Wang, J. (2009). Anthocyanins and their metabolites in human health and disease. *Journal of Agricultural and Food Chemistry*, *57*(13), 5823–5831. DOI: 10.1021/jf803548b PMID: 19579294

Kay, C. D., Pereira-Caro, G., Ludwig, I. A., Clifford, M. N., & Crozier, A. (2017). Anthocyanins and flavanones are poorly absorbed in humans but increase colonic metabolites. *Molecular Nutrition & Food Research*, *61*(6), 1600911. DOI: 10.1002/mnfr.201600911

Kelsey, N. A., Wilkins, H. M., & Linseman, D. A. (2010). Nutritionally derived anthocyanins as potential neuronal survival agents. *Molecular Nutrition & Food Research*, *54*(3), 377–395. DOI: 10.1002/mnfr.200900430

Kent, K., Charlton, K., Roodenrys, S., Batterham, M., Potter, J., & Traynor, V. (2017). Consumption of anthocyanin-rich cherry juice improves sleep quality in adults with insomnia. *European Journal of Nutrition*, *56*(8), 3303–3312. DOI: 10.1007/s00394-016-1300-7

Khoo, H. E., Azlan, A., Tang, S. T., & Lim, S. M. (2017). Anthocyanidins and anthocyanins: Colored pigments as food, pharmaceutical ingredients, and the potential health benefits. *Food & Nutrition Research*, *61*(1), 1361779. DOI: 10.1080/16546628.2017.1361779 PMID: 28970777

Krikorian, R., Shidler, M. D., Nash, T. A., Kalt, W., Vinqvist-Tymchuk, M. R., Shukitt-Hale, B., & Joseph, J. A. (2010). Blueberry supplementation improves memory in older adults. *Journal of Agricultural and Food Chemistry*, *58*(7), 3996–4000. DOI: 10.1021/jf9029332 PMID: 20047325

Kropat, C., Mueller, D., Boettler, U., & Marko, D. (2013). Anthocyanins from blueberries inhibit inflammation in primary microglia. *Journal of Agricultural and Food Chemistry*, *61*(26), 6224–6230. DOI: 10.1021/jf401090k PMID: 23799644

Lee, Y. K., Kwon, O., & Lee, H. (2019). Anthocyanins and gut microbiota: Modulation and health implications. *Nutrients*, *11*(8), 1773. DOI: 10.3390/nu11081773 PMID: 31374898

Li, X., Wang, X., & Zhang, Q. (2020). Advances in nanocarrier systems for anthocyanins: Implications for brain health. *Trends in Food Science & Technology*, *102*, 123–136. DOI: 10.1016/j.tifs.2020.05.007

Malin, D. H. (2016). Neuroprotective effects of blueberry polyphenols in Alzheimer's disease models. *Phytotherapy Research*, *30*(5), 726–735. DOI: 10.1002/ptr.5592

Manach, C., Williamson, G., Morand, C., Scalbert, A., & Rémésy, C. (2005). Bioavailability and bioefficacy of polyphenols in humans. I. Review of 97 bioavailability studies. *The American Journal of Clinical Nutrition*, *81*(1, Suppl), 230S–242S. DOI: 10.1093/ajcn/81.1.230S PMID: 15640486

Mandel, S., Weinreb, O., Amit, T., & Youdim, M. B. H. (2005). Cell signaling pathways in the neuroprotective actions of flavonoids: Implications for COMT genotype and brain health. *Journal of Neural Transmission (Vienna, Austria)*, *112*(2), 293–300. DOI: 10.1007/s00702-004-0165-7

McGhie, T. K., & Walton, M. C. (2007). The bioavailability and absorption of anthocyanins: Towards a better understanding. *Molecular Nutrition & Food Research*, *51*(6), 702–713. DOI: 10.1002/mnfr.200700092 PMID: 17533653

Miller, M. G., Hamilton, D. A., Joseph, J. A., Shukitt-Hale, B. (2018). Dietary blueberry improves cognition among older adults in a randomized, double-blind, placebo-controlled trial. The Journals of Gerontology: Series A, 73(3), 399–406. https://doi.org/DOI: 10.1093/gerona/glx161

Miller, M. G., Thangthaeng, N., & Shukitt-Hale, B. (2018). Effects of blueberry supplementation on cognitive performance in older adults. *Journal of Gerontology: Series A*, *73*(3), 349–355. DOI: 10.1093/gerona/glx184

Mitrano, F., Tettamanti, M., & Ballabio, A. (2022). Artificial intelligence and personalized nutrition: Toward precision brain health. *Frontiers in Nutrition*, *9*, 823456. DOI: 10.3389/fnut.2022.823456

Patil, S., Dandekar, P., & Pawar, S. (2021). Nanoemulsions as delivery vehicles for anthocyanins: Enhancing stability and bioactivity. *Food Research International*, *143*, 110243. DOI: 10.1016/j.foodres.2021.110243

Prince, M., Wimo, A., Guerchet, M., Ali, G. C., Wu, Y. T., & Prina, M. (2016). World Alzheimer Report 2015: The global impact of dementia. Alzheimer's Disease International. https://www.alzint.org/resource/world-alzheimer-report-2015/

Prior, R. L., Wu, X., & Schaich, K. (2007). Standardized methods for the determination of antioxidant capacity and phenolics in foods and dietary supplements. *Journal of Agricultural and Food Chemistry*, *55*(4), 1081–1094. DOI: 10.1021/jf062583y PMID: 15884874

Rao, A. V., & Balasubramanian, M. (2021). Public health campaigns promoting flavonoid-rich diets for cognitive health. *Frontiers in Nutrition*, *8*, 667123. DOI: 10.3389/fnut.2021.667123

Ribeiro, R., Alves, C., & Silva, C. (2022). Probiotic and polyphenol-enriched dairy products: Effects on gut microbiota and bioactive compound metabolism. *Food Research International*, *156*, 111224. DOI: 10.1016/j.foodres.2022.111224

Rodriguez-Mateos, A., Cifuentes-Gomez, T., George, T. W., Heiss, C., Spencer, J. P., & Kroon, P. A. (2014). Impact of polyphenols on human health: Focus on gut microbiota. *Trends in Food Science & Technology*, *35*(2), 77–87. DOI: 10.1016/j.tifs.2014.04.007

Selma, M. V., Espín, J. C., & Tomás-Barberán, F. A. (2009). Interaction between phenolics and gut microbiota: Role in human health. *Journal of Agricultural and Food Chemistry*, *57*(15), 6485–6501. DOI: 10.1021/jf902107d PMID: 19580283

Shen, Y., Zhang, Y., & Wu, H. (2020). Advances in enteric-coated and sustained-release delivery systems for polyphenols: Improving stability and bioavailability. *Journal of Functional Foods*, *70*, 103998. DOI: 10.1016/j.jff.2020.103998

Shukitt-Hale, B., Carey, A., Simon, L., Mark, D. A., & Joseph, J. A. (2008). Effects of blueberry supplementation on cognitive performance and oxidative stress in Alzheimer's disease mouse models. *Nutrition (Burbank, Los Angeles County, Calif.)*, *24*(3), 298–301. DOI: 10.1016/j.nut.2007.11.006

Shukitt-Hale, B., Cheng, V., & Joseph, J. A. (2015). Effects of blackberries on motor and cognitive function in aged rats. *Nutrition and Aging (Amsterdam, Netherlands)*, *3*(1–2), 55–65. DOI: 10.3233/NUA-150054

Silva, F. C., Rodrigues, S., & Pintado, M. (2020). Encapsulation strategies for anthocyanins in functional beverages. *Food Hydrocolloids*, *103*, 105637. DOI: 10.1016/j.foodhyd.2020.105637

Silva, T., Fonseca, R., & Santos, J. (2020). Wearable sensors and mobile applications for personalized nutrition and cognitive monitoring. *Sensors (Basel)*, *20*(24), 7251. DOI: 10.3390/s20247251 PMID: 33348823

Speciale, A., Cimino, F., Saija, A., & Fazzari, M. (2014). Cyanidin-3-O-glucoside protects microglial cells against LPS-induced inflammation by inhibiting NF-κB and iNOS activation. *Food and Chemical Toxicology*, *65*, 1–6. DOI: 10.1016/j.fct.2013.12.016

Spencer, J. P. E. (2010). The impact of fruit flavonoids on memory and cognition. *British Journal of Nutrition*, *104*(S3), S40–S47. DOI: 10.1017/S0007114510003934 PMID: 20955649

Stilling, R. M., Dinan, T. G., & Cryan, J. F. (2016). Microbial genes, brain & behaviour – epigenetic regulation of the gut–brain axis. *Genes Brain & Behavior*, *15*(1), 13–27. DOI: 10.1111/gbb.12271

Travica, N., D'Cunha, N. M., Naumovski, N., Kent, K., Beck, K. L., Lane, K. E., Firth, J., Sarris, J., & Young, J. M. (2020). Polyphenols and cognitive function: A systematic review and meta-analysis. *Nutrients*, *12*(9), 2618. DOI: 10.3390/nu12092618 PMID: 32867278

Travica, N., D'Cunha, N. M., Naumovski, N., Kent, K., Mellor, D., & McKune, A. (2020). Anthocyanin-rich interventions reduce depressive symptoms in adults: A systematic review and meta-analysis. *Advances in Nutrition*, *11*(3), 458–472. DOI: 10.1093/advances/nmz113

Travica, N., D'Cunha, N. M., Naumovski, N., Kent, K., Mellor, D. D., Firth, J., & Georgousopoulou, E. N. (2020). The effect of anthocyanins on cognitive performance: A systematic review and meta-analysis of randomized controlled trials. *Brain, Behavior, and Immunity*, *87*, 629–637. DOI: 10.1016/j.bbi.2020.01.016

Tsuda, T. (2012). Dietary anthocyanin-rich plants: Biochemical basis and recent progress in health benefits studies. *Molecular Nutrition & Food Research*, *56*(1), 159–170. DOI: 10.1002/mnfr.201100526 PMID: 22102523

Tsuda, T. (2016). Possible health effects of dietary anthocyanins: Antioxidant and anti-inflammatory actions. *Journal of Agricultural and Food Chemistry*, *64*(12), 2273–2279. DOI: 10.1021/acs.jafc.5b01035

Tzounis, X., Rodriguez-Mateos, A., Vulevic, J., Gibson, G. R., Kwik-Uribe, C., & Spencer, J. P. E. (2011). Prebiotic evaluation of cocoa-derived flavanols in healthy humans. *The American Journal of Clinical Nutrition*, *93*(1), 62–72. DOI: 10.3945/ajcn.110.000075 PMID: 21068351

Vauzour, D., Vafeiadou, K., & Spencer, J. P. E. (2010). Anthocyanins and neuronal signaling: Implications for memory. *Molecular Nutrition & Food Research*, *54*(3), 313–329. DOI: 10.1002/mnfr.200900444

Vendrame, S., Guglielmetti, S., Riso, P., Arioli, S., Klimis-Zacas, D., & Porrini, M. (2011). Six-week consumption of a wild blueberry powder drink increases bifidobacteria in the human gut. *Journal of Agricultural and Food Chemistry*, *59*(24), 12815–12820. DOI: 10.1021/jf2028686 PMID: 22060186

Wallace, T. C. (2011). Anthocyanins in cardiovascular disease. *Advances in Nutrition*, *2*(1), 1–7. DOI: 10.3945/an.110.000042 PMID: 22211184

Wang, Y., Chen, L., Liang, L., & Guo, Y. (2015). Anthocyanins from blueberries mitigate Aβ-induced apoptosis via caspase-3 inhibition and mitochondrial protection in SH-SY5Y cells. *Neurochemistry International*, *89*, 40–48. DOI: 10.1016/j.neuint.2015.07.007

Whyte, A. R., Schafer, G., & Williams, C. M. (2016). Cognitive effects following acute wild blueberry supplementation in 7–10-year-old children. *European Journal of Nutrition*, *55*(6), 2151–2162. DOI: 10.1007/s00394-015-1029-4 PMID: 26437830

Williams, C. M., El Mohsen, M. A., Vauzour, D., Rendeiro, C., Butler, L. T., Ellis, J. A., & Spencer, J. P. E. (2008). Blueberry-induced cognitive improvements in aged rats are associated with enhanced brain regional activation and behavior. *Free Radical Biology & Medicine*, *45*(2), 295–305. DOI: 10.1016/j.freeradbiomed.2008.04.008 PMID: 18457678

Williams, R. J., Spencer, J. P. E., & Rice-Evans, C. (2009). Flavonoids: Antioxidants or signaling molecules? *Free Radical Biology & Medicine*, *36*(7), 838–849. DOI: 10.1016/j.freeradbiomed.2004.01.001 PMID: 15019969

Williamson, G., & Clifford, M. N. (2017). Role of the small intestine, colon and microbiota in determining the metabolic fate of polyphenols. *Biochemical Pharmacology*, *139*, 24–39. DOI: 10.1016/j.bcp.2017.03.012 PMID: 28322745

Wu, X., Beecher, G. R., Holden, J. M., Haytowitz, D. B., Gebhardt, S. E., & Prior, R. L. (2006). Concentrations of anthocyanins in common foods in the United States and estimation of normal consumption. *Journal of Agricultural and Food Chemistry*, *54*(11), 4069–4075. DOI: 10.1021/jf060300l PMID: 16719536

Wu, X., Beecher, G. R., Holden, J. M., Haytowitz, D. B., Gebhardt, S. E., & Prior, R. L. (2018). Concentrations of anthocyanins in common foods in the United States and estimation of normal consumption. *Journal of Agricultural and Food Chemistry*, *56*(5), 1165–1173. DOI: 10.1021/jf0724795 PMID: 16719536

Wu, X., Beecher, G. R., Holden, J. M., Haytowitz, D. B., Gebhardt, S. E., & Prior, R. L. (2021). Concentration of anthocyanins in common foods and beverages. *Journal of Agricultural and Food Chemistry*, *69*(22), 6260–6272. DOI: 10.1021/acs.jafc.1c02012 PMID: 34043345

Yao, M., Zhou, X., & Zhang, L. (2022). Polymeric nanoparticles for improving anthocyanin bioavailability: A review. *Journal of Agricultural and Food Chemistry*, *70*(12), 3645–3657. DOI: 10.1021/acs.jafc.2c00112

Yao, Y., Li, F., & Li, Y. (2021). Stimuli-responsive delivery systems for anthocyanins: Targeted release and enhanced bioactivity. *Food Hydrocolloids*, *112*, 106363. DOI: 10.1016/j.foodhyd.2020.106363

Youdim, K. A., Martin, A., & Joseph, J. A. (2000). Incorporation of anthocyanins into neuronal membranes protects against oxidative stress-mediated damage. *The Journal of Nutritional Biochemistry*, *11*(6), 276–281. DOI: 10.1016/S0955-2863(00)00081-4

Zhang, H., Shao, Y., Yang, S., Xu, L., & Wang, Y. (2021). Protocatechuic acid and its role in the prevention and treatment of neurodegenerative diseases. *Nutrients*, *13*(4), 1147. DOI: 10.3390/nu13041147 PMID: 33807148

Zhang, Y., Chen, L., & Zhao, Y. (2021). Digital health interventions in personalized brain nutrition and cognitive support. *Nutrients, 13*(10), 3572. DOI: 10.3390/nu13103572 PMID: 34684573

Zhou, S., Yu, L., Zhang, Z., & Zhang, H. (2016). Gut microbiota and polyphenols: From absorption to bioactivity. *Food Research International, 89*, 253–261. DOI: 10.1016/j.foodres.2016.09.029

Chapter 5
Gut–Brain Axis and Microbiome Modulation by Dietary Bioactives

Sahar Fazal
https://orcid.org/0009-0007-5839-3002
Capital University of Science and Technology, Islamabad, Pakistan

Shah Gull
Capital University of Science and Technology, Islamabad, Pakistan

Saher Sajjad
Capital University of Science and Technology, Islamabad, Pakistan

Nazmeen Ismail
Capital University of Science and Technology, Islamabad, Pakistan

Alishba Sarwat
Capital University of Science and Technology, Islamabad, Pakistan

Nasreen Noor
Capital University of Science and Technology, Islamabad, Pakistan

Hafsa Arshad
https://orcid.org/0009-0001-3162-0646
Capital University of Science and Technology, Islamabad, Pakistan

Rao Mubashar Ali
https://orcid.org/0009-0009-9904-4911
Capital University of Science and Technology, Islamabad, Pakistan

Aqsa Qasim Zafar
https://orcid.org/0009-0000-3680-8210
Capital University of Science and Technology, Islamabad, Pakistan

Anushay Shahab
https://orcid.org/0009-0006-7004-9880
Capital University of Science and Technology, Islamabad, Pakistan

Emaan Fatima
Capital University of Science and Technology, Rawalpindi, Pakistan

ABSTRACT

The gut microbiota is a key component of the complex, bidirectional communication system that connects the central nervous system with the gastrointestinal tract. This chapter examines the way in which dietary bioactive, including short-chain fatty acids, polyphenols, probiotics, and prebiotics, alter the gut microbiota and have an impact on the gut-brain axis's neurological, hormonal, immunological, and

DOI: 10.4018/979-8-3373-4139-2.ch005

metabolic pathways. Vagal and endocrine signaling immunological modulation, and gut barrier integrity maintenance are important mechanisms. The chapter also covers the impact of dysbiosis on mental and neurological illnesses, the role of microbial metabolites in brain function, and the therapeutic potential of certain dietary strategies. The standardization, bioavailability, and personalization of these analyses continue to present difficulties despite encouraging preclinical and clinical results. Future research should concentrate on customized diet based on microbiota profiles and comprehensive clinical trials to verify efficacy and safety.

DEFINITION AND OVERVIEW OF GUT BRAIN AXIS

The gut-brain axis (GBA) is the connection between the human brain, with its different cognitive and emotional functions, and the gastrointestinal (GI) system, that involves the enteric nervous system and the varied microbiome that lives in the gut lumen (Diaz et al. 2011). GBA functions in a bidirectional way by combining fast-acting neuronal and delayed immune-mediated processes, with feedback from both systems (Bruce et al.'2018). In addition to their direct impacts on brain pathways and immunological regulation, gut microbiota plays an important role in the creation of several vitamins, cofactors, and minerals necessary for good health and metabolism (Patterson E et al.'2014). The gut-brain axis (GBA) existed for over three decades. The gastrointestinal motor and sensory components convey information to the central nervous system (CNS), and the GBA is defined as the return reaction to the gut (Jones et al.' 2006). Nutrition influences microbial colonization and gut metabolites, which can impact brain development and function via neuronal, immunological, and endocrine mechanisms (Deverman et al.'2009). The brain is the major component of the GBA, with connections to the cerebral cortex, limbic system, hypothalamic-pituitary axis, and brain system. The GBA regulates intestinal immune cells through norepinephrine and neuropeptide messengers, such as vasoactive intestinal peptide, which modulate the activity of dendritic cells and T cells found throughout the intestinal wall and in secondary lymphoid organs such as Peyer's patches. In terms of the gut microbiome, the immune system's potential to regulate brain development has been established (Mayer, 2011).

The gastrointestinal tract's positive influence on brain function became recognized since the middle of the nineteenth century by pioneering researchers like as Claude Bernard, Ivan Pavlov, William Beaumont, William James, and Carl Lange. Even Charles Darwin recognized the importance of this interaction in his classic The Expression of the Emotions in Man and Animals (1872), in which he wrote: "The manner in which the secretions of the alimentary canal and of certain other organs are affected by strong emotions, is another excellent instance of the direct action of the sensorium on these organs, independently of the will or of any serviceable associated habit." (Al Omran Y et al.'2014).

GUT-BRAIN AXIS STRUCTURE AND FUNCTION

The gut–brain axis (GBA) represents a dynamic, bidirectional communication network between the crucial nervous machine (CNS) and the gastrointestinal (GI) tract. This complicated device coordinates neural, hormonal, immune, and metabolic indicators, permitting the mind and intestine to influence each other's features. Central to this communication are the enteric anxious gadget (ENS) and the vagus nerve, both of which play critical roles in transmitting signals among the intestine and the mind. The ENS, regularly referred to as the "2nd mind," operates largely independently in the partitions of the GI tract but additionally interacts with the CNS. In parallel, the vagus nerve acts as a primary neural dual carriageway, linking the intestine and brain through afferent and efferent pathways. These components form the inspiration of gut–brain interplay and could be mentioned in detail to highlight their structure, functions, and relevance in maintaining physiological and emotional homeostasis (Carabotti *et al.*, 2015).

Enteric Nervous System (ENS): An Autonomous Integrator of Gut Function and Microbial Signals

The enteric nervous gadget (ENS) is a dense, specialized network of neurons and glial cells embedded within the gastrointestinal (GI) tract. Functioning in large part independently of the valuable fearful gadget (CNS), the ENS orchestrates important digestive features which include motility, secretion, nutrient absorption, and blood float regulation. Its neuronal population, anticipated at over 600 million, competitors that of the spinal twine, underlining its complexity and useful autonomy. Derived from migrating neural crest cells for the duration of embryogenesis, the ENS forms interconnected plexuses dispensed throughout the GI tract: the myenteric (Auerbach's) plexus and the submucosal (Meissner's) plexus (Furness, 2013). The myenteric plexus, situated between the longitudinal and circular muscle layers, mainly modulates peristaltic moves. The submucosal plexus, located towards the mucosal floor, regulates fluid secretion and local blood drift. In large mammals, the submucosal plexus can also similarly differentiate into wonderful sublayers. Although able to self-reliant reflexes, the ENS keeps consistent communication with the CNS thru sympathetic and parasympathetic pathways of the autonomic anxious system (ANS), bearing in mind bidirectional integration of visceral indicators (Makris *et al.*, 2021). ENS neurons are categorized both functionally and morphologically. Functionally, three major types exist:

- **Intrinsic primary afferent neurons (IPANs)**: Serve as sensory detectors, perceiving mechanical and chemical stimuli from the gut lumen and initiating local reflexes
- **Interneurons**: Coordinate signal transmission between sensory and motor pathways within and across the plexuses.
- **Motor neurons**: Control smooth muscle contraction, glandular secretion, and vasodilation.

From a structural and electrophysiological perspective, ENS neurons are generally classified as:

- **AH neurons (Dogiel Type II)**: These are large, slow-adapting neurons with afterhyperpolarization properties, typically involved in intrinsic sensory signaling (Brookes *et al.*, 2013).
- **S neurons (Dogiel Type I)**: Fast-conducting and often responsible for motor or interneuron activity, characterized by smaller, angular cell bodies (Nurgali *et al.*, 2004).

The enteric nervous system, while physically insulated from the gut lumen by epithelial and mucus barriers, is functionally responsive to microbial cues. Metabolites produced by gut microbiota such as short-chain fatty acids (SCFAs), polysaccharide A, and microbial indoles can reach ENS neurons either directly or via intermediary cells like enteroendocrine cells (EECs), modulating excitability and neurotransmitter expression. Intrinsic primary afferent neurons (IPANs) in particular are sensitive to microbial-derived signals. *Ex vivo* studies have shown that *Lactobacillus rhamnosus JB-1* and *Bacteroides fragilis* can directly or indirectly activate IPANs, potentially through EC cell-mediated release of gut peptides. Evidence from germ-free mice suggests that microbial exposure is essential for normal IPAN excitability. Among microbial metabolites, **butyrate** a key Short-Chain Fatty Acid has shown the most direct neuro modulatory effects. It can depolarize AH neurons in the myenteric plexus, enhance neuronal excitability, and even influence neurotransmitter biosynthesis by upregulating choline acetyltransferase through FFA3 receptor activation. Butyrate absorption occurs via transporters like SMCT-1, linking epithelial nutrient transport to ENS signaling. However, species- and region-specific responses have been observed, highlighting the context-dependent effects of SCFAs (Brookes *et al.*, 2013). In parallel, the serotonergic system plays a central role in ENS signaling. The gut houses extensive 5-HT (serotonin) activity, with several receptor subtypes including excitatory 5-HT3 receptors located on intrinsic and extrinsic nerve terminals. Notably, enterochromaffin (EC) cells form synapse-like junctions with ENS neurons, allowing for rapid activation in response to microbial or luminal stimuli. Serotonin release influenced by microbiota is a key mechanism through which microbial composition shapes neural activity in the gut brain axis. In addition to neurons, enteric glial cells contribute to intestinal homeostasis. Although less well-characterized, this glia engages in bidirectional communication with neurons and epithelial cells, modulating gut motility, barrier integrity, and inflammation (Nurgali *et al.*, 2014).

Visceral Afferent Signaling: Bridging the ENS and CNS

The ENS communicates significantly with the imperative apprehensive machine (CNS) thru a community of visceral afferent pathways which might be vital to the microbiota–gut–mind axis. These neural connections are mostly mediated by way of branches of the autonomic apprehensive gadget (ANS), which include the vagal, pelvic, and splanchnic nerves, each playing a specialised position in relaying intestine-derived signals to the brain. Non-painful indicators, together with gut motility and satiety, are typically transmitted via vagal and pelvic afferents, while nociceptive or discomfort-associated inputs are carried by using splanchnic pathways. Visceral afferents come across loads of signals in the gut surroundings ranging from microbial metabolites and intestine peptides to mechanical stretch thru specialised sensory endings embedded within the intestine wall, especially within the subepithelial layer (Blackshaw *et al.*, 2007). These afferent fibers originate within the nodose ganglia (vagal afferents) and dorsal root ganglia (spinal afferents), projecting to the brainstem and spinal twine, respectively. Notably, those afferents regularly branch into the ENS, permitting bidirectional interactions wherein CNS hobby can modulate ENS characteristic and vice versa (Berthoud *et al.*, 2004).

Emerging evidence indicates that intestine microbiota and their metabolites can have an effect on ENS and CNS signalling through those afferent fibres. Short-chain fatty acids (SCFAs) which include butyrate and acetate, produced by using microbial fermentation, had been shown to modulate vagal nerve pastime. For instance, butyrate turns on vagal afferents in the rat jejunum and may exert its consequences through loose fatty acid receptor 3 (FFA3) expressed on sensory neurons. Similarly, oral management of butyrate has been related to reduced neuronal pastime within the brainstem and hypothalamus, suggesting a satiety-modulating function via vagal signaling. Other microbial metabolites like acetate and propionate have also verified roles in CNS electricity regulation and urge for food manipulate. Acetate complements parasympathetic hobby and hormone launch, at the same time as propionate stimulates intestinal gluconeogenesis through vagal afferent input to the dorsal vagal complicated. Indole, another metabolite, activates vagal afferents and alters brain hobby, highlighting the vagus nerve as a key conversation channel among gut microbiota and better brain features (Li *et al.*, 2018; Kim *et al.*, 2013).

Vagus Nerve: A Bidirectional Mediator in the Microbiota–Gut–Brain Axis

The vagus nerve (VN) plays a pivotal role in gut–brain communication, functioning as a critical bi-directional pathway that relays microbial, hormonal, and neural signals between the gastrointestinal (GI) tract and the central nervous system (CNS). Approximately 80% of its fibers are afferent, transmitting sensory information from the gut to the brain, while 20% are efferent, mediating brain-to-gut regulatory feedback. Although vagal afferents do not directly penetrate the epithelial layer, they can detect microbial activity via indirect sensing of bacterial metabolites and signaling molecules. This transduction often occurs through intermediary epithelial cells especially enteroendocrine cells (EECs) that relay luminal cues through serotonin (5-HT), peptide YY (PYY), cholecystokinin (CCK), and glucagon-like peptide-1 (GLP-1), which activate vagal receptors such as 5-HT_3R and CCKR. EECs themselves recognize microbial ligands through toll-like receptors (TLRs) and sense short-chain fatty acids (SCFAs) via G-protein coupled receptors like FFA2 and FFA3 (Li *et al.*, 2018; Kim *et al.*, 2013). These mechanisms place EECs at the centre of microbiota-to-vagus signaling. Evidence also supports direct sensing of microbial components by vagal afferents. SCFAs like butyrate and isobutyrate can stimulate vagal fibers, either via epithelial intermediaries or direct depolarization of nerve terminals a component of gram-negative bacteria, has been shown to activate vagal neurons at the nodose ganglia, implicating TLR4-mediated signaling. Gut microbiota-derived molecules also modulate vagal activity and brain signaling. Administration of probiotics such as *Lactobacillus johnsonii* and *L. rhamnosus* JB1 enhances vagal activity and alters neurotransmitter expression in the brain. Notably, *L. rhamnosus* treatment increased GABA levels in the cingulate cortex and decreased levels in stress-associated brain regions. These changes were abolished in vagotomized animals, confirming vagus-dependency. Similarly, *Bifidobacterium longum* showed anxiolytic effects during gut inflammation, but these effects required intact vagal innervation (Berthoud *et al.*, 2004).

Vagal Efferents: Anti-Inflammatory Reflex and Barrier Protection

Beyond afferent sensing, vagal efferents exert regulatory control over irritation and intestinal barrier feature. The cholinergic anti-inflammatory pathway (CAP) involves ACh launch from efferent vagal terminals that suppresses seasoned-inflammatory cytokine (e.G., TNF-α) manufacturing via α7-nicotinic acetylcholine receptors (α7nAChRs) on macrophages (Wang *et al.*, 2007). Parallel sympathetic pathways might also supplement this effect, along with splanchnic innervation of the spleen.

Vagal hobby additionally modulates epithelial permeability. Vagal nerve stimulation (VNS) will increase tight junction proteins like occludin and zonula occludens-1 (ZO-1), decreasing intestinal permeability in models of burn damage, hemorrhagic shock, and endotoxemia . This shielding effect is frequently depending on α7nAChR signaling and is mimicked with the aid of marketers like nicotine. Interestingly, vagal alerts may also affect enteric glial cells, which make a contribution to tight junction protein expression and barrier protection thru S-nitrosoglutathione secretion. Although recent findings advise enteric glia may not be important for epithelial maintenance their interaction with vagal and ENS networks can also nevertheless modulate mucosal integrity. Brain-to-gut signaling through the vagus also includes effects on macrophage polarization. Central vagal stimulation increases anti-inflammatory M2 macrophages and suppresses M1 pro-inflammatory populations. Additionally, nutrient-derived signals like CCK can engage vagal pathways to regulate immune tone and barrier function, effects abolished by vagotomy or receptor antagonism (Levy *et al.*, 2013).

Endocrine Signaling in the Microbiota–Gut–Brain Axis

Enteroendocrine cells (EECs) form a specialized subset of the intestinal epithelium, constituting approximately 1% of epithelial cells yet representing the largest endocrine system in mammals. Interspersed among absorptive enterocytes, goblet cells, and Paneth cells, EECs exhibit region-specific distribution along the gastrointestinal (GI) tract, reflecting their diverse hormonal outputs. Contrary to earlier hypotheses suggesting neural crest origin, EECs arise from the same pluripotent intestinal stem cells that give rise to other epithelial lineages. EECs play a pivotal sensory function in gut–brain signaling. Positioned at the interface between luminal contents and underlying neural circuits, they stumble on microbial metabolites and dietary additives through specialized receptors and release bioactive peptides and amines. These secretions modulate the interest of intrinsic number one afferent neurons (IPANs), enteric circuits, and extrinsic sensory fibers, regularly functioning in a synapse-like way. Notably, EECs percentage chemosensory equipment with lingual flavor cells, along with expression of odorant and G-protein–coupled receptors (Rao *et al.*, 2017). Functions of enteroendocrine cells are enlisted in Table 1.

Table 1. Functional Traits of Enteroendocrine Cell (EEC) Subtypes and Their Functions in Gut–Brain Axis Communication (Rao et al., 2017)

EEC Subtype	Key Hormones / Neurotransmitters	Receptors Involved	Function / Role in Gut–Brain Axis
L-cells	GLP-1, PYY	FFA2, FFA3 (on L-cells); GLP-1R, Y1 (on enteric neurons)	Appetite regulation, insulin secretion, modulation of gut motility, ENS-to-CNS signalling
EC cells	Serotonin (5-HT)	5-HT receptors on IPANs; TLRs; GABA receptors	Regulate peristalsis and secretion; mediate microbiota-induced serotonin signaling to CNS
EC cells	5-HT (via GABA interaction)	GABAA, GABAB	Modulate 5-HT release and ENS activity in response to microbial GABA
L-cells	GLP-1, PYY (via bile acids)	TGR5 (on L-cells and enteric neurons), FXR	Mediate bile acid–induced hormonal responses; link gut microbiota metabolism to host energy regulation
Various EECs	Cholecystokinin (CCK)	TLR4	Triggered by LPS via TLR activation; influence ENS signalling
L-cells	GLP-1, PYY (via SCFAs and odorants)	Olfr78 / OR51E2 (odorant receptors)	SCFA-sensing and chemical monitoring of luminal environment
Various EECs	Diverse gut peptides	TLR1, TLR2, TLR4	Detect microbial components, modulate immune–neural signalling

Hormone-Microbiota Interactions: SCFAs, Bile Acids, and Neuromodulators

Among the EEC subtypes, L-cells and enterochromaffin (EC) cells are the most prominent. L-cells secrete glucagon-like peptide-1 (GLP-1) and peptide YY (PYY), both of which regulate appetite, insulin secretion, and GI motility (Mace *et al.*, 2015). EC cells, by contrast, produce approximately 95% of the body's serotonin (5-HT), which modulates GI reflexes and motility. Gut microbiota significantly influence EEC function. Short-chain fatty acids (SCFAs) notably butyrate, acetate, and propionate induce the release of GLP-1 and PYY from L-cells via activation of FFA2 and FFA3 receptors. GLP-1 receptor (GLP-1R)-positive neurons have been localized near L-cells within the submucosal plexus, suggesting that gut peptides may exert paracrine effects on enteric neurons before systemic degradation by DPP-4. Likewise, PYY signals through Y1 receptors expressed on enteric neurons further reinforcing the role of gut hormones in local circuit modulation. Bile acids, transformed by the microbiota into secondary metabolites, additionally affect EEC hobby. The G-protein coupled bile acid receptor TGR5, expressed on L-cells and enteric neurons, mediates GLP-1 and PYY launch in response to those bile acids (. FXR, a nuclear receptor found in enterocytes and hepatocytes, regulates genes worried in bile acid and electricity metabolism, although its direct position in intestine–brain communication remains much less clear. Other bile acid receptors including PXR, CAR, and VDR may also make a contribution, however their roles aren't completely elucidated (Richards *et al.*, 2014).

EECs combine extra microbial cues through a number of molecular sensors. Odorant receptors, inclusive of Olfr78 and its human ortholog OR51E2, are expressed in colonic L-cells and activated by way of SCFAs like acetate and propionate. These receptors may function chemical sentinels, transducing luminal cues into hormonal alerts, though their useful roles within the intestine–mind axis remain underneath research (Fleischer *et al.*, 2015).

Moreover, GABA produced with the aid of the microbiota can modulate EEC characteristic. EC cells explicit GABA_B receptors alongside the GI tract, and activation of these receptors modulates 5-HT release and subsequent ENS hobby. While GABA-B receptor-mediated inhibition is tetrodotoxin (TTX)-insensitive, GABA-A receptor activation promotes muscarinic-dependent, TTX-sensitive five-HT launch, indicating wonderful regulatory outcomes on gut neurotransmission (Breton *et al.*, 2016).

Finally, EECs explicit toll-like receptors (TLRs), which includes TLR1, TLR2, and TLR4, which locate microbial-associated molecular patterns like lipopolysaccharides (LPS). TLR activation on EECs can cause peptide launch, such as cholecystokinin or GLP-1, serving as some other path for microbial–neural communication. This receptor-mediated sensing positions EECs as integrative hubs, able to relaying microbial statistics to IPANs and extrinsic afferents.

Immune Signaling in the Gut-Brain Axis: Microbial Modulation of Mucosal and Neural Immunity

The gastrointestinal (GI) tract represents one of the body's maximum dynamic immunological environments, where the intestine microbiota plays a pivotal position in shaping host immune responses and keeping mucosal homeostasis. The enteric immune system, specifically the intestine-associated lymphoid tissue (GALT), continuously video display units and adapts to luminal contents via exceedingly specialised and context-based signaling mechanisms. This immunological pastime is deeply included into the microbiota–gut–mind axis, influencing now not best local defenses however also systemic and important worried system (CNS) responses. In a nation of microbial balance (eubiosis), intestinal epithelial cells and resident immune populations work synergistically to keep barrier integrity and prevent useless inflammation (Luu *et al.*, 2017). Specialized epithelial cells produce anti-inflammatory cytokines consisting of TGF-β, which, alongside microbial metabolites like quick-chain fatty acids (SCFAs), direct dendritic cells (DCs) inside the lamina propria to adopt a tolerogenic phenotype. These $CD103^+$ DCs migrate to mesenteric lymph nodes where they facilitate the differentiation of naive T cells into regulatory T cells (Tregs), which in flip produce IL-10 and enhance immune tolerance to dietary and microbial antigens (Fleischer *et al.*, 2015).

Simultaneously, B cells activated inside those lymphoid regions go through magnificence-transfer recombination to supply IgA, which is secreted into the gut lumen to neutralize microbial components and prevent bacterial adherence and translocation. Notably, SCFAs also beautify retinoic acid (RA) synthesis in DCs, further selling T_{reg} induction and oral tolerance. This sensitive immunological equilibrium ensures selective reactivity toward pathogens while retaining tolerance to symbionts and food-derived antigens.

Microbial Triggers of Innate Immune Activation and Inflammation

Disruption of microbial balance, or dysbiosis, provokes a shift in immune popularity styles. Pathogen-related molecular styles (PAMPs), inclusive of lipopolysaccharide (LPS), are detected with the aid of host pattern reputation receptors (PRRs), together with toll-like receptors (TLRs) and NOD-like receptors. Activation of these receptors initiates intracellular cascades related to NF-κB and results in the transcription of proinflammatory cytokines like IL-1β, TNF-α, and IFN-γ. These immune mediators compromise epithelial integrity, growth intestine permeability, and recruit additional immune cells to the mucosa. Environmental factors together with high-fats diets, antibiotics, strain, or mucosal damage may also in addition exacerbate epithelial dysfunction and permit microbial translocation, contributing to systemic inflammation . Phagocytic cells along with macrophages and DCs try to incorporate microbial invasion, however the systemic spillover of inflammatory cytokines and microbial fragments (e.G., LPS, flagellin) can also set off neuroimmune responses in the CNS. Interestingly, PRRs consisting of TLRs can distinguish between microbe-related molecular patterns (MAMPs) of commensals and PAMPs of pathogens, contributing to immune tolerance (Hooper *et al.*, 2012; Zhang *et al.*, 2021).

Immune-to-Brain Communication and Neurotransmitter Interplay

Immune-derived alerts can interface with neural circuits via a couple of pathways. Cytokines launched all through mucosal inflammation can go the blood–brain barrier (BBB) or sign via afferent vagal fibers, modulating the hypothalamic–pituitary–adrenal (HPA) axis and CNS characteristic. Conversely, pressure-induced cortisol secretion can impair epithelial barrier function and shift intestine immune responses. Moreover, intestine microbes themselves can synthesize neuroactive substances such as serotonin, dopamine, norepinephrine, and gamma-aminobutyric acid (GABA), influencing each immune and neuronal pastime. Serotonin, predominantly produced by using enterochromaffin (EC) cells of the intestine epithelium, modulates GI motility, immune mobile signaling, and neuroimmune verbal exchange. Bacteria such as Candida, Streptococcus, Escherichia, and Enterococcus immediately contribute to serotonin production, whilst microbial stimulation of TPH1 in addition increases local synthesis (Makris *et al.*, 2021). Other taxa, which includes Bacillus, Klebsiella, Morganella, and Proteus, synthesize catecholamines like dopamine and norepinephrine, that can have an effect on immune cellular recruitment and inflammation. GABA, a microbial metabolite in large part produced through Bifidobacterium and Lactobacillus species, exerts immunosuppressive consequences and modulates cytokine expression in dendritic and T cells. Together, these interactions shape a dynamic immune neural community through which the microbiota regulates now not best mucosal immunity however additionally mind feature. Disruptions in this balance can also underlie quite a number neuroimmune and inflammatory disorders (Sharma *et al.*, 2018).

MICROBIAL METABOLITES

The human body contains many commensal microorganisms, collectively known as the microbiota. These microorganisms live on the skin and in various mucosal areas, such as the nasal, oral, pulmonary, and vaginal cavities. However, they are found throughout the body, with the majority residing in the stomach and intestines, where millions of bacteria, viruses, and fungi exist in a symbiotic relationship with the host, often providing mutual benefits. Over the past 20 years, gut microbiota has been studied more closely, and its essential roles are becoming increasingly clear. It helps protect against pathogens, regulates immune systems, endocrine, and metabolic functions, and influences drug absorption and metabolism. It is now understood that the microbiota's role is not limited to the gastrointestinal (GI) tract. It also communicates with the central nervous system (CNS) via the stomach and intestines (Silva *et al.*, 2020).

Research has shown that gut microbiota can significantly influence brain function, giving rise to the idea that gut microbes affect mood and mental processes are a major focus of current research. Microorganisms can influence the CNS bidirectionally, primarily through the vagus nerve. This interaction impacts the immune system, stress hormones, and the metabolism of brain-related nutrients such as tryptophan. In addition, gut microbes produce several neuroactive chemicals and beneficial substances like short-chain fatty acids (SCFAs), which influence brain function. The main SCFAs acetate, propionate, and butyrate are produced in the colon during bacterial fermentation of dietary fibers and resistant starch. Besides supporting energy metabolism and regulating T regulatory (Treg) cells, SCFAs appear to have important effects on various organs, including the brain. Studies in both humans and animals have linked gut microbiota imbalances to neurological and behavioral disorders, such as depression, Alzheimer's disease, Parkinson's disease, and autism.

Short Chain Fatty Acids

Short-chain fatty acids (SCFAs) are among the most extensively researched metabolites derived from gut microbes. These fatty acids contain 1 to 6 carbon atoms. They are a special type of compound produced by gut bacteria through the fermentation of complex fibres that the human body cannot digest on its own. More than 95% of SCFAs are produced in the colon. Foods such as butter and other dairy products also contain SCFAs in high concentrations (O'Riordan *et al.*, 2022). These fatty acids are linked to various physiological functions in the host, including gastrointestinal activity, blood pressure regulation, circadian rhythm maintenance, and both innate and adaptive immune responses.

In the brain, SCFAs influence the maturation of microglia (the brain's immune cells) and affect astrocyte gene expression in a sex-specific manner. Interestingly, changes in SCFA levels in fecal matter have been observed in conditions where neural behavior and physiology are altered. Lower SCFA levels have been found in disorders such as anorexia nervosa and Parkinson's disease, while higher SCFA levels are seen in obesity, in children exposed to long-term psychosocial stress, and in individuals with autism spectrum disorder (Pluznick, 2017). Recent studies have shown that children with autism have reduced levels of acetate and butyrate in their stool. In animal studies, reduced SCFA levels have been linked to Alzheimer's disease and chronic stress. Additionally, mice with depression showed altered SCFA levels compared to healthy controls, and these changes were associated with specific gut bacteria. When gut bacteria break down dietary fiber from food, they primarily produce three main short-chain fatty acids (SCFAs): acetate, propionate, and butyrate, which together account for about 95% of all SCFAs. These

SCFAs serve as a crucial energy source for colon cells (Pluznick, 2017). In addition to this role, they help regulate cholesterol production and how the body utilizes glucose for energy. SCFAs also support the function of the intestinal barrier through several mechanisms. They help reduce the growth of pathogens, lower gut inflammation, and influence the structure of tight junctions, which are essential for maintaining the integrity of the gut lining. SCFAs promote the production of proteins that strengthen and maintain proper gut barrier function (Fu *et al.*, 2023).

The body can detect SCFAs using specific G protein-coupled receptors (GPCRs), such as:

- GPR41 (FFAR3)
- GPR42
- GPR43 (FFAR2)
- GPR109A (HCAR2)
- GPR164 (OR51E1)
- OR51E2

These receptors play key roles in reducing inflammation. For instance, they inhibit the activation of NF-κB, a protein that triggers immune and inflammatory responses in gut cells. Among these, FFAR2 is involved in a signaling pathway that includes the protein β-arrestin-2, which helps block NF-κB activation and reduce inflammation. Additionally, GPR41 and GPR43 contribute to immune surveillance in the intestinal mucosa by helping the body detect and respond to microbial threats. GPR109A, known for its tumor-suppressing properties, also inhibits NF-κB activity and supports anti-inflammatory responses (Parker *et al.*, 2020).In ulcerative colitis patients, a decrease in butyric acid-producing bacteria and reduced expression of the MCT1 transporter in intestinal tissue have been observed. This may impair the uptake of butyric acid and reduce its anti-inflammatory effect, such as the inhibition of IL-8 (interleukin-8) production via GPR109A signaling (Ghosh *et al.*, 2021). GPR43 also plays a role in regulating immune cells. Inflammatory complexes called NLRP3 inflammasomes can be activated by signals through GPR41 and GPR43, leading to the release of IL-1β and IL-18, which are cytokines that modulate inflammation. Neutrophils, a type of immune cell particularly rich in FFAR2 expression, are implicated in the development of inflammatory bowel disease (IBD) and contribute to its progression (Fu *et al.*, 2023).

SCFA Production by Microbes in the Gut

Acetate ($C_2H_4O_2$)

Acetate is the simplest short-chain fatty acid (SCFA). It has two carbon atoms: one is a methyl group ($-CH_3$), and the other is a carboxyl group (–COOH). This SCFA is primarily produced from pyruvate through the acetyl-CoA pathway (Fusco *et al.*, 2023). Acetate is mainly produced by gut bacteria such as *Akkermansia muciniphila*, *Bacteroides*, *Bifidobacterium*, *Prevotella*, and *Ruminococcus*. It can also be synthesized through another route called the Wood-Ljungdahl pathway, used by microbes like *Blautia hydrogenotrophica*, *Clostridium*, and *Streptococcus*.

Propionate ($C_3H_6O_2$)

Propionate contains three carbon atoms two in the chain and one in the carboxyl group. It is slightly larger than the acetate. The main pathway for its formation is the succinate pathway, involving microbes such as *Bacteroides* species, *Phascolarctobacterium succinatutens*, *Dialister* species, and *Veillonella* species (Ghosh *et al.*, 2021).

Butyrate ($C_4H_8O_2$)

Butyrate consists of four carbon atoms three in the chain and one in the carboxyl group. It is more hydrophobic than both acetate and propionate (Chen *et al.*, 2021). It is produced via two main pathways:

- One pathway involves bacteria such as *Coprococcus comes* and *Coprococcus eutactus*.
- The other is the butyryl-CoA:acetate CoA-transferase route, used by microbes like *Anaerostipes* species, *Coprococcus catus*, *Eubacterium rectale*, *Eubacterium hallii*, *Faecalibacterium prausnitzii*, and *Roseburia* species. These microbes contribute significantly to butyrate production in the human gut .

ROLE OF PROBIOTICS AND PREBIOTICS IN MICROBIOME MODULATION

Probiotics are the bacteria which are commonly consumed and refer to as the health beneficial microorganisms. Because of this property of the probiotics or the health beneficial microorganisms the different food authorities define the probiotics in different ways: Probiotics are defined by the International Scientific Association for Probiotics and Prebiotics as "live microorganisms that, when given in adequate amounts, provide a health benefit on the host" primarily by influencing the intestinal flora and improving immune responses (Montalto et al., 2002) (Kalkan et al., 2018). Composed primarily of bacteria but also including yeasts, these microorganisms are found in some fermented foods (e.g., yoghurt), are added to some food products, and are sold as dietary supplements; however, not all foods and dietary supplements labelled as probiotics have been shown to have health benefits. Probiotics' ability to withstand severe gastrointestinal conditions, like bile and acid resistance, as well as their potential for transient colonization and attachment to intestinal mucosa, are among their key attributes. Effective probiotics should also be safe for human consumption, generate antimicrobial compounds, and retain their advantageous qualities throughout storage and processing (Dao et al., 2024). Lactic acid bacteria and yeasts such as *Saccharomyces cerevisiae var. boulardii* are among the strains that have been identified for their probiotic potential. They have been shown to have a variety of health benefits, including enhancing digestion, preventing infections, and potentially lowering the risk of cancer (Staniszewski & Kordowska-Wiater, 2021). Probiotic bacteria, especially Lactobacillus strains, colonize human mucosa, including the oral cavity and digestive system, by producing biofilms that are stabilized by surface structures such proteins, lipoteichoic acids, and capsular polysaccharides. During this colonization process, specialized binding to host receptors is followed by early physical interactions (van der Waals forces) (Molska & Regula, 2019). Strains of Bifidobacterium and Lactobacillus attach to intestinal epithelial cells and create biofilms to colonize the gut (Dahiya & Nigam, 2022). Probiotics can compete with pathogens for mucosal binding sites thanks to surface features such sticky proteins and capsular polysaccharides.

Mechanisms of Gut-Brain Communication Influenced by Probiotics

Probiotics improve mental health and cognitive abilities by influencing gut-brain interactions through a variety of methods. Probiotics can alter neurotransmitter synthesis, immunological responses, and hormonal signaling in the gut-brain axis (GBA), a bidirectional communication pathway that affects the central nervous system (CNS) and emotional regulation (Alzubide & Alhalafi, 2024; Chowdhury, 2024; Misra et al., 2023). This axis is specifically targeted by psych biotics, a subtype of probiotics, to enhance mood, cognitive function, and lessen anxiety and depression (Chowdhury, 2024). Probiotics have been shown to improve hippocampus function, which is important for memory and learning, by controlling neurotrophins and neurotransmitters and preserving the balance of inflammation (Salami & Soheili, 2022). Additionally, probiotics have demonstrated promise in the treatment of a number of mental health conditions, such as anxiety and depression, underscoring their potential as therapeutic agents in mental health interventions (Mishra et al., 2023).

PREBIOTICS

Prebiotics are non-digestible food components that selectively stimulate the growth and activity of beneficial gut microbes, promoting host health (Gibson et al., 2017). Two well-studied prebiotics, inulin and fructooligosaccharides (FOS), are fructans that resist digestion in the upper gastrointestinal tract and are fermented by gut microbiota in the colon, producing short-chain fatty acids (SCFAs) like butyrate, acetate, and propionate These SCFAs lower colonic pH, inhibit pathogens, and provide energy to colonocytes, contributing to gut and systemic health. Inulin and FOS are naturally found in foods such as chicory root, garlic, onions, asparagus, and bananas, and are also used in functional foods and supplements (Shoaib et al., 2016). Their distinct structural properties, particularly chain length, influence their fermentation rate and microbial selectivity, making them effective in targeting beneficial bacteria like Bifidobacteria and Lactobacilli (Vandeputte et al., 2021).

MOLECULAR AND CELLULAR PATHWAYS UNDERLYING MODULATION BY DIETARY BIOACTIVES

Gut Barrier Integrity and Tight Junction Regulation

Dietary bioactives, including fermentable fibers, polyphenols, polysaccharides, and bioactive peptides, preserve gut barrier integrity through molecular modulation:

1. **Short-Chain Fatty Acids (SCFAs):** Produced from fiber fermentation, acetate, propionate, and butyrate are key regulators. Butyrate activates AMP-activated protein kinase (AMPK), enhancing transcription of *claudin1* and facilitating ZO1 repositioning, thus tightening junctions and reducing permeability (Peng et al., 2022; Ma et al., 2021). SCFAs also signal via G-protein–coupled receptors (GPR41/FFAR3, GPR43/FFAR2, GPR109A), suppressing NFκB activation and reinforcing epithelial resilience (Ma et al., 2021).
2. **Polysaccharides and Polyphenols:** Prebiotic dietary polysaccharides boost beneficial SCFA-producing bacteria, indirectly reinforcing the barrier. Polyphenols like caffeoyltyramine protect against TNFα-

induced barrier loss by enhancing *occludin* and *claudin* expression and reducing oxidative stress (Timewell et al., 2023).

3. **Bioactive Peptides:** Milk- and soy-derived peptides modulate the microbiome, enhancing beneficial microbial populations that support barrier repair and tight junction protein expression.
4. **High-Fat Diet (HFD) Protection:** While HFD can damage junctions via LPS and bile acid dysregulation, specific microbiota-modified bile acids activate FXR and GPBAR1, countering leakiness (Lee et al., 2021).

Effect of different bioactive compounds on barrier and their mechanism is given in table 2.

Table 2. Dietary Bioactives and Barrier Effects

Bioactive Type	Examples	Effects on Barrier	Mechanism
Butyrate (SCFA)	Fiber fermentation	↑ Occludin, claudin-1, ZO-1	AMPK activation; HDAC inhibition; GPCR signaling (FFARs)
Acetate/Propionate	SCFAs from complex carbs	↓ Permeability; ↑ IL-18	NLRP3/IL-18 axis; GPCR pathways (2,3)
Dietary Polysaccharides	Lycium barbarum polysaccharides	↑ SCFAs; microbial balance	Prebiotic → increased barrier support
Polyphenols	Caffeoyltyramine	Barrier restoration	Antioxidant; NF-κB inhibition (5)
Bile Acids*	Secondary bile acids	Jnk/FXR/GPBAR1-mediated defense	Microbiota transformation protects tight junctions (6)

These convergent pathways help maintain mucosal integrity, reduce endotoxemia, and preserve gut homeostasis. Dietary bioactives modulate gut immunity at cellular and molecular levels and include SCFAs via GPCRs and epigenetic pathways. SCFAs activate FFAR2/3 on epithelial and immune cells, leading to IL10 production, downregulation of TNFα/IL6 secretion, and histone acetylation through HDAC inhibition (Wenzel et al., 2022). In vitro, SCFA mixtures reduced monocyte TNFα and IL1β in LPS-stimulated cultures.

T-cell and Macrophage Reprogramming: Butyrate promotes regulatory T-cell differentiation via Foxp3 acetylation, while reprogramming macrophages from pro-inflammatory (M1) to regulatory phenotypes (Chang et al., 2021).

Polyphenols and Peptides: These compounds attenuate NFκB and JAK/STAT activation in colitis models, reducing mucosal inflammation and oxidative damage.

Translational Evidence: Dietary fiber supplementation in critically ill humans was shown to reduce intestinal permeability (lactulose/rhamnose ratio) and systemic inflammation (CRP), highlighting translational relevance (Liu et al., 2022).

SCIENTIFIC AND CLINICAL CHALLENGES IN MODULATING THE GUT–BRAIN AXIS WITH DIETARY BIOACTIVES

Efforts to influence the gut–brain axis (GBA) through dietary bioactives such as polyphenols, probiotics, and phytochemicals face a range of scientific and clinical challenges. One of the primary issues is low bioavailability—many plant-derived compounds like flavonoids and curcumin are poorly absorbed in the gut and require transformation by gut microbes to become effective. This microbial processing varies widely across individuals, making it difficult to ensure consistent outcomes. For example, compounds like quercetin and resveratrol only exhibit anti-inflammatory or neuroprotective effects after being metabolized by specific microbes, and this transformation capacity differs from person to person (Jaberi et al., 2024; Nisha & Elora, 2025; Pizarroso et al., 2021). Another significant hurdle is the inter-individual variability in gut microbiome composition. The effectiveness of compounds such as ellagitannins and isoflavones depends on the presence of particular microbial strains capable of converting them into biologically active metabolites like urolithins or equol, which possess neuroprotective and estrogenic properties. However, not everyone harbors these microbial converters, limiting the universal applicability of such interventions (Nisha & Elora, 2025; Pizarroso et al., 2021). Furthermore, the translation of preclinical findings into reliable human therapies remains a challenge. While animal studies have shown promising results—for example, improvements in stress response and neurotransmitter balance following prebiotic or phytochemical treatment—these effects are not always replicated in human trials. The lack of standardized biomarkers, inconsistent dosage protocols, and limited long-term studies reduce the strength of the evidence supporting these interventions (Petrella et al., 2021; Meher et al., 2024).

Regulatory and Safety Considerations for Probiotics and Nutraceuticals Targeting the Gut–Brain Axis

As dietary supplements and nutraceuticals gain popularity for managing mental and metabolic health through the GBA, questions about safety and regulation have come to the forefront. Many of these products are marketed as food supplements rather than therapeutic agents, allowing them to bypass rigorous clinical evaluations typically required for pharmaceuticals (Petrella et al., 2021). This is especially concerning for live biotherapeutic products (LBPs), which include genetically engineered or next-generation probiotics designed to influence brain function or inflammation. Regulatory frameworks—such as those of the FDA in the U.S. or the EMA in Europe—are still adapting to address the complexity of these microbiome-targeted therapies (Meher et al., 2024). There are also concerns about the long-term safety of these interventions. Probiotics may pose risks for immunocompromised individuals, including the potential for systemic infections or the spread of antibiotic resistance genes. Additionally, inconsistencies in product labeling—such as inaccurate strain identification, unverified colony-forming units (CFUs), and poor shelf stability—raise doubts about product reliability and efficacy (Liu et al., 2025). These inconsistencies make it difficult for consumers and clinicians to evaluate or trust over-the-counter products Moreover, individual responses to probiotics are highly variable depending on host microbiota composition, making it difficult to establish universal guidelines for effectiveness or safety (Cryan et al., 2019).. Another significant issue is the lack of standardized international guidelines for probiotic labeling, clinical claims, and manufacturing quality. For instance, the European Union categorizes probiotics as functional foods, allowing certain health claims, while the United States classifies them under dietary supplements, which restricts what manufacturers can claim about product benefits. Meanwhile, personalized nutrition approaches that use genomic or microbiome data face ethical and data privacy concerns, particularly when such data guide individualized supplement recommendations (Nisha & Elora, 2025; Liu et al., 2025; Jaberi et al., 2024).

To address these challenges, there's a growing consensus on the need for multidisciplinary collaboration—bringing together clinicians, regulatory agencies, microbiologists, and nutrition experts to create clear safety standards, validated clinical endpoints, and algorithms for personalized dosing based on microbiome profiling.

CONCLUSION

The gut-brain axis and microbiome modulation by dietary bioactives offer exciting potential for improving mental health, with evidence linking dysbiosis to disorders and bioactives like probiotics, prebiotics, and polyphenols showing promise through reducing inflammation, enhancing neurotransmitter production, and preventing leaky gut. However, gaps in understanding mechanisms, long-term effects, and individual variability limit their application. Personalized nutrition, leveraging microbiome and genetic data, could transform mental health interventions by tailoring bioactives to each person's unique profile. Future research should prioritize mechanistic clarity, large-scale trials, and scalable personalized approaches to bridge these gaps and make dietary bioactives a cornerstone of mental health care.

REFERENCES

Al Omran, Y., & Aziz, Q. (2014). The brain-gut axis in health and disease. *Microbial endocrinology: the microbiota-gut-brain axis in health and disease*, 135-153.

Alzubide, A., & Alhalafi, M. (2024). Gut-brain axis modulation: The role of probiotics in neurological health. *Journal of Neurogastroenterology*, *12*(3), 45–59.

Berthoud, H. R., Blackshaw, L. A., Brookes, S. J., & Grundy, D. (2004). Neuroanatomy of extrinsic afferents supplying the gastrointestinal tract. *Neurogastroenterology and Motility*, *16*(s1), 28–33. DOI: 10.1111/j.1743-3150.2004.00471.x PMID: 15066001

Blackshaw, L. A., Brookes, S. J. H., Grundy, D., & Schemann, M. (2007). Sensory transmission in the gastrointestinal tract. *Neurogastroenterology and Motility*, *19*(s1), 1–19. DOI: 10.1111/j.1365-2982.2006.00871.x PMID: 17280582

Breton, J., Tennoune, N., Lucas, N., Francois, M., Legrand, R., Jacquemot, J., & Fetissov, S. O. (2016). Gut commensal E. coli proteins activate host satiety pathways following nutrient-induced bacterial growth. *Cell Metabolism*, *23*(2), 324–334. DOI: 10.1016/j.cmet.2015.10.017 PMID: 26621107

Brookes, S. J., Spencer, N. J., Costa, M., & Zagorodnyuk, V. P. (2013). Extrinsic primary afferent signalling in the gut. *Nature Reviews. Gastroenterology & Hepatology*, *10*(5), 286–296. DOI: 10.1038/nrgastro.2013.29 PMID: 23438947

Bruce-Keller, A. J., Salbaum, J. M., & Berthoud, H. R. (2018). Harnessing gut microbes for mental health: Getting from here to there. [DOI] [PMC free article] [PubMed] [Google Scholar]. *Biological Psychiatry*, *83*(3), 214–223. DOI: 10.1016/j.biopsych.2017.08.014 PMID: 29031410

Carabotti, M., Scirocco, A., Maselli, M. A., & Severi, C. (2015). The gut–brain axis: Interactions between enteric microbiota, central and enteric nervous systems. *Annals of Gastroenterology*, *28*(2), 203–209. https://www.ncbi.nlm.nih.gov/pmc/articles/PMC4367209/ PMID: 25830558

Chang, P. V., Hao, L., Offermanns, S., & Medzhitov, R. (2021). The microbial metabolite butyrate regulates intestinal macrophage function via histone deacetylase inhibition. *Proceedings of the National Academy of Sciences of the United States of America*, *118*(1), e2017225118. DOI: 10.1073/pnas.2017225118 PMID: 24390544

Chen, Y., Xu, J., & Chen, Y. (2021). Regulation of neurotransmitters by the gut microbiota and effects on cognition in neurological disorders. *Nutrients*, *13*(6), 2099. DOI: 10.3390/nu13062099 PMID: 34205336

Chowdhury, T. (2024). Psychobiotics: Microbial interventions for mental health. *Advances in Microbial Physiology*, *84*, 112–145.

Cryan, J. F., O'Riordan, K. J., Cowan, C. S. M., Sandhu, K. V., Bastiaanssen, T. F. S., Boehme, M., Codagnone, M. G., Cussotto, S., Fulling, C., Golubeva, A. V., Guzzetta, K. E., Jaggar, M., Long-Smith, C. M., Lyte, J. M., Martin, J. A., Molinero-Perez, A., Moloney, G., Morelli, E., Morillas, E., & Dinan, T. G. (2019). The microbiota-gut-brain axis. *Physiological Reviews*, *99*(4), 1877–2013. DOI: 10.1152/physrev.00018.2018 PMID: 31460832

Dahiya, D., & Nigam, P. S. (2022). Probiotic biofilm formation in the gut: Mechanisms and implications. *Applied Microbiology and Biotechnology, 106*(5), 1985–2002. DOI: 10.1007/s00253-022-11821-5

Dao, M. C., Everard, A., Aron-Wisnewsky, J., Sokolovska, N., Prifti, E., Verger, E. O., Kayser, B. D., Levenez, F., Chilloux, J., Hoyles, L., Dumas, M.-E., Rizkalla, S. W., Doré, J., Cani, P. D., & Clément, K. (2024). Akkermansia muciniphila and improved metabolic health during a dietary intervention in obesity: Relationship with gut microbiome richness and ecology. *Gut Microbes, 16*(1), 1–17. DOI: 10.1080/19490976.2023.2297866

Deverman, B. E., & Patterson, P. H. (2009, October). Cytokines and CNS development. *Neuron, 64*(1), 61–78. DOI: 10.1016/j.neuron.2009.09.002 PMID: 19840550

Diaz Heijtz, R., Wang, S., & Anuar, F.. (2011). Normal gut microbiota modulates brain development and behavior. [DOI] [PMC free article] [PubMed] [Google Scholar]. *Proceedings of the National Academy of Sciences of the United States of America, 108*(7), 3047–3052. DOI: 10.1073/pnas.1010529108 PMID: 21282636

Fleischer, J., Bumbalo, R., Bautze, V., Strotmann, J., & Breer, H. (2015). Expression of odorant receptor Olfr78 in enteroendocrine cells of the colon. *Cell and Tissue Research, 361*(3), 697–710. DOI: 10.1007/s00441-015-2165-0 PMID: 25813788

Furness, J. B., Rivera, L. R., Cho, H. J., Bravo, D. M., & Callaghan, B. (2013). The gut as a sensory organ. *Nature Reviews. Gastroenterology & Hepatology, 10*(12), 729–740. DOI: 10.1038/nrgastro.2013.180 PMID: 24061204

Fusco, W., Lorenzo, M. B., Cintoni, M., Porcari, S., Rinninella, E., Kaitsas, F., Lener, E., Mele, M. C., Gasbarrini, A., Collado, M. C., Cammarota, G., & Ianiro, G. (2023). Short-chain fatty-acid-producing bacteria: Key components of the human gut microbiota. *Nutrients, 15*(9), 2211. DOI: 10.3390/nu15092211 PMID: 37432351

Ghosh, S., Whitley, C. S., Haribabu, B., & Jala, V. R. (2021). Regulation of intestinal barrier function by microbial metabolites. *Cellular and Molecular Gastroenterology and Hepatology, 11*(5), 1463–1482. DOI: 10.1016/j.jcmgh.2021.02.007 PMID: 33610769

Gibson, G. R., Hutkins, R., Sanders, M. E., Prescott, S. L., Reimer, R. A., Salminen, S. J., Scott, K., Stanton, C., Swanson, K. S., Cani, P. D., Verbeke, K., & Reid, G. (2017). Expert consensus document: The International Scientific Association for Probiotics and Prebiotics (ISAPP) consensus statement on the definition and scope of prebiotics. *Nature Reviews. Gastroenterology & Hepatology, 14*(8), 491–502. DOI: 10.1038/nrgastro.2017.75 PMID: 28611480

Hooper, L. V., Littman, D. R., and Macpherson, A. J. (2012). Interactions between the microbiota and the immune system. *science, 336*(6086), 1268-1273.

Jaberi, K. R., Alamdari-Palangi, V., Savardashtaki, A., Vatankhah, P., Jamialahmadi, T., Tajbakhsh, A., & Sahebkar, A. (2024). Modulatory effects of phytochemicals on gut–brain axis: Therapeutic implication. *Current Developments in Nutrition, 8*(6), 103785. DOI: 10.1016/j.cdnut.2024.103785 PMID: 38939650

Jones, M. P., Dilley, J. B., Drossman, D., & Crowell, M. D. (2006, February). Brain-gut connections in functional GI disorders: Anatomic and physiologic relationships. *Neurogastroenterology and Motility, 18*(2), 91–103. DOI: 10.1111/j.1365-2982.2005.00730.x PMID: 16420287

Kalkan, Ç., Soykan, I., & Soydal, Ç. (2018). The role of probiotics in gut barrier function. *The Turkish Journal of Gastroenterology : the Official Journal of Turkish Society of Gastroenterology, 29*(2), 131–140. DOI: 10.5152/tjg.2018.18232

Kim, S. W., Hooker, J. M., Otto, N., Win, K., Muench, L., Shea, C., & Fowler, J. S. (2013). Whole-body pharmacokinetics of HDAC inhibitor drugs, butyric acid, valproic acid and 4-phenylbutyric acid measured with carbon-11 labeled analogs by PET. *Nuclear Medicine and Biology, 40*(7), 912–918. DOI: 10.1016/j.nucmedbio.2013.06.007 PMID: 23906667

Lee, J., Li, Y., Jang, J. H., Park, S., Lee, H. Y., & Jeong, Y. (2021). Microbial transformation of bile acids: Effects on host physiology and intestinal integrity. *International Journal of Molecular Sciences, 22*(14), 7613. DOI: 10.3390/ijms22147613 PMID: 34299233

Levy, G., Fishman, J. E., Xu, D., Chandler, B. T., Feketova, E., Dong, W., Qin, Y., Alli, V., Ulloa, L., & Deitch, E. A. (2013). Parasympathetic stimulation via the vagus nerve prevents systemic organ dysfunction by abrogating gut injury and lymph toxicity in trauma and hemorrhagic shock. *Shock (Augusta, Ga.), 39*(1), 39–44. DOI: 10.1097/SHK.0b013e31827b450d PMID: 23247120

Li, Z., Yi, C. X., Katiraei, S., Kooijman, S., Zhou, E., Chung, C. K., & Wang, Y. (2018). Butyrate reduces appetite and activates brown adipose tissue via the gut–brain neural circuit. *Gut, 67*(7), 1269–1279. DOI: 10.1136/gutjnl-2017-314050 PMID: 29101261

Liu, L., Qi, W., Zhang, N., Zhang, J., Liu, S., Wang, H., Jiang, L., & Sun, Y. (2025). Nutraceuticals for gut–brain axis health: A novel approach to combat malnutrition and future personalised nutraceutical interventions. *Nutrients, 17*(9), 1551. DOI: 10.3390/nu17091551 PMID: 40362863

Liu, T., Li, X., Wang, Y., Luo, C., & Yin, Y. (2022). Dietary fiber supplementation improves gut barrier function and clinical outcomes in critically ill patients: A randomized clinical trial. *Clinical Nutrition (Edinburgh, Lothian), 41*(5), 1234–1241. DOI: 10.1016/j.clnu.2021.09.022

Luu, M., Steinhoff, U., & Visekruna, A. (2017). Functional heterogeneity of gut-resident regulatory T cells. *Clinical & Translational Immunology, 6*(9), e156. DOI: 10.1038/cti.2017.39 PMID: 28983404

Ma, J., Li, Z., Zhang, W., Zhang, C., & Zhang, Y. (2021). Role of short-chain fatty acid–G-protein-coupled receptors in intestinal barrier function regulation. *Animal Nutrition, 7*(1), 259–265. DOI: 10.1016/j.aninu.2020.08.005 PMID: 33997349

Makris, A. P., Karianaki, M., Tsamis, K. I., & Paschou, S. A. (2021). The role of the gut–brain axis in depression: Endocrine, neural, and immune pathways. *Hormones (Athens, Greece), 20*(1), 1–12. DOI: 10.1007/s42000-020-00236-4 PMID: 32827123

Mayer, E. A. (2011). Gut feelings: The emerging biology of gut–brain communication. [A comprehensive recent review of the underlying neurobiology and bidirectional nature of the gut brain axis.]. *Nature Reviews. Neuroscience, 12*(8), 453–466. DOI: 10.1038/nrn3071 PMID: 21750565

Meher, A. K., Acharya, B., & Sahu, P. K. (2024). Probiotics: Bridging the interplay of a healthy gut and psychoneurological well-being. *Food Bioengineering*, *3*(1), 126–147. DOI: 10.1002/fbe2.12081

Mishra, S., Wang, S., Nagpal, R., Miller, B., Singh, R., Taraphder, S., & Yadav, H. (2023). Probiotics and prebiotics for mental health: Focus on depression and anxiety disorders. *Current Nutrition Reports*, *12*(1), 66–81. DOI: 10.1007/s13668-023-00457-0 PMID: 36790719

Molska, M., & Reguła, J. (2019). Potential mechanisms of probiotic action in the prevention and treatment of colorectal cancer. *Nutrients*, *11*(10), 2453. DOI: 10.3390/nu11102453 PMID: 31615096

Montalto, M., Curigliano, V., Santoro, L., Vastola, M., Cammarota, G., Manna, R., Gasbarrini, A., & Gasbarrini, G. (2002). Management and treatment of lactose malabsorption. *World Journal of Gastroenterology*, *12*(2), 187–191. DOI: 10.3748/wjg.v12.i2.187 PMID: 16482616

Nisha, I. J., & Elora, E. J. (2025). Natural bioactives and microbiome-based precision therapeutics for cancer, chronic, and neurodegenerative diseases. *PrimeAsia*, *6*(1), 1–8. DOI: 10.25163/primeasia.6110264

O'Riordan, K. J., Collins, M. K., Moloney, G. M., Knox, E. G., Aburto, M. R., Fülling, C., Morley, S. J., Clarke, G., Schellekens, H., & Cryan, J. F. (2022). Short chain fatty acids: Microbial metabolites for gut-brain axis signalling. *Molecular and Cellular Endocrinology*, *546*, 111572. DOI: 10.1016/j.mce.2022.111572 PMID: 35066114

Parker, A., Fonseca, S., & Carding, S. R. (2020). Gut microbes and metabolites as modulators of blood-brain barrier integrity and brain health. *Gut Microbes*, *11*(2), 135–157. DOI: 10.1080/19490976.2019.1638722 PMID: 31368397

Patterson, E., Cryan, J. F., Fitzgerald, G. F., Ross, R. P., Dinan, T. G., & Stanton, C. (2014). Gut microbiota, the pharmabiotics they produce and host health. [DOI] [PubMed] [Google Scholar]. *The Proceedings of the Nutrition Society*, *73*(4), 477–489. DOI: 10.1017/S0029665114001426 PMID: 25196939

Peng, L., Li, Z.-R., Green, R. S., Holzman, I. R., & Lin, J. (2022). Butyrate enhances the intestinal barrier by facilitating tight junction assembly via activation of AMP-activated protein kinase in Caco-2 cell monolayers. *The Journal of Nutrition*, *152*(7), 1877–1885. DOI: 10.1093/jn/nxac069 PMID: 19625695

Petrella, C., Farioli-Vecchioli, S., Cisale, G. Y., Strimpakos, G., Borg, J. J., Ceccanti, M., Fiore, M., Monteleone, G., & Nisticò, R. (2021). A healthy gut for a healthy brain: Preclinical, clinical and regulatory aspects. *Current Neuropharmacology*, *19*(5), 610–628. DOI: 10.2174/1570159X18666200730111528 PMID: 32744976

Pizarroso, N. A., Fuciños, P., Gonçalves, C., Pastrana, L., & Amado, I. (2021). A review on the role of food-derived bioactive molecules and the microbiota–gut–brain axis in satiety regulation. *Nutrients*, *13*(2), 632. DOI: 10.3390/nu13020632 PMID: 33669189

Pluznick, J. L. (2017). Microbial short-chain fatty acids and blood pressure regulation. *Current Hypertension Reports*, *19*(4), 1–5. DOI: 10.1007/s11906-017-0722-5 PMID: 28315048

Richards, P., Parker, H. E., Adriaenssens, A. E., Hodgson, J. M., Cork, S. C., Trapp, S., & Reimann, F. (2014). Identification and characterization of GLP-1 receptor–expressing cells using a new transgenic mouse model. *Diabetes*, *63*(4), 1224–1233. DOI: 10.2337/db13-1440 PMID: 24296712

Salami, M., & Soheili, M. (2022). Neuroprotective effects of probiotics via gut-brain axis signaling. *Journal of Functional Foods*, *88*, 104887. DOI: 10.1016/j.jff.2021.104887

Sharma, K. K., Singh, D., & Rawat, S. (2018). Molecular dynamics simulation studies suggests unconventional roles of non-secretary laccases from enteropathogenic gut bacteria and Cryptococcus neoformans serotype D. *Computational Biology and Chemistry*, *73*, 41–48. DOI: 10.1016/j.compbiolchem.2018.01.010 PMID: 29425935

Shoaib, M., Shehzad, A., Omar, M., Rakha, A., Raza, H., Sharif, H. R., Shakeel, A., Ansari, A., & Niazi, S. (2016). Inulin: Properties, health benefits and food applications. *Carbohydrate Polymers*, *147*, 444–454. DOI: 10.1016/j.carbpol.2016.04.020 PMID: 27178951

Silva, Y. P., Bernardi, A., & Frozza, R. L. (2020). The role of short-chain fatty acids from gut microbiota in gut-brain communication. *Frontiers in Endocrinology*, *11*, 508738. DOI: 10.3389/fendo.2020.00025 PMID: 32082260

Staniszewski, A., & Kordowska-Wiater, M. (2021). Probiotic and anticancer properties of Lactobacillus spp. *Applied Sciences (Basel, Switzerland)*, *11*(8), 3702. DOI: 10.3390/app11083702

Timewell, M., Butikofer, L., & Garcia-Larsen, V. (2023). Polyphenols as modulators of gut barrier function: Caffeoyltyramine and feruloyltyramine restore TNF-α–induced epithelial damage. *Food & Function*, *14*(4), 7613–7624.

Vandeputte, D., Falony, G., Vieira-Silva, S., Tito, R. Y., Joossens, M., & Raes, J. (2021). Prebiotic inulin-type fructans induce specific changes in the human gut microbiota. *Gut*, *66*(11), 1968–1974. DOI: 10.1136/gutjnl-2016-313271 PMID: 28213610

Wang, F. B., & Powley, T. L. (2007). Vagal innervation of intestines: Afferent pathways mapped with new en bloc horseradish peroxidase adaptation. *Cell and Tissue Research*, *329*(2), 221–230. DOI: 10.1007/s00441-007-0413-7 PMID: 17453246

Wenzel, T. J., Gates, E. J., Ranger, A. L., & Klein, J. R. (2022). Short-chain fatty acids reduce inflammation in LPS-stimulated human monocytes via histone deacetylase inhibition. *Journal of Immunology Research*, *2022*, 8299834.

Zhang, Y., Wang, Z., Peng, J., Gerner, S. T., Yin, S., & Jiang, Y. (2021). Gut microbiota-brain interaction: An emerging immunotherapy for traumatic brain *injury. Experimental Neurology*, *337*, 113585. DOI: 10.1016/j.expneurol.2020.113585 PMID: 33370556

Chapter 6
Mechanisms of Neuroprotection by Flavonoids, Phenolic Acids, Stilbenes, and Lignans

Rabia Akram
https://orcid.org/0000-0001-9416-6391
Independent Researcher, Pakistan

Shahid Mahmood Baig
Aga Khan University, Karachi, Pakistan

Ghulam Hussain
Independent Researcher, Pakistan

Tehreem Iman
https://orcid.org/0009-0002-5992-6791
Independent Researcher, Pakistan

Humaira Muzaffar
https://orcid.org/0000-0003-2172-8474
Independent Researcher, Pakistan

Arslan Iftikhar
https://orcid.org/0000-0002-7113-3455
Government College University, Faisalabad, Pakistan

Ammara Tehreem
Independent Researcher, Pakistan

ABSTRACT

Neurodegenerative and neurological disorders act as a serious global public health problem. These conditions result in oxidative stress, neuronal loss, neuroinflammation, synaptic dysfunction, and mitochondrial impairment. Polyphenolic compounds such as flavonoids, phenolic acids, stilbenes, and lignans have gained attention due to their antioxidant, antinflammatory, cardioprotective, and neuroprotective properties. The underlying mechanism of neuroprotection of these compounds involves inhibiting the activation of cyclin-dependent kinase-5 and GSK3β, activating the PI3K/AKT pathway, increasing antioxidant enzymes, reducing the generation of oxidative species, inhibiting the NF-κB, and activating the NLRP3 pathway. Additionally, they regulate neurotransmitters (dopamine, glutamate), increase neurotrophic factors to support neuronal growth and plasticity, prevent misfolded proteins, and neurotoxicity for neuronal survival. This chapter will discuss the neuroprotective potential of these polyphenols and the current understanding of their neuroprotective mechanism.

DOI: 10.4018/979-8-3373-4139-2.ch006

INTRODUCTION

Neurological and neurodegenerative diseases are a major health concern in the modern day. These conditions include Nervous depression, Alzheimer's disease (AD; 83.6%), Parkinson's disease (PD; 12.3%), Huntington's disease (HD; 0.7%), and multiple sclerosis (MS; 2.6%). Neurological diseases associated with ageing, immunological, and metabolic issues result in memory loss, chronic inflammation, and the accumulation of toxic substances in addition to progressive neuronal damage. Although pharmacology has made significant progress in these areas, there are currently no proven therapies for these ailments. Considerable attention is focused on the possibility that natural medicines may help to prevent them. Food polyphenols serve as secondary metabolites defend themselves from viral, fungal and bacterial infections as well as, to combat oxidative stress (OS), solar heat, and UV light. They naturally occur in a wide variety of fruits, vegetables, herbs, cereals, and drinks derived from them. Often conjugated with sugars, their phenolic structure can vary. These natural compounds, commonly used as dietary supplements, offer several health benefits because of their antioxidant properties. Additionally, they regulate antibacterial and inflammatory processes, which extend to the prevention of malignancies, cardiovascular disorders (CVD), metabolic syndrome, and nervous system disorders. The main pathophysiological changes in brain disorders include inappropriate inflammatory responses, structural abnormalities and oxidative metabolic modifications. Polyphenols exhibit the potential to stop or delay these changes due to their antioxidant and antiinflammatory characteristics. Although each molecule has distinct properties, all polyphenols have comparable effects on the body (Akram, Muzaffar, et al., 2023; Bellavite, 2023).

Vegetables, fruits, cereals, and dry legumes contain polyphenols with at least one phenol group in their structure. According to chemical structure, polyphenols are further classified into a number of subclasses including flavonoids, phenolic acids, coumarins, tannins, quinones, lignans, curcuminoids and stilbenes. These polyhydroxy phytochemicals produced via the phenylpropanoid and shikimic acid pathways are thought to help plants adapt to stressful conditions brought on by environmental changes (Imran et al., 2021; Mutha et al., 2021). This study overviews the neuroprotective potential of polyphenols, such as flavonoids, phenolic acids, stilbenes, and lignans (Figure 1).

Figure 1. Chemical structures of Polyphenols including phenolic acids, lignans, stilbenes and flavonoids.

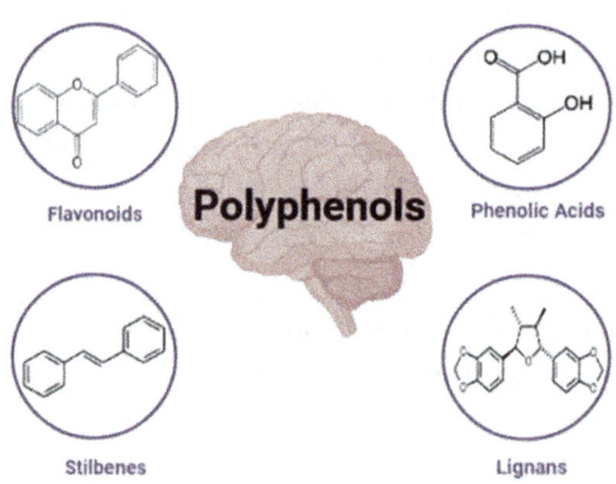

MECHANISMS OF ACTION OF BIOACTIVE COMPOUNDS ON THE NERVOUS SYSTEM

Gut-brain axis

Bidirectional communication between the gastrointestinal (GI) and central nervous systems (CNS) is known as the gut-brain axis (GBA). This axis plays a role in the regulation of gut microbiota, neurotransmitter synthesis and immune system function and improves health status related to nervous system disorders.

Gut microbiota modulation: Polyphenols maintain gut flora and have an indirect impact on brain function. The gut microbiota secretes a variety of neuroactive metabolites, including as gamma-aminobutyric acid (GABA), serotonin, and dopamine precursors, which can have an impact on behavior and mood. They support populations of beneficial bacteria such as Lactobacillus and Bifidobacterium and in turn alleviate symptoms of depression. Polyphenols modify the makeup of the gut microbiota and aid in the ideal balance of gut-derived metabolites which regulate stress reactions and mental health.

Antioxidant effects: Antioxidant-rich diets are linked to beneficial gut microbiota variety, which can produce neuroactive chemicals and change the health of the brain. For instance, polyphenols boost the number of good bacteria that produce short-chain fatty acids (SCFAs). These bacteria penetrate the blood brain barrier (BBB) and possess antioxidant and antiinflammatory properties in the brain.

Immune modulation and antiinflammatory effects: The immune system depends on the gut microbiota, and dysbiosis can cause systemic inflammation, which is linked to mental health conditions. Bioactive substances such as omega-3 fatty acids produce a proinflammatory environment in the stomach and lessened inflammation. SCFAs have been shown to penetrate the BBB and provide neuroprotective benefits via reducing neuroinflammation. Butyrate helps to stabilize mental health, lower CNS inflammation, and enhance the GI tract's epithelial barrier, which guards against pro-inflammatory microorganisms that diffuse from the stomach into the bloodstream.

BBB permeability and neuroprotection: Since the BBB protects the brain from infections and neurotoxins, it is essential for brain health. Various bioactive substances such as flavonoids and polyunsaturated fatty acids (PUFAs) play a role in maintaining the BBB integrity, which is commonly compromised by oxidative stress and chronic inflammation. They safeguard brain health by enhancing the function of the BBB, which prevents harmful substances from entering the CNS. Green tea contains polyphenols, such as epigallocatechin gallate (EGCG), which have a neuroprotective effect by reducing inflammation and oxidative stress. This can help protect the BBB and make individuals less susceptible to mental health issues.

HPA axis regulation: The hypothalamic-pituitary-adrenal (HPA) axis, the primary axis of the GBA, is crucial for regulating cortisol levels and the stress response. Polyphenols like resveratrol (RESV) decrease cortisol release, which in turn downregulates HPA axis response to stress to alleviate anxiety. This lowers inflammation in response to stressors, which may be reflected in mood disorder sufferers' lower cortisol levels.

Modulation of neurotransmitters: Neurotransmitter synthesis and regulation, which form the foundation of neurotransmitter signaling in the CNS, are influenced by the human gut microbiota. They affect the production of neurotransmitters through their effects on precursors or cofactors like tryptophan, phenylalanine, and tyrosine. For instance, strains of Bifidobacterium and Lactobacillus can increase serotonin levels in addition to producing GABA. Some substances, like curcumin have been shown to raise levels of brain-derived neurotrophic factor (BDNF), which is crucial for synaptic plasticity and neuronal growth. This can have an impact on neuronal mechanisms and brain health (Banc et al., 2023).

FLAVONOIDS AND NEUROPROTECTION

Flavonoids are present in leaves, stems, roots, fruits and flowers. Chrysin (5,7-dihydroxyflavone), the first flavonoid, was discovered in 1814. After that, more than 10,000 compounds have been isolated and characterized. They occur naturally as glycosides, glycosidic ligands, and methylated derivatives (Hao et al., 2024). Flavanols, Flavanones, flavones, isoflavones, and anthocyanidins are produced by modifications at methoxy or hydroxy substituents on the phenylchromen-4-one and phenyl rings, C2–C3 (saturated or olefinic), and C4 (a ketone group) (Safe et al., 2021). The antioxidant, antiinflammatory, anticarcinogenic, neuroprotective, antidiabetic, antibacterial, and antithrombotic properties of flavonoids are widely recognized (Abdi Syahputra et al., 2024).

The pharmacokinetics of flavonoids and their ability to cross the BBB are crucial when discussing about brain damage and therapy. Research has shown that human brain endothelial cells as well as endothelial cell lines from rats and mice, absorb flavonoids, indicating that flavonoids possess a favorable interaction with the endothelium BBB. Additionally, this interaction has been seen at the microvasculature level and indicates downstream therapeutic effects of flavonoids. However, the extent to which flavonoids may readily penetrate the BBB and reach various parts of the brain may vary among different flavonoids (Hasan et al., 2023). In this context, the flavonoids' lipid solubility and conjugation ability are two crucial factors to determine their ability to cross the BBB. The methylation of metabolites in the liver and small intestine results in increased lipophilicity and decreased polarity (Oirdi, 2024).

Quercetin (3,3′,4′,5,7-Pentahydroxyflavone) is found in a broad variety of vegetables, with an average daily intake of 0.025–0.050 g estimated and up to around 0.25 g in people who consume excessive fruits and vegetables. Although capers are the richest food source, it is also plentiful in berries, grapes, chocolate, onions, kale, broccoli, and leeks. Quercetin's bioavailability varies depending on intestinal metabolism and food type. Interestingly, its plasma concentration can be raised through dietary supplementation. It protects neurons from OS-induced damage by reducing neuroinflammation and modifying the production of Nuclear factor erythroid 2-related factor 2 (Nrf2) dependent antioxidant reactive components.

Hesperetin together with Hesperidin, Citrus fruits are especially rich in hesperidin (3,5,7-trihydroxyflavanone 7-rhamnoglucoside). It is a glycosylated form of the flavanone hesperetin, which is present in citrus fruits including lemons, oranges, and tangerines. Propolis, grapes, and other crops like dandelion (*Taraxacum officinale*) contain traces of hesperidin. It possesses antidepressant, neuroprotective, and readily penetrates the BBB (Bellavite, 2023).

Anti-Anxiety activity

Flavonoids possess anti-anxiety properties and it has been shown in animal anxiety models. The anxiolytics interact with GABA type A receptors on both benzodiazepine (BZD) and non-BZD sites to perform their action. Chrysin binds with BZDs as a ligand and exhibits anxiolytic efficacy. Similarly, apigenin attaches to the BDZ receptor in the GABA-BDZ-Cl receptor complex and boosts GABAergic neurotransmission. Wogonin's benzodiazepine receptor (BDZ-S) binding degree was 0.92 μmol/L. Wogonin has the potential to raise aminobutyric acid activation, and the BDZ-S antagonist Ro15-1788 counteracts its anxiolytic action. Flumazenil, a GABA receptor antagonist, counteracts the anxiolytic effect of 6-hydroxyflavonoids (6HF). It has been shown to increase aminobutyric acid activation through BZD receptors in rat cortical neurons. Hesperitin and hesperidin treatment were shown to reduce the length of time that mice remained immobile in the open field without influencing the mice's voluntary activity behavior.

Analgesic activity

Through the activation of the Nrf2/Heme Oxygenase-1 (HO-1) pathway, which inhibits the nuclear factor kappa b (NF-κB) pathway, activates the Nitric Oxide/cyclic GMP/Protein Kinase G/ATP sensitive Potassium channel (NO/cGMP/PKG/KATP) pathway, increases the expression of cytokine IL-10, and stops the production of tumour necrosis factor alpha (TNF-α) and interleukin (IL) 1β. Proinflammatory cytokines are activated, and extracellular matrix proteins are cleaved by matrix metalloproteinases (MMPs), which cause neuroinflammation. Through the inhibition of MMP expression and the reduction of TNF-α and Transforming Growth Factor 1 beta (TGF-1β) production, naringenin cures neuropathic pain. Hesperidin has both analgesic and antiinflammatory properties, so it can work in concert with NSAIDs to lessen pain perception and lower Nonsteroidal Antiinflammatory Drug (NSAID) dosages. Hesperidin methylation also produces hesperidin methyl chalcone (HMC), which directly interacts with Ser276 on p65 to reduce macrophage activation of the NF-κB pathway. This alleviates the decrease in arthritic pain brought on by yeast polysaccharides and decreases the levels of NF-κB-dependent IL-33, IL-6and TNF-α. Through the downregulation of the cyclic GMP-AMP synthase/stimulator of interferon genes/NOD-like receptor family pyrin domain containing 3 (cGAS /Sting/NLRP3) pathway, EGCG therapy showed strong protective benefits in cell cycle arrest, cell viability, apoptosis, and inflammatory state. Hydrogen peroxide (H_2O_2) also dramatically elevated the expression of NLRP3, Sting, and cGAS protein levels. Therefore, the main contributing mechanisms to the analgesic activity of flavonoids include lowering the generation of inflammatory indicators and preventing the activation of inflammatory pathways (Hao et al., 2024).

Anti-Parkinson activity

The 6-hydroxydopamine (6-OHDA) induced model of PD was used to assess quercetin. To induce PD, male Sprague-Dawley rats received a single intracisternal injection of 6-OHDA. Quercetin administration for 14 consecutive days substantially increased striatal dopamine and antioxidant enzyme levels as compared to the group treated with 6-OHDA alone. Apart from mitigating memory impairment, quercetin also lowered malondialdehyde (MDA) levels in the hippocampus area and increased antioxidant catalase (CAT), superoxide dismutase (SOD), and total glutathione. Quercetin therapy enhanced mitochondrial activity, lowered OS, and decreased α-synuclein expression in the 6-OHDA-treated pheochromocytoma of the rat (PC12) neuronal line. It decreased mitochondrial damage and α-synuclein accumulation, and delayed neuronal death (Bellavite, 2023).

Anti-Alzheimer's activity

Foods high in flavonoids and polyphenols can help prevent neurodegenerative symptoms and indicators, such as AD and other dementias. For instance, eating strawberries and flavonoids in general was linked to a lower risk of developing Alzheimer's dementia. According to changes in MRI markers, people who consumed the most flavonoid polymers, anthocyanins, and flavonols showed the lowest risk of AD and associated dementias in the prospective Framingham Offspring Cohort research (Safe et al., 2021). Flavonoid-rich drink considerably improved performance on the executive function and psychomotor speed tests in people with cognitive impairment when compared to the placebo (Alharbi et al., 2016).

An investigation into the relationship between dietary flavonoid consumption and cognitive health involved 808 Italian participants. Higher consumption of flavonoids (catechins, anthocyanins, and flavonols) in the diet was significantly associated with a lower risk of cognitive impairment (Godos et al., 2020).

Some flavonoids help in promoting antioxidant processes, preventing tau from aggregating, and interfering with the production and assembly of Amyloid beta (Aβ) peptides in in AD model. Among these are interactions with mitogen-activated protein kinase (MAPK) pathways and phosphatidylinositol 3-kinase/AKT (PI3K/AKT), which control transcription factors and gene expression. Additional mechanisms include the glycogen synthase kinase-3beta (GSK3β) activation and inhibition of cyclin-dependent kinase-5, which prevents aberrant tau phosphorylation, the disruption of Aβ aggregation. Additionally, alpha-secretase activation or beta-secretase inhibition results in changes in the processing of amyloid-precursor protein. Flavonoids' therapeutic potential to halt the development and progression of AD, as well as to enhance cognitive function is highlighted by their interactions with many signaling pathways.

For six weeks, laboratory mice were fed diets supplemented with quercetin (2 mg/g food) alongside a control group. The levels of antioxidant enzymes and alpha or beta-secretase in the cortex did not alter, but the supplementation markedly raised the plasma levels of isorhamnetin (methylated metabolite of quercetin) and quercetin. Preclinical research enables us to view hesperidin, quercetin, and other flavonoids as highly intriguing and safe dietary compounds that should be further explored as supplemental treatments to either prevent or lessen the harmful effects of neurodegenerative diseases, despite the paucity of human clinical trials (Bellavite, 2023).

Anti-diabetic peripheral neuropathy activity

Diabetic peripheral neuropathy (DPN) is characterized by progressive nerve damage caused by OS, chronic inflammation, and persistent hyperglycemia (Akram, Anwar, et al., 2023). NLRP3 inflammasome plays a crucial role in the pathophysiology of DPN by linking metabolic failure to neuroinflammation and neuronal injury. A large family of naturally occurring polyphenols, flavonoids, has garnered attention as a possible therapy for DPN due to their substantial antiinflammatory and antioxidant properties. Flavonoids inhibit NLRP3 activation, which results in neuroinflammation and a decrease in the production of IL-1β and IL-18. The mechanisms by which flavonoids function include lowering OS, changing significant signaling pathways, and inhibiting the actions of caspase-1 and NF-κB, both of which are necessary for the NLRP3 inflammasome to be activated (Figure 2). Clinical data suggests that flavonoids may help diabetic patients with neuropathic pain and enhance nerve function (Khanna et al., 2025).

However, flavonoids' relatively poor bioavailability currently limits their therapeutic utility, despite their wide range of neuroprotective benefits and good safety profile. Future studies should focus on creating novel formulations that increase flavonoid's bioavailability in order to overcome this restriction and fully use its medicinal potential. Since the majority of the flavonoid preparations that are sold as dietary supplements or herbal medicines with varying purported therapeutic effects have not yet undergone thorough controlled clinical trials to prove their effectiveness. To prove the health advantages of natural flavonoids, more thorough *in vitro* and *in vivo* research must be combined with clinical investigations

(Hasan et al., 2023). The role of phenolic acids has been elaborated in the next section of the chapter to better understand their bioavailability and neuroprotective sight for the neurological diseases.

Figure 2. Neuroprotective effects of flavonoids. GABA: Gabba aminobutyric acid; (IL)-10, 1β: Interleukin; TNF-α: Tumor necrosis factor α; TGF-1β: Transforming growth factor 1β; NF-κB: Nuclear factor kappa b; GSK-3β: Glycogen synthase kinase-3β; SOD: Superoxide dismutase; MDA: malondialdehyde; NLRP3: NOD-like receptor family pyrin domain containing-3; ROS: Reactive oxygen species

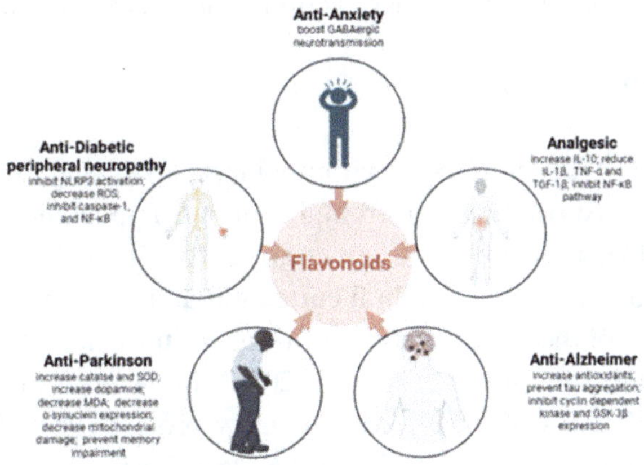

PHENOLIC ACIDS AND NEUROPROTECTION

Phenolic acids are essential parts of food and found in substances that have a single carbohydrate acid group. Plants comprised of free as well as bound forms of phenolic acids. There are two subclasses of phenolic acids including hydroxybenzoic acid (HBA) and hydroxycinnamic acid (HCA). Protocatechuic, P-HBA, gallic, syringic and vanillic acids are examples of HBAs, which have a C6–C1 structure. On the other hand, caffeic, coumaric, sinapic and ferulic acids are categorized in HCAs having a C6–C3 structure. Cereals, Oilseeds, cowpeas, coffee, raspberries, blackberries, squash shells, seeds and black currants are rich sources of HBAs. The primary sources of HCAs include cereals, spinach, peaches, tomatoes, citrus fruits and juices, potatoes, almonds, plums, cherries, and coffee. Chlorogenic acid and caffeic acid are two significant components of HCA with inherent cardioprotective and antioxidant characteristics. The phenylpropanoid route is used to produce phenolic acids, which are available in free, soluble, conjugated, and insoluble forms. Following their absorption in the gastrointestinal system, ferulic and caffeic acids undergo extensive metabolism in the human body. Numerous cancer types, diabetes, obesity, and metabolic syndrome have all been shown to benefit from caffeine. Phenolic acid combinations are effective bioactive dietary components, according to a recent assessment of the literature. Plant phenolics have a variety of uses that can help individuals in tissue damage control, disease management, and skin anti-aging (Mihaylova et al., 2024; Zhang et al., 2022).

Phenolic acids possess strong antioxidant characteristics against reactive oxygen species (ROS) and reactive nitrogen species (RNS). The anti-mutagenic properties of potatoes, lettuce, apples, coffee beans, and berries can help prevent CVD and cancer. Seaweeds are abundant in phenolic chemicals that are good for your health. Seaweeds have a broad and varied set of phenolic acids that are separated into four main groups and many terpenoids (Bento-Silva et al., 2019; Oluwole et al., 2022).

Chlorogenic acid, cyclosporine, and caffeic acid lower blood pressure and stop key enzymes involved in the progression of hypertension. Gallic acid selectively targets adipose tissue in order to minimize excessive lipid storage in obese patients, enhance insulin signaling, limit lipogenesis, and concurrently battle elevated proinflammatory response and OS. In a xenografted tumor model, vanillic acid may significantly inhibit Hypoxia-inducible factor 1 (HIF-1) and reduce tumor development, offering fresh insights into the mechanism behind its anticancer action (Zhang et al., 2022).

Anti-depressant activity

Ferulic acid reduces immobility in forced swimming test (FST), and tail suspension test (TST), increases monoamine oxidase (MAO) levels in the frontal brain and hippocampus, and raises norepinephrine and serotonin levels in the hypothalamus (J. Chen et al., 2015). It also improves FST and TST scores (without influencing locomotor activity), cerebral cortex glutathione peroxidase (GPx), blood levels of SOD and CAT, and reduction of the concentrations of thiobarbituric acid-reactive compounds in mice's blood, hippocampal, and cerebral cortex (Lenzi et al., 2015). In the prefrontal cortex and hippocampus, it shows markedly increased levels of postsynaptic density protein (PSD95), synapsin I, and BDNF, which reverses TST and FST scores and significantly lowers depressive-like behaviors brought on by chronic unpredictable mild stress (CUMS) in the sucrose preference test and FST (Y. M. Liu et al., 2017). Ferulic acid enhances mechanical allodynia, thermal hyperalgesia, and the descending monoaminergic system in conjunction with spinal β2 and 5-hydroxytryptamine 1A (5HTA-1A) receptors, as well as the downstream of mu and δ-opioid receptors, in an animal model of neuropathic pain caused by chronic constriction injury (CCI) (Xu et al., 2016).

Rosmarinic acid downregulates MAPK phosphatase-1, upregulates BDNF expression, and modulates corticosterone and dopamine production in TST in mice with depression (Kondo et al., 2015). **Cholorogenic acid** stimulates the formation of axons and dendrites and promoting the release of serotonin by increasing the expression of synapsin-I (via 5HTA receptors) in the cells of fetal raphe neurons (Wu et al., 2016). **Gallic acid** improves anxiety and depression and lowers DG hippocampal subdivisions and cell densities in the CA1, CA2, and CA3 when trimethyltin is given to Sprague-Dawley rats (assessed in the novelty suppressed feeding test, TST and elevated plus maze) (Moghadas et al., 2016). In a 4-week, randomized, placebo-controlled, double-blind study, **Homovanillic acid** reduces depressed symptoms in 22 men and 25 women. This was due to the fact that after co-supplementing with 1.4 g of cocoa extract daily, which is equivalent to 0.645 mg of total polyphenols daily, the patients' peripheral dopaminergic activity improved and their plasma concentration of homovanillic acid increased by 11.5% (Ibero-Baraibar et al., 2016).

Anti-Schizophrenia activity

Homovanillic acid results in lower rates of suicide among people with schizophrenia (Neider et al., 2016). In those with psychotic spectrum disorders, such as schizophrenia, it restores normal dopaminergic activity by assuming some of the dopaminergic metabolic processes of the central nervous system (Van De Kerkhof et al., 2015).

Anti-Parkinson activity

Ferulic acid reduces nociceptive thresholds, tactile allodynia, mechanical and thermal hyperalgesia, total protein, IL-6, IL-1β, lipid peroxidase, and nitric oxide (NO), and it also improves antiinflammatory and antioxidant activity. In rats, it lessens neuropathic pain caused by CCI (in the left sciatic nerve) (Aswar & Patil, 2016). It causes increased expression of anti-apoptotic protein Bcl2, reduction of the pro-apoptotic protein Bax levels, and Bcl2-associated X protein/ B-cell lymphoma 2 (Bax/Bcl2) ratio in the PD model. It also ameliorates 1-methyl-4-phenyl-1,2,3,6-tetrahydropyridine (MPTP) induced loss of nigro-striatal dopaminergic neurons (Nagarajan et al., 2015).

Caffeic acid causes escitalopram to decrease α-synuclein fibrillation (dose-dependent) (Fazili & Naeem, 2015). **Ellagic acid** increases mobility of the striatum and hippocampus and lowers the neuroinflammatory biomarkers IL-1β and TNF-α in PD model caused by 6-OHDA (Farbood et al., 2015). It lessens the impaired locomotor activity in rotenone-induced *Drosophila melanogaster* and zebrafish (Khatri & Juvekar, 2016).

Antiinflammatory and antioxidant activities

Ferulic acid causes recovery of dopamine neurons in the substantia nigra pars compacta (SNc) region and striatal nerve terminals, restoration of glutamate, CAT, and SOD levels, inhibition of lipid oxidation, reduction in Glial Fibrillary Acidic Protein (GFAP) hyperactivity, pro-inflammatory cytokines and Cyclooxygenase-2 (COX-2), ionized calcium-binding adapter molecule (Iba-1), and inducible nitric oxide synthase (iNOS) activities in PD model (Ojha et al., 2015). **Chlorogenic acid** causes a reduction of ROS and MDA levels in rat cerebral slices following OS caused by H_2O_2 (Gul et al., 2016).

Anti-epilepsy activity

Cholorogenic acid reduces nitrite content, lipid peroxidation, and the messenger ribonucleic acid (mRNA) expressions of metabotropic glutamate receptor 1,5, and N-methyl-d-aspartate receptor, therefore mitigating pilocarpine-induced epilepsy (seizures) (Aseervatham et al., 2016). **Caffeic acid** reduces DNA damage and free radical levels in the kindling Carworth Farms (CF) model of pentylenotetrazole (PTZ) induced epilepsy at different doses (Coelho et al., 2015). It decreases the diazepam-induced sleeping time test latency to sleep and the genotoxic damage caused by pilocarpine in mice with acute seizure models (Coelho et al., 2016).

Rosmarinic acid lowers the amount of deoxyribonucleic acid (DNA) damage and free radicals in the kindling CF, lengthening the latency, and lowers the chances of seizure occurrences in PTZ-induced epilepsy in male mice (Coelho et al., 2015). It prevents the loss of hippocampus neurons in the Cornu ammonis 1 (CA1) and CA3 areas, increases CAT activity, decreases OS, increases defense system activity, prevents mossy fiber sprouting, and attenuates seizures in rats with kainite-model temporal lobe epilepsy (Khamse et al., 2015). In the C57BL/6 female mouse model of PTZ-induced epilepsy, it has an immediate anticonvulsant-like effect on seizures and lengthens the time it takes for generalized seizures and myoclonic twitch responses (Grigoletto et al., 2016). By lowering pilocarpine levels and a sleeping duration test in a mouse model of PTZ-induced seizures, it improves the response time to first seizures and lowers the time interval to sleep when taken with diazepam (Coelho et al., 2016).

Anti-Huntington activity

Rosmarinic acid administered via the nose reduces OS and improves behavioral abnormalities in rats given 3-nitropropionic acid (3NP) (Bhatt et al., 2015).

Anti-ischemic injury activity

P-coumaric acid increases nuclear respiratory factor-1 (NRF-1) and SOD activity while decreasing MDA levels, Aβ protein expressions, and ischemic fiber degeneration in sciatic nerve segments (Guven, Yuksel, et al., 2015). In rat brains exposed to cerebral ischemia, it reduces neurological impairment scores, localized ischemia, and oxidative damage because of its antiapoptotic and antioxidant properties (Guven, Bozkurt Aras, et al., 2015). It also causes a decrease in MDA, NF-κB immunopositive neurons, and HIF-1α level and an increase in SOD activity, NRF1 level, and the count of healthy neurons after ischemia in rats with spinal cord perfusion damage (Guven, Sehitoglu, et al., 2015).

Ellagic acid reduces volume and weight of the brain, the infarct size, and inhibits apoptosis in the ischemic brains of rat by lowering the levels of inflammatory mediators NF-κB and its inhibitory regulators (phosphorylated IK-Ba), as well as MAPK proteins, apoptotic pathway proteins, and caspases (S. Y. Chen et al., 2016). As a result, the morphology of the brain alters, the rats' nerve-related function improves, and the sizes of infarcts decrease. Additionally, in a model of oxygen/glucose deprivation and reperfusion, it increases the expression of the cyclin D1 and β-catenin genes (Q. S. Liu et al., 2017; Szwajgier et al., 2017).

Anti-Diabetes activity

Gallic acid causes reduction of the oxidative damage caused by intracerebroventricular streptozotocin (STZ) by restoring normal levels of total thiol contents, thiobarbituric acid-reactive compounds, SOD, GPx, and CAT activity in the rat striatum (Naghizadeh & Mansouri, 2014). It improves sciatic nerve function during peripheral nerve degeneration and motor coordination and crush velocity in sciatic nerve-affected rats (Hajimoradi et al., 2015).

Interestingly, phenolic acids can be regarded as promising compounds for the future combination therapy of neurological disorders because of their wide distribution in natural sources, significant daily intake, relatively high stability in foods, high intestinal absorption (compared to more complex polyphenols), and effective brain absorption (Szwajgier et al., 2017). The flavonoids and phenolic acids are well studied polypheonls, however, other phenols such as stilbenes and lignans can also contribute to the neuroprotection. Thereby, the mechanistic sight of these bioactive compounds and their treatment approach in animal model versus humans has been elaborated in this chapter in next sections.

STILBENES AND NEUROPROTECTION

Stilbenes are characterized by a range of chemical components and can be found in their methoxylated, prenylated, glycosylated and free forms. Plant tissues and organs contain trace levels of stilbenes. There are several representatives with diverse biological functions. Stilbenes are a type of phenolic chemical that are naturally present in many different foods, including various medicinal plants, berries, red wine, grapes, and peanuts. 3′-hydroxypterostilbene, pterostilbene (PTS), and RESV are a few of the familiar stilbenes. The most widely distributed and well-known forms of RESV are most likely the isomeric, trans, and glycosylated forms. Trans-resveratrol-3-O-β-d glycoside is the predominant type found in plants. A structural analogue of RESV, piceatannol is present in berries, teas, almonds, peanuts, and grapes. It has several biological benefits that promote health, including antiinflammatory, anti-cancer, and antioxidant effects (Mihaylova et al., 2024; Zhang et al., 2022). RESV restores multi-drug resistance in a range of tumor cells by making them more sensitive to chemotherapeutic treatments (Zhang et al., 2022). Stilbenes have anticancer, antibacterial, antioxidant, anti-degenerative disease, antiinflammatory, neuroprotective, cardioprotective, anti-diabetic, and anti-aging properties. Neurotrophic factors, such as BDNF, can be increased by dietary stilbenes (Oluwole et al., 2022; Teka et al., 2022). Pinosylvin (3,5-dihydroxy-trans-stilbene) is another intriguing stilbene compound with possible health benefits. The heartwood of Pinus species contains the majority of this naturally occurring trans-stilbenoid, which is also found in large proportions in bark debris (Reinisalo et al., 2015).

Anti-Alzheimer activity

Using a proteasome-based mechanism, RESV encouraged intracellular breakdown of Aβ peptides generated from cell lines that expressed either Swedish mutant Aβ precursor protein 695 or wild-type (Marambaud et al., 2005). Additionally, it lessens learning impairment and hippocampal neurodegeneration by reducing the acetylation of sirtuin 1 (SIRT1) substrates in an animal model of transgenic AD (D. Kim et al., 2007).

In a randomized, placebo-controlled clinical study, patients with AD showed a favorable response to RESV at dosages up to 1 g, two time a day for 52 weeks. The trial's RESV did not always impact clinical outcomes or other biomarker trajectories, but it did stop a drop in Aβ-40 levels in the cerebrospinal fluid and patients' blood when compared to the placebo (Turner et al., 2015).

It enhances cognition and memory in diabetic patients and healthy people with subclinical cognitive impairment. It improves cognitive function and increases the level of antioxidant enzyme in the hippocampus and cerebral cortex while decreasing MDA levels in an animal model of vascular dementia (a lipid peroxidation result). PTS-based therapy alleviated the STZ-induced memory loss in rats and increased cholinergic transmission by blocking cholinesterase (Naik et al., 2017). By reducing ROS generation, inhibiting glutamate release, and halting a rise in cytosolic calcium levels, hydroxystilbene shielded rat cortical neurons against Aβ (25–35)-induced damage. Additionally, in a mouse model of transient middle cerebral artery occlusion, oxyresveratrol significantly decreased the amount of brain infarct by inhibiting both caspase-3 activation and cytochrome c release (Mendonça et al., 2024).

The little amount of polyphenols in brain tissue after oral administration limits their use against AD because they have trouble crossing the BBB. RESV, and especially its compounds can cross the brain and result in positive physiological changes. This increased transfer of RESV was observed to shield tissue against the harmful effects of Aβ1-42 and the ensuing impairment of memory skills. Quercetin-3-O-glucuronide, a polyphenol metabolite, was also shown to dramatically lower the production of Aβ peptides by primary neuron cells derived from the Tg2576 AD mouse model. Tau peptides are prevented from forming neurotoxic aggregates by polyphenol-rich grape seed extract (GSE), suggesting that polyphenols such as stilbenes have the ability to directly affect the tau aggregation process.

Rats given RESV for 3, 10, and 30 days showed a substantial and dose-dependent increase in BDNF mRNA level in their hippocampus tissue, indicating that RESV may be linked to BDNF pathway activation. By controlling brain CREB and BDNF levels, grape powder extract protects rats against OS-induced anxiety, memory loss, and hypertension. Additionally, compared to rats fed the control diet, animals fed blueberries had improved spatial memory function and a faster rate of learning via BDNF expression. The monomeric proanthocyanidin (PAC) metabolites appear to be the main PAC metabolite that may reach concentrations of around 400 nM in the brain and enhance cognitive performance. 3′-O-methyl-epicatechin-5-O-β-glucuronide, one of the epicatechin metabolites, improves synaptic plasticity via CREB signaling-related pathways.

It was discovered that long-term RESV administration might lessen cognitive impairment by lowering tau hyper-phosphorylation and amyloid load in Senescence-accelerated murine mice (SAMP8). It was also shown to stimulate prosurvival pathways, including the SIRT1 and AMP-activated protein kinase (AMPK) pathways. Interestingly, RESV followed AMPK pathway to improve mitochondrial biogenesis activation and activate this pathway in neural cells both in the brain and in vitro (Reinisalo et al., 2015).

Human amyloid precursor protein (APP) and PS1 genes are expressed by RESV. In hippocampal areas, it results in decreased levels of Aβ42 and Aβ40 and a notable decrease in amyloid plaques. The underlying mechanism revealed a substantial drop in the expression levels of BACE1 (β-secretase), sAPPα, APP, and sAPPβ. It's interesting to note that there was no change in SIRT1 levels. However, RESV has been shown to activate SIRT1 in the SAMP8-related AD mouse model. The mice in the RESV group had greater levels of SIRT1 than the animals in the control group. Additionally, the acetylation of p53, the substrate of SIRT1, was reduced in the cortex and hippocampus. Higher amounts of phosphorylated AMPK (p-AMPK) were seen in the cortical region of the RESV group, whereas AMPK levels were constant. The identification of a small number of Aβ42 and Aβ40 granule clusters in the control group and the almost nonexistent presence of Aβ granules in the RESV group are two of the most significant findings. The α-secretase enzyme was elevated in the cortex and hippocampal regions of the RESV group, although BACE1 remained unchanged. The study's increase in GSK3β phosphorylated at Ser9 was another signal, as seen by the RESV-fed group's lower pTau levels. Decreased acetylation of the recognized SIRT1 substrates peroxisome proliferator-activated receptor gamma co-activator-1 alpha (PGC-1α) and p53 is associated with neuroprotection of RESV. It reduced hippocampal neurodegeneration in the CA1 and CA3 regions. Gnetol competitively and reversibly inhibits BChE, which might be helpful in treating AD. Concentrated gnetol stopped cells from entering apoptosis via regulating the production of vascular endothelial growth factor (VEGF) and heart fatty acid binding protein 3 (hFABP3). This enhanced malathion-induced changes in cellular and nuclear morphology. Furthermore, gnetol treatment raised neurite outgrowth and NGF levels (Küpeli Akkol et al., 2025).

Anti-Parkinson activity

In a mouse model of PD produced by MPTP, RESV improved motor coordination and shielded neurons from ROS (Lu et al., 2008) via hydroxyl radical scavenging. Additionally, it prevented microglial activation and nicotinamide adenine dinucleotide phosphate (NADPH) oxidase inhibition, shielding the evaluated model from dopaminergic neurotoxicity brought on by lipopolysaccharide (LPS) (F. Zhang et al., 2010).

The neuroprotective potential of oligomeric stilbenes (OSs) produced from Alpha grape stems including ε-viniferin, melanoxylin A, vitisin A, vitisin D, trans-vitisin B, α-viniferin, ampelopsin A and cis-vitisin B, was assessed against models induced by neurotoxins, such as rotenone, 6-OHDA, paraquat (PQ), and 1-methyl-4-phenylpyridine (MPP+). The results show that certain stilbenes dramatically increased cell survival and decreased ROS levels. Trans-vitisin B, and Vitisin A reduced cardiolipin peroxidation and mitochondrial ROS and showed promising neuroprotective effects (Pislyagin et al., 2025).

Anti-Huntington activity

RESV, SRT501-M, increased the expression of PGC-1α in brown adipose tissue. It led to a decrease in high blood glucose levels and a reduction in brown adipose tissue (BAT) vacuolation. However, in spite of these benefits, the therapy had no discernible impact on the mice's striatal atrophy, weight loss, survival rates, or motor function (Begh et al., 2024).

Anti-Multiple sclerosis activity

The immune system attacks and destroys the myelin sheath that surrounds the brain and spinal cord in MS. Communication between the brain and other bodily components is disrupted by this impairment. In mice suffering from recurrent encephalomyelitis, a form of MS, RESV prevents neuronal degeneration and reduce inflammation in cases of chronic encephalomyelitis. RESV successfully delayed the onset of encephalomyelitis, but it had no influence on the features of spinal cord inflammation. It increased number of retinal ganglion cells in the neuroinflamed mice's eyes during encephalomyelitis. For seven days, the mice were fed a chow containing 0.7 percent cuprizone; for three weeks, they were then fed a diet containing 0.2 percent cuprizone. The next week saw the start of the RESV administration, which lasted for three weeks. RESV improved mitochondrial function, alleviated OS, decreased NF-κB activation, repaired demyelination, and restored motor balance and coordination. Additionally, it resulted in an increase in Olig1 expression, which is positively correlated with the active remyelination process. This suggests that RESV may enhance the remyelination process, increasing its efficacy in the treatment of MS.

RESV-generated miRNA decreased the severity of encephalomyelitis. Treatment with RESV decreased the severity of the illness, including the inflammatory and immune cell invasion levels. The miRNA microarray analysis showed that encephalitogenic mice receiving RESV treatment had a changed miRNA profile. In activated T cells, especially those in the brain, RESV affected the progression of the cell phase and the incidence of apoptosis. RESV treatment during encephalomyelitis markedly increased miR-124 expression while concurrently suppressing Sphingosine Kinase 1 (SK1) target gene expression. Hence, RESV treatment improves the development of encephalomyelitis by reducing neurological inflammation by altering the miR-124/SK1 axis, halting the cell cycle, and promoting triggered encephalitogenic T cell death (Begh et al., 2024).

Anti-ischemic injury activity

In a rat model of global cerebral ischemia, RESV decreased neuronal mortality via activating PI3K-AKT signaling, lowering GSK-3β and cyclic adenosine monophosphate (cAMP) response element-binding protein (CREB) levels, and other processes. Additionally, in a mouse model of transient middle cerebral artery occlusion, oxyresveratrol significantly decreased the amount of brain infarct by inhibiting both caspase-3 activation and cytochrome c release (Mendonça et al., 2024).

The low bioavailability of natural stilbenes, like RESV, is one of their main drawbacks. Following oral treatment, stilbenes have low systemic concentrations due to their quick metabolism in the liver and intestines. Furthermore, because they are hydrophobic, they are less soluble in water, which decreases their absorption in the digestive system. These elements reduce their therapeutic effectiveness in clinical settings. In order to improve the stability and bioavailability of stilbenes, researchers are creating a variety of delivery methods, such as cyclodextrin complexes, liposomal formulations, and nano-encapsulation. By shielding these substances from quick deterioration, encapsulation methods enable longer-lasting release and better absorption. Significant progress has been made in the pharmaceutical development of stilbene-based medications, with an emphasis on improving their therapeutic effectiveness and bioavailability. Although RESV is still the most researched substance, scientists are increasingly interested in synthetic derivatives like pterostilbene and modified analogues because of its poor bioavailability (Raees et al., 2025).

To improve the stability, potency, and specificity of stilbene molecules for the targeted treatment of illness, new derivatives must be created using medicinal chemistry techniques. More potent and bioavailable stilbene-based medications may result from structural alterations and synthetic analogues. Furthermore, improvements in drug delivery technologies might greatly enhance stilbenes' therapeutic efficacy and absorption while resolving their present bioavailability and metabolism-related drawbacks.

Figure 3. Neuroprotective effects of RESV. AD: Alzheimer's Disease; Aβ: Amyloid beta; SIRT1: Sirtuin 1; ROS: Reactive oxygen species; PD: Parkinson's Disease; NADPH: nicotinamide adenine dinucleotide phosphate; HD: Huntington's Disease; PGC-1α: peroxisome proliferator-activated receptor gamma co-activator-1alpha; BAT: Brown adipose tissue; MS: Multiple Sclerosis;OS: Oxidative stress; NF-κB: Nuclear factor kappa b; IS: Ischemic Stroke.

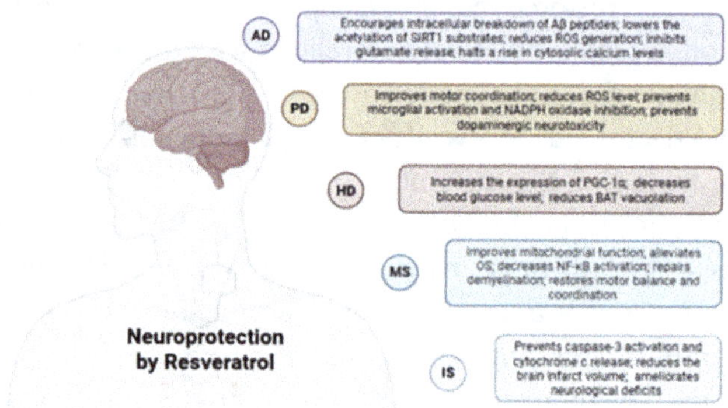

LIGNANS AND NEUROPROTECTION

Oxidative dimerization of two phenyl-propane units results in the production of lignans, a large family of naturally occurring chemicals. Wine, tea, beer, coffee, fruits, nuts, vegetables, olive oil, oilseeds, linseed and garlic are among their dietary sources. They possess biological activities, which include antiviral, anticancer, antiinflammatory, and antioxidant properties. They have been used for a long time in conventional and ethnic medicine. One important source of the complex structure of lignans, which may have biological activity for various biological functions, is *Solanum melongena* L (Zhang et al., 2022).

Two units of phenylpropane combine to generate lignans, a class of naturally occurring chemicals found in plants. They are crucial to the defensive mechanisms of plants. They are found in many different parts of the plant world, primarily in their free form instead of as glycosides. Eight subtypes of lignans have been identified, and they are further divided according to the existence of oxygen. Although lignans build up in every organ of the plant, the majority are found in the roots, seeds, barks and fruits. The most abundant source of lignan in the human diet is flaxseed. Lentils, wheat, prunes, pears, asparagus, garlic, sesame, carrots and so on are other sources (Mihaylova et al., 2024). Lariciresinol, secoisolariciresinol, medioresinol, syringaresinol, and pinoresinol are the most prevalent lignans. Plant sources like flaxseeds and sesame seeds include lignans, which are phenolic chemicals that are neither caloric nor nutritional. Lignans are abundant in flaxseeds, up to 8000 μg/g. Due to their antioxidant qualities, flaxseed lignans offer protection against CVD and cancer. Lignans can be joined to other molecules like lactone, carbon, or ether and include two phenyl-propanoid units at the β1 and β carbons (Oluwole et al., 2022).

More than 70 plant categories contain lignans. The groups Lauraceae, Annonaceae, Orchidaceae, Berberidaceae, and Schisandraceae include a variety of lignan components. Lignans come from the production of shikimic acid and are categorised as diphenolic substances. Lignan and neolignan are the two varieties of lignans. In addition to having a lower number and a more constrained evolutionary range of lignans, they differ in whether or not their monomers have a phenyl-propanoid 8,8′-bond. The additional side groups that lignans contain which might be either aromatic or aliphatic are used to classify them. Furthermore, the four groups of linear lignans—butanolide-derived lignans, butane, tetrahydrofuran derivatives, bisepoxylignans, monoepoxylignans, and derivatives of 3,7-dioxabicyclo (3.3.0)-octane derivatives—can be identified by their capacity to integrate oxygen into the framework. A large family of compounds known as cyclolignans is created when further cyclization is made possible by a C-7/C-6" connection. Among the diphenolic chemical compounds that produce lignans are 7′-hydroxymatairesinol, isolariciresinol, medioresinol (Med), sesamin (Ses), lariciresinol (Lari), syringaresinol (Syr), matairesinol (Mat), secoisolariciresinol (Seco), and pinoresinol (Pin) (Hassanein et al., 2024).

Lignan rich fraction (LF) at both 200 and 400 mg/kg might markedly boost antioxidant activity in the brain by preventing lipid peroxidation. Despite the reduction in superoxide dismutase activity, LF may also raise catalase. Only the LF-200 group had a decrease in NO, the LF-400 group's findings were negligible in comparison to the control group (Dwita et al., 2023).

Nrf2 activation is seen as essential for lignans' neuroprotective properties. Magnolol has a neuroprotective effect on mice with MS in an experimental autoimmune encephalomyelitis model. By lowering MDA, NO, and Myeloperoxidase (MPO), increasing antioxidant enzyme levels in the spinal cord and brain, and decreasing proinflammatory cytokines and inflammatory cell infiltration by upregulating Nrf2 expression, magnesiumolol therapy dramatically reduced OS. Similarly, in chronic unpredictable moderate stress-induced depression, magnolol boosted antiinflammatory cytokines, decreased depressive-like behaviors, and suppressed proinflammatory cytokines by boosting HO-1 and Nrf2 and enhancing Nrf2 nuclear translocation. Moreover, sesamin improved Nrf2 nuclear translocation and upregulated HO-1 expression to protect PC12 cells from OS-related neurodegenerative disorders. Furthermore, sesamol has a positive impact on the liver-brain axis, which suggests that it might be used as a neuroprotective medication as people age. By activating Nrf2 and upregulating HO-1 and NAD(P)H dehydrogenase quinone 1 (NQO1), sesamol therapy improved antioxidant enzymes, prevented mitochondrial dysfunction, and restored cellular redox equilibrium in H_2O_2-treated SH-SY5Y cells (Hassanein et al., 2024).

Pharmacokinetics of lignans: Plants contain lignans as glycosides and aglycones. The gastrointestinal tract absorbs enterolignans, which are mostly glucuronidated and sulfated in the liver and enterocytes after being transformed into lignan glycosides by gut flora. Most of the lignan metabolites were discovered in the portal vein in their conjugated form. Upon conjugation, lignans are mostly carried by the blood and stored in the brain, lung, liver, prostate, and stomach. Moreover, enterolignans are present in plasma and other bodily fluids. Urine and faeces are the two main ways that lignan metabolites are removed from various animals. Either inadequate absorption or enterohepatic circulation might be the cause of the elevated enterolignans in faeces. Bacterial-glucuronidase activity then deglucuronidates the lignan glucuronides that are expelled by the bile. However, only a tiny amount of the lignans are removed as sulphates and aglycones, but the bulk are eliminated as enterolignan glucuronide when expelled in the urine. The hydrolysis needed to separate lignans from sugars, create enterolignans, and make these substances accessible to people is influenced by a number of variables, including antibiotic usage, changed microbiota, and eating habits (Hassanein et al., 2024).

Anti-Alzheimer activity

Vegetables, legumes, seeds, nuts, and whole grains all contain lignan (a class of PPH). It was demonstrated that lignans extracted from *Schisandra chinensis* fruits decreased levels of Aβ peptide1-42 and attenuated Aβ1-42-induced neuronal cell damage by increasing GPx and SOD activity and decreasing ROS, MDA, and lactate dehydrogenase levels (Giuliano et al., 2020). When exposed to LPS, eleven dibenzocyclooctene lignans (1–11) present in *Schisandra chinensis* fruits prevented mouse BV2 microglial cells from producing NO. Interestingly, schisandrin had the highest level of activity, providing a highly effective therapy alternative for AD. It considerably lessens the impairment of short-term and spatial reference memory caused by Aβ. Schisandrin may reduce cognitive impairment through antioxidation, as evidenced by the increased GPx activity and GSH content in the rats' hippocampus and cerebral cortex and decreased GSSG and MDA contents (Han et al., 2022).

Gomisin A decreases mRNA expression and TNF-α, IL-1β, and IL-6 production (D. H. Kim et al., 2006). Neurodegenerative disorders brought on by aberrant OS can be treated with schisantherin. In the hippocampus and cerebral cortex, it lowers Aβ and MDA levels while raising SOD and GPx activity. It significantly lessens the histological changes in the hippocampus (Li et al., 2014).

Schisandrin B (SchB) significantly increases Ach levels and decreases acetylcholine esterase (AChE) activity in mice (Giridharan et al., 2011). Additionally, it exhibits an antiapoptotic effect on Aβ-stimulated rat cortical neurons in vitro (B. Wang & Wang, 2009).

Anti-Parkinson activity

Lignans prevented cell death by downregulating Bcl-2-like protein 4 and caspase-3/8 and upregulating Bcl-2. Sesamin, a lignan obtained from Asiasari radix, shown neuroprotective qualities by preventing the sustained phosphorylation of p38 mitogen-activated protein and c-Jun N-terminal kinases and by momentarily phosphorylating Bad and extracellular signal-regulated kinase 1/2.

Plant lignan enterolactone is derived from lignan 7-hydroxymatairesinol (HMR/lignin). It has potent antiinflammatory and antioxidant qualities and is extracted from the heartwood of Norway spruce. The effects of a 4-week oral treatment with HMR/lignan on neuroinflammatory processes, neurodegenerative illnesses, and motor deficits were evaluated using a traditional cytotoxic model of PD produced by unilateral intrastriatal injection of 6-OHDA in rats. A neuroinflammatory reaction is triggered, and nigrostriatal dopaminergic neurons are gradually damaged, starting at the terminal level and progressing back toward the SNc cell bodies. The treatment of HMR/lignan results in a notable improvement in behavioural assessments as well as a notable reduction in the SNc's striatal terminal loss and associated neuroinflammatory response (Giuliano et al., 2020).

It inhibits the overproduction of NO, iNOS expression, and ROS accumulation in SH-SY5Y cells generated by 6-OHDA. The results indicated that schisantherin A could possess neroprotective qualities by means of anti-oxidation and antiinflammatory processes (L. Q. Zhang et al., 2015).

Anti-Huntington activity

By inhibiting JNK-mediated pyruvate dehydrogenase (PDH) inhibition, SchB improves cellular glutathione redox state and can stop necrosis and death in differentiated neuronal PC12 cells exposed to 3NP (Lam & Ko, 2012).

Antiinflammatory activity

Gomisin A found in *Schisandra chinensis* fruits has been shown to reduce the expression of iNOS and COX-2 in LPS-stimulated N9 microglia in addition to inhibiting the generation of NO and prostaglandin E2 (PGE2). Gomisin A inhibits toll like receptors (TLR-4) mediated NF-κB and MAPK signaling pathways, therefore reducing microglia-mediated neuroinflammation. It significantly lowers ROS produced by LPS, NADPH oxidase activation, and microglia production of glycoprotein (GP91phox) (X. Wang et al., 2014).

In neuron and microglia cocultures, SchB has antiinflammatory effects by blocking the Myeloid differentiation primary response-88/IκB kinase/NF-κB signaling pathway. This leads to a reduction in NO, TNF-α, PGE2, IL-6, and IL-1β production. Furthermore, by significantly lowering NADPH oxidase activity and ROS production in microglia, SchB had a protective impact on neurons (Zeng et al., 2012).

Anti-ischemic injury activity

In the rat model of cerebral ischemia-reperfusion, SchB may lessen the harm caused by the inflammatory response to nerve cells via regulating the Heat Shock Protein Family A (Hsp70) Member 12B/PI3K/AKT (HSPA12B/PI3K/AKT) signaling pathway. This implies that after cerebral ischemia-reperfusion, SchB may also inhibit the subsequent inflammatory response. This implies that AD that is made worse by cerebral ischemia could be treatable with SchB (Jiang et al., 2016).

The protective benefits of lignans in preventive chronic diseases have been discussed in the literature. However, further human research is necessary to establish a strong foundation for the positive effects of lignans. Further clinical and epidemiological studies will serve as the primary source of conclusive data. Numerous plant species have been identified to contain lignans that may aid in combating cancer, hepatitis, microbial, and fungal infections. On the other hand, certain lignans were found to be beneficial for neurological disorders. Through their antiinflammatory and anti-oxidative properties, and interference with the molecular pathways and DNA methylation, lignans can function as effective neuroprotective compounds in animal and cell line models; however, more thorough clinical research is required to confirm this (Das & Kasi, 2019).

CONCLUSION

Polyphenols are naturally occurring secondary metabolites and one of the most well-established types of bioactive chemicals. They present a wide range of potential effects to modulate ROS level, enzymes, proteins, metal toxicity, inflammation, receptors, signal transduction, apoptosis, neurotransmitters, and ion channels. Thus, through regulating neurotransmitters, upregulating neurotrophic factors, modifying cell signaling pathways, inhibiting protein aggregation, and exhibiting antioxidant, antiapoptotic, and antiinflammatory properties, they serve as neuroprotective agents in the treatment of neurodegenerative diseases. Flavonoids, phenolic acids, stilbenes, and lignans are examples of polyphenols that regulate both neuronal survival and function. This study suggests that these compounds could serve as alternative therapies to alleviate the symptoms of neurodegenerative diseases. However, the limited solubility and bioavailability of specialized metabolites pose significant challenges to their medicinal use. Future research should concentrate on metabolite profiling, innovative formulations, advanced omics approaches, as well as preclinical and clinical, and combination and adjunct therapies.

REFERENCES

Abdi Syahputra, R., Dalimunthe, A., Utari, Z. D., Halim, P., Sukarno, M. A., Zainalabidin, S., Salim, E., Gunawan, M., Nurkolis, F., Park, M. N., Luckanagul, J. A., Bangun, H., Kim, B., & Harahap, U. (2024). Nanotechnology and flavonoids: Current research and future perspectives on cardiovascular health. *Journal of Functional Foods*, *120*, 106355. DOI: 10.1016/j.jff.2024.106355

Akram, R., Anwar, H., Javed, M. S., Imran, A., Rasul, A., Malik, S. A., Manzoor, M., Islam, F., Khan, I. U., Sajid, F., Iman, T., Shah, M. A., Sun, T., Hussain, G., & Shah, M. A. (2023). Natural molecules as promising players against diabetic peripheral neuropathy: an emerging nutraceutical approach. Https://Doi.Org/10.1080/10942912.2023.2189569, *26*(1), 894–914. DOI: 10.1080/10942912.2023.2189569

Akram, R., Muzaffar, H., Anwar, H., Malik, S. A., Sajid, F., Iman, T., Ahsan, A., Manzoor, M., Khan, I. U., & Hussain, G. (2023). *Role of Endogenous and Dietary Antioxidants in Brain Disorders*. 171–214. DOI: 10.1007/978-3-031-41188-5_8

Alharbi, M. H., Lamport, D. J., Dodd, G. F., Saunders, C., Harkness, L., Butler, L. T., & Spencer, J. P. E. (2016). Flavonoid-rich orange juice is associated with acute improvements in cognitive function in healthy middle-aged males. *European Journal of Nutrition*, *55*(6), 2021–2029. DOI: 10.1007/s00394-015-1016-9 PMID: 26280945

Aseervatham, G. S. B., Suryakala, U., Doulethunisha, , Sundaram, S., Bose, P. C., & Sivasudha, T. (2016). Expression pattern of NMDA receptors reveals antiepileptic potential of apigenin 8-C-glucoside and chlorogenic acid in pilocarpine induced epileptic mice. *Biomedicine and Pharmacotherapy*, *82*, 54–64. DOI: 10.1016/j.biopha.2016.04.066 PMID: 27470339

Aswar, M., & Patil, V. (2016). Ferulic acid ameliorates chronic constriction injury induced painful neuropathy in rats. *Inflammopharmacology*, *24*(4), 181–188. DOI: 10.1007/s10787-016-0272-5 PMID: 27372349

Banc, R., Rusu, M. E., Filip, L., & Popa, D. S. (2023). The Impact of Ellagitannins and Their Metabolites through Gut Microbiome on the Gut Health and Brain Wellness within the Gut–Brain Axis. *Foods*, *12*(2), 270. Advance online publication. DOI: 10.3390/foods12020270 PMID: 36673365

Begh, M. Z. A., Khan, J., Zehravi, M., Sweilam, S. H., Dinesh Raja, A., Muthukumar, A., Akiful Haque, M., Kar, N. R., Singh, L. P., Dharani Priya, B., Alshehri, M. A., Ahmad, I., Kang, S., Moon, S., Park, M. N., & Emran, T. (2024). Targeting Neurological Disorders with Stilbenes: Bridging the Preclinical-Clinical Gap. *International Journal of Biological Sciences*, *20*(14), 5474–5494. DOI: 10.7150/ijbs.102032 PMID: 39494329

Bellavite, P. (2023). Neuroprotective Potentials of Flavonoids: Experimental Studies and Mechanisms of Action. *Antioxidants*, *12*(2), 280. DOI: 10.3390/antiox12020280 PMID: 36829840

Bento-Silva, A., Koistinen, V. M., Mena, P., Bronze, M. R., Hanhineva, K., Sahlstrøm, S., Kitrytė, V., Moco, S., & Aura, A. M. (2019). Factors affecting intake, metabolism and health benefits of phenolic acids: do we understand individual variability? *European Journal of Nutrition 2019 59:4*, *59*(4), 1275–1293. DOI: 10.1007/s00394-019-01987-6

Bhatt, R., Singh, D., Prakash, A., & Mishra, N. (2015). Development, characterization and nasal delivery of rosmarinic acid-loaded solid lipid nanoparticles for the effective management of Huntingtons disease. *Drug Delivery*, *22*(7), 931–939. DOI: 10.3109/10717544.2014.880860 PMID: 24512295

Chen, J., Lin, D., Zhang, C., Li, G., Zhang, N., Ruan, L., Yan, Q., Li, J., Yu, X., Xie, X., Pang, C., Cao, L., Pan, J., & Xu, Y. (2015). Antidepressant-like effects of ferulic acid: Involvement of serotonergic and norepinergic systems. *Metabolic Brain Disease*, *30*(1), 129–136. DOI: 10.1007/s11011-014-9635-z PMID: 25483788

Chen, S. Y., Zheng, K., & Wang, Z. Q. (2016). Neuroprotective effects of ellagic acid on neonatal hypoxic brain injury via inhibition of inflammatory mediators and down-regulation of JNK/p38 MAPK activation. *Tropical Journal of Pharmaceutical Research*, *15*(2), 241–251. DOI: 10.4314/tjpr.v15i2.4

Coelho, V. R., Vieira, C. G., de Souza, L. P., da Silva, L. L., Pflüger, P., Regner, G. G., Papke, D. K. M., Picada, J. N., & Pereira, P. (2016). Behavioral and genotoxic evaluation of rosmarinic and caffeic acid in acute seizure models induced by pentylenetetrazole and pilocarpine in mice. *Naunyn-Schmiedeberg's Archives of Pharmacology*, *389*(11), 1195–1203. DOI: 10.1007/s00210-016-1281-z PMID: 27476160

Coelho, V. R., Vieira, C. G., De Souza, L. P., Moysés, F., Basso, C., Papke, D. K. M., Pires, T. R., Siqueira, I. R., Picada, J. N., & Pereira, P. (2015). Antiepileptogenic, antioxidant and genotoxic evaluation of rosmarinic acid and its metabolite caffeic acid in mice. *Life Sciences*, *122*, 65–71. DOI: 10.1016/j.lfs.2014.11.009 PMID: 25498895

Das, M., & Kasi, P. D. (2019). A Mini Review on the Protective Effect of Lignans for the Treatment of Neurodegenerative Disorders. *Journal of Nutrition. Food and Lipid Science*, *2019*(01), 40–53. DOI: 10.33513/NFLS/1901-06

Dwita, L. P., Iwo, M. I., Mauludin, R., & Elfahmi, . (2023). Neuroprotective potential of lignan-rich fraction of *Piper cubeba* L. by improving antioxidant capacity in the rat's brain. *Brazilian Journal of Biology*, *82*, e266573. DOI: 10.1590/1519-6984.266573 PMID: 36629543

Farbood, Y., Sarkaki, A., Dolatshahi, M., Mansouri, S. M. T., & Khodadadi, A. (2015). Ellagic Acid Protects the Brain Against 6-Hydroxydopamine Induced Neuroinflammation in a Rat Model of Parkinson's Disease. *Basic and Clinical Neuroscience*, *6*(2), 83. https://pmc.ncbi.nlm.nih.gov/articles/PMC4636882/ PMID: 27307952

Fazili, N. A., & Naeem, A. (2015). Anti-fibrillation potency of caffeic acid against an antidepressant induced fibrillogenesis of human α-synuclein: Implications for Parkinson's disease. *Biochimie*, *108*, 178–185. DOI: 10.1016/j.biochi.2014.11.011 PMID: 25461276

Giridharan, V. V., Thandavarayan, R. A., Sato, S., Ko, K. M., & Konishi, T. (2011). Prevention of scopolamine-induced memory deficits by schisandrin B, an antioxidant lignan from Schisandra chinensis in mice. *Free Radical Research*, *45*(8), 950–958. DOI: 10.3109/10715762.2011.571682 PMID: 21615274

Giuliano, C., Siani, F., Mus, L., Ghezzi, C., Cerri, S., Pacchetti, B., Bigogno, C., & Blandini, F. (2020). Neuroprotective effects of lignan 7-hydroxymatairesinol (HMR/lignan) in a rodent model of Parkinson's disease. *Nutrition (Burbank, Los Angeles County, Calif.)*, *69*, 110494. DOI: 10.1016/j.nut.2019.04.006 PMID: 31586482

Godos, J., Caraci, F., Castellano, S., Currenti, W., Galvano, F., Ferri, R., & Grosso, G. (2020). Association Between Dietary Flavonoids Intake and Cognitive Function in an Italian Cohort. *Biomolecules 2020, Vol. 10, Page 1300, 10*(9), 1300. DOI: 10.3390/biom10091300

Grigoletto, J., de Oliveira, C. V., Grauncke, A. C. B., de Souza, T. L., Souto, N. S., de Freitas, M. L., Furian, A. F., Santos, A. R. S., & Oliveira, M. S. (2016). Rosmarinic acid is anticonvulsant against seizures induced by pentylenetetrazol and pilocarpine in mice. *Epilepsy & Behavior, 62*, 27–34. DOI: 10.1016/j.yebeh.2016.06.037 PMID: 27448240

Gul, Z., Demircan, C., Bagdas, D., & Buyukuysal, R. L. (2016). Protective Effects of Chlorogenic Acid and its Metabolites on Hydrogen Peroxide-Induced Alterations in Rat Brain Slices: A Comparative Study with Resveratrol. *Neurochemical Research, 41*(8), 2075–2085. DOI: 10.1007/s11064-016-1919-8 PMID: 27161374

Guven, M., Bozkurt Aras, A., Akman, T., Murat Sen, H., Ozkan, A., Salis, O., Sehitoglu, I., Kalkan, Y., Silan, C., Deniz, M., & Cosar, M. (2015). Neuroprotective effect of p-coumaric acid in rat model of embolic cerebral ischemia. *Iranian Journal of Basic Medical Sciences., 18*(4), 356. https://pmc.ncbi .nlm.nih.gov/articles/PMC4439450/ PMID: 26019798

Guven, M., Sehitoglu, M. H., Yuksel, Y., Tokmak, M., Aras, A. B., Akman, T., Golge, U. H., Karavelioglu, E., Bal, E., & Cosar, M. (2015). The Neuroprotective Effect of Coumaric Acid on Spinal Cord Ischemia/Reperfusion Injury in Rats. *Inflammation, 38*(5), 1986–1995. DOI: 10.1007/s10753-015-0179-0 PMID: 25943038

Guven, M., Yuksel, Y., Sehitoglu, M. H., Tokmak, M., Aras, A. B., Akman, T., Golge, U. H., Goksel, F., Karavelioglu, E., & Cosar, M. (2015). The Effect of Coumaric Acid on Ischemia–Reperfusion Injury of Sciatic Nerve in Rats. *Inflammation, 38*(6), 2124–2132. DOI: 10.1007/s10753-015-0195-0 PMID: 26049410

Hajimoradi, M., Fazilati, M., Gharib-Naseri, M. K., & Sarkaki, A. (2015). Gallic acid and exercise training improve motor function, nerve conduction velocity but not pain sense reflex after experimental sciatic nerve crush in male rats. *Avicenna Journal of Phytomedicine, 5*(4), 288. /pmc/articles/PMC4587608/

Han, N., Wen, Y., Liu, Z., Zhai, J., Li, S., & Yin, J. (2022). Advances in the roles and mechanisms of lignans against Alzheimer's disease. *Frontiers in Pharmacology, 13*, 960112. DOI: 10.3389/fphar.2022.960112 PMID: 36313287

Hao, B., Yang, Z., Liu, H., Liu, Y., & Wang, S. (2024). Advances in Flavonoid Research: Sources, Biological Activities, and Developmental Prospectives. *Current Issues in Molecular Biology 2024, Vol. 46, Pages 2884-2925, 46*(4), 2884–2925. DOI: 10.3390/cimb46040181

Hasan, S., Khatri, N., Rahman, Z. N., Menezes, A. A., Martini, J., Shehjar, F., Mujeeb, N., & Shah, Z. A. (2023). Neuroprotective Potential of Flavonoids in Brain Disorders. *Brain Sciences 2023, Vol. 13, Page 1258, 13*(9), 1258. DOI: 10.3390/brainsci13091258

Hassanein, E. H. M., Althagafy, H. S., Baraka, M. A., Abd-alhameed, E. K., Ibrahim, I. M., Abd El-Maksoud, M. S., Mohamed, N. M., & Ross, S. A. (2024). The promising antioxidant effects of lignans: Nrf2 activation comes into view. *Naunyn-Schmiedeberg's Archives of Pharmacology*, *397*(9), 6439–6458. DOI: 10.1007/s00210-024-03102-x PMID: 38695909

Ibero-Baraibar, I., Perez-Cornago, A., Ramirez, M. J., Martínez, J. A., & Zulet, M. A. (2016). An increase in plasma homovanillic acid with cocoa extract consumption is associated with the alleviation of depressive symptoms in overweight or obese adults on an energy restricted diet in a randomized controlled trial. *The Journal of Nutrition*, *146*(4), 897S–904S. DOI: 10.3945/jn.115.222828 PMID: 26962189

Imran, A., Maqbool, J., Rasul, A., Anwar, H., Malik, S. A., Akram, R., Sajid, F., Tariq, A. R., & Arshad, M. U. (2021). *Hussain, and G.* Neuroprotective Effects of Polyphenols., DOI: 10.2174/9781681089379121080006

Jiang, E., Tang, Z., Yu, C., Yu, C., & Zhu, W. (2016). Protective effect of schisandrin B on cerebral ischemia reperfusion injury of rats and its mechanisms. *Journal of Jilin University Medicine Edition*, *42*(5), 860–865. DOI: 10.13481/J.1671-587X.20160504

Khamse, S., Sadr, S. S., Roghani, M., Hasanzadeh, G., & Mohammadian, M. (2015). Rosmarinic acid exerts a neuroprotective effect in the kainate rat model of temporal lobe epilepsy: Underlying mechanisms. *Pharmaceutical Biology*, *53*(12), 1818–1825. DOI: 10.3109/13880209.2015.1010738 PMID: 25874386

Khanna, S., Kumar, S., Sharma, P., Daksh, R., Nandakumar, K., & Shenoy, R. R. (2025). Flavonoids regulating NLRP3 inflammasome: A promising approach in alleviating diabetic peripheral neuropathy. *Inflammopharmacology*, *2025*(5), 1–32. DOI: 10.1007/s10787-025-01729-7 PMID: 40205269

Khatri, D., & Juvekar, A. (2016). Abrogation of locomotor impairment in a rotenone-induced Drosophila melanogaster and zebrafish model of Parkinson's disease by ellagic acid and curcumin. *International Journal of Nutrition, Pharmacology, Neurological Diseases*, *6*(2), 90–96. DOI: 10.4103/2231-0738.179969

Kim, D., Nguyen, M. D., Dobbin, M. M., Fischer, A., Sananbenesi, F., Rodgers, J. T., Delalle, I., Baur, J. A., Sui, G., Armour, S. M., Puigserver, P., Sinclair, D. A., & Tsai, L. H. (2007). SIRT1 deacetylase protects against neurodegeneration in models for Alzheimer's disease and amyotrophic lateral sclerosis. *The EMBO Journal*, *26*(13), 3169–3179. DOI: 10.1038/sj.emboj.7601758 PMID: 17581637

Kim, D. H., Hung, T. M., Bae, K. H., Jung, J. W., Lee, S., Yoon, B. H., Cheong, J. H., Ko, K. H., & Ryu, J. H. (2006). Gomisin A improves scopolamine-induced memory impairment in mice. *European Journal of Pharmacology*, *542*(1–3), 129–135. DOI: 10.1016/j.ejphar.2006.06.015 PMID: 16824513

Kondo, S., El Omri, A., Han, J., & Isoda, H. (2015). Antidepressant-like effects of rosmarinic acid through mitogen-activated protein kinase phosphatase-1 and brain-derived neurotrophic factor modulation. *Journal of Functional Foods*, *14*, 758–766. DOI: 10.1016/j.jff.2015.03.001

Küpeli Akkol, E., Karatoprak, G. Ş., Dumlupınar, B., Bahadır Acıkara, Ö., Arıcı, R., Yücel, Ç., Aynal, L. C., & Sobarzo Sánchez, E. (2025). Stilbenes Against Alzheimer's Disease: A Comprehensive Review of Preclinical Studies of Natural and Synthetic Compounds Combined with the Contributions of Developed Nanodrug Delivery Systems. *Molecules 2025, Vol. 30, Page 1982*, *30*(9), 1982. DOI: 10.3390/molecules30091982

Lam, P. Y., & Ko, K. M. (2012). Beneficial effect of (-)Schisandrin B against 3-nitropropionic acid-induced cell death in PC12 cells. *BioFactors (Oxford, England)*, *38*(3), 219–225. DOI: 10.1002/biof.1009 PMID: 22488872

Lenzi, J., Rodrigues, A. F., Rós, A. de S., de Castro, B. B., de Lima, D. D., Magro, D. D. D., & Zeni, A. L. B. (2015). Ferulic acid chronic treatment exerts antidepressant-like effect: Role of antioxidant defense system. *Metabolic Brain Disease*, *30*(6), 1453–1463. DOI: 10.1007/s11011-015-9725-6 PMID: 26340979

Li, X., Zhao, X., Xu, X., Mao, X., Liu, Z., Li, H., Guo, L., Bi, K., & Jia, Y. (2014). Schisantherin A recovers Aβ-induced neurodegeneration with cognitive decline in mice. *Physiology & Behavior*, *132*, 10–16. DOI: 10.1016/j.physbeh.2014.04.046 PMID: 24813830

Liu, Q. S., Li, S. R., Li, K., Li, X., Yin, X., & Pang, Z. (2017). Ellagic acid improves endogenous neural stem cells proliferation and neurorestoration through Wnt/β-catenin signaling in vivo and in vitro. *Molecular Nutrition & Food Research*, *61*(3), 1600587. Advance online publication. DOI: 10.1002/mnfr.201600587 PMID: 27794200

Liu, Y. M., Hu, C. Y., Shen, J. D., Wu, S. H., Li, Y. C., & Yi, L. T. (2017). Elevation of synaptic protein is associated with the antidepressant-like effects of ferulic acid in a chronic model of depression. *Physiology & Behavior*, *169*, 184–188. DOI: 10.1016/j.physbeh.2016.12.003 PMID: 27940143

Lu, K. T., Ko, M. C., Chen, B. Y., Huang, J. C., Hsieh, C. W., Lee, M. C., Chiou, R. Y. Y., Wung, B. S., Peng, C. H., & Yang, Y. L. (2008). Neuroprotective effects of resveratrol on MPTP-induced neuron loss mediated by free radical scavenging. *Journal of Agricultural and Food Chemistry*, *56*(16), 6910–6913. DOI: 10.1021/jf8007212 PMID: 18616261

Marambaud, P., Zhao, H., & Davies, P. (2005). Resveratrol promotes clearance of Alzheimer's disease amyloid-β peptides. *The Journal of Biological Chemistry*, *280*(45), 37377–37382. DOI: 10.1074/jbc.M508246200 PMID: 16162502

Mendonça, E. L. S. S., Xavier, J. A., Fragoso, M. B. T., Silva, M. O., Escodro, P. B., Oliveira, A. C. M., Tucci, P., Saso, L., & Goulart, M. O. F. (2024). E-Stilbenes: General Chemical and Biological Aspects, Potential Pharmacological Activity Based on the Nrf2 Pathway. *Pharmaceuticals 2024, Vol. 17, Page 232*, *17*(2), 232. DOI: 10.3390/ph17020232

Mihaylova, D., Dimitrova-Dimova, M., & Popova, A. (2024). Dietary Phenolic Compounds—Wellbeing and Perspective Applications. *International Journal of Molecular Sciences*, *25*(9), 4769. DOI: 10.3390/ijms25094769 PMID: 38731987

Moghadas, M., Edalatmanesh, M. A., & Robati, R. (2016). Histopathological Analysis from Gallic Acid Administration on Hippocampal Cell Density, Depression, and Anxiety Related Behaviors in A Trimethyltin Intoxication Model. *Cell Journal*, *17*(4), 659–667. DOI: 10.22074/CELLJ.2016.3838 PMID: 26862525

Mutha, R. E., Tatiya, A. U., & Surana, S. J. (2021). Flavonoids as natural phenolic compounds and their role in therapeutics: an overview. *Future Journal of Pharmaceutical Sciences 2021 7:1*, *7*(1), 1–13. DOI: 10.1186/s43094-020-00161-8

Nagarajan, S., Chellappan, D. R., Chinnaswamy, P., & Thulasingam, S. (2015). Ferulic acid pretreatment mitigates MPTP-induced motor impairment and histopathological alterations in C57Bl/6 mice. *Pharmaceutical Biology*, *53*(11), 1591–1601. DOI: 10.3109/13880209.2014.993041 PMID: 25857436

Naghizadeh, B., & Mansouri, M. T. (2014). Protective effects of gallic acid against streptozotocin-induced oxidative damage in rat striatum. *Drug Research*, *65*(10), 515–520. DOI: 10.1055/s-0034-1377012 PMID: 24941087

Naik, B., Nirwane, A., & Majumdar, A. (2017). Pterostilbene ameliorates intracerebroventricular streptozotocin induced memory decline in rats. *Cognitive Neurodynamics*, *11*(1), 35–49. DOI: 10.1007/s11571-016-9413-1 PMID: 28174611

Neider, D., Lindström, L. H., & Bodén, R. (2016). Risk factors for suicide among patients with schizophrenia: A cohort study focused on cerebrospinal fluid levels of homovanillic acid and 5-hydroxyindoleacetic acid. *Neuropsychiatric Disease and Treatment*, *12*, 1711–1714. DOI: 10.2147/NDT.S107178 PMID: 27468235

Oirdi, E. (2024). Harnessing the Power of Polyphenols: A New Frontier in Disease Prevention and Therapy. *Pharmaceuticals (Basel, Switzerland)*, *17*(6), 692. https://www.mdpi.com/1424-8247/17/6/692#B197-pharmaceuticals-17-00692. DOI: 10.3390/ph17060692 PMID: 38931359

Ojha, S., Javed, H., Azimullah, S., Khair, S. B. A., & Haque, M. E. (2015). Neuroprotective potential of ferulic acid in the rotenone model of Parkinson's disease. *Drug Design, Development and Therapy*, *9*, 5499–5510. DOI: 10.2147/DDDT.S90616 PMID: 26504373

Oluwole, O., Fernando, W. B., Lumanlan, J., Ademuyiwa, O., & Jayasena, V. (2022). Role of phenolic acid, tannins, stilbenes, lignans and flavonoids in human health – a review. *International Journal of Food Science & Technology*, *57*(10), 6326–6335. DOI: 10.1111/ijfs.15936

Pislyagin, E. A., Tarbeeva, D. V., Yurchenko, E. A., Menchinskaya, E. S., Gorpenchenko, T. Y., Pokhilo, N. D., Kalinovskiy, A. I., Aminin, D. L., & Fedoreyev, S. A. (2025). Neuroprotective Activity of Oligomeric Stilbenes from Alpha Grape Stems in In Vitro Models of Parkinson's Disease. *International Journal of Molecular Sciences*, *26*(6), 2411. DOI: 10.3390/ijms26062411 PMID: 40141054

Raees, S., Shah, S. A., Karim, A., Raees, S., Shah, S. A., & Karim, A. (2025). *Stilbenes: Emerging Applications in Health*. Agriculture, and Industry., DOI: 10.5772/intechopen.1010264

Reinisalo, M., Kårlund, A., Koskela, A., Kaarniranta, K., & Karjalainen, R. O. (2015). Polyphenol Stilbenes: Molecular Mechanisms of Defence against Oxidative Stress and Aging-Related Diseases. *Oxidative Medicine and Cellular Longevity*, *340520*, 1–24. Advance online publication. DOI: 10.1155/2015/340520 PMID: 26180583

Safe, S., Jayaraman, A., Chapkin, R. S., Howard, M., Mohankumar, K., & Shrestha, R. (2021). Flavonoids: Structure–function and mechanisms of action and opportunities for drug development. *Toxicological Research*, *37*(2), 147–162. DOI: 10.1007/s43188-020-00080-z PMID: 33868973

Szwajgier, D., Borowiec, K., & Pustelniak, K. (2017). The Neuroprotective Effects of Phenolic Acids: Molecular Mechanism of Action. *Nutrients*, *9*(5), 477. DOI: 10.3390/nu9050477 PMID: 28489058

Teka, T., Zhang, L., Ge, X., Li, Y., Han, L., & Yan, X. (2022). Stilbenes: Source plants, chemistry, biosynthesis, pharmacology, application and problems related to their clinical Application-A comprehensive review. *Phytochemistry*, *197*, 113128. DOI: 10.1016/j.phytochem.2022.113128 PMID: 35183567

Turner, R. S., Thomas, R. G., Craft, S., Van Dyck, C. H., Mintzer, J., Reynolds, B. A., Brewer, J. B., Rissman, R. A., Raman, R., Aisen, P. S., Mintzer, J., Reynolds, B. A., Karlawish, J., Galasko, D., Heidebrink, J., Aggarwal, N., Graff-Radford, N., Sano, M., Petersen, R., & Alexopoulis, G. (2015). A randomized, double-blind, placebo-controlled trial of resveratrol for Alzheimer disease. *Neurology*, *85*(16), 1383–1391. DOI: 10.1212/WNL.0000000000002035 PMID: 26362286

Van De Kerkhof, N. W. A., Fekkes, D., Van Der Heijden, F. M. M. A., Egger, J. I. M., & Verhoeven, W. M. A. (2015). Relationship between Plasma Homovanillic Acid and Outcome in Patients with Psychosis Spectrum Disorders. *Neuropsychobiology*, *71*(4), 212–217. DOI: 10.1159/000431095 PMID: 26279280

Wang, B., & Wang, X. M. (2009). Schisandrin B protects rat cortical neurons against Aβ1-42- induced neurotoxicity. *Die Pharmazie*, *64*(7), 450–454. DOI: 10.1691/PH.2009.9552 PMID: 19694182

Wang, X., Hu, D., Zhang, L., Lian, G., Zhao, S., Wang, C., Yin, J., Wu, C., & Yang, J. (2014). Gomisin A inhibits lipopolysaccharide-induced inflammatory responses in N9 microglia via blocking the NF-κB/MAPKs pathway. *Food and Chemical Toxicology*, *63*, 119–127. DOI: 10.1016/j.fct.2013.10.048 PMID: 24211520

Wu, J., Chen, H., Li, H., Tang, Y., Yang, L., Cao, S., & Qin, D. (2016). Antidepressant potential of chlorogenic acid-enriched extract from Eucommia ulmoides Oliver bark with neuron protection and promotion of serotonin release through enhancing synapsin I expression. *Molecules (Basel, Switzerland)*, *21*(3), 260. Advance online publication. DOI: 10.3390/molecules21030260 PMID: 26927040

Xu, Y., Lin, D., Yu, X., Xie, X., Wang, L., Lian, L., Fei, N., Chen, J., Zhu, N., Wang, G., Huang, X., & Pan, J. (2016). The antinociceptive effects of ferulic acid on neuropathic pain: Involvement of descending monoaminergic system and opioid receptors. *Oncotarget*, *7*(15), 20455–20468. DOI: 10.18632/oncotarget.7973 PMID: 26967251

Zeng, K. W., Zhang, T., Fu, H., Liu, G. X., & Wang, X. M. (2012). Schisandrin B exerts anti-neuroinflammatory activity by inhibiting the Toll-like receptor 4-dependent MyD88/IKK/NF-κB signaling pathway in lipopolysaccharide-induced microglia. *European Journal of Pharmacology*, *692*(1–3), 29–37. DOI: 10.1016/j.ejphar.2012.05.030 PMID: 22698579

Zhang, F., Shi, J. S., Zhou, H., Wilson, B., Hong, J. S., & Gao, H. M. (2010). Resveratrol protects dopamine neurons against lipopolysaccharide-induced neurotoxicity through its antiinflammatory actions. *Molecular Pharmacology*, *78*(3), 466–477. DOI: 10.1124/mol.110.064535 PMID: 20554604

Zhang, L. Q., Sa, F., Chong, C. M., Wang, Y., Zhou, Z. Y., Chang, R. C. C., Chan, S. W., Hoi, P. M., & Yuen Lee, S. M. (2015). Schisantherin A protects against 6-OHDA-induced dopaminergic neuron damage in zebrafish and cytotoxicity in SH-SY5Y cells through the ROS/NO and AKT/GSK3β pathways. *Journal of Ethnopharmacology*, *170*, 8–15. DOI: 10.1016/j.jep.2015.04.040 PMID: 25934514

Zhang, Y., Cai, P., Cheng, G., & Zhang, Y. (2022). A Brief Review of Phenolic Compounds Identified from Plants: Their Extraction, Analysis, and Biological Activity. *Natural Product Communications*, *17*(1), 1934578X211069721. Advance online publication. DOI: 10.1177/1934578X211069721

Chapter 7
Neuroprotective Potential of Flavonoids:
Mechanisms and Delivery Strategies

Adnan Afzal
https://orcid.org/0009-0001-4359-5695
Department of Zoology, Government College University, Faisalabad, Pakistan

Arifa Mehreen
University of Agriculture, Faisalabad, Pakistan

Hafiz Aamir Ali Kharl
https://orcid.org/0000-0002-7365-2937
University of Agriculture, Faisalabad, Pakistan

Qurat-ul-Ane Gillani
https://orcid.org/0000-0002-1068-4472
Women University, Multan, Pakistan

Faiza Shakoor
Comstas University, Vehari, Pakistan

Aiman Siddique
National University of Sciences and Technology, Islamabad, Pakistan

Muhammad Hussain Taqi
https://orcid.org/0009-0002-5868-0929
Cholistan University of Veterinary and Animal Sciences, Bahawalpur, Pakistan

Ayesha Bintay Bintay Farooq
Government College University, Faisalabad, Pakistan

ABSTRACT

Neurodegenerative diseases like Alzheimer's, Parkinson's, stroke, and multiple sclerosis pose a major global health challenge due to progressive neuronal damage. Flavonoids, abundant in plant-based foods, offer neuroprotective potential through antioxidant, anti-inflammatory, and signaling-modulatory effects. This chapter reviews their structural diversity, mechanisms of action—including regulation of oxidative stress, inflammation, and cell survival—and their role in preclinical and clinical models. It also discusses nanotechnology-based delivery systems that improve flavonoid bioavailability and brain targeting. Despite promising effects, challenges like poor bioavailability and variable clinical outcomes highlight the need for further research and innovative therapeutic strategies.

DOI: 10.4018/979-8-3373-4139-2.ch007

1. INTRODUCTION

Neurodegenerative diseases are progressive medical conditions characterized by the gradual decline in the structure and function of the nervous system (Teleanu et al., 2022). Prominent examples include Alzheimer's disease, Parkinson's disease, Huntington's disease, and amyotrophic lateral sclerosis, which predominantly impair the brain and spinal cord, leading to cognitive decline, motor dysfunction, and disruption of vital physiological processes. The global burden of these disorders continues to rise, while effective treatment options remain limited, thereby motivating extensive research into mechanisms of neurodegeneration and potential therapeutic strategies (Wareham et al., 2022).

Although the etiologies of neurodegenerative diseases are multifactorial, oxidative stress, chronic inflammation, mitochondrial dysfunction, and impaired protein homeostasis are recognized as central contributors to neuronal injury and death. These interconnected processes accelerate neuronal degeneration and highlight the urgent need for innovative, disease-modifying interventions. In recent years, natural bioactive compounds derived from plants, herbs, and other organisms have gained attention for their ability to modulate these pathways and potentially delay or mitigate neurodegeneration. Polyphenols, alkaloids, terpenoids, and flavonoids are among the most studied due to their antioxidant, anti-inflammatory, and neuroprotective properties (Spagnuolo, Moccia, & Russo, 2018).

Among these, flavonoids stand out as particularly promising neuroprotective agents. Widely distributed in fruits, vegetables, nuts, seeds, tea, and wine, flavonoids are polyphenolic compounds with diverse structures and biological functions. Their health-promoting roles have long been acknowledged, but their capacity to influence neurological processes has attracted increasing scientific focus (Wróbel-Biedrawa, Grabowska, Galanty, Sobolewska, & Podolak, 2022). Structurally, flavonoids are composed of 15 carbon atoms arranged in a C6-C3-C6 skeleton, forming three rings that allow for extensive substitution patterns (Karak, 2019). This structural diversity gives rise to multiple subclasses, including flavones, flavonols, isoflavones, anthocyanins, and flavan-3-ols. Despite differences across subclasses, all flavonoids share the capacity to scavenge free radicals, regulate inflammatory signaling pathways, and modulate processes critical to neuronal survival and function.

The neuroprotective mechanisms of flavonoids are multifaceted and involve modulation of major brain-related molecular systems (Calderaro et al., 2022). One of the primary protective roles is their strong antioxidant activity. The central nervous system is highly vulnerable to oxidative damage due to its elevated metabolic rate and the abundance of polyunsaturated fatty acids in neuronal membranes. Oxidative stress-induced neuronal injury has been implicated as a key factor in the onset and progression of many neurodegenerative diseases (Singh, Kukreti, Saso, & Kukreti, 2019). Compounds such as quercetin, catechins, and anthocyanins directly neutralize reactive oxygen and nitrogen species (ROS and RNS), thereby reducing oxidative stress and preventing neuronal damage. Furthermore, flavonoids enhance the activity of endogenous antioxidant enzymes, including superoxide dismutase (SOD) and catalase, strengthening intrinsic defense mechanisms against oxidative insults.

In addition to combating oxidative stress, flavonoids exert potent anti-inflammatory effects that are critical in neurodegenerative contexts. Persistent neuroinflammation, primarily mediated by activated microglia, is a hallmark of Alzheimer's disease, Parkinson's disease, and related conditions (Borst et al., 2021). While microglia maintain homeostasis under physiological conditions, in neurodegenerative states they release pro-inflammatory cytokines, chemokines, and neurotoxic mediators that exacerbate neuronal injury and disease progression. Flavonoids such as curcumin, resveratrol, and apigenin have demonstrated the ability to modulate microglial activation and suppress the excessive release of inflammatory mediators, thereby preserving neuronal integrity and maintaining balanced immune function within the brain (Calis et al., 2020).

Taken together, these mechanisms highlight the potential of flavonoids as either adjunctive therapies alongside conventional treatments or as independent neuroprotective agents. Their ability to simultaneously target oxidative stress, neuroinflammation, and other molecular pathways positions flavonoids as valuable candidates in efforts to slow or prevent neurodegenerative processes.

Another important way in which flavonoids protect nerve cells is by modulating cellular pathways that are essential for neuronal survival. A key example is the activation of the PI3K/Akt pathway, which plays central roles in promoting cell survival, growth, and nutrient metabolism (Zughaibi et al., 2021). Activation of this pathway enhances neuronal resistance to oxidative stress and excitotoxicity, both of which are central contributors to neurodegeneration. Flavonoids are also capable of stimulating other protective signaling pathways, including Nrf2 and AMPK, which regulate antioxidant defenses, energy metabolism, and overall neuronal function. Because signaling pathways are vital for maintaining neural health, their modulation by flavonoids highlights the compounds' therapeutic promise for neurodegenerative disorders.

Despite the strong laboratory evidence supporting flavonoids' neuroprotective properties, their clinical application remains challenging (Bellavite, 2023). A major limitation is their poor bioavailability. Flavonoids tend to have low absorption in the digestive tract, are rapidly metabolized, and often fail to accumulate at effective levels in the brain. Their relatively large size and water-soluble nature further limit their ability to cross biological membranes and the blood–brain barrier (BBB). To overcome these barriers, researchers are developing innovative delivery systems designed to improve flavonoid stability, absorption, and targeted delivery (Chen et al., 2022). Nanotechnology-based approaches such as nanoparticles, liposomes, and micelles have shown particular promise. These delivery vehicles not only enhance bioavailability but also improve the capacity of flavonoids to cross the BBB, thereby increasing their therapeutic potential (Furtado et al., 2018).

While preclinical studies in animal models have consistently demonstrated the neuroprotective effects of flavonoids, results from human clinical studies have been less clear and sometimes contradictory (Naomi et al., 2023). Some trials suggest cognitive and neuroprotective benefits, whereas others report little or no significant effect. Such discrepancies likely reflect differences in study design, sample size, dosage, duration of treatment, and individual variation in flavonoid metabolism. Furthermore, the long-term safety profile of flavonoids, including possible side effects from high-dose or chronic use, requires further evaluation before these compounds can be widely recommended in clinical settings (Safe et al., 2021).

Taken together, flavonoids represent a highly promising strategy for preventing or slowing neurodegenerative diseases due to their antioxidant, anti-inflammatory, and neuroprotective properties. Because they act through natural biological pathways, flavonoids may complement existing therapies and offer a safer, integrative approach to managing neurological decline. However, to realize their full therapeutic potential, ongoing research must focus on improving their bioavailability, clarifying their mechanisms of action in humans, and establishing evidence from large-scale clinical trials. Advances in nanotechnology and drug delivery systems are expected to accelerate this process, paving the way for flavonoid-based therapies to play a meaningful role in both the prevention and management of neurodegenerative diseases worldwide (Sciences et al., 2019).

2. FLAVONOIDS: STRUCTURE AND CLASSIFICATION

Flavonoids are plant substances known for their wide variety of health benefits. You can find these compounds which help promote good health, in fruits, vegetables, nuts, seeds and beverages such as tea and wine. A 15-carbon foundation, present in three sets of rings, is the basic structure of flavonoids and many hydroxyl, methoxy and related substituent groups are attached to the main rings (Chen et al., 2023). Thanks to their many forms, flavonoids are sorted into several subclasses, each offering special benefits to health. This section focuses on the structure of flavonoids, their various classes and the main sources from which flavonoids are commonly found in food.

Figure 1. Protective mechanism of flavonoids

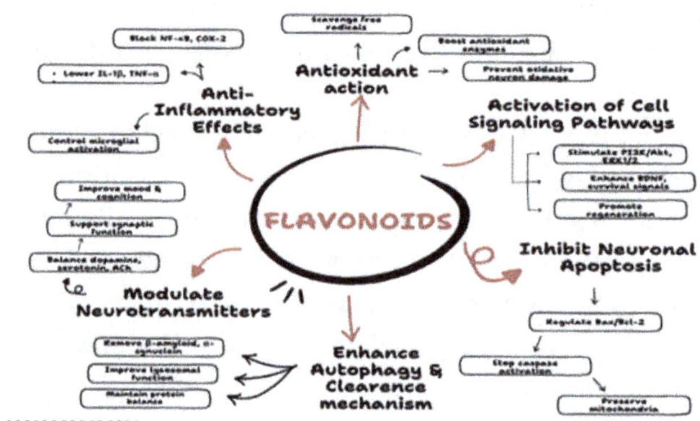

Flavonoids are composed of a flavan backbone, which is formed when a pyran-fused ring (C) links two phenyl rings (A and B). This characteristic structure is often referred to as the **"C6–C3–C6" skeleton**. In their phenolic framework, both the A-ring and B-ring typically carry hydroxyl (-OH) groups on their benzene rings, while the C-ring forms a pyran that bridges the two prenylated phenolic rings (van Dinteren, 2024). The flexibility of this structure allows for a wide variety of substitutions on each ring, leading to the formation of numerous subclasses with distinct chemical features. Many biological activities of flavonoids are directly related to this chemical structure.

Hydroxyl groups on the phenolic rings donate electrons, neutralizing free radicals and reducing oxidative stress (Vuolo, Lima, & Junior, 2019). In addition, the structure of flavonoids enables them to interact with enzymes, receptors, and cellular signaling pathways, thereby enhancing their antioxidant, anti-inflammatory, anticancer, and neuroprotective functions (Al-Khayri et al., 2022). Based on their structural variations, flavonoids are classified into several subclasses, the most studied of which include flavones, flavonols, isoflavones, flavan-3-ols (catechins), anthocyanins, and flavanones (Burton-Freeman et al., 2019).

Flavones

Flavones are characterized by a double bond between the second and third carbons of the C-ring. This structural feature influences their biological roles. Common dietary sources include parsley, celery, and green peppers (Dias et al., 2021). Apigenin, a well-known flavone found in parsley and chamomile, exhibits anti-inflammatory, antioxidant, and anticancer activities and is being investigated for potential therapeutic applications.

Flavonols

Flavonols differ from flavones by having a hydroxyl group at position 3 of the C-ring (Lu, Qi, Zhao, & Jin, 2020). They are abundant in apples, onions, grapes, and kale, and key examples include quercetin, kaempferol, and myricetin. Quercetin, in particular, has been widely studied for its ability to combat oxidative stress, reduce inflammation, and protect against cancer. Evidence also suggests quercetin supports cardiovascular health, boosts immune function, and protects the brain from damage (Azeem et al., 2023).

Isoflavones

Isoflavones are mainly found in legumes, especially soybeans, and are structurally unique because the B-ring is attached differently compared with flavones and flavonols. Genistein and daidzein are notable examples that mimic estrogen activity in the body (Sarao, Kaur, Malik, & Singh, 2022). Due to these estrogen-like effects, isoflavones may benefit individuals with breast or prostate cancer and help alleviate menopausal symptoms.

Flavan-3-ols (Catechins)

Flavan-3-ols, often referred to as catechins, are primarily found in green tea, cocoa, and red wine. They contain a hydroxyl group at the third position of the C-ring (Spiegel et al., 2020). Catechins such as epicatechin, epicatechin gallate, and epigallocatechin gallate (EGCG) are extensively studied, with EGCG gaining particular attention for its anticancer and antioxidant effects. Green tea, which is rich in catechins, has been associated with a reduced risk of cancer, improved cardiovascular health, and enhanced metabolism (Xing et al., 2019).

Anthocyanins

Anthocyanins are pigments responsible for the red, purple, and blue colors of many fruits and vegetables. These flavonoids are often modified with sugar molecules, which influence their solubility and biological activity. Rich sources include berries, cherries, grapes, and eggplants (Deis et al., 2021). Studies suggest that anthocyanins reduce oxidative stress, lower inflammation, and improve the health of the heart, brain, and eyes. They may also decrease the risk of cardiovascular disease, diabetes, and cancer.

Flavanones

Flavanones are abundant in citrus fruits such as oranges, lemons, and grapefruits (Barreca et al., 2020). They are structurally distinct due to a hydroxyl group at the 5th position of the C-ring. Important examples include hesperidin and naringenin, which possess antioxidant, anti-inflammatory, and anticancer properties. Hesperidin, in particular, has been shown to enhance blood circulation, reduce hypertension, and support heart health.

Biological and Nutritional Importance

Beyond structural classification, flavonoids are often grouped by their biological activities. They act as **antioxidants**, reduce inflammation, protect against cancer, regulate glucose metabolism in diabetes, and play a protective role in neurodegenerative diseases (Ginwala et al., 2019). These diverse activities explain their importance in preventing and managing chronic diseases.

Dietary sources of flavonoids are widespread. Fruits, vegetables, herbs, nuts, seeds, tea, coffee, and wine all contribute to daily intake (Guven et al., 2019). For instance, citrus fruits provide flavanones, berries are rich in anthocyanins, green tea supplies catechins, while apples, kale, and onions are high in flavonols. Herbs such as parsley, thyme, and dill, and legumes like soybeans, are additional sources.

The **bioavailability of flavonoids** varies depending on the food matrix, the type of flavonoid, and preparation methods. Whole-food consumption often enhances absorption, as accompanying nutrients such as fiber and fats aid digestion (Shahidi & Pan, 2022). Other factors—such as gut microbiota composition, genetic variability, and overall health status—also influence metabolism and uptake.

In summary, flavonoids are a diverse group of plant-derived compounds with significant health-promoting properties. Their structural variations underpin a wide range of biological functions, from antioxidant and anti-inflammatory activity to cancer and neuroprotection. Among them, anthocyanins, flavonols, flavones, flavanones, catechins, and isoflavones stand out as dietary compounds with broad therapeutic potential. Ongoing research continues to highlight their role in preventing cardiovascular, metabolic, and neurodegenerative disorders, making flavonoids an essential component of both nutrition and therapeutic discovery (Dangles & Fenger, 2018).

3. MECHANISMS OF ACTION IN NEUROPROTECTION

Neuroprotection refers to safeguarding neurons from illness and injury, including harmful effects such as oxidative stress, inflammation, excitotoxicity, and metabolic changes. Without these protective mechanisms, neurodegenerative diseases progress, leading to serious consequences such as loss of memory, impaired motor control, and overall decline in brain function (Armstrong, 2020). Various strategies, including the use of antioxidants and anti-inflammatory agents, can help defend neurons against damage. Flavonoids, because of their antioxidant and anti-inflammatory properties, are being investigated as potential treatments to protect the brain (Kruszka et al., 2025).

Chapter 3 highlights the neuroprotective pathways influenced by flavonoids, with a particular focus on their ability to scavenge reactive oxygen species (ROS) and inhibit NF-κB and COX-2 activity. Oxidative stress is a major contributor to neuronal damage in neurodegenerative disorders. The brain is especially vulnerable to oxidative injury because of its high metabolic activity and the abundance of polyunsaturated fatty acids in neurons, which are highly susceptible to oxidative damage (Schönfeld & Reiser, 2021). Compounding this vulnerability, the brain naturally produces fewer protective enzymes, leaving it more exposed to damage from free radicals.

ROS, such as superoxide anion, hydrogen peroxide, and hydroxyl radicals, can damage lipids, proteins, and DNA, causing widespread cellular injury. When ROS accumulate excessively, they overwhelm natural antioxidant defenses, leading to oxidative stress. This imbalance activates immune responses and can ultimately trigger neuronal cell death (Volpe, Villar-Delfino, Dos Anjos, & Nogueira-Machado, 2018). The body's primary antioxidant defenses include enzymes such as superoxide dismutase (SOD), catalase, and glutathione peroxidase. However, if ROS levels surpass the protective capacity of these enzymes, oxidative stress contributes to the onset and progression of neurodegenerative diseases, including Alzheimer's, Parkinson's, and Huntington's disease.

Flavonoids combat this problem by neutralizing free radicals and preventing their accumulation. Their chemical structure, particularly hydroxyl groups on aromatic rings, allows them to scavenge ROS and protect cellular integrity (Wang, Li, & Bi, 2018). This antioxidant role is especially important in the brain, where oxidative damage is a key driver of neurodegeneration. Examples of flavonoids with strong neuroprotective effects include quercetin, catechins, and anthocyanins. These compounds directly remove ROS and also enhance the activity of endogenous antioxidant enzymes such as SOD and catalase (Hassan et al., 2022).

Quercetin, found in foods like apples, onions, and tea, has been shown in experiments to reduce ROS levels and protect lipids, proteins, and DNA in neuronal cells from oxidative harm (Grewal et al., 2021). Similarly, epigallocatechin gallate (EGCG) and other catechins present in green tea act as potent antioxidants, inhibiting lipid peroxidation and boosting the brain's own antioxidant enzyme systems (Payne, Nahashon, Taka, Adinew, & Soliman, 2022). Through these mechanisms, flavonoids not only scavenge harmful ROS but also strengthen the brain's defenses against oxidative stress, thereby reducing the risk of neurodegenerative disease.

In addition to their antioxidant effects, flavonoids also exert neuroprotection by regulating inflammation. Chronic inflammation in the brain is a major contributor to neuronal injury. Normally, microglial cells serve a protective role by clearing debris and pathogens. However, in many neurodegenerative conditions, microglia remain persistently overactivated, releasing inflammatory molecules such as TNF-α, IL-1β, IL-6, and prostaglandins that damage neurons (Yu et al., 2022). This prolonged neuroinflammation exacerbates oxidative stress and disrupts neuronal survival signaling, accelerating disease progression. Therefore, controlling inflammation is essential for preserving neuronal health and slowing neurodegeneration (Henek et al., 2018).

Figure 2. Advanced delivery methods

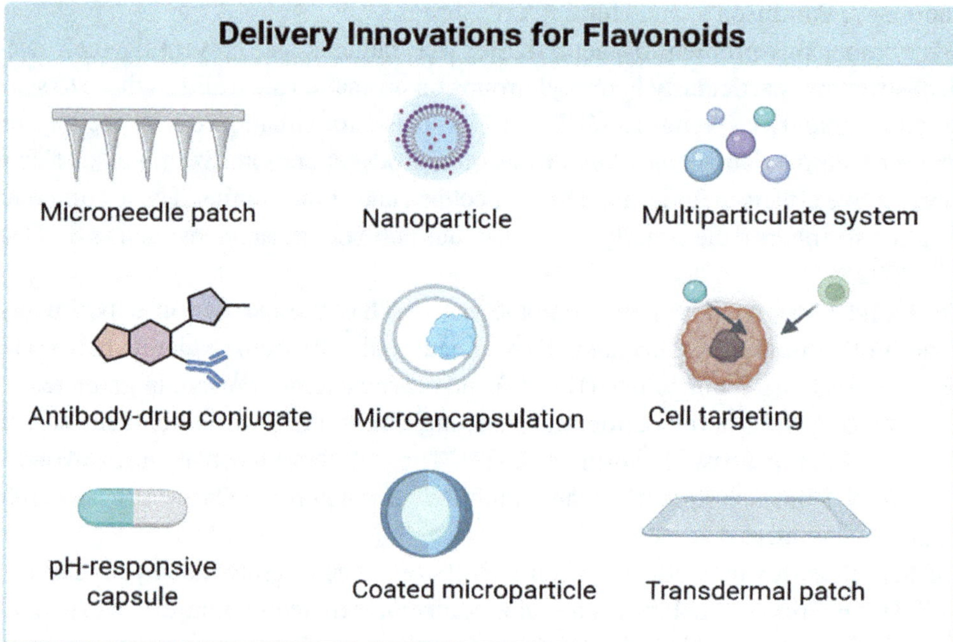

Table 1. Therapeutic roles and mechanisms of flavonoids in neurodegenerative diseases

Neurodegenerative Disease	Flavonoids Involved	Key Mechanisms	Effects/Outcomes
Alzheimer's Disease (AD)	Quercetin, EGCG, Curcumin	- Reduces amyloid-beta plaque formation- Inhibits tau hyperphosphorylation- Reduces oxidative stress and neuroinflammation- Activates neuroprotective enzymes- Modulates NF-κB, NLRP3 inflammasome pathways	- Improves cognitive function- Enhances memory and learning in animal models- Promotes clearance of toxic protein aggregates- Reduces inflammation and plaque buildup
Parkinson's Disease (PD)	Quercetin, EGCG	- Antioxidant action against ROS- Enhances mitochondrial function- Inhibits α-synuclein aggregation- Protects dopaminergic neurons	- Preserves dopamine-producing neurons- Improves motor function in PD models- May slow disease progression
Stroke / Cerebral Ischemia	Quercetin, EGCG, Resveratrol	- Reduces oxidative stress- Attenuates inflammation- Activates PI3K-Akt signaling pathway- Promotes neurogenesis and synaptic plasticity	- Decreases neuronal damage- Enhances functional recovery post-stroke- Supports neuroprotection and healing
Multiple Sclerosis (MS)	Various (esp. Quercetin, EGCG)	- Immunomodulatory effects- Reduces neuroinflammation- Regulates microglia and astrocyte activation	- Supports immune system balance- Alleviates inflammation-related symptoms- Potential improvement in cognitive function

Neuroprotection by flavonoids in stroke may take place through differing molecular routes (Xu, Wang, Chen, Xiao, & Wang, 2021). Consuming flavonoids can stop ROS and pro-inflammatory cytokines from being made. Without these, ischemia does less damage to the brain. Flavonoids manage apoptosis by blocking caspase activation and causing Bcl-2 expression which prevents cells from dying (Rahman et al., 2021). In addition, flavonoids enhance levels of brain-derived neurotrophic factor (BDNF) and other growth factors needed for surviving and reviving neurons after a stroke. Multiple sclerosis (MS) is a disease in which myelin, the protective covering of nerve fibers in the central nervous system, is destroyed by the body's immune system (Fymat, 2023). MS is linked to neuroinflammation, increased oxidative stress and problems with the immune system which in turn damages nerves and causes disabilities to keep getting worse. Researchers believe that flavonoids may be effective in treating MS because they control immune responses and decrease inflammation (Martínez et al., 2019). Flavonoids affect MS by controlling T cell function, microglia and how pro-inflammatory cytokines are made. TNF-α and IL-6 are important cytokines involved in MS and quercetin and curcumin help to prevent their release when T cells are activated (Qureshi, Al-Suhaimi, Wahid, Shehzad, & Shehzad, 2018). Flavonoids have been found to improve disease symptoms and limit harm to myelin in mice with MS. Flavonoids do this by decreasing the activity and inflammation levels of microglia which live in the central nervous system. Published results indicate flavonoids may be effective in treating MS (Maher, 2019). It has been shown that increasing your intake of flavonoids could slow the myelin loss and make it easier to move for animals with MS. In addition, resveratrol and other flavonoids seem to prevent brain deterioration and help regenerate the covering of damaged axons. While research on people with MS is rare, preclinical evidence suggests that flavonoids could strengthen the immune system and lower inflammation in the brain.

In short, flavonoids have shown strong neuroprotective results in managing Alzheimer's disease, Parkinson's disease, stroke and multiple sclerosis. The antioxidant, anti-inflammatory and cell-protective abilities of flavonoids help improve brain function, shield dopaminergic neurons, boost healing in the brain after injury and influence immune reactions found in some autoimmune illnesses. Whereas studies done in labs show that flavonoids work in these diseases, additional clinical testing is needed to reveal their value and help guide their use by doctors.

Table 2. Modern delivery strategies for enhancing flavonoid efficacy in neurodegenerative diseases

Delivery Method	Description	Key Mechanisms/Advantages	Impact on Flavonoid Therapy
Liposomes	Lipid bilayer vesicles	- Enhance solubility of hydrophobic flavonoids- Protect against degradation- Targeted release with surface modification	- Improved bioavailability- Increased brain delivery- Enhanced therapeutic effectiveness
Nanoparticles (PLGA-based)	Biodegradable polymeric carriers	- Controlled release- Improved stability and solubility- Ligand attachment for BBB targeting	- Prolonged circulation time- Site-specific delivery to neurons- Enhanced neuroprotection
Polymeric Micelles	Amphiphilic copolymer carriers	- Solubilize hydrophobic flavonoids- Stabilize drug in aqueous environments	- Enhanced absorption- Better drug retention in the body
Dendrimers	Branched synthetic polymers	- High drug loading- Surface modification for targeted delivery	- Extended half-life- Specific brain region targeting
Nanostructured Lipid Carriers (NLCs)	Lipid matrix-based nanoparticles	- High drug loading- Sustained release- BBB penetration using targeting ligands	- Superior bioavailability- Increased brain cell protection
Solid Lipid Nanoparticles (SLNs)	Solid lipid-based nanoformulations	- Controlled and sustained release- Improved stability and protection from degradation	- Enhanced efficacy in animal models- Better brain-targeting options

continued on following page

Table 2. Continued

Delivery Method	Description	Key Mechanisms/Advantages	Impact on Flavonoid Therapy
Microneedles	Minimally invasive transdermal devices	- Direct bloodstream access- Avoids first-pass metabolism- Painless delivery	- Rapid absorption- Suitable for chronic therapy
Oral Systems (Enteric-coated capsules/soft gels)	Protected oral formulations	- Withstand stomach acid- Release in small intestine for absorption	- Improved gastrointestinal stability- Enhanced intestinal absorption
Advanced Nanocarriers (Gold NPs, Carbon Nanotubes, MSNs)	Innovative nanomaterials with tunable properties	- Controlled release- High surface area- Ligand-targeted delivery	- Long-lasting therapeutic action- Enhanced targeting and safety profile
Personalized Delivery Systems	Tailored to patient-specific factors	- Biomarker-based targeting- Dose and release customization	- Individualized treatment response- Optimized clinical outcomes

4. CLINICAL AND PRECLINICAL EVIDENCE

There is much evidence suggesting that flavonoids are helpful for brain health, in tests on animals as well as on humans. There are many of these compounds in fruits, vegetables and similar edibles. They boast antioxidant, anti-inflammatory and protect the nervous system. As they can impact neuromodulators, reduce cellular damage and control parts of the immune system, antioxidants may be beneficial in approaches to Alzheimer's disease, Parkinson's disease and stroke. However, despite the promising results from preclinical research, clinical outcomes are still unclear, so more study is needed to fully understand their uses. Scientists have found conflicting results when testing flavonoids for their effects on neurological diseases in clinical trials. Several researchers have studied how flavonoids might help manage Alzheimer's disease, Parkinson's disease and cognitive problems (Melrose, 2023). Curcumin which comes from turmeric, is a popular choice for study in clinical trials because it displays strong antioxidant and anti-inflammatory effects. It seems that curcumin may be helpful for Alzheimer's patients, as it lowers amyloid-beta plaque formation and controls neuroinflammation. Taking curcumin supplements in an RCT improved brain function and lowered signs of oxidative stress in Alzheimer's patients, suggesting it may delay the course of the disease (Chico et al., 2018). It remains difficult for much of the curcumin we eat to be absorbed and stay intact, since both digestion and the liver break it down quickly. Consequently, approaches that help the body take in curcumin more efficiently such using nanoparticles, are being studied in clinical settings. Researchers have examined quercetin as a possible treatment for age-related memory problems and Parkinson's disease. In older people taking part in a clinical study, quercetin supplementation was associated with improvements in cognitive function, perhaps because of its antioxidant and anti-inflammatory properties (Li, Chen, Yang, Qiao, & Zhang, 2021). Researchers have found that quercetin might protect dopamine-releasing neurons and help improve motor performance in rats with Parkinson's disease. Yet, the studies using quercetin for Parkinson's disease are not common. Similarly, since many people have a hard time absorbing quercetin, finding new ways to deliver it may be necessary for clinical use.

Another study is looking at EGCG which is one of the main polyphenols in green tea, for its ability to preserve brain cells. Many studies have looked into how it may be helpful in Alzheimer's and Parkinson's diseases. Research has found that EGCG, when included in trials, can control the inflammation in the brain and help by improving thinking and memory skills, especially for people with early signs of dementia. Though EGCG has been helpful as a treatment for problems in animals, further studies on humans are needed since clinical findings are mixed (Eng, Thanikachalam, & Ramamurthy, 2018). The outcomes from these studies may not be used widely, as they rely on few participants and have several different approaches. Although these findings look promising, certain factors keep these therapies from being applied in clinical settings. A key problem is understanding how well nutrients are available to the body (Whitney, Rolfes, Crowe, & Walsh, 2019). Most flavonoids, including quercetin, curcumin and EGCG, do not get absorbed well because they are rapidly used by the body and quickly eliminated. Therefore, very little of the active compound enters the brain and only modest medical effects are possible. In order to improve flavonoids' absorption by the body, researchers are using new methods such as nanoparticles, liposomes and encapsulation, to stabilize and improve their solubility.

5. CONCLUSION

Because flavonoids act as antioxidants, fight inflammation and control cell activity, they have shown great potential in protecting the brain from harm. Because of the widespread research that has proved they can reduce oxidative stress, lower neuroinflammation and help save neurons, these natural compounds are considered promising medicines for the treatment of Alzheimer's, Parkinson's, stroke and multiple sclerosis. Substances like flavonoids are believed to look after the health of neurons by influencing the action of neurotransmitters, supporting cell survival and shielding the blood-brain barrier. Thanks to their ability to scavenge ROS, flavonoids prevent oxidative damage leading to brain disorders. In addition, these antioxidants hinder neuroinflammation by controlling important pro-inflammatory pathways, making it less likely for brain cells to become damaged and better support brain wellness. Flavonoids are known to interfere with neurotransmitter pathways, supporting brain performance, positive feelings and trouble-free body movements. Especially in Alzheimer's and Parkinson's, loss of neurons is connected to problems with thinking and movement. Testing flavonoids in preclinical neurodegeneration models has demonstrated how they save neurons and stimulate recovery. According to studies, flavonoids can protect neurons in the brain of people with Parkinson's, lower signs of amyloid-beta in those with Alzheimer's and assist those who suffer from stroke and ischemic injuries. Still, findings from studies are mixed with some showing mental function and disease symptoms improve and others stating that effects are not significant. A main difficulty with clinical translation is that flavonoids are not easily absorbed by the body, making their effects in trials less noticeable. Still, a number of issues remain when we look at the findings from preclinical studies. The difficulties in their clinical application are greatly caused by the rapid metabolism and low solubility of flavonoids. Changes in the design of the study and how the flavonoids are taken may be the source of variability in clinical trials. Sufficient long-term safety information has not yet been found, namely for taking large doses of flavonoids. Still, even though preclinical research has shown flavonoids may be good for the brain, further research with people is needed to prove the point.

To continue using flavonoids to support neurological health, it is necessary to address the specific problems found here. Today's innovations in nanoparticles, liposomes and micelles can improve how flavonoids are delivered to the brain. A therapy approach that uses patients' genetic and molecular profiles might make flavonoids safer and more beneficial for different people. Doing further experiments and better clinical studies will be crucial for knowing whether flavonoids can help improve neurodegenerative diseases. All in all, flavonoids may serve as a natural approach to protecting the brain and studies suggest they can help retard the growth of neurodegenerative disorders. Although the research available is encouraging, the use of these medicines can be improved, mainly by exploring how to best give them, what the right doses are and how they should be used longer term. Further progress in methods of delivery and research could make flavonoids an important part of future neuroprotection treatments.

REFERENCES

Al-Khayri, J. M., Sahana, G. R., Nagella, P., Joseph, B. V., Alessa, F. M., & Al-Mssallem, M. Q. (2022). Flavonoids as potential anti-inflammatory molecules: A review. *Molecules (Basel, Switzerland)*, *27*(9), 2901. DOI: 10.3390/molecules27092901 PMID: 35566252

Armstrong, R. (2020). What causes neurodegenerative disease? *Folia Neuropathologica*, *58*(2), 93–112. DOI: 10.5114/fn.2020.96707 PMID: 32729289

Azeem, M., Hanif, M., Mahmood, K., Ameer, N., Chughtai, F. R. S., & Abid, U. (2023). An insight into anticancer, antioxidant, antimicrobial, antidiabetic and anti-inflammatory effects of quercetin: A review. *Polymer Bulletin*, *80*(1), 241–262. DOI: 10.1007/s00289-022-04091-8 PMID: 35125574

Barreca, D., Mandalari, G., Calderaro, A., Smeriglio, A., Trombetta, D., Felice, M. R., & Gattuso, G. (2020). Citrus flavones: An update on sources, biological functions, and health promoting properties. *Plants*, *9*(3), 288. DOI: 10.3390/plants9030288 PMID: 32110931

Bellavite, P. (2023). Neuroprotective potentials of flavonoids: Experimental studies and mechanisms of action. *Antioxidants*, *12*(2), 280. DOI: 10.3390/antiox12020280 PMID: 36829840

Borst, K., Dumas, A. A., & Prinz, M. (2021). Microglia: Immune and non-immune functions. *Immunity*, *54*(10), 2194–2208. DOI: 10.1016/j.immuni.2021.09.014 PMID: 34644556

Burton-Freeman, B., Brzeziński, M., Park, E., Sandhu, A., Xiao, D., & Edirisinghe, I. (2019). A selective role of dietary anthocyanins and flavan-3-ols in reducing the risk of type 2 diabetes mellitus: A review of recent evidence. *Nutrients*, *11*(4), 841. DOI: 10.3390/nu11040841 PMID: 31013914

Calderaro, A., Patanè, G. T., Tellone, E., Barreca, D., Ficarra, S., Misiti, F., & Laganà, G. (2022). The neuroprotective potentiality of flavonoids on Alzheimer's disease. *International Journal of Molecular Sciences*, *23*(23), 14835. DOI: 10.3390/ijms232314835 PMID: 36499159

Calis, Z., Mogulkoc, R., & Baltaci, A. K. (2020). The roles of flavonols/flavonoids in neurodegeneration and neuroinflammation. *Mini-Reviews in Medicinal Chemistry*, *20*(15), 1475–1488. DOI: 10.2174/1389557519666190617150051 PMID: 31288717

Chen, B., Zhao, J., Zhang, R., Zhang, L., Zhang, Q., Yang, H., & An, J. (2022). Neuroprotective effects of natural compounds on neurotoxin-induced oxidative stress and cell apoptosis. *Nutritional Neuroscience*, *25*(5), 1078–1099. DOI: 10.1080/1028415X.2020.1840035 PMID: 33164705

Chen, L., Cao, H., Huang, Q., Xiao, J., & Teng, H. (2022). Absorption, metabolism and bioavailability of flavonoids: A review. *Critical Reviews in Food Science and Nutrition*, *62*(28), 7730–7742. DOI: 10.1080/10408398.2021.1917508 PMID: 34078189

Chen, S., Wang, X., Cheng, Y., Gao, H., & Chen, X. (2023). A review of classification, biosynthesis, biological activities and potential applications of flavonoids. *Molecules (Basel, Switzerland)*, *28*(13), 4982. DOI: 10.3390/molecules28134982 PMID: 37446644

Chico, L., Ienco, E. C., Bisordi, C., Lo Gerfo, A., Petrozzi, L., Petrucci, A., Mancuso, M., & Siciliano, G. (2018). Amyotrophic lateral sclerosis and oxidative stress: A double-blind therapeutic trial after curcumin supplementation. *CNS & Neurological Disorders - Drug Targets*, *17*(10), 767–779. DOI: 10.2174/1871527317666180720162029 PMID: 30033879

Dangles, O., & Fenger, J.-A. (2018). The chemical reactivity of anthocyanins and its consequences in food science and nutrition. *Molecules (Basel, Switzerland)*, *23*(8), 1970. DOI: 10.3390/molecules23081970 PMID: 30087225

Deis, L., Quiroga, A. M., & De Rosas, M. I. (2021). Coloured compounds in fruits and vegetables and health. *Psychiatry and Neuroscience Update: From Epistemology to Clinical Psychiatry–Vol., IV*, 343–358. DOI: 10.1007/978-3-030-61721-9_25

Dias, M. C., Pinto, D. C., & Silva, A. M. (2021). Plant flavonoids: Chemical characteristics and biological activity. *Molecules (Basel, Switzerland)*, *26*(17), 5377. DOI: 10.3390/molecules26175377 PMID: 34500810

Eng, Q. Y., Thanikachalam, P. V., & Ramamurthy, S. (2018). Molecular understanding of Epigallocatechin gallate (EGCG) in cardiovascular and metabolic diseases. *Journal of Ethnopharmacology*, *210*, 296–310. DOI: 10.1016/j.jep.2017.08.035 PMID: 28864169

Furtado, D., Björnmalm, M., Ayton, S., Bush, A. I., Kempe, K., & Caruso, F. (2018). Overcoming the blood–brain barrier: The role of nanomaterials in treating neurological diseases. *Advanced Materials*, *30*(46), 1801362. DOI: 10.1002/adma.201801362 PMID: 30066406

Fymat, A. L. (2023). Multiple sclerosis: I. Symptomatology and etiology. *J Neurol Psychol Res*, *4*(1).

Ginwala, R., Bhavsar, R., Chigbu, D. G. I., Jain, P., & Khan, Z. K. (2019). Potential role of flavonoids in treating chronic inflammatory diseases with a special focus on the anti-inflammatory activity of apigenin. *Antioxidants*, *8*(2), 35. DOI: 10.3390/antiox8020035 PMID: 30764536

Grewal, A. K., Singh, T. G., Sharma, D., Sharma, V., Singh, M., Rahman, M. H., & Albadrani, G. M. (2021). Mechanistic insights and perspectives involved in neuroprotective action of quercetin. *Biomedicine and Pharmacotherapy*, *140*, 111729. DOI: 10.1016/j.biopha.2021.111729 PMID: 34044274

Guven, H., Arici, A., & Simsek, O. (2019). Flavonoids in our foods: A short review. *Journal of Basic and Clinical Health Sciences*, *3*(2), 96–106. DOI: 10.30621/jbachs.2019.555

Karak, P. (2019). Biological activities of flavonoids: An overview. *International Journal of Pharmaceutical Sciences and Research*, *10*(4), 1567–1574.

Kruszka, J., Martyński, J., Szewczyk-Golec, K., Woźniak, A., & Nuszkiewicz, J. (2025). The Role of Selected Flavonoids in Modulating Neuroinflammation in Alzheimer's Disease: Mechanisms and Therapeutic Potential. *Brain Sciences*, *15*(5), 485. DOI: 10.3390/brainsci15050485 PMID: 40426656

Li, H., Chen, F.-J., Yang, W.-L., Qiao, H.-Z., & Zhang, S.-J. (2021). Quercetin improves cognitive disorder in aging mice by inhibiting NLRP3 inflammasome activation. *Food & Function*, *12*(2), 717–725. DOI: 10.1039/D0FO01900C PMID: 33338087

Lu, H., Qi, Y., Zhao, Y., & Jin, N. (2020). Effects of hydroxyl group on the interaction of carboxylated flavonoid derivatives with S. Cerevisiae α-glucosidase. *Current Computer-Aided Drug Design, 16*(1), 31–44. DOI: 10.2174/1573409914666181022142553 PMID: 30345924

Maher, P. (2019). The potential of flavonoids for the treatment of neurodegenerative diseases. *International Journal of Molecular Sciences, 20*(12), 3056. DOI: 10.3390/ijms20123056 PMID: 31234550

Martínez, G., Mijares, M. R., & De Sanctis, J. B. (2019). Effects of flavonoids and its derivatives on immune cell responses. *Recent Patents on Inflammation & Allergy Drug Discovery, 13*(2), 84–104. DOI: 10.2174/1872213X13666190426164124 PMID: 31814545

Melrose, J. (2023). The potential of flavonoids and flavonoid metabolites in the treatment of neurodegenerative pathology in disorders of cognitive decline. *Antioxidants, 12*(3), 663. DOI: 10.3390/antiox12030663 PMID: 36978911

Naomi, R., Yazid, M. D., Teoh, S. H., Balan, S. S., Shariff, H., Kumar, J., Bahari, H., & Embong, H. (2023). Dietary polyphenols as a protection against cognitive decline: Evidence from animal experiments; mechanisms and limitations. *Antioxidants, 12*(5), 1054. DOI: 10.3390/antiox12051054 PMID: 37237920

Payne, A., Nahashon, S., Taka, E., Adinew, G. M., & Soliman, K. F. (2022). Epigallocatechin-3-Gallate (EGCG): New therapeutic perspectives for neuroprotection, aging, and neuroinflammation for the modern age. *Biomolecules, 12*(3), 371. DOI: 10.3390/biom12030371 PMID: 35327563

Qureshi, M., Al-Suhaimi, E. A., Wahid, F., Shehzad, O., & Shehzad, A. (2018). Therapeutic potential of curcumin for multiple sclerosis. *Neurological Sciences, 39*(2), 207–214. DOI: 10.1007/s10072-017-3149-5 PMID: 29079885

Rahman, N., Khan, H., Zia, A., Khan, A., Fakhri, S., Aschner, M., Gul, K., & Saso, L. (2021). Bcl-2 modulation in p53 signaling pathway by flavonoids: A potential strategy towards the treatment of cancer. *International Journal of Molecular Sciences, 22*(21), 11315. DOI: 10.3390/ijms222111315 PMID: 34768743

Safe, S., Jayaraman, A., Chapkin, R. S., Howard, M., Mohankumar, K., & Shrestha, R. (2021). Flavonoids: Structure–function and mechanisms of action and opportunities for drug development. *Toxicological Research, 37*(2), 147–162. DOI: 10.1007/s43188-020-00080-z PMID: 33868973

Sarao, L., Kaur, S., Malik, T., & Singh, A. (2022). Genistein and daidzein. In *Nutraceuticals and health care* (pp. 331–341). Elsevier. DOI: 10.1016/B978-0-323-89779-2.00016-8

Schönfeld, P., & Reiser, G. (2021). How the brain fights fatty acids' toxicity. *Neurochemistry International, 148*, 105050. DOI: 10.1016/j.neuint.2021.105050 PMID: 33945834

Sciences, N. A. o., Medicine, Policy, Affairs, G., Data, B. o. R., Information, . . . Statistics, T. (2019). *Reproducibility and replicability in science*: National Academies Press.

Shahidi, F., & Pan, Y. (2022). Influence of food matrix and food processing on the chemical interaction and bioaccessibility of dietary phytochemicals: A review. *Critical Reviews in Food Science and Nutrition, 62*(23), 6421–6445. DOI: 10.1080/10408398.2021.1901650 PMID: 33787422

Singh, A., Kukreti, R., Saso, L., & Kukreti, S. (2019). Oxidative stress: A key modulator in neurodegenerative diseases. *Molecules (Basel, Switzerland), 24*(8), 1583. DOI: 10.3390/molecules24081583 PMID: 31013638

Spagnuolo, C., Moccia, S., & Russo, G. L. (2018). Anti-inflammatory effects of flavonoids in neurodegenerative disorders. *European Journal of Medicinal Chemistry, 153*, 105–115. DOI: 10.1016/j.ejmech.2017.09.001 PMID: 28923363

Spiegel, M., Andruniów, T., & Sroka, Z. (2020). Flavones' and flavonols' antiradical structure–activity relationship—A quantum chemical study. *Antioxidants, 9*(6), 461. DOI: 10.3390/antiox9060461 PMID: 32471289

Teleanu, R. I., Niculescu, A.-G., Roza, E., Vladâcenco, O., Grumezescu, A. M., & Teleanu, D. M. (2022). Neurotransmitters—Key factors in neurological and neurodegenerative disorders of the central nervous system. *International Journal of Molecular Sciences, 23*(11), 5954. DOI: 10.3390/ijms23115954 PMID: 35682631

van Dinteren, S. (2024). *Defining the window of opportunity of prenylated phenolics as natural antimicrobials: Extraction, antimicrobial activity, and impact on the intestinal epithelium.* Wageningen University and Research, Volpe, C. M. O., Villar-Delfino, P. H., Dos Anjos, P. M. F., & Nogueira-Machado, J. A. (2018). Cellular death, reactive oxygen species (ROS) and diabetic complications. *Cell Death & Disease, 9*(2), 119. PMID: 38331884

Vuolo, M. M., Lima, V. S., & Junior, M. R. M. (2019). Phenolic compounds: Structure, classification, and antioxidant power. In *Bioactive compounds* (pp. 33–50). Elsevier. DOI: 10.1016/B978-0-12-814774-0.00002-5

Wang, T.-y., Li, Q., & Bi, K. (2018). Bioactive flavonoids in medicinal plants: Structure, activity and biological fate. *Asian Journal of Pharmaceutical Sciences, 13*(1), 12–23. DOI: 10.1016/j.ajps.2017.08.004 PMID: 32104374

Wareham, L. K., Liddelow, S. A., Temple, S., Benowitz, L. I., Di Polo, A., Wellington, C., & Bu, G. (2022). Solving neurodegeneration: Common mechanisms and strategies for new treatments. *Molecular Neurodegeneration, 17*(1), 23. DOI: 10.1186/s13024-022-00524-0 PMID: 35313950

Whitney, E. N., Rolfes, S. R., Crowe, T., & Walsh, A. (2019). *Understanding nutrition.* Cengage AU.

Wróbel-Biedrawa, D., Grabowska, K., Galanty, A., Sobolewska, D., & Podolak, I. (2022). A flavonoid on the brain: Quercetin as a potential therapeutic agent in central nervous system disorders. *Life (Chicago, Ill.), 12*(4), 591. DOI: 10.3390/life12040591 PMID: 35455082

Xing, L., Zhang, H., Qi, R., Tsao, R., & Mine, Y. (2019). Recent advances in the understanding of the health benefits and molecular mechanisms associated with green tea polyphenols. *Journal of Agricultural and Food Chemistry, 67*(4), 1029–1043. DOI: 10.1021/acs.jafc.8b06146 PMID: 30653316

Xu, H., Wang, E., Chen, F., Xiao, J., & Wang, M. (2021). Neuroprotective phytochemicals in experimental ischemic stroke: Mechanisms and potential clinical applications. *Oxidative Medicine and Cellular Longevity, 2021*(1), 6687386. DOI: 10.1155/2021/6687386 PMID: 34007405

Yu, F., Wang, Y., Stetler, A. R., Leak, R. K., Hu, X., & Chen, J. (2022). Phagocytic microglia and macrophages in brain injury and repair. *CNS Neuroscience & Therapeutics*, *28*(9), 1279–1293. DOI: 10.1111/cns.13899 PMID: 35751629

Zughaibi, T. A., Suhail, M., Tarique, M., & Tabrez, S. (2021). Targeting PI3K/Akt/mTOR pathway by different flavonoids: A cancer chemopreventive approach. *International Journal of Molecular Sciences*, *22*(22), 12455. DOI: 10.3390/ijms222212455 PMID: 34830339

Chapter 8 Phytopharmaceuticals and the Gut–Brain Axis in Neurodegenerative Therapy

Hafiz Aamir Ali Kharl
https://orcid.org/0000-0002-7365-2937
University of Agriculture, Faisalabad, Pakistan

Arifa Mehreen
University of Agriculture, Faisalabad, Pakistan

Qurat-ul-Ane Gillani
https://orcid.org/0000-0002-1068-4472
Women University, Multan, Pakistan

Adnan Afzal
https://orcid.org/0009-0001-4359-5695
Cholistan University of Veterinary and Animal Sciences, Pakistan

Faiza Shakoor
Comsats University Islamabad, Vehari, Pakistan

Aiman Siddique
National University of Sciences and Technology, Islamabad, Pakistan

Muhammad Hussain Taqi
https://orcid.org/0009-0002-5868-0929
Cholistan University of Veterinary and Animal Sciences, Pakistan

Ayesha Bintay Farooq
Allied Health Sciences, Government College University, Faisalabad, Pakistan

ABSTRACT

The rising incidence of Alzheimer's and Parkinson's diseases has prompted the exploration of treatment options beyond conventional medicine. Increasing attention is being given to plant-based polyphenols for their role in modulating gut microbiota to support brain health. This chapter integrates pharmacy, microbiology, and neuroscience to explore polyphenols like curcumin, resveratrol, and quercetin, focusing on their anti-inflammatory and neuroprotective effects. It highlights how gut microbiota influences the activation and effectiveness of these compounds through short-chain fatty acids, neurotransmitters, and immune responses. Challenges such as poor bioavailability are addressed through advanced drug delivery systems including nanoformulations, phytosomes, and colon-targeted drugs. The chapter also evaluates the combined use of phytopharmaceuticals, functional foods, and probiotics in neurodegenerative disease management.

DOI: 10.4018/979-8-3373-4139-2.ch008

1. INTRODUCTION

These diseases Alzheimer's, Parkinson's and Huntington's pose a major challenge to people's health around the world and have big social and economic effects . Gradual injury to central nervous system neurons causes these diseases which result in poorer thinking, movement control and daily functions (Vinters & Kleinschmidt-DeMasters, 2018). The number of people affected by neurodegenerative diseases is rising because of an aging population and adjustments to people's lifestyles. Based on data from WHO, a current total of 50 million people have dementia and this number is expected to nearly triple by 2050 (Nichols et al., 2022). Because of the growing number of neurodegenerative disorders, many healthcare systems and caregivers struggle and the money needed to manage these conditions can easily run into billions of dollars each year. In fact, about 60-70% of all dementia cases are due to Alzheimer's disease (Association, 2019). With symptoms such as forgetfulness, confusion and a drop in thinking skills, AD often leads to disability in seniors. Another important neurodegenerative disease is Parkinson's which impacts millions of people worldwide by leading to tremors, stiff muscles and slow movements as the disease develops, often happening along with memory and thinking problems. Huntington's disease which is caused by heredity, causes people to find difficulty with movement, lose brain function and experience mental health problems (Stoker et al., 2022). While we have come to better understand these diseases, there are not many ways to treat them and a cure is still missing for all the main neurodegenerative conditions. Typically, drugs used to treat neurodegenerative diseases control symptoms but seldom stop the disease from getting worse. For Alzheimer's disease, doctors often prescribe acetylcholinesterase inhibitors (like donepezil, rivastigmine or galantamine) to address symptoms by raising brain acetylcholine levels which are short in people with AD. Still, the benefits these drugs give for thought processes are modest and do not impact the main brain damage in these disorders (Drossman et al., 2018). In Parkinson's disease, the drug levodopa is recommended because it is converted to dopamine in the brain and helps relieve movement problems. In the first years of the disease, levodopa is very effective, but over time, symptoms return and it may no longer help, so long-term use can also lead to serious side effects. Besides, available treatments for Parkinson's and Huntington's disease do not modify the condition, so it still gets worse over time.

Since conventional medicines have their own limits, other treatments are wanted as quickly as possible. Since neurodegenerative diseases are complicated, involving problems such as oxidative stress, neuroinflammation, protein aggregation and mitochondrial dysfunction, using a general approach to therapy is essential (Teleanu et al., 2022). Researchers are considering medicinal plants and substances from plants as possible treatments for brain health disorders. Phytopharmaceuticals have been around since ancient medicine and modern science is starting to recognize that they may also care for the brain. These natural substances include polyphenols, alkaloids, flavonoids and terpenoids–all of which influence certain pathways connected to neurodegeneration by lowering stress, controlling inflammation, supporting the growth of brain cells and boosting function of mitochondria. Because they act on many processes that can damage the brain, phytopharmaceuticals are becoming more widely studied (Patil et al., 2019). Unlike traditional drugs which usually target just one mechanism in the body, phytopharmaceuticals act on several cellular systems involved in causing disease. For instance, green tea, Ginkgo biloba and turmeric all possess strong antioxidant and anti-inflammatory properties that can keep neurons safe from being damaged (Ara, Maqbool, & Gani, 2022). They help get rid of damaging free radicals, reduce swelling and protect the brain from problems caused by amyloid-beta, all of which are important for Alzheimer's disease. In addition, particular phytopharmaceuticals support neuron well-being, promote the growth of new brain cells and promote synaptic plasticity which may assist in better brain function in diseases that result in brain cell loss. The bidirectional communication between the gut and the brain, known as the gut-brain axis, is an area of increasing investigation in neurodegenerative therapy (Loh et al., 2024). Most functions of the immune system, metabolism and mood depend on the gut-brain axis. Researchers have found that microorganisms in the gut microbiota can influence brain function and their alterations may lead to the development of neurodegenerative illnesses. A loss of balance in the gut microorganisms, called dysbiosis, may play a role in the development of Alzheimer's, Parkinson's and multiple sclerosis. By producing SCFAs, the gut microbiota can improve brain function by lowering inflammation and protecting brain cells (Y. P. Silva, Bernardi, & Frozza, 2020). Besides, gut microbes can manage the immune system which then links to both neuroinflammation and issues that lead to brain degeneration.

These phytopharmaceuticals can shape the connection between the gut and the brain by influencing gut bacteria and increasing healthy gut function. Some compounds that come from plants, including polyphenols, flavonoids and terpenoids, appear to encourage a variety of helpful bacteria in the gut and reduce growth of harmful bacteria. In fact, studies suggest that polyphenols in green tea, red wine and berries can help grow Bifidobacteria and Lactobacilli, important gut microbes linked to both healthier cognition and less inflammation in the brain (Filosa, Di Meo, & Crispi, 2018). They can also increase SCFAs which fuel the cells in your gut and protect the blood-brain barrier. Promoting a healthy gut microbiome with phytopharmaceuticals can modulate both neuroinflammation and oxidative stress, reducing the chance of neurodegenerative diseases. Little is still understood about how phytopharmaceuticals affect the gut-brain axis, so greater research needs to be done to figure it out. Still, research into how phytopharmaceuticals can boost brain health through the gut microbiota is very interesting. The latest research points to uses of plant-based compounds in stopping or easing neurodegenerative diseases by restoring the correct fit of the gut microbiota and stimulating helpful metabolites (Tan et al., 2021). It may offer a new, organic way to complement current medicines, letting doctors use a whole-body approach for handling Alzheimer's, Parkinson's and similar health problems.

Figure 1. Microbiota and the Gut-Brain Axis

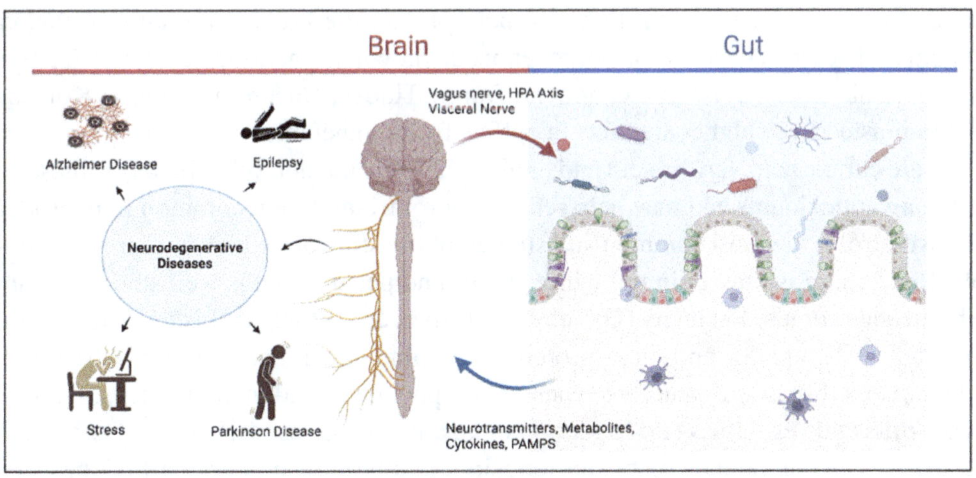

All in all, the rising number of neurodegenerative diseases globally means that new and combined treatment strategies are needed. While traditional drugs for mental health work well on the symptoms, they rarely solve the root problems and are usually not very effective (Sarris, 2018). Phytopharmaceuticals have numerous benefits because of their active ingredients and may supplement or replace traditional medicine. Phytopharmaceuticals could help brain health by targeting oxidative stress, neuroinflammation, clumps of abnormal proteins and issues in the mitochondria (S.-Y. Chen et al., 2020). Researchers are now exploring that the gut-brain axis may explain how plant compounds work on the brain by adjusting the level of bacteria in our digestive tract. With on-going studies into phytopharmaceuticals, they could become a key tool in managing and preventing neurodegenerative diseases for patients and their families.

2. POLYPHENOLS AS NEUROPROTECTIVE AGENTS

Teas, fruits, vegetables and herbs are rich in a family of chemicals called polyphenols. They have attracted interest due to their protection against brain damage, especially in diseases like Alzheimer's disease, Parkinson's disease and Huntington's disease. They cause neurons to degenerate which results in mental decline, problems with movement and many additional unpleasant symptoms (Fatuzzo et al., 2023). Their influence on several processes associated with Alzheimer's—such as oxidative stress, neuroinflammation, amyloid formation and damaged neurons—makes them attractive choices for therapy. This chapter discusses brain active polyphenols, the materials they are found in, their role in the brain and their pathways in the body.

Polyphenols are organized into various groups based on what they are chemically made of. Classes of polyphenols such as flavonoids, phenolic acids, stilbenes and lignans, are among the most familiar (Oluwole, Fernando, Lumanlan, Ademuyiwa, & Jayasena, 2022). Natural sources of food are frequently rich in the various subtypes of these chemical classes. The study of flavonoids among polyphenolic compounds has focused greatly on their ability to protect the brain (Bellavite, 2023). For example, this type includes flavonols, flavones, flavanones and isoflavones which you will find in apples, onions, citrus fruits and berries. Because flavonoids like quercetin, kaempferol and anthocyanins are strong antioxidants and reduce inflammation, they can help protect the brain from damage that can lead to neurodegenerative disease. This class of polyphenols is known as phenolic acids which are made from benzoic and cinnamic acids (Belščak-Cvitanović, Durgo, Huđek, Bačun-Družina, & Komes, 2018). In food, these vitamins occur in higher amounts in coffee, fruits, vegetables and whole grains. Examples of phenolic acids are caffeic acid, ferulic acid and gallic acid. Beyond their other benefits, these compounds function as strong antioxidants and may help reduce neuroinflammation common in neurodegenerative disorders. Resveratrol is the most common and studied of the stilbenes which comprise a group of lesser polyphenols. Resveratrol is present in red wine, grapes and peanuts and is well studied for its ability to protect the brain and promote longevity (Galiniak, Aebisher, & Bartusik-Aebisher, 2019). As it can fight oxidative damage, inflammation and sticky protein accumulation, it is being considered a good option for stopping Alzheimer's disease. Lignans are types of polyphenols found in many seeds, whole grains and some types of fruits and vegetables (Rodríguez-García, Sánchez-Quesada, Toledo, Delgado-Rodríguez, & Gaforio, 2019). The compounds secoisolariciresinol and lariciresinol are part of these compounds. Research has found that lignans might defend the brain against harmful effects by decreasing oxidative stress and inflammation. Most of the protection from neurodegenerative diseases given by polyphenols is due to their effect on several brain pathways involved in disease development. One way polyphenols influence health is by helping to clear free radicals from your body. Oxidative stress greatly increases the risk of damage and death to neurons found in neurodegenerative diseases (Singh, Kukreti, Saso, & Kukreti, 2019). These free radicals, especially reactive oxygen species (ROS), react strongly with lipids, proteins and DNA to cause problems and death in cells. Polyphenols help clear out free radicals so they can no longer damage neurons or cause oxidative injury to brain cells (Rudrapal et al., 2022). Quercetin and kaempferol are among the flavonoids that lower ROS in the body and caffeic acid belongs to the phenolic family that does the same. With a lowering of oxidative stress, polyphenols support cellular balance and protect the brain from cell loss.

3. MICROBIOTA AND THE GUT-BRAIN AXIS

The gut-brain axis describes how the GI tract connects with the brain by sharing information. The system helps support a healthy gut and brain (Chakrabarti et al., 2022). It means there are signals passing between the gut microbiota, enteric nervous system, vagus nerve and the central nervous system (CNS). Digestion, managing the immune system and feelings all depend on a healthy gut-brain axis (Miller, 2018). Studies in recent times have shown new links between gut health and the functioning of the brain, suggesting there is a role for treating brain illnesses with drugs for the gut. Because the gut microbiota interacts closely with the brain, it sheds light on why diseases like Alzheimer's disease, Parkinson's disease and multiple sclerosis begin. Many important structures are part of the gut-brain axis to support their active interaction. The "second brain" is the term given to the enteric nervous system, a huge system of neurons within the GI tract's lining (Sharkey & Mawe, 2023). It doesn't rely only on the central nervous system and is able to oversee functions in the intestines such as contraction, fluid release and blood activity. Messages are passed from the enteric nervous system to the brain by the vagus nerve. Part of the way our cells talking to each other involves the microbes in our gut. Trillions of bacteria, viruses, fungi and archaea, known as the human gut microbiota, live mainly in the large intestine (Matijašić et al., 2020). They help the body digest, metabolize food and defend against germs. Among the things it produces, the microbiota in the gut creates short-chain fatty acids, neurotransmitters and hormones that affect brain function. People are paying more attention these days to how the bacteria in our gut affect our brain and body inflammation. It is now accepted that our gut bacteria have several ways of changing how the brain functions. An important means by which gut microbes influence the brain is by emitting metabolites that serve as messaging molecules (Parker, Fonseca, & Carding, 2020). SCFAs,, butyrate and propionate, are formed in the gut from the bacteria-induced fermentation of fiber from foods. It has been found that these SCFAS may affect brain function by entering the brain via the blood-brain barrier. Butyrate seems to promote brain health by encouraging neurogenesis, reducing brain inflammation and enhancing the barrier between the blood and the brain (Alpino, Pereira-Sol, Dias, Aguiar, & Peluzio, 2024). The microorganisms in our gut also make neurotransmitters that help regulate our mood, thinking and movements. Most of the serotonin our bodies make comes from the gut and upsets in gut microbiota can lead to changes in serotonin production and brain activity. How the immune system works is impacted by the gut microbiota and in turn affects the brain. The gut's immune system detects and controls reactions to both food and microbes (Yoo, Groer, Dutra, Sarkar, & McSkimming, 2020). Proper gut bacteria help reduce chances of harmful inflammation in the body. Still, when the microbes in the gut get out of sync, too many pro-inflammatory chemicals can get into the brain through the blood and the vagus nerve. Cytokines may impact how the brain works and are often linked to neuroinflammation which plays a key role in developing many neurodegenerative diseases (W. Zhang, Xiao, Mao, & Xia, 2023). Microglia which are part of the immune system in the nervous system, can respond to infection elsewhere in the body. This results in their release of chemicals that can harm brain and nerve cells. For this reason, looking after healthy gut microbiota is necessary to manage immune functions and stop neuroinflammation.

Figure 2. Microbial Metabolism of Polyphenols

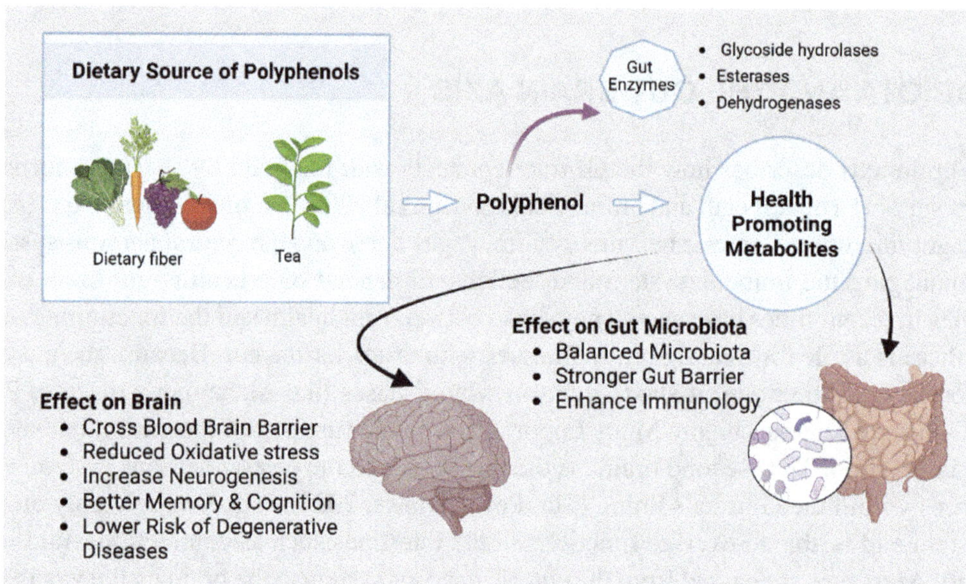

Dysbiosis of the gut microbiota plays a major role in how neurodegenerative diseases advance (H. Zhang et al., 2022). Many things such as diet, medicines, diseases, stress and age can affect how the gut microbiota changes. Disturbances in the microorganisms in the intestines may result in fewer types of microbes and have been linked to different health issues, for instance, neurodegenerative ones (Sochocka et al., 2019). If dysbiosis occurs, the body may not produce probiotic materials called SCFAs, the blood-brain barrier may become damaged and overall inflammation might rise, all adding to the risk of neurodegenerative diseases. Alzheimer's disease is made worse by changes in gut microbiota because it causes greater levels of amyloid-beta plaques and tau tangles, the disease's main symptoms. Alzheimer's disease has been linked to key changes in the gut microbiota, research has found (Mancuso & Santangelo, 2018). Because of these shifts, there are more pro-inflammatory cytokines around and lower SCFAs which can raise inflammation and the level of 'free radicals' in the brain. Moreover, issues with gut dysbiosis can weaken the gut-brain barrier so that dangerous substances can get into the brain. As a result of this disruption, toxins and immune cells are able to reach the brain and add to the damage to brain cells (Al-Obaidi & Desa, 2018). Our knowledge indicates that effective changes to gut microbiota may result in less inflammation and fewer amyloid-beta plaques in people with Alzheimer's disease. Before motor symptoms show up in Parkinson's disease, there is a change in the gut microbiota. In people with Parkinson's disease, abnormalities in their gut microbiota have been linked to increased levels of pro-inflammatory cytokines that can cause both bodywide and brain inflammation (Lin et al., 2019). Gut microbial activity may also affect the creation of alpha-synuclein aggregates which are a sign of Parkinson's disease. Research indicates that certain bacteria in the gut make chemicals that can change alpha-synuclein's structure and lead to its misfolding and buildup in the brain (Anis, Xie, Brundin, & Brundin, 2022). Finding this connection shows how important gut bacteria are in starting and worsening Parkinson's disease which suggests new approaches for treatment. Dysbiosis or a shift in the gut microbiome, is detectable in multiple sclerosis (MS), a condition involving immune-based neuroinflammation and loss of myelin in the brain (Sittipo, Choi, Lee, & Lee, 2022). Research has found that changes in gut microbes in people with MS result in the overactivity of immune cells which make T-cells target the myelin covering the nerves. A study has proven that Firmicutes and Bacteroidetes bacteria are likely to be involved in immunological reactions and might affect the development of MS in some people. Recent research revealed that imbalanced gut bacteria in MS patients can cause increased pro-inflammatory immune cells and a reduction in anti-inflammatory regulatory T-cells (Morshedi, Hashemi, Moazzen, Sahebkar, & Hosseinifard, 2019). Changing the gut microbes through what we eat or with special therapies might restore the immune balance and ease the symptoms of MS. Since the role of the gut-brain axis in neurodegenerative diseases is relatively new, researchers have already uncovered a relationship between problems in the gut and diseases such as Alzheimer's, Parkinson's and multiple sclerosis. It's essential to understand how the gut microbiota can play a role in brain function (by making neurotransmitters, influencing the immune system and controlling inflammation) to understand how the disease develops (Rutsch, Kantsjö, & Ronchi, 2020). Knowing the role the gut-brain axis plays in neurodegenerative diseases can guide treatments meant to improve gut health and slow the progress of these diseases. A number of approaches such as food modifications and gut microbiome therapies, are being tested to improve the health of your gut and brain. In short, the gut-brain axis links the gut and brain, helping to affect the brain's function and health. Samples have shown that dysbiosis or problems in the gut microbiota, may result in the development of neurodegenerative diseases. A shift in microbes, lower levels of positive metabolites such as short-chain fatty acids and more general inflammation lead to inflammation in the brain and damage to its cells in Alzheimer's, Parkinson's and multiple sclerosis (Missiego-Beltrán & Beltrán-Velasco, 2024). Bringing the microbes in the gut back into balance through changes in diet or probiotics could potentially help lower the risk of or slow down, neurodegenerative diseases and therefore deserves more attention by researchers.

4. MICROBIAL METABOLISM OF POLYPHENOLS

Because polyphenols are greatly altered in the gut by microbes, their potential health and therapeutic benefits are receiving more study. Many plant-based foods such as fruits, vegetables, tea and red wine contain lots of polyphenols that are valued for their antioxidant, anti-inflammatory and neuroprotective effects (Rojas-García, Fernández-Ochoa, Cádiz-Gurrea, Arráez-Román, & Segura-Carretero, 2023). Gut bacteria have to transform many polyphenols into bioactive substances for them to have a full impact on your health. Because gut bacteria break down polyphenols, they provide the body with helpful metabolites that may act on the brain and other areas. Polyphenols become more easily used for health when they are converted into urolithins, phenolic acids and similar bioactive compounds and this change greatly supports a healthy nervous system (Fraga, Croft, Kennedy, & Tomás-Barberán, 2019).

Table 1. Microbial Metabolism of Polyphenols and Their Impact on Brain Health

Polyphenol Source	Microbial Metabolites	Key Microbial Enzymes	Health Benefits of Metabolites	Impact on Brain Health
Ellagic acid (e.g., pomegranates, berries)	Urolithins (e.g., Urolithin A)	Glycoside hydrolases, dehydrogenases	Antioxidant, anti-inflammatory, enhances mitochondrial function	Crosses BBB; protects neurons; reduces neuroinflammation
Flavonoids (e.g., quercetin, kaempferol in onions, kale, apples)	Phenolic acids (e.g., 3,4-DHPAA, 3-HPAA)	Esterases, glycosidases	Antioxidant, anti-inflammatory	Reduces NF-κB activation, protects against oxidative stress and inflammation
Green tea polyphenols (EGCG)	Phenolic acids and smaller flavonoids	Polyphenol oxidases	Reduces microglial activity, cytokine production	Supports brain antioxidant systems; reduces neuroinflammation
Resveratrol (red wine, grapes)	SCFAs via microbial modulation	Fermentative bacteria (e.g., Bifidobacteria, Lactobacilli)	Modulates microbiota, increases SCFA production	Enhances serotonin/ dopamine levels, reduces neuroinflammation, supports neurogenesis
General Polyphenol Intake	Various bioactives (phenolics, SCFAs)	Depends on gut microbial composition	Promotes beneficial bacteria, inhibits pathogens	Improves gut-brain axis, strengthens blood-brain barrier, prevents neurodegenerative damage

5. FORMULATION CHALLENGES AND PHARMACEUTICAL INNOVATIONS

How much polyphenols can treat neurodegenerative illnesses like Alzheimer's disease and Parkinson's disease is largely based on their ability to accumulate enough in the proper tissues to work effectively. Despite their importance, polyphenols suffer from tough challenges when being combined and delivered because they are hardly soluble, not well absorbed and are quickly metabolized by the body (Aatif, 2023). Due to these challenges, they are not very effective in clinical use and fixing them will make polyphenols much better as neuroprotective substances. Various new approaches in drug development are now in use to make polyphenols more soluble, less likely to break down and easier for our bodies to take up. These approaches feature nanoemulsions, phytosomes, liposomes and systems that are designed to work well for the colon and respond to changes in pH. The goal is to link polyphenols with delivery vehicles so they end up where they are needed, in the brain, to help them protect the nervous system (Grabska-Kobyłecka et al., 2023). Poor solubility of polyphenols stands out as a major challenge when trying to use them effectively. Most polyphenols prefer fats and are not easily absorbed into water, making their uptake through the gastrointestinal tract limited (Jakobek, 2015). Because it takes a lot of work for the substance to dissolve, the amount entering the blood to reach the target brain is often small. In addition, because polyphenols have poor stability in water, they break down quickly as food moves through the digestive system. For example, curcumin is a polyphenol that is not easily absorbed by water and is rapidly processed in the liver which lessens its opportunities for use in medicine. As with other polyphenols, resveratrol and EGCG often encounter solubility and stability problems which hinders how effective they are when given by mouth (Yin et al., 2022). Among other problems, low bioavailability makes it difficult for polyphenols to be used effectively in treatments for neurological diseases. Most polyphenols are metabolized by the liver right after being absorbed in the gastrointestinal tract into their inactive form. The large metabolic processing of tea decreases the active polyphenol you absorb and so reduces its ability to help treat various conditions (Yan, Zhong, Duan, Chen, & Li, 2020). Moreover, since polyphenols are poorly absorbed into the brain through the blood-brain barrier, they face many difficulties in the treatment of neurodegenerative diseases. A number of strategies have been invented to solve these formulation problems and increase the solubility, stability and bioavailability of polyphenols. Using nanoemulsions is a highly effective way to improve how much polyphenols dissolve in water. Being so small, nanoemulsions improve the ability of polyphenols to be absorbed by the body and dissolve in water. Furthermore, nanotechnology helps protect polyphenols so they remain stable and can be better used by the body (Milinčić et al., 2019). Such formulations help move polyphenols past biologic barriers, including the one that protects the brain, making them more accessible to the tissues where they are needed. Using nanoemulsions with curcumin in animals with neurodegenerative conditions has improved both how readily curcumin enters the body and its effectiveness as a treatment. Phytosomes are another new method for improving the body's absorption of polyphenols. Phytosomes include both polyphenols and phospholipids which make up the main part of cell membranes (Dutt et al., 2023). As a result, complexation makes polyphenols more able to dissolve, keep stable and be used by the body. The absorption of polyphenols in the gut and their bioavailability improve when phytosomes are used rather than unformulated polyphenols. For example, curcumin attached to phosphatidylcholine, forming a phytosome, improves how the body takes in curcumin and makes it more useful for treating different diseases, including those of the brain. They boost the body's absorption of polyphenols and also deliver them to particular tissues, including the brain (Figueira, Menezes, Macedo, Costa, & Nunes dos Santos, 2017). Because phytosomes encourage polyphenols to be better absorbed and concentrated in the blood, they boost the protection for the brain and have greater potential as therapies against Alzheimer's and similar conditions.

6. POLYPHENOL-PROBIOTIC SYNERGY

People are showing growing interest in combining polyphenols and probiotics to help manage several illnesses, especially neurodegenerative diseases such as Alzheimer's disease, Parkinson's disease and other types of brain disorders. Both polyphenols and probiotics have complementary effects on our health, mainly helping brain function and the digestive system. Antioxidant and neuroprotective polyphenols that come from plants can greatly affect cellular signaling, oxidative stress and inflammation (Fakhri et al., 2022). Unlike prebiotics, probiotics are living organisms that, given in the appropriate dose, help maintain health mostly by guiding the activity of gut bacteria and boosting the health of the intestinal tract. Because the gut microbiota is key to good general health and handles neuroinflammatory, oxidative and brain responses, using polyphenols together with probiotics is seen as effective (Shabbir, Tyagi, Elahi, Aloo, & Oh, 2021). It is thought that when polyphenols and probiotics are combined, their therapeutic effects on the brain account for a broader range of benefits in people who could develop neurodegenerative conditions. The gut-brain axis makes sure the brain maintains good health and manages disease development. Neurological problems including neurodegenerative diseases can occur when the collection of gut microorganisms is upset (Makdissi, Parsons, & Di Cara, 2023). A lack of healthy microbes in the gut can increase the opening between our intestines and blood which allows harmful materials like inflammatory cytokines, toxins and amyloid-beta to enter the blood and move to the brain, supporting both neuroinflammation and brain damage. Because of this, probiotics may support a healthy balance of gut bacteria and lower inflammation, strengthen the gut barrier and boost brain health (Gou, Zhang, Ren, Li, & Zhang, 2022). Those polyphenols that are antioxidant and anti-inflammatory can further boost this support by changing the microbiota, controlling oxidative stress and defending nerve cells. In combination, polyphenols and probiotics help control the microbes in the gut, reduce body-wide inflammation and support the brain, making them a useful strategy against neurodegenerative diseases. One important way polyphenols and probiotics act together is by helping to control the makeup of gut bacteria. Many types of microorganisms called the gut microbiota work together in digesting food, making vitamins and taking care of the immune system (Rowland et al., 2018). Scientists have learned that fruits, vegetables, tea and red wine contain polyphenols which can boost the growth of helpful bacteria and limit the growth of hazardous bacteria in the gut. As an example, some evidence shows that polyphenols help to grow useful bacteria in the gut such as Lactobacilli and Bifidobacteria which are related to boosting the gut barrier and managing immune activity. Moreover, polyphenols affect the way metabolites like short-chain fatty acids (SCFAs), made by gut bacteria from fiber in the diet, are produced. SCFA metabolites are known to reduce inflammation and protect neurons (O'Riordan et al., 2022). Polyphenols are believed to grow more good bacteria in the gut and control the production of SCFAs, supporting the gut's health and helping it to talk to the brain. With probiotics, the same thing occurs in the gut, encouraging the development of healthy bacteria while slowing the growth of harmful ones. Probiotics appear to help increase the number of bacteria in the gut, thicken the lining of the stomach and control immune reactions (Xinzhou Wang, Zhang, & Zhang, 2021). By teaming probiotics with polyphenols, the good bacteria can thrive in the gut, reduce the openings in the intestines and guard against toxins getting into the blood. As a result of these effects working together, neurodegenerative diseases—where intestinal dysfunction and inflammation are often seen—can worsen. Research in animals and in people shows that combining polyphenols and probiotics may have additive effects on brain health. Tests on animals suggest that polyphenols like curcumin, resveratrol and EGCG may enhance brain function, lower brain inflammation and guard neurons from being damaged by diseases of the brain (Maqbool & Zehravi, 2021). Researchers found that polyphenols can support the brain by acting directly and also indirectly through their interaction with the gut microbiota. In animals with Alzheimer's disease models, using curcumin and probiotics together is able to reduce the buildup of amyloid-beta, boost memory and control brain inflammation (Asl, Sepehri, & Salami, 2019). Researchers have also measured how using probiotics with resveratrol in rodent models of Parkinson's disease can help dopaminergic neurons, reduce excessive cellular damage and improve movement. The results from human clinical trials show that the combination of polyphenols and probiotics may bring benefits (Ma & Chen, 2020). Research on mild cognitive impairment (MCI), a step before Alzheimer's disease, found that taking polyphenols from berries and a probiotic supplement at the same time improved memory test results and lowered signs of neuroinflammation. Researchers observed that taking a probiotic together with pomegranate extract helped elderly individuals remember things better and enhance other brain functions (Siddarth et al., 2020). It appears that using polyphenols and probiotics together might benefit cognitive function and help lower the risk of neurodegenerative diseases in people.

Table 2. Synergistic Effects of Polyphenols and Probiotics on Gut-Brain Axis and Neurodegenerative Diseases

Component	Primary Actions	Synergistic Effects	Impact on Neurodegenerative Diseases
Polyphenols	Antioxidant, anti-inflammatory, modulate gut microbiota	Promote growth of beneficial gut bacteria (e.g., *Bifidobacteria*, *Lactobacilli*), enhance SCFA production	Reduce neuroinflammation, oxidative stress, and amyloid-β accumulation; protect neurons
Probiotics	Enhance gut microbiota diversity and gut barrier integrity	Improve microbial balance, reduce gut permeability, and control immune responses	Prevent gut-derived inflammation from reaching brain; support dopaminergic neurons and cognitive function
Combined Intake	Dual modulation of gut microbiota and immune pathways	Greater SCFA production (e.g., butyrate), improved neurotransmitter levels (e.g., serotonin, dopamine), enhanced intestinal barrier function	Improve memory, reduce β-amyloid plaques, and boost cognition in Alzheimer's and Parkinson's models
Gut Microbiota	Metabolize polyphenols, produce SCFAs, maintain immune and gut health	Act as a mediator between polyphenol-probiotic actions and brain response	Healthier microbiota supports brain homeostasis via the gut-brain axis
Functional Foods/ Supplements	Contain both polyphenols and probiotics (e.g., yogurt + berries)	Convenient, natural strategy to deliver brain-protective and gut-supportive bioactives	Potential long-term neuroprotection, easier incorporation into diet and lifestyle

7. REGULATORY AND CLINICAL CONSIDERATIONS

Because more people are turning to phytopharmaceuticals, especially those rich in polyphenols, to treat and prevent neurodegenerative diseases, new concerns have arisen about their clinical development. The issues include ensuring safety, standard development, the setup of clinical trials, government regulation and possibilities for translating phytopharmaceuticals into pharmacy and nutraceuticals (Pavithra & Manimaran, 2024). Gaining knowledge about these factors is vital to making sure that polyphenol-based treatments are effective and safe enough to be used broadly for diseases like Alzheimer's disease, Parkinson's disease and others. Since more people are turning to plant-based treatments, it's important to learn how to deal with the rules so these natural medicines can be marketed and put to safe and effective medical use.

Safety plays a big role when it comes to developing treatments based on polyphenols. Even though polyphenols are generally safe when we get them from foods, their use in clinical treatments must be checked closely (R. F. Silva & Pogačnik, 2020). It means looking into side effects, toxic levels and how one drug might affect another. Even though high concentrations of things such as curcumin, resveratrol and EGCG found in food are risk-free, using these polyphenols as supplements in bigger amounts might result in adverse effects in a few people. Curcumin, for example, may result in upset stomach or interact with medicines that thin the blood (Bahramsoltani, Rahimi, & Farzaei, 2017). In addition, if you take more than the recommended amount, polyphenols could reduce how well your body takes in certain minerals or change liver activity. No polyphenol-based therapies can be given permission for use without first being properly studied and considered for safety. Polyphenol safety depends on comprehensive tests on animals, followed by testing them on humans. Most of these studies aim to figure out MTD and determine possible harmful effects (Navas, Cuppens, Cuppens, Toutain, & Papadopoulos, 2020). During Phase I, healthy people take the polyphenol-based treatment to assess its safety and to see how it is processed by the body. Patients with conditions such as Alzheimer's disease are invited to take part in Phase II and III clinical trials to assess how the treatment works and is safe. Long-term safety of polyphenol therapy should be checked, especially in people with other health problems or those using different medicines.

One more major obstacle in phytopharmaceutical clinical development is the need for standardization (Parveen, Parveen, Parveen, & Ahmad, 2015). Conventional pharmaceuticals are created in laboratories to ensure they are always the same, but polyphenols come from plants which contain different chemicals. Polyphenols may be concentrated differently and their bioactivity can change with the type of plant, where it is grown, the way it is produced and used and how it is grown and harvested (Brglez Mojzer, Knez Hrnčič, Škerget, Knez, & Bren, 2016). Because polyphenols can vary so much, it's challenging to guarantee their quality in clinical research or products sold to consumers. Making polyphenol extracts standard means setting rules for each batch to guarantee the same amount of active ingredients and fulfill safety and purity requirements set by the industry.

8. CONCLUSION AND FUTURE DIRECTIONS

Although polyphenols have vast applications in neurodegenerative disease therapy, using them fully as medicines is hindered by their weak absorption and quick breakdown by the body. Known antioxidants, anti-inflammatory agents and neuro-protectors, polyphenols are compounds naturally found in fruits, vegetables and tea. Experts believe they can alter major cellular processes linked to Alzheimer's, Parkinson's and related conditions. Still, using polyphenols as treatments in humans is problematic due to their inability to pass through the body quickly, be absorbed by the body well and their susceptibility to breakdown. Research has found that polyphenols do not normally get to sufficient concentrations in the brain when used medicinally. Because of their large molecules, poor solubility in water and first-pass breakdown by the liver, this issue occurs in these medications. Since it is hard for most drugs to get through the blood-brain barrier, treating neurodegenerative diseases with them is still difficult. Nanoemulsions, liposomes and phytosomes are new methods being used to improve the solubility, stability and how much of polyphenols the body can use from foods and dietary supplements. Thanks to their advanced release system, the polyphenols used in these foods can avoid destruction, pass well through our digestive system and enter the brain. As a result, better treatment options for neurodegenerative diseases may become available, since the polyphenols can get to the brain in enough quantity. In addition, new discoveries about the connection between the gut and brain have led to different kinds of medical treatments. Scientists have discovered that trillions of gut microorganisms play a key part in overseeing brain health. Polyphenols are thought to boost the development of beneficial bacteria and reduce those that are harmful to our bodies. It can also reduce swelling, create a stronger gut lining and improve how your brain works. Studies indicate that polyphenols can influence microbiota, reducing the risk of diseases affecting the brain and sometimes lessening the symptoms of people with these illnesses. Researchers are also investing in how polyphenol compounds can be used in customized medicine. It may be possible to make the use of polyphenols more effective by designing treatments that suit a person's genetic profile and gut microbiome. Different people metabolize polyphenols differently due to genetic differences and figuring out those differences can help find better ways to use them. Thanks to gut microbiome profiling, doctors may be able to choose the best polyphenol-based approach for each patient which would help ensure safety and success in treating them. Both the growth of nutraceuticals and the creation of polyphenol forms of therapy tend to support the brain and its functions. Working together, polyphenols and probiotics, prebiotics or similar substances may offer increased brain working and a lower likelihood of neurodegenerative diseases. Functional foods made from these special formulas provide a natural solution for protecting against and managing diseases.

Although it is apparent that polyphenols can aid in treating neurodegenerative diseases, some barriers exist about their bioavailability, ways of delivering them and how they are metabolized. New drug delivery solutions, customized medicine and controlling communication between the stomach and brain will help us make the most of polyphenols. Advanced studies and technological progress suggest that polyphenol-based treatments may become important in both preventing and controlling neurodegenerative diseases, providing a less complicated solution for brain health.

REFERENCES

Aatif, M. (2023). Current understanding of polyphenols to enhance bioavailability for better therapies. *Biomedicines, 11*(7), 2078. DOI: 10.3390/biomedicines11072078 PMID: 37509717

Al-Obaidi, M. M. J., & Desa, M. N. M. (2018). Mechanisms of blood brain barrier disruption by different types of bacteria, and bacterial–host interactions facilitate the bacterial pathogen invading the brain. *Cellular and Molecular Neurobiology, 38*(7), 1349–1368. DOI: 10.1007/s10571-018-0609-2 PMID: 30117097

Alpino, G. C. Á., Pereira-Sol, G. A., Dias, M. M. e., Aguiar, A. S., & Peluzio, M. C. G. (2024). Beneficial effects of butyrate on brain functions: A view of epigenetic. *Critical Reviews in Food Science and Nutrition, 64*(12), 3961–3970. DOI: 10.1080/10408398.2022.2137776 PMID: 36287024

Anis, E., Xie, A., Brundin, L., & Brundin, P. (2022). Digesting recent findings: Gut alpha-synuclein, microbiome changes in Parkinson's disease. *Trends in Endocrinology and Metabolism, 33*(2), 147–157. DOI: 10.1016/j.tem.2021.11.005 PMID: 34949514

Ara, I., Maqbool, M., & Gani, I. (2022). Neuroprotective Activity of Herbal Medicinal Products: A Review. *International Journal of Current Research in Physiology and Pharmacology*, 1-10.

Asl, Z. R., Sepehri, G., & Salami, M. (2019). Probiotic treatment improves the impaired spatial cognitive performance and restores synaptic plasticity in an animal model of Alzheimer's disease. *Behavioural Brain Research, 376*, 112183. DOI: 10.1016/j.bbr.2019.112183 PMID: 31472194

Association, A. s. (2019). 2019 Alzheimer's disease facts and figures. *Alzheimer's & dementia, 15*(3), 321-387.

Bahramsoltani, R., Rahimi, R., & Farzaei, M. H. (2017). Pharmacokinetic interactions of curcuminoids with conventional drugs: A review. *Journal of Ethnopharmacology, 209*, 1–12. DOI: 10.1016/j.jep.2017.07.022 PMID: 28734960

Bellavite, P. (2023). Neuroprotective potentials of flavonoids: Experimental studies and mechanisms of action. *Antioxidants, 12*(2), 280. DOI: 10.3390/antiox12020280 PMID: 36829840

Belščak-Cvitanović, A., Durgo, K., Huđek, A., Bačun-Družina, V., & Komes, D. (2018). Overview of polyphenols and their properties. In *Polyphenols: Properties, recovery, and applications* (pp. 3–44). Elsevier.

Brglez Mojzer, E., Knez Hrnčič, M., Škerget, M., Knez, Ž., & Bren, U. (2016). Polyphenols: Extraction methods, antioxidative action, bioavailability and anticarcinogenic effects. *Molecules (Basel, Switzerland), 21*(7), 901. DOI: 10.3390/molecules21070901 PMID: 27409600

Chakrabarti, A., Geurts, L., Hoyles, L., Iozzo, P., Kraneveld, A. D., La Fata, G., & Shortt, C. (2022). The microbiota–gut–brain axis: Pathways to better brain health. Perspectives on what we know, what we need to investigate and how to put knowledge into practice. *Cellular and Molecular Life Sciences : CMLS, 79*(2), 80. DOI: 10.1007/s00018-021-04060-w PMID: 35044528

Chen, S.-Y., Gao, Y., Sun, J.-Y., Meng, X.-L., Yang, D., Fan, L.-H., Xiang, L., & Wang, P. (2020). Traditional Chinese medicine: Role in reducing β-amyloid, apoptosis, autophagy, neuroinflammation, oxidative stress, and mitochondrial dysfunction of Alzheimer's disease. *Frontiers in Pharmacology*, *11*, 497. DOI: 10.3389/fphar.2020.00497 PMID: 32390843

Drossman, D. A., Tack, J., Ford, A. C., Szigethy, E., Törnblom, H., & Van Oudenhove, L. (2018). Neuromodulators for functional gastrointestinal disorders (disorders of gut– brain interaction): a Rome foundation working team report. *Gastroenterology, 154*(4), 1140-1171. e1141.

Dutt, Y., Pandey, R. P., Dutt, M., Gupta, A., Vibhuti, A., Raj, V. S., Chang, C.-M., & Priyadarshini, A. (2023). Liposomes and phytosomes: Nanocarrier systems and their applications for the delivery of phytoconstituents. *Coordination Chemistry Reviews*, *491*, 215251. DOI: 10.1016/j.ccr.2023.215251

Fakhri, S., Abbaszadeh, F., Moradi, S. Z., Cao, H., Khan, H., & Xiao, J. (2022). Effects of polyphenols on oxidative stress, inflammation, and interconnected pathways during spinal cord injury. *Oxidative Medicine and Cellular Longevity*, *2022*(1), 8100195. DOI: 10.1155/2022/8100195 PMID: 35035667

Fatuzzo, I., Niccolini, G. F., Zoccali, F., Cavalcanti, L., Bellizzi, M. G., Riccardi, G., & Minni, A. (2023). Neurons, nose, and neurodegenerative diseases: Olfactory function and cognitive impairment. *International Journal of Molecular Sciences*, *24*(3), 2117. DOI: 10.3390/ijms24032117 PMID: 36768440

Figueira, I., Menezes, R., Macedo, D., Costa, I., & Nunes dos Santos, C. (2017). Polyphenols beyond barriers: A glimpse into the brain. *Current Neuropharmacology*, *15*(4), 562–594. DOI: 10.2174/1570159X14666161026151545 PMID: 27784225

Filosa, S., Di Meo, F., & Crispi, S. (2018). Polyphenols-gut microbiota interplay and brain neuromodulation. *Neural Regeneration Research*, *13*(12), 2055–2059. DOI: 10.4103/1673-5374.241429 PMID: 30323120

Fraga, C. G., Croft, K. D., Kennedy, D. O., & Tomás-Barberán, F. A. (2019). The effects of polyphenols and other bioactives on human health. *Food & Function*, *10*(2), 514–528. DOI: 10.1039/C8FO01997E PMID: 30746536

Galiniak, S., Aebisher, D., & Bartusik-Aebisher, D. (2019). Health benefits of resveratrol administration. *Acta Biochimica Polonica*, *66*(1), 13–21. PMID: 30816367

Gou, H.-Z., Zhang, Y.-L., Ren, L.-F., Li, Z.-J., & Zhang, L. (2022). How do intestinal probiotics restore the intestinal barrier? *Frontiers in Microbiology*, *13*, 929346. DOI: 10.3389/fmicb.2022.929346 PMID: 35910620

Grabska-Kobyłecka, I., Szpakowski, P., Król, A., Książek-Winiarek, D., Kobyłecki, A., Głąbiński, A., & Nowak, D. (2023). Polyphenols and their impact on the prevention of neurodegenerative diseases and development. *Nutrients*, *15*(15), 3454. DOI: 10.3390/nu15153454 PMID: 37571391

Jakobek, L. (2015). Interactions of polyphenols with carbohydrates, lipids and proteins. *Food Chemistry*, *175*, 556–567. DOI: 10.1016/j.foodchem.2014.12.013 PMID: 25577120

Lin, C.-H., Chen, C.-C., Chiang, H.-L., Liou, J.-M., Chang, C.-M., Lu, T.-P., & Lin, H.-Y. (2019). Altered gut microbiota and inflammatory cytokine responses in patients with Parkinson's disease. *Journal of Neuroinflammation*, *16*(1), 1–9. DOI: 10.1186/s12974-019-1528-y PMID: 31248424

Loh, J. S., Mak, W. Q., Tan, L. K. S., Ng, C. X., Chan, H. H., Yeow, S. H., Foo, J. B., Ong, Y. S., How, C. W., & Khaw, K. Y. (2024). Microbiota–gut–brain axis and its therapeutic applications in neurodegenerative diseases. *Signal Transduction and Targeted Therapy*, *9*(1), 37. DOI: 10.1038/s41392-024-01743-1 PMID: 38360862

Ma, G., & Chen, Y. (2020). Polyphenol supplementation benefits human health via gut microbiota: A systematic review via meta-analysis. *Journal of Functional Foods*, *66*, 103829. DOI: 10.1016/j.jff.2020.103829

Makdissi, S., Parsons, B. D., & Di Cara, F. (2023). Towards early detection of neurodegenerative diseases: A gut feeling. *Frontiers in Cell and Developmental Biology*, *11*, 1087091. DOI: 10.3389/fcell.2023.1087091 PMID: 36824371

Mancuso, C., & Santangelo, R. (2018). Alzheimer's disease and gut microbiota modifications: The long way between preclinical studies and clinical evidence. *Pharmacological Research*, *129*, 329–336. DOI: 10.1016/j.phrs.2017.12.009 PMID: 29233677

Maqbool, M., & Zehravi, M. (2021). Neuroprotective role of polyphenols in treatment of neurological disorders: A review. *Interventional Pain Medicine and Neuromodulation, 1*(1).

Matijašić, M., Meštrović, T., Čipčić Paljetak, H., Perić, M., Barešić, A., & Verbanac, D. (2020). Gut microbiota beyond bacteria—Mycobiome, virome, archaeome, and eukaryotic parasites in IBD. *International Journal of Molecular Sciences*, *21*(8), 2668. DOI: 10.3390/ijms21082668 PMID: 32290414

Milinčić, D. D., Popović, D. A., Lević, S. M., Kostić, A. Ž., Tešić, Ž. L., Nedović, V. A., & Pešić, M. B. (2019). Application of polyphenol-loaded nanoparticles in food industry. *Nanomaterials (Basel, Switzerland)*, *9*(11), 1629. DOI: 10.3390/nano9111629 PMID: 31744091

Miller, I. (2018). The gut–brain axis: Historical reflections. *Microbial Ecology in Health and Disease*, *29*(2), 1542921. DOI: 10.1080/16512235.2018.1542921 PMID: 30425612

Missiego-Beltrán, J., & Beltrán-Velasco, A. I. (2024). The Role of Microbial Metabolites in the Progression of Neurodegenerative Diseases—Therapeutic Approaches: A Comprehensive Review. *International Journal of Molecular Sciences*, *25*(18), 10041. DOI: 10.3390/ijms251810041 PMID: 39337526

Morshedi, M., Hashemi, R., Moazzen, S., Sahebkar, A., & Hosseinifard, E.-S. (2019). Immunomodulatory and anti-inflammatory effects of probiotics in multiple sclerosis: A systematic review. *Journal of Neuroinflammation*, *16*(1), 1–11. DOI: 10.1186/s12974-019-1611-4 PMID: 31752913

Navas, R. E., Cuppens, F., Cuppens, N. B., Toutain, L., & Papadopoulos, G. Z. (2020). MTD, where art thou? A systematic review of moving target defense techniques for IoT. *IEEE Internet of Things Journal*, *8*(10), 7818–7832. DOI: 10.1109/JIOT.2020.3040358

Nichols, E., Steinmetz, J. D., Vollset, S. E., Fukutaki, K., Chalek, J., Abd-Allah, F., & Akram, T. T. (2022). Estimation of the global prevalence of dementia in 2019 and forecasted prevalence in 2050: An analysis for the Global Burden of Disease Study 2019. *The Lancet. Public Health*, *7*(2), e105–e125. DOI: 10.1016/S2468-2667(21)00249-8 PMID: 34998485

O'Riordan, K. J., Collins, M. K., Moloney, G. M., Knox, E. G., Aburto, M. R., Fülling, C., Morley, S. J., Clarke, G., Schellekens, H., & Cryan, J. F. (2022). Short chain fatty acids: Microbial metabolites for gut-brain axis signalling. *Molecular and Cellular Endocrinology*, *546*, 111572. DOI: 10.1016/j.mce.2022.111572 PMID: 35066114

Oluwole, O., Fernando, W. B., Lumanlan, J., Ademuyiwa, O., & Jayasena, V. (2022). Role of phenolic acid, tannins, stilbenes, lignans and flavonoids in human health–a review. *International Journal of Food Science & Technology*, *57*(10), 6326–6335. DOI: 10.1111/ijfs.15936

Parker, A., Fonseca, S., & Carding, S. R. (2020). Gut microbes and metabolites as modulators of blood-brain barrier integrity and brain health. *Gut Microbes*, *11*(2), 135–157. DOI: 10.1080/19490976.2019.1638722 PMID: 31368397

Parveen, A., Parveen, B., Parveen, R., & Ahmad, S. (2015). Challenges and guidelines for clinical trial of herbal drugs. *Journal of Pharmacy & Bioallied Sciences*, *7*(4), 329–333. DOI: 10.4103/0975-7406.168035 PMID: 26681895

Patil, K. R., Mahajan, U. B., Unger, B. S., Goyal, S. N., Belemkar, S., Surana, S. J., Ojha, S., & Patil, C. R. (2019). Animal models of inflammation for screening of anti-inflammatory drugs: Implications for the discovery and development of phytopharmaceuticals. *International Journal of Molecular Sciences*, *20*(18), 4367. DOI: 10.3390/ijms20184367 PMID: 31491986

Pavithra, K., & Manimaran, V. (2024). A Review of Safety, Quality, Regulation, and Delivery Approaches for Phytopharmaceuticals. *Jordan Journal of Pharmaceutical Sciences*, *17*(2), 316–332. DOI: 10.35516/jjps.v17i2.1768

Rodríguez-García, C., Sánchez-Quesada, C., Toledo, E., Delgado-Rodríguez, M., & Gaforio, J. J. (2019). Naturally lignan-rich foods: A dietary tool for health promotion? *Molecules (Basel, Switzerland)*, *24*(5), 917. DOI: 10.3390/molecules24050917 PMID: 30845651

Rojas-García, A., Fernández-Ochoa, Á., Cádiz-Gurrea, M. L., Arráez-Román, D., & Segura-Carretero, A. (2023). Neuroprotective effects of agri-food by-products rich in phenolic compounds. *Nutrients*, *15*(2), 449. DOI: 10.3390/nu15020449 PMID: 36678322

Rowland, I., Gibson, G., Heinken, A., Scott, K., Swann, J., Thiele, I., & Tuohy, K. (2018). Gut microbiota functions: Metabolism of nutrients and other food components. *European Journal of Nutrition*, *57*(1), 1–24. DOI: 10.1007/s00394-017-1445-8 PMID: 28393285

Rudrapal, M., Khairnar, S. J., Khan, J., Dukhyil, A. B., Ansari, M. A., Alomary, M. N., Alshabrmi, F. M., Palai, S., Deb, P. K., & Devi, R. (2022). Dietary polyphenols and their role in oxidative stress-induced human diseases: Insights into protective effects, antioxidant potentials and mechanism (s) of action. *Frontiers in Pharmacology*, *13*, 806470. DOI: 10.3389/fphar.2022.806470 PMID: 35237163

Rutsch, A., Kantsjö, J. B., & Ronchi, F. (2020). The gut-brain axis: How microbiota and host inflammasome influence brain physiology and pathology. *Frontiers in Immunology*, *11*, 604179. DOI: 10.3389/fimmu.2020.604179 PMID: 33362788

Sarris, J. (2018). Herbal medicines in the treatment of psychiatric disorders: 10-year updated review. *Phytotherapy Research*, *32*(7), 1147–1162. DOI: 10.1002/ptr.6055 PMID: 29575228

Shabbir, U., Tyagi, A., Elahi, F., Aloo, S. O., & Oh, D.-H. (2021). The potential role of polyphenols in oxidative stress and inflammation induced by gut microbiota in alzheimer's disease. *Antioxidants*, *10*(9), 1370. DOI: 10.3390/antiox10091370 PMID: 34573002

Sharkey, K. A., & Mawe, G. M. (2023). The enteric nervous system. *Physiological Reviews*, *103*(2), 1487–1564. DOI: 10.1152/physrev.00018.2022 PMID: 36521049

Siddarth, P., Li, Z., Miller, K. J., Ercoli, L. M., Merril, D. A., Henning, S. M., Heber, D., & Small, G. W. (2020). Randomized placebo-controlled study of the memory effects of pomegranate juice in middle-aged and older adults. *The American Journal of Clinical Nutrition*, *111*(1), 170–177. DOI: 10.1093/ajcn/nqz241 PMID: 31711104

Silva, R. F., & Pogačnik, L. (2020). Polyphenols from food and natural products: Neuroprotection and safety. *Antioxidants*, *9*(1), 61. DOI: 10.3390/antiox9010061 PMID: 31936711

Silva, Y. P., Bernardi, A., & Frozza, R. L. (2020). The role of short-chain fatty acids from gut microbiota in gut-brain communication. *Frontiers in Endocrinology*, *11*, 508738. DOI: 10.3389/fendo.2020.00025 PMID: 32082260

Singh, A., Kukreti, R., Saso, L., & Kukreti, S. (2019). Oxidative stress: A key modulator in neurodegenerative diseases. *Molecules (Basel, Switzerland)*, *24*(8), 1583. DOI: 10.3390/molecules24081583 PMID: 31013638

Sittipo, P., Choi, J., Lee, S., & Lee, Y. K. (2022). The function of gut microbiota in immune-related neurological disorders: A review. *Journal of Neuroinflammation*, *19*(1), 154. DOI: 10.1186/s12974-022-02510-1 PMID: 35706008

Sochocka, M., Donskow-Łysoniewska, K., Diniz, B. S., Kurpas, D., Brzozowska, E., & Leszek, J. (2019). The gut microbiome alterations and inflammation-driven pathogenesis of Alzheimer's disease—A critical review. *Molecular Neurobiology*, *56*(3), 1841–1851. DOI: 10.1007/s12035-018-1188-4 PMID: 29936690

Stoker, T. B., Mason, S. L., Greenland, J. C., Holden, S. T., Santini, H., & Barker, R. A. (2022). Huntington's disease: Diagnosis and management. *Practical Neurology*, *22*(1), 32–41. DOI: 10.1136/practneurol-2021-003074 PMID: 34413240

Tan, L. Y., Yeo, X. Y., Bae, H.-G., Lee, D. P. S., Ho, R. C., Kim, J. E., Jo, D.-G., & Jung, S. (2021). Association of gut microbiome dysbiosis with neurodegeneration: Can gut microbe-modifying diet prevent or alleviate the symptoms of neurodegenerative diseases? *Life (Chicago, Ill.)*, *11*(7), 698. DOI: 10.3390/life11070698 PMID: 34357070

Teleanu, D. M., Niculescu, A.-G., Lungu, I. I., Radu, C. I., Vladâcenco, O., Roza, E., Costăchescu, B., Grumezescu, A. M., & Teleanu, R. I. (2022). An overview of oxidative stress, neuroinflammation, and neurodegenerative diseases. *International Journal of Molecular Sciences*, *23*(11), 5938. DOI: 10.3390/ijms23115938 PMID: 35682615

Vinters, H. V., & Kleinschmidt-DeMasters, B. (2018). General pathology of the central nervous system. In *Greenfield's Neuropathology-Two Volume Set* (pp. 25–82). CRC Press.

Wang, X., Qi, Y., & Zheng, H. (2022). Dietary polyphenol, gut microbiota, and health benefits. *Antioxidants*, *11*(6), 1212. DOI: 10.3390/antiox11061212 PMID: 35740109

Wang, X., Zhang, P., & Zhang, X. (2021). Probiotics regulate gut microbiota: An effective method to improve immunity. *Molecules (Basel, Switzerland)*, *26*(19), 6076. DOI: 10.3390/molecules26196076 PMID: 34641619

Yan, Z., Zhong, Y., Duan, Y., Chen, Q., & Li, F. (2020). Antioxidant mechanism of tea polyphenols and its impact on health benefits. *Animal Nutrition*, *6*(2), 115–123. DOI: 10.1016/j.aninu.2020.01.001 PMID: 32542190

Yin, Z., Zheng, T., Ho, C.-T., Huang, Q., Wu, Q., & Zhang, M. (2022). Improving the stability and bioavailability of tea polyphenols by encapsulations: A review. *Food Science and Human Wellness*, *11*(3), 537–556. DOI: 10.1016/j.fshw.2021.12.011

Yoo, J. Y., Groer, M., Dutra, S. V. O., Sarkar, A., & McSkimming, D. I. (2020). Gut microbiota and immune system interactions. *Microorganisms*, *8*(10), 1587. DOI: 10.3390/microorganisms8101587 PMID: 33076307

Zhang, H., Chen, Y., Wang, Z., Xie, G., Liu, M., Yuan, B., Chai, H., Wang, W., & Cheng, P. (2022). Implications of gut microbiota in neurodegenerative diseases. *Frontiers in Immunology*, *13*, 785644. DOI: 10.3389/fimmu.2022.785644 PMID: 35237258

Zhang, W., Xiao, D., Mao, Q., & Xia, H. (2023). Role of neuroinflammation in neurodegeneration development. *Signal Transduction and Targeted Therapy*, *8*(1), 267. DOI: 10.1038/s41392-023-01486-5 PMID: 37433768

Zhang, Y., Yu, W., Zhang, L., Wang, M., & Chang, W. (2022). The interaction of polyphenols and the gut microbiota in neurodegenerative diseases. *Nutrients*, *14*(24), 5373. DOI: 10.3390/nu14245373 PMID: 36558531

Chapter 9

Polyphenol-Mediated Neuroprotection Against Traumatic Injury to the Nervous System

Hamna Khan
https://orcid.org/0009-0004-5200-9610
Government College University, Faisalabad, Pakistan

Bilal Waheed
https://orcid.org/0009-0000-5352-2126
Faisalabad Medical University, Pakistan

Maida Asif
Governement College University, Faisalabad, Pakistan

Usra Khan
https://orcid.org/0009-0006-3155-6011
Government College University, Faisalabad, Pakistan

Syeda Meryam Waseem
https://orcid.org/0009-0006-4169-5437
Governement College University, Faisalabad, Pakistan

Muhammad Hannan
https://orcid.org/0009-0006-4169-5437
Governement College University, Faisalabad, Pakistan

Hira Javaid
Governement College University, Faisalabad, Pakistan

ABSTRACT

Polyphenols are a broad class of chemicals that come from plants that have been shown to have the ability to defend the central and peripheral nervous systems against damage. Flavonoids, lignans, curcuminoids, phenolic acids, and stilbenes are members of this class and have potent anti-inflammatory, anti-apoptotic, and antioxidant qualities. Through influencing signaling pathways like NF-κB, Nrf2, PI3K/Akt, and MAPK, maintaining mitochondrial function, and lowering oxidative stress, polyphenols help neuronal survival and functional recovery in traumatic brain injury (TBI) and spinal cord injury (SCI). They increase Schwann cell activity, decrease neuroinflammation, and encourage axonal regeneration in peripheral nerve injuries (PNIs). Strong neuroprotective effects are demonstrated by substances such

DOI: 10.4018/979-8-3373-4139-2.ch009

as gallic acid, EGCG, resveratrol, hesperidin, curcumin, and quercetin. The classification, origins, and processes of polyphenols are highlighted in this chapter, with an emphasis on how they help to promote repair and lessen secondary damage in neurotrauma.

1. INTRODUCTION

Phytochemicals are substances derived from plants that are widely distributed in nature and have long been used to treat a wide range of illnesses. Because they have fewer or no threatening effects, they are attracting attention. Since ancient times, they have been utilized for therapeutic purposes. The beneficial effects of many phytochemicals, including alkaloids, flavonoids, and tannins, against brain disorders have recently been thoroughly demonstrated by Hussain et al (Hussain et al., 2018). In the kingdom of plants, phenols are one of the biggest and most prevalent classes of secondary metabolites. They are involved in several vital physiological processes. They are considered healthful to consume and play a significant role in the daily diet of humans as microconstituents. Phenolic compounds' physical and chemical characteristics make them adaptable ligands that can interact with a variety of targets, including the carbonic anhydrases (Karioti et al., 2016). Plants produce nonnutritional natural compounds called polyphenols, which are secondary metabolites that can be found in fruits, seeds, and vegetables. The phenylpropanoid pathway converts l-phenylalanine into the majority of polyphenols. These are distinguished by the presence of one or more phenolic groups with a wide variety of structures; the main structural types of polyphenols are flavonoids, phenolic acids, lignans, stilbenes, tannins, and coumarins (Quideau et al., 2011). Polyphenols are becoming more and more important, especially due to their health benefits. They are increasingly being used as antioxidants to prevent and treat a variety of illnesses, especially those related to oxidative stress, cancer, cardiovascular disease, inflammation, neurodegenerative diseases, and age-related degenerative diseases (Hano & Tungmunnithum, 2020). They can be used as dietary supplements, cosmetic additives, and medications, among other things. More than 8000 polyphenols have been identified from plants, and hundreds more have been described from fruits, grains, and vegetables (Berndtsson et al., 2019). Numerous studies back up polyphenolic compounds' neuroprotective properties, which include the capacity to reduce neuroinflammation, protect neurons from damage caused by neurotoxins, and maybe enhance memory, learning, and cognitive processes (Spagnuolo et al., 2016). The purpose of this study is to investigate the neuroprotective role of polyphenols in neurotrauma, with a focus on their therapeutic potential in central and peripheral nerve injuries. Figure 1 below describes the classification of poly-phenolic compounds.

Figure 1. Classification of poly-phenolic compounds

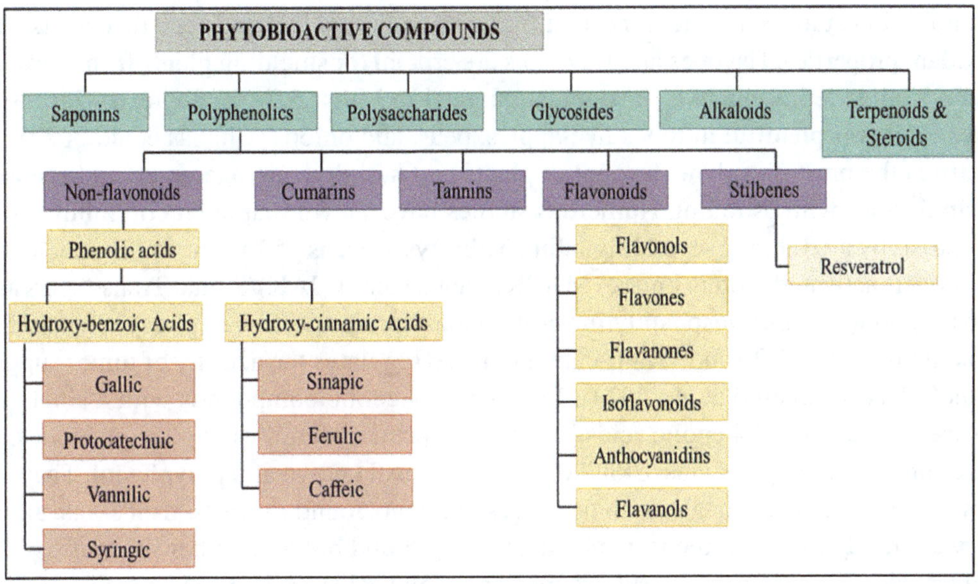

2 POLYPHENOLS: CLASSIFICATION AND SOURCES

Plant-derived compounds known as polyphenols help plants fend off a variety of environmental stresses. The largest class of phytochemicals with aromatic rings and ligand groups (primarily hydroxyl joined with various carbon sites) is called polyphenols. More than 10,000 natural polyphenols have been found and assessed so far. Natural polyphenols are secondary metabolites. Polyphenols can be categorized as either flavonoids or nonflavonoids based on their structural characteristics. A heterogeneous pyrone C ring connects the two aromatic rings that make up the 15-carbon skeleton that all flavonoids share. The structure of this carbon bridge is represented by the notation C6-C3-C6.

Flavonols, flavones, isoflavones, flavanones, anthocyanidins, flavan-3-ols, and minor subclasses of flavonoids, such chalcones, are among the various subgroups that flavonoids, which are heterocyclic ring-linked groups, contain.

Conversely, phenolic acids, hydroxycinnamic acids, stilbenes, lignans, and coumarins are referred to as non-flavonoids (Liu *et al.*, 2019). One of the most significant classes of naturally occurring antioxidants and chemopreventive agents is dietary polyphenols, which are present in fruits, vegetables, grains, tea, essential oils, and the foods and drinks that are made from them. Dietary phenolic compounds improve human health by reducing risk and delaying the onset of degenerative diseases like cancer, cardiovascular disease, and metabolic disorders, according to epidemiological, clinical, and nutritional research (Zhang and Tsao, 2016). These molecules may have anti-inflammatory qualities in addition to being strong antioxidants. Their potential involvement in preventing and treating a variety of clinical illnesses linked to inflammation and oxidative stress has been emphasized in numerous research (e.g., cancer, cardiovascular and neurological disorders) (Silva & Pogačnik, 2020). In recent decades, polyphenols—

chemical compounds abundant in plants—have gained attention in the field of nutrition (Biao, 2022). Fruits and vegetables like apples, berries, grapes, tomatoes, and onions contain flavonols, which are major flavonoids in nature. They are important for drawing pollinators and seed dispersers. Because of their antioxidant properties, flavones and flavonols are crucial for shielding plants from ultraviolet light. One of nature's most representative flavonoids, quercetin (3,3′,4′,5,7-pentahydroxyflavone) is also a dietary flavonol that is plentiful in foods like apples, peas, and onions (Chagas et al., 2022).

Quercetin is the most prevalent flavonol in plants and has phytoestrogenic effect because it shares structural similarities with estradiol. Numerous studies have shown that quercetin inhibits growth and induces apoptosis by modulating several signaling pathways, such as p53 a gene family, nuclear factor-κB (NF-κB), mitogen-activated protein kinase (MAPK), Janus Kinase(JAK)/Signal Transducers and Activators of Transcription(STAT), phosphatidylinositol 3-kinase/protein kinase B, phosphoinositide-3-kinase (PI3K)/protein kinase B(AKT) and Wnt/β-catenin. It also regulates the activity of tumor suppressor and oncogenic ncRNAs (Chimento et al., 2023). In general, phenolic compounds with a single carboxylic acid group are referred to as "phenolic acids." One of the primary families of plant phenolic compounds is phenolic or phenolcarboxylic acids, a kind of phytochemical known as a polyphenol. They are present in a wide range of plant-based meals, with the largest amounts found in seeds, fruit skins, and vegetable leaves. They are rarely found in free form and are usually found bound to other substances like amides, esters, or glycosides. The two primary subgroups of phenolic acids are hydroxybenzoic and hydroxycinnamic acid (Kumar and Goel, 2019). A diet high in antioxidant fruits and vegetables dramatically lowers the incidence of numerous oxidative stress-related diseases, such as cancer, diabetes, and cardiovascular disorders, according to epidemiological studies. A phenolic compound's antioxidant potential varies depending on the quantity and location of its hydroxyl groups (Kumar & Goel, 2019). Moreover, the methoxy and carboxylic acid groups also have important effects on the antioxidant ability of phenolic acids (Chen et al., 2020). Several unrelated plant species produce stilbenes, a small class of secondary metabolites that are formed from the phenylpropanoid pathway. These substances have a wide range of effects on human health and plant disease resistance (Chong et al., 2009).

Stilbene compounds are found in many plant species and belong to a large group of natural defense polyphenols. Found mostly in grape skin, resveratrol (3,5,4′-trihydroxy-trans-stilbene) is a well-known polyphenol phytoalexin that has garnered a lot of scientific interest because of its potential health benefits, including cardiovascular (French paradox), chemopreventive, antiobesity, antidiabetic, and neuroprotective qualities. Other stilbene compounds, such as pterostilbene (3,5-dimethyl ether derivative of resveratrol), may, however, have greater bioavailability and neuroprotective activity against AD than resveratrol itself, according to recent findings (Reinisalo et al., 2015). By coupling two coniferyl alcohol residues via the phenylpropanoid or shikimic acid pathway, lignans—structurally complex, bioactive polyphenolic compounds—are created. They are generally divided into two categories based on where they come from: plant lignans and mammalian lignans. Whereas mammalian lignans have hydroxyl groups in the meta position, plant lignans primarily have oxygenated substituents in the para locations. They are also divided into a number of classes according to their structure, including arylnaphtalene, aryltetraline lactone derivatives, dibenzylfuran, dibenzylbutyrolactol, and dihydroxybenzylbutane (Soleymani et al., 2020). The highest quantities of lignans, which are bioactive, non-nutritional, non-caloric phenolic plant chemicals, are found in flax and sesame seeds, while smaller concentrations are found in grains, other seeds, fruits, and vegetables (Peterson et al., 2010).

Non-flavonoids called lignan polyphenols have been shown to have anti-inflammatory, anti-cancer, and heart disease risk-lowering properties. Non-flavonoids called lignan polyphenols have been shown to have anti-inflammatory, anti-cancer, and heart disease risk-lowering properties (Xu et al., 2022). A class of naturally occurring compounds with a variety of phenolic structures, flavonoids can be found in tea, wine, fruits, vegetables, grains, bark, roots, stems, and flowers. The health benefits of these natural products are widely recognized, and attempts are underway to separate the components known as flavonoids (Panche et al., 2016). By scavenging free radicals and ROS, chelating metals, and inhibiting the oxidation of low-density lipoproteins (LDLs), flavonoids can demonstrate their exceptional antioxidant qualities. The most fundamental forms of flavonoids are flavones. All flavonoids have a C6-C3-C6 carbon skeleton from a structural perspective. The carbon atoms are arranged in three phenolic rings, A, B, and C, with oxygen typically found in the C ring (Hassanpour & Doroudi, 2023). Flavonoids, another name for hydroxylated polyphenols, are abundant in fruits, vegetables, cereals, nuts, herbs, seeds, stems, and flowers of many different plants. They have many therapeutic qualities, including anti-inflammatory, anti-cancer, anti-microbial, and neuroprotective (Al-Khayri et al., 2022).

Table 1. Role of Poly-phenolic compounds in Nervous system disorders

Compounds	Assay	Model	Disease	References
Fisetin	In vitro	Mouse motor neuron (NSC-34) mutant hSOD1G93A gene Male wildtype	Amyotrophic Lateral Sclerosis	(T. Wang et al., 2018)
	In vivo	C57BL/6N mice	Alzheimer's disease	(Ahmad et al., 2017)
Luteolin	In vitro	HT-22 Mouse Hippocampal Neuronal Cell Line	Alzheimer's disease	(Ahmad et al., 2021)
	In vitro	PC-12 cells (6-hydroxy-dopamine) (6-OHDA	Parkinson's disease	(Guo et al., 2013)
Resveratrol	In vitro	Neuro2a neuroblastoma cells	Caloric restriction associated with neurodegenerative diseases in general	(Dasgupta & Milbrandt, 2007)
	In vivo	Neonatal mice	Sevoflurane-induced cognitive impairment	(le Tang et al., 2021)
	In vivo	Gerbils	Global cerebral ischemia	(Wang et al., 2002)
Rosmarinic acid	In vitro	PC-12 cells	β-amyloid-induced oxidative stress	(Rong et al., 2018)
Ferulic acid	In vivo	Male Wistar rats—using rotenone (ROT)-induced rat model of PD	Parkinson's disease	(Ojha et al., 2015)
	In vivo	Male Sprague–Dawley rats and pheochromocytoma (PC-12) cells	Hypoxia/ischemia	(Ren et al., 2017)

continued on following page

Table 1. Continued

Compounds	Assay	Model	Disease	References
Quercetin	In vivo	Male Wistar rat	Parkinson's disease	(Karuppagounder et al., 2013)
Caffeic Acid	In vivo	Mice	Focal cerebral ischemia	(Fernandes et al., 2014)
Flavonoids	In vivo	Aged male rats	Age-related cognitive decline	(Williams et al., 2008)
Catechin	In vivo	6-OHDA lesioned rats	Parkinson's disease	(Teixeira et al., 2013)
Astilbin	In vivo	Male C57BL/6 mice (Parkinson's disease model)	Parkinson's disease	(Zhu et al., 2019)
Curcumin	Clinical trial/ Randomized Controlled Trial	Idiopathic PD patients aged ≥30	Parkinson's disease	(Ghodsi et al., 2022)
	Clinical trial/ Randomized Controlled Trial	Patients at high risk of developing Alzheimer's disease	Alzheimer's disease	(Thota et al., 2020)
	In vivo	Adult male ICR mice	Traumatic brain injury	(Dai et al., 2018)
	In vivo	Nine-week-old Sprague–Dawley male rats/culture primary cortical neurons	Ischemic	(Wu et al., 2013)
Genistein	In vivo	Adult male C57/BL6J mice	Ischemic	(Zhou et al., 2020)
	Clinical trial/ Randomized Controlled Trial	Alzheimer's disease patients	Alzheimer's disease	(Viña et al., 2022)

One of the leading causes of death and morbidity among children and young people is accidents that result in head injuries (Patten, 1996). When the central nervous system (CNS) or peripheral nervous system (PNS) is traumatized, a series of processes are set off that lead to a strong inflammatory response. It is unclear exactly what part inflammation plays in degeneration and regeneration. Inflammation from CNS injuries is frequently linked to poor recovery, although in peripheral nerves, the inflammatory response is thought to be necessary for the proper course of Wallerian degeneration and regeneration (Mietto et al., 2015). A broad class of naturally occurring substances present in plants; polyphenols have attracted a lot of interest due to their possible use as therapeutics in the treatment of neurodegenerative diseases (NDs). This study investigates the possible protective effects of several polyphenolic chemicals, including phenolic acids, lignans, flavonoids, and stilbenes. The primary focus is on the therapeutic potential of molecular mechanisms that enable polyphenolic compounds to improve neuron survival and function by lowering oxidative stress, inflammation, and apoptotic processes (Utpal et al., 2025). The neuroprotective properties of polyphenolic compounds have been demonstrated in numerous studies to protect neurons from damage caused by neurotoxins, reduce neuroinflammation, and possibly enhance memory, learning, and cognitive processes (Spagnuolo et al., 2016; Vauzour et al., 2008).

3 TRAUMATIC INJURIES TO THE NERVOUS SYSTEM

3.1 Central Nervous System (CNS)

3.1.1 Traumatic Brain injury (TBI)

Traumatic brain injury (TBI) is a type of acquired brain injury that happens when the brain is damaged by an abrupt trauma. TBI is the world's biggest cause of death and disability among young adults, despite its historical disregard as a serious health concern (Williamson & Rajajee, 2021). In the United States, there are ~1.5 to 2 million persons that incur a TBI each year (Campbell et al., 2018). Since TBI lasts far longer than the actual physical impact and is divided into two phases—primary and subsequent brain injury—it is actually a process. The term "primary injury" describes an abrupt and severe brain injury that results in mechanical damage to the tissue at the site of the impact (Mustafa & Alshboul, 2013). The term "secondary brain injury" describes the range of alterations that occur several hours, days, weeks, months, and occasionally even years following the insult and result in biochemical, cellular, and physiological changes, including the BBB being further disrupted, neuroinflammation, excitotoxicity, axonal damage, necrosis, apoptosis, mitochondrial dysfunction, and oxidative stress production (Kaur & Sharma, 2018). The activation of cysteine proteases like calpain and caspase-3, which break important cellular substrates and result in cell death, is the result of destructive numerous mediators of the secondary damage process eventually taking precedence over a few fundamental protective mechanisms. As of right now, there is no exact treatment plan in place to stop the pathogenesis and neurodegeneration that occur after traumatic brain injury in people. New therapeutic targets and the creation of potential medications for the treatment of TBI are still being sought after (Ray et al., 2002).

3.1.1.1 Pathophysiology of TBI

Excess neurotransmitters are released and glutamate transporters that are typically engaged in the reuptake of glutamate fail as a result of TBI's breakdown of the blood–brain barrier (BBB) (Chamoun et al., 2010). Cell death results from the activation of apoptotic proteins including calcineurin, caspases, and calpain caused by an excess of $Ca2+$ in the cytosol (Aarts & Tymianski, 2005). TBI is characterized by mitochondrial dysfunction, which changes physiological and metabolic processes and ultimately results in cell death. When too many ions enter the mitochondria, ROS are produced, the mitochondrial membrane depolarizes, and ATP synthesis is suppressed (Susin et al., 1998). Additional mitochondrial proteins that are essential for apoptotic cell death include cytochrome c and apoptosis-inducing factor (AIF) (Tsujimoto & Shimizu, 2007). Secondary cell death and oxidative stress can lead to elevated quantities of free radicals, including reactive nitrogen species (RNS) and reactive oxygen species (ROS). Excessive ROS production disrupts mitochondrial activity and damages mitochondrial membranes through lipid peroxidation (Aarts & Tymianski, 2005). Specifically, the constant release of ROS and lipid peroxidation has an adverse effect on cerebral blood flow, which causes immunosuppression and brain plasticity (Ansari et al., 2008). TBI induces a multifaceted range of immunological/inflammatory tissue responses with similarities to ischemia/reperfusion injury (Werner & Engelhard, 2007). The prolonged release of cytokines showed that the permeability of the BBB is altered, which leads to edema formation and neurological defects (Aarts & Tymianski, 2005). Chemokines, for example IL-8 (CXCL8), MIP-α, and MCP-1, have been observed in traumatized sites that recruit leukocytes to the site of injury (Frugier et al., 2010).

Figure 2. Pathophysiology of traumatic brain injury

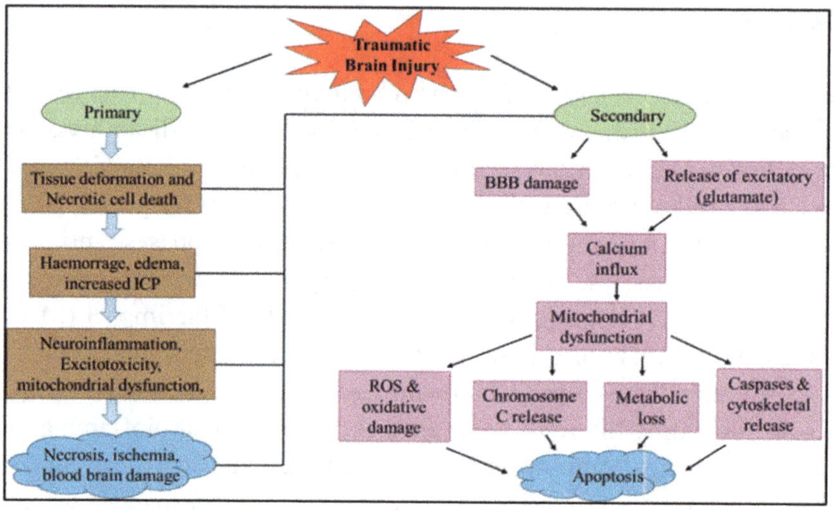

Neuronal cell death is visible in the human hippocampus up to a year after traumatic brain injury (TBI), and neuronal apoptosis is the primary feature of subsequent brain injury. (Smith et al., 1997). Two pathways mediate apoptosis, the extrinsic pathways (EP) and the intrinsic pathways (IP) (Rossi & Gaidano, 2003). However, in TBI, caspase-independent apoptosis causes mitochondrial proteins such as AIF, Smac/DIABLO, endonuclease G, and polymerase-1 to be released into the nucleus (Hasegawa et al., 2011).

3.1.2 Spinal Cord Injury

Damage or trauma to the spinal cord that causes paralysis and loss of sensation is known as spinal cord injury (SCI). In the United States, there are over 400 000 people with spinal injuries, and more than 14,000 new injuries are reported year (Sekhon & Fehlings, 2001). Oxidative stress (OS) and neuroinflammation are caused by an imbalance between the production of reactive oxygen species and antioxidant defense levels. OS and inflammatory pathways are important, demanding drivers of cross-linked dysregulated pathways following SCI (Islam et al., 2022).

3.1.2.1 Pathophysiology of SCI

Axonal connection disruption and neural cell death are the main pathophysiological events of spinal cord injury (SCI). While the proximal stump of the long axons retracts, the distal stump experiences a degenerative process known as WD, increasing the susceptibility of the remaining cell bodies to later events (Ruff et al., 2008). A secondary cascade of events characterized by disruption of normal blood flow, followed by hemorrhage, edema, inflammation, release of extracellular matrix molecules and pro- and anti-regenerative factors, and activation of the cell death machinery occurs within a few hours following SCI (Harkey III et al., 2003). SCI is categorized into primary, secondary, and chronic (McDonald & Sadowsky, 2002). Compression, shearing, laceration, and severe stretch/distraction are examples of these forces (Ackery et al., 2004). After the first injury, secondary SCI is a gradual and progressive injury Figure 3 showing the critical role of inflammation, oxidative stress, and interconnected pathways during SCI.

Figure 3. Pathophysiology of SCI

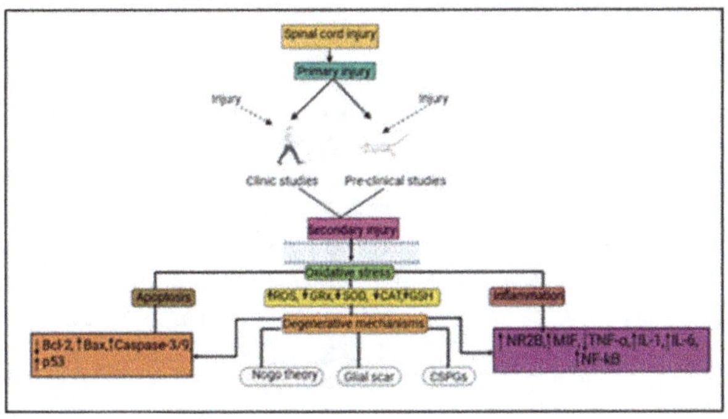

The phagocytic inflammatory cells work as reactive oxygen species (ROS) producers (Jamme et al., 1995). During the chronic phase, glial scar formation, reactive astrocytes, microglia/macrophages, and extracellular matrix molecules—especially chondroitin sulfate proteoglycans—all play a critical role in inhibiting axon growth by serving as a protective barrier (Silva et al., 2014) and stop the healing axons from passing through the wound (Rolls et al., 2009). After injury, other kinds of inhibitory molecules, like Nogo-A, which is derived from oligodendrocytes, are highly expressed and inhibit the growth cone's ability to elongate (Yiu & He, 2006).

3.2 Peripheral Nervous System (PNS)

Peripheral nerve injuries (PNIs), resulting from trauma or iatrogenic damage (Lee & Wolfe, 2000) can be severe and permanently debilitating, affecting over one million people every year worldwide (Daly et al., 2012). About one-third of all PNIs still have incomplete functional recovery, which might include complete or partial recovery of motor and/or sensory function, even with careful microsurgical repair procedures, such as direct repair and auto- or allografting. In addition to causing severe weakness, muscle atrophy, and chronic pain, these disorders can cause permanent impairments (Wang et al., 2019).

3.2.1 Pathophysiology of PNI

A peripheral nerve injury results in several alterations to the soma of the injured neuron as well as other physiological and metabolic changes close to the wound site (Razzaq, Hussain, et al., 2020). There is a noticeable difference between the two segments of a neuron that comprise the proximal and distal halves of the injury location (Doron-Mandel et al., 2015). The two sections of a neuron that make up the proximal and distal parts of the damage site differ noticeably from one another (Carroll et al., 1997a). When pro-inflammatory and anti-inflammatory factors in the body rise sharply following an injury, neuroglial cells like microglia and astrocytes are activated to trigger an immune response, which leads to symptoms like pain and inflammation. This condition is known as neuroinflammation (Matsuda et al., 2019). There are two stages of neuroinflammation: acute and chronic. While inflammation aids in nerve healing and the phagocytosis of undesirable cells during the acute phase, neuroinflammation is more likely to cause damaged neurons to deteriorate during the chronic phase (Mishra et al., 2021).

Figure 4. Wallerian degeneration and axonal regeneration following peripheral nerve injury

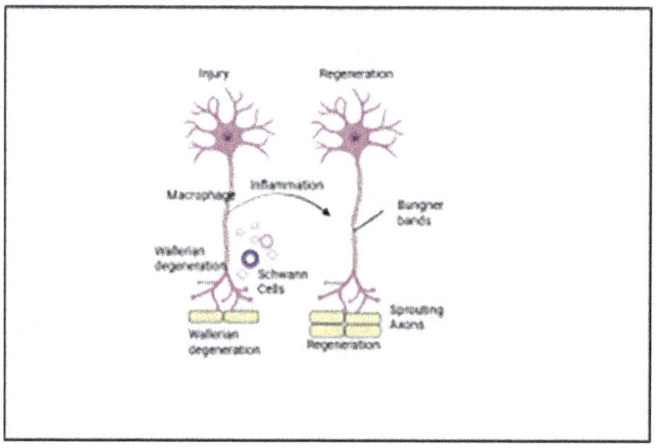

The regeneration process starts when all retrograde degenerative changes are reversed. When the Schwann cells stop producing myelin, macrophages phagocytose the myelin sheath deposits, which also stimulates macrophages (Hu et al., 2015). Interleukin-6 is produced by Schwann cells and macrophages to both eliminate myelin waste and promote axonal growth. Starting at the proximal end of the injured area, regeneration progresses to the distal end as the debris is removed. As the axon passes between the membranes of two nerve endings, the Schwann cells direct its cytoplasmic sprout. A clean endoneurial tube is used to guide sprouting axons toward the desired area for re-innervation, and Schwann cells are arranged in longitudinal columns to form Bungner's bands. These bands then release growth factors, including as fibroblast growth factor, interleukin growth factor, nerve growth factor, brain-derived neurotrophic factor, and vascular endothelial growth factor (Hussain et al., 2020).

Since polyphenols have positive effects as antioxidants and anti-inflammatory chemicals, their involvement in promoting and regulating the healing process following damage to various tissues has been thoroughly studied recently. The plant kingdom has a large class of bioactive substances called polyphenols that are part of the food chain and have the capacity to interact with living tissues (García-Conesa & Larrosa, 2020). Therefore, polyphenols could be good candidates as complementary diet substitutes for supporting peripheral nerve regeneration (Kumar et al., 2022) (Proestos & Varzakas, 2017).

4 EFFECT OF POLY-PHENOLIC COMPOUNDS ON TRAUMATIC INJURIES TO CNS AND PNS

4.1 Flavonoids

4.1.1 Flavanols

Only three flavanols—catechin, (-)-epicatechin (EC), and (-)-epigallocatechin-3-gallate (EGCG)—were examined in TBI models that showed pleiotropic effects influencing many pathogenic features of TBI. After TBI, both catechin and EC therapy dramatically decreased the development of brain edema (Jiang et al., 2017). as well as brain tissue damage (Cheng et al., 2016). Moreover, EGCG has been shown to induce an increase in the number of neural stem cells (nestin-positive cells) in the peri-lesion area (Itoh et al., 2012). Despite the fact that Itoh et al. (2013) showed that EGCG therapy dramatically increases the quantity of B-cell lymphoma 2 (Bcl2)-positive neuronal cells surrounding the damaged area (Itoh et al., 2013). By inhibiting IL-1β, iNOS, and IL-6 and increasing arginase-1, catechin indirectly improved survival and decreased neuronal death following TBI, hence reducing oxidative stress and neuroinflammation.

(Jiang et al., 2017). Furthermore, following a TBI, EC therapy significantly decreased neutrophil infiltration into the damaged tissue. By inhibiting microglial and astrocyte activation and decreasing Iba-1 and GFAP expression, EGCG decreased neuroinflammation following TBI. Additionally, it decreased TNF-α, IL-1β, and IL-6, most likely by preventing the activation of the NF-κB pathway by lowering the phosphorylation of IκBα, IKKα/β, and p65 translocation (Wu & Cui, 2020). All flavanol studies consistently demonstrated that they improve functional outcomes following traumatic brain injury (TBI) by mitigating important secondary injury processes, including neuroinflammation, oxidative stress, and neurodegeneration, regardless of the timing of treatment (Itoh et al., 2013). EGCG also demonstrated faster motor and sensory functioning, which was significant in sciatic nerve crush injury. By upregulating neurotrophic factors and their receptors, EGCG demonstrated neuroprotective and pro-regenerative effects following sciatic nerve crush injury. In the L3–L6 spinal cord, it markedly elevated the expression of BDNF, GDNF, NT-3, TrkB, Trk-C, and p75 receptors (Renno et al., 2013). By decreasing the expression of caspase-3, COX-2, and TUNEL-positive cells and boosting the expression of S100B, EGCG shielded neurons from retrograde apoptosis following sciatic nerve transection, suggesting improved neuronal survival (Kian et al., 2019).

4.1.2 Flavanones

Only hesperidin and naringenin were examined in the TBI paradigm, and several experimental strategies were used to treat the condition. According to a study by Kosari-Nasab and colleagues, hesperidin increased BDNF in the hippocampus while decreasing IL-1β and MDA levels. This suggests that hesperidin's antidepressant-like effect may be due to decreased neuroinflammation, oxidative stress, and improved neurotrophic support (Deng et al., 2021). Hesperidin-loaded alginate/chitosan hydrogels (0.1%, 1%, and 10% w/v) were assessed for nerve regeneration in a recent study. The hydrogels' overall suitability was confirmed by in vitro evaluations that revealed good porosity, swelling, degradation, controlled hesperidin release, hemocompatibility, antibacterial activity, and no cytotoxicity. In vivo, functional recovery was enhanced by injecting hesperidin-loaded hydrogel at the site of the sciatic crush injury. The group that received 1% hesperidin demonstrated the best sciatic functional index, sensory response, hotplate test outcomes, and decreased muscle atrophy. Healthy morphology of reinnervated muscles and regenerated nerves was verified by H&E staining. Thus, the authors came to the conclusion that hesperidin in an alginate/chitosan hydrogel may be used to treat damage to peripheral nerves (Bagher et al., 2021). Because of its potent anti-inflammatory and antioxidant qualities, naringenin also had a positive impact on nerve regeneration. Being a non-enzymatic antioxidant, it lessens cellular damage after nerve injury by preserving redox equilibrium (Basu & Basu, 2020; Rao et al., 2021). By decreasing mechanical hyperalgesia and improving the sciatic functional index, systemic administration of naringenin, either by alone or in combination with hydroxypropyl-β-cyclodextrin, promoted nerve regeneration following crush injury. It reduced apoptosis and caspase-3 levels by inhibiting the p75NTR and JNK pathways, raising IL-10, and decreasing TNF-α and IL-1β. Improved nerve regeneration was demonstrated by morphological examination, which was further boosted by the cyclodextrin complex (Oliveira et al., 2020). By modifying TGF-β1/Smad3, Nrf2/ARE/HO-1, and peroxisome proliferator-activated receptor γ (PPARγ) signaling, hesperidin reduced oxidative stress and inflammation (Mahmoud et al., 2017). An *in vitro* study based on oxidative stress showed that activation of the ERK/MAPK signaling by hesperidin is involved in the expression of HO-1 and Nrf2 (Chen et al., 2010). According to a recent study by Heo et al., hesperidin (100 mg/kg, i.p.) for 14 days (from 7 days before SCI to 7 days after SCI) reduced inflammation and oxidative stress by upregulating the HO-1/Nrf2 pathway (Heo et al., 2020). At the same dose, hesperidin led to a decrease in TNF-α, IL-1β, and IL-6 in the spinal cord tissue (Carballo-Villalobos et al., 2017).

4.1.3 Flavanols

Four flavonols—quercetin, fisetin, rutin, and kaempferol—have been studied in TBI models. (Almeida et al., 2018). By decreasing pro-inflammatory cytokines and increasing IL-10, quercetin reduces TBI-induced inflammation while also reestablishing mitochondrial function and antioxidant enzymes. In order to fight oxidative stress, it stimulates the Nrf2 pathway, increasing the expression of HO-1 and NQO-1 (Song et al., 2021; Zhang et al., 2018). Fisetin has been shown to protect the integrity of the BBB and reduce the brain lesion size (Zhang et al., 2018). By raising the number of live neurons, lowering the number of apoptotic cells, and decreasing the levels of active caspase-3, quercetin, fisetin, and rutin decreased neurodegeneration and apoptosis in the peri-lesioned region (Du et al., 2018; Li et al., 2018; Yang et al., 2014; Zhai et al., 2016; Zhang et al., 2018). The neuroprotective effects of quercetin may include decreased ERK½ phosphorylation, decreased Bax, increased Bcl-2 expression, increased Akt phosphorylation, and activation of the PI3K/Akt pathway (Du et al., 2018; Li et al., 2018; Yang et al., 2014; Zhai et al., 2016). After traumatic brain injury (TBI), kaempferol (KMF) enhances behavior and stimulus-induced brain activation in the early stages of development. It improved brain vitality, maintained mitochondrial functional integrity, and raised TCA cycle flow. The way it affected the metabolic profile greatly improved sensorimotor behavior (Chitturi et al., 2019). After traumatic brain injury, KMF enhances rat behaviour and energy metabolism. It influences the development of TBI by affecting neuronal, vascular, and synaptic/axonal compartments in diverse ways and targeting mitochondrial ion homeostasis. With improvements in parenchymal microstructure, neurovascular coupling, and neuronal connection in particular brain regions, early KMF treatment dramatically improved the prognosis for TBI in adolescents (Parent et al., 2020). In addition, numerous studies have documented the positive effects of myricetin and quercetin on peripheral nervous system regeneration. Following peripheral nerve damage,

EGCG also reduced the expression of NADPH and neuronal nitric oxide synthase (nNOS) (Wei et al., 2011) . In several in vitro, in vivo, and clinical investigations, quercetin has demonstrated antioxidative, anti-inflammatory, immunoprotective, and anticarcinogenic properties, improving conditions like cancer, diabetes, and cardiovascular disorders (Shah et al., 2016). In vitro, it enhances neuronal survival (Ansari et al., 2009; Jiang et al., 2016) and reduces neurotoxicity and neuroinflammation. Additionally, quercetin's anti-inflammatory and antioxidant properties shown beneficial effects against oxidative damage caused by SCI (Çevik et al., 2013). Quercetin (7.5 mg/kg, i.p.) suppressed the STAT1 and NF-κB pathway after SCI, preventing necroptosis of oligodendrocytes, according to an in vitro and in vivo investigation (Fan et al., 2019). According to Jiang et al., 100 mg/kg of quercetin significantly reduced the production of ROS, TNF-α, IL-1β, and IL-18 in female rats following SCI (Jiang et al., 2016). According to a number of experimental investigations, EGCG provides neuroprotection against spinal cord damage (Urdzikova et al., 2017). Rutin is known for various pharmacological activities, like neuroprotection (Javed et al., 2012) and post-SCI neuroinflammation (Wu et al., 2016) activities. Additionally, Zhang and Ma demonstrated that in a rat model of SCI, 10 μmol/kg of rutin (i.p.) decreased the expression levels of MIP-2 and p-Akt (Zhang & Ma, 2015). Additionally, by inhibiting the p38MAPK pathway following SCI, rutin (30 mg/kg) reduced oxidative stress-related markers and inflammatory cytokines (such as IL-1β, IL-6, and TNF-α) (Song et al., 2018). In an in vitro investigation, rutin shielded cells from H2O2-induced oxidative stress and apoptosis by controlling ROS, inhibiting LPO, preserving the intracellular antioxidant system, and controlling the NF-κB/p65 signaling pathway and Bax/Bcl-2 ratio (Zhou et al., 2016). Additionally, it was reported that by suppressing the TGF-β/Smad pathway in the SCI model, an intraperitoneal injection of 30 mg/kg of rutin for three days following SCI plus mild hypothermia decreased inflammatory factors (Yao et al., 2021). Quercetin (7.5 mg/kg) inhibited oligodendrocyte necroptosis following SCI by modifying the STAT1 and NF-κB pathways, according to in vitro and in vivo studies (Fan et al., 2019). A prior study found that nuclear factor-κB (NF-κB) was crucial for neuroinflammation caused by activated microglia following SCI and that ROS and the Mitogen-activated protein kinases (MAPK) signaling pathway may activate it. The prognosis of secondary damage brought on by inflammation mediated by microglia is improved when the NF-κB signaling pathway is inhibited in the spinal cord because it directly reduces the expression of several inflammatory factors like IL-6, TNF-α, or IL-1β. Because of its anti-oxidant and anti-inflammatory qualities, kaempferol may have neuroprotective effects in a range of neurological conditions. In a model of neuroinflammation, kaempferol was shown to suppress the inflammatory response mediated by microglia (Liu et al., 2021). The anti-inflammatory action of KPF stands out among its many pharmacological effects, and it is primarily ascribed to its capacity to inhibit the enzymes phospholipase A2, lipoxygenase, cyclooxygenase, and nitric oxide by modulating the enzyme nitric oxide synthase (iNOS). Furthermore, an earlier in vivo study demonstrated that KPF could cross the blood-brain barrier (BBB) in rats when given a single dose of 600 mg/kg (Silva dos Santos et al., 2021).

4.1.4 Flavones

One of the most abundant dietary flavones, Luteolin and chrysin were the only flavones exploited in TBI settings so far (Dos Santos et al., 2020). By decreasing TNF-α, IL-1β, and NF-κB p65 nuclear accumulation, luteolin, when given intraperitoneally (IP) before and after traumatic brain injury (TBI), decreased brain inflammation (Sawmiller et al., 2014). After TBI, luteolin failed to protect $Nrf2^{-/-}$ mice, indicating that its neuroprotective benefits depend on Nrf2-ARE pathway activation (Xu et al., 2014). After the TBI, luteolin had a significant impact on the level of oxidative stress by raising the levels of SOD, CAT, GPx, and GSH and lowering the levels of MDA three days later. In all tested doses, the levels of these proteins were fully restored fourteen days later. Fourteen days after TBI, chyrsin dose-dependently decreased Bax and apoptotic cells while raising Bcl-2 in the brain and hippocampus. It enhanced motor coordination, memory, and learning (Rashno et al., 2019). The effect of the flavone apigenin on nerve regeneration in vitro has only been investigated once in relation to peripheral nerve damage. This scant data indicates that more investigation is necessary to fully understand its possible neuroregenerative qualities. Additionally, it has been reported that apigenin treatment inhibits the proliferation of Schwann cells and changes their phenotype into repairing Schwann cells without the need for ERK signaling. According to reports, apigenin solution (0.1, 1, or 10 mM) inhibits myelin fragmentation in a way that is independent of Krox20, which in turn inhibits axonal deterioration following nerve transection (Kim et al., 2019).

4.1.5 Isoflavones

The effects of the isoflavone family members in the TBI mode have only been described in one study thus far. According to a study by Soltani and colleagues, genistein can reduce brain edema by significantly lowering the water content of the brain when administered at a low dose (15 mg/kg) following a traumatic brain injury. Additionally, genistein improved the integrity of the blood-brain barrier and decreased intracranial pressure. Furthermore, genistein enhanced motor function as a functional consequence (Soltani et al., 2015). Furthermore, the impact of genistein on peripheral nerve regeneration following traumatic injury was only reported in one study (Özbek et al., 2017). The authors observed improved paw mechanical withdrawal threshold and sciatic function index four weeks following end-to-end repair or sciatic nerve crush. Through nerve tissue analysis, they also found that the regenerated sciatic nerve had lower levels of proinflammatory cytokines (TNF-α and IL-1β) and higher GAP-43 and myelin basic protein (MBP) immunoreactivity than the untreated animals. This suggested that genistein enhanced nerve regeneration following traumatic nerve injury and repair (Zhang et al., 2021).

Table 2 showing the list of poly-phenolic compounds their experimental methods, the dose and mechanism of action against TBI, SCI & PNI.

Table 2. List of poly-phenolic compounds

Traumatic Brain Injury (TBI)	Compound	Assay/ Model	Dose	Mechanism of Action	Reference
	Ellagic Acid	Diffuse TBI in rats	100 mg/kg every hour for 48 h	Decrease TNF-α in brain, restore BBB integrity, improves hippocampal LTP and cognitive memory performance	(Mashhadizadeh et al., 2017)
	Resveratrol	Controlled cortical impact (CCI) in rats	100mg/kg daily for 3 days	Suppresses TLR4/NF-κB pathway, Reduces autophagy, Decreases inflammation, Improves brain edema, Enhances cognitive outcomes	(Feng et al., 2016)
		Blast-induced brain injury in rodents	Various (shown) effective in preclinical review)	Activates Nrf2/ Keap1, inhibits NF-KB: reduces oxidative stress, inflammation and apoptosis in injured brain tissue	(Nath et al., 2022) (Cong et al., 2021)
	Quercetin	Weight-drop TBI in rats	10 mg /kg, daily 7 days post-injury	Reduce oxidative stress; increase SOD and catalase; inhibits ERK1/2 phosphorylation; activates Akt; reduces apoptosis, edema, and neurobehavioral deficits	(Yavtushenko et al., 2020)
	Epigallocatechin-3-Gallate EGCG	Fenny weight-drop TBI in wistar or sprouge-dawley rats	1OO mg/kg immediately post TBI	Brain edema, decrease vascular permeability, decrease TNF-a mRNA, decrease microglial activation (CD68), reduce AQP4 and GFAP, increase SOD and GSH-Px, reduce MDA, inhibits NADPH oxidase activation	(Zhang et al., 2015)
Spinal Cord Injury (SCI)	Baicalin	Rat compression SCI using weight-drop	100mg/kg daily in first 24h	Decrease spinal cord water content and BSCB permeability, decrease MDA, TNF-a, NF-KB, decrease Bax and caspase-3, increase BCI-2, improve locomotor recovery	(Zhao et al., 2022) (Cao et al., 2010)
	Curcumin	T8-T9 clip compression in rats	200mg/kg daily for 7 days post-SCI	Decrease MDA, increase SOD, decrease cavity volume, high BBB score, decrease macrophage infiltration	(Cao et al., 2010) (Jin et al., 2014)

continued on following page

Table 2. Continued

Traumatic Brain Injury (TBI)	Compound	Assay/ Model	Dose	Mechanism of Action	Reference
	Epigallocatechin-3-Gallate EGCG	Sprauge-Dawley rata	One time 10 or 20mg/ kg EGCG injected intrathecally immediately after SCI	Up-regulates neurotrophic factor BDNF and GDNF Preserves myelin and improves locomotors recovery in open field and inclined plane tests	(Tian et al., 2013)
	Resveratrol	Thorasic (T8-T9) clip compression SCI in Sprague-dawley rats	100mg/kg, administered once, one day after injury	Decrease oxidative stress markers (decrease MDA, NO, XO and increase SOD, GSH) Decrease apoptosis (decrease Bax and caspase-3 increase Bcl-2)	(Şentürk et al., 2018)
Peripheral Nerve Injury (PNI)	Epigallocatechin-3-Gallate EGCG	Sciatic nerve crush injury in rats	50mg/kg, daily, pre-injury for 3 days	Decrease lipid peroxidation (MDA), increase antioxidant enzymes (SOD, CAT), promotes axon and myelin regeneration via neurotrophic support	(Kian et al., 2019)
	Resveratrol	Rat sciatic nerve crush injury model	Administered daily for 10-day, post injury, 50mg/ kg	Enhance p300 acetyltransferase-mediated VEGH expression via Akt signaling	(Ding et al., 2018)
	Quercetin	Sciatic nerve crush in mice	High dose of quercetin (20 mg/kg/day)	Significant upregulation of genes linked to axonal growth, cAMP signaling pathway & myelin function.	(Chen et al., 2017)

4.2 Non-flavonoids

4.2.1 Phenolic acids

Phenolic acids, (Hydroxybenzoic Derivatives) a major subfamily of polyphenols, have shown strong effects in TBI models by reducing neuroinflammation, oxidative stress, and neurodegeneration, while also improving functional outcomes.

Gallic acid reduced TNF-α, IL-1β, and IL-6 levels after TBI (Gatson et al., 2013; Itoh et al., 2012; Itoh et al., 2013; Sawmiller et al., 2014), while protocatechuic acid attenuated microglial activation and prevented GSH depletion (Lee SangHwon et al., 2017). Both ferulic and protocatechuic acids restored antioxidant levels, while ferulic acid reduced neuronal apoptosis and raised SOD and GSH (Dong et al., 2023). According to Nasution et al., CAPE completely restored GPx, SOD, and MDA levels following TBI and decreased oxidative damage (Kerman et al., 2012). Protocatechuic acid's effects on the peripheral nervous system have only been examined in vitro on the RSC96 Schwann cell line, where it improved cell motility, proliferation, and survival (Ju, Kuo, et al., 2015; Ju, Liao, et al., 2015). In fact, the treatment markedly increased the expression of the proliferating cell nuclear antigen (PCNA), suppressed the proapoptotic protein Bad, upregulated the pro-survival proteins p-Bad and Bcl-xL, and phosphorylated the insulin-like growth factor-I (IGF-I)-mediated phosphatidylinositol 3 kinase/serine–threonine kinase (PI3K/Akt) pathway by inhibiting Bad. The expression of the cell cycle regulators cyclins D1, E, and A increased, indicating an increase in the G1–S-phase cell cycle transition as well. Additionally, following protocatechuic acid treatment, there was an increase in the phosphorylation of MEK1/2, ERK1/2, JNK1/2, and p38, which resulted in the production of proteolytic enzymes of tissue type (tPA) and urokinase type (uPA). Lastly, zymography data showed that protocatechuic acid treatment of RSC96 Schwann cells increased the activity of metallopeptidase 9 (MMP9) and MMP2 (Ju, Liao, et al., 2015). Additionally, studies have shown that gallic acid has neuroprotective, anti-inflammatory, anti-cancer, anti-hyperglycemic, anti-hyperlipidemic, and antioxidant properties (Latha & Daisy, 2011; Mansouri et al., 2013). It has also been shown that gallic acid reverses cyclophosphamide- and paclitaxel-induced neurotoxicity (Oyagbemi et al., 2016) and that it has antinociceptive and antiedematogenic effects in a chronic constriction injury neuropathic pain mice model (Kaur et al., 2012). Since the activities of the Na+/K+, Ca2+, and Mg2+ ATPases were shown to be significantly decreased in SCI, gallic acid restored the activity of the Na+/K+, Ca2+, and Mg2+ atpases during SCI. When compared to the SCI group, the activities of the Na+/K+ ($P<0.01$), Ca2+ ($P<0.01$), and Mg2+ ($P<0.001$) ATPases increased statistically significantly after treatment with gallic acid (Yang et al., 2015).

4.2.2 Curcumins

Zhu et al. demonstrated that curcumin decreased pro-inflammatory cytokines and glial activation following TBI, and they connected these effects to the inhibition of the TLR4/MyD88/NF-κB pathway (Zhu et al., 2014). In addition to restoring BDNF, synapsin I, and CREB levels and reducing oxidative damage, Wu et al. found that dietary curcumin enhanced energy balance by raising AMPK, uMtCK, and cytochrome c oxidase II (Wu et al., 2006). In the injured cortex, Dong et al. demonstrated that post-injury curcumin treatment increased Bcl-2, decreased cleaved caspase-3, and decreased apoptotic neurons. Increased Nrf2 nuclear translocation and the overexpression of antioxidant enzymes such as HO-1, NQO-1, GCLC, and GCLM were associated with these effects, suggesting involvement of the Nrf2-ARE pathway (Dong et al., 2018). Furthermore, a recent study by Sun and coworkers found that curcumin administration causes clear neurogenesis in the treated animals' hippocampal regions in addition to suppressing neuronal cell death following TBI (Sun et al., 2020). Curcumin reversed the induction of aquaporin-4 (AQP4), an astrocytic water channel implicated in the development of cellular edema following head trauma, according to a study by Laird and colleagues. This suggests a potential mechanism of action by which curcumin reduces the development of brain edema following traumatic brain injury (Laird et al., 2010). The impact of curcumin on autophagy, myelination, and associated signal transduction pathways, such as Akt and Erk1/2, was also examined by Zhao et al. They showed that during the healing of damaged sciatic nerves following crush lesions, an Erk1/2 pathway inhibitor and an Akt pathway agonist abolished the beneficial effect of curcumin on remyelination and axon regeneration (Z. Zhao et al., 2017). Flaxseed oil, which is high in lignans and α-linolenic acid, an important ω-3 PUFA, helped rats recover functionally from sciatic nerve crush injury quicker. Compared to untreated controls, treated mice displayed regenerated myelinated nerve fibers with thicker myelin (Che Ramli et al., 2020). Sesame oil has demonstrated potent antioxidative and anti-inflammatory properties due to its abundance of lignans, different fatty acids, and antioxidants including α-tocopherol. Sesame oil treatment enhanced functional recovery in a rat sciatic nerve crush model by lowering lipid peroxidation in serum and nerve tissue and markedly upregulating nuclear Nrf2 expression, suggesting improved antioxidant defense (Hsu et al., 2016). The anti-inflammatory effects of curcumin following SCI have been associated with an increase in Nrf2 and a decrease of NF-κB, IL-1β, IL-6, and TNF-α activity (Jin et al., 2014). NF-κB stimulation caused curcumin to produce antioxidative preservation through Nrf2 pathways and a decrease in ROS (Daverey & Agrawal, 2020).

4.2.3 Stilbenes

Resveratrol, a member of the stilbenes, a large subfamily of polyphenolic compounds, has anti-inflammatory qualities and can raise the amount of the anti-inflammatory cytokine IL-10 in damaged brain tissue while lowering the levels of TNF-α, IL-1β, IL-6, IL-4, IL-12, high mobility group box 1 (HMGB1), and NF-kB p65 (Feng et al., 2016; Gatson et al., 2013; Zou et al., 2018) and significantly decreasing microglial activation after the TBI (Gatson et al., 2013). Resveratrol significantly raised the levels of GPx and 8-hydroxy-2′-deoxyguanosine (8-OhdG) protein in the wounded tissue following the TBI (ATAlAY et al., 2017). Furthermore, it has been demonstrated to reduce ROS generation when taken orally in a low dose following a TBI. In the damaged brain, resveratrol raised Nrf2 levels and, consequently, HO1 levels, as was previously demonstrated for a number of other (poly)phenols (Shi et al., 2018). Additionally, resveratrol may help promote peripheral nerve regeneration, as evidenced by in vivo experiments that demonstrated daily systemic therapy reduced motor impairments brought on by sciatic nerve compression injury (Ding et al., 2018). Resveratrol treatment increased axon numbers and improved myelin morphology to near-normal levels in a rat model of chronic constriction injury to the sciatic nerve, promoting nerve regeneration. It averted motor impairment by increasing walking distance and open field activity, indicating functional recovery, and it restored IGF-1 expression, which had decreased in rats given saline (Bagriyanik et al., 2014). Resveratrol has the potential to lessen the severity of SCI health consequences by overriding the NF-κB signaling pathway (Xu et al., 2018). Following SCI, resveratrol (100 mg/kg, i.p.) decreased the transcription of p62, caspase-3, caspase-9, and Bax while increasing the activity of p-AMPK, Bcl-2, and SIRT1. It has also been observed that resveratrol protects neurons by inhibiting the SIRT1/AMPK signaling pathway (Yan et al., 2017; H. Zhao et al., 2017). Zhang et al. found that resveratrol has anti-apoptotic effects following SCI by lowering related p53, caspase-3, and cytochrome C (Zhang et al., 2019). Additionally, via inducing autophagy via the AMPK/mTOR pathway, resveratrol reduced neuroinflammation after SCI (Meng et al., 2018). In rats with SCI, resveratrol dramatically increased neuronal autophagic flux to reduce programmed cell death and promote functional repair (P. Wang et al., 2018).

4.2.4 Ellagitannins

Ellagic acid, the sole representative of the ellagitannin subfamily in the TBI model, has been demonstrated to dramatically lower the levels of TNF-α, IL-1β, and IL-6 proteins in the brain after a TBI (Farbood et al., 2015; Mashhadizadeh et al., 2017; Mirshekar et al., 2018; Sarkaki et al., 2015). A study by Mashhadizadeh and colleagues also showed that ellagic acid can reduce BBB permeability after the TBI (Mashhadizadeh et al., 2017). In one study, TBI rats fed a 5% w/w blueberry-enriched diet had restored CREB and CaMKII phosphorylation and elevated BDNF. Additionally, it decreased 4-HNE levels, indicating that blueberries can help maintain neuronal survival and combat oxidative stress brought on by TBI (Krishna et al., 2019). In neuropathic pain caused by tibial and sural nerve transection, ellagic acid demonstrated positive effects, while oral pomegranate extract decreased oxidative stress, nitric oxide generation, and inflammation (Jain et al., 2013). In rats with STZ-induced diabetes, ellagic acid had a neuroprotective effect by reducing oxidative stress indicators such as MDA, TOS, OSI, and NO in the sciatic nerves (Uzar et al., 2012).

5. CONCLUSION

The treatment of traumatic injuries to the central and peripheral nerve systems may benefit greatly from the use of polyphenolic chemicals, which are found in many plants and common food sources. The ability of polyphenolic chemicals, such as flavonoids, tannins, curcumins, stilbenes, phenolic acids and non-flavonoids, to regulate oxidative stress, lower inflammation, and influence apoptotic signalling pathways underlies their neuroprotective and regenerative qualities. Experimental evidence from in vitro, in vivo, and a few clinical studies supports the efficacy of these compounds in lowering secondary damage cascades following traumatic nervous system injuries. Because of their wide range of biological action and low toxicity, polyphenols are an intriguing class of natural compounds that may enhance existing neurotrauma treatment approaches. To turn these results into successful, evidence-based treatments, more investigation and clinical validation are necessary.

REFERENCES

Aarts, M. M., & Tymianski, M. (2005). TRPMs and neuronal cell death. *Pflügers Archiv, 451*(1), 243–249. DOI: 10.1007/s00424-005-1439-x PMID: 16044308

Ackery, A., Tator, C., & Krassioukov, A. (2004). A global perspective on spinal cord injury epidemiology. *Journal of Neurotrauma, 21*(10), 1355–1370. DOI: 10.1089/neu.2004.21.1355 PMID: 15672627

Ahmad, A., Ali, T., Park, H. Y., Badshah, H., Rehman, S. U., & Kim, M. O. (2017). Neuroprotective effect of fisetin against amyloid-beta-induced cognitive/synaptic dysfunction, neuroinflammation, and neurodegeneration in adult mice. *Molecular Neurobiology, 54*(3), 2269–2285. DOI: 10.1007/s12035-016-9795-4 PMID: 26944285

Ahmad, S., Jo, M. H., Ikram, M., Khan, A., & Kim, M. O. (2021). Deciphering the potential neuroprotective effects of luteolin against Aβ1–42-Induced alzheimer's disease. *International Journal of Molecular Sciences, 22*(17), 9583. DOI: 10.3390/ijms22179583 PMID: 34502488

Al-Khayri, J. M., Sahana, G. R., Nagella, P., Joseph, B. V., Alessa, F. M., & Al-Mssallem, M. Q. (2022). Flavonoids as potential anti-inflammatory molecules: A review. *Molecules (Basel, Switzerland), 27*(9), 2901. DOI: 10.3390/molecules27092901 PMID: 35566252

Almeida, A. F., Borge, G. I. A., Piskula, M., Tudose, A., Tudoreanu, L., Valentová, K., Williamson, G., & Santos, C. N. (2018). Bioavailability of quercetin in humans with a focus on interindividual variation. *Comprehensive Reviews in Food Science and Food Safety, 17*(3), 714–731. DOI: 10.1111/1541-4337.12342 PMID: 33350133

Ansari, M. A., Abdul, H. M., Joshi, G., Opii, W. O., & Butterfield, D. A. (2009). Protective effect of quercetin in primary neurons against Aβ (1–42): Relevance to Alzheimer's disease. *The Journal of Nutritional Biochemistry, 20*(4), 269–275. DOI: 10.1016/j.jnutbio.2008.03.002 PMID: 18602817

Ansari, M. A., Roberts, K. N., & Scheff, S. W. (2008). Oxidative stress and modification of synaptic proteins in hippocampus after traumatic brain injury. *Free Radical Biology & Medicine, 45*(4), 443–452. DOI: 10.1016/j.freeradbiomed.2008.04.038 PMID: 18501200

ATAlAY, T., Gülşen, İ., Çölçimen, N., Alp, H., SOSuNCu, E., AlACA, I., Ak, H., & Rağbetli, M. (2017). Resveratrol treatment prevents hippocampal neurodegeneration in a rodent model of traumatic brain injury. *Turkish Neurosurgery, 27*(6). PMID: 27509464

Bagher, Z., Ehterami, A., Nasrolahi, M., Azimi, M., & Salehi, M. (2021). Hesperidin promotes peripheral nerve regeneration based on tissue engineering strategy using alginate/chitosan hydrogel: In vitro and in vivo study. *International Journal of Polymeric Materials, 70*(5), 299–308. DOI: 10.1080/00914037.2020.1713781

Bagriyanik, H., Ersoy, N., Cetinkaya, C., Ikizoglu, E., Kutri, D., Ozcana, T., Kamanga, L., & Kiray, M. (2014). The effects of resveratrol on chronic constriction injury of sciatic nerve in rats. *Neuroscience Letters, 561*, 123–127. DOI: 10.1016/j.neulet.2013.12.056 PMID: 24394911

Basu, P., & Basu, A. (2020). In vitro and in vivo effects of flavonoids on peripheral neuropathic pain. *Molecules (Basel, Switzerland), 25*(5), 1171. DOI: 10.3390/molecules25051171 PMID: 32150953

Berndtsson, E., Nynäs, A.-L., Newson, W., Langton, M., Andersson, R., Johansson, E., & Olsson, M. (2019). 21. The underutilised side streams of broccoli and kale–valorisation via proteins and phenols. In *Sustainable governance and management of food systems* (pp. 153–159). Academic. DOI: 10.3920/978-90-8686-892-6_21

Campbell, J. C., Anderson, J. C., McFadgion, A., Gill, J., Zink, E., Patch, M., Callwood, G., & Campbell, D. (2018). The effects of intimate partner violence and probable traumatic brain injury on central nervous system symptoms. *Journal of Women's Health, 27*(6), 761–767. DOI: 10.1089/jwh.2016.6311 PMID: 29083256

Cao, Y., Li, G., Wang, Y., Fan, Z., Yu, D., Wang, Z., & Bi, Y. (2010). Neuroprotective effect of baicalin on compression spinal cord injury in rats. *Brain Research, 1357*, 115–123. DOI: 10.1016/j.brainres.2010.07.108 PMID: 20708606

Carballo-Villalobos, A., González-Trujano, M., Alvarado-Vázquez, N., & López-Muñoz, F. (2017). Pro-inflammatory cytokines involvement in the hesperidin antihyperalgesic effects at peripheral and central levels in a neuropathic pain model. *Inflammopharmacology, 25*(2), 265–269. DOI: 10.1007/s10787-017-0326-3 PMID: 28265836

Çevik, Ö., Erşahin, M., Şener, T. E., Tinay, İ., Tarcan, T., Çetinel, Ş., Şener, A., Toklu, H. Z., & Şener, G. (2013). Beneficial effects of quercetin on rat urinary bladder after spinal cord injury. *The Journal of Surgical Research, 183*(2), 695–703. DOI: 10.1016/j.jss.2013.02.016 PMID: 23490140

Chagas, M. S. S., Behrens, M. D., Moragas-Tellis, C. J., Penedo, G. X., Silva, A. R., & Gonçalves-de-Albuquerque, C. F. (2022). Flavonols and flavones as potential anti-inflammatory, antioxidant, and antibacterial compounds. *Oxidative Medicine and Cellular Longevity, 2022*(1), 9966750. DOI: 10.1155/2022/9966750 PMID: 36111166

Chamoun, R., Suki, D., Gopinath, S. P., Goodman, J. C., & Robertson, C. (2010). Role of extracellular glutamate measured by cerebral microdialysis in severe traumatic brain injury. *Journal of Neurosurgery, 113*(3), 564–570. DOI: 10.3171/2009.12.JNS09689 PMID: 20113156

Che Ramli, M. D., Sumari, N. A., Selamat, N. W., Muhammad, H., & Sanusi, J. (2020). Effectiveness of Flaxseed Oil on Peripheral Nerve Regeneration Following Crush Injury of Sciatic Nerve in Rat: Behavioural Analysis and an Electron Microscopic Study. *Malaysian Journal of Medicine & Health Sciences, 16*.

Chen, J., Yang, J., Ma, L., Li, J., Shahzad, N., & Kim, C. K. (2020). Structure-antioxidant activity relationship of methoxy, phenolic hydroxyl, and carboxylic acid groups of phenolic acids. *Scientific Reports, 10*(1), 2611. DOI: 10.1038/s41598-020-59451-z PMID: 32054964

Chen, M., Gu, H., Ye, Y., Lin, B., Sun, L., Deng, W., Zhang, J., & Liu, J. (2010). Protective effects of hesperidin against oxidative stress of tert-butyl hydroperoxide in human hepatocytes. *Food and Chemical Toxicology, 48*(10), 2980–2987. DOI: 10.1016/j.fct.2010.07.037 PMID: 20678535

Chen, M.-M., Qin, J., Chen, S.-J., Yao, L.-M., Zhang, L.-Y., Yin, Z.-Q., & Liao, H. (2017). Quercetin promotes motor and sensory function recovery following sciatic nerve-crush injury in C57BL/6J mice. *The Journal of Nutritional Biochemistry, 46*, 57–67. DOI: 10.1016/j.jnutbio.2017.04.006 PMID: 28458138

Cheng, T., Wang, W., Li, Q., Han, X., Xing, J., Qi, C., Lan, X., Wan, J., Potts, A., Guan, F., & Wang, J. (2016). Cerebroprotection of flavanol (-)-epicatechin after traumatic brain injury via Nrf2-dependent and-independent pathways. *Free Radical Biology & Medicine*, *92*, 15–28. DOI: 10.1016/j.freeradbiomed.2015.12.027 PMID: 26724590

Chimento, A., De Luca, A., D'Amico, M., De Amicis, F., & Pezzi, V. (2023). The involvement of natural polyphenols in molecular mechanisms inducing apoptosis in tumor cells: A promising adjuvant in cancer therapy. *International Journal of Molecular Sciences*, *24*(2), 1680. DOI: 10.3390/ijms24021680 PMID: 36675194

Chitturi, J., Santhakumar, V., & Kannurpatti, S. S. (2019). Beneficial effects of kaempferol after developmental traumatic brain injury is through protection of mitochondrial function, oxidative metabolism, and neural viability. *Journal of Neurotrauma*, *36*(8), 1264–1278. DOI: 10.1089/neu.2018.6100 PMID: 30430900

Chong, J., Poutaraud, A., & Hugueney, P. (2009). Metabolism and roles of stilbenes in plants. *Plant Science*, *177*(3), 143–155. DOI: 10.1016/j.plantsci.2009.05.012

Cong, P., Wang, T., Tong, C., Liu, Y., Shi, L., Mao, S., Shi, X., Jin, H., Liu, Y., & Hou, M. (2021). Resveratrol ameliorates thoracic blast exposure-induced inflammation, endoplasmic reticulum stress and apoptosis in the brain through the Nrf2/Keap1 and NF-κB signaling pathway. *Injury*, *52*(10), 2795–2802. DOI: 10.1016/j.injury.2021.08.019 PMID: 34454721

Dai, W., Wang, H., Fang, J., Zhu, Y., Zhou, J., Wang, X., Zhou, Y., & Zhou, M. (2018). Curcumin provides neuroprotection in model of traumatic brain injury via the Nrf2-ARE signaling pathway. *Brain Research Bulletin*, *140*, 65–71. DOI: 10.1016/j.brainresbull.2018.03.020 PMID: 29626606

Daly, W., Yao, L., Zeugolis, D., Windebank, A., & Pandit, A. (2012). A biomaterials approach to peripheral nerve regeneration: Bridging the peripheral nerve gap and enhancing functional recovery. *Journal of the Royal Society, Interface*, *9*(67), 202–221. DOI: 10.1098/rsif.2011.0438 PMID: 22090283

Dasgupta, B., & Milbrandt, J. (2007). Resveratrol stimulates AMP kinase activity in neurons. *Proceedings of the National Academy of Sciences of the United States of America*, *104*(17), 7217–7222. DOI: 10.1073/pnas.0610068104 PMID: 17438283

Daverey, A., & Agrawal, S. K. (2020). Curcumin protects against white matter injury through NF-κB and Nrf2 cross talk. *Journal of Neurotrauma*, *37*(10), 1255–1265. DOI: 10.1089/neu.2019.6749 PMID: 31914858

Deng, C., Yi, R., Fei, M., Li, T., Han, Y., & Wang, H. (2021). Naringenin attenuates endoplasmic reticulum stress, reduces apoptosis, and improves functional recovery in experimental traumatic brain injury. *Brain Research*, *1769*, 147591. DOI: 10.1016/j.brainres.2021.147591 PMID: 34324877

Ding, Z., Cao, J., Shen, Y., Zou, Y., Yang, X., Zhou, W., Guo, Q., & Huang, C. (2018). Resveratrol promotes nerve regeneration via activation of p300 acetyltransferase-mediated VEGF signaling in a rat model of sciatic nerve crush injury. *Frontiers in Neuroscience*, *12*, 341. DOI: 10.3389/fnins.2018.00341 PMID: 29875625

Dong, Q., Yang, S., Liao, H., He, Q., & Xiao, J. (2023). Bioinformatics findings reveal the pharmacological properties of ferulic acid treating traumatic brain injury via targeting of ferroptosis. *International Journal of Food Properties*, *26*(1), 778–786. DOI: 10.1080/10942912.2023.2185178

Dong, W., Yang, B., Wang, L., Li, B., Guo, X., Zhang, M., Jiang, Z., Fu, J., Pi, J., Guan, D., & Zhao, R. (2018). Curcumin plays neuroprotective roles against traumatic brain injury partly via Nrf2 signaling. *Toxicology and Applied Pharmacology*, *346*, 28–36. DOI: 10.1016/j.taap.2018.03.020 PMID: 29571711

Dos Santos, C. N., Menezes, R., Carregosa, D., Valentova, K., Foito, A., McDougall, G., & Stewart, D. (2020). Flavonols and flavones. *Dietary Polyphenols: Their Metabolism and Health Effects*, 163-198.

Du, G., Zhao, Z., Chen, Y., Li, Z., Tian, Y., Liu, Z., Liu, B., & Song, J. (2018). Quercetin protects rat cortical neurons against traumatic brain injury. *Molecular Medicine Reports*, *17*(6), 7859–7865. DOI: 10.3892/mmr.2018.8801 PMID: 29620218

Fan, H., Tang, H.-B., Shan, L.-Q., Liu, S.-C., Huang, D.-G., Chen, X., Chen, Z., Yang, M., Yin, X.-H., Yang, H., & Hao, D.-J. (2019). Quercetin prevents necroptosis of oligodendrocytes by inhibiting macrophages/microglia polarization to M1 phenotype after spinal cord injury in rats. *Journal of Neuroinflammation*, *16*(1), 206. DOI: 10.1186/s12974-019-1613-2 PMID: 31699098

Farbood, Y., Sarkaki, A., Dianat, M., Khodadadi, A., Haddad, M. K., & Mashhadizadeh, S. (2015). Ellagic acid prevents cognitive and hippocampal long-term potentiation deficits and brain inflammation in rat with traumatic brain injury. *Life Sciences*, *124*, 120–127. DOI: 10.1016/j.lfs.2015.01.013 PMID: 25637685

Feng, Y., Cui, Y., Gao, J.-L., Li, M.-H., Li, R., Jiang, X.-H., Tian, Y.-X., Wang, K.-J., Cui, C.-M., & Cui, J.-Z. (2016). Resveratrol attenuates neuronal autophagy and inflammatory injury by inhibiting the TLR4/NF-κB signaling pathway in experimental traumatic brain injury. *International Journal of Molecular Medicine*, *37*(4), 921–930. DOI: 10.3892/ijmm.2016.2495 PMID: 26936125

Fernandes, F. D. P., Menezes, A. P. F., de Sousa Neves, J. C., Fonteles, A. A., da Silva, A. T. A., de Araújo Rodrigues, P., do Carmo, M. R. S., de Souza, C. M., & de Andrade, G. M. (2014). Caffeic acid protects mice from memory deficits induced by focal cerebral ischemia. *Behavioural Pharmacology*, *25*(7), 637–647. DOI: 10.1097/FBP.0000000000000076 PMID: 25171077

Frugier, T., Morganti-Kossmann, M. C., O'Reilly, D., & McLean, C. A. (2010). In situ detection of inflammatory mediators in post mortem human brain tissue after traumatic injury. *Journal of Neurotrauma*, *27*(3), 497–507. DOI: 10.1089/neu.2009.1120 PMID: 20030565

García-Conesa, M.-T., & Larrosa, M. (2020). *Polyphenol-rich foods for human health and disease* (Vol. 12). MDPI.

Gatson, J. W., Liu, M.-M., Abdelfattah, K., Wigginton, J. G., Smith, S., Wolf, S., & Minei, J. P. (2013). Resveratrol decreases inflammation in the brain of mice with mild traumatic brain injury. *The Journal of Trauma and Acute Care Surgery*, *74*(2), 470–475. DOI: 10.1097/TA.0b013e31827e1f51 PMID: 23354240

Ghodsi, H., Rahimi, H. R., Aghili, S. M., Saberi, A., & Shoeibi, A. (2022). Evaluation of curcumin as add-on therapy in patients with Parkinson's disease: A pilot randomized, triple-blind, placebo-controlled trial. *Clinical Neurology and Neurosurgery*, *218*, 107300. DOI: 10.1016/j.clineuro.2022.107300 PMID: 35636380

Guo, D.-J., Li, F., Yu, P. H.-F., & Chan, S.-W. (2013). Neuroprotective effects of luteolin against apoptosis induced by 6-hydroxydopamine on rat pheochromocytoma PC12 cells. *Pharmaceutical Biology*, *51*(2), 190–196. DOI: 10.3109/13880209.2012.716852 PMID: 23035972

Hano, C., & Tungmunnithum, D. (2020). *Plant polyphenols, more than just simple natural antioxidants: Oxidative stress, aging and age-related diseases* (Vol. 7). MDPI.

Harkey, H. L.III, White, E. A.IV, Tibbs, R. E.Jr, & Haines, D. E. (2003). A clinician's view of spinal cord injury. *Anatomical Record. Part B, New Anatomist*, *271*(1), 41–48. DOI: 10.1002/ar.b.10012 PMID: 12619085

Hasegawa, Y., Suzuki, H., Sozen, T., Altay, O., & Zhang, J. H. (2011). Apoptotic mechanisms for neuronal cells in early brain injury after subarachnoid hemorrhage. *Early Brain Injury or Cerebral Vasospasm: Volume 1. Pathophysiology : the Official Journal of the International Society for Pathophysiology*, •••, 43–48.

Hassanpour, S. H., & Doroudi, A. (2023). Review of the antioxidant potential of flavonoids as a subgroup of polyphenols and partial substitute for synthetic antioxidants. *Avicenna Journal of Phytomedicine*, *13*(4), 354. PMID: 37663389

Heo, S.-D., Kim, J., Choi, Y., Ekanayake, P., Ahn, M., & Shin, T. (2020). Hesperidin improves motor disability in rat spinal cord injury through anti-inflammatory and antioxidant mechanism via Nrf-2/HO-1 pathway. *Neuroscience Letters*, *715*, 134619. DOI: 10.1016/j.neulet.2019.134619 PMID: 31715292

Hsu, C.-C., Huang, H.-C., Wu, P.-T., Tai, T.-W., & Jou, I.-M. (2016). Sesame oil improves functional recovery by attenuating nerve oxidative stress in a mouse model of acute peripheral nerve injury: Role of Nrf-2. *The Journal of Nutritional Biochemistry*, *38*, 102–106. DOI: 10.1016/j.jnutbio.2016.09.003 PMID: 27732910

Hu, X., Leak, R. K., Shi, Y., Suenaga, J., Gao, Y., Zheng, P., & Chen, J. (2015). Microglial and macrophage polarization—New prospects for brain repair. *Nature Reviews. Neurology*, *11*(1), 56–64. DOI: 10.1038/nrneurol.2014.207 PMID: 25385337

Hussain, G., Wang, J., Rasul, A., Anwar, H., Qasim, M., Zafar, S., Aziz, N., Razzaq, A., Hussain, R., De Aguilar, J.-L. G., & Sun, T. (2020). Current status of therapeutic approaches against peripheral nerve injuries: A detailed story from injury to recovery. *International Journal of Biological Sciences*, *16*(1), 116–134. DOI: 10.7150/ijbs.35653 PMID: 31892850

Hussain, G., Zhang, L., Rasul, A., Anwar, H., Sohail, M. U., Razzaq, A., Aziz, N., Shabbir, A., Ali, M., & Sun, T. (2018). Role of plant-derived flavonoids and their mechanism in attenuation of Alzheimer's and Parkinson's diseases: An update of recent data. *Molecules (Basel, Switzerland)*, *23*(4), 814. DOI: 10.3390/molecules23040814 PMID: 29614843

Islam, F., Bepary, S., Nafady, M. H., Islam, M. R., Emran, T. B., Sultana, S., Huq, M. A., Mitra, S., Chopra, H., Sharma, R., Sweilam, S. H., Khandaker, M. U., & Idris, A. M. (2022). Polyphenols targeting oxidative stress in spinal cord injury: Current status and future vision. *Oxidative Medicine and Cellular Longevity*, *2022*(1), 8741787. DOI: 10.1155/2022/8741787 PMID: 36046682

Itoh, T., Imano, M., Nishida, S., Tsubaki, M., Mizuguchi, N., Hashimoto, S., Ito, A., & Satou, T. (2012). (−)-Epigallocatechin-3-gallate increases the number of neural stem cells around the damaged area after rat traumatic brain injury. *Journal of Neural Transmission (Vienna, Austria)*, *119*(8), 877–890. DOI: 10.1007/s00702-011-0764-9 PMID: 22212485

Itoh, T., Tabuchi, M., Mizuguchi, N., Imano, M., Tsubaki, M., Nishida, S., Hashimoto, S., Matsuo, K., Nakayama, T., Ito, A., Munakata, H., & Satou, T. (2013). Neuroprotective effect of (−)-epigallocatechin-3-gallate in rats when administered pre-or post-traumatic brain injury. *Journal of Neural Transmission (Vienna, Austria)*, *120*(5), 767–783. DOI: 10.1007/s00702-012-0918-4 PMID: 23180302

Jain, V., Pareek, A., Bhardwaj, Y. R., & Singh, N. (2013). Attenuating effect of standardized fruit extract of Punica granatum L in rat model of tibial and sural nerve transection induced neuropathic pain. *BMC Complementary and Alternative Medicine*, *13*(1), 274. DOI: 10.1186/1472-6882-13-274 PMID: 24499201

Jamme, I., Petit, E., Divoux, D., Gerbi, A., Maixent, J.-M., & Nouvelot, A. (1995). Modulation of mouse cerebral Na+, K+-ATPase activity by oxygen free radicals. *Neuroreport*, *7*(1), 333–337. PMID: 8742483

Javed, H., Khan, M., Ahmad, A., Vaibhav, K., Ahmad, M., Khan, A., Ashafaq, M., Islam, F., Siddiqui, M., Safhi, M., & Islam, F. (2012). Rutin prevents cognitive impairments by ameliorating oxidative stress and neuroinflammation in rat model of sporadic dementia of Alzheimer type. *Neuroscience*, *210*, 340–352. DOI: 10.1016/j.neuroscience.2012.02.046 PMID: 22441036

Jiang, W., Huang, Y., Han, N., He, F., Li, M., Bian, Z., Liu, J., Sun, T., & Zhu, L. (2016). Quercetin suppresses NLRP3 inflammasome activation and attenuates histopathology in a rat model of spinal cord injury. *Spinal Cord*, *54*(8), 592–596. DOI: 10.1038/sc.2015.227 PMID: 26754474

Jiang, Z., Zhang, J., Cai, Y., Huang, J., & You, L. (2017). Catechin attenuates traumatic brain injury-induced blood–brain barrier damage and improves longer-term neurological outcomes in rats. *Experimental Physiology*, *102*(10), 1269–1277. DOI: 10.1113/EP086520 PMID: 28678393

Jin, W., Wang, J., Zhu, T., Yuan, B., Ni, H., Jiang, J., Wang, H., & Liang, W. (2014). Anti-inflammatory effects of curcumin in experimental spinal cord injury in rats. *Inflammation Research*, *63*(5), 381–387. DOI: 10.1007/s00011-014-0710-z PMID: 24468890

Ju, D.-T., Kuo, W.-W., Ho, T.-J., Paul, C. R., Kuo, C.-H., Viswanadha, V. P., Lin, C.-C., Chen, Y.-S., Chang, Y.-M., & Huang, C.-Y. (2015). Protocatechuic acid from Alpinia oxyphylla induces Schwann cell migration via ERK1/2, JNK and p38 activation. *The American Journal of Chinese Medicine*, *43*(04), 653–665. DOI: 10.1142/S0192415X15500408 PMID: 26119854

Ju, D.-T., Liao, H.-E., Shibu, M. A., Ho, T.-J., Padma, V. V., Tsai, F.-J., Chung, L.-C., Day, C. H., Lin, C.-C., & Huang, C.-Y. (2015). Nerve regeneration potential of protocatechuic acid in RSC96 Schwann cells by induction of cellular proliferation and migration through IGF-IR-PI3K-Akt signaling. *The Chinese Journal of Physiology*, *58*(6), 412–419. DOI: 10.4077/CJP.2015.BAD340 PMID: 26717920

Karioti, A., Carta, F., & Supuran, C. T. (2016). Phenols and polyphenols as carbonic anhydrase inhibitors. *Molecules (Basel, Switzerland), 21*(12), 1649. DOI: 10.3390/molecules21121649 PMID: 27918439

Karuppagounder, S., Madathil, S., Pandey, M., Haobam, R., Rajamma, U., & Mohanakumar, K. (2013). Quercetin up-regulates mitochondrial complex-I activity to protect against programmed cell death in rotenone model of Parkinson's disease in rats. *Neuroscience, 236*, 136–148. DOI: 10.1016/j.neuroscience.2013.01.032 PMID: 23357119

Kaur, P., & Sharma, S. (2018). Recent advances in pathophysiology of traumatic brain injury. *Current Neuropharmacology, 16*(8), 1224–1238. DOI: 10.2174/1570159X15666170613083606 PMID: 28606040

Kaur, S., Rana, A., Gangwani, S., & Sharma, R. (2012). Punica granatum attenuates sciatic nerve ligation induced-neuropathic pain. *International Journal of Pharmaceutical Sciences and Research, 3*(2), 509.

Kerman, M., Kanter, M., Coşkun, K. K., Erboga, M., & Gurel, A. (2012). Neuroprotective effects of caffeic acid phenethyl ester on experimental traumatic brain injury in rats. *Journal of Molecular Histology, 43*(1), 49–57. DOI: 10.1007/s10735-011-9376-9 PMID: 22124729

Kian, K., Khalatbary, A. R., Ahmadvand, H., Karimpour Malekshah, A., & Shams, Z. (2019). Neuroprotective effects of (−)-epigallocatechin-3-gallate (EGCG) against peripheral nerve transection-induced apoptosis. *Nutritional Neuroscience, 22*(8), 578–586. DOI: 10.1080/1028415X.2017.1419542 PMID: 29292676

Kim, M., Jung, J., Jeong, N. Y., & Chung, H.-J. (2019). The natural plant flavonoid apigenin is a strong antioxidant that effectively delays peripheral neurodegenerative processes. *Anatomical Science International, 94*(4), 285–294. DOI: 10.1007/s12565-019-00486-2 PMID: 30949912

Krishna, G., Ying, Z., & Gomez-Pinilla, F. (2019). Blueberry supplementation mitigates altered brain plasticity and behavior after traumatic brain injury in rats. *Molecular Nutrition & Food Research, 63*(15), 1801055. DOI: 10.1002/mnfr.201801055 PMID: 31115168

Kumar, R., Kumar, R., Khurana, N., Singh, S. K., Khurana, S., Verma, S., Sharma, N., Vyas, M., Dua, K., Khursheed, R., Awasthi, A., & Vishwas, S. (2022). Improved neuroprotective activity of Fisetin through SNEDDS in ameliorating the behavioral alterations produced in rotenone-induced Parkinson's model. *Environmental Science and Pollution Research International, 29*(33), 50488–50499. DOI: 10.1007/s11356-022-19428-z PMID: 35230633

Laird, M. D., Sukumari-Ramesh, S., Swift, A. E., Meiler, S. E., Vender, J. R., & Dhandapani, K. M. (2010). Curcumin attenuates cerebral edema following traumatic brain injury in mice: A possible role for aquaporin-4? *Journal of Neurochemistry, 113*(3), 637–648. DOI: 10.1111/j.1471-4159.2010.06630.x PMID: 20132469

Latha, R. C. R., & Daisy, P. (2011). Insulin-secretagogue, antihyperlipidemic and other protective effects of gallic acid isolated from Terminalia bellerica Roxb. in streptozotocin-induced diabetic rats. *Chemico-Biological Interactions, 189*(1-2), 112–118. DOI: 10.1016/j.cbi.2010.11.005 PMID: 21078310

le Tang, X., Wang, X., & Fang, G. (2021). Resveratrol ameliorates sevoflurane-induced cognitive impairment by activating the SIRT1/NF-κB pathway in neonatal mice. *The Journal of Nutritional Biochemistry, 90*, 108579. DOI: 10.1016/j.jnutbio.2020.108579 PMID: 33388350

Lee, S. K., & Wolfe, S. W. (2000). Peripheral nerve injury and repair. *The Journal of the American Academy of Orthopaedic Surgeons*, *8*(4), 243–252. DOI: 10.5435/00124635-200007000-00005 PMID: 10951113

SangHwon, L. S., Choi BoYoung, C. B., Lee SongHee, L. S., Kho, A., Jeong JeongHyun, J. J., Hong DaeKi, H. D., & Suh SangWon, S. S. (2017). Administration of protocatechuic acid reduces traumatic brain injury-induced neuronal death.

Li, X., Wang, H., Wen, G., Li, L., Gao, Y., Zhuang, Z., Zhou, M., Mao, L., & Fan, Y. (2018). Neuroprotection by quercetin via mitochondrial function adaptation in traumatic brain injury: PGC-1α pathway as a potential mechanism. *Journal of Cellular and Molecular Medicine*, *22*(2), 883–891. DOI: 10.1111/jcmm.13313 PMID: 29205806

Liu, Z., Yao, X., Sun, B., Jiang, W., Liao, C., Dai, X., Chen, Y., Chen, J., & Ding, R. (2021). Pretreatment with kaempferol attenuates microglia-mediate neuroinflammation by inhibiting MAPKs–NF–κB signaling pathway and pyroptosis after secondary spinal cord injury. *Free Radical Biology & Medicine*, *168*, 142–154. DOI: 10.1016/j.freeradbiomed.2021.03.037 PMID: 33823244

Mahmoud, A. M., Mohammed, H. M., Khadrawy, S. M., & Galaly, S. R. (2017). Hesperidin protects against chemically induced hepatocarcinogenesis via modulation of Nrf2/ARE/HO-1, PPARγ and TGF-β1/Smad3 signaling, and amelioration of oxidative stress and inflammation. *Chemico-Biological Interactions*, *277*, 146–158. DOI: 10.1016/j.cbi.2017.09.015 PMID: 28935427

Mansouri, M. T., Naghizadeh, B., Ghorbanzadeh, B., Farbood, Y., Sarkaki, A., & Bavarsad, K. (2013). Gallic acid prevents memory deficits and oxidative stress induced by intracerebroventricular injection of streptozotocin in rats. *Pharmacology, Biochemistry, and Behavior*, *111*, 90–96. DOI: 10.1016/j.pbb.2013.09.002 PMID: 24036471

Mashhadizadeh, S., Farbood, Y., Dianat, M., Khodadadi, A., & Sarkaki, A. (2017). Therapeutic effects of ellagic acid on memory, hippocampus electrophysiology deficits, and elevated TNF-α level in brain due to experimental traumatic brain injury. *Iranian Journal of Basic Medical Sciences.*, *20*(4), 399. PMID: 28804609

Matsuda, M., Huh, Y., & Ji, R.-R. (2019). Roles of inflammation, neurogenic inflammation, and neuroinflammation in pain. *Journal of Anesthesia*, *33*(1), 131–139. DOI: 10.1007/s00540-018-2579-4 PMID: 30448975

McDonald, J. W., & Sadowsky, C. (2002). Spinal-cord injury. *Lancet*, *359*(9304), 417–425. DOI: 10.1016/S0140-6736(02)07603-1 PMID: 11844532

Meng, H. Y., Shao, D. C., Li, H., Huang, X. D., Yang, G., Xu, B., & Niu, H. Y. (2018). Resveratrol improves neurological outcome and neuroinflammation following spinal cord injury through enhancing autophagy involving the AMPK/mTOR pathway. *Molecular Medicine Reports*, *18*(2), 2237–2244. DOI: 10.3892/mmr.2018.9194 PMID: 29956767

Mietto, B. S., Mostacada, K., & Martinez, A. M. B. (2015). Neurotrauma and inflammation: CNS and PNS responses. *Mediators of Inflammation*, *2015*(1), 251204. DOI: 10.1155/2015/251204 PMID: 25918475

Mirshekar, M. A., Sarkaki, A., Farbood, Y., Naseri, M. K. G., Badavi, M., Mansouri, M. T., & Haghparast, A. (2018). Neuroprotective effects of gallic acid in a rat model of traumatic brain injury: Behavioral, electrophysiological, and molecular studies. *Iranian Journal of Basic Medical Sciences.*, *21*(10), 1056. PMID: 30524680

Mishra, A., Bandopadhyay, R., Singh, P. K., Mishra, P. S., Sharma, N., & Khurana, N. (2021). Neuroinflammation in neurological disorders: Pharmacotherapeutic targets from bench to bedside. *Metabolic Brain Disease*, *36*(7), 1591–1626. DOI: 10.1007/s11011-021-00806-4 PMID: 34387831

Mustafa, A. G., & Alshboul, O. A. (2013). Pathophysiology of traumatic brain injury. *Neuroscience Journal*, *18*(3), 222–235. PMID: 23887212

Nath, J., Roy, R., Sathyamoorthy, Y. K., Paul, S., Goswami, S., Chakravarty, H., Paul, R., & Borah, A. (2022). Resveratrol as a therapeutic choice for traumatic brain injury: An insight into its molecular mechanism of action. *Brain Disorders (Amsterdam, Netherlands)*, *6*, 100038. DOI: 10.1016/j.dscb.2022.100038

Ojha, S., Javed, H., Azimullah, S., Abul Khair, S. B., & Haque, M. E. (2015). Neuroprotective potential of ferulic acid in the rotenone model of Parkinson's disease. *Drug Design, Development and Therapy*, •••, 5499–5510. PMID: 26504373

Oliveira, M. A., Heimfarth, L., Passos, F. R. S., Miguel-dos-Santos, R., Mingori, M. R., Moreira, J. C. F., Lauton, S. S., Barreto, R. S., Araújo, A. A., Oliveira, A. P., Oliveira, J. T., Baptista, A. F., Martinez, A. M. B., Quintans-Júnior, L. J., & Quintans, J. S. S. (2020). Naringenin complexed with hydroxypropyl-β-cyclodextrin improves the sciatic nerve regeneration through inhibition of p75NTR and JNK pathway. *Life Sciences*, *241*, 117102. DOI: 10.1016/j.lfs.2019.117102 PMID: 31790691

Oyagbemi, A. A., Omobowale, T. O., Saba, A. B., Olowu, E. R., Dada, R. O., & Akinrinde, A. S. (2016). Gallic acid ameliorates cyclophosphamide-induced neurotoxicity in Wistar rats through free radical scavenging activity and improvement in antioxidant defense system. *Journal of Dietary Supplements*, *13*(4), 402–419. DOI: 10.3109/19390211.2015.1103827 PMID: 26716793

Özbek, Z., Aydin, H., Kocman, A., Özkara, E., Şahin, E., Bektur, E., Vural, M., Köse, A., Arslantaş, A., & Baycu, C. (2017). Neuroprotective effect of genistein in peripheral nerve injury. *Turkish Neurosurgery*, *27*(5). PMID: 27759874

Panche, A. N., Diwan, A. D., & Chandra, S. R. (2016). Flavonoids: An overview. *Journal of Nutritional Science*, *5*, e47. DOI: 10.1017/jns.2016.41 PMID: 28620474

Parent, M., Chitturi, J., Santhakumar, V., Hyder, F., Sanganahalli, B. G., & Kannurpatti, S. S. (2020). Kaempferol treatment after traumatic brain injury during early development mitigates brain parenchymal microstructure and neural functional connectivity deterioration at adolescence. *Journal of Neurotrauma*, *37*(7), 966–974. DOI: 10.1089/neu.2019.6486 PMID: 31830867

Patten, J. P. (1996). Trauma and the nervous system. In *Neurological Differential Diagnosis* (pp. 399–417). Springer. DOI: 10.1007/978-3-642-58981-2_23

Peterson, J., Dwyer, J., Adlercreutz, H., Scalbert, A., Jacques, P., & McCullough, M. L. (2010). Dietary lignans: Physiology and potential for cardiovascular disease risk reduction. *Nutrition Reviews*, *68*(10), 571–603. DOI: 10.1111/j.1753-4887.2010.00319.x PMID: 20883417

Proestos, C., & Varzakas, T. (2017). Aromatic plants: Antioxidant capacity and polyphenol characterisation. *Foods*, *6*(4), 28. DOI: 10.3390/foods6040028 PMID: 28375185

Quideau, S., Deffieux, D., Douat-Casassus, C., & Pouységu, L. (2011). Plant polyphenols: Chemical properties, biological activities, and synthesis. *Angewandte Chemie International Edition*, *50*(3), 586–621. DOI: 10.1002/anie.201000044 PMID: 21226137

Rao, P. N., Mainkar, O., Bansal, N., Rakesh, N., Haffey, P., Urits, I., Orhurhu, V., Kaye, A. D., Urman, R. D., Gulati, A., & Jones, M. (2021). Flavonoids in the treatment of neuropathic pain. *Current Pain and Headache Reports*, *25*(7), 43. DOI: 10.1007/s11916-021-00959-y PMID: 33961144

Rashno, M., Sarkaki, A., Farbood, Y., Rashno, M., Khorsandi, L., Naseri, M. K. G., & Dianat, M. (2019). *RETRACTED: Therapeutic effects of chrysin in a rat model of traumatic brain injury: A behavioral, biochemical, and histological study*. Elsevier. DOI: 10.1016/j.lfs.2019.05.007

Ray, S., Dixon, C., & Banik, N. (2002). Molecular mechanisms in the pathogenesis of traumatic brain injury. *Histology and Histopathology*. PMID: 12371142

Reinisalo, M., Kårlund, A., Koskela, A., Kaarniranta, K., & Karjalainen, R. O. (2015). Polyphenol stilbenes: Molecular mechanisms of defence against oxidative stress and aging-related diseases. *Oxidative Medicine and Cellular Longevity*, *2015*(1), 340520. DOI: 10.1155/2015/340520 PMID: 26180583

Ren, Z., Zhang, R., Li, Y., Li, Y., Yang, Z., & Yang, H. (2017). Ferulic acid exerts neuroprotective effects against cerebral ischemia/reperfusion-induced injury via antioxidant and anti-apoptotic mechanisms in vitro and in vivo. *International Journal of Molecular Medicine*, *40*(5), 1444–1456. DOI: 10.3892/ijmm.2017.3127 PMID: 28901374

Renno, W. M., Al-Maghrebi, M., AlShammari, A., & George, P. (2013). (−)-Epigallocatechin-3-gallate (EGCG) attenuates peripheral nerve degeneration in rat sciatic nerve crush injury. *Neurochemistry International*, *62*(3), 221–231. DOI: 10.1016/j.neuint.2012.12.018 PMID: 23313191

Rolls, A., Shechter, R., & Schwartz, M. (2009). The bright side of the glial scar in CNS repair. *Nature Reviews. Neuroscience*, *10*(3), 235–241. DOI: 10.1038/nrn2591 PMID: 19229242

Rong, H., Liang, Y., & Niu, Y. (2018). Rosmarinic acid attenuates β-amyloid-induced oxidative stress via Akt/GSK-3β/Fyn-mediated Nrf2 activation in PC12 cells. *Free Radical Biology & Medicine*, *120*, 114–123. DOI: 10.1016/j.freeradbiomed.2018.03.028 PMID: 29555592

Rossi, D., & Gaidano, G. (2003). Messengers of cell death: Apoptotic signaling in health and disease. *Haematologica*, *88*(2), 212–218. PMID: 12604411

Ruff, R. L., McKerracher, L., & Selzer, M. E. (2008). Repair and neurorehabilitation strategies for spinal cord injury. *Annals of the New York Academy of Sciences*, *1142*(1), 1–20. DOI: 10.1196/annals.1444.004 PMID: 18990118

Sarkaki, A., Farbood, Y., Gharib-Naseri, M. K., Badavi, M., Mansouri, M. T., Haghparast, A., & Mirshekar, M. A. (2015). Gallic acid improved behavior, brain electrophysiology, and inflammation in a rat model of traumatic brain injury. *Canadian Journal of Physiology and Pharmacology*, *93*(8), 687–694. DOI: 10.1139/cjpp-2014-0546 PMID: 26222320

Sawmiller, D., Li, S., Shahaduzzaman, M., Smith, A. J., Obregon, D., Giunta, B., Borlongan, C. V., Sanberg, P. R., & Tan, J. (2014). Luteolin reduces Alzheimer's disease pathologies induced by traumatic brain injury. *International Journal of Molecular Sciences*, *15*(1), 895–904. DOI: 10.3390/ijms15010895 PMID: 24413756

Sekhon, L. H., & Fehlings, M. G. (2001). Epidemiology, demographics, and pathophysiology of acute spinal cord injury. *Spine*, *26*(24S), S2–S12. DOI: 10.1097/00007632-200112151-00002 PMID: 11805601

Şentürk, S., Yaman, M. E., Aydın, H. E., Güney, G., Bozkurt, I., Paksoy, K., & Abdioğlu, A. A. (2018). Effects of resveratrol on inflammation and apoptosis after experimental spinal cord injury.

Shah, P. M., Priya, V. V., & Gayathri, R. (2016). Quercetin-a flavonoid: A systematic review. *Journal of Pharmaceutical Sciences and Research*, *8*(8), 878.

Shi, Z., Qiu, W., Xiao, G., Cheng, J., & Zhang, N. (2018). Resveratrol attenuates cognitive deficits of traumatic brain injury by activating p38 signaling in the brain. *Medical Science Monitor*, *24*, 1097–1103. DOI: 10.12659/MSM.909042 PMID: 29467361

Silva, N. A., Sousa, N., Reis, R. L., & Salgado, A. J. (2014). From basics to clinical: A comprehensive review on spinal cord injury. *Progress in Neurobiology*, *114*, 25–57. DOI: 10.1016/j.pneurobio.2013.11.002 PMID: 24269804

Silva dos Santos, J., Goncalves Cirino, J. P., de Oliveira Carvalho, P., & Ortega, M. M. (2021). The pharmacological action of kaempferol in central nervous system diseases: A review. *Frontiers in Pharmacology*, *11*, 565700. DOI: 10.3389/fphar.2020.565700 PMID: 33519431

Smith, D. H., Chen, X.-H., Pierce, J. E., Wolf, J. A., Trojanowski, J. Q., Graham, D. I., & Mcintosh, T. K. (1997). Progressive atrophy and neuron death for one year following brain trauma in the rat. *Journal of Neurotrauma*, *14*(10), 715–727. DOI: 10.1089/neu.1997.14.715 PMID: 9383090

Soleymani, S., Habtemariam, S., Rahimi, R., & Nabavi, S. M. (2020). The what and who of dietary lignans in human health: Special focus on prooxidant and antioxidant effects. *Trends in Food Science & Technology*, *106*, 382–390. DOI: 10.1016/j.tifs.2020.10.015

Soltani, Z., Khaksari, M., Jafari, E., Iranpour, M., & Shahrokhi, N. (2015). Is genistein neuroprotective in traumatic brain injury? *Physiology & Behavior*, *152*, 26–31. DOI: 10.1016/j.physbeh.2015.08.037 PMID: 26367454

Song, H., Zhang, X., Wang, W., Liu, R., Zhao, K., Liu, M.-y., Gong, W., & Ning, B. (2018). Neuroprotective mechanisms of rutin for spinal cord injury through anti-oxidation and anti-inflammation and inhibition of p38 mitogen activated protein kinase pathway. *Neural Regeneration Research*, *13*(1), 128–134. DOI: 10.4103/1673-5374.217349 PMID: 29451217

Song, J., Du, G., Wu, H., Gao, X., Yang, Z., Liu, B., & Cui, S. (2021). RETRACTED: Protective effects of quercetin on traumatic brain injury induced inflammation and oxidative stress in cortex through activating Nrf2/HO-1 pathway. *Restorative Neurology and Neuroscience*, *39*(1), 73–84. DOI: 10.3233/RNN-201119 PMID: 33612499

Spagnuolo, C., Napolitano, M., Tedesco, I., Moccia, S., Milito, A., & Luigi Russo, G. (2016). Neuroprotective role of natural polyphenols. *Current Topics in Medicinal Chemistry*, *16*(17), 1943–1950. DOI: 10.2174/1568026616666160204122449 PMID: 26845551

Sun, G., Miao, Z., Ye, Y., Zhao, P., Fan, L., Bao, Z., Tu, Y., Li, C., Chao, H., Xu, X., & Ji, J. (2020). Curcumin alleviates neuroinflammation, enhances hippocampal neurogenesis, and improves spatial memory after traumatic brain injury. *Brain Research Bulletin*, *162*, 84–93. DOI: 10.1016/j.brainresbull.2020.05.009 PMID: 32502596

Susin, S. A., Zamzami, N., & Kroemer, G. (1998). Mitochondria as regulators of apoptosis: Doubt no more. *Biochimica et Biophysica Acta (BBA)-. Biochimica et Biophysica Acta. Bioenergetics*, *1366*(1-2), 151–165. DOI: 10.1016/S0005-2728(98)00110-8

Teixeira, M., Souza, C., Menezes, A., Carmo, M., Fonteles, A., Gurgel, J., Lima, F., Viana, G., & Andrade, G. (2013). Catechin attenuates behavioral neurotoxicity induced by 6-OHDA in rats. *Pharmacology, Biochemistry, and Behavior*, *110*, 1–7. DOI: 10.1016/j.pbb.2013.05.012 PMID: 23714698

Thota, R. N., Rosato, J. I., Dias, C. B., Burrows, T. L., Martins, R. N., & Garg, M. L. (2020). Dietary supplementation with curcumin reduce circulating levels of glycogen synthase kinase-3β and islet amyloid polypeptide in adults with high risk of type 2 diabetes and Alzheimer's disease. *Nutrients*, *12*(4), 1032. DOI: 10.3390/nu12041032 PMID: 32283762

Tian, W., Han, X.-G., Liu, Y.-J., Tang, G.-Q., Liu, B., Wang, Y.-Q., Xiao, B., & Xu, Y.-F. (2013). Intrathecal epigallocatechin gallate treatment improves functional recovery after spinal cord injury by upregulating the expression of BDNF and GDNF. *Neurochemical Research*, *38*(4), 772–779. DOI: 10.1007/s11064-013-0976-5 PMID: 23344852

Tsujimoto, Y., & Shimizu, S. (2007). Role of the mitochondrial membrane permeability transition in cell death. *Apoptosis*, *12*(5), 835–840. DOI: 10.1007/s10495-006-0525-7 PMID: 17136322

Urdzikova, L. M., Ruzicka, J., Karova, K., Kloudova, A., Svobodova, B., Amin, A., Dubisova, J., Schmidt, M., Kubinova, S., & Jhanwar-Uniyal, M. (2017). A green tea polyphenol epigallocatechin-3-gallate enhances neuroregeneration after spinal cord injury by altering levels of inflammatory cytokines. *Neuropharmacology*, *126*, 213–223. DOI: 10.1016/j.neuropharm.2017.09.006 PMID: 28899730

Utpal, B. K., Roy, S. C., Zehravi, M., Sweilam, S. H., Raja, A. D., Haque, M. A., Nayak, C., Balakrishnan, S., Singh, L. P., Panigrahi, S., Alshehri, M. A., Rab, S. O., Minhaj, N. S., & Emran, T. B. (2025). Polyphenols as Wnt/β-catenin pathway modulators: A promising strategy in clinical neurodegeneration. *Animal Models and Experimental Medicine*, *8*(2), 266–286. DOI: 10.1002/ame2.12525 PMID: 39808166

Uzar, E., Alp, H., Cevik, M. U., Fırat, U., Evliyaoglu, O., Tufek, A., & Altun, Y. (2012). Ellagic acid attenuates oxidative stress on brain and sciatic nerve and improves histopathology of brain in streptozotocin-induced diabetic rats. *Neurological Sciences*, *33*(3), 567–574. DOI: 10.1007/s10072-011-0775-1 PMID: 21922312

Vauzour, D., Vafeiadou, K., Rodriguez-Mateos, A., Rendeiro, C., & Spencer, J. P. (2008). The neuroprotective potential of flavonoids: A multiplicity of effects. *Genes & Nutrition*, *3*(3), 115–126. DOI: 10.1007/s12263-008-0091-4 PMID: 18937002

Viña, J., Escudero, J., Baquero, M., Cebrián, M., Carbonell-Asíns, J. A., Muñoz, J. E., Satorres, E., Meléndez, J. C., Ferrer-Rebolleda, J., Cózar-Santiago, M. P., Santabárbara-Gómez, J. M., Jové, M., Pamplona, R., Tarazona-Santabalbina, F. J., & Borrás, C. (2022). Genistein effect on cognition in prodromal Alzheimer's disease patients. The GENIAL clinical trial. *Alzheimer's Research & Therapy*, *14*(1), 164. DOI: 10.1186/s13195-022-01097-2 PMID: 36329553

Wang, M. L., Rivlin, M., Graham, J. G., & Beredjiklian, P. K. (2019). Peripheral nerve injury, scarring, and recovery. *Connective Tissue Research*, *60*(1), 3–9. DOI: 10.1080/03008207.2018.1489381 PMID: 30187777

Wang, P., Jiang, L., Zhou, N., Zhou, H., Liu, H., Zhao, W., Zhang, H., Zhang, X., & Hu, Z. (2018). Resveratrol ameliorates autophagic flux to promote functional recovery in rats after spinal cord injury. *Oncotarget*, *9*(9), 8427–8440. DOI: 10.18632/oncotarget.23877 PMID: 29492205

Wang, Q., Xu, J., Rottinghaus, G. E., Simonyi, A., Lubahn, D., Sun, G. Y., & Sun, A. Y. (2002). Resveratrol protects against global cerebral ischemic injury in gerbils. *Brain Research*, *958*(2), 439–447. DOI: 10.1016/S0006-8993(02)03543-6 PMID: 12470882

Wang, T., Wang, S., Wang, X., Jiang, H., Yang, Y., Wang, Y., Cheng, J., Zhang, C., Liang, W., & Feng, H. (2018). Fisetin exerts antioxidant and neuroprotective effects in multiple mutant hSOD1 models of amyotrophic lateral sclerosis by activating ERK. *Neuroscience*, *379*, 152–166. DOI: 10.1016/j.neuroscience.2018.03.008 PMID: 29559385

Wei, I.-H., Tu, H.-C., Huang, C.-C., Tsai, M.-H., Tseng, C.-Y., & Shieh, J.-Y. (2011). (-)-Epigallocatechin gallate attenuates NADPH-d/nNOS expression in motor neurons of rats following peripheral nerve injury. *BMC Neuroscience*, *12*(1), 52. DOI: 10.1186/1471-2202-12-52 PMID: 21627848

Werner, C., & Engelhard, K. (2007). Pathophysiology of traumatic brain injury. *British Journal of Anaesthesia*, *99*(1), 4–9. DOI: 10.1093/bja/aem131 PMID: 17573392

Williams, C. M., Abd El Mohsen, M., Vauzour, D., Rendeiro, C., Butler, L. T., Ellis, J. A., Whiteman, M., & Spencer, J. P. (2008). Blueberry-induced changes in spatial working memory correlate with changes in hippocampal CREB phosphorylation and brain-derived neurotrophic factor (BDNF) levels. *Free Radical Biology & Medicine*, *45*(3), 295–305. DOI: 10.1016/j.freeradbiomed.2008.04.008 PMID: 18457678

Williamson, C., & Rajajee, V. (2021). Traumatic brain injury: Epidemiology, classification, and pathophysiology. *Waltham: UpToDate*.

Wu, A., Ying, Z., & Gomez-Pinilla, F. (2006). Dietary curcumin counteracts the outcome of traumatic brain injury on oxidative stress, synaptic plasticity, and cognition. *Experimental Neurology*, *197*(2), 309–317. DOI: 10.1016/j.expneurol.2005.09.004 PMID: 16364299

Wu, J., Li, Q., Wang, X., Yu, S., Li, L., Wu, X., Chen, Y., Zhao, J., & Zhao, Y. (2013). Neuroprotection by curcumin in ischemic brain injury involves the Akt/Nrf2 pathway. *PLoS One*, *8*(3), e59843. DOI: 10.1371/journal.pone.0059843 PMID: 23555802

Wu, J., Maoqiang, L., Fan, H., Zhenyu, B., Qifang, H., Xuepeng, W., & Liulong, Z. (2016). Rutin attenuates neuroinflammation in spinal cord injury rats. *The Journal of Surgical Research*, *203*(2), 331–337. DOI: 10.1016/j.jss.2016.02.041 PMID: 27363641

Wu, Y., & Cui, J. (2020). (-)-Epigallocatechin-3-gallate provides neuroprotection via AMPK activation against traumatic brain injury in a mouse model. *Naunyn-Schmiedeberg's Archives of Pharmacology, 393*(11), 2209–2220. DOI: 10.1007/s00210-020-01841-1 PMID: 32062732

Xu, J., Wang, H., Ding, K., Zhang, L., Wang, C., Li, T., Wei, W., & Lu, X. (2014). Luteolin provides neuroprotection in models of traumatic brain injury via the Nrf2–ARE pathway. *Free Radical Biology & Medicine, 71*, 186–195. DOI: 10.1016/j.freeradbiomed.2014.03.009 PMID: 24642087

Xu, L., Botchway, B. O., Zhang, S., Zhou, J., & Liu, X. (2018). Inhibition of NF-κB signaling pathway by resveratrol improves spinal cord injury. *Frontiers in Neuroscience, 12*, 690. DOI: 10.3389/fnins.2018.00690 PMID: 30337851

Xu, X.-Y., Wang, D.-Y., Li, Y.-P., Deyrup, S. T., & Zhang, H.-J. (2022). Plant-derived lignans as potential antiviral agents: A systematic review. *Phytochemistry Reviews, 21*(1), 239–289. DOI: 10.1007/s11101-021-09758-0 PMID: 34093097

Yan, P., Bai, L., Lu, W., Gao, Y., Bi, Y., & Lv, G. (2017). Regulation of autophagy by AMP-activated protein kinase/sirtuin 1 pathway reduces spinal cord neurons damage. *Iranian Journal of Basic Medical Sciences., 20*(9), 1029. PMID: 29085598

Yang, T., Kong, B., Gu, J.-W., Kuang, Y.-Q., Cheng, L., Yang, W.-T., Xia, X., & Shu, H.-F. (2014). Anti-apoptotic and anti-oxidative roles of quercetin after traumatic brain injury. *Cellular and Molecular Neurobiology, 34*(6), 797–804. DOI: 10.1007/s10571-014-0070-9 PMID: 24846663

Yang, Y. H., Wang, Z., Zheng, J., & Wang, R. (2015). Protective effects of gallic acid against spinal cord injury-induced oxidative stress. *Molecular Medicine Reports, 12*(2), 3017–3024. DOI: 10.3892/mmr.2015.3738 PMID: 25955644

Yao, S., Wang, L., Chen, Q., Lu, T., Pu, X., & Luo, C. (2021). The effect of mild hypothermia plus rutin on the treatment of spinal cord injury and inflammatory factors by repressing TGF-β/smad pathway. *Acta Cirurgica Brasileira, 36*(3), e360307. DOI: 10.1590/acb360307 PMID: 33978063

Yavtushenko, I. V., Nazarenko, S. M., Katrushov, O. V., & Kostenko, V. O. (2020). Quercetin limits the progression of oxidative and nitrosative stress in the rats' tissues after experimental traumatic brain injury. *Wiadomosci Lekarskie (Warsaw, Poland), 73*(10), 2127–2132. DOI: 10.36740/WLek202010104 PMID: 33310934

Yiu, G., & He, Z. (2006). Glial inhibition of CNS axon regeneration. *Nature Reviews. Neuroscience, 7*(8), 617–627. DOI: 10.1038/nrn1956 PMID: 16858390

Zhai, X., Ding, Y., Wang, Q., Zhang, H., & Li, F. (2016). Rutin acid ameliorates neural apoptosis induced by traumatic brain injury via mitochondrial pathways in mice. *Neuroimmunomodulation, 23*(3), 179–187. DOI: 10.1159/000448716 PMID: 27644033

Zhang, B., Wang, B., Cao, S., & Wang, Y. (2015). Epigallocatechin-3-gallate (EGCG) attenuates traumatic brain injury by inhibition of edema formation and oxidative stress. *The Korean Journal of Physiology & Pharmacology; Official Journal of the Korean Physiological Society and the Korean Society of Pharmacology, 19*(6), 491. DOI: 10.4196/kjpp.2015.19.6.491 PMID: 26557015

Zhang, G., Liu, Y., Xu, L., Sha, C., Zhang, H., & Xu, W. (2019). Resveratrol alleviates lipopolysaccharide-induced inflammation in PC-12 cells and in rat model. *BMC Biotechnology*, *19*(1), 10. DOI: 10.1186/s12896-019-0502-1 PMID: 30691440

Zhang, L., Wang, H., Zhou, Y., Zhu, Y., & Fei, M. (2018). Fisetin alleviates oxidative stress after traumatic brain injury via the Nrf2-ARE pathway. *Neurochemistry International*, *118*, 304–313. DOI: 10.1016/j.neuint.2018.05.011 PMID: 29792955

Zhang, P., & Ma, X. (2015). Effect of rutin on spinal cord injury through inhibition of the expression of MIP-2 and activation of MMP-9, and downregulation of Akt phosphorylation. *Molecular Medicine Reports*, *12*(5), 7554–7560. DOI: 10.3892/mmr.2015.4357 PMID: 26502930

Zhang, S., Gai, Z., Gui, T., Chen, J., Chen, Q., & Li, Y. (2021). Antioxidant effects of protocatechuic acid and protocatechuic aldehyde: Old wine in a new bottle. *Evidence-Based Complementary and Alternative Medicine : eCAM*, *2021*(1), 6139308. DOI: 10.1155/2021/6139308 PMID: 34790246

Zhao, H., Chen, S., Gao, K., Zhou, Z., Wang, C., Shen, Z., Guo, Y., Li, Z., Wan, Z., Liu, C., & Mei, X. (2017). Resveratrol protects against spinal cord injury by activating autophagy and inhibiting apoptosis mediated by the SIRT1/AMPK signaling pathway. *Neuroscience*, *348*, 241–251. DOI: 10.1016/j.neuroscience.2017.02.027 PMID: 28238848

Zhao, R., Wu, X., Bi, X.-Y., Yang, H., & Zhang, Q. (2022). Baicalin attenuates blood-spinal cord barrier disruption and apoptosis through PI3K/Akt signaling pathway after spinal cord injury. *Neural Regeneration Research*, *17*(5), 1080–1087. DOI: 10.4103/1673-5374.324857 PMID: 34558536

Zhao, Z., Li, X., & Li, Q. (2017). Curcumin accelerates the repair of sciatic nerve injury in rats through reducing Schwann cells apoptosis and promoting myelinization. *Biomedicine and Pharmacotherapy*, *92*, 1103–1110. DOI: 10.1016/j.biopha.2017.05.099 PMID: 28622711

Zhou, L., Wu, Z., Wang, G., Xiao, L., Wang, H., Sun, L., & Xie, Y. (2020). Long-term maternal separation potentiates depressive-like behaviours and neuroinflammation in adult male C57/BL6J mice. *Pharmacology, Biochemistry, and Behavior*, *196*, 172953. DOI: 10.1016/j.pbb.2020.172953 PMID: 32450088

Zhou, Y.-F., Guo, B., Ye, M.-J., Liao, R.-F., & Li, S.-L. (2016). Protective effect of rutin against H2O2-induced oxidative stress and apoptosis in human lens epithelial cells. *Current Eye Research*, *41*(7), 933–942. DOI: 10.3109/02713683.2015.1082186 PMID: 26576853

Zhu, H., Bian, C., Yuan, J., Chu, W., Xiang, X., Chen, F., Wang, C., Feng, H., & Lin, J. (2014). Curcumin attenuates acute inflammatory injury by inhibiting the TLR4/MyD88/NF-κB signaling pathway in experimental traumatic brain injury. *Journal of Neuroinflammation*, *11*(1), 59. DOI: 10.1186/1742-2094-11-59 PMID: 24669820

Zhu, Y.-L., Sun, M.-F., Jia, X.-B., Cheng, K., Xu, Y.-D., Zhou, Z.-L., Zhang, P.-H., Qiao, C.-M., Cui, C., Chen, X., Yang, X.-S., & Shen, Y.-Q. (2019). Neuroprotective effects of Astilbin on MPTP-induced Parkinson's disease mice: Glial reaction, α-synuclein expression and oxidative stress. *International Immunopharmacology*, *66*, 19–27. DOI: 10.1016/j.intimp.2018.11.004 PMID: 30419450

Zou, P., Liu, X., Li, G., & Wang, Y. (2018). Resveratrol pretreatment attenuates traumatic brain injury in rats by suppressing NLRP3 inflammasome activation via SIRT1. *Molecular Medicine Reports*, *17*(2), 3212–3217. PMID: 29257276

Chapter 10
Polyphenol-Rich Medicinal Plants:
Mechanistic Insights and Therapeutic Potentials Against Alzheimer's Disease

Qurat-ul-Ane Gillani
https://orcid.org/0000-0002-1068-4472
The Women University, Multan, Pakistan

Arifa Mehreen
University of Agriculture, Faisalabad, Pakistan

Hafiz Aamir Ali Kharl
https://orcid.org/0000-0002-7365-2937
University of Agriculture, Faisalabad, Pakistan

Adnan Afzal
https://orcid.org/0009-0001-4359-5695
Cholistan University of Veterinary and Animal Sciences, Pakistan

Faiza Shakoor
Comstas University, Vehari, Pakistan

Aiman Siddique
National University of Sciences and Technology, Pakistan

Muhammad Hussain Taqi
https://orcid.org/0009-0002-5868-0929
Cholistan University of Veterinary and Animal Sciences, Pakistan

Ayesha Bintay Bintay Farooq
Government College University, Faisalabad, Pakistan

ABSTRACT

Alzheimer's disease (AD) is a leading neurodegenerative disorder marked by memory loss, cognitive decline, and β-amyloid plaques in the brain. Current treatments offer only short-term relief. This chapter explores the role of plant-derived polyphenols—flavonoids, phenolic acids, and lignans from Ginkgo biloba, Curcuma longa, Camellia sinensis, and Withania somnifera—in managing AD. It highlights their antioxidant, anti-inflammatory, and neuroprotective actions through mechanisms like free radical scavenging, acetylcholinesterase inhibition, and modulation of Nrf2 and NF-κB pathways. Laboratory, animal, and clinical studies are reviewed, along with bioavailability challenges and advanced drug delivery approaches. The chapter offers insight into polyphenol-based strategies for AD prevention and treatment.

DOI: 10.4018/979-8-3373-4139-2.ch010

1. INTRODUCTION

Alzheimer's Disease (AD) causes the gradual loss of mental abilities in older people. Alzheimer's disease is the leading cause of dementia and its symptoms are memory problems, difficulties with learning or remembering, slowed thinking ability and changes in mood and behavior (Zvěřová, 2019). With more people experiencing aging worldwide, AD is estimated to become a much greater threat to public health. According to the World Health Organization (WHO), about 50 million people around the world currently have dementia and Alzheimer's disease accounts for much of these cases. By 2050, it is predicted that this number will triple as the result of older people living longer. AD impacts health and economies, as well as hurting patients and those close to them (Castro, Dillon, Machnicki, & Allegri, 2010). Long-term health care is expensive, losing time at work and caring for an ill person causes emotional hardship for caregivers. Living with Alzheimer's and related dementias in the US is expected to cost over $350 billion each year and this total is likely to increase as the public ages. Alzheimer's disease is believed to have a variety of causes, including the buildup of amyloid plaques, abnormal tau proteins, stress related to oxygen and brain inflammation (W. J. Huang, Zhang, & Chen, 2016). Such processes, when repeated, cause both damage to nerves and impairment of mental function. It has been very hard to find effective ways to treat AD. Today's medication for Alzheimer's disease merely treats the symptoms, without addressing the real causes of the disease. Even though these medicines might improve how people think, they do not stop or undo the changes caused by Alzheimer's disease (Moser, 2008). What's more, some of their side effects such as stomach problems, tiredness and dizziness may make patients less likely to stick to their treatment plans. In addition, the high prices of these prescriptions can make things very tough for both patients and healthcare providers.Since the effectiveness of existing treatments is limited, more and more people are interested in trying new approaches that include natural compounds from medicinal plants. Another type of compound gaining interest is polyphenols which are found in many fruits, vegetables, teas and plant food (Ignat, Volf, & Popa, 2011). It is known that polyphenols are antioxidants, fight inflammation and help protect the nervous system and they have been shown to be beneficial for Alzheimer's disease and other neurodegenerative illnesses. They are naturally present and have been examined to find out if they reduce oxidative stress and inflammation, important causes of AD (Christen, 2000). If the creation of ROS exceeds the power of the body to tackle them, oxidative stress will create harm to the lipids, proteins and DNA of the cell. Oxidative stress is important in Alzheimer's disease by helping cause the buildup of amyloid-beta, tau hyperphosphorylation and cell death among neurons. Also, the brain's immune cells (microglia and astrocytes) are activated during neuroinflammation and contribute greatly to AD. Neuronal damage and poor cognitive function are made worse by continuous neuroinflammation (Ransohoff, 2016).

2. PATHOPHYSIOLOGY OF ALZHEIMER'S DISEASE

Most of the global population diagnosed with Alzheimer's Disease is elderly, since this condition largely affects seniors. It is usually noticeable by a worsening of mental skills, forgetfulness and shifts in how people behave (Budson & Solomon, 2021). Many factors such as family history, environmental conditions and molecules, lead to the brain shrinkage and death of neurons found in the pathophysiology of AD. Four important problems occur in the disease: accumulation of β-amyloid plaques, buildup of tau protein tangles, increased oxidative stress and neuroinflammation (Teleanu et al., 2022). When the cholinergic system degenerates, these events happen together and lead to problems with cognition. Learning how AD happens is vital for creating successful ways to treat it.

Figure 1. Pathophysiology of Alzheimer's disease

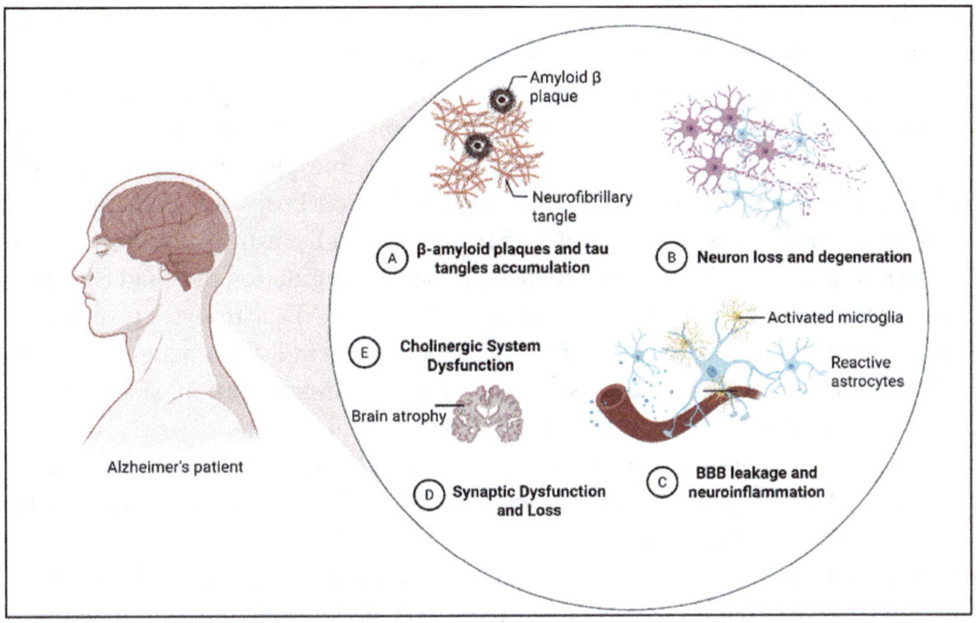

Two main features found in Alzheimer's disease, β-amyloid plaques and tau protein tangles, are often studied in depth (Villemagne et al., 2018). Many experts think the growth of β-amyloid mainly triggers what happens in Alzheimer's disease. APP, a protein on the cell membrane, is cut apart by enzymes and creates a number of peptides, including β-amyloid. In regular brain health, β-amyloid is easily removed however in AD, the brain can't get rid of it, causing β-amyloid plaques to build up. The plaques in Alzheimer's disease are made up of insoluble clumps of β-amyloid found outside neurons (Walker, 2020). They interfere with cell messaging and may help create oxidative stress, inflammation and problems in nerve signaling, as well as interact with tau to affect AD. In most cases, Tau helps stabilize microtubules in neurons, so they work normally. Unfortunately, normal changes to tau in AD involve hyperphosphorylation which makes it separate from microtubules. Because of this, tau tangles form and damage the normal work of cells, contributing to the death of neurons (Chi, Chang, & Sang, 2018). Both β-amyloid plaques and tau tangles cause the gradual cognitive problems seen in Alzheimer's patients. Oxidative stress is a major factor contributing to the development of AD. The problem happens when the production of reactive oxygen species (ROS) exceeds the body's antioxidant level (Bardaweel et al., 2018). Reactive oxygen species in the brain have the ability to injure neurons by harming lipids, proteins and DNA. Gathering of β-amyloid plaques has proven to trigger oxidative stress by causing a rise in ROS. Because of this oxidative damage, the disease starts to more rapidly spread due to the additional deposition of amyloid and tau (Buccellato, D'Anca, Fenoglio, Scarpini, & Galimberti, 2021). In Alzheimer's disease, problems with mitochondria are closely related to oxidative stress. In cells, mitochondria are responsible for generating energy and keeping the cell's balance. As a result, cells in

AD have mitochondrial problems, struggle to produce energy and the extra ROS damage neuron cells. Reduced ATP and disrupted calcium management by mitochondria also support the development of AD (Calvo-Rodriguez, Kharitonova, & Bacskai, 2020). Problems in mitochondria cause them to produce factors that encourage cells to die which makes the disease worse.

Alzheimer's disease is caused partly by neuroinflammation and the activation of microglial cells. Neuroinflammation is defined as when the brain's immune system, made up of microglia and astrocytes, turns on. Microglia are present throughout the brain and, when at rest, maintain constant balance and guard neurons against damag (Augusto-Oliveira et al., 2019). Yet, microglia get activated when there are more beta-amyloid plaques in AD. The result of activation is more cytokines and chemokines being released which cause extra damage to neurons and spark amyloid plaque growth. At first, microglial activation helps defend against damage, but continued activation keeps inflammation going, helping neurodegeneration to spread. Besides becoming activated, microglia, astrocytes also have a role in neuroinflammation (Kwon & Koh, 2020). The blood-brain barrier, regulation of blood flow and metabolic support for neurons are all functions of astrocytes in the brain. In AD, astrocytes respond by producing inflammatory substances which damage the neurons. The persistent inflammation in the brains of AD patients is an important reason for the disease's development and progression, since it raises oxidative stress and leads to increased β-amyloid formation (Tamagno et al., 2021).

Experts also find that changes to the cholinergic system are a major aspect of Alzheimer's disease. Memory, attention and learning depend on the acetylcholine, produced and controlled by the cholinergic system of nerve fibers (Huang et al., 2022). The area in the basal forebrain responsible for making acetylcholine loses many cholinergic neurons in AD. A lowering of acetylcholine in the brain due to this degeneration is one reason behind cognitive impairment. A lack of the chemical choline is detected early and repeatedly in AD and has been connected to most cognitive changes seen in patients. Learning and memory are made possible by synaptic plasticity which depends on the brain chemical acetylcholine (Abraham et al., 2019). A drop in acetylcholine can disrupt how our brains store new information. As cholinergic neurons are damaged in AD, communication between different brain regions becomes harder which worsens cognitive decline. In addition, existing medications for Alzheimer's disease focus on raising acetylcholine levels by targeted acetylcholinesterase inhibition (Akıncıoğlu & Gülçin, 2020). The drugs help ease the symptoms but do less to slow down the brain's degeneration over time. β-amyloid plaques, tau tangles, oxidative stress, neuroinflammation and cholinergic degeneration are all part of the complicated network of Alzheimer's disease (Chen et al., 2022). If β-amyloid plaques and tau tangles disrupt neuronal activity, microglia and astrocytes become active and that can cause neuroinflammation and damage due to oxidative stress. A build-up of ROS only increases neuronal damage, leading to problems with mitochondria and a loss of synapses. When the cholinergic system breaks down, it leads to a loss of thinking ability and problems with memory. The continuous web of chemical and cellular events causes the AD patients' brain function to get worse and cannot be reversed (Decourt et al., 2022).

All told, the development of Alzheimer's disease relies on several factors such as accumulation of β-amyloid plaques, buildup of tau tangles, stress from oxidation, inflammation and loss of cholinergic systems. Such negative characteristics disrupt nerve cell functioning and lead to mental decline, so knowing more about their relationships is important for successful treatments (Reynolds et al., 2022). Although the current treatments do not cure AD, research into how AD functions will help develop better treatments in the future.

3. POLYPHENOLS: CLASSIFICATION AND MECHANISMS OF NEUROPROTECTION

A look at the types and families of polyphenols

These polyphenols are compounds from plants known for carrying multiple phenolic groups or aromatic rings with at least one hydroxyl group (Tsimogiannis & Oreopoulou, 2019). Classification of these compounds depends on their chemical structure. Flavonoids, phenolic acids, lignans and stilbenes are the principal classes of polyphenols. Many different subtypes of these classes have been investigated to see if they can help treat neurodegenerative diseases (Kovacs, 2018).

1. Flavonoids

Flavonoids are a group of polyphenolic chemicals present in many kinds of plant-derived foods and beverages (Mutha et al., 2021). The group includes many subclasses, among them are flavonols, flavones, flavanones, isoflavones and anthocyanins. Flavonoids are well-known for protecting neurons from both oxidation and inflammation. Examples of flavonoids are quercetin, catechins, kaempferol and anthocyanins. Research shows that they can lower the accumulation of amyloid-beta, protect neurons and help improve thinking abilities.

2. Phenolic Acids

A major group of polyphenols, phenolic acids, comes from benzoic acid and cinnamic acid. Many of these nutrients are at their highest levels in foods such as coffee, a variety of fruits and vegetables and whole grains (Roman et al., 2019). Phenolic acids are mostly grouped into hydroxybenzoic acids and hydroxycinnamic acids. Examples commonly found are caffeic acid, ferulic acid and gallic acid. Because of their strong antioxidant properties, phenolic acids are able to remove harmful free radicals and lower stress in the brain. They work in part by decreasing inflammation in the brain, a key symptom of Alzheimer's disease.

3. Lignans

The main sources of lignans are seeds, whole grains and certain vegetables and fruit. The molecule of a lignan is formed through the linking of two phenylpropanoid units (Rodríguez-García et al., 2019). Flaxseeds, sesame seeds and, of course, whole grains are all sugar-free foods rich in lignans. Lignans contain strong antioxidants and anti-inflammatories which help protect the nervous system. Studies have found that secoisolariciresinol is connected to better memory and brain function (Rom et al., 2018).

4. Stilbenes

Stilbenes include a small number of polyphenols and resveratrol is its best-known member. Red wine, grapes, peanuts and berries all contain resveratrol. It has been widely examined for its abilities to fight antioxidants, reduce inflammation and protect the brain (Lee et al., 2020). Resveratrol is known to lower oxidative stress, enhance mitochondrial work and influence brain-derived neurotrophic factor (BDNF), an important factor for neurons.

Ways in which polyphenols protect the brain.

The reason polyphenols protect the brain has much to do with their actions against free radicals, adjustments to immune function and influence on neuronal signaling pathways (Vauzour, 2012). Because of these mechanisms, polyphenols are good candidates for helping prevent or lessen neurodegenerative diseases such as Alzheimer's.

1. Active against free radicals and excessive inflammation

Oxidative stress and inflammation help bring about Alzheimer's disease. Consuming polyphenols can help stop these negative processes by giving the body antioxidant and anti-inflammatory support (Joseph et al., 2016). Polyphenols help get rid of harmful free radicals that can harm our cells. Eliminating them, polyphenols help prevent free radicals from damaging lipids, proteins and DNA in the brain and causing damage. It is especially significant in people with Alzheimer's disease, because plaques and tangles made of amyloid-beta and tau lead to increased ROS and result in oxidative damage to the brain (Simunkova et al., 2019).

In addition to being antioxidants, polyphenols control inflammation in the brain by influencing the actions of microglia and astrocytes. These cells are very important for the body’s response to inflammation during Alzheimer’s. The appearance of amyloid-beta plaques in the brain turns microglia on and they then release pro-inflammatory chemicals such as cytokines and chemokines (W.-Y. Wang, Tan, Yu, & Tan, 2015). Continuous brain inflammation speeds up the deterioration of nerves and their function. Quercetin, curcumin and resveratrol, considered polyphenols, have been found to decrease microglial activation and lower the amount of pro-inflammatory substances, helping to reduce neuroinflammation and guard neurons from damage.

2. Change in the release of Neurotrophic Factors

Neurotrophic factors help neurons stay healthy, live longer and multiply. One important neurotrophic factor, brain-derived neurotrophic factor (BDNF), is key to changes in our synapses and supports both learning and memory (Miranda, Morici, Zanoni, & Bekinschtein, 2019). Because BDNF levels tend to be reduced in Alzheimer's disease, patients experience difficulties with thinking and memory. Studies suggest polyphenols help increase the expression of BDNF and similar neurotrophic factors which safeguard neurons, encourage their growth and enhance our ability to think (Moosavi et al., 2015).

In animal studies of Alzheimer's disease, resveratrol has boosted the expression of BDNF and resulted in improved memory and learning (R. Wang, Zhang, Li, & Zhang, 2017). For example, green tea has epigallocatechin gallate (EGCG) while turmeric has curcumin and it was found that both help to control BDNF levels and help produce new brain cells.

3. Enzyme Inhibition

Polyphenols are able to help protect the nervous system by stopping certain enzymes from contributing to Alzheimer's disease. Polyphenols mainly act on acetylcholinesterase (AChE) which helps to break down the brain messenger acetylcholine important for memory and thinking (Roseiro et al., 2012). Because acetylcholine-producing neurons are lost in Alzheimer's, it also causes less of the acetylcholine chemical in the brain. Such a reduction may lead to the cognitive problems seen in AD.

Some polyphenols, including curcumin, EGCG and quercetin, seem to stop acetylcholinesterase from working and help acetylcholine build up in the brain (Jabir et al., 2018). Because of its ability to manage this enzyme, galantamine is associated with better function of cholinergic systems and has shown improvement in memory and thinking in both lab and medical studies. Besides, various polyphenols have been proven to stop or slow other enzymes involved in Alzheimer's, including BACE1 which creates amyloid-beta (William Raja et al., 2023). Because they suppress these enzymes, polyphenols may prevent the plaques and slow the advancement of Alzheimer's disease.

4. These pathways are involved: Nrf2, NF-κB and MAPK

Polyphenols help direct important cellular signals related to oxidative stress, inflammation and the life span of neurons (J. A. Joseph, SHUKITT-HALE, & Lau, 2007). One good example is the Nrf2 pathway which is important for protecting cells against adverse effects due to oxidative stress. Nrf2 is a way for the body to increase the activity of antioxidant enzymes, among them superoxide dismutase (SOD) and glutathione peroxidase (GPx), when it is activated. Coenzyme Q10 acts as a shield against cell harm from oxidation. Polyphenols such as resveratrol and curcumin have been discovered to activate the Nrf2 pathway which increases the body's defense against oxidative stress (Thiruvengadam et al., 2021). Inflammation is also affected by the important signaling pathway, known as the NF-κB (nuclear factor kappa-light-chain-enhancer of activated B cells) pathway (Dąbek, Kułach, & Gąsior, 2010). When Alzheimer's disease occurs, NF-κB is frequently activated which results in increased levels of inflammatory cytokines and injury to nerve cells. Scientists have discovered that quercetin and EGCG help lower NF-κB activation and as a result, lessen inflammation and save neurons from injury (Grewal et al., 2021).

Figure 2. Top 8 polyphenolic-rich plants for AD

The MAPK (mitogen-activated protein kinase) pathway is also a major way cells communicate, taking part in inflammation, the ability of cells to live and apoptosis. In Alzheimer's disease, a problem with the MAPK pathway may be responsible for both neuron death and declining memory. Resveratrol and curcumin are polyphonols that can activate the MAPK pathway to help keep cells alive and lower neuronal apoptosis (Islam et al., 2024). Polyphenols provide effective protection for the brain because they possess antioxidant, anti-inflammatory and enzyme-inhibiting abilities. Because they change important pathways linked to oxidative stress, inflammation and brain health, they can help prevent and address neurodegenerative diseases such as Alzheimer's. The most important types of polyphenols, flavonoids, phenolic acids, lignans and stilbenes, work differently to help protect the brain. Through these actions, polyphenols are thought to slow or help reduce the harmful symptoms of Alzheimer's disease. Much more investigation is needed to discover how effective these drugs are in treating Alzheimer's and other neurological conditions.

4. MEDICINAL PLANTS RICH IN POLYPHENOLS AND THEIR ANTI-ALZHEIMER'S EFFECTS

Alzheimer's disease (AD) causes the gradual loss of mental skills for the elderly. The causes of its pathophysiology are complex and include amyloid-beta plaques, tau protein tangles, high levels of oxidative stress, brain inflammation and the degeneration of nerve cells (Sehar, Rawat, Reddy, Kopel, & Reddy, 2022). Although there are medicines to help the symptoms of the disease, they do not always tackle the reasons it occurs. For this reason, seeking out alternative remedies found in nature, especially ones full of polyphenols from medicinal plants, is becoming more popular because they have beneficial antioxidant, anti-inflammatory and brain protectors. Numerous medicinal plants have been discovered that might provide polyphenols useful for both relieving the symptoms of AD and slowing its progress (Yan et al., 2022). Plants that belong in this group are Ginkgo biloba, Curcuma longa, Camellia sinensis, Withania somnifera and Bacopa monnieri. Every one of these plants includes polyphenolic substances that tackle the important causes of Alzheimer's. People have long investigated ginkgo biloba, a well-known medicinal plant that protects the brain. Both flavonoid glycosides and terpene lactones are the main types of active compounds present (Y. Zhang et al., 2017). Flavonol glycosides in turmeric such as quercetin, kaempferol and isorhamnetin, are effective at removing free radicals and reducing stress in the brain. When oxidative stress is high, it helps cause Alzheimer's. Decreasing it can help neurons escape the damage from reactive oxygen species (Tauffenberger & Magistretti, 2021). Ginkgolides and bilobalide which are terpene lactones, increase brain blood flow and help protect the brain. These drugs cause blood vessels to relax and reduce how easily platelets clump which helps improve oxygen delivery to the brain. In addition, both ginkgolides and bilobalide prevent the clumping of amyloid-beta protein, a main feature of Alzheimer's disease. Such an effect may prevent toxic plaques from forming and disturbing communication between brain cells which can result in memory loss (Singhrao et al., 2014). Ginkgo biloba is said by some studies to improve mental function, memory and possibly stop the advances of Alzheimer's disease. It should be noted that anti-inflammatory effects of Ginkgo biloba are also beneficial since they change the work of the resident brain immune cells called microglia, resulting in less neuroinflammation. Recurring brain inflammation also contributes to Alzheimer's disease and taking Ginkgo biloba may alleviate some of the symptoms (Nowak et al., 2021).

Table 1. Medicinal plants rich in polyphenols and their reported anti-Alzheimer's effects

Plant Name	Key Polyphenolic Compounds	Primary Anti-Alzheimer's Actions	Notable Challenges
Ginkgo biloba	Flavonoid glycosides (quercetin, kaempferol), terpene lactones (ginkgolides, bilobalide)	Antioxidant, anti-inflammatory, prevents amyloid-beta aggregation, improves blood flow to brain	Variability in response; dosage standardization needed
Curcuma longa (Turmeric)	Curcumin	Reduces oxidative stress, suppresses NF-κB pathway, prevents tau and amyloid-beta aggregation	Poor bioavailability; delivery enhancement required
Camellia sinensis (Green tea)	Catechins (EGCG)	Antioxidant, reduces amyloid-beta/ tau aggregation, enhances antioxidant enzymes (SOD, catalase), controls inflammation	Limited brain bioavailability of EGCG

continued on following page

Table 1. Continued

Plant Name	Key Polyphenolic Compounds	Primary Anti-Alzheimer's Actions	Notable Challenges
Withania somnifera (Ashwagandha)	Withanolides, flavonoids	Antioxidant, anti-inflammatory, increases acetylcholine levels, promotes neurogenesis	Further human trials needed
Bacopa monnieri	Bacosides, phenolic acids	Enhances cholinergic transmission, reduces oxidative stress and inflammation, decreases amyloid-beta plaques	Clinical efficacy still under investigation

5. PRECLINICAL AND CLINICAL EVIDENCE

Evidence showing that polyphenols from medicinal plants help protect the brain in AD has advanced considerably over recent years. Most of this research has looked into what these natural compounds can do to help slow or prevent the development of Alzheimer's. Experiments done in vitro provide controlled conditions for researchers to test the direct impact of polyphenolic compounds on neurons and in vivo experiments are used to evaluate these compounds in entire organisms. Models, however, can be limited, but clinical trials on humans reveal the real world benefits and risks of using these compounds for AD (Scearce-Levie et al., 2020). Together, the findings from these studies help us better understand how to use polyphenols to treat Alzheimer's disease.

Scientists have often used in vitro models, with cells, to investigate the ways in which polyphenols protect the nervous system (Kujawska & Jodynis-Liebert, 2018). Often, neuronal cell lines grown in the lab are exposed to amyloid-beta peptides or other harmful agents that resemble the disease process of Alzheimer's. With these models, researchers can test how polyphenols can either stop or lessen the harm of amyloid-beta and additional neurotoxic chemical compounds. Scientists have proven that some naturally occurring polyphenols in turmeric, green tea and grape products can inhibit the accumulation of amyloid-beta in the brain and therefore prevent the creation of amyloid plaques. By binding with amyloid-beta, polyphenols can keep it from self-assembling and forming the plaque seen in people with Alzheimer's disease (Giorgetti et al., 2018). In addition to controlling amyloid-beta, polyphenols also reduce the tau protein formation seen in AD. It has been discovered in laboratories that some polyphenols block tau hyperphosphorylation, the main event causing tau tangles. As an example, curcumin has been proven to lower tau phosphorylation in brain cells, suggesting it might be able to slow or stop the formation of tau tangles in the brain. Apart from assisting in amyloid-beta and tau clearance, polyphenols also have antioxidant and inflammatory suppression properties that aid nerves in coping with stress from free radicals and harm from inflammation (Shabbir et al., 2021). In Alzheimer's disease, oxidative stress is important and various polyphenols, including resveratrol, curcumin and quercetin, have been found to work as antioxidants and save neurons from harm. The activities of SOD and catalase have been seen to increase and ROS levels reduced by these polyphenols, thanks to studies conducted under laboratory conditions. As well as being antioxidants, polyphenols also control inflammation by impacting how microglia are turned on or off (Hornedo-Ortega et al., 2018). Studies in a laboratory have found that polyphenols such as EGCG from green tea can stop NF-κB from becoming active and reduce the release of inflammatory proteins such as TNF-α and IL-1β. Neurons may be safeguarded from chronic inflammatory damage in Alzheimer's disease because polyphenols lessen neuroinflammation.

More findings show that polyphenols can prevent damage to the brain in animal studies. Researchers have depended on animal models to learn how polyphenols influence memory, heart cells in the brain and the development of Alzheimer's (Naomi et al., 2023). Most studies use mice that are transgenic for amyloid precursor protein or tau to show the disease-like changes found in Alzheimer's disease. In these animal models, polyphenols have improved how well animals perform mentally by reducing the presence of amyloid-beta plaques and abnormal tau tangles, both important signs in Alzheimer's disease. As a result, the researchers found that when animals were treated with curcumin, there was a lower amount of amyloid-beta plaque and better scores during the Morris water maze, a test of memory and learning. EGCG and resveratrol, like Catechin, have both been found in studies to reduce the clusters in the brain linked to Alzheimer's disease and to improve cognitive abilities in animal tests (Farzaei et al., 2019). In the body, polyphenols work to aid brain health by encouraging new brain cell growth, better synaptic activity and preventing brain cells from dying. In some studies, curcumin has been shown to encourage the release of the protein BDNF which supports healthy brain cells by promoting their survival, change in synapses and growth of new ones (Jin et al., 2022). In experiments on rodents, researchers saw that curcumin increased the formation of new brain cells in the hippocampus, a sign that it might aid in repairing brain cells damaged by disease. Just like curcumin, resveratrol and quercetin increase brain-derived neurotrophic factor and boost neurons' life span in experiments done with animals having Alzheimer's disease. Improved memory and flexibility to new inputs depend on changes in both neurogenesis and synaptic plasticity in the brain.

Besides what polyphenols do to amyloid-beta, tau and neurogenesis, they may also secure and protect the blood vessels of the brain. It has been researched that resveratrol and EGCG polyphenols can help increase blood flow in the brain by support of endothelial function and vasodilation. Alzheimer's disease patients suffer from poor blood circulation to the brain which often speeds up memory loss and may enhance the risk of vascular dementia. According to animal studies, polyphenols improve the strength of the blood-brain barrier by allowing more oxygen and nutrients to reach the brain, helping these substances protect the brain.

Experiments with people have shown that polyphenols may help with Alzheimer's disease. Although clinical trials have shown various outcomes, before polyphenols can be widely used in medicine, some problems must be solved. In studies on people, polyphenols including curcumin, EGCG, resveratrol and Ginkgo biloba have proven to help improve thinking skills and ease symptoms of Alzheimer's disease. Some studies reveal that AD can significantly improve recall, concentration and mental control, but other studies do not find such benefits (Liss et al., 2021). A research study on curcumin and people with mild to moderate Alzheimer's disease showed that those given curcumin performed better in several tasks and experienced better well-being. Exactly like the previous studies, EGCG has been found to help patients with early Alzheimer's disease improve their thinking skills. Several studies have not found meaningful benefits and the differences in results might be explained by changes in study methods, the participants involved and how much the polyphenol can be absorbed.

Using phytosomes is a successful way to carry polyphenols like quercetin and silymarin to the body. The results demonstrate that these phytosome formulations provide better absorption, improved pharmacokinetics and better health benefits than the free polyphenols (Riva, Ronchi, Petrangolini, Bosisio, & Allegrini, 2019). Compared to standard versions, phytosome formulations of silymarin improve liver function and minimize oxidative stress, while phytosomes of quercetin help the brain, reduce inflammation and improve cognition.

All in all, the substantial challenge with polyphenol bioavailability prevents them from being effective for Alzheimer's disease and similar situations. Because certain polyphenols are easily broken down, absorbed in low amounts and struggle to get into the brain, their bioavailability is low (Di Lorenzo, Colombo, Biella, Stockley, & Restani, 2021). So, different methods such as nanoencapsulation, phytosomes and liposomes, have been made to help polyphenols be absorbed and delivered well for maximum benefit. Preliminary research and clinical studies have confirmed that nanoencapsulated curcumin and phytosome formulation of quercetin are effective when studied in the lab and in people. Advances in how polyphenols are delivered will be important for raising their ability to treat Alzheimer's disease and similar conditions

6. FUTURE PERSPECTIVES

As scientists learn more about polyphenols in AD, some promising ways to make them work better are starting to appear. Synergistic combinations between polyphenols and regular medicines seem like an interesting direction to take. The main AD treatments affect only the symptoms and have only mild effects on the disease itself. Taking these conventional drugs in combination with polyphenols may mean we can both enjoy more effective treatment and better safety (Rejhová, Opattová, Čumová, Slíva, & Vodička, 2018). Because of their multiple actions as antioxidants, anti-inflammatories and neuroprotectants, polyphenols can possibly strengthen the effects of drugs now given to AD patients. A combination of polyphenols, along with acetylcholinesterase inhibitors or glutamate regulators, might effectively deal with the main problems and signs of Alzheimer's disease (Rebas, Rzajew, Radzik, & Zylinska, 2020). Since current approaches struggle to stop disease progression, this method could solve this problem by dealing with several aspects of the disease at the same time. Blending polyphenols with regular drugs could lead to better health results for patients. Polyphenols appear to impact the important signals that control oxidative stress, inflammation in the brain and the lives of brain cells, all linked to Alzheimer's disease. Using natural substances together with regular drugs might allow us to influence a number of pathological processes at the same time. Polyphenols such as curcumin or EGCG could stop the misfolding of amyloid-beta proteins and acetylcholinesterase inhibitors such as donepezil or rivastigmine help keep cholinergic function at an improved level (Y. Wang, Wang, Yan, Zhou, & Wang, 2022). Used together, these methods might support patients in Alzheimer's research by boosting brain function and living quality. Besides, because polyphenols reduce inflammation and provide antioxidants, they might protect the body from some of the side effects linked to standard drugs.

Personalized neuro-nutrition is another area that looks at a person's DNA and unique traits to suggest the best diet for their health (Novikov, Badaeva, Danilov, & Vorobyeva, 2023). The amount and effectiveness of polyphenols depends on genetic shifts in body processes. For instance, variations in genes encoding polyphenol enzymes such as cytochrome P450 may affect how much of these chemicals we can use. If we use genetic data, we may be able to improve the way polyphenols help in Alzheimer's disease. Neuro-nutrition is personalized when the diet is adjusted to a person's genetics, habits and disease state so they are supported effectively.

7. CONCLUSION

In short, ongoing studies suggest that polyphenols from plants could be important medicines for treating Alzheimer's disease (AD). Plants with medical uses have been found to have naturally occurring compounds that can act as antioxidants, help control inflammation and hinder certain enzymes. According to research, certain polyphenols and other substances found in Ginkgo biloba and Bacopa monnieri work by: (a) breaking up the amyloid-beta protein, (b) preventing tau proteins from changing to harmful forms and (c) easing oxidative stress and inflammation in the brain—events that take place in AD. What's more, these substances are known to protect brain cells, increase brain cell growth and improve thinking in both test animals and humans. The value of plant-derived polyphenols in AD comes from their various actions which let them address various pathways that contribute to the disease. Because they help decrease amyloid plaques, block tau tangles and modulate neuroinflammation, this new approach is promising for AD treatment when combined with standard drugs. Besides, since polyphenols are easy to obtain and safe, using them with other approaches could support therapy for Alzheimer's diseases. Yet, despite showing promise, polyphenols have not been fully applied in AD because of low bioavailability, ineffective delivery approaches and differences in study results. Much potential in polyphenols remains and more collaboration between specialists is necessary to explore it. In this research, input from pharmacology, molecular biology and biotechnology is needed to develop and enhance novel methods of drug administration and their chemicals. Research that moves from lab science to medical practice is needed to prove the usefulness of polyphenols in health care. Going forward, teamwork between various professionals will play a crucial role in improving treatments which should help patients live better and respond effectively to Alzheimer's disease.

REFERENCES

Abraham, W. C., Jones, O. D., & Glanzman, D. L. (2019). Is plasticity of synapses the mechanism of long-term memory storage? *NPJ Science of Learning*, *4*(1), 9. DOI: 10.1038/s41539-019-0048-y PMID: 31285847

Akıncıoğlu, H., & Gülçin, İ. (2020). Potent acetylcholinesterase inhibitors: Potential drugs for Alzheimer's disease. *Mini-Reviews in Medicinal Chemistry*, *20*(8), 703–715. DOI: 10.2174/1389557520666200103100521 PMID: 31902355

Augusto-Oliveira, M., Arrifano, G. P., Lopes-Araújo, A., Santos-Sacramento, L., Takeda, P. Y., Anthony, D. C., Malva, J. O., & Crespo-Lopez, M. E. (2019). What do microglia really do in healthy adult brain? *Cells*, *8*(10), 1293. DOI: 10.3390/cells8101293 PMID: 31652490

Bardaweel, S. K., Gul, M., Alzweiri, M., Ishaqat, A., ALSalamat, H. A., & Bashatwah, R. M. (2018). Reactive oxygen species: The dual role in physiological and pathological conditions of the human body. *The Eurasian Journal of Medicine*, *50*(3), 193–201. DOI: 10.5152/eurasianjmed.2018.17397 PMID: 30515042

Buccellato, F. R., D'Anca, M., Fenoglio, C., Scarpini, E., & Galimberti, D. (2021). Role of oxidative damage in Alzheimer's disease and neurodegeneration: From pathogenic mechanisms to biomarker discovery. *Antioxidants*, *10*(9), 1353. DOI: 10.3390/antiox10091353 PMID: 34572985

Budson, A. E., & Solomon, P. R. (2021). *Memory loss, Alzheimer's disease, and dementia-e-book: a practical guide for clinicians*. Elsevier Health Sciences.

Calvo-Rodriguez, M., Kharitonova, E. K., & Bacskai, B. J. (2020). Therapeutic strategies to target calcium dysregulation in Alzheimer's disease. *Cells*, *9*(11), 2513. DOI: 10.3390/cells9112513 PMID: 33233678

Castro, D. M., Dillon, C., Machnicki, G., & Allegri, R. F. (2010). The economic cost of Alzheimer's disease: Family or public-health burden? *Dementia & Neuropsychologia*, *4*(4), 262–267. DOI: 10.1590/S1980-57642010DN40400003 PMID: 29213697

Chen, Z.-R., Huang, J.-B., Yang, S.-L., & Hong, F.-F. (2022). Role of cholinergic signaling in Alzheimer's disease. *Molecules (Basel, Switzerland)*, *27*(6), 1816. DOI: 10.3390/molecules27061816 PMID: 35335180

Chi, H., Chang, H.-Y., & Sang, T.-K. (2018). Neuronal cell death mechanisms in major neurodegenerative diseases. *International Journal of Molecular Sciences*, *19*(10), 3082. DOI: 10.3390/ijms19103082 PMID: 30304824

Christen, Y. (2000). Oxidative stress and Alzheimer disease. *The American Journal of Clinical Nutrition*, *71*(2), 621S–629S. DOI: 10.1093/ajcn/71.2.621s PMID: 10681270

Dąbek, J., Kułach, A., & Gąsior, Z. (2010). Nuclear factor kappa-light-chain-enhancer of activated B cells (NF-κB): A new potential therapeutic target in atherosclerosis? *Pharmacological Reports*, *62*(5), 778–783. DOI: 10.1016/S1734-1140(10)70338-8 PMID: 21098861

Decourt, B., D'Souza, G. X., Shi, J., Ritter, A., Suazo, J., & Sabbagh, M. N. (2022). The cause of Alzheimer's disease: The theory of multipathology convergence to chronic neuronal stress. *Aging and Disease*, *13*(1), 37. DOI: 10.14336/AD.2021.0529 PMID: 35111361

Di Lorenzo, C., Colombo, F., Biella, S., Stockley, C., & Restani, P. (2021). Polyphenols and human health: The role of bioavailability. *Nutrients*, *13*(1), 273. DOI: 10.3390/nu13010273 PMID: 33477894

Farzaei, M. H., Bahramsoltani, R., Abbasabadi, Z., Braidy, N., & Nabavi, S. M. (2019). Role of green tea catechins in prevention of age-related cognitive decline: Pharmacological targets and clinical perspective. *Journal of Cellular Physiology*, *234*(3), 2447–2459. DOI: 10.1002/jcp.27289 PMID: 30187490

Giorgetti, S., Greco, C., Tortora, P., & Aprile, F. A. (2018). Targeting amyloid aggregation: An overview of strategies and mechanisms. *International Journal of Molecular Sciences*, *19*(9), 2677. DOI: 10.3390/ijms19092677 PMID: 30205618

Grewal, A. K., Singh, T. G., Sharma, D., Sharma, V., Singh, M., Rahman, M. H., & Albadrani, G. M. (2021). Mechanistic insights and perspectives involved in neuroprotective action of quercetin. *Biomedicine and Pharmacotherapy*, *140*, 111729. DOI: 10.1016/j.biopha.2021.111729 PMID: 34044274

Hornedo-Ortega, R., Cerezo, A. B., De Pablos, R. M., Krisa, S., Richard, T., García-Parrilla, M. C., & Troncoso, A. M. (2018). Phenolic compounds characteristic of the mediterranean diet in mitigating microglia-mediated neuroinflammation. *Frontiers in Cellular Neuroscience*, *12*, 373. DOI: 10.3389/fncel.2018.00373 PMID: 30405355

Huang, Q., Liao, C., Ge, F., Ao, J., & Liu, T. (2022). Acetylcholine bidirectionally regulates learning and memory. *Journal of Neurorestoratology*, *10*(2), 100002. DOI: 10.1016/j.jnrt.2022.100002

Huang, W. J., Zhang, X., & Chen, W. W. (2016). Role of oxidative stress in Alzheimer's disease. *Biomedical Reports*, *4*(5), 519–522. DOI: 10.3892/br.2016.630 PMID: 27123241

Ignat, I., Volf, I., & Popa, V. I. (2011). A critical review of methods for characterisation of polyphenolic compounds in fruits and vegetables. *Food Chemistry*, *126*(4), 1821–1835. DOI: 10.1016/j.foodchem.2010.12.026 PMID: 25213963

Islam, F., Roy, S., Zehravi, M., Paul, S., Sutradhar, H., Yaidikar, L., & Nainu, F. (2024). Polyphenols targeting MAP kinase signaling pathway in neurological diseases: Understanding molecular mechanisms and therapeutic targets. *Molecular Neurobiology*, *61*(5), 2686–2706. DOI: 10.1007/s12035-023-03706-z PMID: 37922063

Jabir, N. R., Khan, F. R., & Tabrez, S. (2018). Cholinesterase targeting by polyphenols: A therapeutic approach for the treatment of Alzheimer's disease. *CNS Neuroscience & Therapeutics*, *24*(9), 753–762. DOI: 10.1111/cns.12971 PMID: 29770579

Jin, T., Zhang, Y., Botchway, B. O., Zhang, J., Fan, R., Zhang, Y., & Liu, X. (2022). Curcumin can improve Parkinson's disease via activating BDNF/PI3k/Akt signaling pathways. *Food and Chemical Toxicology*, *164*, 113091. DOI: 10.1016/j.fct.2022.113091 PMID: 35526734

Joseph, J. A., Shukitt-Hale, B., & Lau, F. C. (2007). Fruit polyphenols and their effects on neuronal signaling and behavior in senescence. *Annals of the New York Academy of Sciences*, *1100*(1), 470–485. DOI: 10.1196/annals.1395.052 PMID: 17460212

Joseph, S. V., Edirisinghe, I., & Burton-Freeman, B. M. (2016). Fruit polyphenols: A review of anti-inflammatory effects in humans. *Critical Reviews in Food Science and Nutrition*, *56*(3), 419–444. DOI: 10.1080/10408398.2013.767221 PMID: 25616409

Kovacs, G. G. (2018). Concepts and classification of neurodegenerative diseases. [): Elsevier.]. *Handbook of Clinical Neurology*, *145*, 301–307. DOI: 10.1016/B978-0-12-802395-2.00021-3 PMID: 28987178

Kujawska, M., & Jodynis-Liebert, J. (2018). Polyphenols in Parkinson's disease: A systematic review of in vivo studies. *Nutrients*, *10*(5), 642. DOI: 10.3390/nu10050642 PMID: 29783725

Kwon, H. S., & Koh, S.-H. (2020). Neuroinflammation in neurodegenerative disorders: The roles of microglia and astrocytes. *Translational Neurodegeneration*, *9*(1), 42. DOI: 10.1186/s40035-020-00221-2 PMID: 33239064

Lee, K. H., Cha, M., & Lee, B. H. (2020). Neuroprotective effect of antioxidants in the brain. *International Journal of Molecular Sciences*, *21*(19), 7152. DOI: 10.3390/ijms21197152 PMID: 32998277

Liss, J., Seleri Assunção, S., Cummings, J., Atri, A., Geldmacher, D., Candela, S., & Mintzer, J. (2021). Practical recommendations for timely, accurate diagnosis of symptomatic Alzheimer's disease (MCI and dementia) in primary care: A review and synthesis. *Journal of Internal Medicine*, *290*(2), 310–334. DOI: 10.1111/joim.13244 PMID: 33458891

Miranda, M., Morici, J. F., Zanoni, M. B., & Bekinschtein, P. (2019). Brain-derived neurotrophic factor: A key molecule for memory in the healthy and the pathological brain. *Frontiers in Cellular Neuroscience*, *13*, 472800. DOI: 10.3389/fncel.2019.00363 PMID: 31440144

Moosavi, F., Hosseini, R., Saso, L., & Firuzi, O. (2015). Modulation of neurotrophic signaling pathways by polyphenols. *Drug Design, Development and Therapy*, •••, 23–42. PMID: 26730179

Moser, I. (2008). Making Alzheimer's disease matter. Enacting, interfering and doing politics of nature. *Geoforum*, *39*(1), 98–110. DOI: 10.1016/j.geoforum.2006.12.007

Mutha, R. E., Tatiya, A. U., & Surana, S. J. (2021). Flavonoids as natural phenolic compounds and their role in therapeutics: An overview. *Future Journal of Pharmaceutical Sciences*, *7*(1), 1–13. DOI: 10.1186/s43094-020-00161-8 PMID: 33495733

Naomi, R., Yazid, M. D., Teoh, S. H., Balan, S. S., Shariff, H., Kumar, J., Bahari, H., & Embong, H. (2023). Dietary polyphenols as a protection against cognitive decline: Evidence from animal experiments; mechanisms and limitations. *Antioxidants*, *12*(5), 1054. DOI: 10.3390/antiox12051054 PMID: 37237920

Novikov, V. N., Badaeva, A. V., Danilov, A. B., & Vorobyeva, Y. D. (2023). *Precision Neuronutrition: Personalized Approaches for Optimizing Brain Health.* Paper presented at the Biology and Life Sciences Forum. DOI: 10.3390/IECN2023-15524

Nowak, A., Kojder, K., Zielonka-Brzezicka, J., Wróbel, J., Bosiacki, M., Fabiańska, M., Wróbel, M., Sołek-Pastuszka, J., & Klimowicz, A. (2021). The use of Ginkgo biloba L. as a neuroprotective agent in the Alzheimer's disease. *Frontiers in Pharmacology*, *12*, 775034. DOI: 10.3389/fphar.2021.775034 PMID: 34803717

Ransohoff, R. M. (2016). How neuroinflammation contributes to neurodegeneration. *Science*, *353*(6301), 777–783. DOI: 10.1126/science.aag2590 PMID: 27540165

Rebas, E., Rzajew, J., Radzik, T., & Zylinska, L. (2020). Neuroprotective polyphenols: A modulatory action on neurotransmitter pathways. *Current Neuropharmacology*, *18*(5), 431–445. DOI: 10.2174/1570159X18666200106155127 PMID: 31903883

Rejhová, A., Opattová, A., Čumová, A., Slíva, D., & Vodička, P. (2018). Natural compounds and combination therapy in colorectal cancer treatment. *European Journal of Medicinal Chemistry*, *144*, 582–594. DOI: 10.1016/j.ejmech.2017.12.039 PMID: 29289883

Reynolds, C. F.III, Jeste, D. V., Sachdev, P. S., & Blazer, D. G. (2022). Mental health care for older adults: Recent advances and new directions in clinical practice and research. *World Psychiatry; Official Journal of the World Psychiatric Association (WPA)*, *21*(3), 336–363. DOI: 10.1002/wps.20996 PMID: 36073714

Riva, A., Ronchi, M., Petrangolini, G., Bosisio, S., & Allegrini, P. (2019). Improved oral absorption of quercetin from quercetin phytosome®, a new delivery system based on food grade lecithin. *European Journal of Drug Metabolism and Pharmacokinetics*, *44*(2), 169–177. DOI: 10.1007/s13318-018-0517-3 PMID: 30328058

Rodríguez-García, C., Sánchez-Quesada, C., Toledo, E., Delgado-Rodríguez, M., & Gaforio, J. J. (2019). Naturally lignan-rich foods: A dietary tool for health promotion? *Molecules (Basel, Switzerland)*, *24*(5), 917. DOI: 10.3390/molecules24050917 PMID: 30845651

Rom, S., Zuluaga-Ramirez, V., Reichenbach, N. L., Erickson, M. A., Winfield, M., Gajghate, S., Christofidou-Solomidou, M., Jordan-Sciutto, K. L., & Persidsky, Y. (2018). Secoisolariciresinol diglucoside is a blood-brain barrier protective and anti-inflammatory agent: Implications for neuroinflammation. *Journal of Neuroinflammation*, *15*(1), 1–10. DOI: 10.1186/s12974-018-1065-0 PMID: 29373982

Roman, G. C., Jackson, R., Gadhia, R., Román, A., & Reis, J. (2019). Mediterranean diet: The role of long-chain ω-3 fatty acids in fish; polyphenols in fruits, vegetables, cereals, coffee, tea, cacao and wine; probiotics and vitamins in prevention of stroke, age-related cognitive decline, and Alzheimer disease. *Revue Neurologique*, *175*(10), 724–741. DOI: 10.1016/j.neurol.2019.08.005 PMID: 31521398

Roseiro, L. B., Rauter, A. P., & Serralheiro, M. L. M. (2012). Polyphenols as acetylcholinesterase inhibitors: Structural specificity and impact on human disease. *Nutrition and Aging (Amsterdam, Netherlands)*, *1*(2), 99–111. DOI: 10.3233/NUA-2012-0006

Scearce-Levie, K., Sanchez, P. E., & Lewcock, J. W. (2020). Leveraging preclinical models for the development of Alzheimer disease therapeutics. *Nature Reviews. Drug Discovery*, *19*(7), 447–462. DOI: 10.1038/s41573-020-0065-9 PMID: 32612262

Sehar, U., Rawat, P., Reddy, A. P., Kopel, J., & Reddy, P. H. (2022). Amyloid beta in aging and Alzheimer's disease. *International Journal of Molecular Sciences*, *23*(21), 12924. DOI: 10.3390/ijms232112924 PMID: 36361714

Shabbir, U., Tyagi, A., Elahi, F., Aloo, S. O., & Oh, D.-H. (2021). The potential role of polyphenols in oxidative stress and inflammation induced by gut microbiota in alzheimer's disease. *Antioxidants*, *10*(9), 1370. DOI: 10.3390/antiox10091370 PMID: 34573002

Simunkova, M., Alwasel, S. H., Alhazza, I. M., Jomova, K., Kollar, V., Rusko, M., & Valko, M. (2019). Management of oxidative stress and other pathologies in Alzheimer's disease. *Archives of Toxicology*, *93*(9), 2491–2513. DOI: 10.1007/s00204-019-02538-y PMID: 31440798

Singhrao, S. K., Harding, A., Simmons, T., Robinson, S., Kesavalu, L., & Crean, S. (2014). Oral inflammation, tooth loss, risk factors, and association with progression of Alzheimer's disease. *Journal of Alzheimer's Disease*, *42*(3), 723–737. DOI: 10.3233/JAD-140387 PMID: 24946875

Tamagno, E., Guglielmotto, M., Vasciaveo, V., & Tabaton, M. (2021). Oxidative stress and beta amyloid in Alzheimer's disease. Which comes first: The chicken or the egg? *Antioxidants*, *10*(9), 1479. DOI: 10.3390/antiox10091479 PMID: 34573112

Tauffenberger, A., & Magistretti, P. J. (2021). Reactive oxygen species: Beyond their reactive behavior. *Neurochemical Research*, *46*(1), 77–87. DOI: 10.1007/s11064-020-03208-7 PMID: 33439432

Teleanu, D. M., Niculescu, A.-G., Lungu, I. I., Radu, C. I., Vladâcenco, O., Roza, E., Costăchescu, B., Grumezescu, A. M., & Teleanu, R. I. (2022). An overview of oxidative stress, neuroinflammation, and neurodegenerative diseases. *International Journal of Molecular Sciences*, *23*(11), 5938. DOI: 10.3390/ijms23115938 PMID: 35682615

Thiruvengadam, M., Venkidasamy, B., Subramanian, U., Samynathan, R., Ali Shariati, M., Rebezov, M., & Fedoseeva, N. (2021). Bioactive compounds in oxidative stress-mediated diseases: Targeting the NRF2/ARE signaling pathway and epigenetic regulation. *Antioxidants*, *10*(12), 1859. DOI: 10.3390/antiox10121859 PMID: 34942962

Tsimogiannis, D., & Oreopoulou, V. (2019). Classification of phenolic compounds in plants. In *Polyphenols in plants* (pp. 263–284). Elsevier. DOI: 10.1016/B978-0-12-813768-0.00026-8

Vauzour, D. (2012). Dietary polyphenols as modulators of brain functions: Biological actions and molecular mechanisms underpinning their beneficial effects. *Oxidative Medicine and Cellular Longevity*, *2012*(1), 914273. DOI: 10.1155/2012/914273 PMID: 22701758

Villemagne, V. L., Doré, V., Burnham, S. C., Masters, C. L., & Rowe, C. C. (2018). Imaging tau and amyloid-β proteinopathies in Alzheimer disease and other conditions. *Nature Reviews. Neurology*, *14*(4), 225–236. DOI: 10.1038/nrneurol.2018.9 PMID: 29449700

Walker, L. C. (2020). Aβ plaques. *Free Neuropathology*, *1*, 1–31. PMID: 33345256

Wang, R., Zhang, Y., Li, J., & Zhang, C. (2017). Resveratrol ameliorates spatial learning memory impairment induced by Aβ1–42 in rats. *Neuroscience*, *344*, 39–47. DOI: 10.1016/j.neuroscience.2016.08.051 PMID: 27600946

Wang, W.-Y., Tan, M.-S., Yu, J.-T., & Tan, L. (2015). Role of pro-inflammatory cytokines released from microglia in Alzheimer's disease. *Annals of Translational Medicine*, *3*(10), 136. PMID: 26207229

Wang, Y., Wang, K., Yan, J., Zhou, Q., & Wang, X. (2022). Recent progress in research on mechanisms of action of natural products against Alzheimer's disease: Dietary plant polyphenols. *International Journal of Molecular Sciences*, *23*(22), 13886. DOI: 10.3390/ijms232213886 PMID: 36430365

William Raja, T. R., Duraipandiyan, V., Ignacimuthu, S., Janakiraman, U., & Packiam, S. M. (2023). Role of polyphenols in alleviating alzheimer's disease: A review. *Current Medicinal Chemistry*, *30*(35), 4032–4047. DOI: 10.2174/0929867330666221202152540 PMID: 36476438

Yan, L., Guo, M.-S., Zhang, Y., Yu, L., Wu, J.-M., Tang, Y., & Chen, Q. (2022). Dietary plant polyphenols as the potential drugs in neurodegenerative diseases: Current evidence, advances, and opportunities. *Oxidative Medicine and Cellular Longevity*, *2022*(1), 5288698. DOI: 10.1155/2022/5288698 PMID: 35237381

Zhang, Y., Peiying, L., Qianqian, K., Shiyan, X., Xiaojing, L., & Limin, S. (2017). The contents of terpene trilactone and flavonoid in leaves of seedlings from ancient female ginkgo trees in China. *Horticultural Plant Journal*, *3*(4), 165–171. DOI: 10.1016/j.hpj.2017.06.002

Zvěřová, M. (2019). Clinical aspects of Alzheimer's disease. *Clinical Biochemistry*, *72*, 3–6. DOI: 10.1016/j.clinbiochem.2019.04.015 PMID: 31034802

Chapter 11
Polyphenols Against Oxidative Stress, Neuroinflammation, and Mitochondrial Dysfunction

Nimra Aziz
https://orcid.org/0009-0006-3565-6853
The University of Lahore, Pakistan

Asra Khan
https://orcid.org/0009-0005-4239-6140
The University of Lahore, Pakistan

Qammar Abbas
The University of Lahore, Pakistan

Shazia Batool
The University of Lahore, Pakistan

Laaraib Nawaz
https://orcid.org/0000-0002-9420-632X
The University of Lahore, Pakistan

ABSTRACT

Neurodegeneration is driven by complex interplay of oxidative stress, chronic neuroinflammation, and mitochondrial dysfunction. These interconnected mechanisms not only exacerbate neuronal damage but also challenge the effectiveness of single-target therapies. Polyphenols, naturally occurring compounds, have significant multi-target therapeutic potential to modulate these pathological mechanisms. This chapter gives a comprehensive overview of the therapeutic potentials of polyphenols against neuroprotection. It highlights how polyphenols counteract neuroinflammation by focusing on their roles in limiting glial activation and downregulating pro-inflammatory molecules and enzymes via targeting multiple associated pathways. In addition, polyphenols increase mitochondrial integrity by reducing reactive oxygen species, preserving membrane potential, and promoting biogenesis. By targeting various neurodegenerative hallmarks, polyphenols signify the compelling strategies for disease modification, paving the way for innovative therapeutic interventions in neurodegenerative disorders.

DOI: 10.4018/979-8-3373-4139-2.ch011

1. INTRODUCTION

1.1 Overview of Neurodegenerative Diseases

"Neurodegenerative diseases (NDs)" are a large range of illnesses characterized by the slow loss of neurons, which results in movement dysfunction, cognitive decline, and other neurological symptoms. Neurodegenerative illnesses encompass a wide range of conditions, including amyotrophic lateral sclerosis (ALS), Huntington's disease (HD), Parkinson's disease (PD), and Alzheimer's disease (AD). Nonetheless, the unique clinical manifestations, affected brain regions, and underlying pathogenic processes of these NDs set them apart (Kormas & Moutzouri, 2023). For example, tau tangles and beta-amyloid plaques are linked to memory loss and cognitive deterioration in AD. Tremors, bradykinesia, and stiffness are symptoms of Parkinson's disease (PD) brought on by the loss of dopamine-producing neurons. Due to the increase of CAG trinucleotide repeats in the HTT gene, HD presents as uncontrollable movements and cognitive impairment. While, motor neuron loss in ALS causes gradual muscular weakening and paralysis (Ahmad, Zeyaullah, Khan, et al., 2024). The number of fatalities and disabilities brought on by neurological illnesses is becoming more widely acknowledged as a global public health concern, and it is expected to increase over the coming decades. Disability-adjusted life-years (DALYs), which are the total of years of life lost and years lived with a disability, have climbed by 15% during the last 30 years, while the number of deaths has increased by 39% in absolute terms (Feigin et al., 2020).

1.2 Common Pathophysiological Hallmarks:

The genesis and progression of neurodegenerative illnesses are known to be accelerated by: oxidative stress, chronic neuroinflammation, and mitochondrial dysfunction (Grabska-Kobyłecka et al., 2023). It has been suggested that oxidative stress regulates aging and a number of neurological conditions. Oxidative stress is caused by a lack of antioxidants and is typified by high quantities of reactive species, such as oxygen and hydroxyl free radicals (Olufunmilayo et al., 2023). The inherent and adaptive reactions to dangerous conditions such as infections, ischemia, stress, and trauma are referred to as neuroinflammation. Although neuroinflammation starts off as a protective response, research has shown that chronic or severe inflammation plays a major role in the emergence of a number of neurological conditions, particularly degenerative diseases (Adamu et al., 2024). The pathophysiology of neurodegenerative diseases is significantly influenced by disruptions of mitochondrial homeostasis and the resulting dysfunction of the mitochondria, so maintaining the integrity of the mitochondria through various surveillance mechanisms is essential for neuron survival (Mangrulkar et al., 2023). These mechanisms not only contribute individually to neuronal injury but also interact synergistically, amplifying neurodegenerative cascades.

2. UNDERSTANDING THE TARGETS

2.1 Neuroinflammation:

Neuroinflammation is a well-orchestrated and complex process involving glial cells, particularly microglia and astrocytes (Aziz et al., 2024).

2.1.1 Activation of Astrocytes and Microglia

Microglia are the resident macrophages within the central nervous system (CNS) and are first responders to infection of injury. Once activated, they switch to an amoeboid morphology and increase their antigen presentation, phagocytic activity, and proinflammatory cytokines secretion. This contributes to the neuroinflammation and influence the adjacent neuronal and glial cells (Zhao et al., 2019). This process is initiated via pattern recognition receptors (PRRs) – such as Toll-like receptors (TLRs) – that identify molecular patterns associated with infection or cell injury, including pathogen-associated molecular patterns (PAMPs) and damage-associated molecular patterns (DAMPs), leading to the activation of downstream signaling pathways. In addition, astrocytes are the most abundant glia in the CNS and their activation, known as astrogliosis, features elevated glial fibrillary acidic protein (GFAP) levels, hypertrophy, and secretion of neurotoxic and neuroinflammatory mediators (Lahooti et al., 2021; Wang et al., 2024).

In physiological states, microglia and astrocytes maintain the homeostasis of CNS by clearing debris, neurovascular coupling, and regulating synaptic activity. Conversely, under the pathological conditions, the glial cells undergo functional and morphological alterations, leading to persistent neuroinflammation. In the early stages of neuroinflammation, microglia tend to adopt a phagocytic and anti-inflammatory M2 state, but if the inflammation is not resolved, microglial cells proliferate and swap to their reactive phenotype M1, termed microgliosis (Cherry et al., 2014; Li et al., 2019). Astrocytes also exhibit a biphasic response. Initially supportive, but later astrocytes undergo a reactive transformation A1 phenotype, marked by the secretion of reactive oxygen species, cytokines, and chemokines, thereby aggravating neuronal dysfunction (Zhang et al., 2025). Notably, systemic inflammation and peripheral immune signals can also worsen the central glial activation by disrupting the blood brain barrier (BBB).

Typically, microglia are activated before astrocytes in response to brain injury, contributing to the subsequent activation of astrocytes. The reactive state of microglial cells then releases proinflammatory cytokines (proteins that modulate inflammatory processes) and other inflammatory molecules in the brain. The astrocytes sense these molecules as potent neuroinflammation signals, switch to astrogliosis (a reactive astrocyte state A1), and ultimately secrete more pro-inflammatory stimuli. In fact, microglial activation can also be modulated by astrocytes, either inhibiting or promoting it, based on the context and signals. This two-way communication is crucial for the maintenance of homeostasis and managing pathological alterations (L. Liu et al., 2020). Sustained glial activation results in self-perpetuating neuroinflammation cycle, which is characterized by continuous proinflammatory cytokines production, oxidative stress, and BBB dysfunction (Escartin et al., 2021).

Chronic astrocytic and microglial activation may contribute to various neurological and behavioral disorders, including PD, AD, ALS, multiple sclerosis (MS) and epilepsy (Ahmad, Zeyaullah, AlShahrani, et al., 2024). In AD, for instance, astrocytes alter the calcium signaling and alleviate glutamate uptake, resulting in excitotoxicity. Likewise, microglial cells surround amyloid-beta (Aβ) deposits in an attempt to clear them which is also accompanied by release of neurotoxic agents that exacerbate neuronal damage (Leng & Edison, 2021).

2.1.2 Role of proinflammatory cytokines

Neuroinflammation is predominately driven by proinflammatory cytokines that significantly contribute to the pathogenesis of various CNS disorders, including AD and PD. Activated astrocytes and microglial cells are the main sources of cytokines production, including interleukin-1 beta (IL-1β), interleukin-6 (IL-6), and tumor necrosis factor-alpha (TNF-α) in response to CNS injury, such as trauma, ischemia, or infection (Monkkonen & Debnath, 2018). In addition to cellular signaling, these mediators regulate BBB permeability, modulate peripheral immune cell recruitment, and directly affect neuronal functions and synaptic activity (Besedovsky & del Rey, 2011).

Proinflammatory cytokines are distinctly known for their ability to disrupt integrity of BBB. TNF-α and IL-1β promote the disruption of tight junction proteins, namely occludin and claudin-5, resulting in elevated BBB permeability and facilitating filtration of T-cells, neutrophils, and monocytes into the CNS. The subsequent influx of immune cells perpetuates a chronic inflammatory state, contributing to a vicious cycle of neuronal injury repair failure (Gryka-Marton et al., 2025). In addition, cytokines also play a significant role in altering synaptic function and neurotransmitter balance. TNF-α influences neuronal activity by regulating the trafficking of α-Amino-3-hydroxy-5-methyl-4-isoxazolepropionic acid (AMPA) and gamma-aminobutyric acid (GABA) receptors, promoting excitatory transmission and suppressing inhibitory input. As a result, neurons become more susceptible to glutamate-mediated excitotoxicity, a central factor in neuronal death during neuroinflammatory states (Muhammad, 2019). IL-1β modulates N-Methyl-D-Aspartate (NMDA) receptor signaling, which triggers calcium overload and mitochondrial damage in neuronal cells (Dong et al., 2017).

Glial–cytokine crosstalk contributes significantly to the complexity of neuroinflammatory signaling. Under the influence of IL-1β and TNF-α, the activated astrocytes undergo hypertrophy and exhibit upregulated levels of GFAP. They concurrently release chemokines and cytokines, such as Granulocyte-Macrophage Colony-Stimulating Factor (GM-CSF), C-X-C Motif Chemokine Ligand 10 (CXCL10), and C-C Motif Chemokine Ligand 2 (CCL2) that perpetuates immune cells' recruitment and activation (Mohammad et al., 2025).

2.1.3 NF-κB and NLRP3 inflammasome pathways

The NOD-like receptor family pyrin domain containing 3 (NLRP3) inflammasome and nuclear factor kappa-light-chain-enhancer of activated B cells (NF-κB) signaling pathway are crucial molecular mechanisms that regulate the inflammatory response within CNS. These pathways are pivotal drivers of neuroinflammation, promoting the maturation and expression of proinflammatory cytokines, like IL-1β, thereby promoting cellular stress. The activation of these pathways in astrocytes, microglia, and even neuronal cells facilitates the pathogenesis and progression of the neuroinflammatory and neurodegenerative diseases, including AD, PD, and MS (Holbrook et al., 2021; S. Singh & Singh, 2020).

In the CNS, NF-κB is rapidly activated in microglia following exposure to lipopolysaccharide (LPS), Aβ, or misfolded proteins. This activation leads to a proinflammatory transcriptional profile that supports the sustained production of cytokines and chemokines. For instance, in AD, NF-κB activation by Aβ aggregates upregulates inducible nitric oxide synthase (iNOS), cyclooxygenase-2 (COX-2), and IL-1β, thereby exacerbating oxidative stress and synaptic dysfunction (G. Singh et al., 2020). In astrocytes, NF-κB contributes to astrogliosis and the release of neurotoxic mediators that impairs neuronal viability (Hung et al., 2016).

Closely linked to NF-κB signaling is the NLRP3 inflammasome, a multiprotein cytosolic complex that mediates the maturation and secretion of IL-1β and IL-18. The NLRP3 inflammasome consists of the sensor protein NLRP3, the adaptor protein ASC (apoptosis-associated speck-like protein containing a CARD), and pro-caspase-1. Activation of the NLRP3 inflammasome is a two-step process: the first "priming" signal—often mediated by NF-κB—upregulates the transcription of NLRP3 and pro-IL-1β, while the second signal triggers assembly of the inflammasome complex and cleavage of pro-caspase-1 into its active form. Active caspase-1 then processes pro-IL-1β and pro-IL-18 into their mature, secreted forms, initiating a potent inflammatory response (Vezzani et al., 2020). The pathological consequences of sustained NF-κB and NLRP3 activation include disruption of the BBB, synaptic loss, oxidative damage, and recruitment of peripheral immune cells. For example, in stroke models, NF-κB activation in endothelial cells leads to upregulation of adhesion molecules and MMPs, contributing to BBB breakdown (Huang et al., 2020).

2.2 Mitochondrial Dysfunction

2.2.1 Importance of mitochondria in neurons

Neurons need a constant supply of energy, and because of their intricate structure and high energy metabolic needs, they are especially susceptible to mitochondrial malfunction. Some of the most crucial organelles in cells, mitochondria, contribute to the stability and development of cells by maintaining the homeostasis of the mitochondrial network through energy metabolism, mitophagy, and mitochondrial dynamics (Rummel & Butterfield, 2022). Abnormal mitochondrial energy metabolism can increase reactive oxygen species (ROS) production and induce oxidative stress. Thus, mitochondria are regarded as a key target for drugs to cure ND and as one of the key probable pathways of disease development.

In addition, abnormalities in mitochondrial homeostasis play a significant role in the development of ND (K. Meng et al., 2025). Mitochondrial dynamics actively regulates both physiological and pathological processes while preserving equilibrium through constant fusion and division (Bereiter-Hahn & Vöth, 1994). Cell signaling pathways become disrupted, synaptic function deteriorates, and neuronal cell death results from aberrant mitochondrial fusion and fission, which also causes disorganized distribution, restricted neural transport, decreased energy efficiency, and excessive ROS production.

2.2.2 Mitochondrial ROS generation, membrane depolarization, ATP depletion

The main location of cellular ROS formation is in the mitochondria, specifically in complexes I and III of the electron transport chain (ETC). While low amounts of ROS operate as signaling molecules in healthy situations, excessive ROS synthesis causes oxidative damage and cellular malfunction in pathological states (Peggion et al., 2024). Electron leakage, particularly from complex I, results from disruption of the ETC and lowers molecular oxygen to produce superoxide anions. These reactive species intensify oxidative stress in the mitochondrial matrix by quickly converting to hydroxyl radicals and hydrogen peroxide (Alqahtani et al., 2023).

ROS buildup destroys mitochondrial DNA, proteins, and lipids, impairing membrane integrity and starting the opening of the mitochondrial permeability transition pore (mPTP). The mitochondrial membrane potential, a crucial gradient that propels ATP synthesis through oxidative phosphorylation, is lost when the mPTP is opened for an extended period of time. Depolarization of the membrane significantly impairs ATP synthesis by ATP synthase, which results in the depletion of cellular energy (Henrich et al., 2023). By affecting ion pump performance, vesicle transport, and synaptic activity, ATP depletion jeopardizes neuronal survival and leaves neurons more susceptible to additional insults. In neurodegenerative disorders like Alzheimer's, Parkinson's, and ALS, this series of events, which starts with the production of ROS and ends with energy failure, represents a key mechanism of neuronal damage (Wang et al., 2022).

2.2.3 Links to apoptosis and neuronal loss

Intrinsic apoptotic pathways, which are essential to the development of neurodegenerative diseases, are implicated in mitochondrial malfunction. They release important pro-apoptotic molecules into the cytosol, including cytochrome c, apoptosis-inducing factor (AIF), and second mitochondria-derived activator of caspases (Smac/DIABLO), in response to oxidative stress, calcium overload, or prolonged permeability transition pore (mPTP) opening (Wang et al., 2022). When cytochrome c binds to procaspase-9 and apoptotic protease activating factor-1 (Apaf-1), the apoptosome is formed. This triggers caspase-9 and then caspase-3, which coordinates DNA breakage, cytoskeletal disintegration, and cell death. AIF also moves from the mitochondria to the nucleus without the help of caspase, where it causes extensive chromatin condensation and DNA deterioration (Klemmensen et al., 2024).

Neurons are especially vulnerable to these apoptotic events because they are post-mitotic and cannot be replenished once they are destroyed. In conditions AD, PD, and ALS, a chronic stimulation of mitochondrial apoptosis in response to prolonged stress, such as protein aggregation, excitotoxicity, and defective mitophagy, causes gradual loss of neurons. In AD, amyloid-beta causes mitochondrial fragmentation and the release of cytochrome c, whereas in PD, α-synuclein aggregation damages the integrity of the mitochondrial membrane and encourages the activation of caspase (Alqahtani et al., 2023). Furthermore, calcium buffering and ATP supply both necessary for preserving synaptic transmission and plasticity are compromised by malfunctioning mitochondria. Apoptosis is facilitated by this bioenergetic failure, which also plays a role in the early synaptic malfunction and cognitive decline seen in neurodegenerative diseases (Klemmensen et al., 2024).

2.3 Oxidative Stress

Oxidative stress is an imbalance between two different types of molecules in human body: free radicals and antioxidants. Specifically, there are too many free radicals and not enough antioxidants. As a result, the excess free radicals start to harm body's own cells and tissues (Martemucci et al., 2022). They damage the different parts of cells, including lipids and proteins that allow them to work normally. Unlike free radicals, antioxidants don't snatch electrons from healthy cells in the body. Instead, they donate one of their electrons to a free radical (Chandimali et al., 2025). The brain is particularly vulnerable to oxidative stress because it consumes a large amount of oxygen, contains high levels of polyunsaturated fatty acids (which are prone to oxidation), and has relatively low antioxidant defenses compared to other organs (Franzoni et al., 2021). It leads to a harmful condition caused by an imbalance between the production of reactive oxygen species (ROS) and reactive nitrogen species (RNS) and the brain's ability to detoxify them with antioxidants.

2.3.1 Mechanisms of Neural Damage by ROS and RNS

- **Lipid Peroxidation:** Neurons contain high levels of polyunsaturated fatty acids (PUFAs) in their membranes. ROS such as hydroxyl radicals initiate lipid peroxidation, damaging cell membranes, disrupting ion channels, and altering membrane fluidity. This leads to neuronal dysfunction and death (Zheng et al., 2024).

- **Protein Oxidation and Nitration:** ROS/RNS oxidize proteins or nitrate tyrosine residues, altering their structure and function. This affects enzymes, receptors, and transport proteins, impairing cell signaling and repair mechanisms (Campolo et al., 2014).
- **DNA Damage:** Hydroxyl radicals and peroxynitrite cause strand breaks and base modifications in neuronal DNA. This activates cell cycle arrest, mitochondrial dysfunction, and apoptosis pathways (programmed cell death) (Sadiq, 2023).
- **Mitochondrial Dysfunction:** ROS are both generated by and damage mitochondria. This impairs ATP production, increases ROS leakage, and triggers the release of pro-apoptotic proteins like cytochrome c (Zong et al., 2024).
- **Inflammatory Response:** Oxidative stress activates microglia (brain immune cells), which release more ROS/RNS and pro-inflammatory cytokines. This creates a vicious cycle of oxidative and inflammatory damage (Zhu et al., 2022).

2.3.2 Disruption of Redox Homeostasis in Neurodegeneration

Redox homeostasis refers to the balanced state between the productions of ROS/RNS and the action of antioxidant defense systems in cells. This balance is crucial for normal cellular function, especially in the brain, which is highly sensitive to oxidative damage due to its high oxygen consumption and lipid-rich environment (Goldsteins et al., 2022). When redox balance is disrupted often due to excessive ROS/RNS or insufficient antioxidant response it leads to oxidative stress. This stress results in damage to vital biomolecules such as lipids, proteins, and deoxyribonucleic acid (DNA). Over time, such damage accumulates in neuronal cells and impairs cellular signaling, mitochondrial function, and ion homeostasis (Selvaraj et al., 2025). One key mechanism through which redox imbalance contributes to neurodegeneration is via mitochondrial dysfunction. Mitochondria are both a major source and target of ROS, and their impairment leads to decreased ATP production, release of pro-apoptotic factors, and further ROS generation, establishing a vicious cycle of damage (Guo et al., 2013). In diseases like AD, PD, and ALS, oxidative stress has been shown to impair synaptic function, promote tau protein hyperphosphorylation, increase β-amyloid aggregation, leading to neuronal apoptosis. These effects are exacerbated by redox-sensitive pathways like NF-κB and mitogen-activated protein kinase (MAPK), which become dysregulated under oxidative conditions (Perluigi et al., 2024). Additionally, the failure of glutathione-dependent systems one of the brain's primary antioxidant defenses is commonly observed in neurodegenerative conditions. Reduced levels of glutathione (GSH) impair detoxification of ROS and disturb redox-sensitive transcription factors, further worsening neuronal survival (Iskusnykh et al., 2022). Disruption in redox homeostasis also activates microglial cells, the resident immune cells of the brain. Activated microglia release large quantities of ROS and pro-inflammatory cytokines, leading to chronic neuroinflammation, which contributes significantly to the progression of neurodegenerative disorders (Bellanti et al., 2025).

2.3.4 Biomarkers of Oxidative Stress in Neurodegenerative Conditions

Biomarkers of oxidative stress are measurable indicators that reflect the level of oxidative damage occurring in cells and tissues. In neurodegenerative diseases, these biomarkers help in diagnosis of the stage and severity of the disease and understanding its underlying mechanisms. The brain's vulnerability to oxidative stress makes these biomarkers especially valuable in conditions like AD, PD, HD, and ALS (A. Singh et al., 2019).

- **Lipid Peroxidation Products:** These include malondialdehyde (MDA) and 4-hydroxynonenal (4-HNE), which result from the oxidative degradation of PUFAs in neuronal membranes. Elevated levels of these markers are consistently reported in the cerebrospinal fluid (CSF), plasma, and post-mortem brain tissues of patients with AD and PD. They contribute to membrane dysfunction, inflammation, and synaptic impairment (Mas-Bargues et al., 2021).
- **Protein Oxidation Markers:** Protein carbonylation and 3-nitrotyrosine are key indicators of oxidative modifications to proteins. These changes can affect enzyme activity, ion transport, and receptor binding in neurons. In AD brains, increased carbonyl content has been associated with tau hyperphosphorylation and β-amyloid toxicity (Pluta et al., 2022).
- **DNA Oxidation Products:** The most widely studied DNA oxidative marker is 8-hydroxy-2′-deoxyguanosine (8-OHdG). It signifies ROS-induced damage to guanine bases and is linked to mitochondrial DNA mutations. Elevated CSF levels of 8-OHdG are seen in early AD and correlate with memory impairment (Lovell & Markesbery, 2007).
- **Antioxidant Enzyme Levels:** Measuring the activity of antioxidant enzymes such as superoxide dismutase (SOD), catalase (CAT), and glutathione peroxidase (GPx) provides insight into the brain's defensive status. Decreased levels of SOD and GPx have been observed in PD and ALS, highlighting redox system failure (Niedzielska et al., 2016).

Figure 1 highlights the key mediators for neurodegeneration:

Figure 1. In neurodegenerative diseases (NDDs), activated microglia and astrocytes release proinflammatory cytokines. Astrocytes also exhibit upregulated expression of glial fibrillary acidic protein (GFAP). The activation of nuclear factor kappa B (NF-κB) and the NLRP3 inflammasome further contributes to cytokine release. Additionally, mitochondrial dysfunction and the production of reactive oxygen species (ROS) are observed, ultimately leading to blood-brain barrier (BBB) disruption and neuronal cell death

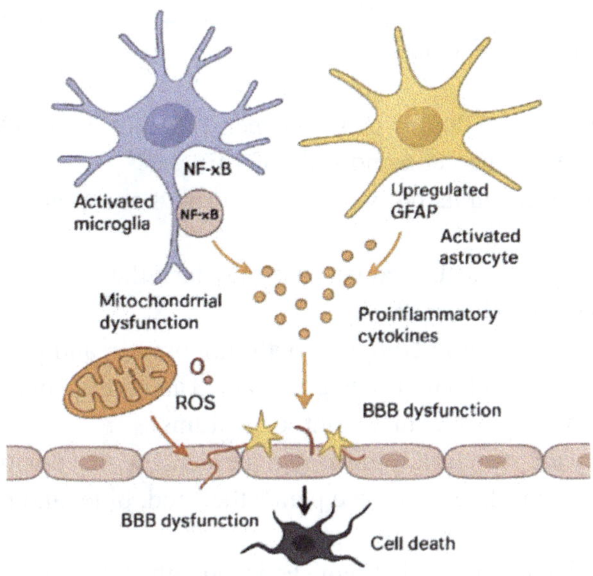

3. OVERVIEW OF POLYPHENOLS

3.1. Definition and classification of Polyphenols

The plant kingdom comprises an extensive and structurally varied family of secondary metabolites known as polyphenols. They are characterized chemically by the presence of a variety of phenol units, which are aromatic rings with one or more hydroxyl (-OH) groups that contribute to their biological activity and redox characteristics. Potent antioxidants, polyphenols can scavenge reactive oxygen and nitrogen species, chelate metal ions, and regulate signaling pathways that have implications in the proliferation of cells, inflammation, and apoptosis (Vauzour et al., 2010).

With respect to organization, polyphenols may vary from fundamental compounds like phenolic acids to intricate polymers like tannins. The degree of methylation, glycosylation, hydroxylation, and the location of functional groups all have a significant impact on their biological activity. Additionally, the foundation for their classification is the quantity and interconnectivity of aromatic rings. Even though there are thousands of distinct polyphenolic chemicals known to exist, they are often categorized into four major groups: lignans, stilbenes, phenolic acids, and flavonoids. Their primary carbon skeletons and biochemical origins in the phenylpropanoid pathway are reflected in these groups (Crozier et al., 2006).

3.1.1 Flavonoids

The most meticulously researched and prevalent subgroup of polyphenols are flavonoids. Two aromatic rings (referred to as A and B rings) joined by a three-carbon bridge which frequently creates a closed heterocyclic ring (C-ring) makes up their shared backbone, C6-C3-C6. Through hydroxylation, methoxylation, glycosylation, acylation, and prenylation, the fundamental flavonoid structure permits great diversity, giving rise to several subclasses with unique biological characteristics (Panche et al., 2016).

The following are subclasses of flavonoids:

- **Flavonols**: Those that have a double bond between C2 and C3 and a hydroxyl group at the third position of the C-ring (e.g., quercetin and kaempferol).
- **Flavones:** Identical to flavonols but without a hydroxyl group at the 3-position (e.g., luteolin, apigenin).
- **Flavanones:** They are more saturated because they lack the double bond that links C2 and C3. These include (naringenin and hesperetin).
- **Flavanols (flavan-3-ols):** These comprise both monomeric and polymeric catechins; they are distinguished by the presence of a hydroxyl group at C3 and the absence of a double bond between C2 and C3 (e.g., catechin, epicatechin, proanthocyanidins).
- **Anthocyanins:** They are structurally flavonoid pigments with a positively charged oxygen atom in the C-ring that are glycosylated and give plants their red, blue, and purple hues (e.g., cyanidin, delphinidin).
- **Isoflavones:** These flavonoids are distinguished from others by the location of the B-ring, which is joined to the C-ring at position three (genistein, daidzein, etc.). Often, these substances possess estrogenic properties.

Flavonoids have a variety of biological implications, such as modifying important enzymes and transcription factors that are involved in inflammation (NF-κB), mitochondrial function, and the oxidative stress response (Nrf2 pathway) (Fraga et al., 2019)(Ghosh et al., 2011).

3.1.2 Phenolic Acids

Low-molecular-weight polyphenols known as phenolic acids are distinguished by a carboxylic acid moiety and a single aromatic ring that has been replaced with hydroxyl groups. Based on their carbon side chain, they can be categorized into two main structural types:

- **Hydroxybenzoic acids (C6-C1):** These comprise substances like syringic acid, gallic acid, and protocatechuic acid. Plant cell walls frequently include them in bound forms or as constituents of hydrolyzable tannins.
- **Hydroxycinnamic acids (C6-C3):** Comprising p-coumaric acid, caffeic acid, ferulic acid, and sinapic acid, have a longer carbon chain. Most often, they are found as quinic acid esters or as conjugates with other plant metabolites.

By scavenging free radicals and inhibiting pro-inflammatory signaling pathways, phenolic acids are known to have antibacterial, anti-inflammatory, and antioxidant properties. They can also serve as precursors to more complex polyphenols (El-Saadony et al., 2025).

3.1.3 Stilbenes

A relatively small but pharmacologically pertinent class of polyphenols, stilbenes are identified by their C6-C2-C6 structure, which consists of two phenyl rings connected by an ethylene bridge. The most thoroughly investigated stilbene is resveratrol (3,5,4'-trihydroxy-trans-stilbene), which has been explored for its ability to improve metabolic health, lower oxidative stress, and lengthen lifespan (Bhaskara et al., 2020). Stilbenes' geometric structure (cis vs. trans isomers) and hydroxylation pattern affect their biological action. For instance, among other mechanisms, resveratrol works via modifying SIRT1, AMPK, and mTOR signaling. The anticancer and neuroprotective effects of synthetic and natural stilbene derivatives are being additionally explored (Baur & Sinclair, 2006).

3.1.4 Lignans

Phenylpropanoid dimers called lignans are produced when two C6-C3 units oxidatively couple. Lactone rings, methoxyls, hydroxyls, and other functional groups can all be found in their basic configuration. Enterolignans, such as enterodiol and enterolactone, which have modest estrogenic activity and affect hormonal balance, are produced when intestinal microbes metabolize lignans.

Numerous bioactivities, such as anticancer, anti-inflammatory, and antioxidant properties, have been linked to lignans. Research on hormone-related malignancies and cardiovascular disorders has focused on them because of their potential for modifying enzymes involved in the metabolism of estrogen and affect the cell signaling cascades (Landete, 2012).

3.1.5 Other Polyphenols

In addition to the four main classes, polyphenols consist of:

- **Coumarins:** These compounds have anticoagulant and anti-inflammatory properties and have a benzopyrone backbone.
- **Curcuminoids:** Turmeric-derived polyphenolic diketones, including curcumin, that influence inflammation and oxidative stress via the Nrf2 and NF-κB pathways.
- **Tannins:** High molecular weight substances that are separated into condensed tannins (proanthocyanidins) and hydrolysable tannins (based on gallic acid and ellagic acid esters). They have the ability to bind proteins and influence how nutrients are absorbed.

The pharmacological potency of these "non-classical" polyphenols is being intensively studied in clinical settings. The major classes of polyphenols along with their key examples, based on their biological relevance and structural characteristics are illustrated in Figure 2.

Figure 2. Classification of polyphenols

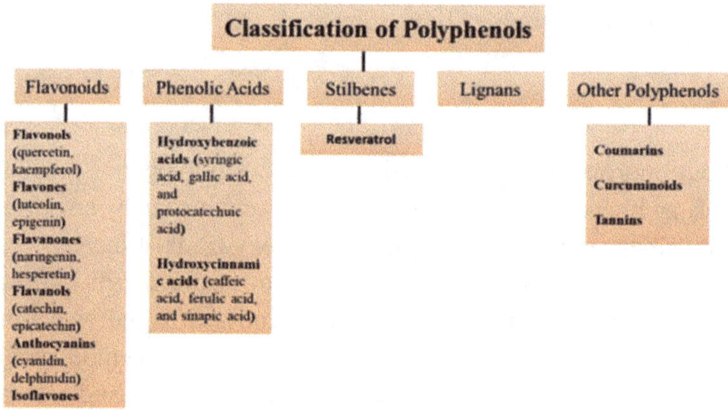

3.2. Pharmacokinetics and Bioavailability Considerations of Polyphenols

The pharmacokinetic characteristics and bioavailability of dietary polyphenols continue to be important barriers to attaining therapeutic efficacy, despite the widely recognized health advantages of these compounds. The degree and speed at which an active ingredient enters the bloodstream and becomes accessible at the site of action is referred to as bioavailability. The chemical structure, solubility, food matrix, metabolic changes, and interactions with gut bacteria of polyphenols are some of the elements that affect their pharmacokinetics.

3.2.1 Absorption

In the small intestine, polyphenols are mostly absorbed as conjugated compounds (such as glycosides) or aglycones. Large molecular size, low permeability, or fast metabolism, however, cause many polyphenols to have poor intestinal absorption. For instance, before being absorbed, flavonoid glycosides need to be digested by microbial enzymes or intestinal β-glucosidases (Ciupei et al., 2024).

3.2.2 Distribution and Metabolism

Following absorption, polyphenols are extensively metabolized in the liver and intestinal cells throughout phases I and II. Polyphenols are quickly changed into glucuronidated, sulfated, and methylated conjugates by phase II enzymes like catechol-O-methyltransferases (COMTs), sulfotransferases (SULTs), and UDP-glucuronosyltransferases (UGTs).(Nemzer et al., 2025) These metabolites may nonetheless have biological impacts even though they frequently have lower bioactivity than the parent chemicals. Polyphenols are carried by the bloodstream attached to lipoproteins or albumin, and they can build up in particular organs such the brain, liver, kidneys, and adipose tissue, but usually in trace amounts (Del Rio et al., 2012).

3.2.3 Role of Gut Microbiota

When unabsorbed polyphenols make it to the colon, the gut bacteria further breaks them down into low-molecular-weight phenolic acids, which are easier to absorb and may play a major role in systemic bioactivity. Individual differences in gut microbiota composition have a significant impact on polyphenols' biological impact and metabolic destiny. For example, individual differences in the metabolism of ellagitannins into urolithins or isoflavones into equal may impact the capacity to modulate disease (Ozdal et al., 2016).

3.2.4 Excretion

The main excretion routes for polyphenol metabolites are bile and urine. Their half-life and systemic availability are prolonged by biliary excretion, which may result in enterohepatic recirculation, whereas renal excretion includes conjugated forms (Williamson, 2025). Their neuroprotective, cardioprotective, and anti-aging qualities are largely attributed to their mitochondrial-targeted activities. Polyphenols provide protection in Alzheimer's and Parkinson's disease models by reducing mitochondrial dysfunction and neuronal death in neurodegenerative illnesses. Reduction of oxidative damage, maintenance of cardiac mitochondrial function, and avoidance of ischemia-reperfusion injury are the mechanisms by which cardioprotective benefits are achieved (Ha et al., 2018). Moreover, polyphenols prolong lifespan in a variety of model organisms and postpone cellular senescence by modifying mitochondrial biosynthesis and dynamics. Many of the therapeutic effects of polyphenols in a variety of oxidative stress-related diseases are based on their capacity to target mitochondria (Baur & Sinclair, 2006).

4. ANTI-INFLAMMATORY EFFECTS OF POLYPHENOLS

As mentioned earlier, the glial cells produce a variety of proinflammatory mediators that aggravate neuronal dysfunction. Moreover, chronic activation of the NLRP3 inflammasome and NF-κB exacerbates the inflammatory milieu, triggering BBB dysfunction, synaptic loss, and neuronal damage. Polyphenols counteract these pathological processes via targeting multiple pathways, including antioxidant scavenging, inflammasome suppression, gut-brain axis modulation, and downregulation of transcription factors. Details are as follows:

4.1 Downregulation of NLRP3 Inflammasome Signaling

Polyphenols play significant anti-inflammatory role via downregulation of the NLRP3 inflammasome, a key regulator of innate immune activities. Upon activation, the NLRP3 inflammasome triggers the cleavage of pro-caspase-1 into its active form, resulting in maturation of IL-18 and IL-1β, contributing to chronic neuroinflammation and cellular damage. Polyphenols interfere with the inflammasome-mediated production of IL-18 and IL-1β, thereby countering the cycle of self-perpetuating neuroinflammation (Winiarska-Mieczan et al., 2023). For instance, resveratrol suppresses both the activation and priming (via NF-κB inhibition) of the TXNIP/TRX/NLRP3 axis Aβ-stimulated microglia, leading to decreased caspase-1 cleavage as well as IL-1β release. Moreover, it also blocks the NLRP3 activation by inhibiting the generation of ROS and stimulating Sirtuin-1 (SIRT1) (He et al., 2017). Likewise, Curcumin selectively blocks the activation of NLRP3 inflammasome through the blockage of K^+ efflux and disruption of mitochondrial arrangement (Yin et al., 2018).

4.2 Inhibition of NF-κB Signaling

Polyphenols like curcumin, epigallocatechin-3-gallate (EGCG), and resveratrol significantly contribute to mitigate neuroinflammation by suppressing NF-κB signaling. These compounds exert significant anti-inflammatory effects by preventing the NF-κB pathway and promote cell survival. Resveratrol suppress the activities of p65 and IkappaB kinase to prevent the IκBα degradation, which is important for the nuclear translocation of NF-κB and its activation. It also blocks the IκBα phosphorylation and reduce the generation of ROS that stimulates NF-κB activity (Ren et al., 2013). EGCG also inhibits the activation of NF-κB by stabilizing IκBα and blocking its degradation as well activate Nrf-2/HO-1 to protect against inflammation induced by hypoxia. (S.-R. Kim et al., 2022).

4.3 Supression of Pro-Inflammatory Enzymes

In addition to other mechanisms, polyphenols attenuate neuroinflammation by inhibiting essential pro-inflammatory enzymes, including inducible nitric oxide synthase (iNOS), lipoxygenase (LOX), and cyclooxygenase-2 (COX-2). These enzymes are key contributors in production of inflammatory mediators, such as nitric oxide (NO), leukotrienes, and prostaglandins that promote oxidative stress and neuroinflammation. Resveratrol inhibits iNOS and COX-2 at both protein and mRNA levels by modulating AP-1 and NF-κB pathways (T. Meng et al., 2021). Similarly, flavonoids like catechin, EGCG, quercetin, and genistein have been detected to downregulate the expression of iNOS and COX-2 as part of their anti-neuroinflammatory activities (Calis et al., 2020).

4.4 Inhibition of Proinflammatory Cytokines Production

Polyphenols directly alleviate the proinflammatory cytokines, including IL-1β, IL-6, and TNF-α. For example, flavonoids like quercetin, myricetin, and EGCG inhibit the iNOS expression and production of cytokines production within microglia via downregulation of MAPK phosphorylation and NF-κB signaling pathways (Mokra et al., 2023; Vauzour, 2012).

4.5 Modulation of Glial Phenotypic Polarization

Polyphenols significantly modulate glial cell polarization by promoting the shift of astrocytes and microglia from pro-inflammatory (A1/M1) to anti-inflammatory (A2/M2) phenotypes. Resveratrol stimulates polarization of M2 microglia by Peroxisome proliferator-activated receptor gamma coactivator 1-alpha (PGC-1α) signaling and decreasing inflammatory marker(X. Liu et al., 2023; Yang et al., 2017). Flavonoids like luteolin, quercetin, baicalein, and baicalin decrease astrogliosis (A1 reactive astrocytes), reduce GFAP expression, and limit proinflammatory cytokines production via modulating Nuclear Factor Erythroid 2–Related Factor 2 (Nrf2) and Signal Transducer and Activator of Transcription 3 (STAT3) pathways (Calis et al., 2020) to restore homeostasis and support neuronal survival the CNS. The details of anti-inflammatory effects of polyphenols are summarized in table 4.1.

Table 1. Summary of anti-inflammatory mechanisms of polyphenols

Mechanism	Polyphenols	Molecular targets	References
Downregulation of NLRP3 Inflammasome	Resveratrol, Curcumin	TXNIP/TRX/NLRP3, Caspase-1, SIRT1	(Q. He et al., 2017; Winiarska-Mieczan et al., 2023; Yin et al., 2018)
Inhibition of NF-κB Signaling	Resveratrol, Curcumin, EGCG	NF-κB (p65), IKK, IκBα	(S.-R. Kim et al., 2022; Ren et al., 2013)
Inhibition of Pro-Inflammatory Enzymes	EGCG, Resveratrol, Quercetin, Catechin, Genistein	COX-2, iNOS, LOX	(Calis et al., 2020; S. Li et al., 2024; T. Meng et al., 2021)
Downregulation of Pro-Inflammatory Cytokines	Hesperidin, Quercetin, Myricetin, EGCG	TNF-α, IL-1β, IL-6, MAPKs	(H.-W. Ding et al., 2017; Mokra et al., 2023; Vauzour, 2012)
Modulation of Glial Phenotypic Polarization	Resveratrol, Quercetin, Luteolin, Baicalein	PGC-1α, STAT3, Nrf2, GFAP	(Calis et al., 2020; X. Liu et al., 2023; X. Yang et al., 2017)

5. POLYPHENOLS AND MITOCHONDRIAL PROTECTION

Under oxidative stress and inflammatory circumstances, polyphenols maintain mitochondrial integrity and function, resulting in strong cytoprotective benefits. The stability of the mitochondrial membrane potential, suppression of mitochondrial-mediated apoptosis, blockage of the opening of the mitochondrial permeability transition pore (mPTP), and mitigation of the formation of ROS within the mitochondria are some of the processes that facilitate this preservation.(Kung et al., 2021) Polyphenols such as resveratrol, epigallocatechin gallate (EGCG), curcumin, and hydroxytyrosol also enhance mitochondrial biogenesis through upregulation of PGC-1α and support mitophagy for the removal of damaged mitochondria, thereby maintaining mitochondrial quality control (Dong et al., 2022)

5.1. Preservation of Mitochondrial Membrane Potential ΔΨm

For cells to thrive and generate energy, the potential of the mitochondrial membrane must be maintained. Polyphenols like EGCG and resveratrol prevent mitochondrial depolarization in a number of ways:

- **Prevention of membrane leakage and mPTP opening:** By suppressing voltage dependent anion channel 1 (VDAC1) expression and preventing mPTP opening through the activation of the AMPK–Mitofusin-2 pathway in stressed cardiomyocytes, resveratrol helps maintaining mitochondrial membrane potential (Ahmed, 2025).

- **Enhancement of antioxidant defenses and scavenging of ROS:** By decreasing ROS, downregulating pro-apoptotic DRP1/BAX pathways, and raising anti-apoptotic BCL-2 levels, EGCG restores $\Delta\Psi m$ in pancreatic β-cells subjected to high-glucose-induced oxidative stress (Jia et al., 2024).
- **Targeted mitochondrial accumulation of catechins**: EGCG and its mitochondrial-directed derivatives show both direct membrane-protective activity and antioxidant benefits by reducing H_2O_2-induced mitochondrial depolarization and death in cardiac cells (Sahadevan et al., 2023).
- **Improvement of mitochondrial biogenesis and functional stability**: Resveratrol upregulates genes necessary for mitochondrial integrity by activating the SIRT1–AMPK–PGC-1α signaling axis, which indirectly aids in the stabilization of $\Delta\Psi m$ (Wu et al., 2024).

When combined, these polyphenolic activities stimulate ATP synthesis, inhibiting the release of apoptotic agents, and ultimately enhancing cellular viability by maintaining $\Delta\Psi m$ in the face of oxidative or metabolic stress.

5.2. Stimulation of mitochondrial biogenesis

Peroxisome proliferator-activated receptor γ coactivator-1 alpha (PGC-1α) signaling axis, a master regulator of mitochondrial gene expression, is the primary mechanism by which polyphenols stimulate mitochondrial biogenesis. Resveratrol increases SIRT1's deacetylation and PGC-1α activity while also activating AMPK, which increases PGC-1α expression. Resveratrol enhanced the expression of PGC-1α, NRF-1, and TFAM in models of metabolic stress and brain injury (such as subarachnoid hemorrhage), enhanced mitochondrial DNA content, membrane potential, and ATP generation, and decreased ROS and apoptosis (Zhou et al., 2021).

Quercetin has been demonstrated to increase PGC-1α levels both in vitro and in vivo, through SIRT1 activation. It also increases the expression of downstream effectors NRF-1 and TFAM, as well as mitochondrial DNA copy number, respiratory enzyme activity, and ATP generation. In models of the heart, neurons, and pancreatic β-cells, it also decreased ROS and apoptotic markers (Peng et al., 2023). In addition, Hydroxytyrosol, which is found in skeletal muscle, adipocytes, and retinal epithelial cells, stimulates PGC-1α through SIRT1 deacetylation. better mtDNA, elevated OXPHOS complex I–II expression, better endurance, and improved mitochondrial function are all correlated with these benefits (Hao et al., 2010).

The combined action of these polyphenols via SIRT1–AMPK–PGC-1α not only boosts mitochondrial biogenesis but also contributes to enhanced bioenergetics and redox homeostasis in diverse cell types.

5.3. Reduction of mitochondrial ROS Generation

One of the main causes of mitochondrial malfunction and cell death is the overproduction of mitochondrial reactive oxygen species (mtROS). In order to protect mitochondrial integrity, polyphenols including EGCG, resveratrol, and myricetin lower mtROS through both direct and indirect processes.

- **Direct mtROS scavenging:** It has been demonstrated that EGCG, a key green tea catechin, dramatically lowers mitochondrial superoxide levels in hepatocytes under lipotoxic stress, as revealed by decreased lipid peroxidation, an elevated GSH/GSSG ratio, and a lessened MitoSOX signal.

EGCG also inhibited the production of mtROS in skin fibroblast models, increasing the efficiency of mitochondrial respiration and reversing malfunction linked to senescence (Ding et al., 2023).

- **Delivery targeted to the mitochondria improves efficacy:** MitoEGCG$_4$ and MitoEGCG$_6$ are examples of synthetic mitochondria-directed EGCG derivatives that efficiently accumulate in mitochondria to counteract oxidative damage caused by H_2O_2 in cardiomyocytes. These compounds were far more effective than EGCG alone at preventing mitochondrial depolarization and apoptosis in addition to lowering mtROS (Sahadevan et al., 2023).
- **Activation of endogenous antioxidant defenses:** In brain damage models, EGCG reduces mtROS and restores mitochondrial dynamics by upregulating antioxidant enzymes such as GPx1 and mitochondrial SOD. EGCG further decreased oxidative stress in epilepsy models by preserving DRP1-mediated mitochondrial fission and increasing GPx1 activity (J.-E. Kim et al., 2023).
- **Metal chelation and polyphenol synergy:** Flavonoids like myricetin maintain mitochondrial redox equilibrium by chelating transition metals and directly scavenging ROS, which prevents ROS from spreading and strengthens endogenous antioxidant systems like glutathione-S-transferase (J.-E. Kim et al., 2023).

These multifaceted actions direct mtROS neutralization, mitochondrial targeting, enzyme induction, and metal chelation underscore the robust potential of polyphenols to preserve mitochondrial function under oxidative stress.

5.4. Prevention of Cytochrome c Release and Caspase Activation

Resveratrol, EGCG, and hydroxytyrosol are examples of polyphenols that stabilize mitochondrial membranes and alter the dynamics of apoptotic proteins to block the intrinsic mitochondrial apoptotic pathway. Resveratrol and quercetin block cytochrome c release and downstream activation of caspase-9 and caspase-3 by preventing mitochondrial membrane permeabilization in neuronal and cancer models. When used together, they show synergistic inhibition of apoptosis through this route, which lowers cell death in PC12 and cancer cell lines (Mouria et al., 2002). EGCG preserves mitochondrial integrity by attenuating mitochondrial permeabilization, maintaining $\Delta\Psi m$, and reducing activation of pro-apoptotic proteins such as Bax, while maintaining Bcl-2 expression in models of ischemia-reperfusion and β-cell damage (Sirangelo et al., 2022).

Hydroxytyrosol exerts anti-apoptotic effects in cardiac and colon cells by inhibiting cytochrome c release and caspase-3/-9 activation, modulating Bax/Bcl-2 balance, and preventing mitochondrial outer membrane permeabilization (Batarfi et al., 2024). Broadly, flavonoids including hesperidin, myricetin, rosmarinic acid, theaflavin, and curcumin inhibit mPTP opening and downstream caspase signaling under stress conditions such as Aβ exposure, ischemia, or neurotoxin insult (Naoi et al., 2019). By stabilizing mitochondrial membranes and regulating Bcl-2 family proteins, polyphenols effectively block cytochrome c efflux and caspase cascade activation—thereby safeguarding cells against intrinsic apoptosis. The details of Mitochondrial Protective Mechanisms of polyphenols are summarized in Table 2.

Table 2. Summary of mitochondrial protective mechanisms of polyphenols

Mechanism	Polyphenols	Molecular Targets	References
Preservation of mitochondrial membrane potential	EGCG, Resveratrol	Inhibits VDAC1, increases AMPK-Mitofusin2 signaling, stabilizes $\Delta\Psi m$, inhibits mPTP opening, and lowers ROS and apoptosis.	(Ahmed, 2025),(Jia et al., 2024),(Sahadevan et al., 2023),(Wu et al., 2024)
Stimulation of mitochondrial biogenesis	Hydroxytyrosol, Quercetin, Resveratrol	The SIRT1–AMPK–PGC-1α pathway is activated. Increases in mtDNA content, ATP generation, and OXPHOS complex expression are caused by PGC-1α, NRF1, and TFAM.	(Zhou et al., 2021),(Peng et al., 2023), (Hao et al., 2010)
Reduction of mitochondrial ROS generation	EGCG, Hydroxytyrosol, Myricetin	increases GPx1/SOD, scavenges mtROS, and enhances redox balance; mitochondria-targeted administration increases effectiveness.	(Sahadevan et al., 2023), (S. Bin Ding et al., 2023), (J.-E. Kim et al., 2023)
Preservation of cytochrome c release and caspase activation	EGCG, Hydroxytyrosol, Resveratrol	Suppresses cytochrome c release, stabilizes mitochondrial membranes, modifies the Bax/Bcl-2 ratio, and prevents the activation of caspase-3 and -9.	(Mouria et al., 2002), (Sirangelo et al., 2022), (Batarfi et al., 2024), (Naoi et al., 2019)
Integrated mitochondrial protection	Curcumin, EGCG, Resveratrol	Uses several mitochondrial mechanisms to combine bioenergetic restoration, anti-apoptotic, and antioxidant effects.	(Kung et al., 2021), (Dong et al., 2022), (Ha et al., 2018)

6. POLYPHENOLS AND OXIDATIVE STRESS

Polyphenols not only scavenge free radicals but also exert regulatory effects on multiple cell-signaling pathways. Their neuroprotective actions are categorized into:

- **Direct Mechanisms:** Free radical scavenging: Polyphenols like quercetin, EGCG, and resveratrol donate electrons to neutralize ROS (e.g., •OH, O_2^-, NO) and terminate lipid peroxidation chain reactions .Metal chelation, Curcumin and quercetin chelate redox-active metals such as Fe^{2+} and Cu^{2+}, preventing Fenton-type ROS generation (Chu, 2022).
- **Indirect Mechanisms:** Polyphenols disrupt the interaction between Keap1 and Nrf2—either by competing with Nrf2 for Keap1 binding or modifying Keap1 cysteine residues—leading to Nrf2 stabilization and nuclear translocation. In the nucleus, Nrf2 binds to the Antioxidant Response Element (ARE), initiating transcription of key antioxidant genes such as HO-1, SOD, (CAT, and GPx (Suraweera et al., 2020).
- **SIRT1 modulation:** Resveratrol activates SIRT1, which deacetylates and activates PGC-1α to promote mitochondrial biogenesis, and collaborates with Nrf2 to induce antioxidant enzymes. Concurrently, SIRT1 suppresses pro-inflammatory NF-κB signaling by deacetylating its p65 subunit, thus reducing inflammation and oxidative stress (Shaito et al., 2022).

• **Anti-inflammatory effects:** Polyphenols inhibit activation of microglia and astrocytes, reduce NF-κB and MAPK pathway activity, and suppress production of pro-inflammatory mediators including IL-1β, TNF-α, and iNOS (Jalouli et al., 2025).

The details of antioxidative efficacy of polyphenols are summarized in Table 3.

Table 3. Summary of antioxidant efficacy of polyphenols

Mechanisms	Polyphenols	Molecular targets	References
ROS scavenging, metal chelation, Protects against Aβ toxicity and neuroinflammation	Quercetin	Nrf2 activation	(Costa et al., 2016).
Anti-apoptotic, mitochondrial protection	EGCG	Nrf2/HO-1 axis	(F. He et al., 2018)
Improves cognition and reduces amyloid burden in AD	Resveratrol	SIRT1 activation, anti-inflammatory, Nrf2 modulation	(M. Yang et al., 2023)

7. CLINICAL TRIALS AND TRANSLATIONAL CHALLENGES

While polyphenols exhibit strong potential in preclinical models of neurodegeneration, metabolism, inflammation, and cardiovascular disease, but their translation into clinical outcomes remain inconsistent. One of the main challenges lies in variable and poor bioavailability of polyphenol compounds. For instance, resveratrol and many flavonoids are rapidly metabolized and conjugated, resulting in minimal systemic absorption. So, far below effective plasma and tissue concentrations in humans are identified in both in vitro and in vivo models. This pharmacokinetic limitation or gap contributes to the failure in identifying dose–response relationships in clinical trial (Salla et al., 2024). Moreover, Safety, regulatory, and quality-control issues add practical barriers. Since, botanical extracts and nutraceuticals often circumvent the drug-approval process, their quality is inconsistent and regulatory guidance for claims remain unclear. Clinical translation of polyphenols as potential therapeutics requires compliance with GMP guidelines, establish dose–toxicity relationships, and ensure reliable product manufacturing (Rudrapal et al., 2024).

Poor clinical translation of polyphenols is hindered by its extensive metabolism; once ingested, they are rapidly conjugated and microbiota transformed, leaving circulating metabolites—rather than parent compounds—as the primary bioactive agents, with marked inter-individual variability. Their CNS penetration is also constrained by the BBB, where efflux pumps and metabolic enzymes restrict CNS penetration. While approaches such as nano-formulations, liposomes, and intranasal delivery enhance brain exposure in preclinical studies, human data remain scarce, underscoring the need for trials incorporating PK/PD analysis, microbiome stratification, and CNS biomarkers (Marques et al., 2024; Quesada-Vázquez et al., 2024).

CONCLUSION

Polyphenols exhibit promising neuroprotective effects as they stand out multi-targeted actions against the major pathological mechanisms supporting neurodegeneration. The interplay of persistent oxidative stress, chromic neuroinflammation, and mitochondrial dysfunction accelerates neuronal injury and polyphenols counteract these processes through multi-level intervention. They modulate glial activation by promoting the shift of astrocytes and microglia from pro-inflammatory to anti-inflammatory states. This concurrently downregulate proinflammatory cytokines, such as TNF-α, IL-1β, and IL-6 and inhibit key signaling pathways like NLRP3 inflammasome and NF-κB. Simultaneously, polyphenols protect mitochondrial function by reducing mitochondrial ROS, stabilizing the membrane potential, and increasing biogenesis via the SIRT1–AMPK–PGC-1α axis. As a result, neuronal energy metabolism is restored, apoptosis is suppressed, and overall cell viability is improved. In addition, the Nrf2 pathway is activated by polyphenols that boost endogenous antioxidant mechanism and enhance redox homeostasis. Collectively, these pleiotropic actions of polyphenols offer a robust therapeutic strategy that addresses the multifactorial nature of neurodegeneration.

ABBREVIATIONS

Neurodegenerative diseases (ND); Parkinson's disease (PD); Huntington's disease (HD); Amyotrophic lateral sclerosis (ALS); Alzheimer's disease (AD); Huntingtin (HTT gene); Disability-adjusted life-years (DALYs); Reactive oxygen species (ROS); Central nervous system (CNS); Pattern recognition receptors (PRRs); Toll-like receptors (TLRs); Pathogen-associated molecular patterns (PAMPs); Damage-associated molecular patterns (DAMPs); Glial fibrillary acidic protein (GFAP); Blood brain barrier (BBB); Amyloid-beta (Aβ); Interleukin-1 beta (IL-1β); Interleukin-6 (IL-6); Tumor necrosis factor-alpha (TNF-α); Amino-3-hydroxy-5-methyl-4-isoxazolepropionic acid (AMPA); Gamma-aminobutyric acid (GABA); N-Methyl-D-Aspartate (NMDA); Granulocyte-Macrophage Colony-Stimulating Factor (GM-CSF); C-X-C Motif Chemokine Ligand 10 (CXCL10); C-C Motif Chemokine Ligand 2 (CCL2); NOD-like receptor family pyrin domain containing 3 (NLRP3); Nuclear factor kappa-light-chain-enhancer of activated B cells (NF-κB); IκB kinase (IKK); Lipopolysaccharide (LPS); Inducible nitric oxide synthase (iNOS); Cyclooxygenase-2 (COX-2); Apoptosis-associated speck-like protein containing a CARD (ASC); Experimental autoimmune encephalomyelitis (EAE); Electron transport chain (ETC); Mitochondrial permeability transition pore (mPTP); Apoptosis-inducing factor (AIF); Second mitochondria-derived activator of caspases (SMAC); Apoptotic protease activating factor-1 (Apaf-1); Reactive nitrogen species (RNS); Polyunsaturated fatty acids (PUFAs); Deoxyribonucleic acid (DNA); Mitogen-activated protein kinase (MAPK); Glutathione (GSH); malondialdehyde (MDA); 4-hydroxynonenal (4-HNE); Cerebrospinal fluid (CSF); 8-hydroxy-2′-deoxyguanosine (8-OHdG); superoxide dismutase (SOD); catalase (CAT); Glutathione peroxidase (GPx); Sirtuin-1 (SIRT1); Epigallocatechin gallate (EGCG); Lipoxygenase (LOX), Cyclooxygenase-2 (COX-2); Nuclear Factor Erythroid 2–Related Factor 2 (Nrf2); Signal Transducer and Activator of Transcription 3 (STAT3); Mitochondrial permeability transition pore (mPTP); Catechol-O-methyltransferases (COMTs); Sulfotransferases (SULTs); UDP-glucuronosyltransferases (UGTs); Mitochondrial Membrane Potential (ΔΨm); Voltage-Dependent Anion Channel 1 (VDAC1); AMPK (AMP-activated protein kinase); Mitofusin 2 (Mfn2); Peroxisome proliferator-activated receptor γ coactivator-1 alpha (PGC-1α); Antioxidant Response Element (ARE).

REFERENCES

Adamu, A., Li, S., Gao, F., & Xue, G. (2024). The role of neuroinflammation in neurodegenerative diseases: Current understanding and future therapeutic targets. *Frontiers in Aging Neuroscience*, *16*(April), 1–16. DOI: 10.3389/fnagi.2024.1347987 PMID: 38681666

Ahmad, S. R., Zeyaullah, M., AlShahrani, A. M., Dawria, A., Ali, H., Mohieldin, A., Altijani, A. A. G., Razi, U., Mehdi, M., Akram, S., & Hussain, E. R. (2024). Deciphering the enigma of neuron-glial interactions in neurological disorders. *Frontiers in Bioscience (Landmark Edition)*, *29*(4), 142. DOI: 10.31083/j.fbl2904142 PMID: 38682185

Ahmad, S. R., Zeyaullah, M., Khan, M. S., AlShahrani, A. M., Altijani, A. A. G., Ali, H., Dawria, A., Mohieldin, A., Alam, M. S., & Mohamed, A. O. A. (2024). Pharmacogenomics for neurodegenerative disorders - a focused review. *Frontiers in Pharmacology*, *15*(December), 1–11. DOI: 10.3389/fphar.2024.1478964 PMID: 39759457

Ahmed, M. (2025). Targeting aging pathways with natural compounds: a review of curcumin, epigallocatechin gallate, thymoquinone, and resveratrol. *Immunity & Ageing: I & A, 22*, 28.

Alqahtani, T., Deore, S. L., Kide, A. A., Shende, B. A., Sharma, R., Chakole, R. D., Nemade, L. S., Kale, N. K., Borah, S., & Deokar, S. S. (2023). Mitochondrial dysfunction and oxidative stress in Alzheimer's disease, and Parkinson's disease, Huntington's disease and amyotrophic lateral sclerosis-an updated review. *Mitochondrion*, *71*, 83–92. DOI: 10.1016/j.mito.2023.05.007 PMID: 37269968

Aziz, N., Ruzza, C., Falcicchia, C., Guarino, A., Soukupova, M., Asth, L., Aleotti, V., Bettegazzi, B., Simonato, M., & Zucchini, S. (2024). Lack of Direct Effects of Neurotrophic Factors in an In Vitro Model of Neuroinflammation. *International Journal of Molecular Sciences*, *25*(8), 4160. Advance online publication. DOI: 10.3390/ijms25084160 PMID: 38673746

Batarfi, W. A., Yunus, M. H. M., Hamid, A. A., Lee, Y. T., & Maarof, M. (2024). Hydroxytyrosol: A promising therapeutic agent for mitigating inflammation and apoptosis. *Pharmaceutics*, *16*(12), 1504. DOI: 10.3390/pharmaceutics16121504 PMID: 39771483

Baur, J. A., & Sinclair, D. A. (2006). Therapeutic potential of resveratrol: The in vivo evidence. *Nature Reviews. Drug Discovery*, *5*(6), 493–506. DOI: 10.1038/nrd2060 PMID: 16732220

Bellanti, F., Coda, A. R. D., Trecca, M. I., Lo Buglio, A., Serviddio, G., & Vendemiale, G. (2025). Redox Imbalance in Inflammation: The Interplay of Oxidative and Reductive Stress. In *Antioxidants* (Vol. 14, Issue 6). Multidisciplinary Digital Publishing Institute (MDPI). DOI: 10.3390/antiox14060656

Bereiter-Hahn, J., & Vöth, M. (1994). Dynamics of mitochondria in living cells: Shape changes, dislocations, fusion, and fission of mitochondria. *Microscopy Research and Technique*, *27*(3), 198–219. DOI: 10.1002/jemt.1070270303 PMID: 8204911

Besedovsky, H. O., & del Rey, A. (2011). Central and peripheral cytokines mediate immune-brain connectivity. *Neurochemical Research*, *36*(1), 1–6. DOI: 10.1007/s11064-010-0252-x PMID: 20820913

Bhaskara, V. K., Mittal, B., Mysorekar, V. V., Amaresh, N., & Simal-Gandara, J. (2020). Resveratrol, cancer and cancer stem cells: A review on past to future. *Current Research in Food Science*, *3*(August), 284–295. DOI: 10.1016/j.crfs.2020.10.004 PMID: 33305295

Calis, Z., Mogulkoc, R., & Baltaci, A. K. (2020). The Roles of Flavonols/Flavonoids in Neurodegeneration and Neuroinflammation. *Mini-Reviews in Medicinal Chemistry*, *20*(15), 1475–1488. DOI: 10.2174/1389557519666190617150051 PMID: 31288717

Campolo, N., Bartesaghi, S., & Radi, R. (2014). Metal-catalyzed protein tyrosine nitration in biological systems. *Redox Report : Communications in Free Radical Research*, *19*(6), 221–231. DOI: 10.1179/1351000214Y.0000000099 PMID: 24977336

Chandimali, N., Bak, S. G., Park, E. H., Lim, H. J., Won, Y. S., Kim, E. K., Park, S. I., & Lee, S. J. (2025). Free radicals and their impact on health and antioxidant defenses: a review. In *Cell Death Discovery* (Vol. 11, Issue 1). Springer Nature. DOI: 10.1038/s41420-024-02278-8

Cherry, J. D., Olschowka, J. A., & O'Banion, M. K. (2014). Neuroinflammation and M2 microglia: The good, the bad, and the inflamed. *Journal of Neuroinflammation*, *11*(1), 98. DOI: 10.1186/1742-2094-11-98 PMID: 24889886

Chu, A. J. (2022). Quarter-Century Explorations of Bioactive Polyphenols: Diverse Health Benefits. In *Frontiers in Bioscience - Landmark* (Vol. 27, Issue 4). Bioscience Research Institute. DOI: 10.31083/j.fbl2704134

Ciupei, D., Colişar, A., Leopold, L., Stănilă, A., & Diaconeasa, Z. M. (2024). Polyphenols: From classification to therapeutic potential and bioavailability. *Foods*, *13*(24), 4131. DOI: 10.3390/foods13244131 PMID: 39767073

Crozier, A., Clifford, M. N., & Ashihara, H. (2006). *Plant secondary metabolites. Occurrence, Structure and Role in the Human Diet*. Blackwell-Publishers. DOI: 10.1002/9780470988558

Del Rio, D., Rodriguez-Mateos, A., Spencer, J. P. E., Tognolini, M., & Borges, G. (2012). phenolics in human health: Structures, bioavailability, and evidence of protective effects against chronic diseases., 2013, 18. *DOI:* Https://Doi. Org/10.1089/Ars, 1818–1892.

Ding, S., Chu, X.-L., Jin, Y.-X., Jiang, J.-J., Zhao, X., & Yu, M. (2023). Epigallocatechin gallate alleviates high-fat diet-induced hepatic lipotoxicity by targeting mitochondrial ROS-mediated ferroptosis. *Frontiers in Pharmacology*, *14*(March), 1–11. DOI: 10.3389/fphar.2023.1148814 PMID: 37025486

Dong, Y., Kalueff, A. V., & Song, C. (2017). N-methyl-d-aspartate receptor-mediated calcium overload and endoplasmic reticulum stress are involved in interleukin-1beta-induced neuronal apoptosis in rat hippocampus. *Journal of Neuroimmunology*, *307*, 7–13. https://doi.org/https://doi.org/10.1016/j.jneuroim.2017.03.005. DOI: 10.1016/j.jneuroim.2017.03.005 PMID: 28495142

Dong, Y., Yu, M., Wu, Y., Xia, T., Wang, L., Song, K., Zhang, C., Lu, K., & Rahimnejad, S. (2022). Hydroxytyrosol Promotes the Mitochondrial Function through Activating Mitophagy. *Antioxidants*, *11*(5), 1–15. DOI: 10.3390/antiox11050893 PMID: 35624756

El-Saadony, M. T., Saad, A. M., Mohammed, D. M., Korma, S. A., Alshahrani, M. Y., Ahmed, A. E., Ibrahim, E. H., Salem, H. M., Alkafaas, S. S., Saif, A. M., Elkafas, S. S., Fahmy, M. A., Abd El-Mageed, T. A., Abady, M. M., Assal, H. Y., El-Tarabily, M. K., Mathew, B. T., AbuQamar, S. F., El-Tarabily, K. A., & Ibrahim, S. A. (2025). Medicinal plants: Bioactive compounds, biological activities, combating multidrug-resistant microorganisms, and human health benefits - a comprehensive review. *Frontiers in Immunology*, *16*(April), 1–37. DOI: 10.3389/fimmu.2025.1491777 PMID: 40375989

Escartin, C., Galea, E., Lakatos, A., O'Callaghan, J. P., Petzold, G. C., Serrano-Pozo, A., Steinhäuser, C., Volterra, A., Carmignoto, G., Agarwal, A., Allen, N. J., Araque, A., Barbeito, L., Barzilai, A., Bergles, D. E., Bonvento, G., Butt, A. M., Chen, W.-T., Cohen-Salmon, M., & Verkhratsky, A. (2021). Reactive astrocyte nomenclature, definitions, and future directions. *Nature Neuroscience*, *24*(3), 312–325. DOI: 10.1038/s41593-020-00783-4 PMID: 33589835

Feigin, V. L., Vos, T., Nichols, E., Owolabi, M. O., Carroll, W. M., Dichgans, M., Deuschl, G., Parmar, P., Brainin, M., & Murray, C. (2020). The global burden of neurological disorders: Translating evidence into policy. *Lancet Neurology*, *19*(3), 255–265. DOI: 10.1016/S1474-4422(19)30411-9 PMID: 31813850

Fraga, C. G., Croft, K. D., Kennedy, D. O., & Tomás-Barberán, F. A. (2019). The effects of polyphenols and other bioactives on human health. *Food & Function*, *10*(2), 514–528. DOI: 10.1039/C8FO01997E PMID: 30746536

Franzoni, F., Scarfò, G., Guidotti, S., Fusi, J., Asomov, M., & Pruneti, C. (2021). Oxidative Stress and Cognitive Decline: The Neuroprotective Role of Natural Antioxidants. In *Frontiers in Neuroscience* (Vol. 15). Frontiers Media S.A. DOI: 10.3389/fnins.2021.729757

Ghosh, N., Ghosh, R., Mandal, V., & Mandal, S. C. (2011). Recent advances in herbal medicine for treatment of liver diseases. *Pharmaceutical Biology*, *49*(9), 970–988. DOI: 10.3109/13880209.2011.558515 PMID: 21595500

Goldsteins, G., Hakosalo, V., Jaronen, M., Keuters, M. H., Lehtonen, Š., & Koistinaho, J. (2022). CNS Redox Homeostasis and Dysfunction in Neurodegenerative Diseases. In *Antioxidants* (Vol. 11, Issue 2). DOI: 10.3390/antiox11020405

Grabska-Kobyłecka, I., Szpakowski, P., Król, A., Książek-Winiarek, D., Kobyłecki, A., Głąbiński, A., & Nowak, D. (2023). Polyphenols and Their Impact on the Prevention of Neurodegenerative Diseases and Development. *Nutrients*, *15*(15), 1–34. DOI: 10.3390/nu15153454 PMID: 37571391

Gryka-Marton, M., Grabowska, A. D., & Szukiewicz, D. (2025). Breaking the Barrier: The Role of Proinflammatory Cytokines in BBB Dysfunction. *International Journal of Molecular Sciences*, *26*(8), 3532. Advance online publication. DOI: 10.3390/ijms26083532 PMID: 40331982

Guo, C. Y., Sun, L., Chen, X. P., & Zhang, D. S. (2013). Oxidative stress, mitochondrial damage and neurodegenerative diseases. *Neural Regeneration Research*, *8*(21), 2003–2014. DOI: 10.3969/j.issn.1673-5374.2013.21.009 PMID: 25206509

Ha, T., Kim, M. K., Park, K.-S., Jung, W., Choo, H., & Chong, Y. (2018). Structural Modification of (-)-Epigallocatechin Gallate (EGCG) Shows Significant Enhancement in Mitochondrial Biogenesis. *Journal of Agricultural and Food Chemistry*, *66*(15), 3850–3859. DOI: 10.1021/acs.jafc.8b00364 PMID: 29514455

Hao, J., Shen, W., Yu, G., Jia, H., Li, X., Feng, Z., Wang, Y., Weber, P., Wertz, K., Sharman, E., & Liu, J. (2010). Hydroxytyrosol promotes mitochondrial biogenesis and mitochondrial function in 3T3-L1 adipocytes. *The Journal of Nutritional Biochemistry*, *21*(7), 634–644. DOI: 10.1016/j.jnutbio.2009.03.012 PMID: 19576748

He, Q., Li, Z., Wang, Y., Hou, Y., Li, L., & Zhao, J. (2017). Resveratrol alleviates cerebral ischemia/reperfusion injury in rats by inhibiting NLRP3 inflammasome activation through Sirt1-dependent autophagy induction. *International Immunopharmacology*, *50*, 208–215. DOI: 10.1016/j.intimp.2017.06.029 PMID: 28683365

Henrich, M. T., Oertel, W. H., Surmeier, D. J., & Geibl, F. F. (2023). Mitochondrial dysfunction in Parkinson's disease–a key disease hallmark with therapeutic potential. *Molecular Neurodegeneration*, *18*(1), 83. DOI: 10.1186/s13024-023-00676-7 PMID: 37951933

Holbrook, J. A., Jarosz-Griffiths, H. H., Caseley, E., Lara-Reyna, S., Poulter, J. A., Williams-Gray, C. H., Peckham, D., & McDermott, M. F. (2021). Neurodegenerative disease and the NLRP3 inflammasome. *Frontiers in Pharmacology*, *12*, 643254. DOI: 10.3389/fphar.2021.643254 PMID: 33776778

Huang, Y., Chen, S., Luo, Y., & Han, Z. (2020). Crosstalk between inflammation and the BBB in stroke. *Current Neuropharmacology*, *18*(12), 1227–1236. DOI: 10.2174/1570159X18666200620230321 PMID: 32562523

Hung, C.-C., Lin, C.-H., Chang, H., Wang, C.-Y., Lin, S.-H., Hsu, P.-C., Sun, Y.-Y., Lin, T.-N., Shie, F.-S., Kao, L.-S., Chou, C.-M., & Lee, Y.-H. (2016). Astrocytic GAP43 induced by the TLR4/NF-κB/STAT3 axis attenuates astrogliosis-mediated microglial activation and neurotoxicity. *The Journal of Neuroscience : The Official Journal of the Society for Neuroscience*, *36*(6), 2027–2043. DOI: 10.1523/JNEUROSCI.3457-15.2016 PMID: 26865625

Iskusnykh, I. Y., Zakharova, A. A., & Pathak, D. (2022). Glutathione in Brain Disorders and Aging. In *Molecules* (Vol. 27, Issue 1). MDPI. DOI: 10.3390/molecules27010324

Jalouli, M., Rahman, M. A., Biswas, P., Rahman, H., Harrath, A. H., Lee, I. S., Kang, S., Choi, J., Park, M. N., & Kim, B. (2025). Targeting natural antioxidant polyphenols to protect neuroinflammation and neurodegenerative diseases: a comprehensive review. In *Frontiers in Pharmacology* (Vol. 16). Frontiers Media SA. DOI: 10.3389/fphar.2025.1492517

Jia, X., Mao, D., Guo, J., Ke, J., Zhu, Y., Zhao, X., Luo, Z., Liu, X., Tang, R., Hou, R., Lan, H., & zheng, Q. (2024). Epigallocatechin gallate attenuated high glucose-induced pancreatic beta cell dysfunction by modulating DRP1-mediated mitochondrial apoptosis pathways. *Scientific Reports*, *14*(1), 16809. DOI: 10.1038/s41598-024-67867-0 PMID: 39039202

Kim, J.-E., Kim, T.-H., & Kang, T.-C. (2023). EGCG Attenuates CA1 Neuronal Death by Regulating GPx1, NF-κB S536 Phosphorylation and Mitochondrial Dynamics in the Rat Hippocampus following Status Epilepticus. *Antioxidants*, *12*(4), 966. Advance online publication. DOI: 10.3390/antiox12040966 PMID: 37107343

Kim, S.-R., Seong, K.-J., Kim, W.-J., & Jung, J.-Y. (2022). Epigallocatechin Gallate Protects against Hypoxia-Induced Inflammation in Microglia via NF-κB Suppression and Nrf-2/HO-1 Activation. *International Journal of Molecular Sciences*, *23*(7), 4004. Advance online publication. DOI: 10.3390/ijms23074004 PMID: 35409364

Klemmensen, M. M., Borrowman, S. H., Pearce, C., Pyles, B., & Chandra, B. (2024). Mitochondrial dysfunction in neurodegenerative disorders. *Neurotherapeutics; the Journal of the American Society for Experimental NeuroTherapeutics*, *21*(1), e00292. DOI: 10.1016/j.neurot.2023.10.002 PMID: 38241161

Kormas, P., & Moutzouri, A. (2023). Current psychological approaches in neurodegenerative diseases. In *Handbook of computational neurodegeneration* (pp. 261–289). Springer. DOI: 10.1007/978-3-319-75922-7_10

Kung, H.-C., Lin, K.-J., Kung, C.-T., & Lin, T.-K. (2021). Oxidative stress, mitochondrial dysfunction, and neuroprotection of polyphenols with respect to resveratrol in Parkinson's disease. *Biomedicines*, *9*(8), 918. DOI: 10.3390/biomedicines9080918 PMID: 34440122

Lahooti, B., Chhibber, T., Bagchi, S., Varahachalam, S. P., & Jayant, R. D. (2021). Therapeutic role of inflammasome inhibitors in neurodegenerative disorders. *Brain, Behavior, and Immunity*, *91*, 771–783. DOI: 10.1016/j.bbi.2020.11.004 PMID: 33157255

Landete, J. M. (2012). Plant and mammalian lignans: A review of source, intake, metabolism, intestinal bacteria and health. *Food Research International*, *46*(1), 410–424. DOI: 10.1016/j.foodres.2011.12.023

Leng, F., & Edison, P. (2021). Neuroinflammation and microglial activation in Alzheimer disease: Where do we go from here? *Nature Reviews. Neurology*, *17*(3), 157–172. DOI: 10.1038/s41582-020-00435-y PMID: 33318676

Li, K., Li, J., Zheng, J., & Qin, S. (2019). Reactive astrocytes in neurodegenerative diseases. In *Aging and Disease*. DOI: 10.14336/AD.2018.0720

Liu, L., Liu, J., Bao, J., Bai, Q., & Wang, G. (2020). Interaction of microglia and astrocytes in the neurovascular unit. *Frontiers in Immunology*, *11*, 1024. DOI: 10.3389/fimmu.2020.01024 PMID: 32733433

Liu, X., Tang, M., He, T.-Y., Zhao, S., Li, H.-Z., Li, Z., Guo, Y.-X., & Wang, X.-L. (2023). Resveratrol Improves Paclitaxel-Induced Cognitive Impairment in Mice by Activating SIRT1/PGC-1α Pathway to Regulate Neuronal State and Microglia Cell Polarization. *Drug Design, Development and Therapy*, *17*, 1125–1138. DOI: 10.2147/DDDT.S400936 PMID: 37077409

Lovell, M. A., & Markesbery, W. R. (2007). Oxidative DNA damage in mild cognitive impairment and late-stage Alzheimer's disease. *Nucleic Acids Research*, *35*(22), 7497–7504. DOI: 10.1093/nar/gkm821 PMID: 17947327

Mangrulkar, S. V., Wankhede, N. L., Kale, M. B., Upaganlawar, A. B., Taksande, B. G., Umekar, M. J., Anwer, M. K., Dailah, H. G., Mohan, S., & Behl, T. (2023). Mitochondrial dysfunction as a signaling target for therapeutic intervention in major neurodegenerative disease. *Neurotoxicity Research*, *41*(6), 708–729. DOI: 10.1007/s12640-023-00647-2 PMID: 37162686

Marques, D., Moura-Louro, D., Silva, I. P., Matos, S., Dos Santos, C. N., & Figueira, I. (2024). Unlocking the potential of low-molecular-weight (Poly)phenol metabolites: Protectors at the blood-brain barrier frontier. *Neurochemistry International, 179*, 105836. DOI: 10.1016/j.neuint.2024.105836 PMID: 39151552

Martemucci, G., Costagliola, C., Mariano, M., D'andrea, L., Napolitano, P., & D'Alessandro, A. G. (2022). Free Radical Properties, Source and Targets, Antioxidant Consumption and Health. In *Oxygen* (Vol. 2, Issue 2, pp. 48–78). Multidisciplinary Digital Publishing Institute (MDPI). DOI: 10.3390/oxygen2020006

Mas-Bargues, C., Escrivá, C., Dromant, M., Borrás, C., & Viña, J. (2021). Lipid peroxidation as measured by chromatographic determination of malondialdehyde. Human plasma reference values in health and disease. In *Archives of Biochemistry and Biophysics* (Vol. 709). Academic Press Inc., DOI: 10.1016/j.abb.2021.108941

Meng, K., Jia, H., Hou, X., Zhu, Z., Lu, Y., Feng, Y., Feng, J., Xia, Y., Tan, R., Cui, F., & Yuan, J. (2025). Mitochondrial Dysfunction in Neurodegenerative Diseases: Mechanisms and Corresponding Therapeutic Strategies. *Biomedicines, 13*(2), 1–30. DOI: 10.3390/biomedicines13020327 PMID: 40002740

Meng, T., Xiao, D., Muhammed, A., Deng, J., Chen, L., & He, J. (2021). Anti-Inflammatory Action and Mechanisms of Resveratrol. *Molecules (Basel, Switzerland), 26*(1), 229. Advance online publication. DOI: 10.3390/molecules26010229 PMID: 33466247

Mohammad, Z. B., Yudin, S. C. Y., Goldberg, B. J., Serra, K. L., & Klegeris, A. (2025). Exploring neuroglial signaling: Diversity of molecules implicated in microglia-to-astrocyte neuroimmune communication. *Reviews in the Neurosciences, 36*(1), 91–117. DOI: 10.1515/revneuro-2024-0081 PMID: 39240134

Mokra, D., Joskova, M., & Mokry, J. (2023). Therapeutic Effects of Green Tea Polyphenol (–)-Epigallocatechin-3-Gallate (EGCG) in Relation to Molecular Pathways Controlling Inflammation, Oxidative Stress, and Apoptosis. *International Journal of Molecular Sciences, 24*(1), 340. Advance online publication. DOI: 10.3390/ijms24010340 PMID: 36613784

Monkkonen, T., & Debnath, J. (2018). Inflammatory signaling cascades and autophagy in cancer. In *Autophagy*. DOI: 10.1080/15548627.2017.1345412

Mouria, M., Gukovskaya, A. S., Jung, Y., Buechler, P., Hines, O. J., Reber, H. A., & Pandol, S. J. (2002). Food-derived polyphenols inhibit pancreatic cancer growth through mitochondrial cytochrome C release and apoptosis. *International Journal of Cancer, 98*(5), 761–769. DOI: 10.1002/ijc.10202 PMID: 11920648

Muhammad, M. (2019). Tumor Necrosis Factor Alpha: A Major Cytokine of Brain Neuroinflammation. In Behzadi, P. (Ed.), *Cytokines*. IntechOpen., DOI: 10.5772/intechopen.85476

Naoi, M., Wu, Y., Shamoto-Nagai, M., & Maruyama, W. (2019). Mitochondria in Neuroprotection by Phytochemicals: Bioactive Polyphenols Modulate Mitochondrial Apoptosis System, Function and Structure. *International Journal of Molecular Sciences, 20*(10), 2451. Advance online publication. DOI: 10.3390/ijms20102451 PMID: 31108962

Nemzer, B. V., Al-Taher, F., Kalita, D., Yashin, A. Y., & Yashin, Y. I. (2025). Health-improving effects of polyphenols on the human intestinal microbiota: A review. *International Journal of Molecular Sciences, 26*(3), 1335. DOI: 10.3390/ijms26031335 PMID: 39941107

Niedzielska, E., Smaga, I., Gawlik, M., Moniczewski, A., Stankowicz, P., Pera, J., & Filip, M. (2016). Oxidative Stress in Neurodegenerative Diseases. In *Molecular Neurobiology* (Vol. 53, Issue 6, pp. 4094–4125). Humana Press Inc. DOI: 10.1007/s12035-015-9337-5

Olufunmilayo, E. O., Gerke-Duncan, M. B., & Holsinger, R. M. D. (2023). Oxidative stress and antioxidants in neurodegenerative disorders. *Antioxidants*, *12*(2), 517. DOI: 10.3390/antiox12020517 PMID: 36830075

Ozdal, T., Sela, D. A., Xiao, J., Boyacioglu, D., Chen, F., & Capanoglu, E. (2016). The reciprocal interactions between polyphenols and gut microbiota and effects on bioaccessibility. *Nutrients*, *8*(2), 78. DOI: 10.3390/nu8020078 PMID: 26861391

Panche, A. N., Diwan, A. D., & Chandra, S. R. (2016). Flavonoids: An overview. *Journal of Nutritional Science*, *5*, e47. DOI: 10.1017/jns.2016.41 PMID: 28620474

Peggion, C., Calì, T., & Brini, M. (2024). Mitochondria dysfunction and neuroinflammation in neurodegeneration: Who comes first? *Antioxidants*, *13*(2), 240. DOI: 10.3390/antiox13020240 PMID: 38397838

Peng, J., Yang, Z., Li, H., Hao, B., Cui, D., Shang, R., Lv, Y., Liu, Y., Pu, W., Zhang, H., He, J., Wang, X., & Wang, S. (2023). Quercetin Reprograms Immunometabolism of Macrophages via the SIRT1/PGC-1α Signaling Pathway to Ameliorate Lipopolysaccharide-Induced Oxidative Damage. *International Journal of Molecular Sciences*, *24*(6), 5542. Advance online publication. DOI: 10.3390/ijms24065542 PMID: 36982615

Perluigi, M., Di Domenico, F., & Butterfield, D. A. (2024). OXIDATIVE DAMAGE IN NEURODEGENERATION: ROLES IN THE PATHOGENESIS AND PROGRESSION OF ALZHEIMER DISEASE. In *Physiological Reviews* (Vol. 104, Issue 1, pp. 103–197). American Physiological Society. DOI: 10.1152/physrev.00030.2022

Pluta, R., Kiś, J., Januszewski, S., Jabłoński, M., & Czuczwar, S. J. (2022). Cross-talk between amyloid, tau protein and free radicals in post-ischemic brain neurodegeneration in the form of Alzheimer's disease proteinopathy. *Antioxidants*, *11*(1), 146. DOI: 10.3390/antiox11010146 PMID: 35052650

Quesada-Vázquez, S., Eseberri, I., Les, F., Pérez-Matute, P., Herranz-López, M., Atgié, C., Lopez-Yus, M., Aranaz, P., Oteo, J. A., Escoté, X., Lorente-Cebrian, S., Roche, E., Courtois, A., López, V., Portillo, M. P., Milagro, F. I., & Carpéné, C. (2024). Polyphenols and metabolism: From present knowledge to future challenges. *Journal of Physiology and Biochemistry*, *80*(3), 603–625. DOI: 10.1007/s13105-024-01046-7 PMID: 39377969

Ren, Z., Wang, L., Cui, J., Huoc, Z., Xue, J., Cui, H., Mao, Q., & Yang, R. (2013). Resveratrol inhibits NF-kB signaling through suppression of p65 and IkappaB kinase activities. *Die Pharmazie*, *68*(8), 689–694. PMID: 24020126

Rudrapal, M., Rakshit, G., Singh, R. P., Garse, S., Khan, J., & Chakraborty, S. (2024). Dietary Polyphenols: Review on Chemistry/Sources, Bioavailability/Metabolism, Antioxidant Effects, and Their Role in Disease Management. In *Antioxidants* (Vol. 13, Issue 4, p. 429). DOI: 10.3390/antiox13040429

Rummel, N. G., & Butterfield, D. A. (2022). Altered metabolism in Alzheimer disease brain: Role of oxidative stress. *Antioxidants & Redox Signaling*, *36*(16–18), 1289–1305. DOI: 10.1089/ars.2021.0177 PMID: 34416829

Sadiq, I. Z. (2023). Free radicals and oxidative stress: Signaling mechanisms, redox basis for human diseases, and cell cycle regulation. *Current Molecular Medicine*, *23*(1), 13–35. DOI: 10.2174/1566524022666211222161637 PMID: 34951363

Sahadevan, R., Binoy, A., Shajan, I., & Sadhukhan, S. (2023). Mitochondria-targeting EGCG derivatives protect H9c2 cardiomyocytes from H2O2-induced apoptosis: Design, synthesis and biological evaluation. *RSC Advances*, *13*(42), 29477–29488. DOI: 10.1039/D3RA04527G PMID: 37818277

Salla, M., Karaki, N., El Kaderi, B., Ayoub, A. J., Younes, S., Abou Chahla, M. N., Baksh, S., & El Khatib, S. (2024). Enhancing the Bioavailability of Resveratrol: Combine It, Derivatize It, or Encapsulate It? *Pharmaceutics*, *16*(4), 569. DOI: 10.3390/pharmaceutics16040569 PMID: 38675230

Selvaraj, N. R., Nandan, D., Nair, B. G., Nair, V. A., Venugopal, P., & Aradhya, R. (2025). Oxidative Stress and Redox Imbalance: Common Mechanisms in Cancer Stem Cells and Neurodegenerative Diseases. In *Cells* (Vol. 14, Issue 7). Multidisciplinary Digital Publishing Institute (MDPI). DOI: 10.3390/cells14070511

Shaito, A., Al-Mansoob, M., Ahmad, S. M. S., Haider, M. Z., Eid, A. H., Posadino, A. M., Pintus, G., & Giordo, R. (2022). Resveratrol-Mediated Regulation of Mitochondria Biogenesis-associated Pathways in Neurodegenerative Diseases: Molecular Insights and Potential Therapeutic Applications. *Current Neuropharmacology*, *21*(5), 1184–1201. DOI: 10.2174/1570159X20666221012122855 PMID: 36237161

Singh, A., Kukreti, R., Saso, L., & Kukreti, S. (2019). Oxidative stress: A key modulator in neurodegenerative diseases. In *Molecules* (Vol. 24, Issue 8). MDPI AG. DOI: 10.3390/molecules24081583

Singh, G., Kaur, J., Kaur, M., Singh, P., & Bhatti, R. (2020). Anti-nociceptive and anti-inflammatory effect of imperatorin: Evidences for involvement of COX-2, iNOS, NFκB and inflammatory cytokines. *The International Journal of Neuroscience*, *130*(2), 176–185. DOI: 10.1080/00207454.2019.1667789 PMID: 31524564

Singh, S., & Singh, T. G. (2020). Role of nuclear factor kappa B (NF-κB) signalling in neurodegenerative diseases: An mechanistic approach. *Current Neuropharmacology*, *18*(10), 918–935. DOI: 10.2174/1570159X18666200207120949 PMID: 32031074

Sirangelo, I., Liccardo, M., & Iannuzzi, C. (2022). Hydroxytyrosol Prevents Doxorubicin-Induced Oxidative Stress and Apoptosis in Cardiomyocytes. *Antioxidants*, *11*(6), 1087. Advance online publication. DOI: 10.3390/antiox11061087 PMID: 35739984

Suraweera, T. L., Vasantha Rupasinghe, H. P., Dellaire, G., & Xu, Z. (2020). Regulation of Nrf2/are pathway by dietary flavonoids: A friend or foe for cancer management? In *Antioxidants* (Vol. 9, Issue 10, pp. 1–44). MDPI. DOI: 10.3390/antiox9100973

Vauzour, D. (2012). Dietary polyphenols as modulators of brain functions: Biological actions and molecular mechanisms underpinning their beneficial effects. *Oxidative Medicine and Cellular Longevity*, *914273*, 1–16. Advance online publication. DOI: 10.1155/2012/914273 PMID: 22701758

Vauzour, D., Rodriguez-Mateos, A., Corona, G., Oruna-Concha, M. J., & Spencer, J. P. E. (2010). Polyphenols and human health: Prevention of disease and mechanisms of action. *Nutrients*, *2*(11), 1106–1131. DOI: 10.3390/nu2111106 PMID: 22254000

Vezzani, B., Carinci, M., Patergnani, S., Pasquin, M. P., Guarino, A., Aziz, N., Pinton, P., Simonato, M., & Giorgi, C. (2020). The dichotomous role of inflammation in the CNS: A mitochondrial point of view. *Biomolecules*, *10*(10), 1–36. DOI: 10.3390/biom10101437 PMID: 33066071

Wang, Q., Xue, H., Yue, Y., Hao, S., Huang, S.-H., & Zhang, Z. (2022). Role of mitophagy in the neurodegenerative diseases and its pharmacological advances: A review. *Frontiers in Molecular Neuroscience*, *15*, 1014251. DOI: 10.3389/fnmol.2022.1014251 PMID: 36267702

Wang, Q., Yang, S., Zhang, X., Zhang, S., Chen, L., Wang, W., Chen, N., & Yan, J. (2024). Inflammasomes in neurodegenerative diseases. *Translational Neurodegeneration*, *13*(1), 65. DOI: 10.1186/s40035-024-00459-0 PMID: 39710713

Williamson, G. (2025). Bioavailability of Food Polyphenols: Current State of Knowledge. *Annual Review of Food Science and Technology*, *16*(1), 315–332. DOI: 10.1146/annurev-food-060721-023817 PMID: 39899845

Winiarska-Mieczan, A., Kwiecień, M., Jachimowicz-Rogowska, K., Donaldson, J., Tomaszewska, E., & Baranowska-Wójcik, E. (2023). Anti-Inflammatory, Antioxidant, and Neuroprotective Effects of Polyphenols-Polyphenols as an Element of Diet Therapy in Depressive Disorders. *International Journal of Molecular Sciences*, *24*(3), 2258. Advance online publication. DOI: 10.3390/ijms24032258 PMID: 36768580

Wu, S. K., Wang, L., Wang, F., & Zhang, J. (2024). Resveratrol improved mitochondrial biogenesis by activating SIRT1/PGC-1α signal pathway in SAP. *Scientific Reports*, *14*(1), 1–15. DOI: 10.1038/s41598-024-76825-9 PMID: 38167627

Yang, X., Xu, S., Qian, Y., & Xiao, Q. (2017). Resveratrol regulates microglia M1/M2 polarization via PGC-1α in conditions of neuroinflammatory injury. *Brain, Behavior, and Immunity*, *64*, 162–172. DOI: 10.1016/j.bbi.2017.03.003 PMID: 28268115

Yin, H., Guo, Q., Li, X., Tang, T., Li, C., Wang, H., Sun, Y., Feng, Q., Ma, C., Gao, C., Yi, F., & Peng, J. (2018). Curcumin Suppresses IL-1β Secretion and Prevents Inflammation through Inhibition of the NLRP3 Inflammasome. *Journal of Immunology (Baltimore, Md. : 1950)*, *200*(8), 2835–2846. DOI: 10.4049/jimmunol.1701495

Zhang, H., Zhang, X., Chai, Y., Wang, Y., Zhang, J., & Chen, X. (2025). Astrocyte-mediated inflammatory responses in traumatic brain injury: Mechanisms and potential interventions. *Frontiers in Immunology*, *16*, 1584577. DOI: 10.3389/fimmu.2025.1584577 PMID: 40406119

Zhao, J., Bi, W., Xiao, S., Lan, X., Cheng, X., Zhang, J., Lu, D., Wei, W., Wang, Y., Li, H., Fu, Y., & Zhu, L. (2019). Neuroinflammation induced by lipopolysaccharide causes cognitive impairment in mice. *Scientific Reports*, *9*(1), 5790. DOI: 10.1038/s41598-019-42286-8 PMID: 30962497

Zheng, Y., Sun, J., Luo, Z., Li, Y., & Huang, Y. (2024). Emerging mechanisms of lipid peroxidation in regulated cell death and its physiological implications. In *Cell Death and Disease* (Vol. 15, Issue 11). Springer Nature. DOI: 10.1038/s41419-024-07244-x

Zhou, J., Yang, Z., Shen, R., Zhong, W., Zheng, H., Chen, Z., Tang, J., & Zhu, J. (2021). Resveratrol Improves Mitochondrial Biogenesis Function and Activates PGC-1α Pathway in a Preclinical Model of Early Brain Injury Following Subarachnoid Hemorrhage. *Frontiers in Molecular Biosciences*, *8*, 620683. DOI: 10.3389/fmolb.2021.620683 PMID: 33968980

Zhu, G., Wang, X., Chen, L., Lenahan, C., Fu, Z., Fang, Y., & Yu, W. (2022). Crosstalk between the oxidative stress and glia cells after stroke: From mechanism to therapies. *Frontiers in Immunology*, *13*, 852416. DOI: 10.3389/fimmu.2022.852416 PMID: 35281064

Zong, Y., Li, H., Liao, P., Chen, L., Pan, Y., Zheng, Y., Zhang, C., Liu, D., Zheng, M., & Gao, J. (2024). Mitochondrial dysfunction: Mechanisms and advances in therapy. *Signal Transduction and Targeted Therapy*, *9*(1), 124. DOI: 10.1038/s41392-024-01839-8 PMID: 38744846

Chapter 12
Probiotics and Human Microbiome:
Therapeutic Potentials in Cancer, Cardiovascular Disease, and Cognitive Disorders

Saba Mehboob
https://orcid.org/0009-0006-3598-8516
Government College University, Faisalabad, Pakistan

Muhammad Awais
Government College University, Faisalabad, Pakistan

Ayesha Shafqat
Government College University, Faisalabad, Pakistan

Arslan Iftikhar
https://orcid.org/0000-0002-7113-3455
Government College University, Faisalabad, Pakistan

Humaira Muzaffar
https://orcid.org/0000-0003-2172-8474
Government College University, Faisalabad, Pakistan

Maham Fatima
Government College University, Faisalabad, Pakistan

Moin Akhtar
Government College University, Faisalabad, Pakistan

Bushra Basharat
Government College University, Faisalabad, Pakistan

Maryam Mustafa
Department of Physiology, Government College University, Faisalabad, Pakistan

ABSTRACT

Human microbiota is a diverse and rich ecosystem composed of trillion microorganisms, such as bacteria, fungi, and viruses. These microbial populations reside at different locations of the body including the skin, mouth, and more importantly, gastrointestinal (GI) tract. The gut microbiota is among them and serves the main physiologic homeostasis, supportive metabolic processes, immune-modulatory and defensive functions against the invasion of pathogenic organisms. The term probiotics refers to live

DOI: 10.4018/979-8-3373-4139-2.ch012

microorganisms which when taken in sufficient quantities exert a number of health benefits on the host. They are important to the gut health, as they provide necessary micronutrients, enzymes, and vitamins (this includes B complex and vitamin K that is vital to digestive processes and mucosal integrity). In this chapter, the association of human microbiota with probiotics is highlighted and the use of probiotics to rebalance the microbe in the host is demonstrated as contributing to good health.

1. INTRODUCTION

In the fast-hustled life today, a lot of people are found to lack sufficient sleep because their mind is engaged in office work or business related duties till late evening hours. This causes a great imbalance in their sleep and typically causes poor diet habits. Being limited by time, people are prone to excessive consumption of fast food and ready-to-eat dishes rich in harmful components, and poor in such essential nutrients. These packaged foods that have extended shelf life and improved taste are generally deep fried and heavily loaded with salt and sugar. The processed foods, therefore, lead to malnutrition, but are still preferred because of their availability in supermarkets. Such a breakdown in life-style – manifested by lack of adequate sleep and poor nutrition has emerged as one of the primary causes of many health conditions including chronic diseases like cancer. The presence of probiotics, beneficial microorganisms, is a strong indicator of digestive wellness, a state which reflects physical and mental wellbeing overall. The rising awareness of the lifestyle-based health issues has fueled a greater demand for healthier food products such as yogurt, a probiotic rich product, that is nutrient dense and easily digestible. By microbiota we mean the beneficial microorganisms that reside in different sites in the human body, such as the colon, skin, lungs, and oral cavity. These may be bacteria, viruses, yeast or a combination of those. Human body has two genomes. the human genome, and the microbial genome. Together, all the genetic material of such microorganisms is called the microbiome that possesses much more genes than human genome itself and is critical for sustaining health and physiologic balance (Grice & Segre, 2012). A collection of microbial populations which form an assemblage around the host are commonly known as the microbiota. Microorganisms synthesis various minerals like Nitrogen [N], phosphorus [P], and metabolize in the hosts shortly after birth (Peterson et al., 2009). In the gut, the resident bacteria have become suited to their local environment through developing complex interrelations with the host and other microbial communities, that allow them to gain the nutrients they need to survive (Clemente et al., 2012). Further, microorganisms within the gut not only compete with each other for space and nutrients but also interfere with the colonization of pathogenic or foreign microbes thus maintaining the microbial balance as well as host protection (Kamada et al., 2013). The gut microbiota of man consists of more than 100 trillion cells of microorganisms, which are important for the breakdown of complex carbohydrates like plant polysaccharides, starch. The human intestine is deficient in the endogenous enzymes needed for starch digestion, so these microbes play a critical compensatory role. The production of a range of short-chain fatty acids (SCFAs) including butyrate, acetate and propionate, which are main microbial metabolite by-products, promotes the proliferation and metabolic activity of gut microbiota specifically within the large intestine. These SCFAs not only behave as the source of energy for colonocytes but also demonstrate the anti-inflammatory functions and immune modulation.

As a whole, the gut microbiota plays a key role in intestinal as well as overall health maintenance. This microbiota imbalance, termed as dysbiosis, has been closely associated with a variety of gastrointestinal disorders such as inflammatory bowel disease (IBD), colon cancer, Crohn's disease, and ulcerative colitis (Qin, J., et al. 2010). According to the World Health Organization (WHO), probiotics are live microorganisms, mostly bacteria and yeast, which, when given in adequate amounts, provide health benefits to the host (Sanders et al., 2011). These beneficial microbes are now better understood for their multi-faceted roles in human health, going far beyond the gut to affect cognitive functions. New studies indicate that probiotics can help maintain brain functions by improving cognitive performance, contributing to the gut-brain axis, a two-way communication system between the gastrointestinal tract and the central nervous system (Kelly et al., 2017). Probiotics are especially good at ensuring digestive system health because they participate in the breakdown of food and prevent the pathogenic microorganisms from colonizing the gastrointestinal tract. Besides, they have been proven to relieve some skin conditions like atopic dermatitis, an inflammatory condition that involves dry, itchy skin (Avena-Woods, 2017). Probiotic positive effects are mediated through various mechanisms. Examples include the suppression of harmful bacterial growth in the gut, the creation of vital short-chain acids and vitamins as well as an improvement of the intestinal barrier. Additionally, certain probiotic strains induce the immune system by releasing interferons and interleukins as immunomodulatory agents in assisting to control immune responses and achieve immune homeostasis (Hill et al., 2014).

1.1 Probiotics, Prebiotic, and Postbiotic

Probiotic bacteria include *Lactobacillus, Bifidobacterium, Enterococcus*, and *Streptococcus* (Hill et al., 2014) as shown in Figure 1. Beneficial yeast species such as Saccharomyces cerevisiae are also classified under probiotic. Probiotics affect gut microbiota in a positive way as they elevate the populations of beneficial microbes that consume available nutrients more productively. This inhibits the well-being of, and the growth of pathogenic micro-organisms consequently reducing their occurrence in the intestines. In addition, these favorable microbes assist in fortifying the immunity of the intestines as well as regulating the responses of the immune systems (Mary Ellen, S., et al. 2013).

Prebiotics are non digestible food components that are usually fibers (e.g. Inulin or Fructooligosaccharides) that are a nutrient fuel for probiotics. Prebiotics play an important role in the maintenance and promotion of gut health due to their stimulation effects on beneficial gut bacteria proliferation and activity (Gibson et al., 2017).

Postbiotics or metabiotics as they are also known are the bioactive, soluble compounds formed when probiotics undergo metabolic processes and fermentation. These compounds - short chain fatty acids, enzymes, peptides and vitamins – are critical in sustaining the integrity of the gut as well as modulating immunity and general health of the gastrointestinal tract (Salminen et al., 2021).

Probiotic strains exhibit distinct physiological effects; therefore, strain-specific assessment is essential in both research and clinical practice. As an example, *Lactobacillus rhamnosus* GG was repeatedly identified in the context of the reduction of gastrointestinal infections, whereas *Bifidobacterium longum* was proved effective in the symptoms of irritable bowel syndrome (Steele et al., 2023; Zhang et al., 2023). Recent clinical findings emphasize that the effects across strains, dosage and populations of patients are very different and it is important to be precise in probiotic recommendations (Huang et al., 2024).

Although probiotics are generally considered safe, recent studies have shown negative effects in susceptible groups. Probiotics can also cause bacteremia or fungemia as a result of immunocompromised patients (Chen et al., 2023). The most frequent DSS symptoms include gastrointestinal discomfort (bloating and gas), and infrequent but severe infections are associated with improper use of probiotics

(Martínez et al., 2022). The regulatory bodies have begun to emphasize strain-specific safety testing before therapeutic use is approved.

Figure 1. Probiotic bacteria include lactobacillus, bifidobacterium, enterococcus, and streptococcus

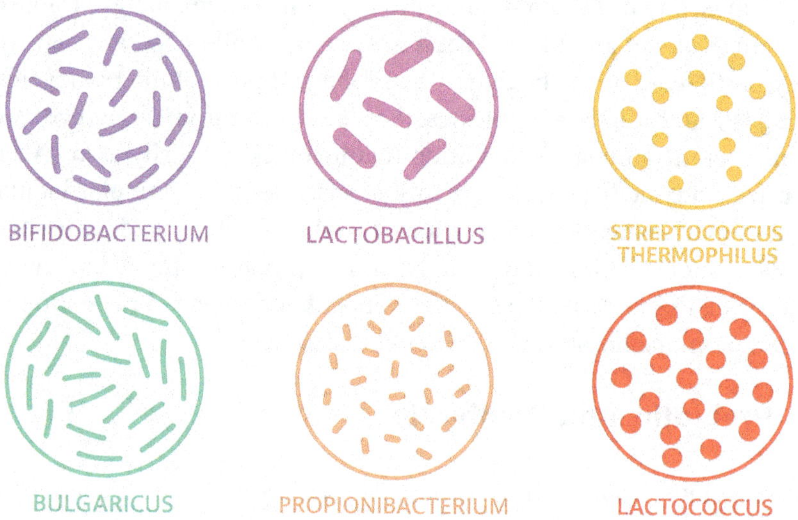

Table 1. Classification of probiotics and their health benefits

Probiotic Strain	Type	Key Health Benefits	References
Lactobacillus & Bifidobacterium	Bacteria	Enhance gut barrier function, aid in metabolite production, modulate immune responses, and exhibit anti-inflammatory effects.	Vogt et al., 2013
Lactobacillus rhamnosus GG (LGG)	Bacteria	Known for its therapeutic potential in treating gastrointestinal disorders.	(Steele et al., 2023)
Saccharomyces boulardii	Yeast	Commonly used to manage and relieve gastrointestinal tract disorders.	(Taj et al., 2022)
Streptococcus thermophilus	Bacteria	Used in combination with other probiotics; supports fermentation and enhances overall probiotic effectiveness.	(Sharma et al., 2014)
Escherichia coli (E. coli)	Bacteria	Certain strains demonstrate probiotic properties with potential health benefits.	(Secher et al., 2017)
Faecalibacterium prausnitzii	Bacteria	Lower abundance linked with IBD; contributes to intestinal health through anti-inflammatory activity.	(Rowland et al., 2018)
Akkermansia muciniphila	Bacteria	Promotes gut barrier function and reduces the risk of metabolic diseases; supports a balanced microbiome.	(Hansen et al., 2012)

1.2 Origin of Probiotic Food

Probiotic foods are those that contain live microorganisms beneficial to human health, particularly in maintaining gut microbiota. These foods have been part of traditional diets across various cultures for centuries. Some of the most common sources include dairy products, fermented foods, and certain plant-based items. (Liu et al., 2020)

Table 2. Major probiotic food sources and their benefits

Probiotic Food	Main Microorganisms / Components	Health Benefits
Yogurt	*Lactobacillus, Bifidobacterium*	Supports gut health, improves digestion
Kefir	Yeasts and carboxylic acid bacteria	Enhances immune system, aids lactose digestion
Curd (Dahi)	Contains ~250 strains of *Lactobacillus*	Promotes intestinal health
Paneer (Cottage Cheese)	Mild source	Good for heart, bones, and teeth
Buttermilk	Carboxylic and lactic acid bacteria	Improves digestion, reduces acidity
Raw Cheese	Natural bacterial cultures	Enhances gut microbiota
Processed Cheese	Often lacks live cultures	Not considered a probiotic source
Idli & Dosa	Fermented with lactic acid bacteria	Easily digestible due to fiber; contains antioxidants
Apple Cider Vinegar	Acetic acid bacteria	Detoxification, improves gut health
Green Peas & Apples	Natural fibers & prebiotics	Support growth of probiotic bacteria
Whole Wheat Bread	Contains prebiotics (if fermented)	Aids digestion and supports gut health
Smoothies (with probiotics)	Varies	Improves nutrient absorption, boosts energy (Zhang et al., 2023)

Probiotics also produce **bioactive metabolites** that contribute to various physiological processes:

Table 3. Probiotic metabolites and their functions

Metabolite Type	Examples	Functions
Proteins	Secreted enzymes	Aid in nutrient breakdown and absorption
Indole	Derived from tryptophan	Regulates intestinal barrier function
Extracellular Vesicles	Nano-carriers of bacterial molecules	Communication with host cells, immune modulation
Bacteriocins	Antimicrobial peptides	Fight pathogens
Fatty Acids & Short-Chain FAs	Acetate, Propionate, Butyrate	Energy source for colon cells, anti-inflammatory actions Secher et al., 2017)

1.3 Mechanism of Action of Probiotics

Probiotic activity on host health is moderated by a variety of mechanisms, most of these effects are associated with the gastrointestinal (GI) tract and the immune system. These probiotics are important in the homeostasis in the intestines and in the improvement of mucosal immunity. Probiotics, after consumed, can temporarily colonize the human gut mucosa. Nevertheless, such colonization is exceptionally species specific, with a great degree of reliance on the former microbiome structure of the host and the area of the GI tract, which will be colonized. The transient nature of the presence of probiotics in the gut is enough to create favorable biological effects though probiotics fail to establish permanent residence in the gut. It has been demonstrated that such effects depend on the individual pre-existing body of microbes therefore implying a more individualized trend of probiotic gut functionality (Zmora et al., 2018)

Activation of intestinal immune system is one of the most important immunological effects of probiotics. Namely, probiotics boost the mucosal immunity by regulating Immunoglobulin A (IgA) production which is a critical antibody in defending mucosal surface. This is done by using B lymphocytes to develop clonal expansion and differentiation to a type of plasma cells, IgA, to secrete. The surge of the IgA-producing cells is an aid to better immovable surveillance and safety against pathogens in the intestinal microbiome.Remarkably, on the one hand, the probiotics induce a strong stimulus of proliferation in the B cells and the production of IgA, on the other hand, the probiotics do not seem to influence the number of CD4+ T cells. Such optimized modulation enables an increase in the desired mucosal immune response without impairing on a systemic immune equilibrium (Vitini et al., 2000).

In addition, the high levels of IgA increases the resistance of the gut mucosal barrier, especially the part in the luminal area. With the strengthening of this initial defensive line, probiotics contribute to the prevention of intrusion of harmful microorganisms and minimizing inflammation, which lead to the overall health of the gastrointestinal tract as illustrated in Figure 2.

Figure 2. Mechanism of action of probiotics

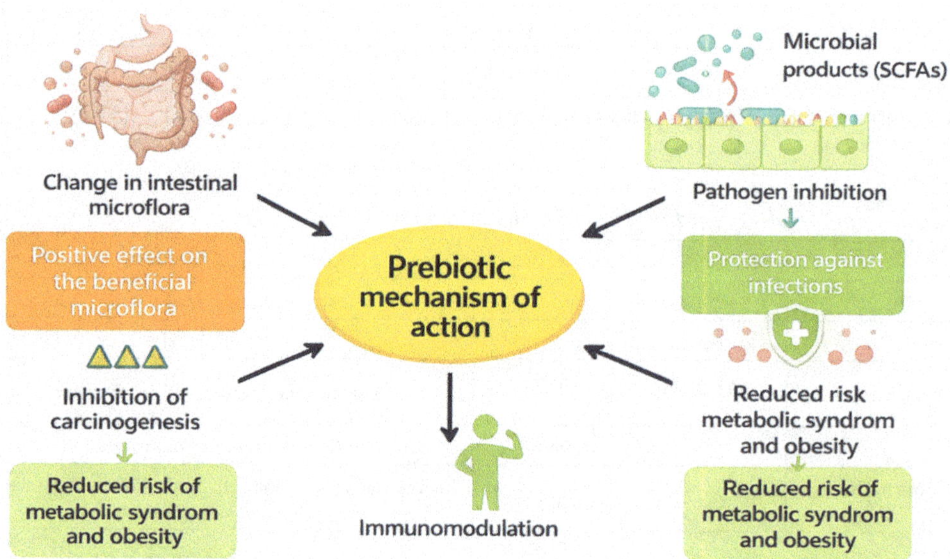

Table 4. Advantages and disadvantages of probiotics

Advantages	Disadvantages
Supports proper digestive system function	May cause gas, bloating, or diarrhoea, especially during initial use (Chen et al., 2021)
Maintains balance of beneficial gut bacteria	Can lead to skin-related issues such as rashes
Protects against diarrhoea, including antibiotic-associated diarrhoea	May trigger allergic reactions in sensitive individuals (Boleij et al., 2015)
Enhances mental health via the gut-brain axis	Risk of Small Intestinal Bacterial Overgrowth (SIBO)
Lowers LDL (bad cholesterol) and regulates blood pressure	Some strains produce amines that may cause headaches
Reduces severity of allergies and eczema	May not be suitable for immunocompromised individuals without medical advice
Helps manage digestive disorders like ulcerative colitis and Crohn's disease	Effectiveness may vary between individuals due to unique gut microbiota
Boosts immune system response	Overuse may disrupt natural gut flora balance
Aids in weight loss and reduction of belly fat	Some products may contain additives or allergens causing adverse effects

2 CLASSIFICATION AND ROLE OF HUMAN MICROBIOTA

Microbiota refers to the collective population of microorganisms, including bacteria, fungi, viruses, and archaea, that reside in and on the human body. The human microbiota comprises approximately **10 to 100 trillion** symbiotic microbial cells. In contrast, the term microbiome refers to the combined genetic material (genomes) of these microbes, along with their associated activities and functions. (Chen et al., 2021)

2.1 Metagenomic Analysis

Modern techniques such as metagenomic sequencing allow for the detailed analysis of microbial communities by studying their DNA directly from environmental samples, without the need for culturing. This provides insights into the genetic makeup, diversity, and functional roles of microbes within a specific site, such as the gut.

2.2 Common Microbial Members

Table 5. Bacterium functions

Domain	Microbial Groups	Examples
Bacteria	**Bacteroides**	Among the most abundant bacterial genera in the human gut; essential for carbohydrate metabolism and immune regulation (García-Mantrana et al. 2023).
	Firmicutes	Includes *Clostridium*, *Lactobacillus*, and *Ruminococcus*; major contributors to fermentation and short-chain fatty acid production (Al-Obaid et al. 2020).
	Actinobacteria	Includes *Bifidobacterium*; supports gut barrier integrity and metabolic balance, particularly in infants (García-Mantrana et al. 2023).
	Proteobacteria	Contains *Escherichia* and *Helicobacter*; often elevated in dysbiosis and inflammatory conditions.
Viruses	**Bacteriophages**	Infect bacterial cells and shape gut bacterial diversity; key in microbial community regulation.
Fungi	**Candida species**	Common members of the gut mycobiome; overgrowth associated with gastrointestinal and systemic disease.
Archaea	**Methanobrevibacter species** (*M. smithii, M. intestini*)	Methanogenic archaea involved in hydrogen consumption and intestinal gas metabolism.

2.3 Role in the Homeostatic and Hedonic Control of Appetite

The gut-brain axis refers to the bidirectional communication pathway between the gastrointestinal tract and the brain. This complex system allows the brain to regulate gut function and, conversely, enables the gut to influence brain activity. It plays a crucial role in the regulation of appetite, integrating both homeostatic and hedonic signals. (Molska & Reguła, 2019)

Homeostatic appetite regulation involves the maintenance of energy balance. It is governed by hormonal signals such as **ghrelin**, which stimulates hunger, and **leptin**, which promotes a sense of fullness. These hormones are secreted in response to nutritional status and are regulated by interactions between the gut and brain.

On the other hand, hedonic regulation of appetite pertains to the pleasure and reward aspects of eating. It is influenced by the delight, utility, and emotional satisfaction derived from food intake.

The gut microbiota contributes to both regulatory systems by influencing the production of neurotransmitters and signaling molecules such as serotonin, dopamine, and gamma-aminobutyric acid (GABA). These chemicals not only affect mood and emotional well-being but also shape food preferences, cravings, and eating behavior. Thus, the microbiota–gut–brain axis plays a central role in balancing physiological hunger cues with psychological factors, impacting both nutritional intake and mental health (Triantafillidis et al., 2009).

3 FACTORS INFLUENCING MICROBIOTA BALANCE

The composition and balance of the human gut microbiota are shaped by multiple factors throughout life. Early-life exposures and physiological conditions play a particularly important role in establishing and modulating microbial communities. Key influences include:

3.1 Birth Method

The mode of delivery significantly affects the initial microbial colonization in infants. Babies delivered vaginally are exposed to the mother's vaginal and intestinal microbiota, which helps establish a foundational community of beneficial bacteria. In contrast, infants born via cesarean section are often colonized by skin-associated bacteria, leading to a microbiota composition that differs from those born vaginally. These differences may influence immune development and health outcomes later in life (Dominguez-Bello et al., 2016).

3.2 Infant Feeding

Breast milk is a vital source of early-life microbiota, containing over 700 different bacterial species. Contrary to earlier beliefs, breast milk is not sterile; it plays a crucial role in seeding the infant gut with health-promoting bacteria. These microbial communities support the development of the immune system, in part by contributing to a proliferative pool of Immunoglobulin A (IgA) antibodies, which protect against pathogens and regulate immune responses (Boleij et al., 2015).

3.3. Genetics

The genetic makeup of the host significantly influences the composition of gut microbiota. Certain genes regulate immune responses, mucosal properties, and other physiological functions that create a specific environment for microbial colonization and growth. Thus, the host genome can modulate microbial diversity and stability (Ambalam et al., 2016).

3.4 Infections

Exposure to viral or bacterial infections can disrupt the normal microbial community in the gut. Pathogenic invasions may alter the abundance and function of resident bacteria, leading to microbial imbalance (dysbiosis) and potentially contributing to digestive or immune-related disorders.

4 DISEASE AND THE GUT MICROBIOTA

The gut microbiota plays a crucial role in maintaining gastrointestinal (GI) health, and disruptions to this microbial balance can contribute to or worsen several diseases. Below are examples of conditions closely linked to gut microbiota imbalance:

4.1 Necrotizing Enterocolitis (NEC)

Necrotizing enterocolitis is a serious gastrointestinal complication, primarily affecting premature infants. The condition typically develops within the first 2–3 weeks of life and is especially common in infants with very low birth weight (less than 2 pounds). These infants often have underdeveloped immune systems, making them more susceptible to severe infections.NEC causes the formation of holes in the intestinal wall, allowing bacteria to leak into the abdominal cavity or bloodstream, leading to life-threatening infection (Dominguez-Bello et al., 2016).

4.2 Rotavirus Infection

Rotavirus is a highly contagious viral infection that targets the gastrointestinal tract, especially in young children. The virus causes inflammation of the stomach's inner lining, leading to symptoms that usually appear 2–3 days after infection. It spreads primarily through hand-to-mouth contact, especially in settings with inadequate hygiene. There is currently no specific antiviral treatment for rotavirus, but symptoms can be managed. Common clinical features include:

- Vomiting
- Watery diarrhea
- Fever
- Loss of appetite
- Dehydration (e.g., dry mouth, sunken eyes)

Preventive measures such as rotavirus vaccination are recommended for children to reduce disease severity.

4.3 Enteric Bacterial Infections

Enteric bacterial infections affect the gastrointestinal tract, typically resulting from the ingestion of contaminated food or water. These infections are highly contagious and lead to inflammation of the digestive system, impairing its ability to properly process and absorb nutrients.

Symptoms may include:

- Nausea and vomiting
- Diarrhea
- Abdominal cramps and pain
- Fever

Treatment usually involves Rehydration therapy to prevent dehydration. Antimicrobial or antibiotic medications as prescribed, depending on the causative bacteria as shown in Figure 3.

Figure 3. Microbiome impact on gastrointestinal infections

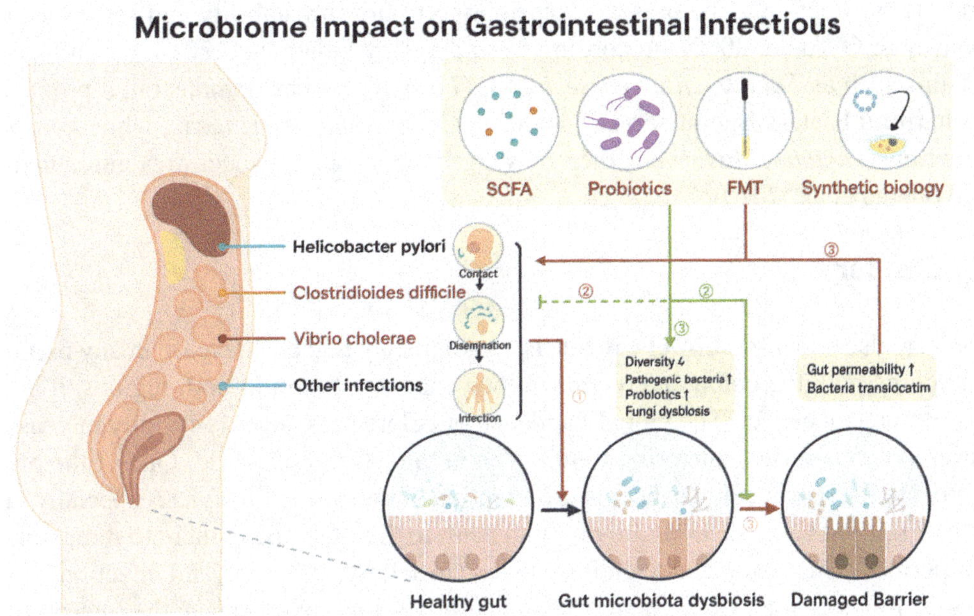

5 THE ROLE OF PROBIOTICS IN CANCER

The disruption of gut microbiota, particularly with elevated levels of certain bacterial genera such as *Bacteroides* and *Prevotella*, has been identified as a contributing factor in the development of cancers, notably colorectal cancer as shown in Figure 4. These bacterial genera can secrete carcinogenic metabolites and toxins that induce chronic inflammation, thereby promoting tumorigenesis. Among cancer patients, especially those with colorectal malignancies, there is often a notable deficiency of beneficial bacteria such as *Lactobacillus*, while pathogenic genera like *Salmonella* and *Clostridium* are more commonly found. In particular, *Bacteroides fragilis* produces a toxin known as fragilysin, which contributes to intestinal inflammation and cancer progression (Boleij et al., 2015). Furthermore, the presence of *Streptococcus gallolyticus* has been associated with colorectal cancer, as has *Enterococcus faecalis*, both of which are considered cancer-related bacterial species (Molska & Reguła, 2019).

Probiotic strains can play a protective role in this context. They form biofilms that adhere to the host cell membrane through a matrix, which serves as a barrier and communication system. The composition of this matrix varies depending on the microbial species and typically includes:

- Polysaccharides
- Proteins
- Phospholipids
- DNA and RNA

One key element is the structural polysaccharide, which is functionally significant. It aids in adhesion through glyosidic bonds and supports the integrity of the biofilm. Additionally, components such as collagen and elastin fibers may be present, contributing to biofilm stability and resistance to external stressors (Davey & O'Toole, 2000). According to findings by Ohara et al.2010, the intake of specific probiotic strains such as *Lactobacillus gasseri* OLL2716 (LG21) has demonstrated promising results in reducing harmful bacterial populations, including *Clostridium perfringens*, while concurrently increasing beneficial *Lactobacillus* levels, thereby suggesting a role in cancer prevention and gut health maintenance (Ohara et al., 2010).

5.1 Colon Cancer

Colon cancer, also known as **bowel cancer**, is a malignancy that can develop in any part of the large intestine, including both the colon and rectum. When cancer is found in both regions, it is commonly referred to as colorectal cancer. The global incidence of colorectal cancer is steadily increasing, particularly in developing countries, mirroring trends seen in the developed world. One of the primary contributing factors to the rise in colon cancer is the adoption of unhealthy lifestyles, especially the frequent consumption of fast food. These foods are often difficult to digest, causing them to remain in the bowel for extended periods. This prolonged retention can lead to acidity, irritation, and infection. Furthermore, many processed foods contain toxic chemicals and additives that may damage the intestinal lining and increase cancer risk. In addition to lifestyle factors, inflammatory bowel diseases (IBD) such as Crohn's disease and ulcerative colitis, as well as genetic predispositions, are known contributors to colon cancer development (Triantafillidis et al., 2009).

Figure 4. The disruption of gut microbiota with elevated levels of certain bacterial genera such as bacteroides and bifidobacterium, as a contributing factor in the development of cancers, notably colorectal cancer

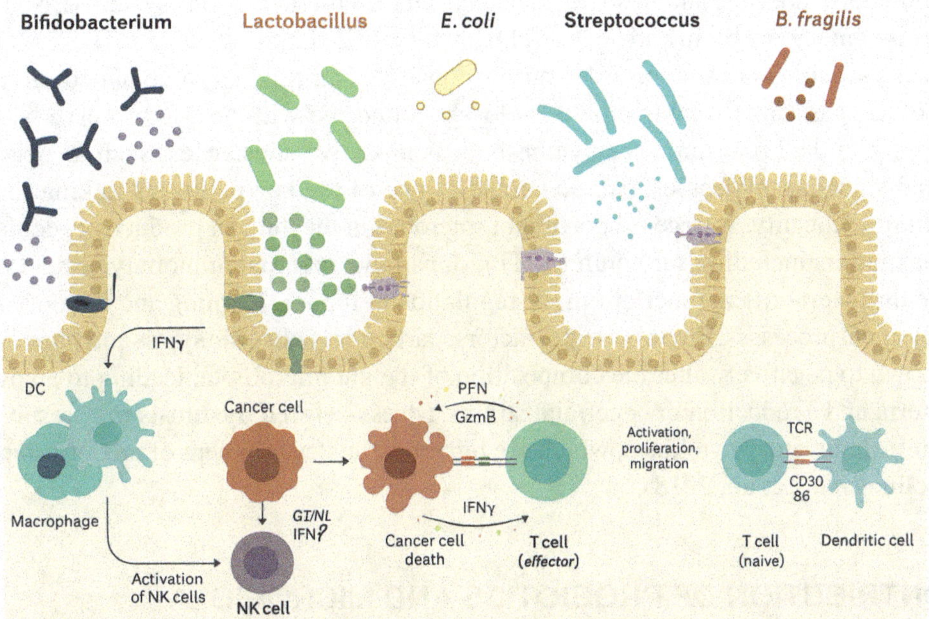

Recent research highlights the potential therapeutic role of probiotics in the management of colon cancer. Studies suggest that probiotics may enhance the population of beneficial **gut** microbiota, and contribute to cancer prevention or treatment by secreting immunomodulatory substances such as cytokines. However, further studies are needed to fully understand the mechanisms by which probiotics influence the efficacy of colorectal cancer therapies. Moreover, the consumption of synbiotics, a combination of probiotics and prebiotics has been shown to increase the population of beneficial bacteria, such as ***Lactobacillus***, supporting gut health and potentially offering protective effects against colorectal cancer (Krebs, 2016).

6 PROBIOTICS, MICROBIOTA, AND THEIR ROLE IN COGNITION

Emerging research has established that the gut microbiota plays a significant role in influencing cognitive functions through the microbiota–gut–brain axis. This axis represents a bidirectional communication pathway that links the central nervous system (CNS) with the enteric nervous system (ENS), allowing gut microorganisms to affect brain function and behavior. The human large intestinal microbiota produces a wide range of metabolites and signaling molecules that can alter physiological processes in the brain. These include amino acids, neurotransmitters (such as serotonin), short-chain fatty acids

(SCFAs), and essential fatty acids, all of which can cross the blood-brain barrier and impact cognitive and emotional regulation (Bagga et al., 2018).

Microbiota in the gut has a drastic effect on the central nervous system (CNS) that runs through a complex and two-way communication line known as the gut-brain axis. This communication takes place on neural, immune, endocrine, and metabolic routes, and the enteric nervous system (ENS) is a decisive connection point among the brain and the bowel (Cawthon & de La Serre, 2018). The ENS is frequently referred to as the second brain and this is because it consists of an expert system of neurons incorporated in the walls of the gastrointestinal tract and is closely connected with the vagus nerve which connects the gut promptly to the brainstem. The synthesis of neuroactive substances, such as neurotransmitter precursors and signaling molecules, can be named as one of the key pathways linking the microbiota with the brain functionality. A large proportion of microbes in the gut can produce or regulate supply of key neurotransmitters including serotonin (5-HT), dopamine, gamma-aminobutyric acid (GABA) and acetylcholine that exert critical functions in the regulation of moods, learning and memory among other neuropsychological processes. Environmental factors, particularly **chronic stress** and **poor sleep quality**, have been shown to negatively alter the composition of the gut microbiota, leading to an overgrowth of harmful bacteria and a reduction in beneficial strains. **Stress-related dysbiosis** may result in increased production of inflammatory cytokines, which are linked to the development of anxiety, depression, and cognitive decline (Jiang et al., 2018).

7 THE CONTRIBUTION OF PROBIOTICS AND MICROBIOTA TO OBESITY AND CARDIOVASCULAR DISEASE

Gut microbiota constitution is a limiting factor in the control of host metabolism but it has also been found to be a determinant of the occurrence of obesity and CVDs. There is growing evidence to support this claim that certain sets of micro-organisms, commonly referred to as obesogenic microbiota are linked to metabolic derailment and the excessive body fat storage. The most relevant bacteria genera in this process are Clostridium, Firmicutes, and Bacteroidetes that, in dysbiotic conditions, may drive energy harvesting of the diet and reshaping host metabolic routes (Cao et al., 2019). Short-chain fatty acids (SCFAs), including acetate, propionate and butyrate, are among the key processes in which these microbes have their effect on a number of substrates that occur in the fermentation of dietary fibers. SCFAs may exert a favorable influence on gut health and inflammation in usual physiological circumstances, but its overproduction or signaling mechanisms can contribute to poor metabolic consequences. Specifically, overproduction of SCFAs has been described as a factor in inhibiting the activity of adenosine-monophosphate-activated protein kinase (AMPK). The AMPK is a very important cellular energy-sensing mechanism that facilitates fatty acid oxidation and, subsequently, ensures the maintenance of ATP levels under the attack of energy stress. It acts so in such a way that it stimulates the pathways of catabolism producing ATP and prevents the activity of anabolic processes consuming it. When the activities of AMPK are blocked (in part due to microbial metabolites), the result is the subsequent reduction in fatty acid oxidation and the subsequent accumulation of lipids in both adipose tissues and the liver. This impairment of the lipid metabolism plays a major role in weight gain and the emergence of obesity-dominant complications, such as insulin resistance and cardiovascular dysfunction (Lopez, 2017). These discoveries emphasize the complex interaction between intestinal microbiome and metabolic homeostasis of the host, which stresses the possibility to use altering microbial composition and functionality as an intervention towards obesity and related cardiovascular disease, prevention, or mitigation.

Being involved in energy regulation and lipid metabolism, short-chain fatty acids (SCFAs) are very important in regulation of the blood pressure as well. Found evidence has shown that SCFA could also affect cardiovascular health by impacting the host signaling pathways with the renal angiotensin system (RAS) first coming to mind. The SCFA, including acetate and propionate, have been identified to activate G-protein-coupled receptors (e.g., GPR41, GPR43) so as to influence vascular transient and systemic blood pressure homeostasis. More importantly, they could induce the upregulation of angiotensin pathways, causing vasoconstriction and higher sodium retention, the two crucial factors in the pathogenesis of high blood pressure (Kang & Cai, 2018).

In addition, an imbalance of microbial community termed as gut dysbiosis has been imminently associated with the increase in blood pressure and development of cardiovascular diseases. Alteration of the equilibrium of the regular makeup of the gut microbiota is able to result in excess production of some metabolites, the systemic inflammation, and the deficient endothelial functionality, which are all known to contribute to the hypertensive pathology. The assumption that patients with hypertension tend to have less microbial diversity and changes in the number of bacteria to produce SCFA is supported by a study conducted by Yang et al. (2015).

A growing amount of evidence has been accumulated to suggest the gut microbiota involvement in the pathophysiology of hypertension and cardiovascular diseases (CVDs). Specifically, the excessive proliferation of some bacterial phyla, like Firmicutes and Bacteroidetes, were linked to a higher blood pressure and the high level of occurrence of hypertension. Such microbial imbalances alter the vascular response and systemic inflammation, two important predisposing cardiovascular risks in addition to host metabolism (Pluznick et al., 2013). Another key point of cardiovascular health is the integrity of the intestinal barrier which is mostly maintained by a heterogeneous and well-balanced microbiome of the intestine. Disruption of this barrier could be caused by dysbiosis an adverse microbial change which may lead to elevated intestinal permeability commonly known as a leaky gut. This defective barrier permits the passage of microbial elements e.g. lipopolysaccharides (LPS) into systemic circulation, which causes chronic low grade inflammation which is a risk factor in diseases like stroke, atherosclerosis and heart failure. The fact that the microbial-cardiovascular relationship exists is further touted by the fact that recent literature identified increased presence of pathogenic genus of bacteria like Streptococcus and members of Enterobacteriaceae family in the intestinal microbiome of coronary artery disease (CAD) patients. These microbial species have been linked to increase pro-inflammatory reactions and endothelial dysfunction, thereby depicting the existence of a microbial factor in the development and advancement of atherosclerotic disease (Jie et al., 2017). On the other hand, probiotics, especially the ones that are enriched with strains of genus Lactobacillus, have demonstrated beneficial effect in lessening cardiovascular risk. It has been reported that these positive microbes reduce adverse blood pressure, circulating pro-inflammatory cytokine levels, enhance lipid metabolism. The underlying mechanisms might include alteration of the makeup of the gut microbiome, increased production of short-chain fatty acids and repair of gut barrier. The potential of probiotics as supplemental measures of hypertension prevention and a reduction of the cardiovascular risk is stressed in studies like the one by Gomez-Guzman et al. (2015).

CONCLUSION

Probiotics referred to as live microorganisms which render health benefits to the host at a sufficient level are vital in defining and sustaining the composition of the human gut microbiota. Gut microbiota is a pertinent and dynamic community of trillions of microorganisms such as fungi, viruses, and bacteria living in the gastrointestinal (GI) tract. The microbial populations play a vital role in the host physiology which comprises a significant role in digestion, nutrient absorption, immune response and protection against pathogens. A healthy microbiota of the gut is essential in homeostasis and one of the main contributors of homeostasis is probiotics. Scientific research has revealed that probiotics aids in keeping the guts in their optimal atmosphere by competing against pathogenic bacteria in the use of attachment points and nutrients, synthesizing anti-microbial substances, and controlling the immune reactions of the host. They impact positively on the stable and diverse structure of the gut microbiome, as well as keep the advantage of healthy bacteria over pathogenic ones. The strains which are well documented and extensively studied on their impact on the positive health of the gut are probiotics including, Lactobacillus acidophilus, Bifidobacterium bifidum and Lactobacillus rhamnosus. Probably the greatest advantage of probiotics is that it improves the digestion process. These probiotic strains synthesize some enzymes like lactase that helps in lactose digestion and hence reduce lactose intolerance syndrome symptoms. They also help to digest the complex carbohydrates, proteins, and fats, thus leading to ease in absorption of the nutrients and general digestive purposes. Besides the stated digestive results, probiotics are significant in immunomodulation. They bring about both innate and adaptative immunity by stimulating the synthesis of the immunoglobulin A (IgA), boosting the activity of macrophages as well as modulation of cytokine secretion. Such measures will help people have a stronger immune system, which can react to infections well and lower inflammation.

Dysbiosis or an imbalance in the microbiota of the gut has been linked in many health disorders such as inflammatory bowel diseases (IBD), irritable bowel syndrome (IBS), metabolic syndrome, obesity, diabetes type 2, allergies, and even to conditions of the neuropsychiatric axis like anxiety and depression. This knowledge has created greater interest in the use of probiotics supplementation as prophylaxis and management of different illnesses. An increasingly large literature underlies the finding that probiotics may also be effective in the treatment of gastrointestinal conditions and even as potential systemic agents.

The lifestyle including diet, sleep, hydration, stress and physical activity is also important in promoting a healthy gut microbiota. Probiotic organisms can be promoted by following a diet that includes fermented food products in large quantities, e.g. yogurt, kefir, sauerkraut, kimchi, and miso, as well as prebiotics (fiber that nourishes probiotics). Other components of gut health include appropriate levels of water intake, sleep, and stress. Poor lifestyle choices and chronic stress can cause dysbiosis as a gut-brain axis, which causes physical and mental health issues. New scientific knowledge has opened up different meanings of probiotics other than digestive health. Tooley (2020) adds that probiotics can also act as mental health supplements by affecting the gut-brain axis, a highly communicable network that connects the gut and the central nervous system. Probiotics have also been found to regulate mood, anxiety and cognitive performance through changes in neurotransmitter user-productions (e.g. serotonin and GABA), changes in systemic inflammation and tissue barrier integrity (gut). The latter has led to the notion of psychobiotics-probiotic strains which keep a person mentally in good health. Also, probiotics have been encouraging towards prevention of some types of cancer specifically colon cancer. The possible mechanisms are changing a composition of the gut microbiota towards the favorable bacteria, binding and inactivating carcinogens, stimulating the activity of detoxifying enzymes and immunomodulation of the host immune response by sclerosing cytokines. Research has also shown that certain types of probiotics might prevent incidence of precancerous lesions and cancerous growth of cells in vivo and in vitro in animals.

REFERENCES

Al-Obaid, S., Al-Khafaji, Z., Al-Jebori, S., & et al. (2020). Gut microbiome of native Arab Kuwaitis reveals diet-dependent variation. [Note: Journal title, volume, and pages/DOI missing in original – please complete if available]

Ambalam, P., Raman, M., Purama, R. K., & Doble, M. (2016). Probiotics, prebiotics and colorectal cancer prevention. *Best Practice & Research. Clinical Gastroenterology*, *30*(1), 119–131. DOI: 10.1016/j.bpg.2016.02.009 PMID: 27048903

Avena-Woods, C. (2017). Overview of atopic dermatitis. *The American Journal of Managed Care*, *23*(6, Suppl), S115–S123. PMID: 28978208

Bagga, D., Reichert, J. L., Koschutnig, K., Aigner, C. S., Holzer, P., Koskinen, K., Moissl-Eichinger, C., & Schöpf, V. (2018). Probiotics drive gut microbiome triggering emotional brain signatures. *Gut Microbes*, *9*(6), 486–496. DOI: 10.1080/19490976.2018.1460015 PMID: 29723105

Boleij, A., Hechenbleikner, E. M., Goodwin, A. C., Badani, R., Stein, E. M., Lazarev, M. G., Ellis, B., Carroll, K. C., Albesiano, E., Wick, E. C., Platz, E. A., Pardoll, D. M., & Sears, C. L. (2015). The *Bacteroides fragilis* toxin gene is prevalent in the colon mucosa of colorectal cancer patients. *Clinical Infectious Diseases*, *60*(2), 208–215. DOI: 10.1093/cid/ciu787 PMID: 25305284

Cao, S. Y., Zhao, C. N., Xu, X. Y., Tang, G. Y., Corke, H., Gan, R. Y., & Li, H. B. (2019). Dietary plants, gut microbiota, and obesity: Effects and mechanisms. *Trends in Food Science & Technology*, *92*, 194–204. DOI: 10.1016/j.tifs.2019.08.004

Cawthon, C. R., & de La Serre, C. B. (2018). Gut bacteria interaction with vagal afferents. *Brain Research, 1693*(Pt B), 134–139. https://doi.org/DOI: 10.1016/j.brainres.2018.03.021

Chen, Y., Zhou, J., & Wang, L. (2021). Role and mechanism of gut microbiota in human disease. * linical and Infection Microbiology, 11*. *Frontiers in Cellular and Infection Microbiology*, *625913*, 625913. Advance online publication. DOI: 10.3389/fcimb.2021.625913 PMID: 33816335

Clemente, J. C., Ursell, L. K., Parfrey, L. W., & Knight, R. (2012). The impact of the gut microbiota on human health: An integrative view. *Cell*, *148*(6), 1258–1270. DOI: 10.1016/j.cell.2012.01.035 PMID: 22424233

Davey, M. E., & O'Toole, G. A. (2000). Microbial biofilms: From ecology to molecular genetics. *Microbiology and Molecular Biology Reviews : MMBR*, *64*(4), 847–867. DOI: 10.1128/MMBR.64.4.847-867.2000 PMID: 11104821

Dominguez-Bello, M. G., De Jesus-Laboy, K. M., Shen, N., Cox, L. M., Amir, A., Gonzalez, A., Bokulich, N. A., Song, S. J., Hoashi, M., Rivera-Vinas, J. I., Mendez, K., Knight, R., & Clemente, J. C. (2016). Partial restoration of the microbiota of cesarean-born infants via vaginal microbial transfer. *Nature Medicine*, *22*(3), 250–253. DOI: 10.1038/nm.4039 PMID: 26828196

García-Mantrana, I., Selma-Royo, M., Alcántara, C., & et al. (2023). Review of the relationships between human gut microbiome, diet, and obesity. *Nutrients, 16*(23), Article 3996. https://doi.org/ (Note: Volume 16 is unusually high – double-check year/journal)DOI: 10.3390/nu16233996

Gibson, G. R., Hutkins, R., Sanders, M. E., Prescott, S. L., Reimer, R. A., Salminen, S. J., Scott, K., Stanton, C., Swanson, K. S., Cani, P. D., Verbeke, K., & Reid, G. (2017). Expert consensus document: The International Scientific Association for Probiotics and Prebiotics (ISAPP) consensus statement on the definition and scope of prebiotics. *Nature Reviews. Gastroenterology & Hepatology*, *14*(8), 491–502. DOI: 10.1038/nrgastro.2017.75 PMID: 28611480

Gómez-Guzmán, M., Toral, M., Romero, M., Jiménez, R., Galindo, P., Sánchez, M., Zarzuelo, M. J., Olivares, M., Gálvez, J., & Duarte, J. (2015). Antihypertensive effects of probiotics *Lactobacillus* strains in spontaneously hypertensive rats. *Molecular Nutrition & Food Research*, *59*(11), 2326–2336. DOI: 10.1002/mnfr.201500290 PMID: 26255877

Grice, E. A., & Segre, J. A. (2012). The human microbiome: Our second genome. *Annual Review of Genomics and Human Genetics*, *13*(1), 151–170. DOI: 10.1146/annurev-genom-090711-163814 PMID: 22703178

Hansen, C. H. F., Krych, L., Nielsen, D. S., Vogensen, F. K., Hansen, L. H., Sørensen, S. J., Buschard, K., & Hansen, A. K. (2012). Early life treatment with vancomycin propagates *Akkermansia muciniphila* and reduces diabetes incidence in the NOD mouse. *Diabetologia*, *55*(8), 2285–2294. DOI: 10.1007/s00125-012-2564-7 PMID: 22572803

Hill, C., Guarner, F., Reid, G., Gibson, G. R., Merenstein, D. J., Pot, B., Morelli, L., Canani, R. B., Flint, H. J., Salminen, S., Calder, P. C., & Sanders, M. E. (2014). The International Scientific Association for Probiotics and Prebiotics consensus statement on the scope and appropriate use of the term probiotic. *Nature Reviews. Gastroenterology & Hepatology*, *11*(8), 506–514. DOI: 10.1038/nrgastro.2014.66 PMID: 24912386

Jiang, H., Zhang, X., Yu, Z., Zhang, Z., Deng, M., Zhao, J., & Ruan, B. (2018). Altered gut microbiota profile in patients with generalized anxiety disorder. *Journal of Psychiatric Research*, *104*, 130–136. DOI: 10.1016/j.jpsychires.2018.07.007 PMID: 30029052

Jie, Z., Xia, H., Zhong, S. L., Feng, Q., Li, S., Liang, S., Zhong, H., Liu, Z., Gao, Y., Zhao, H., Zhang, D., Su, Z., Fang, Z., Lan, Z., Li, J., Xiao, L., Li, J., Li, R., Li, X., & Kristiansen, K. (2017). The gut microbiome in atherosclerotic cardiovascular disease. *Nature Communications*, *8*(1), 845. Advance online publication. DOI: 10.1038/s41467-017-00900-1 PMID: 29018189

Kamada, N., Chen, G. Y., Inohara, N., & Núñez, G. (2013). Control of pathogens and pathobionts by the gut microbiota. *Nature Immunology*, *14*(7), 685–690. DOI: 10.1038/ni.2608 PMID: 23778796

Kang, Y., & Cai, Y. (2018). Gut microbiota and hypertension: From pathogenesis to new therapeutic strategies. *Clinics and Research in Hepatology and Gastroenterology*, *42*(2), 110–117. DOI: 10.1016/j.clinre.2017.09.006 PMID: 29102544

Kelly, J. R., Allen, A. P., Temko, A., Hutch, W., Kennedy, P. J., Farid, N., Murphy, E., Boylan, G., Bienenstock, J., Cryan, J. F., Clarke, G., & Dinan, T. G. (2017). Lost in translation? The potential psychobiotic *Lactobacillus rhamnosus* (JB-1) fails to modulate stress or cognitive performance in healthy male subjects. *Brain, Behavior, and Immunity*, *61*, 50–59. DOI: 10.1016/j.bbi.2016.11.018 PMID: 27865949

Krebs, B. (2016). Prebiotic and synbiotic treatment before colorectal surgery—Randomised double-blind trial. *Collegium Antropologicum, 40*(1), 35–40. PMID: 27301235

Liu, Q., Yu, Z., Tian, F., Zhao, J., Zhang, H., Zhai, Q., & Chen, W. (2020). Surface components and metabolites of probiotics for regulation of intestinal epithelial barrier. *Microbial Cell Factories, 19*(1), 23. Advance online publication. DOI: 10.1186/s12934-020-1289-4 PMID: 32024520

López, M. (2017). EJE PRIZE 2017: Hypothalamic AMPK: A golden target against obesity? *European Journal of Endocrinology, 176*(6), R235–R246. DOI: 10.1530/EJE-16-0927 PMID: 28232370

Mary Ellen, S., Francisco, G., & Richard, G.. (2013). An update on the use and investigation of probiotics in health and disease. *Gut, 62*(5), 787–796. DOI: 10.1136/gutjnl-2012-302504 PMID: 23474420

Molska, M., & Reguła, J. (2019). Potential mechanisms of probiotics action in the prevention and treatment of colorectal cancer. *Nutrients, 11*(10), 2453. Advance online publication. DOI: 10.3390/nu11102453 PMID: 31615096

Ohara, T., Yoshino, K., & Kitajima, M. (2010). Possibility of preventing colorectal carcinogenesis with probiotics. *Hepato-Gastroenterology, 57*(101), 1411–1415. PMID: 21443095

Peterson, J., Garges, S., Giovanni, M., McInnes, P., Wang, L., Schloss, J. A., Bonazzi, V., McEwen, J. E., Wetterstrand, K. A., Deal, C., Baker, C. C., Di Francesco, V., Howcroft, T. K., Karp, R. W., Lunsford, R. D., Wellington, C. R., Belachew, T., Wright, M., Giblin, C., & Guyer, M. (2009). The NIH Human Microbiome Project. *Genome Research, 19*(12), 2317–2323. DOI: 10.1101/gr.096651.109 PMID: 19819907

Pluznick, J. L., Protzko, R. J., Gevorgyan, H., Peterlin, Z., Sipos, A., Han, J., Brunet, I., Wan, L.-X., Rey, F., Wang, T., Firestein, S. J., Yanagisawa, M., Gordon, J. I., Eichmann, A., Peti-Peterdi, J., & Caplan, M. J. (2013). Olfactory receptor responding to gut microbiota-derived signals plays a role in renin secretion and blood pressure regulation. *Proceedings of the National Academy of Sciences of the United States of America, 110*(11), 4410–4415. DOI: 10.1073/pnas.1215927110 PMID: 23401498

Qin, J., Li, R., Raes, J., Arumugam, M., Burgdorf, K. S., Manichanh, C., Nielsen, T., Pons, N., Levenez, F., Yamada, T., Mende, D. R., Li, J., Xu, J., Li, S., Li, D., Cao, J., Wang, B., Liang, H., Zheng, H., & Wang, J. (2010). A human gut microbial gene catalogue established by metagenomic sequencing. *Nature, 464*(7285), 59–65. DOI: 10.1038/nature08821 PMID: 20203603

Rowland, I., Gibson, G., Heinken, A., Scott, K., Swann, J., Thiele, I., & Tuohy, K. (2018). Gut microbiota functions: Metabolism of nutrients and other food components. *European Journal of Nutrition, 57*(1), 1–24. DOI: 10.1007/s00394-017-1445-8 PMID: 28393285

Salminen, S., Collado, M. C., Endo, A., Hill, C., Lebeer, S., & Mättö, J.. (2021). The International Scientific Association of Probiotics and Prebiotics (ISAPP) consensus statement on the definition and scope of postbiotics. *Nature Reviews. Gastroenterology & Hepatology, 18*(9), 649–667. DOI: 10.1038/s41575-021-00440-6 PMID: 33948025

Sanders, M. E., Heimbach, J. T., Pot, B., Tancredi, D. J., Lenoir-Wijnkoop, I., Lähteenmäki-Uutela, A., Gueimonde, M., & Bañares, S. (2011). Health claims substantiation for probiotic and prebiotic products. *Gut Microbes, 2*(3), 127–133. DOI: 10.4161/gmic.2.3.16174 PMID: 21646865

Secher, T., Kassem, S., Benamar, M., Bernard, I., Boury, M., Barreau, F., Oswald, E., & Saoudi, A. (2017). Oral administration of the probiotic strain *Escherichia coli* Nissle 1917 reduces susceptibility to neuroinflammation and repairs experimental autoimmune encephalomyelitis-induced intestinal barrier dysfunction. *Frontiers in Immunology*, *8*, 1096. Advance online publication. DOI: 10.3389/fimmu.2017.01096 PMID: 28959254

Sharma, R., Bhaskar, B., Sanodiya, B. S., Thakur, G. S., Jaiswal, P., Yadav, N., Sharma, A., & Bisen, P. S. (2014). Probiotic efficacy and potential of *Streptococcus thermophilus* modulating human health: A synoptic review. *IOSR Journal of Pharmacy and Biological Sciences*, *9*(3), 52–58. DOI: 10.9790/3008-09325258

Tooley, K. L. (2020). Effects of the human gut microbiota on cognitive performance, brain structure and function: A narrative review. *Nutrients*, *12*(10), 3009. Advance online publication. DOI: 10.3390/nu12103009 PMID: 33007941

Triantafillidis, J. K., Nasioulas, G., & Kosmidis, P. A. (2009). Colorectal cancer and inflammatory bowel disease: Epidemiology, risk factors, mechanisms of carcinogenesis and prevention strategies. *Anticancer Research*, *29*(7), 2727–2737. PMID: 19596953

Vitini, E., Alvarez, S., Medina, M., & Perdigón, G. (2000). Gut mucosal immunostimulation by lactic acid bacteria. *Biocell*, *24*(3), 223–232. PMID: 11201658

Vogt, L., Ramasamy, U., Meyer, D., Pullens, G., Venema, K., Faas, M. M., Schols, H. A., & de Vos, P. (2013). Immune modulation by different types of β2→1-fructans is toll-like receptor dependent. *PLoS One*, *8*(7), e68367. Advance online publication. DOI: 10.1371/journal.pone.0068367 PMID: 23861894

Yang, C., Fei, Y., Qin, Y., Luo, D., Yang, S., Kou, X., Zi, Y., Deng, T., & Jin, M. (2015). Bacterial flora changes in conjunctiva of rats with streptozotocin-induced type I diabetes. *PLoS One*, *10*(7), e0133021. Advance online publication. DOI: 10.1371/journal.pone.0133021 PMID: 26176548

Zhang, Y., Zheng, T., Ma, D., Shi, P., Zhang, H., Li, J., & Sun, Z. (2023). Probiotics *Bifidobacterium lactis* M8 and *Lactobacillus rhamnosus* M9 prevent high blood pressure via modulating the gut microbiota composition and host metabolic products. *mSystems*, *8*(6), e0033123. Advance online publication. DOI: 10.1128/msystems.00331-23 PMID: 37855616

Zmora, N., Zilberman-Schapira, G., Suez, J., Mor, U., Dori-Bachash, M., Bashiardes, S., Kotler, E., Zur, M., Regev-Lehavi, D., Brik, R. B.-Z., Federici, S., Cohen, Y., Linevsky, R., Rothschild, D., Moor, A. E., Ben-Moshe, S., Harmelin, A., Itzkovitz, S., Maharshak, N., & Elinav, E. (2018). Personalized gut mucosal colonization resistance to empiric probiotics is associated with unique host and microbiome features. *Cell*, *174*(6), 1388–1405. DOI: 10.1016/j.cell.2018.08.041 PMID: 30193112

Chapter 13
The Role of Polyphenols and Bioactive Compounds for Combating Neurodegenerative Diseases

P. Jayashree
https://orcid.org/0009-0008-6957-9256
Vels Institute of Science, Technology, and Advanced Studies, India

S. S. Meenambiga
https://orcid.org/0000-0002-5445-525X
Vels Institute of Science, Technology, and Advanced Studies, India

Hanishka Pothiraj
https://orcid.org/0009-0002-5692-0386
Vels Institute of Science, Technology, and Advanced Studies, India

ABSTRACT

Neurodegenerative illnesses afflict hundreds of millions of people globally, resulting in significant healthcare expenses. Conventional treatments have hardly an effect on the disease, which at most are symptomatic, and recent studies have found, that the action of polyphenols and bioactive molecules are beneficial against the disease progression and also in maintaining cognitive function. These phytochemicals abundant in fruits, vegetables, herbs and medicinal plants exhibit a range of beneficial effects which could lead to the transformation of the concept of neuroprotection. Polyphenols such as flavonoids, phenolic acids, and tannins, have gained significant attention due to their potential in antioxidant and anti-inflammatory properties. It can therefore modulate specific signalling pathways implicated in autophagy, amyloid clearance and synaptic plasticity thus, could be potentially beneficial in the treatment of pathologies such Alzheimer's, Parkinson's and Huntington's diseases.

DOI: 10.4018/979-8-3373-4139-2.ch013

INTRODUCTION

Neurodegenerative diseases, which include Alzheimer's, Parkinson's and Huntington's disease, are serious health problems worldwide, and involve complex, multicausal disorders causing oxidative stress, neuroinflammation and disturbance of mitochondrial function and of proteostasis leading to progressive neuronal loss and cognitive decline (Bhullar and Rupasinghe 2013) (Grabska-Kobyłecka, Szpakowski et al., 2023) .The few pharmacological therapies that are currently available only provide temporary symptomatic relief as well as side effects and they are not effective at modifying the progression of the disease. This highlights the importance for new therapeutic approaches (Renaud and Martinoli 2019). Polyphenols, one of the bioactive compounds of plant-based diet and other drugs obtained from such diet have become the agents of choice in fighting oxidative stress and inflammation in neurodegenerative disease, such as Alzheimer's Disease (Uddin et al., 2020). They hit multiple cellular pathways such as Nrf2, MAPK, JAK/STAT and Wnt/β-catenin and are able to produce anti-oxidant, anti-inflammatory actions as well as to enhance neuronal survival and synaptic plasticity (Kooshki et al., 2023). Polyphenol-rich diets are associated with lower risk for neurodegenerative disorders and decreased cognitive decline when compared with other plant extracts (Yan, Guo et al., 2022). Bioactive substances such as alkaloids and terpenoids have some potential: that of modulating neurotransmission, preventing tau protein aggregation, and fostering cognitive resilience. This chapter intends to link, through analysis of the mechanisms of actions and therapeutic relevance, the basic level of knowledge with the applied level of knowledge, thereby emphasizing the potential of such natural compounds for use in the development of novel dietary and pharmacologic therapeutic options for neurodegenerative disorders (Higham, Hidalgo et al., 2019).

MOLECULAR MECHANISMS UNDERLYING NEURODEGENERATION

Neurodegeneration is impelled via molecular mechanisms that interact sophisticatedly, disturbing neuronal survival plus function. Anomalies within protein equilibrium stand as pivotal to these mechanisms, notably by way of diminished autophagy, which elicits the accretion of misshapen proteins and deleterious clusters within neurons (Esmaeili, Yarjanli et al., 2022). Mitochondrial impairment constitutes another vital determinant because it obstructs energy synthesis, elevates reactive oxygen species (ROS) creation, and engenders oxidative pressure, all of which harm and eradicate neurons (Esteras, Rohrer et al., 2017). Aberrations within mitochondrial DNA coupled with its dynamism additionally strengthen oxidative strain which elevates cellular impairment within maladies such as Alzheimer's and Parkinson's (Moradi Vastegani, Nasrolahi et al., 2023). As dysregulated proteins aggregate in the ER, cellular stress responses inducing apoptosis or necroptosis are provoked, therefore endoplasmic reticulum (ER) stress is additionally consequential, as a programmed type of cell demise noticed in maladies like multiple sclerosis (Picon, Jayaraman et al., 2021). Microglia and astrocytes that are activated instigate neuroinflammation. Genetic factors such as TREM2 variants regulate microglial activation and neurodegeneration magnitude since it emits pro-inflammatory cytokines which aggravate neurodegeneration (Leyns, Ulrich et al., 2017). MicroRNAs are dysregulated which additionally disrupts gene expression networks because they affect neural and immune pathways as they contribute to disease progression across various neurodegenerative disorders (Juźwik, Drake et al., 2019). Multi-omics approaches reveal that genetic along with epigenetic factors, inclusive of rare and common variants, interact among environmental influences to mold disease susceptibility with progression. These linked molecular pathways impair autophagy as well as induce mitochondrial and ER stress. They oxidatively impair likewise, incite neurons, modulate genetics, constitute the etiology of neurodegenerative disease, and furnish targets therapeutically (Ellwardt & Zipp 2014).

Figure 1. Molecular mechanisms underlying neurodegeneration

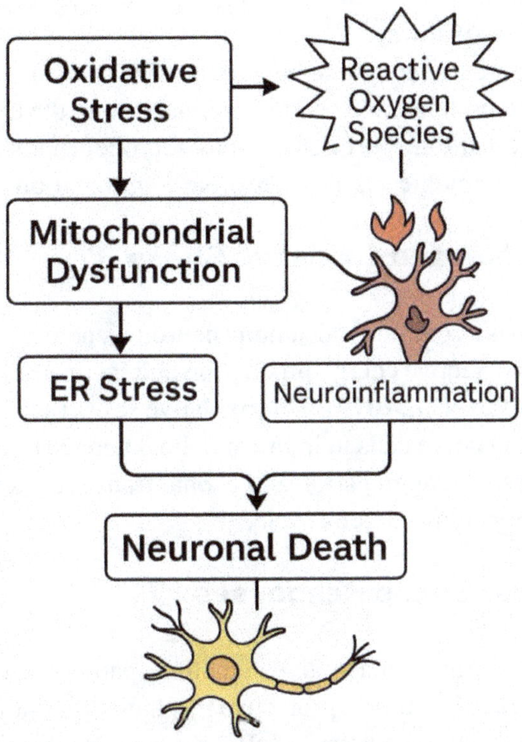

Protein Misfolding and Aggregation

The misfolding and clumping of specific molecules signals several brain diseases. For example, amyloid-beta and tau indicate Alzheimer's, α-synuclein points to Parkinson's, and TDP-43 suggests amyotrophic lateral sclerosis (Tsoi, Quan et al., 2023). These proteins have intrinsically disordered regions that let them to adopt abnormal conformations and aggregates. This causes disruption in cell functioning and neuronal death. A phenomenon called liquid-liquid phase separation happens when proteins bunches up spontaneously on their own into highly dynamic condensates that has recently been implicated in triggering the pathogenic aggregation of proteins and formation of inclusion bodies in neurons (Zbinden et al., 2020). Mutations, post- translational modifications, and truncations further enhance the aggregation propensity of these proteins, exacerbating disease progression.

Impaired Proteostasis and Axonal Transport

Protein homeostasis, or simply proteostasis, refers to the cellular process proteins to maintain proper concentration, conformation, localization, and degradation of proteins that helps ensure their correct function. Neurons require these protein balance to stay healthy. As we get older, our body's ability to maintain this balance declines. This leads to the accumulation of toxic clumps of protein aggregates (Tan et al., 2019). The system that keeps proteins in check includes two main parts: ubiquitin-proteasome system and autophagy-lysosome pathway.

Defects with axonal transport are thought to be an early sign of many brain diseases. This movement is crucial for moving cell parts, proteins, and chemical signals along the neuron (Guo et al., 2020). When genes are mutated, it links axonal transport defects into neurodegeneration. These transport problems often interact alongside other pathological issues, like protein aggregation and mitochondrial dysfunction.

Mitochondrial Dysfunction and Oxidative Stress

Mitochondrial dysfunction is a common trait among neurodegenerative diseases that cause nerve cells to break down. These leads to defective energy production and increased reactive oxygen species (ROS) production which is harmful (Tan et al., 2019). This oxidative stress then damages cellular elements like fats, proteins, and DNA causing nerve cells to injure and die. Damage to mitochondria is linked to other cellular processes including protein aggregation and axonal transport along nerve fibres. These factors make brain cell breakdown worse in a vicious cycle.

Dysregulation of Cellular Stress Responses

Neurons respond to stressors through adaptive signalling pathways, including the integrated stress response (ISR). These stressors are, at times, protective mechanisms, but the prolonged enhancement of phosphorylation of eukaryotic initiation factor 2α (eIF2α) results in exacerbated neurodegeneration due to global translation inhibition, neuroinflammation, and increased cell death. It means the ISR serves several roles: transient ISR activation acts as a protective one, chronic ISR activation can induce pathological characteristics such as memory loss, protein aggregation, and a neurodegenerative phenotype (Bond et al., 2020).

Role of microRNAs and Genetic Factors

Regulators of gene expression post-transcription includes a group of small, non-coding RNAs called MicroRNAs (miRNAs). Previously unrecognized aspects of neurodegeneration are now discovered to involve miRNAs (Juźwik et al., 2019). Dysregulation of certain miRNAs affects both neural and immune systems pathways and thus drives disease progression in several neurodegenerative conditions. Genetic mutations such as those in presenilin's and amyloid precursor protein (APP) in familial Alzheimer's further acknowledge the molecular pathways that bring about the onset and progression of the disease (Maire et al., 2020; Ratan et al., 2023) . Environmental and lifestyle factors, in combination to a genetic susceptibility, determines the risk and course of neurodegenerative disorders (Liu et al., 2023).

POLYPHENOLS: CLASSIFICATION, SOURCES, AND BIOAVAILABILITY

Classification of Polyphenols:

Polyphenols are a heterogeneous class of naturally occurring phytochemicals extensively found in plants, playing a role in colouration, flavour, and defence against infections. They are structurally defined by the presence of many phenol units. Polyphenols are categorically divided into flavonoids, phenolic acids, lignans, and stilbenes, according to their chemical structure and the quantity of phenolic rings.

Flavonoids: Flavonoids constitute the most common class of polyphenols, accounting for more than 60% of total dietary polyphenols. Their basic structure consists of two aromatic rings (A and B) connected by a three-carbon bridge, resulting in a closed pyran ring (C ring). Flavonoids are categorised into:

Flavanols, such as quercetin and kaempferol, are present in onions, kale, and broccoli. **Flavones**, such as apigenin and luteolin, are prevalent in parsley and celery. **Flavanones**, such as hesperidin and naringenin, are prevalent in citrus fruits. **Flavanols,** such as catechins and epicatechins, are prevalent in tea, chocolate, and grapes. **Anthocyanins**, such as cyanidin and delphinidin, are responsible for the red, blue, and purple hues in berries and red cabbage. **Isoflavones,** such as genistein and daidzein, are mostly located in soybeans and legumes.

Phenolic Acids

Phenolic acids are categorised into two subtypes:

Hydroxybenzoic acids, such as gallic acid and protocatechuic acid, are present in tea, berries, and nuts. Hydroxycinnamic acids (e.g., caffeic acid, ferulic acid, chlorogenic acid) are prevalent in coffee, whole grains, and specific fruits. Phenolic acids have an essential part in antioxidant activity and are implicated in the modulation of inflammatory pathways.

Stilbenes

Stilbenes are less prevalent in the plant kingdom. Resveratrol is the most extensively researched chemical, predominantly located in grapes, red wine, and peanuts. Stilbenes have attracted interest for their possible cardioprotective, anticancer, and neuroprotective properties.

Lignans

Lignans are polyphenolic chemicals present in seeds, particularly flaxseeds, as well as in grains and vegetables. Prominent examples encompass secoisolariciresinol and matairesinol. Lignans in humans are transformed into enterolignans, which demonstrate oestrogenic and antioxidant properties.

Additional Polyphenols

This group comprises tannins (both hydrolysable and condensed) and curcuminoids (including curcumin). Tannins, present in tea and wine, are recognised for their astringent flavour and antibacterial characteristics. Curcuminoids, chiefly derived from turmeric, are acknowledged for their anti-inflammatory and antioxidant properties.

Polyphenols have several health advantages, encompassing antioxidant, anti-inflammatory, cardioprotective, neuroprotective, and anticancer properties. Their bioavailability and metabolism are contingent upon the dietary matrix, gut flora, and host variables.

SOURCES OF POLYPHENOLS

Polyphenols are abundantly found in the plant kingdom and are prevalent in numerous foods, playing an essential part in the human diet. Their sources can be defined roughly depending to the elementary dietary kinds.

1. Vegetables and Fruits

Fruits provide the most frequent dietary source of polyphenols. Berries, including blueberries, strawberries, raspberries, and blackberries, are significantly rich in anthocyanins, flavonols, and phenolic acids. Apples, grapes, cherries, and citrus fruits exhibit significant amounts of flavanols, flavanones, and phenolic acids. Vegetables such as onions, spinach, broccoli, kale, and artichokes contain flavonoids, including flavonols (quercetin and kaempferol) and phenolic acids (chlorogenic and caffeic acids) (Manach et al., 2004).

2. Libations

Beverages high in polyphenols significantly contribute to daily polyphenol consumption. Tea, particularly green and black types, is abundant in catechins and theaflavins. Coffee is a prominent source of chlorogenic acid, a principal hydroxycinnamic acid. Red wine contains resveratrol, flavonols, and tannins as a result of the fermentation of grape skins. Cocoa-based beverages are abundant in flavanols, chiefly catechins (Pandey & Rizvi, 2009).

3. Nuts, Seeds, and Pulses

Walnuts, almonds, and hazelnuts possess a variety of polyphenols, notably ellagic acid and proanthocyanidins. Flaxseeds and sesame seeds are superior suppliers of lignans. Legumes, such as soybeans, are abundant in isoflavones, including genistein and daidzein.

4. Entire Grains and Cereals

Whole grains, such as oats, wheat, and rye, are rich in phenolic acids, including ferulic acid and p-coumaric acid. The outer layers (bran) are notably abundant in bound phenolic chemicals, which enhance the antioxidant activity of cereals.

5. Botanicals, Seasonings, and Oils

Herbs and spices, including cloves, oregano, thyme, cinnamon, and turmeric, are rich in polyphenols, particularly flavonoids and curcuminoids. Extra virgin olive oil contains secoiridoids and phenolic alcohols, such as hydroxytyrosol.

6. Additional Sources

Dark chocolate, originating from cocoa, is plentiful in flavanols, particularly epicatechin. Processed foods that include berries, teas, or cocoa may provide polyphenols, however processing can influence their levels. Integrating a variety of polyphenol-rich foods into the diet is linked to many positive health effects, including decreased risks of chronic diseases attributed to their antioxidant and anti-inflammatory properties (Pérez-Jiménezet al., 2010).

Figure 2. Bioactive polyphenols and neurodegeneration

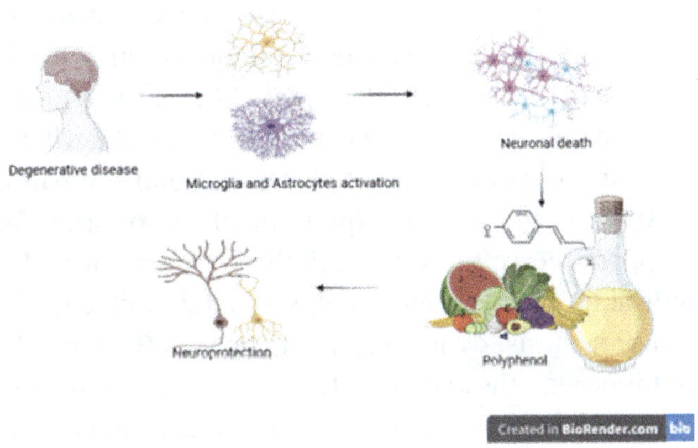

BIOAVAILABILITY OF POLYPHENOLS

The bioavailability of polyphenols indicates the quantity of consumed polyphenols that are absorbed and utilised by the body to produce biological effects. Although prevalent in plant-based meals, polyphenols typically demonstrate low and inconsistent bioavailability owing to their intricate chemical structures and interactions in the digestive system (Scalbert & Williamson 2000).Polyphenols are found in foods in multiple forms, such as glycosides, esters, and polymers, which exhibit limited absorption in the small intestine. Intestinal enzymes or gut microbes frequently require enzymatic hydrolysis to liberate aglycone molecules for subsequent absorption. Upon absorption, polyphenols undergo significant phase II metabolism (glucuronidation, sulfation, and methylation) in the intestinal wall and liver, resulting in metabolites that circulate in the bloodstream (Manach et al., 2005).

Factors affecting polyphenol bioavailability comprise:

Chemical structure: Monomeric forms like catechins and isoflavones are better absorbed than polymeric proanthocyanidins and tannins. **Food matrix:** The presence of fats can enhance the absorption of lipophilic polyphenols. **Gut microbiota:** Microbial fermentation in the colon converts unabsorbed polyphenols into bioactive metabolites. **Individual variability:** Genetics, age, health status, and microbiota composition significantly affect polyphenol metabolism.

Polyphenol metabolites may preserve or even enhance biological activity relative to their parent molecules, so contributing to antioxidant, anti-inflammatory, and cardioprotective properties. Enhancing polyphenol bioavailability is a current research priority, employing techniques such as nanoencapsulation, emulsions, and co-administration with absorption enhancers (Del Rio, Costa et al., 2010).

NEUROPROTECTIVE MECHANISMS OF POLYPHENOLS

Polyphenols protect neurons through several ways that address the main causes of neurodegeneration. They act by scavenging reactive oxygen and nitrogen molecules, which lowers oxidative stress — a big factor in neuronal damage and in diseases like Alzheimer's and Parkinson's (Di Meo et al., 2020, Uddin et al., 2020). These plant-based compounds also chelate metal ions limiting the formation of more harmful free radicals and shielding nerve cell structures (Naoi et al., 2019). Beyond their antioxidant properties, they alter crucial cell signalling pathways, such as those associated with inflammation and cell death such as the NF-κB and Wnt/β-catenin pathways which reduces neuroinflammation and promotes neuronal survival (Al Mamun et al., 2024). They boost the expression of neurotrophic factors and interaction with surface receptors, which supports nerve growth, flexibility, and cognitive function (Arias-Sánchez et al., 2023). In the mitochondrial level, polyphenols helps maintain mitochondrial function and integrity by regulating mitochondrial biogenesis, dynamics, and apoptosis (Campos-Esparza et al., 2009). This is the key supply of energy for nerve cells and helps them resist cell death. Also, polyphenols can modulate neuro-transmitter systems by helping them bring back balance in nerve signals and supports brain health (Rebas et al., 2020). Their ability to cross the blood-brain barrier and safety makes polyphenols suitable for both to prevent and to help treat neurodegenerative diseases alongside other methods (Silva & Pogačnik 2020).

Neuroprotective Mechanisms of Polyphenols

Progressive neuronal loss driven by oxidative stress, chronic inflammation, mitochondrial dysfunction, protein misfolding, and apoptosis is a hallmark of neurodegenerative illnesses like HD, PD, and AD. Many studies have looked at the neuroprotective effects of polyphenols, a broad class of phytochemicals present in many foods including fruits, vegetables, tea, coffee, and wine. Slowing neuronal degeneration and enhancing brain health are considerably aided by their complex mechanisms.

Antioxidant Defense Mechanism:

When there is an accumulation of reactive nitrogen species (RNS) and reactive oxygen species (ROS), it damages the DNA, proteins, and lipids of neurons, a process known as neurodegeneration. They directly scavenging free radicals, polyphenols greatly enhance antioxidant activity, by Preventing Fenton reactions by chlorating redox-active metals such as iron and copper, Enhancing the activity of natural antioxidant enzymes made by the body, including SOD, CAT, and GPx. Furthermore, polyphenols promote cellular defense mechanisms against oxidative damage and activate the Nrf2/ARE pathway, which in turn enhances the transcription of phase II detoxifying enzymes (Scalbert et al., 2005).

Anti-Inflammatory Mechanism:

The production and release of pro-inflammatory cytokines (TNF-α, IL-1β, IL-6) that worsen neuronal injury is caused by activated microglia and astrocytes in chronic neuroinflammation. On one-way polyphenols reduce inflammation is by blocking NF-κB signalling, which is a crucial transcription factor in this process. Another way is by lowering levels of pro-inflammatory mediators such as COX-2 and iNOS, Reduced activation of microglia through modulation of pathways involving Toll-like receptors and To prevent immune-mediated damage to nerve cells, polyphenols inhibit neuroinflammation (Tang & Le, 2016).

Mitochondrial Protection:

Mitochondrial inability leads to reduced ATP synthesis and elevated ROS production, which contribute to neuronal death. Polyphenols maintain mitochondrial integrity by stabilising mitochondrial membranes. Augmenting mitochondrial biogenesis through the activation of PGC-1α (peroxisome proliferator-activated receptor gamma coactivator 1-alpha) and Enhancing ATP synthesis and diminishing mitochondrial oxidative stress. This protection guarantees the preservation of neuronal energy metabolism essential for appropriate synaptic function.

Modulation of Cell Survival Pathways:

Polyphenols modify neuronal survival and death by influencing many cell signalling pathways:

- PI3K/Akt Pathway: Prevents apoptosis and increases neuronal survival.
- MPK Pathways (ERK, JNK, p38): Control reactions to survival cues, inflammation, and oxidative stress.
- Nrf2/ARE Pathway: Promotes the expression of cytoprotective genes.

Polyphenols preserve cellular homeostasis and shield neurones against apoptosis by means of these pathways.

Inhibition of Protein Aggregation: Neurodegenerative diseases often involve the aggregation of misfolded proteins, such as Amyloid-beta plaques in AD, Alpha-synuclein aggregates in PD. Polyphenols like epigallocatechin gallate (EGCG) and resveratrol have demonstrated the ability to: Inhibit protein aggregation, Disaggregate preformed fibrils and Reduce oligomer-induced neurotoxicity. This prevents synaptic dysfunction and neuronal loss caused by toxic protein aggregates.

ENHANCEMENT OF NEUROTROPHIC SUPPORT AND SYNAPTIC PLASTICITY:

Neurotrophins like brain-derived neurotrophic factor (BDNF) are critical for synaptic plasticity, neuronal survival, and cognitive functions. Polyphenols enhance BDNF expression, promoting: Neurogenesis, Synaptic plasticity and Cognitive performance. By supporting neuronal repair and connectivity, polyphenols contribute to cognitive resilience against degenerative processes.

Metal Ion Chelation:

Oxidative damage occurs when metal ion homeostasis is disturbed in the brain. Zinc, iron, and copper are neurotoxic, although certain polyphenols mitigate this effect by acting as natural metal chelators. This effect reduces the neurotoxicity associated with protein Pathy by preventing the aggregation of amyloid-beta and alpha-synuclein, which can be caused by metals.

Polyphenols exhibit promising neuroprotective actions via antioxidant, anti-inflammatory, anti-apoptotic, and anti-aggregation mechanisms. Their ability to modulate multiple pathological processes simultaneously highlights their potential as therapeutic agents in managing neurodegenerative diseases.

Modulation of Mitochondrial Function and Apoptosis:

Polyphenols, which usually targets mitochondria, plays a neuroprotective role due to damaged mitochondria presence in the neurons during neurodegenerative diseases. Understanding their mechanism of action is important as helps to prevent free oxygen and nitrogen radicals, antioxidant gene expression induction and pro- and anti-apoptotic protein regulation (Naoiet al., 2019). Polyphenols also mediate mitochondrial biogenesis, dynamics (fission and fusion), and autophagic degradation into optimal quality and quantity of mitochondria. Moreover, they antagonize membrane potential stabilisation and apoptotic factor release triggered by these stimulants and attenuate cell death in neurons in vitro (Campos-Esparza et al., 2009).

Regulation of Neuroinflammation and Cell Signaling Pathways:

Polyphenols contribute to modifying signal transduction pathways that are implicated in neuroinflammation and neuronal survival besides acting as antioxidants. These polyphenols stop the stimulants enclosing pro-inflammatory mediators like NF-κB pathway and the production of cytokines fostering neurodegeneration (Silva & Pogačnik 2020; Al Mamun et al., 2024). Additionally, polyphenol-induced signaling of the neurotropic factors also contributes to neuronal growth, plasticity and cognitive functions (Arias-Sánchez et al., 2023). Polyphenols exert their influence over pathways such as the Wnt/β-catenin and neurotransmitter systems with the action of sustain neuronal health and perhaps retard the pathology found in neurodegenerative diseases (Rebas, Rzajew et al., 2020). These pleiotropic effects, combined with their capacity to cross the blood-brain barrier and low toxicity profiles, render polyphenols promising agents for neuroprotection. Molecular Mechanisms Underlying Neurodegeneration have shown in Table 1.

Table 1. Molecular mechanisms underlying neurodegeneration

S. No.	Mechanism	Key Features	Associated Diseases	Reference
1	Protein Misfolding & Aggregation	Abnormal accumulation of misfolded proteins like α-synuclein, tau, Aβ	Alzheimer's, Parkinson's, Huntington's	Soto & Pritzkow, 2018
2	Oxidative Stress	Excessive ROS production damages lipids, proteins, and DNA	ALS, Parkinson's, Alzheimer's	Wang et al.,, 2014
3	Mitochondrial Dysfunction	Impaired ATP production, increased ROS, calcium imbalance	Parkinson's, Huntington's, ALS	Booth et al.,, 2017
4	Excitotoxicity	Overactivation of glutamate receptors causes Ca^{2+} influx and neuronal death	ALS, Huntington's, Alzheimer's	Wang et al.,, 2019
5	Neuroinflammation	Chronic activation of microglia and astrocytes releasing cytokines	Alzheimer's, Parkinson's, MS	Heneka et al.,, 2015
6	Impaired Proteostasis	Dysfunction in ubiquitin-proteasome and autophagy-lysosome pathways	Alzheimer's, Parkinson's	Hegde, 2010
7	DNA Damage & Repair Defects	Accumulation of mutations and impaired repair mechanisms	ALS, Ataxias	Madabhushi et al.,, 2014
8	Axonal Transport Defects	Disruption of microtubule-based transport of organelles and proteins	ALS, Huntington's	Millecamps & Julien, 2013
9	Loss of Synaptic Integrity	Synaptic dysfunction and degeneration before neuron loss	Alzheimer's, Parkinson's	Palop & Mucke, 2010
10	Epigenetic Dysregulation	Aberrant methylation, histone modification affecting gene expression	Alzheimer's, Huntington's	De Jager et al.,, 2014

PRINCIPAL CELLULAR SIGNALLING PATHWAYS INVOLVED

Numerous cellular signalling mechanisms modulate neuroinflammation, influencing glial activation and neuronal well-being:

Nuclear Factor-kappa B (NF-κB) Signalling Pathway

Immune responses, inflammation, cell proliferation, and survival are all crucially regulated by the Nuclear Factor-kappa B (NF-κB) signalling system. Inhibitor proteins (IκBs) keep NF-κB transcription factors in the cytoplasm when cells are at rest. When the IκB kinase (IKK) complex is stimulated by signals like cytokines, infections, or stress, IκBs are phosphorylated and degraded. This facilitates the translocation of NF-κB dimers, primarily made up of p65 and p50 subunits, into the nucleus, where they stimulate the transcription of target genes related to cell survival, inflammation, and immunological control.

Because pro-inflammatory genes are spontaneously and persistently activated, dysregulated NF-κB signalling plays a role in cancer, autoimmune illnesses, and chronic inflammatory diseases. NF-κB is a prospective therapeutic target in conditions including rheumatoid arthritis and some kinds of cancer because of its essential involvement in inflammation and immunity (Zhang et al., 2017).

Mitogen-Activated Protein Kinase (MAPK) Routes

Mitogen-Activated Protein Kinase (MAPK) pathways are essential signalling cascades that control cellular functions such as proliferation, differentiation, inflammation, and death. The three principal MAPK pathways are ERK (extracellular signal-regulated kinase), JNK (c-Jun N-terminal kinase), and p38 MAPK (Kyriakis & Avruch 2012). These cascades are generally initiated by growth factors, cytokines, or stressors. Following activation, MAPKs phosphorylate several transcription factors and proteins, hence regulating gene expression and cellular responses. The dysregulation of MAPK signalling is associated with cancer, inflammatory illnesses, and neurological disorders, rendering these pathways relevant therapeutic targets (Huddleston, 2011).

NLRP3 Inflammasome

The NLRP3 inflammasome is a crucial multiprotein complex in innate immunity, tasked with detecting pathogenic microbes and cellular stress. This inflammasome, primarily consisting of the sensor protein NLRP3 (NOD-like receptor family pyrin domain-containing 3), the adaptor protein ASC (apoptosis-associated speck-like protein containing a CARD), and pro-caspase-1, is activated upon the detection of stimuli such as ATP, crystalline substances, or reactive oxygen species. Upon activation, NLRP3 oligomerizes and recruits ASC, subsequently attracting pro-caspase-1, resulting in its activation (Heneka et al., 2014). Active caspase-1 converts pro-inflammatory cytokines such as interleukin-1β (IL-1β) and interleukin-18 (IL-18) into their mature forms, hence enhancing inflammation. Although essential to host defence, dysregulated NLRP3 activation is implicated with chronic inflammatory conditions, including gout, type 2 diabetes, and neurodegenerative illnesses. NLRP3 is an emerging therapeutic target for managing excessive inflammation due to its pathogenic significance (Swanson et al., 2019).

Janus Kinase/Signal Transducer and Activator of Transcription (JAK/STAT) Pathway

The Janus Kinase/Signal Transducer and Activator of Transcription (JAK/STAT) pathway is an essential signalling mechanism that regulates immunological function, cellular proliferation, differentiation, and survival. This pathway, activated mostly by cytokines and growth factors, facilitates communication from extracellular signals to the nucleus, hence affecting gene expression.

The process initiates when cytokines attach to their designated cell surface receptors, resulting in receptor dimerisation. This facilitates the closeness of related Janus Kinases (JAKs)—a family of non-receptor tyrosine kinases—thereby facilitating their activation by trans-phosphorylation (Shuai & Liu, 2003). Upon activation, JAKs phosphorylate particular tyrosine residues on the receptor's intracellular domain, establishing docking sites for Signal Transducers and Activators of Transcription (STAT) proteins (Villarino et al., 2015). Upon binding to these phosphorylated sites, STATs are phosphorylated by JAKs. Phosphorylated STATs dimerise and translocate to the nucleus, where they bind to specific DNA sequences to regulate the transcription of target genes associated with immunity, inflammation, cell proliferation, and death. The JAK/STAT system is stringently controlled by negative feedback mechanisms, including Suppressors of Cytokine Signalling (SOCS) proteins and protein tyrosine phosphatases, to avert excessive or protracted activation (Banerjeeet al., 2017). The dysregulation of this pathway is associated with numerous clinical illnesses, such as autoimmune diseases, malignancies, and inflammatory disorders. The JAK/STAT pathway is a crucial therapeutic target due to its pivotal involvement in immune signalling and disease development. JAK inhibitors, like tofacitinib and ruxolitinib, have been formulated to address illnesses such as rheumatoid arthritis, psoriasis, and certain cancers.(O'Shea & Plenge, 2012).

PI3K/Akt/mTOR Signalling Pathway

The PI3K/Akt/mTOR signalling system is an essential intracellular cascade that regulates cell growth, proliferation, metabolism, and survival. Activation often initiates when growth hormones or insulin attach to receptor tyrosine kinases (RTKs), resulting in the recruitment and activation of phosphoinositide 3-kinase (PI3K). PI3K phosphorylates phosphatidylinositol (4,5)-bisphosphate (PIP2) to produce PIP3, which attracts Akt (protein kinase B) to the plasma membrane (Fruman, Chiu et al., 2017). Akt is triggered by phosphorylation by PDK1 and the mTOR complex 2 (mTORC2). Activated Akt phosphorylates downstream targets that regulate glucose metabolism, protein synthesis, and the prevention of apoptosis (Saxton & Sabatini, 2017). A primary objective is the mechanistic target of rapamycin (mTOR), a serine/threonine kinase that assembles into two separate complexes: mTORC1 and mTORC2. mTORC1 modulates protein synthesis through S6K1 and 4E-BP1, facilitating cellular proliferation. The dysregulation of the PI3K/Akt/mTOR pathway is associated with cancer, diabetes, and neurodegenerative illnesses, positioning it as a significant therapeutic target in oncology and metabolic disorders (Salman et al., 2018). Major Cell Signaling Pathways Regulating Neuroinflammation are shown in Table 2.

Table 2. Major cell signaling pathways regulating neuroinflammation

Pathway	Key Regulators	Effect on Neuro-inflammation	Therapeutic Target Example	Reference
NF-κB	IκB, RelA/p65	Promotes pro-inflammatory cytokines	Curcumin, BAY 11-7082	Liu et al.,, 2017
MAPK (p38, JNK)	p38 MAPK, ERK1/2, JNK	Promotes glial activation and inflammation	SB203580 (p38 inhibitor)	Kim & Choi, 2010
NLRP3 Inflammasome	NLRP3, Caspase-1	Activates IL-1β, IL-18, pyroptosis	MCC950 (NLRP3 inhibitor)	Swanson et al.,, 2019
JAK/STAT	JAK1/2, STAT1/3	Cytokine-mediated pro-inflammatory signaling	Tofacitinib (JAK inhibitor)	Villarino et al.,, 2017
PI3K/Akt/mTOR	PI3K, Akt, mTOR	Anti-inflammatory / pro-survival effects	Rapamycin (mTOR inhibitor)	Huang et al.,, 2020

MODULATION STRATEGIES AND THERAPEUTIC APPROACHES

Pharmacological Inhibitors: Numerous small molecule inhibitors aimed at NF-κB, MAPK, and JAK/STAT pathways have demonstrated effectiveness in preclinical neuroinflammation models. BAY 11-7082 suppresses NF-κB activation, hence diminishing the generation of pro-inflammatory cytokines by microglia (Zhang et al., 2017). MCC950 specifically suppresses NLRP3 inflammasome activation and mitigates neuroinflammation in Alzheimer's disease models (Heneka et al., 2014).

Natural Compounds and Phytochemicals:

Polyphenols, including curcumin, resveratrol, and quercetin, regulate many signalling pathways, such as NF-κB and MAPK, demonstrating anti-inflammatory and antioxidant properties (Stith et al., 2018). These molecules offer a multitarget strategy with reduced adverse effects relative to manufactured pharmaceuticals.

Gene Suppression and RNA-Directed Treatments:

RNA interference aimed at crucial inflammatory mediators (e.g., NLRP3, NF-κB subunits) has surfaced as a viable approach to mitigate neuroinflammation in experimental models. Nonetheless, clinical translation is constrained by delivery challenges.

Modulation of Microbiota:

Recent research linked gut bacteria with central nervous system inflammation through the gut-brain axis. Modulating the microbiome through prebiotics, probiotics, or dietary treatments can indirectly influence neuroinflammatory signalling pathways (Cryan et al., 2019).

Endogenous Cellular Resilience and Alternative Splicing:

Cyto-protection furthers cell survival and tissue resilience from being injured from reactive oxygen species, inflammation, and other toxic stimuli. Neurons possess self-repair pathways that may assist these cells in living through damage or neurodegeneration. Neuroprotection includes increasing expression of protective proteins, activation of autophagy, as well as modulating stress response pathways for the preservation of neuronal integrity and function (Marmolejo-Martínez-Artesero et al., 2021). Another important process is alternative splicing, which enables the cells to produce different protein variants from a single gene. Alternative splicing in response to environmental changes can give rise to protein variants that Favor neuronal survival and adaptation. This thus opens promising means for neuroprotection and provides a therapeutic target for intervention.

Neurogenesis, Glial Modulation, and Lifestyle Factors:

Neuroprotection is perhaps achieved by promoting neurogenesis, the process of generating new neurons, and modifying the activity of other brain cells called glial cells. For instance, stem cell-secreted extracellular vesicles can carry neuroprotective and neurogenic factors to the brain to reduce neuronal injury and to foster memory function (Ma et al., 2020). Another type of CNS cell (glial cells) is astrocytes, that can also acquire neuroprotective phenotypes inhibit harmful inflammation and enhancing neuronal survival under the action of specific molecular switches (Cameron et al., 2024). In addition to that some lifestyle-related factors, such as physical activity, cognitive stimulation, and social engagement, have been demonstrated to exert neuroprotective effects, stimulate neurotrophic signalling and to build cognitive reserve, which reduces the risk of developing dementia in age-related brain disorders.

Comparison with Existing Pharmacological Approaches

Cholinesterase inhibitors (like donepezil and rivastigmine), NMDA receptor antagonists (like memantine), dopamine precursors (like levodopa), and MAO-B inhibitors (like selegiline) are all traditional drugs used to treat neurodegenerative illnesses. They mostly help with symptoms. These medications usually focus on specific pathways and are often linked to limited success in stopping disease progression. Additionally, negative consequences include stomach problems, trouble sleeping, and motor problems are typical (Cummings et al., 2019). Conversely, polyphenolic chemicals and other bioactive sourced from natural origins offer a multi-targeted treatment framework. These substances affect a number of related processes, including as oxidative stress, neuroinflammation, problems with mitochondria, and proteins that don't fold correctly. For example:

Resveratrol - activates SIRT1 and inhibits NF-κB, which helps fight inflammation and amyloid (Porquet et al., 2014).Curcumin - affects MAPK and JAK/STAT signalling, which helps neurones stay alive and stops inflammatory cascades (Gupta et al., 2013).Epigallocatechin gallate (EGCG) - from green tea stops amyloid-beta from clumping together and helps keep mitochondria healthy (Mandel et al.,2008).

These bioactive exhibit enhanced tolerance and reduced toxicity profiles, rendering them appropriate for prolonged administration and preventive measures. its capacity to traverse the blood-brain barrier and modulate neurotrophic signalling pathways amplifies its therapeutic significance (Rege et al., 2015). Conventional pharmacological medicines are limited by their restricted mechanisms and side-effect profiles. In contrast, polyphenols provide a systems-level approach that aligns with personalised therapy. (Wang et al., 2020).

Impact of Lifestyle Variables

Lifestyle factors can profoundly influence neurogenesis by affecting neuroinflammation, oxidative stress, and neurotrophic signalling.

Physical Activity: Aerobic exercise is a highly effective promoter of neurogenesis in the adult hippocampus. It elevates BDNF levels, stimulates NSC growth, and facilitates synaptic plasticity (Van Praag, Fleshner et al., 2014). Exercise modifies glial function by promoting microglial polarisation to the M2 phenotype and diminishing astrocyte reactivity. Dietary Practices and Nutritional Science: Diets abundant in polyphenols, omega-3 fatty acids, and antioxidants enhance neurogenesis. Compounds such as curcumin and resveratrol diminish neuroinflammation and oxidative stress, fostering an environment favourable for neural stem cell survival (Gómez-Pinilla 2008). In contrast, diets rich in fat and sugar hinder neurogenesis by exacerbating systemic and central nervous system inflammation. Cognitive Stimulation: Environmental enrichment, encompassing cognitive training and learning activities, enhances hippocampus neurogenesis by promoting neurotrophic factor expression and improving synaptic connections. Cognitive engagement also inhibits glial overactivation, preserving homeostasis in the neurogenic niche (Sierra, Encinas et al., 2010). Sleep and Stress: Chronic stress and sleep deprivation adversely affect neurogenesis via increasing cortisol levels and facilitating glial-mediated neuroinflammation. In contrast, stress management and sufficient sleep facilitate neurogenic processes.

Neurogenesis is a dynamic process governed by glial cells and affected by lifestyle factors. Regulating glial activation and implementing advantageous lifestyle modifications—such as physical activity, adequate diet, cognitive stimulation, and stress reduction—can cultivate a neurogenic milieu, offering prospects for preventive and therapeutic approaches to neurodegenerative diseases.

OTHER NEUROACTIVE PHYTOCHEMICALS AND BIOACTIVE MOLECULES

Classes and Examples of Neuroactive Phytochemicals:

From terpenoids to alkaloids, carotenoids, essential fatty acids, phytosterols, and even some essential oils, an ample variety of substances represent phytochemicals with possible neuroprotective and neuromodulator activities, derived from plants, fungi, and some animals (Mohd Sairazi and Sirajudeen 2020, Babazadeh, Vahed et al., 2023). Many of these compounds tend to act through a handful of mechanisms-NMDA receptor antagonism, antioxidant power, anti-inflammatory activities, versus reactive oxygen species formation, or apoptotic activation but they can modulate neurotransmitter systems, neurotrophic signaling, and gene expression (Naoi, Inaba-Hasegawa et al., 2017). For instance, they may modulate biosynthesis in serotonin, dopamine, and norepinephrine, paralleled with mood, cognition, and behavior. Some phytochemical activation of Nrf2 signaling-related antioxidant mechanisms in brain cells is exhibited by curcumin, resveratrol, and sulforaphane. These ginsenosides and several flavonoids, work in the help of or by mimicking the action of neurotrophic factors that promote growth and survival of neurons (Bai, Bian et al., 2023).

Neuroactive phytochemicals exhibit structural and functional diversity, often categorised according to their chemical composition and biological functions. The primary categories comprise polyphenols, alkaloids, terpenoids, saponins, sterols, and polysaccharides. Polyphenols: Polyphenols constitute the most extensive category of neuroactive phytochemicals, recognised for their robust antioxidant and anti-inflammatory characteristics. Plant Sources and Therapeutic Applications are shown in Table 3. They are categorised into multiple subgroups: **Flavonoids**

Flavonoids are extensively found in fruits, vegetables, tea, and wine. Quercetin, present in apples, onions, and berries, functions as an antioxidant, regulates cell signaling, suppresses lipid peroxidation, and safeguards neurons from oxidative stress-related damage. Epigallocatechin-3-gallate (EGCG): The predominant catechin in green tea, EGCG safeguards against neuroinflammation, oxidative stress, and amyloid-beta aggregation. Rutin, found in citrus fruits, exhibits neuroprotective properties via its antioxidant capabilities and anti-apoptotic processes. Luteolin, present in celery and green peppers, demonstrates anti-inflammatory and neuroprotective properties through the suppression of pro-inflammatory cytokines. **Phenolic Acids**

Phenolic acids are secondary metabolites present in coffee, fruits, and vegetables. Caffeic acid functions as an antioxidant and mitigates neuroinflammation. Ferulic acid: Demonstrates neuroprotective properties against oxidative stress and safeguards dopaminergic neurones. Rosmarinic acid: Present in rosemary and mint, recognised for its capacity to diminish neuroinflammation and mitigate amyloid-beta toxicity. **Stilbenes**: These compounds possess a 1,2-diphenylethylene core. Resveratrol, present in grapes and red wine, activates SIRT1, demonstrates antioxidant properties, controls neuroinflammation, and diminishes amyloid formation. **Lignans**: Sesamin, present in sesame seeds, possesses antioxidant and anti-inflammatory effects that provide neuroprotection.

Alkaloids: Alkaloids are nitrogenous chemicals that exhibit notable pharmacological effects on the central nervous system.

- Huperzine A: Extracted from Huperzia serrata, it functions as a reversible acetylcholinesterase inhibitor, enhancing cognitive performance in Alzheimer's disease patients.

• Berberine: Present in Berberis species, berberine regulates neurotransmitter systems, diminishes neuroinflammation, and demonstrates anti-amyloid properties.

Caffeine, found in coffee and tea, functions as an adenosine receptor antagonist, promoting alertness and maybe diminishing the risk of neurodegenerative disorders.

Nicotine, present in tobacco leaves, activates nicotinic acetylcholine receptors, enhancing attention and memory at low dosages; yet, its medicinal application is contentious due to the risk of addiction.

Arecoline, derived from Areca catechu, functions as a muscarinic receptor agonist; nevertheless, its neurotoxic and addictive characteristics constrain its therapeutic applicability.

Terpenoids

Terpenoids are lipophilic compounds included in essential oils and numerous medicinal botanicals.

Ginkgolides and Bilobalide: Extracted from Ginkgo biloba, these chemicals demonstrate neuroprotective effects through free radical scavenging, platelet-activating factor reduction, and cerebral blood flow enhancement.

Curcumin: A diferuloylmethane derived from turmeric (Curcuma longa), curcumin exhibits neuroprotective, anti-inflammatory, antioxidant, and anti-amyloid characteristics. Curcumin also regulates other signalling pathways, including NF-κB and MAPK.

Cannabidiol (CBD), derived from Cannabis sativa, interacts with CB1 and CB2 receptors, producing anxiolytic, anti-inflammatory, and neuroprotective benefits without inducing euphoric side effects.

Ginsenosides: Active saponins found in Panax ginseng, including as Rb1 and Rg1, promote neurogenesis, mitigate oxidative stress, and regulate neurotransmitter systems.

Saponins

Saponins are glycosides possessing soap-like characteristics. Ginsenosides: Besides being terpenoids, ginsenosides are categorised as triterpenoid saponins. They improve cognitive function and neuronal viability. Bacosides, present in Bacopa monnieri (Brahmi), facilitate memory enhancement, mitigate oxidative damage, and stimulate synaptic activity.

Sterols

Plant sterols exhibit structural similarity to cholesterol and influence brain health.

Withanolides, derived from Withania somnifera (Ashwagandha), possess anti-inflammatory, antioxidant, and neuroprotective activities. They promote axonal regeneration and synaptic reorganisation. β-Sitosterol: Present in nuts and seeds, β-sitosterol regulates membrane fluidity and may safeguard brain membranes from oxidative damage.

Polysaccharides

Polysaccharides, particularly derived from medicinal fungi, exhibit neuroprotective properties. Ganoderma lucidum polysaccharides demonstrate antioxidant and anti-inflammatory properties that safeguard neurones from oxidative damage. Polysaccharides from Hericium erinaceus (Lion's Mane): Facilitate the synthesis of nerve growth factor (NGF), hence enhancing neurogenesis and myelination.

Table 3. Plant sources and therapeutic applications

Plant Source	Neuroactive Phytochemicals	Therapeutic Potential	Reference
Ginkgo biloba	Ginkgolides, Bilobalide	AD, cognitive enhancement	Smith & Luo, 2004; Mahadevan & Park, 2008
Curcuma longa	Curcumin	AD, PD, anti-amyloid, anti-inflammatory	Gupta et al.,, 2013; Small et al.,, 2018
Camellia sinensis	EGCG, catechins	PD, AD, antioxidant, neuroprotective	Mandel et al.,, 2008; Panickar, 2013
Vitis vinifera	Resveratrol	AD, PD, anti-amyloid, anti-inflammatory	Sawda et al.,, 2017; Turner et al.,, 2015
Bacopa monnieri	Bacosides	Cognitive enhancement, memory support	Calabrese et al.,, 2008; Kongkeaw et al.,, 2014
Panax ginseng	Ginsenosides	Memory, neuroprotection, anti-fatigue	Reay et al.,, 2005; Yuan et al.,, 2021
Withania somnifera	Withanolides	Cognitive function, neuroprotection	Kulkarni & Dhir, 2008; Tiwari et al.,, 2014
Cannabis sativa	Cannabidiol (CBD)	Anxiety, neuroprotection, seizure control	Ibeas Bih et al.,, 2015; Devinsky et al.,, 2014
Huperzia serrata	Huperzine A	Acetylcholinesterase inhibition in AD	Wang et al.,, 2006; Zhou et al.,, 2017
Ganoderma lucidum	Polysaccharides	Antioxidant, neuroprotective	Wachtel-Galor et al.,, 2011; Xu et al.,, 2011
Hericium erinaceus	Polysaccharides, hericenones	Neurogenesis, cognitive support	Wong et al.,, 2016; Nagano et al.,, 2010

Obstacles and Constraints

• Bioavailability: Numerous phytochemicals, including curcumin and resveratrol, exhibit inadequate absorption, fast metabolism, and restricted brain penetration.
• Challenges of Standardisation: The inconsistency in plant extracts and absence of standardised dose impede reproducibility.
• Safety and Toxicity: Elevated dosages or prolonged usage may present hazards; extensive safety investigations over the long term are scarce.
• Regulatory Obstacles: The majority of neuroactive phytochemicals are categorised as supplements rather than pharmaceuticals, constraining their clinical advancement.
Prospective Trajectories
• Nanotechnology: Utilisation of nanoparticles, liposomes, and micelles to enhance delivery and bioavailability.
• Synergistic Formulations: Integrating several phytochemicals or employing them alongside synthetic medications to augment therapeutic effectiveness.

- Network Pharmacology: Comprehending multi-target effects via systems biology to enhance phytochemical-based therapeutics.

Neuroactive phytochemicals offer promising multi-target treatment strategies for the management of neurodegenerative illnesses and the enhancement of cognitive function. Their natural origin, along with various pharmacological actions including antioxidant, anti-inflammatory, and neurotrophic properties, renders them appealing candidates for neuroprotective methods. Ongoing research aimed at addressing issues such as inadequate bioavailability and standardisation, in conjunction with clinical validations, will ascertain their prospective function in neurotherapeutics (Escartin, Galea et al., 2021).

Mechanisms and Therapeutic Potential:

Plant chemicals that act on the nervous system impact upon more than one pathway associated with brain diseases. These are oxidative cell damage, protein aggregation, mitochondrial dysfunction, and impairment of autophagy (Khan et al., 2020, Limanaqi, Biagioni et al., 2020, Stacchiotti and Corsetti 2020). They can fix mitochondrial balance in cells and help remove misfolded proteins to change inflammatory response. All these slow down neuronal cell death (Naoi, Wu et al., 2019). Some plant extracts and pure compounds can get into the blood-brain barrier to protect neuronal cells in pre-clinical models. This shows better outcomes for diseases like Alzheimer's Parkinson's, and Huntington's (Sánchez-Martínez, Valdés et al., 2022). Provided the very early research looks good but more human studies are needed to prove that they work and determine the best route of administration and safety for use on humans (Khan, Jahan et al., 2020).

Synergistic Effects and Multi-Targeted Therapies

Neuroprotective and promise-altering approaches must have synergistic effects and multi-targeted neuroactive compounds. These methods use combined mechanisms of multiple bioactive agents for better efficacy than single-agent treatments, which commonly involve simultaneous targeting of multiple pathophysiological pathways.

The combination of saikosaponin A and albiflorin showing the synergetic neuroprotective effect against corticosterone-triggered apoptosis and this combination is accomplished via the regulation of metabolic pathways, enhancing of mitochondrial function, decreasing of oxidative stress and inhibition of neuroinflammation. The combination modulates more metabolites and influences metabolic balance and inflammation stronger than the single compounds, thus encouraging multi-target intervention (Li, Hou et al., 2022).

With a similar approach, synergistic neuroprotective effects have been demonstrated with natural products applied in Alzheimer's disease neurodegeneration animal models: docosahexaenoic acid, Ginkgo biloba, D-pinitol, and ursolic acid. The combination benefited survival time and lifespan in animal models, along with significant reductions in amyloid-beta plaque accumulation, and enhanced cognitive performance in animals, outperforming the effects of individual components. These results opens up the possibility of formulating mixtures of either synthetic or bioactive molecules to simultaneously address multiple hallmarks of neurodegeneration (Griñán-Ferré et al., 2021).

These also have been a topic of synergistic benefits when pairing synthetic drugs with natural compounds. For example, neuroprotection from cerebral ischemia/reperfusion injury was best provided by the simultaneous application of rosuvastatin and resveratrol combinations than either on its own. This is due to the anti-apoptotic, anti-inflammatory, and autophagy-activating synergic actions, showing the benefit of simultaneously attacking several neurodegenerative pathways (Liu, Yang et al., 2018). Beta vulgaris juice and DMSA have a synergistic effect on reducing lead-induced neurotoxicity. They also boost antioxidant and anti-inflammatory responses and help regulate enzymes linked to neurotransmitters (Shaban, Abd El-Kader et al., 2021). Likewise, the combination of phytochemical compounds, such as quercetin, folic acid, with conventional drugs poses the possibility of enhanced antioxidant, anti-inflammatory, and acetylcholinesterase-inhibitory activities. Hence, these combinations could prove suitable as dietary supplements or adjuvant therapies for neurological disorders. Synergistic inhibitory effect on acetylcholinesterase was seen from the combination of Safranin O with quercetin or gallic acid, thus propounding the quest for new multi-targeted anti-neurodegenerative drug treatments (Imededdine et al., 2023).

Real-World Translational Barriers

Despite promising preclinical studies, numerous challenges in real-world applications hinder the use of polyphenol-based therapeutics and enhanced delivery systems for neurodegenerative illnesses.

Restricted Clinical Evidence - The majority of studies on polyphenols are confined to in vitro or animal models. Human trials are scarce, and those that exist often lack sufficient power or standardised outcome measures (Vauzour et al., 2010).

Bioavailability and Metabolism - Natural compounds like curcumin and resveratrol demonstrate inadequate absorption, rapid metabolism, and reduced systemic retention, hence influencing dose optimisation and therapeutic reliability (Wang et al., 2020).

Regulatory Complexity - Polyphenols occupy a position between nutraceuticals and medicines, complicating the understanding of applicable regulations. Approval processes vary by jurisdiction and sometimes lack defined criteria for safety and efficacy (Patel et al., 2021).

Manufacturing and Scalability - Nanocarriers and intelligent delivery systems require sophisticated manufacturing techniques that may not be economically viable or easily scalable for mass production (Kakkar et al., 2019).

Patient Adherence and Acceptance - Innovative forms such as microneedle patches or intranasal gels may face resistance due to unfamiliarity, discomfort, or cultural inappropriateness, particularly among elderly individuals.

Intellectual Property and Commercialisation - The prevalence of polyphenols in commonly consumed foods complicates patent acquisition and diminishes the business incentives for pharmaceutical companies to spend (Yang et al., 2022).

To overcome these challenges, collaboration across disciplines is essential, alongside the design of robust clinical trials and the formulation of policies that promote innovation while ensuring safety and accessibility.

Future-Oriented Perspectives on Innovative Delivery Methods

Polyphenols and other bioactive compounds often exhibit restricted therapeutic efficacy due to their poor absorption, rapid degradation, and instability in physiological environments. Innovative delivery systems are being devised to enhance targeted distribution, regulated release, and penetration of the blood-brain barrier.

Nanotechnology based systems - such as liposomes, solid lipid nanoparticles (SLNs), and polymeric nanoparticles, have shown promise in improving the pharmacokinetics of polyphenols. Curcumin-loaded nanoparticles have demonstrated enhanced brain absorption and neuroprotective benefits in preclinical animal studies (Kakkar et al., 2019).

Hydrogel based - carriers and microneedle patches offer less invasive techniques for prolonged release, particularly for transdermal and intranasal applications. These platforms can circumvent first-pass metabolism and deliver drugs directly to the central nervous system (Patel et al., 2021).

Exosome-mediated - delivery is a study domain that utilises natural vesicles for the safe and targeted transport of bioactive compounds. Researchers have engineered exosomes derived from mesenchymal stem cells capable of transporting resveratrol and other neuroprotective pharmaceuticals across the blood-brain barrier (Yang et al., 2022).

Researchers are exploring "smart delivery systems" capable of releasing therapeutic agents exclusively in diseased tissues in response to variations in pH, temperature, or oxidative stress. These solutions align with the objectives of precision medicine and may reduce off-target effects while enhancing efficacy. As research advances, the integration of these innovative delivery vehicles with polyphenol-based therapeutics may revolutionise treatment approaches for neurodegenerative diseases.

CHALLENGES

Most neuroactive substances like natural products, peptides, etc., shows an inability to cross the blood-brain barrier efficiently, thus experiencing reduced bioavailability and hence, having limited therapeutic impact in the brain. Many compounds get metabolized rapidly or degraded before they hit their biological targets and their pharmacokinetics and pharmacodynamics in humans remain poorly understood. Though many compounds show promising results at the preclinical stage, few clinical trials have been either inconclusive or show modest results due to varying compound purities, doses, and patient populations. Safety and tolerability concerns arise, as some injections or neuroactive steroids may have levels of neurotoxicity or addictive potential.

REFERENCES

Al Mamun, A., Nsiah, N. Y., Hossain, M. A., Islam, M. R., Herrera-Calderon, O., Baghel, M., & Hafez, M. M. (2024). Polyphenols targeting NF-κB pathway in neurological disorders: What we know so far? *International Journal of Biological Sciences*, *20*(4), 1332–1355. DOI: 10.7150/ijbs.90982 PMID: 38385077

Arias-Sánchez, R. A., Saralegui, Y., Hernández-Baltazar, D., & Rovirosa-Hernández, M. J. (2023). Polyphenols and neurodegenerative diseases: Potential effects and mechanisms of neuroprotection. *Molecules (Basel, Switzerland)*, *28*(14), 5415. Advance online publication. DOI: 10.3390/molecules28145415 PMID: 37513286

Babazadeh, A., Ghanbarzadeh, B., & Hamishehkar, H. (2023). Natural bioactive molecules as neuromedicines for the treatment/prevention of neurodegenerative diseases. *ACS Omega*, *8*(4), 3667–3683. DOI: 10.1021/acsomega.2c06098 PMID: 36743024

Bai, X., Bian, Z., Zhang, M., & Li, J. (2023). Targeting the Nrf2 signaling pathway using phytochemical ingredients: A novel therapeutic road map to combat neurodegenerative diseases. *Phytomedicine*, *109*, 154582. Advance online publication. DOI: 10.1016/j.phymed.2022.154582 PMID: 36610130

Banerjee, S., Biehl, A., Gadina, M., Hasni, S., & Schwartz, D. M. (2017). JAK–STAT signaling as a target for inflammatory and autoimmune diseases: Current and future prospects. *Drugs*, *77*(5), 521–546. DOI: 10.1007/s40265-017-0701-9 PMID: 28255960

Bhullar, K. S., & Rupasinghe, H. P. V. (2013). Polyphenols: Multipotent therapeutic agents in neurodegenerative diseases. *Oxidative Medicine and Cellular Longevity*, *2013*, 891748. Advance online publication. DOI: 10.1155/2013/891748 PMID: 23840922

Bond, S., Driano, J. M., & Costa, A. S. H. (2020). The integrated stress response and phosphorylated eukaryotic initiation factor 2α in neurodegeneration. *Journal of Neuropathology and Experimental Neurology*, *79*(2), 123–143. DOI: 10.1093/jnen/nlz129 PMID: 31913484

Cameron, E. G., Götz, M., & Goldman, S. A. (2024). A molecular switch for neuroprotective astrocyte reactivity. *Nature*, *626*(7999), 574–582. DOI: 10.1038/s41586-023-06935-3 PMID: 38086421

Campos-Esparza, M. R., Sánchez-Gómez, M. V., & Matute, C. (2009). Molecular mechanisms of neuroprotection by two natural antioxidant polyphenols. *Cell Calcium*, *45*(4), 358–368. DOI: 10.1016/j.ceca.2008.12.007 PMID: 19201465

Cryan, J. F., O'Riordan, K. J., Cowan, C. S. M., Sandhu, K. V., Bastiaanssen, T. F. S., Boehme, M., Codagnone, M. G., Cussotto, S., Fulling, C., Golubeva, A. V., Guzzetta, K. E., Jaggar, M., Long-Smith, C. M., Lyte, J. M., Martin, J. A., Molinero-Pérez, A., Moloney, G., Morelli, E., Morillas, E., & Dinan, T. G. (2019). The microbiota-gut-brain axis. *Physiological Reviews*, *99*(4), 1877–2013. DOI: 10.1152/physrev.00018.2018 PMID: 31460832

Del Rio, D., Borges, G., & Crozier, A. (2010). Polyphenols and health: What compounds are involved? *Nutrition, Metabolism, and Cardiovascular Diseases*, *20*(1), 1–6. DOI: 10.1016/j.numecd.2009.05.015 PMID: 19713090

Di Meo, F., Margarucci, S., Galderisi, U., Crispi, S., & Peluso, G. (2020). Bioactive polyphenols and neuromodulation: Molecular mechanisms in neurodegeneration. *International Journal of Molecular Sciences*, *21*(7), 2564. Advance online publication. DOI: 10.3390/ijms21072564 PMID: 32272735

Ellwardt, E., & Zipp, F. (2014). Molecular mechanisms linking neuroinflammation and neurodegeneration in MS. *Experimental Neurology, 262*(Pt A), 8–17. https://doi.org/DOI: 10.1016/j.expneurol.2014.02.006

Escartin, C., Galea, E., Lakatos, A., O'Callaghan, J. P., Petzold, G. C., Serrano-Pozo, A., Steinhäuser, C., Volterra, A., Carmignoto, G., Agarwal, A., Allen, N. J., Araque, A., Barbeito, L., Barzilai, A., Bergles, D. E., Bonvento, G., Butt, A. M., Chen, W.-T., Cohen-Salmon, M., & Verkhratsky, A. (2021). Reactive astrocyte nomenclature, definitions, and future directions. *Nature Neuroscience*, *24*(3), 312–325. DOI: 10.1038/s41593-020-00783-4 PMID: 33589835

Esmaeili, Y., Yarjanli, Z., Pakniyat, A., Aghajanzadeh, T., Namdar, A., & Kheiry, F. (2022). Targeting autophagy, oxidative stress, and ER stress for neurodegenerative disease treatment. *Journal of Controlled Release*, *345*, 147–175. DOI: 10.1016/j.jconrel.2022.03.001 PMID: 35248646

Esteras, N., Kundel, F., Amodeo, G. F., Pavlov, E. V., Klenerman, D., & Abramov, A. Y. (2017). Mitochondrial hyperpolarization in iPSC-derived neurons from patients of FTDP-17 with 10+16 MAPT mutation leads to oxidative stress and neurodegeneration. *Redox Biology*, *12*, 410–422. DOI: 10.1016/j.redox.2017.03.008 PMID: 28319892

Fruman, D. A., Chiu, H., Hopkins, B. D., Bagrodia, S., Cantley, L. C., & Abraham, R. T. (2017). The PI3K pathway in human disease. *Cell*, *170*(4), 605–635. DOI: 10.1016/j.cell.2017.07.029 PMID: 28802037

Gómez-Pinilla, F. (2008). Brain foods: The effects of nutrients on brain function. *Nature Reviews. Neuroscience*, *9*(7), 568–578. DOI: 10.1038/nrn2421 PMID: 18568016

Grabska-Kobyłecka, I., Szpakowski, P., Król, A., & Kobyłecki, A. (2023). Polyphenols and their impact on the prevention of neurodegenerative diseases and development. *Nutrients*, *15*(15), 3454. Advance online publication. DOI: 10.3390/nu15153454 PMID: 37571391

Griñán-Ferré, C., Corpas, R., Puigoriol-Illamola, D., Palomera-Ávalos, V., Sanfeliu, C., & Pallàs, M. (2021). Synergistic neuroprotective effects of a natural product mixture against AD hallmarks and cognitive decline in *Caenorhabditis elegans* and an SAMP8 mice model. *Nutrients*, *13*(7), 2411. Advance online publication. DOI: 10.3390/nu13072411 PMID: 34371921

Guo, W., Stoklund Dittlau, K., & Van Den Bosch, L. (2020). Axonal transport defects and neurodegeneration: Molecular mechanisms and therapeutic implications. *Seminars in Cell & Developmental Biology*, *99*, 133–150. DOI: 10.1016/j.semcdb.2019.07.010 PMID: 31542222

Heneka, M. T., Kummer, M. P., & Latz, E. (2014). Innate immune activation in neurodegenerative disease. *Nature Reviews. Immunology*, *14*(7), 463–477. DOI: 10.1038/nri3705 PMID: 24962261

Higham, J. P., Hidalgo, A., & Gistelinck, M. (2019). Restoration of olfactory memory in *Drosophila* overexpressing human Alzheimer's disease associated tau by manipulation of L-type Ca^{2+} channels. *Frontiers in Cellular Neuroscience*, *13*, 409. Advance online publication. DOI: 10.3389/fncel.2019.00409 PMID: 31551716

Huddleston, J. E. (2011). Metabolic regulation by ERK. *Nature Reviews. Molecular Cell Biology, 12*(9), 546. DOI: 10.1038/nrm3179 PMID: 21860387

Imededdine, K., Yahya, M. F. Z. R., Bouzina, M. A., & Touil, A. (2023). Molecular interactions, binding stability, and synergistic inhibition on acetylcholinesterase activity of safranin O in combination with quercetin and gallic acid: In vitro and in silico study. *Journal of Molecular Structure, 1286*, 135562. Advance online publication. DOI: 10.1016/j.molstruc.2023.135562

Juźwik, C. A., Drake, S. S., Zhang, Y., Paradis-Isler, N., Sylvester, C., Amar-Zifkin, A., Douglas, C., Morquette, B., Moore, C. S., & Fournier, A. E. (2019). microRNA dysregulation in neurodegenerative diseases: A systematic review. *Progress in Neurobiology, 182*, 101664. Advance online publication. DOI: 10.1016/j.pneurobio.2019.101664 PMID: 31356849

Khan, A., Ali, T., Rehman, S. U., Waheed, A., & Kim, M. O. (2020). Neuroprotection: Targeting multiple pathways by naturally occurring phytochemicals. *Biomedicines, 8*(8), 284. Advance online publication. DOI: 10.3390/biomedicines8080284 PMID: 32806490

Kooshki, L., Ghasemi, N., & Azadbakht, M. (2023). The pivotal role of JAK/STAT and IRS/PI3K signaling pathways in neurodegenerative diseases: Mechanistic approaches to polyphenols and alkaloids. *Phytomedicine, 112*, 154686. Advance online publication. DOI: 10.1016/j.phymed.2023.154686 PMID: 36804755

Kyriakis, J. M., & Avruch, J. (2012). Mammalian MAPK signal transduction pathways activated by stress and inflammation: A 10-year update. *Physiological Reviews, 92*(2), 689–737. DOI: 10.1152/physrev.00028.2011 PMID: 22535895

Leyns, C. E. G., Ulrich, J. D., Finn, M. B., Stewart, F. R., Koscal, L. J., Serrano, J. R., Robinson, G. O., Anderson, E., Colonna, M., & Holtzman, D. M. (2017). TREM2 deficiency attenuates neuroinflammation and protects against neurodegeneration in a mouse model of tauopathy. *Proceedings of the National Academy of Sciences of the United States of America, 114*(43), 11524–11529. DOI: 10.1073/pnas.1710311114 PMID: 29073081

Li, X., Wang, H., Li, J., & Li, Y. (2022). Synergistic neuroprotective effect of saikosaponin A and albiflorin on corticosterone-induced apoptosis in PC12 cells via regulation of metabolic disorders and neuroinflammation. *Molecular Biology Reports, 49*(9), 8801–8813. DOI: 10.1007/s11033-022-07730-5 PMID: 36002654

Limanaqi, F., Biagioni, F., Busceti, C. L., Ryskalin, L., & Fornai, F. (2020). Merging the multi-target effects of phytochemicals in neurodegeneration: From oxidative stress to protein aggregation and inflammation. *Antioxidants, 9*(10), 1022. Advance online publication. DOI: 10.3390/antiox9101022 PMID: 33092300

Liu, J., Zhou, H., Fan, W., & Dong, W. (2023). New insights into molecular mechanisms underlying neurodegenerative disorders. *Journal of Integrative Neuroscience, 22*(3), 58. Advance online publication. DOI: 10.31083/j.jin2203058 PMID: 37258447

Liu, Y., Zhang, Y., Zheng, X., Fang, H., & Li, H. (2018). The synergistic neuroprotective effects of combined rosuvastatin and resveratrol pretreatment against cerebral ischemia/reperfusion injury. *Journal of Stroke and Cerebrovascular Diseases*, *27*(6), 1697–1704. DOI: 10.1016/j.jstrokecerebrovasdis.2018.01.033 PMID: 29525080

Ma, X., Liu, H., Li, J., Wang, Y., Ding, S., Zhang, B., & Wang, R. (2020). ADSCs-derived extracellular vesicles alleviate neuronal damage, promote neurogenesis and rescue memory loss in mice with Alzheimer's disease. *Journal of Controlled Release*, *327*, 688–702. DOI: 10.1016/j.jconrel.2020.09.019 PMID: 32931898

Manach, C., Scalbert, A., Morand, C., Rémésy, C., & Jiménez, L. (2004). Polyphenols: Food sources and bioavailability. *The American Journal of Clinical Nutrition*, *79*(5), 727–747. DOI: 10.1093/ajcn/79.5.727 PMID: 15113710

Manach, C., Williamson, G., Morand, C., Scalbert, A., & Rémésy, C. (2005). Bioavailability and bioefficacy of polyphenols in humans. I. Review of 97 bioavailability studies. *The American Journal of Clinical Nutrition*, *81*(1), 230S–242S. DOI: 10.1093/ajcn/81.1.230S PMID: 15640486

Marmolejo-Martínez-Artesero, S., Aparicio, M., & Navarro, A. (2021). Endogenous mechanisms of neuroprotection: To boost or not to be. *Cells*, *10*(2), 370. Advance online publication. DOI: 10.3390/cells10020370 PMID: 33578870

Mohd Sairazi, N. S., & Sirajudeen, K. N. S. (2020). Natural products and their bioactive compounds: Neuroprotective potentials against neurodegenerative diseases. *Evidence-Based Complementary and Alternative Medicine : eCAM*, *2020*(1), 6565396. Advance online publication. DOI: 10.1155/2020/6565396 PMID: 32148547

Moradi Vastegani, S., Nasrolahi, A., Ghaderi, S., Tavassoly, I., & Ghasemi, A. (2023). Mitochondrial dysfunction and Parkinson's disease: Pathogenesis and therapeutic strategies. *Neurochemical Research*, *48*(8), 2285–2308. DOI: 10.1007/s11064-023-03904-0 PMID: 36943668

Naoi, M., Maruyama, W., & Akao, Y. (2017). Neurotrophic function of phytochemicals for neuroprotection in aging and neurodegenerative disorders: Modulation of intracellular signaling and gene expression. *Journal of Neural Transmission (Vienna, Austria)*, *124*(12), 1515–1527. DOI: 10.1007/s00702-017-1797-5 PMID: 29030688

Naoi, M., Wu, Y., Shamoto-Nagai, M., & Maruyama, W. (2019). Mitochondria in neuroprotection by phytochemicals: Bioactive polyphenols modulate mitochondrial apoptosis system, function and structure. *International Journal of Molecular Sciences*, *20*(10), 2451. Advance online publication. DOI: 10.3390/ijms20102451 PMID: 31108962

O'Shea, J. J., & Plenge, R. (2012). JAK and STAT signaling molecules in immunoregulation and immune-mediated disease. *Immunity*, *36*(4), 542–550. DOI: 10.1016/j.immuni.2012.03.014 PMID: 22520847

Pandey, K. B., & Rizvi, S. I. (2009). Plant polyphenols as dietary antioxidants in human health and disease. *Oxidative Medicine and Cellular Longevity*, *2*(5), 270–278. DOI: 10.4161/oxim.2.5.9498 PMID: 20716914

Pérez-Jiménez, J., Neveu, V., Vos, F., & Scalbert, A. (2010). Identification of the 100 richest dietary sources of polyphenols: An application of the Phenol-Explorer database. *European Journal of Clinical Nutrition*, *64*(S3, Suppl 3), S112–S120. DOI: 10.1038/ejcn.2010.221 PMID: 21045839

Picon, C., Jayaraman, A., Kirkwood, R., Pradeepan, S., Amor, S., & Reynolds, R. (2021). Neuron-specific activation of necroptosis signaling in multiple sclerosis cortical grey matter. *Acta Neuropathologica*, *141*(4), 585–604. DOI: 10.1007/s00401-021-02274-7 PMID: 33569629

Ratan, Y., Rajput, A., & Sharma, A. (2023). An insight into cellular and molecular mechanisms underlying the pathogenesis of neurodegeneration in Alzheimer's disease. *Biomedicines*, *11*(5), 1398. Advance online publication. DOI: 10.3390/biomedicines11051398 PMID: 37239068

Rebas, E., Rorbach-Dolata, A., & Piwowar, A. (2020). Neuroprotective polyphenols: A modulatory action on neurotransmitter pathways. *Current Neuropharmacology*, *18*(5), 431–445. DOI: 10.2174/1570159X18666200106155127 PMID: 31903883

Renaud, J., & Martinoli, M.-G. (2019). Considerations for the use of polyphenols as therapies in neurodegenerative diseases. *International Journal of Molecular Sciences*, *20*(8), 1883. Advance online publication. DOI: 10.3390/ijms20081883 PMID: 30995776

Salman, A., Koparde, R., & Zand, M. S. (2018). Determining the quantitative principles of T cell response to antigenic disparity in stem cell transplantation. *Frontiers in Immunology*, *9*, 2284. Advance online publication. DOI: 10.3389/fimmu.2018.02284 PMID: 30364159

Sánchez-Martínez, J. D., Martínez-Gardeazabal, J., & Fernández-Villamarín, M. (2022). Blood–brain barrier permeability study of potential neuroprotective compounds recovered from plants and agri-food by-products. *Frontiers in Nutrition*, *9*, 924596. Advance online publication. DOI: 10.3389/fnut.2022.924596 PMID: 35782945

Saxton, R. A., & Sabatini, D. M. (2017). mTOR signaling in growth, metabolism, and disease. *Cell*, *168*(6), 960–976. DOI: 10.1016/j.cell.2017.02.004 PMID: 28283069

Scalbert, A., Johnson, I. T., & Saltmarsh, M. (2005). Polyphenols: Antioxidants and beyond. *The American Journal of Clinical Nutrition*, *81*(1), 215S–217S. DOI: 10.1093/ajcn/81.1.215S PMID: 15640483

Scalbert, A., & Williamson, G. (2000). Dietary intake and bioavailability of polyphenols. *The Journal of Nutrition*, *130*(8), 2073S–2085S. DOI: 10.1093/jn/130.8.2073S PMID: 10917926

Shaban, N. Z., Abdelrahman, S. A., & El-Kersh, M. A. (2021). Synergistic protective effect of *Beta vulgaris* with meso-2,3-dimercaptosuccinic acid against lead-induced neurotoxicity in male rats. *Scientific Reports*, *11*(1), 252. Advance online publication. DOI: 10.1038/s41598-020-80669-4 PMID: 33420282

Shuai, K., & Liu, B. (2003). Regulation of JAK–STAT signalling in the immune system. *Nature Reviews. Immunology*, *3*(11), 900–911. DOI: 10.1038/nri1226 PMID: 14668806

Sierra, A., Encinas, J. M., Deudero, J. J., Chancey, J. H., Enikolopov, G., Overstreet-Wadiche, L. S., Tsirka, S. E., & Maletic-Savatic, M. (2010). Microglia shape adult hippocampal neurogenesis through apoptosis-coupled phagocytosis. *Cell Stem Cell*, *7*(4), 483–495. DOI: 10.1016/j.stem.2010.08.014 PMID: 20887954

Silva, R. F., & Pogačnik, L. (2020). Polyphenols from food and natural products: Neuroprotection and safety. *Antioxidants, 9*(1), 61. Advance online publication. DOI: 10.3390/antiox9010061 PMID: 31936711

Stacchiotti, A., & Corsetti, G. (2020). Natural compounds and autophagy: Allies against neurodegeneration. *Frontiers in Cell and Developmental Biology, 8*, 555409. Advance online publication. DOI: 10.3389/fcell.2020.555409 PMID: 33072744

Stith, S. S., Diviant, J. P., Brock, C., & Vigil, J. (2018). Patient-reported symptom relief following medical cannabis consumption. *Frontiers in Pharmacology, 9*, 916. Advance online publication. DOI: 10.3389/fphar.2018.00916 PMID: 30210337

Swanson, K. V., Deng, M., & Ting, J. P.-Y. (2019). The NLRP3 inflammasome: Molecular activation and regulation to therapeutics. *Nature Reviews. Immunology, 19*(8), 477–489. DOI: 10.1038/s41577-019-0165-0 PMID: 31036962

Tan, S. H., Karri, V., Tay, N. W. R., Chang, K. H., & Wong, K. H. (2019). Emerging pathways to neurodegeneration: Dissecting the critical molecular mechanisms in Alzheimer's disease, Parkinson's disease. *Biomedicine and Pharmacotherapy, 111*, 765–777. DOI: 10.1016/j.biopha.2018.12.101 PMID: 30612001

Tang, Y., & Le, W. (2016). Differential roles of M1 and M2 microglia in neurodegenerative diseases. *Molecular Neurobiology, 53*(2), 1181–1194. DOI: 10.1007/s12035-014-9070-5 PMID: 25598354

Tsoi, P. S., Quan, M. D., & Choi, M. L. (2023). Aggregation of disordered proteins associated with neurodegeneration. *International Journal of Molecular Sciences, 24*(4), 3380. Advance online publication. DOI: 10.3390/ijms24043380 PMID: 36834792

Uddin, M. S., Kabir, M. T., Rahman, M. S., Alim, M. A., Rahman, M. M., & Rauf, A. (2020). Neuroprotective role of polyphenols against oxidative stress-mediated neurodegeneration. *European Journal of Pharmacology, 886*, 173412. Advance online publication. DOI: 10.1016/j.ejphar.2020.173412 PMID: 32771668

Van Praag, H., Fleshner, M., Schwartz, M. W., & Mattson, M. P. (2014). Exercise, energy intake, glucose homeostasis, and the brain. *The Journal of Neuroscience : The Official Journal of the Society for Neuroscience, 34*(46), 15139–15149. DOI: 10.1523/JNEUROSCI.2814-14.2014 PMID: 25392482

Villarino, A. V., Kanno, Y., & O'Shea, J. J. (2015). Mechanisms of Jak/STAT signaling in immunity and disease. *The Journal of Immunology : Official Journal of the American Association of Immunologists, 194*(1), 21–27. DOI: 10.4049/jimmunol.1401867 PMID: 25527793

Yan, L., Xie, J., & Wang, X. (2022). Dietary plant polyphenols as the potential drugs in neurodegenerative diseases: Current evidence, advances, and opportunities. *Oxidative Medicine and Cellular Longevity, 2022*(1), 5288698. Advance online publication. DOI: 10.1155/2022/5288698 PMID: 35237381

Zbinden, A., Pérez-Berlanga, M., De Rossi, P., & Polymenidou, M. (2020). Phase separation and neurodegenerative diseases: A disturbance in the force. *Developmental Cell, 55*(1), 45–68. DOI: 10.1016/j.devcel.2020.09.014 PMID: 33049211

Zhang, Q., Lenardo, M. J., & Baltimore, D. (2017). 30 years of NF-κB: A blossoming of relevance to human pathobiology. *Cell, 168*(1-2), 37–57. DOI: 10.1016/j.cell.2016.12.012 PMID: 28086098

Chapter 14
Neurodegenerative Natural Foes:
Polyphenols, Especially Flavonoids, as Therapeutic Agents

Iram Malik
http://orcid.org/0009-0003-0298-7443
University of Sargodha, Sargodha, Pakistan

Sheryar Afzal
Department of Biosciences, College of Veterinary Medicine, King Faisal University, Saudi Arabia

Ali Attiq
Universiti Sains Malaysia, Malaysia

ABSTRACT

Polyphenols, are a diverse group of plant-derived compounds found in various foods and beverages such as fruits, berries, legumes, tea and red wine. These compounds exhibit complex structures and are known for their potent bioactive properties. Among polyphenols, flavonoids are the most abundant and have demonstrated significant therapeutic potential, especially in combating neurodegenerative disorders like Alzheimer's, Parkinson's, Huntington's diseases and amyotrophic lateral sclerosis. The increasing prevalence of NNDs due to aging populations underscores the urgency of exploring effective treatments. Flavonoids exert neuroprotective effects through multiple mechanisms. These involve antioxidant activity, modulation of neuroinflammation, inhibition of protein aggregation, and mitigation of endoplasmic reticulum stress. They also influence the microbe-gut-brain axis and have shown promise in enhancing cognitive function and memory.

1. INTRODUCTION

Medicinal plants have been utilized for centuries in traditional medicine systems across the globe, offering therapeutic benefits for a wide range of human ailments. In recent decades, scientific investigations have increasingly validated these traditional practices, revealing that many phytochemicals derived

DOI: 10.4018/979-8-3373-4139-2.ch014

from medicinal plants possess potent antioxidant and anti-inflammatory properties, which are central to their neuroprotective mechanisms (Moradi, Momtaz, Bayrami, Farzaei, & Abdollahi, 2020).

Parallel to this, nutritional science has emphasized the critical role of diet in the prevention and advancement of non-communicable diseases. Epidemiological evidence consistently shows that diets rich in fruits and vegetables are associated with a reduced incidence of cardiovascular diseases (Estruch et al., 2018), type 2 diabetes (Carter, Gray, Troughton, Khunti, & Davies, 2010), stroke (He, Nowson, & MacGregor, 2006), and various cancers (Masala et al., 2012). Additionally, the consumption of green tea has been inversely associated with cognitive decline, suggesting a protective role against neurodegeneration (Feng, Gwee, Kua, & Ng, 2010).

These health benefits are largely attributed to polyphenols- a diverse class of non-essential plant-derived compounds found abundantly in fruits, vegetables, and beverages like tea and wine. Polyphenols have demonstrated the ability to modulate oxidative stress, neuroinflamation, and protein aggregation, all of which are implicated in the pathogenesis of neurodegenerative disorders such as Parkinson's and Alzheimer's disease (Bhullar & Rupasinghe, 2013). These disorders which lack definitive etiopathogenetic origins, are believed to arise from a complex interplay of aging, environmental exposures, and genetic predispositions. Recent research has also highlighted the ability of polyphenols to influence cellular signaling pathways, including those involved in autophagy, mitochondrial function, and neuronal survival (Renaud & Martinoli, 2019). These findings underscore the therapeutic potential of polyphenols and other phytochemicals not only in mitigating disease progression but also in enhancing cognitive resilience. Given their historical use, low toxicity, and growing body of experimental support, polyphenols represent a promising frontier in the development of novel therapeutic agents.

2. STRUCTURE OF POLYPHENOLS

Polyphenols are a diverse a diverse group of naturally occurring compounds characterized by the presence of multiple phenolic structural units. According to a refined biochemical definition, the term "polyphenols" should be reversed for compounds exclusively derived from phenylpropanoid/shikimate and/or polyketide pathways. These compounds must contain more than one phenolic ring and lack nitrogen –based functional groups. This definition excludes monophenolic structures and their derivatives such as phenyl esters, methyl ethers and O-phenyl glycosides.

The majority of plant polyphenols originate from amino acid phynylalanine, which undergoes deamination to form cinnamic acid. This compound then enters the phenylpropanoid pathway, a central metabolic route in plants responsible for the biosynthesis of a wide array of secondary metabolites (Parr & Bolwell, 2000). The phenylpropanoid unit, characterized by a C6-C3 structure-a phenol ring with a three carbon side chain (pyranic ring) serves as the foundational building block for polyphenol synthesis. A large number of polyphenols have two aromatic rings (A and B) which are linked with pyranic ring (C) (Figure 1).

Polyphenols are classified based on the number of phenolic rings they contain and the structural elements that link these rings. The major subclasses include phenolic acids (C6-C1 and C6-C3), flavonoids (C6-C3-C6), stilbenes (C6-C2-C6), and lignans (C6-C3-C3-C6) (Renaud & Martinoli, 2019). On the basics of the degree of oxidation of C-ring, polyphenols can be classified as: flavonoids (Flavanones, flavonols, isoflavones, flavanols, flavones, anthocyanins), liganans, stilbenes, curcuminoids and phenolic acids (Macready et al., 2009); (Gomez-Pinilla & Nguyen, 2012), (Leclerc, Dudonné, & Calon, 2021), (Lamport & Williams, 2021), (Figueira et al., 2017), (Koudoufio et al., 2020).

Figure 1. General classification of polyphenols with example of each subclass

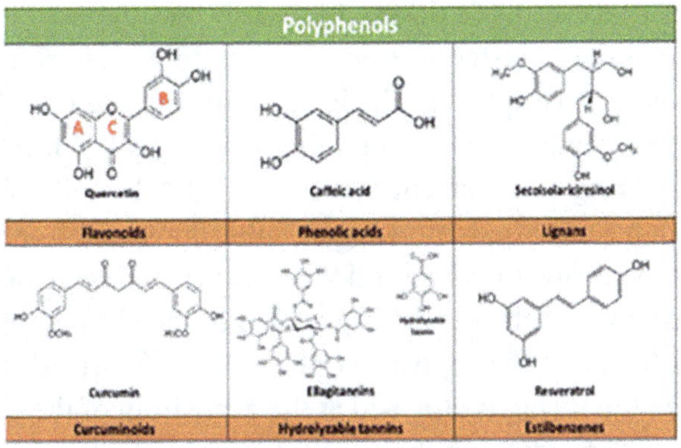

Note: adapted from Arias-Sánchez, R. A., Torner, L., & Fenton Navarro, B. (2023). (Arias-Sánchez, Torner, & Fenton Navarro, 2023).

Within these categories, structural diversity is further enhanced by hydroxylation, O-glycosylation, and cis-trans isomerization, resulting in over 8,000 distinct polyphenolic compounds identified to date (Rothwell et al., 2013), (Naoi, Shamoto-Nagai, & Maruyama, 2019). These compounds are widely distributed in plant-based foods. They are naturally found in fruits, berries, legumes, cereals, spices and variety of beverages like coffee, tea, red wine and dry beans. They are known for their potential pharmacological properties, including antioxidant, anti-inflammatory, inhibition of protein aggregation, mitigation of endoplasmic reticulum stress. And other neuroprotective effects.

3. FLAVONOIDS

Flavonoids constitute a subclass within the extensive group of polyphenolic compounds characterized by a benzo- γ-pyrone structure. It has 6000 types and are characterized by fifteen carbon with two fused rings attached to an aromatic ring (Rana, Samtiya, Dhewa, Mishra, & Aluko, 2022), (B. Singh & Semwal, 2024). These versatile natural substances are recognized as secondary metabolites and have garnered significant scientific interest over the past decade due to their widespread presence in the plant kingdom and their diverse health benefits. In contemporary literature, flavonoids are frequently referred to as "functional ingredients" and "health-promoting biomolecules" because of their potential to enhance well-being and prevent chronic degenerative diseases (S. Kumar & Pandey, 2013) (Karak, P, 2019). Historically and in modern medicine, flavonoids have played pivotal role in therapeutic application in neurodegenerative diseases.

These compounds exhibit a wide range of biochemical and antioxidant activities, contributing to the prevention and management of Alzheimer's disease (AD), Parkinson's disease (PD), and other neurodisease. Their health promoting effects with broad spectrum makes them indispensable in neutraceutical, medicinal, pharmaceutical, and cosmetic industries. This is largely attributed to their potent antioxidant, anti-inflammatory, anti-mutagenic, anti-microbial and anti-carcinogenic properties. Additionally, they also exhibit vascular protective effects, free radical scavenging capabilities, and ability to modulate key cellular enzyme functions. (López-Lázaro, 2009)

3.1 Structure and Classification of Flavonoids:

The fundamental chemical framework of a flavonoid is based on a diphenylpropane skeleton, comprising fifteen carbon atoms arranged in two six-membered aromatic rings connected by a three carbon bridge, which may or may not form part of a third ring. Typically, two benzene rings (designated as rings A and B) are linked via a third heterocyclic oxygen-containing ring (ring C), forming a structure commonly described as C6-C3-C6. These rings are conventionally labeled A,B, and C (Zakaryan, Arabyan, Oo, & Zandi, 2017). Flavonoids can be classified into various subgroups based on two key structural criteria: the position of attachment of the B ring to the central C ring, and the degree of unsaturation and oxidation within C ring. When the B ring is attached at the 3-position of the C ring, the compounds are known as isoflavones. If the B ring is linked at the 4-position, they are categorized as neoflavonoids. The most common flavonoids, in which the B ring is attached at the 2-position of the C ring, are further subdivided into several distinct classes depending on the specific structural characteristics of the C ring. These subclasses include: flavones, flavanones, flavonols, flavanonols, flavanols, chalcones and anthocyanins. Each of these subgroups exhibits unique chemical properties and biological activities, contributing to the wide ranging functionality of flavonoids in human health (Panche, Diwan, & Chandra, 2016), (Figure 2).

Especially cocoa flavanols in diet improves cognition and counteracts different cognitive decline in elderly people (Brickman et al., 2014), (Socci, Tempesta, Desideri, De Gennaro, & Ferrara, 2017). It also enhances cerebral cortical oxygenation and cognition in adults Genistein and daidzein (isoflavones) (Qian et al., 2012), (Aras et al., 2015), luteolin and apigenin (flavones), kaempferol and quercetin (flavonols) (Barreca et al., 2016), cyaniding and delphinidin (anthocyanins) (Strathearn et al., 2014), naringenin and hesperetin (flavanones) (Cirmi et al., 2016), (Heidary Moghaddam et al., 2020) and epigallocatechin-3-gallate (EGCG) (flavanol) (N. A. Singh, Mandal, & Khan, 2016) are some of the famous flavonoids of dietary and medicinal plants with protective roles in various neurodegenerative diseases.

Figure 2. Classification of flavonoids (Adapted from Naoi & Shamoto-Nagai, 2019; Naoi et al., 2019)

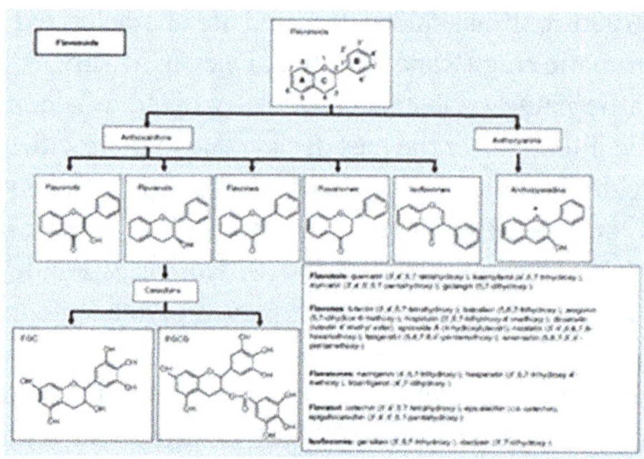

4. NEUROPROTECTION ROLE OF FLAVONOIDS

Neuro-transitions that include synaptic dysfunction, dementia and cognition impairments are more common in aging people and lead to nervous diseases (Bhullar & Rupasinghe, 2013). Multiple sclerosis, amyotrophic lateral sclerosis, Huntington´s disease and Alzheimer´s disease are neurodegenerative disorders (NNDs). (Moradi et al., 2022). As life expectancy in humans increases, the number of aged people also increases thus, become more vulnerable to NNDs. It is predicted that in 2030 there is an increase of around 50% people with age group between 60 to 80 years (Arias-Sánchez et al., 2023). The etiopathogenetic origins of neurodegenerative disorders are not clear. Besides environmental and genetic risk factors, the major causative mechanisms include neuro inflammation, oxidative stress, protein aggregation, mitochondrial dysfunction and metal toxicity (Renaud & Martinoli, 2019). Lower concentration of reactive oxygen and nitrogen species are necessary for maintaining normal brain function. Elevated concentration leads to neural apoptosis, dysfunction of mitochondria, Ca^{+2} stress and excitotoxicity.

Structural variations in polyphenols (flavonoids) play a crucial role in their potential therapeutic applications especially in neuroprotection. Presence of phenol rings, different hydroxylation patterns and conjugation in double bonds enable polyphenols to perform neuroprotective function through various mechanisms like inhibition of the production and aggregation of beta-amyloid, metal chelation, reducing concentration of reactive oxygen and nitrogen species. They can also activate redox responsible transcription factors, regulate the mitochondrial apoptosis and gene expression (Rehman et al., 2019),(Renaud & Martinoli, 2019).

4.1 Flavonoids as Antioxidants:

Oxidative stress results from an imbalance characterized by the excessive production of reactive oxygen species (ROS) and reactive nitrogen species (RNS), primarily generated by mitochondria and NADPH oxidase, alongside a reduction in the cellular antioxidant defense mechanisms. The imbalance constitutes an early and critical pathological event in several neurodegenerative disorders, including PD, AD, amyotrophic lateral sclerosis, and multiple sclerosis (Dias, Junn, & Mouradian, 2013), (Tönnies & Trushina, 2017). The resulting free radicals inflict oxidative damage by targeting proteins, RNA, and DNA, initiating lipid peroxidation in cellular membranes, compromising membrane integrity and promoting protein crosslinking.

Flavonoids, Particularly those with catechol structures, exhibit antioxidant activity that is influenced by their structural features specifically, the number and position of hydroxyl groups on the aromatic A and B rings, the presence of a 2,3 –double bond and a 4-carbonyl group on C ring. Their antioxidant mechanisms encompass several pathways (Zuo et al., 2018).

4.1.1 Direct Scavenging of Reactive Oxygen Species (ROS):

In laboratory (in vivo) studies, the antioxidant activity of flavonoids is closely linked to the arrangement of hydroxyl (-OH) groups in their structure. Specifically, the B-ring of flavonoids, when it contains an ortho-dihydroxy group, allows the resulting flavonoid phenoxyl radicals to undergo electron delocalization. This movement of electrons within the B-ring enhances the molecule's ability to neutralize ROS, thereby increasing its antioxidant potential (Vicente & Boscaiu, 2018).

Additionally, antioxidant activity is significantly improved when flavonoid monomers undergo polymerization. For example, proanthocyanidins (also known as condensed tannins) and catechin polymers show strong antioxidant effects due to higher number of hydroxyl groups in their structures (Kirschweng, Tátraaljai, Földes, & Pukánszky, 2017). Other structural features that boost free radical scavenging include the presence of double bonds in the phenolic ring, hydroxyl side chains, and glycosylation in anthocyanidins (Vicente & Boscaiu, 2018).

4.1.2 Activation of Antioxidant Enzymes:

Flavonoids also enhance antioxidant defense by influencing enzyme activity. Some flavonoids can inhibit enzymes that generate free radicals, such as xanthine oxidase and nitric oxide synthase (Engwa, 2018). At the same time, they can activate antioxidant enzymes by inducing the expression of phase II detoxifying enzymes. These include NADPH-quinone oxidoreductase, glutathione S-transferase, and UDP-glucuronosyltransferase, which play key roles in protecting cells from oxidative stress and harmful chemicals. This activation is regulated through a genetic mechanism involving the electrophile responsive element (EpRE), a DNA sequence found in the promoter regions of genes encoding these phase II enzymes. Flavonoids have shown to trigger EpRE-mediated gene expression, contributing to their antioxidant and protective effects (Lee-Hilz et al., 2006).

4.1.3 Inhibition of Oxidases:

Certain structures in flavonoids, like the ortho-dihydroxy group on the B-ring, help them trap and neutralize ROS. When flavonoids form large molecules (like proanthocyanidins), their antioxidant power increases due to more hydroxyl groups (Kirschweng et al., 2017). Furthermore, they act as buffer and generate flavinic acid by capturing free radicals. It is less reactive because unreactive electrons are delocalized (Figure 3A) (Arias-Sánchez et al., 2023). Some flavonoids reduce the activity of enzymes that produce ROS, such as xanthine oxidase and nitric oxide synthase (Engwa, 2018). They also boost the production of protective enzymes like glutathiones S-transferase and NADPH-quinone oxidoreductase by activating specific genes (Lee-Hilz et al., 2006). Flavonoids can stop enzymes like xanthine oxidase, protein kinase C, and others that contribute to oxidative stress (Zeng, Zhang, Hu, Pan, & Gong, 2019), (Hanáková et al., 2017).

4.1.4 Binding to Metal Ions:

Flavonoids can chelate some metals ions like iron and copper, which helps prevent these metals from producing ROS from Fenton and Haber-Weiss reactions (Figure 3B) (Arias-Sánchez et al., 2023). For example quercetin and catechin can form stable complexes with metals like Fe (III), Cu (II) and others, reducing their harmful effects (Kejík et al., 2021), (Lesjak et al., 2018), (Simunkova et al., 2021), (K. Zhang et al., 2021).

4.1.5 Supporting other antioxidants:

Flavonoids help regenerate α-tocopherol (vitamin E), which protects fats in the body from oxidation. This helps slow down the oxidation of low density lipoprotein (LDL), which linked to heart disease (Torello, Alvarez, & Olalla Saad, 2021).

4.1.6 Reducing nitric oxide-related stress:

While nitric oxide (NO) is important for blood vessel health, too much of it can lead to harmful compounds like peroxynitrite. Flavonoids can reduce NO production by blocking the enzyme that makes it, especially in inflamed cells. Quercetin, for instance lowers both NO and vascular endothelial growth factor (VEGF) levels, which are involved in inflammation and blood vessel growth. (Papi, Ahmadizar, & Hasanvand, 2019), (Terao, 2017).

Figure 3. Flavonoids (polyphenols) from medicinal plant showing antioxidant property (A) by donation of proton from flavonoids to the radicals and generating flavin radical (B) Elimination of ROS and chelation of metal ions by Fenton and Haber Weiss reactions.

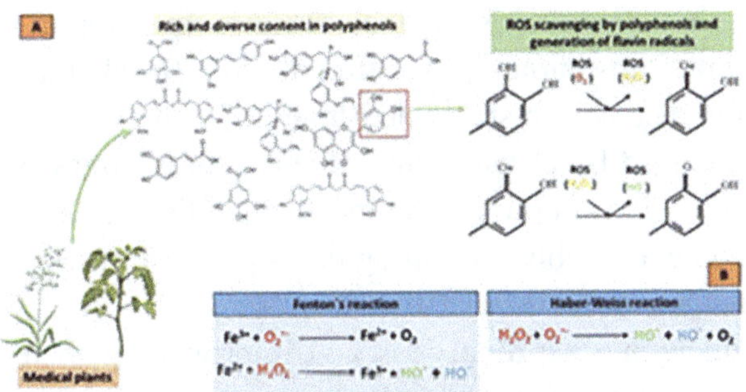

Note: adapted from Arias-Sánchez, R. A., Torner, L., & Fenton Navarro, B. (2023). (Arias-Sánchez et al., 2023)

Although the antioxidant potential of phytochemicals-- especially flavonoids has been widely associated with neuroprotection in both in vitro and in vivo models, effective antioxidant action in vivo typically requires polyphenol concentrations in the 10-100 μM range. Often, the needed concentrations are much higher than what s found in the bloodstream. Interestingly, the structural features of flavonoid molecules that protect brain cells do not always align with those responsible for antioxidant activity. Quercetin's ability to protect neurons depends on specific hydroxyl groups in structure, not just its antioxidant power. This suggests that flavonoids may protect the brain through other mechanisms beyond just fighting oxidative stress. (Echeverry et al., 2010).

4.2 Flavonoids as Neuroinflammation Modulators:

Neuroinflamtion represents a central immune response within the central nervous system (CNS). It activates the glial cell particularly microglia. Upon activation, these cells release a variety of inflammatory mediators, including chemokines, cytokines, and ROS/RNS (Sochocka, Diniz, & Leszek, 2017). ROS contribute inflammation by releasing inflammatory mediators such as peroxiredoxin 2 which activate macrophages to secrete IL-6. This inflammatory cascade is often accompanied by elevated levels of prostaglandins, prostaglandin E2 (PGE2). It is particularly seen in elevated levels in aging brain and cause acute inflammation (Ricciotti & FitzGerald, 2011). Once neuroinflammation surpass a critical threshold, it can lead to cellular damage and neuronal dysfunction. This process is marked by microglial activation, proliferation, and reactive astrogliosis. Which leads to activated microglia promoting astrocyte proliferation (D. Zhang et al., 2009). Chronic inflammation is a pivotal pathogenic factor for both AD and PD (Hirsch & Hunot, 2009).

Flavonoids have demonstrated significant neuroprotective effects in both in vitro and in vivo models by modulating neuroinflammatory responses (Khan, Ikram, Hahm, & Kim, 2020), (Maleki, Crespo, & Cabanillas, 2019). These compounds interact with neuronal receptors and influence transcription factors, kinase signaling pathways, and gene/protein expression. All these are associated with memory and learning in hippocampus region (Rendeiro, Rhodes, & Spencer, 2015). Flavonoids suppress microglial activation and mitigate the neurotoxicity caused by neurotoxins released from these cells. For example, wogonin (plant flavonoid) inhibits nitric oxide (NO) production and prevents activation-induced death in C6 glial cells by suppressing NF-kappaB-mediated nitric oxide synthase expression (Kim, Kim, Kim, & Suk, 2001). Luteolin and apigenin, two well studies flavonoids, protect dopaminergic neurons by reducing oxidative stress, neuroinflamation, and microglial activation. In MPTP-induce Parkinson's disease (PD) mouse models, these flavonoids enhance neurotrophic support by increasing brain-derived neurotrophic factor levels in the substantia nigra region of brain (Patil et al., 2022).

In another study apigenin when administered at 20mg/kg for three weeks, significantly ameliorated behavioral deficits. It was induced by chronic unpredictable mild stress (CUMS), including reduced locomotors activity and sucrose consumption. The mechanism lies in the activity of apigenin is that it disrupts NLRP3 inflammasome assembly and upregulates peroxisome proliferator-activated receptor-gamma (PPARγ) (Li, Wang, Qin, Qu, & Ma, 2016).

It also mitigates dopaminergic neuronal loss and improves biochemical, behavioral and mitochondrial enzyme functions by suppressing oxidative stress and neuroinflammtion (Anusha, Sumathi, & Joseph, 2017). Rutin reduces mRNA expression of cytokines, caspase-1, ASC, and ASC-NLRP3 in male albino Wistar rats (Aruna, Geetha, & Suguna, 2014). Daidazein, another flavonoid, has shown effective in Alzheimer's disease risk by lowering IL-1, TNF-α, and IL-6 expression in primary astrocytes stimulated with amyloid-beta and lipopolysaccharides (M. H. Liu, Lin, Sheu, & Sun, 2009). Catechin protects murine microglial N9 cells from oxidative damage by inhibiting NF-kappaB and p53 activity (Huang et al., 2005).

Blueberry extracts, rich in flavonoids, suppresses iNOS and COX-2 expression and reduce TNF-α, IL-Iβ, NO and ROS levels in microglial cells (Lau, Bielinski, & Joseph, 2007). Naringenin prevents glial cell death by reducing TNF- α levels, iNOS and NO, and inhibiting p38 MAPK and STAT-1 signaling (Vafeiadou et al., 2009). Genistein and daidzein, when used in combination preserved neuronal function and cognitive abilities in HIV-infected individuals via estrogen receptor-meditated mechanisms (Adams, Aksenova, Aksenov, Mactutus, & Booze, 2012).

Figure 4. Flavonoids role in inflammation

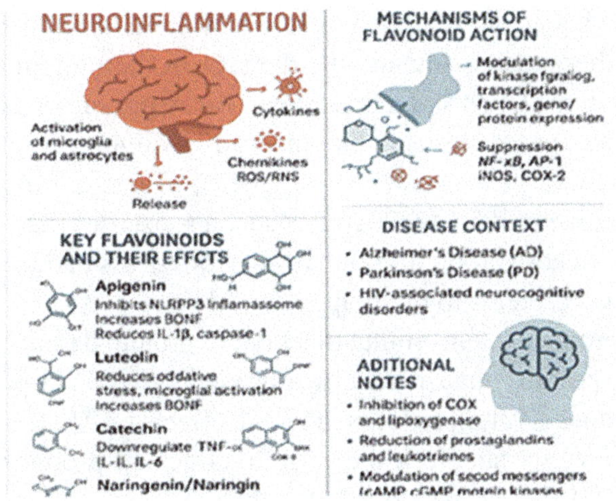

4.3 Flavonoids in Mitigating Endoplasmic Reticulum Stress (ERS):

Endoplasmic reticulum stress occur when misfolded proteins build up in the endoplasmic reticulum (ER) or when calcium levels are disturbed. This effects the ER's normal structure and function. In response, the cell activates a process called the unfolded protein response (UPR), which increases the production of helper proteins (chaperones) and those that remove misfolded proteins. If ER stress continues for too long, it can lead to cell damage and even cell death. ER stress is linked to brain diseases like PD, AD and Huntington's disease (HD). Recently, scientists have been exploring small molecules that can block ER stress and UPR as possible treatments (Uddin et al., 2020), (H. Liu et al., 2016), (Martucciello, Masullo, Cerulli, & Piacente, 2020).

Eating foods rich in flavonoids has shown health benefits. Kaempferol, a natural flavonol, helps protect human neuroblastoma cells (IMR32) from ER stress-related cell death by blocking UPR signals. It lowers the levels of stress markers like glucose regulating proteins (GRP78) and CHOP caused by Brefeldin-A (BFA) (Abdullah & Ravanan, 2018). Luteolin, a flavanol found in celery and broccoli, also reduces GRP78 and CHOP levels in PC12 cells (Hu, Yen, Shen, Wu, & Wu, 2014).

Apigenin, tested on mouse hippocampal cells (HT22), lowers ER stress proteins like CHOP, GRP78, and GRP94. It also effects several signaling proteins, including ATF6α, elF2α and MAPKs like p38, JNK and ERK (Chiang et al., 2011). Epicatechin (EC), a flavan-3-ol with antioxidant and other protective properties, prevents damage in HT22 cells caused by methamphetamine (METH). EC blocks the activation of DR4, ERK, CHOP and p38, helping protect brain cells (Kang et al., 2019).

These findings suggest that flavonoids like kaempferol, luteolin, apigenin and epicatechin may help reduce ER stress and protect brain cells from damage (Devi, Kumar, Singh, Dubey, & Kim, 2021).

4.4 Flavonoids as proteotoxicity modulator:

Another major causative agent of Neurodegenerative diseases are the buildup of harmful protein clumps called amyloids. Other reasons include the damage in the protein-cleaning system known as the ubiquitin-proteasome system (UPS) and a breakdown in the protein quality control (PQC) network. These protein aggregates can appear as insoluble amyloid- β oligomers (AβO), which are especially toxic in Alzheimer's disease (AD) (Sharma & Priya, 2017), (Sontag, Samant, & Frydman, 2017). In AD, amyloid- β is the main toxic protein, while in Parkinson's disease (PD), clumps of α-synuclein and Lewy bodies are key features. These problems often result from stress in the cells, which causes the PQC system to fail and allows toxic proteins to build up.

Flavonoids have shown promise in preventing this kind of protein damage. Studies using cell models suggest that flavan-3-ols, especially their breakdown products may help prevent AD. One such compound, phenyl- γ-valerolactones (PVLs), reduces the toxic effects of Aβ. In mammalian and yeast cells, monohydroxylated form of PVLs - can block the harmful effects of β-oligomers and inhibit cell death. One more PVL, called (4'-OH)-PVL, can stop Aβ from forming toxic clumps. Atomic force microscopy (AFM) images show that it changes toxic AβO into harmless, shapeless forms (Ruotolo et al., 2020). When protein aggregates build up, they interfere with PQC system and disturb the balance of proteins in the cell (Cook, Stetler, & Petrucelli, 2012). Myricetin, a type of flavone, helps stop different harmful proteins from clumping together. It also affects the levels of HSP70 (a helper protein) and E6-AP (a protein involved in quality control). By stabilizing E6-AP, myricetin reduces the buildup of misfolded proteins and their toxic effects (Joshi et al., 2019).

Modified flavonoids may be even more effective. Dihydroquercetin, a changed form of quercetin, works better under stress. While regular quercetin fails to protect cells from heat and chemical stress, dihydroquercetin successfully prevents cell damage. Under heat stress or exposure to sodium arsenite, quercetin also reduces the production of HSP70, which is important for cell protection (Budagova, Zhmaeva, Grigor'ev, Goncharova, & Kabakov, 2003).

Pesticides are linked to many diseases, and studies show they can cause toxic protein buildup, like Aβ amyloids. Silymarin, a flavonolignan from milk thistle seeds, helps reduced Aβ clumps. They are caused by paraquat in *C.elegans* (J. Kumar, Park, Awasthi, & Prasad, 2015). Epimedium, tested in two *C.elegans* models of human protein-related diseases (CL4176 and AM140), showed it can reduce the toxic effects of Aβ1-42 and polyglutamine proteins. It also reduced paralysis in both models (Cai et al., 2019).

Another flavonoid, 6‴- feruloylspinosin (6-FS), found in sour Jujube seeds, was tested on transgenic *C.elegans* (GMC10) and PC12 cells. It delayed aging, reduced paralysis, improved heat resistance, increased movement towards chemical signals, and boosted autophagy (the cell's cleanup process). It occurs through lysosome pathway. 6-FS also reduced Aβ buildup and protein clumping. In PC12 cells, it increased mitophagy (removal of damaged mitochondria) by raising the levels of Pink1 and Parkin, two key proteins in the mitophagy pathway (Yang, Zhao, Zhang, Xie, & Zhou, 2020).

5. BLOOD-BRAIN BARRIER PERMEABILITY OF FLAVONOIDS

There are variety of flavonoids which have therapeutic effects on NNDs. Their trial on cell and animal model has been extensively studied. But utilization of flavonoids in clinical trials has yet to be resolved. One main reason is that they can't pass the Brain Blood Barrier (BBB) (Youdim, Qaiser, Begley, Rice-

Evans, & Abbott, 2004). BBB is a protective structure composed of brain endothelial cells, pericytes and astrocyte extensions. It plays a crucial role in regulating and restricting the movement of chemical substances and immune cells from the bloodstream into the brain tissue. This selective transport by brain endothelial cells is essential for maintaining brain homeostasis. However, the BBB also limits the availability of dietary polyphenols and their metabolites in the brain. Although the exact mechanisms of how polyphenols cross the BBB are not fully understood, it is believed that both passive diffusion and active transport processes may be involved (Grabska-Kobyłecka et al., 2023).

Recent studies have shown that the ability of polyphenol compounds to cross the blood-brain barrier BBB depends largely on their chemical structure and modifications. Figueira et al., highlighted differences in how polyphenol derivatives from berry juice are transported across the BBB. Catechol derivatives and gallic acid which are coupled with methyl or sulphate groups show better transport across the BBB. In contrary, modified pyrogallol derivatives with same groups are less effective. The authors have also described how polyphenols derivatives are not accumulated in brain tissue. The main retarding factor was breast cancer resistance protein (BCRP) a member of the ATP-binding cassette (ABC) transporter family in endothelium cells of human brain (Figueira et al., 2017). The same activity of BCRP transporter was reported fo quercetin buildup (Youdim et al., 2003). Additionally, brain endothelial cells were also considered for lowering of flavonoids concentration due to their metabolizing and transporting from plasma (Figueira et al., 2017).

Larger lipophilic polyphenols (flavonoids) or their metabolites are believed to transport easily through endothelial cells via passive diffusion. However, the presence of effective efflux transporters must always be considered. Flavonoids can cross the BBB through several pathways: lipid mediated diffusion, paracellular diffusion, receptor mediated transcytosis, carrier-mediated transport, active effect transport and absorptive-mediated transcytosis (Abbott, Rönnbäck, & Hansson, 2006).

The permeability of BBB endothelial cells by polyphenols is mainly affected by efflux transporters and by another factor that is solute carrier (Ziberna, Fornasaro, Čvorović, Tramer, & Passamonti, 2014).

Polyphenols can also influence ABC transporters, which act as efflux pumps. Some flavonoids inhibit these transporters (K. W. Tan, Li, Paxton, Birch, & Scheepens, 2013), while others are substrates for them, including P-glycoprotein (P-gp), multidrug-resistant protein 2 (MRP2) and BCRP (Alvarez et al., 2010). These interactions can significantly reduce polyphenol bioavailability. Youdim's team showed that quercetin is able to cross the BBB when given with P-gp or BCRP inhibitor. It likely enters the central nervous system through diffusion due to its hydrophobic nature. It is then actively removed by efflux transporters (Youdim et al., 2003). Despite these challenges, some flavonoids can still reach the brain in meaningful amounts and provide protective effects.

Animal studies have confirmed that flavonoids can enter the brain and accumulate in various regions, regardless of how they are administered (Janle et al., 2010). For example, curcumin has been shown to reach the central nervous system where it provides neuroprotection by preventing cell death and supporting mitochondrial function (Lao et al., 2006). Resveratrol also crosses the BBB and integrates into brain tissue (Q. Wang et al., 2002). Similarly phenolic acids like caffeic acid have been found in the brain and cerebrospinal fluid of patients with neurological disorders, where they help protect the brain during ischemic injury (Pinheiro Fernandes et al., 2014), (Grabska-Kobylecka et al., 2020).

A study utilizing the human brain micro vascular endothelial cell (HBMEC) model demonstrated that among the three flavonoids: epigalloccatechin gallate (EGCG), quercetin and cyaniding-3-glucoside (C3G), EGCG was able to cross the BBB with greater efficiency than C3G. Quercetin did not cross the BBB under the same conditions. However, another experiment involving three month old male

Sprague Dawley rats, eighteen in number, revealed that quercetin can penetrate the BBB when given with tocopherol (vitamin E) (Ferri et al., 2015). These findings support the potential of flavonoids by modifications in treating neurological disorders, as their ability to cross the BBB enables them to reach the site of neuronal damage and exert therapeutic effects.

6. NANO-TECHNOLOGICAL SOLUTIONS TO IMPROVE THE BIOAVAILABILITY OF FLAVONOIDS

To be therapeutically effective, flavonoids must cross the BBB and reach brain tissue in their active form. However, their bioavailability is generally low. Several factors hinder this process and impact therapeutic outcomes, including the selective permeability of the BBB. Other common factors include different stereochemistry, interactions with efflux transporters, lipophilicity, administration method and synergistic interactions. Gastrointestinal transformations, limited absorption, rapid colonic and hepatic metabolism and systematic clearance also decreases the bioavailability of flavonoids (Squillaro, T. et al., 2018).

Oral administration, the most common route, further complicates bioavailability. Consequently, phenolic compounds often exhibit insufficient bioavailability to exert beneficial effects in humans. Moreover, polyphenols are sensitive to various physiological conditions, necessitating enhanced stability for clinical use. In recent years nanformulations of polyphenols have resulted increase in bioavailability. There are several methods in which flavonoids are converted into nanostructures and are evaluated for their neuroprotective roles.

6.1 Nanoencapsulation:

Nanoencapsulation is a technique used to enhance the bioavailability of phenolic compounds. It involves enclosing small active particles (solid, liquid or a gas) in a secondary material matrix or shell (Augustin, M. A., & Hemar, Y., 2009). This process produces capsules ranging from 1 to 1000nm in diameter, increasing the surface to volume ratio. Nanoencapsulation offers protection against harsh conditions, enables controlled release at absorption sites, reduces toxicity to peripheral organs, improves solubility, and minimize degradation in gastrointestinal tract.

Nanoparticles are typically nonimmunogenic, nonthrombogenic and noniflammatory. It is crucial that the encapsulation process does not alter the drug's properties. Various nanoscale delivery systems can be employed for flavonoids encapsulation, including nanocapsules, micelles, nanospheres, solid lipid nanoparticles, liposomes and cyclodextrins (Conte, R.et al., 2016).

6.2 Natural Nano-Carriers:

Natural nano-carriers such as nanocrystals, chitosan, cyclodextrins, and casein can be used to encapsulate phenolic compounds. Quercetin, a dietary bioflavonoid, has demonstrated anti-inflammatory an neuroprotective effects in vivo. Its anti-Alzheimer's properties include inhibition of amyloid-beta aggregation (Caruana, M., 2016). However, quercetin suffers from low solubility, extensive gut metabolism, poor absorption and limited BBB penetration (Cai, X., 2013). To enhance its BBB permeability and bioavailability, Testa and colleagues developed β-cyclodextrin-dodecylcarbonate based nanoparticles containing quercetin (Testa, G., 2014). Cyclodextrin improves the stability and functional properties of the encapsulated compound (Esfanjani, A. F., & Jafari, S. M. 2016). In vitro studies using human neuroblastoma SH-SY5Y cells showed that quercetin loaded nanoparticles were non toxic and exhibited stronger anti-inflammatory effects than free quercetin. This novel therapeutic strategy may enhance the neuroprotective potential of quercetin in preventing or slowing Alzheimer's disease (AD) progression.

In another study quercetin encapsulated in chitosan-alaginate nanoparticles using two in vitro models: neurotoxicity induced by 6-hydroxydopamine (6-OHDA) rat brain synaptosomes and H_2O_2 in SH-SY5Y cells in rat (Aluani, D., 2017). Chitosan is a natural polysaccharide, known for its biocompatibility, biodegradability and low toxicity. Quercetin loaded nanoparticles with chitosan protected against oxidative stress in both models, suggesting that nanoparticle incorporation can overcome limitations and enhance its effects in neurodegenerative diseases.

Rifaai et al., reported the treatment of an animal model of AD by quercetin nanoparticles. AD was induced in rats using oral $AlCl_3$. Which caused hippocampal degenation. Quercetin nanoparticles administered alongside $AlCl_3$, mitigated these effects of AD nearly eliminating neurofibrillary tangles and amyloid plaques. It also preserved hippocampal dentate gyrus proliferation resulting improved bioavailability of quercetin and reduced $AlCl_3$ induced neuronal damage at molecular, cellular and subcellular levels (Rifaai, R. A., 2020).

A very recent research has showed enhanced quercetin's bioavailability and solubility by making magnetic targeting using superparamagnetic nanoparticles. Researchers developed quercetin-conjugated superparamagnetic iron oxide (QT-SPIONs) nanoparticles and tested them in a rat model of AD induced by $AlCl_{3.}$ QT-SPIONs significantly outperformed free quercetin in all experiments. It improved brain tissue penetration and bioavailability (Amanzadeh Jajin, E., 2021).

6.3 Solid Lipid Nanoparticles (SLNs):

Solid lipid nanoparticles (SLNs) are emerging as a promising therapeutic tool for AD treatment. These colloidal carriers consist of a hydrophobic lipid core in an oil-in-water emulsion stabilized by aqueous surfactants. The drug is either dissolved or dispersed within the solid core (Squillaro, T., 2018). For effective brain delivery, SLNs must be smaller than 200 nm to traverse the tight endothelial junctions of the BBB, ensuring efficient tissue penetration.

SLNs show minmal inherent cytotoxicity, enhanced permeability across BBB, notable stability and the ability to evade filtration by liver and spleen (Naqvi, S., 2020). In one study human brain like endothelial cells were used to develop a delivery method aimed at improving the bioavailability of resveratrol (Loureiro, J. A., 2017). The researchers utilized morphologically defined lipid nanoparticles that were functionalized with the OX26 monoclonal antibody. It targeted transferrin receptors (TfR) expressed on BBB cells. These nanoparticles served as carrier for resveratrol. Their findings showed that the encapsulated compounds successfully crossed the BBB and inhibited the aggregation of Aβ peptides responsible for Alzheimer's disease. Furthermore, the study demonstrated that SLNs functionalized with OX26 antibody can effectively deliver polyphenols or other therapeutic agents to the brain.

7. THE GUT-BRAIN AXIS: MODULATORY EFFECTS OF FLAVONOIDS IN NNDS

Flavonoids are highly abundant in diet but their bioavailability is very low in body. They can be classified into two types: free and bound polyphenols. Free polyphenols (flavonoids) are those that do not physically or chemically interact with macromolecules. They can be easily hydrolyzed in small intestinal esterases and get absorbed. In this way keeping humans healthy. However, free phenols constitute just 5-10% of total dietary polyphenols. The remaining 90-95% are bound polyphenols. These bound forms require breakdown into smaller metabolites by intestinal microbiota before they can enter hepatic and intestinal circulation (Y. Liu, Li, Du, et al., 2022).

Upon ingestion, small fraction of polyphenols is absorbed in the stomach. In small intestine, free polyphenols are mainly absorbed, but overall absorption remain limited. These compounds undergo extensive metabolism in enterocytes and the liver through phase I and phase II biotransformation processes. Flavonoids present in food are mostly conjugated with sugars or organic acids. They can also exist as unconjugated oligomers. Certain flavonoids are hydrolyzed by enzymes such as rhamnosidase, glucosidase and glucuronidase, facilitating phase I biotransformation (Zhou et al., 2022).

The resulting aglycones are either passively diffused into enterocytes or transported to the liver via the portal vein. In these tissues, phase II biotransformation occurs. Here polyphenolic aglycones are conjugated with methyl groups,glucuronates, or sulfates. These reactions including methylation, hydroxylation, sulphation and glucosidation convert them into O-glucuronate or O-sulfonate forms. They are then sent to various organs and excreted from body via urine. Due to their different structural properties, most dietary flavonoids bypass phase I and are directly metabolized through phase II biotransformation (Di Lorenzo, Colombo, Biella, Stockley, & Restani, 2021).

The polyphenols which are unabsorbed and are present in bound forms come to the colon, broken down and absorbed through the action of intestinal microbiota. These microbes not only facilitate polyphenol metabolites but also promote the growth of beneficial bacteria such as *Lactobacillus* and *Bifidobacterium*. These support human health. In turn, they help suppress pathogenic species like *Clostridium histolyticum* and *Clostridium perfringens* (Sorrenti et al., 2020). Moreover, gut microbiota can convert polyphenols having large molecular weight into more bioactive metabolites. Research indicates that up to 90% polyphenols taken in body are transformed into improved bioavailable compounds through gut microbiota.

In healthy persons, colon is home to many types of bacteria, including *Firmicutes, Actinobacteria, Bacteroidetes, Verrucomicrobia* and *Proteobacteria.* Mostly *Firmicutes* and *Bacteroidetes* make up over 90% of the gut microbiome. However, factors like diet, illness and medication can significantly alter this balance Patients with neurodegenerative diseases often show signs of gut imbalance, known as intestinal micro ecological dysfunction. This includes reduced diversity and richness of gut bacteria. The drop in beneficial microbes and increase in bacteria increase inflammation (Sochocka et al., 2019). These changes can lower the production of short chain fatty acids (SCFAs) like acetic acid, butyric acid and propionic acid which are important for brain health. The imbalance in gut bacteria and their byproducts plays a key role in the development of NNDs by affecting inflammation and oxidative stress in the brain.

Flavonoids can help restore gut health. They encourage the growth of beneficial bacteria, which then produce SCFAs. They also release important hormones and neurotransmitters like serotonin and gamma-aminobutyric acid (GABA). These substances support brain function and help prevent or slow down the neurodegenerative conditions by acting through the microbiome-gut-brain (MGB) axis (Wu et al., 2021). It solely depends on how gut microbes process flavonoids. These microbes break down polyphenols (flavonoids) into smaller, more active compounds that can reduce brain inflammation and oxidative stress.

Figure 5. Role of polyphenols (flavonoid) in modulating microbiota-gut-brain (MGB) axis to support brain health and slow the progression of neurodegenerative diseases (NNDs)

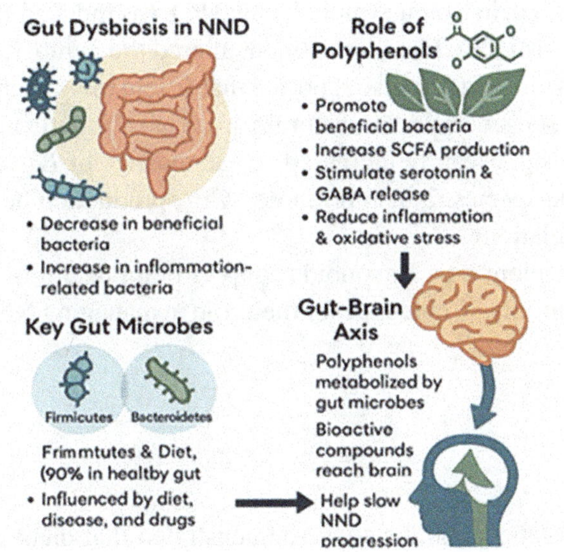

Quercetin has demonstrated potent role in neuroprotection in neurodegenerative diseases, primarily due to its anti-inflammatory as well as antioxidant properties. Despite its relatively low oral bioavailability as previously discussed, it exerts beneficial effects by interacting with gut microbiota. It is reported that quercetin has reduced myelin and axonal damage in rat models with diabetic peripheral neuropathy (DPN). In an early study, they induced an Alzheimer's disease model using streptozotocin, where quercetin treatment reversed ROS levels and enhanced gut microbiota diversity. Quercetin appears to modulate the gut microbiota by decreasing the colonies of potentially pathogenic bacteria like *f_Porphyromonadaceae*, *g_Oxalobacter*, *f_Oxalobacteraceae* and *g_Klebsiella*. These bacteria are certainly associated with DPN phenotypes and oxidative stress. Simultaneously, it enriches beneficial bacteria like *c_Actinobacteria* and *p_Actinobacteria*. They lower oxidative stress and DPN. Therefore, quercetin may support neuroprotection in AD by alleviating DNP-induced gut dysbiosis, particularly through the increase of Actinobacteria at both the phylum and class level (Xie et al., 2020).

Isorientin, a flavone, naturally occurs in corn, bamboo, silk, *Passiflora edulis* and gentian leaves. It possesses strong anti-inflammatory and antioxidant properties. Research has shown that isoorientin reduces the accumulation of Aβ oligomer in the cortex and hippocampus. It alleviates synaptic dysfunction, and improves spatial memory deficits in APP/PSI transgenic mice. This shows its potential therapeutic value in Alzheimer disease (AD) (X. Tan et al., 2021).

Further evidence suggests that isoorientin exerts neuroprotection by modulating the gut microbiota. Zhang and coworkers, demonstrated that isoorientin significantly targeted Aβ plaque, reducing its deposition in the brain's cortex and hippocampus region of AD mice. It altered the gut microbial composition by promoting the abundance of *Mollicutes* class, *Prevotellaceae* (family), *Prevotellaceae UCG*

001 (genus) in fecal samples. *Proteobacteria* was found in the cecum of large intestine of AD mice (Z. Zhang et al., 2022).

Fisetin is a natural flavonoid found in many fruits and vegetables. It has shown significant antioxidant, anti-inflammatory and brain-protective effects, making it a promising potential treatment for neurodegenerative diseases like amyotrophic lateral sclerosis, PD and AD (T. H. Wang et al., 2018). Recent studies suggest that fisetin may exert its neuroprotective effects by modulating the composition and diversity of the gut microbiota. For instance, in a mouse model where PD was induced by giving 1-methyl-4-phenyl-1, 2, 3, 6-tetrahydropyidine (MPTP). Fisetin treatment altered the number, diversity and distribution of intestinal microbes (Chen et al., 2020). It also reported that fistein improved the behavioral problems and reduced dopaminergic neurodegeneration in the brain areas of substantia nigre and striatum in the mice with PD. Moreover, fistein significantly increased the abundance of *Lachnospiraceae* while reducing levels of uncultured bacteria species. It also promote SCFA production hence, supporting brain health through gut microbiota regulation.

Research continues to explore how flavonoids support brain health by regulating gut bacteria and their metabolites, offering promising evidence for their role in managing NNDs (Y. Liu, Li, Zhou, Wang, & Ao, 2022).

8. CONCLUSION

Many clinical and observational studies have demonstrated that diets high in plant polyphenols or supplements containing polyphenol rich extracts can act against neurodegenerative conditions in humans. The mechanisms underlying these benefits are complex encompassing both direct and indirect actions but not yet fully understood. The existing evidence suggests that increased dietary intake of plant polyphenols support central nervous system health. However, when considering individual polyphenols or defined mixtures of isolated polyphenols, several critical issues remain to be addressed in future clinical trials:

1. Establishment of appropriate dosage and treatment duration
2. Enhancement of polyphenol bioavailability
3. Identification of indications and potential contradictions
4. Determination of the optimal composition of polyphenol mixtures
5. Development of nanoformulations to improve delivery to the CNS, including crossing the blood-brain barrier (BBB).
6. Pharmacokinetic profiling of flavonoids in human blood and brain tissue, including the relationship between plasma and CNS concentration.

While current evidence supports the neuroprotective potential of plant flavonoids, further research is essential to optimize their formulation, delivery, and clinical application. Addressing these challenges will be crucial for the development of effective flavonoid based therapies for neurodegenerative diseases.

ACKNOWLEDGEMENT AND FUNDING STATEMENT

This work was supported by the Deanship of Scientific Research, Vice Presidency for Graduate Studies and Scientific Research, King Faisal University, Saudi Arabia (Grant No: KFU KFU253515).

REFERENCES

Abbott, N. J., Rönnbäck, L., & Hansson, E. (2006). Astrocyte-endothelial interactions at the blood-brain barrier. *Nature Reviews. Neuroscience*, *7*(1), 41–53. DOI: 10.1038/nrn1824

Abdullah, A., & Ravanan, P. (2018). Kaempferol mitigates Endoplasmic Reticulum Stress Induced Cell Death by targeting caspase 3/7. *Scientific Reports*, *8*(1), 2189. DOI: 10.1038/s41598-018-20499-7

Adams, S. M., Aksenova, M. V., Aksenov, M. Y., Mactutus, C. F., & Booze, R. M. (2012). Soy isoflavones genistein and daidzein exert anti-apoptotic actions via a selective ER-mediated mechanism in neurons following HIV-1 Tat(1-86) exposure. *PLoS One*, *7*(5), e37540. DOI: 10.1371/journal.pone.0037540

Alvarez, A. I., Real, R., Pérez, M., Mendoza, G., Prieto, J. G., & Merino, G. (2010). Modulation of the activity of ABC transporters (P-glycoprotein, MRP2, BCRP) by flavonoids and drug response. *Journal of Pharmaceutical Sciences*, *99*(2), 598–617. DOI: 10.1002/jps.21851

Anusha, C., Sumathi, T., & Joseph, L. D. (2017). Protective role of apigenin on rotenone induced rat model of Parkinson's disease: Suppression of neuroinflammation and oxidative stress mediated apoptosis. *Chemico-Biological Interactions*, *269*, 67–79. DOI: 10.1016/j.cbi.2017.03.016

Aras, A. B., Guven, M., Akman, T., Ozkan, A., Sen, H. M., Duz, U., & Cosar, M. (2015). Neuroprotective effects of daidzein on focal cerebral ischemia injury in rats. *Neural Regeneration Research*, *10*(1), 146–152. DOI: 10.4103/1673-5374.150724

Arias-Sánchez, R. A., Torner, L., & Fenton Navarro, B. (2023). Polyphenols and Neurodegenerative Diseases: Potential Effects and Mechanisms of Neuroprotection. *Molecules (Basel, Switzerland)*, *28*(14). Advance online publication. DOI: 10.3390/molecules28145415

Aruna, R., Geetha, A., & Suguna, P. (2014). Rutin modulates ASC expression in NLRP3 inflammasome: A study in alcohol and cerulein-induced rat model of pancreatitis. *Molecular and Cellular Biochemistry*, *396*(1-2), 269–280. DOI: 10.1007/s11010-014-2162-8

Barreca, D., Bellocco, E., D'Onofrio, G., Nabavi, S. F., Daglia, M., Rastrelli, L., & Nabavi, S. M. (2016). Neuroprotective Effects of Quercetin: From Chemistry to Medicine. *CNS & Neurological Disorders - Drug Targets*, *15*(8), 964–975. DOI: 10.2174/1871527315666160813175406

Bhullar, K. S., & Rupasinghe, H. P. (2013). Polyphenols: Multipotent therapeutic agents in neurodegenerative diseases. *Oxidative Medicine and Cellular Longevity*, *891748*. Advance online publication. DOI: 10.1155/2013/891748

Brickman, A. M., Khan, U. A., Provenzano, F. A., Yeung, L. K., Suzuki, W., Schroeter, H., & Small, S. A. (2014). Enhancing dentate gyrus function with dietary flavanols improves cognition in older adults. *Nature Neuroscience*, *17*(12), 1798–1803. DOI: 10.1038/nn.3850

Budagova, K. R., Zhmaeva, S. V., Grigor'ev, A. N., Goncharova, A. Y., & Kabakov, A. E. (2003). Flavonoid dihydroquercetin, unlike quercetin, fails to inhibit expression of heat shock proteins under conditions of cellular stress. *Biochemistry (Moscow)*, *68*(9), 1055–1061. DOI: 10.1023/a:1026081016923

Cai, W., Wu, J., Wang, X., Huang, J., Shen, Z., & Chen, X. (2019). Epimedium flavonoids mitigate proteotoxicity and extend healthspan via DAF-16 in C. elegans. *Traditional Medicine and Modern Medicine*, *02*(01), 19–25. DOI: 10.1142/s2575900019500046

Carter, P., Gray, L. J., Troughton, J., Khunti, K., & Davies, M. J. (2010). Fruit and vegetable intake and incidence of type 2 diabetes mellitus: Systematic review and meta-analysis. *BMJ (Clinical Research Ed.)*, *341*, c4229. DOI: 10.1136/bmj.c4229

Chen, T. J., Feng, Y., Liu, T., Wu, T. T., Chen, Y. J., Li, X., & Wu, Y. C. (2020). Fisetin Regulates Gut Microbiota and Exerts Neuroprotective Effect on Mouse Model of Parkinson's Disease. *Frontiers in Neuroscience*, *14*, 549037. DOI: 10.3389/fnins.2020.549037

Chiang, C. K., Hsu, S. P., Wu, C. T., Huang, J. W., Cheng, H. T., Chang, Y. W., & Liu, S. H. (2011). Endoplasmic reticulum stress implicated in the development of renal fibrosis. *Molecular Medicine (Cambridge, Mass.)*, *17*(11-12), 1295–1305. DOI: 10.2119/molmed.2011.00131

Cirmi, S., Ferlazzo, N., Lombardo, G. E., Ventura-Spagnolo, E., Gangemi, S., Calapai, G., & Navarra, M. (2016). Neurodegenerative Diseases: Might Citrus Flavonoids Play a Protective Role? *Molecules (Basel, Switzerland)*, *21*(10). Advance online publication. DOI: 10.3390/molecules21101312

Cook, C., Stetler, C., & Petrucelli, L. (2012). Disruption of protein quality control in Parkinson's disease. *Cold Spring Harbor Perspectives in Medicine*, *2*(5), a009423. DOI: 10.1101/cshperspect.a009423

Devi, S., Kumar, V., Singh, S. K., Dubey, A. K., & Kim, J. J. (2021). Flavonoids: Potential Candidates for the Treatment of Neurodegenerative Disorders. *Biomedicines*, *9*(2). Advance online publication. DOI: 10.3390/biomedicines9020099

Di Lorenzo, C., Colombo, F., Biella, S., Stockley, C., & Restani, P. (2021). Polyphenols and Human Health: The Role of Bioavailability. *Nutrients*, *13*(1). Advance online publication. DOI: 10.3390/nu13010273

Dias, V., Junn, E., & Mouradian, M. M. (2013). The Role of Oxidative Stress in Parkinson's Disease. *Journal of Parkinson's Disease*, *3*(4), 461–491. DOI: 10.3233/jpd-130230

Echeverry, C., Arredondo, F., Abin-Carriquiry, J. A., Midiwo, J. O., Ochieng, C., Kerubo, L., & Dajas, F. (2010). Pretreatment with natural flavones and neuronal cell survival after oxidative stress: A structure-activity relationship study. *Journal of Agricultural and Food Chemistry*, *58*(4), 2111–2115. DOI: 10.1021/jf902951v

Engwa, G. A. (2018). Free Radicals and the Role of Plant Phytochemicals as Antioxidants Against Oxidative Stress-Related Diseases. In Asao, T., & Asaduzzaman, M. (Eds.), *Phytochemicals - Source of Antioxidants and Role in Disease Prevention*. IntechOpen.

Estruch, R., Ros, E., Salas-Salvadó, J., Covas, M. I., Corella, D., Arós, F., & Martínez-González, M. A. (2018). Primary Prevention of Cardiovascular Disease with a Mediterranean Diet Supplemented with Extra-Virgin Olive Oil or Nuts. *The New England Journal of Medicine*, *378*(25), e34. DOI: 10.1056/NEJMoa1800389

Feng, L., Gwee, X., Kua, E. H., & Ng, T. P. (2010). Cognitive function and tea consumption in community dwelling older Chinese in Singapore. *The Journal of Nutrition, Health & Aging*, *14*(6), 433–438. DOI: 10.1007/s12603-010-0095-9

Ferri, P., Angelino, D., Gennari, L., Benedetti, S., Ambrogini, P., Del Grande, P., & Ninfali, P. (2015). Enhancement of flavonoid ability to cross the blood-brain barrier of rats by co-administration with α-tocopherol. *Food & Function*, *6*(2), 394–400. DOI: 10.1039/c4fo00817k

Figueira, I., Garcia, G., Pimpão, R. C., Terrasso, A. P., Costa, I., Almeida, A. F., & Santos, C. N. (2017). Polyphenols journey through blood-brain barrier towards neuronal protection. *Scientific Reports*, *7*(1), 11456. DOI: 10.1038/s41598-017-11512-6

Gomez-Pinilla, F., & Nguyen, T. T. (2012). Natural mood foods: The actions of polyphenols against psychiatric and cognitive disorders. *Nutritional Neuroscience*, *15*(3), 127–133. DOI: 10.1179/1476830511y.0000000035

Grabska-Kobylecka, I., Kaczmarek-Bak, J., Figlus, M., Prymont-Przyminska, A., Zwolinska, A., Sarniak, A., & Nowak, D. (2020). The Presence of Caffeic Acid in Cerebrospinal Fluid: Evidence That Dietary Polyphenols Can Cross the Blood-Brain Barrier in Humans. *Nutrients*, *12*(5). Advance online publication. DOI: 10.3390/nu12051531

Grabska-Kobyłecka, I., Szpakowski, P., Król, A., Książek-Winiarek, D., Kobyłecki, A., Głąbiński, A., & Nowak, D. (2023). Polyphenols and Their Impact on the Prevention of Neurodegenerative Diseases and Development. *Nutrients*, *15*(15). Advance online publication. DOI: 10.3390/nu15153454

Hanáková, Z., Hošek, J., Kutil, Z., Temml, V., Landa, P., Vaněk, T., & Šmejkal, K. (2017). Anti-inflammatory Activity of Natural Geranylated Flavonoids: Cyclooxygenase and Lipoxygenase Inhibitory Properties and Proteomic Analysis. *Journal of Natural Products*, *80*(4), 999–1006. DOI: 10.1021/acs.jnatprod.6b01011

He, F. J., Nowson, C. A., & MacGregor, G. A. (2006). Fruit and vegetable consumption and stroke: Meta-analysis of cohort studies. *Lancet*, *367*(9507), 320–326. DOI: 10.1016/s0140-6736(06)68069-0

Heidary Moghaddam, R., Samimi, Z., Moradi, S. Z., Little, P. J., Xu, S., & Farzaei, M. H. (2020). Naringenin and naringin in cardiovascular disease prevention: A preclinical review. *European Journal of Pharmacology*, *887*, 173535. DOI: 10.1016/j.ejphar.2020.173535

Hirsch, E. C., & Hunot, S. (2009). Neuroinflammation in Parkinson's disease: A target for neuroprotection? *Lancet Neurology*, *8*(4), 382–397. DOI: 10.1016/s1474-4422(09)70062-6

Hu, L. W., Yen, J. H., Shen, Y. T., Wu, K. Y., & Wu, M. J. (2014). Luteolin modulates 6-hydroxydopamine-induced transcriptional changes of stress response pathways in PC12 cells. *PLoS One*, *9*(5), e97880. DOI: 10.1371/journal.pone.0097880

Huang, Q., Wu, L. J., Tashiro, S., Gao, H. Y., Onodera, S., & Ikejima, T. (2005). (+)-Catechin, an ingredient of green tea, protects murine microglia from oxidative stress-induced DNA damage and cell cycle arrest. *Journal of Pharmacological Sciences*, *98*(1), 16–24. DOI: 10.1254/jphs.fpj04053x

Janle, E. M., Lila, M. A., Grannan, M., Wood, L., Higgins, A., Yousef, G. G., & Weaver, C. M. (2010). Pharmacokinetics and tissue distribution of 14C-labeled grape polyphenols in the periphery and the central nervous system following oral administration. *Journal of Medicinal Food*, *13*(4), 926–933. DOI: 10.1089/jmf.2009.0157

Joshi, V., Mishra, R., Upadhyay, A., Amanullah, A., Poluri, K. M., Singh, S., & Mishra, A. (2019). Polyphenolic flavonoid (Myricetin) upregulated proteasomal degradation mechanisms: Eliminates neurodegenerative proteins aggregation. *Journal of Cellular Physiology*, *234*(11), 20900–20914. DOI: 10.1002/jcp.28695

Kang, Y., Lee, J. H., Seo, Y. H., Jang, J. H., Jeong, C. H., Lee, S., & Park, B. (2019). Epicatechin Prevents Methamphetamine-Induced Neuronal Cell Death via Inhibition of ER Stress. *Biomolecules & Therapeutics*, *27*(2), 145–151. DOI: 10.4062/biomolther.2018.092

Kejík, Z., Kaplánek, R., Masařík, M., Babula, P., Matkowski, A., Filipenský, P., & Jakubek, M. (2021). Iron Complexes of Flavonoids-Antioxidant Capacity and Beyond. *International Journal of Molecular Sciences*, *22*(2). Advance online publication. DOI: 10.3390/ijms22020646

Khan, A., Ikram, M., Hahm, J. R., & Kim, M. O. (2020). Antioxidant and Anti-Inflammatory Effects of Citrus Flavonoid Hesperetin: Special Focus on Neurological Disorders. *Antioxidants*, *9*(7). Advance online publication. DOI: 10.3390/antiox9070609

Kim, H., Kim, Y. S., Kim, S. Y., & Suk, K. (2001). The plant flavonoid wogonin suppresses death of activated C6 rat glial cells by inhibiting nitric oxide production. *Neuroscience Letters*, *309*(1), 67–71. DOI: 10.1016/s0304-3940(01)02028-6

Kirschweng, B., Tátraaljai, D., Földes, E., & Pukánszky, B. (2017). Natural antioxidants as stabilizers for polymers. *Polymer Degradation & Stability*, *145*, 25–40. DOI: 10.1016/j.polymdegradstab.2017.07.012

Koudoufio, M., Desjardins, Y., Feldman, F., Spahis, S., Delvin, E., & Levy, E. (2020). Insight into Polyphenol and Gut Microbiota Crosstalk: Are Their Metabolites the Key to Understand Protective Effects against Metabolic Disorders? *Antioxidants*, *9*(10). Advance online publication. DOI: 10.3390/antiox9100982

Kumar, J., Park, K. C., Awasthi, A., & Prasad, B. (2015). Silymarin extends lifespan and reduces proteotoxicity in C. elegans Alzheimer's model. *CNS & Neurological Disorders - Drug Targets*, *14*(2), 295–302. DOI: 10.2174/1871527314666150116110212

Kumar, S., & Pandey, A. K. (2013). Chemistry and biological activities of flavonoids: An overview. *TheScientificWorldJournal*, *162750*. Advance online publication. DOI: 10.1155/2013/162750

Lamport, D. J., & Williams, C. M. (2021). Polyphenols and Cognition In Humans: An Overview of Current Evidence from Recent Systematic Reviews and Meta-Analyses. *Brain Plasticity*, *6*(2), 139–153. DOI: 10.3233/bpl-200111

Lao, C. D., Ruffin, M. T. t., Normolle, D., Heath, D. D., Murray, S. I., Bailey, J. M., & Brenner, D. E. (2006). Dose escalation of a curcuminoid formulation. *BMC Complementary and Alternative Medicine, 6*, 10. DOI: 10.1186/1472-6882-6-10

Lau, F. C., Bielinski, D. F., & Joseph, J. A. (2007). Inhibitory effects of blueberry extract on the production of inflammatory mediators in lipopolysaccharide-activated BV2 microglia. *Journal of Neuroscience Research, 85*(5), 1010–1017. DOI: 10.1002/jnr.21205

Leclerc, M., Dudonné, S., & Calon, F. (2021). Can Natural Products Exert Neuroprotection without Crossing the Blood-Brain Barrier? *International Journal of Molecular Sciences, 22*(7). Advance online publication. DOI: 10.3390/ijms22073356

Lee-Hilz, Y. Y., Boerboom, A. M., Westphal, A. H., Berkel, W. J., Aarts, J. M., & Rietjens, I. M. (2006). Pro-oxidant activity of flavonoids induces EpRE-mediated gene expression. *Chemical Research in Toxicology, 19*(11), 1499–1505. DOI: 10.1021/tx060157q

Lesjak, M., Beara, I., Simin, N., Pintać, D., Majkić, T., Bekvalac, K., & Mimica-Dukić, N. (2018). Antioxidant and anti-inflammatory activities of quercetin and its derivatives. *Journal of Functional Foods, 40*, 68–75. DOI: 10.1016/j.jff.2017.10.047

Li, R., Wang, X., Qin, T., Qu, R., & Ma, S. (2016). Apigenin ameliorates chronic mild stress-induced depressive behavior by inhibiting interleukin-1β production and NLRP3 inflammasome activation in the rat brain. *Behavioural Brain Research, 296*, 318–325. DOI: 10.1016/j.bbr.2015.09.031

Liu, H., Yang, J., Li, L., Shi, W., Yuan, X., & Wu, L. (2016). The Natural Occurring Compounds Targeting Endoplasmic Reticulum Stress. *Evidence-Based Complementary and Alternative Medicine : eCAM, 7831282*. Advance online publication. DOI: 10.1155/2016/7831282

Liu, M. H., Lin, Y. S., Sheu, S. Y., & Sun, J. S. (2009). Anti-inflammatory effects of daidzein on primary astroglial cell culture. *Nutritional Neuroscience, 12*(3), 123–134. DOI: 10.1179/147683009x423274

Liu, Y., Li, X., Zhou, X., Wang, J., & Ao, X. (2022). FADD as a key molecular player in cancer progression. *Molecular Medicine (Cambridge, Mass.), 28*(1), 132. DOI: 10.1186/s10020-022-00560-y

Liu, Y., Li, Y., Du, C., Kuang, S., Zhou, X., Zhang, J., & Ao, X. (2022). Underlying mechanisms of epithelial splicing regulatory proteins in cancer progression. *Journal of Molecular Medicine (Berlin, Germany), 100*(11), 1539–1556. DOI: 10.1007/s00109-022-02257-5

López-Lázaro, M. (2009). Distribution and biological activities of the flavonoid luteolin. *Mini-Reviews in Medicinal Chemistry, 9*(1), 31–59. DOI: 10.2174/138955709787001712

Macready, A. L., Kennedy, O. B., Ellis, J. A., Williams, C. M., Spencer, J. P., & Butler, L. T. (2009). Flavonoids and cognitive function: A review of human randomized controlled trial studies and recommendations for future studies. *Genes & Nutrition, 4*(4), 227–242. DOI: 10.1007/s12263-009-0135-4

Maleki, S. J., Crespo, J. F., & Cabanillas, B. (2019). Anti-inflammatory effects of flavonoids. *Food Chemistry, 299*, 125124. DOI: 10.1016/j.foodchem.2019.125124

Martucciello, S., Masullo, M., Cerulli, A., & Piacente, S. (2020). Natural Products Targeting ER Stress, and the Functional Link to Mitochondria. *International Journal of Molecular Sciences*, *21*(6). Advance online publication. DOI: 10.3390/ijms21061905

Masala, G., Assedi, M., Bendinelli, B., Ermini, I., Sieri, S., Grioni, S., & Palli, D. (2012). Fruit and vegetables consumption and breast cancer risk: The EPIC Italy study. *Breast Cancer Research and Treatment*, *132*(3), 1127–1136. DOI: 10.1007/s10549-011-1939-7

Moradi, S. Z., Jalili, F., Farhadian, N., Joshi, T., Wang, M., Zou, L., & Xiao, J. (2022). Polyphenols and neurodegenerative diseases: Focus on neuronal regeneration. *Critical Reviews in Food Science and Nutrition*, *62*(13), 3421–3436. DOI: 10.1080/10408398.2020.1865870

Moradi, S. Z., Momtaz, S., Bayrami, Z., Farzaei, M. H., & Abdollahi, M. (2020). Nanoformulations of Herbal Extracts in Treatment of Neurodegenerative Disorders. *Frontiers in Bioengineering and Biotechnology*, *8*, 238. DOI: 10.3389/fbioe.2020.00238

Naoi, M., Shamoto-Nagai, M., & Maruyama, W. (2019). Neuroprotection of Multifunctional Phytochemicals as Novel Therapeutic Strategy for Neurodegenerative Disorders: Antiapoptotic and Antiamyloidogenic Activities by Modulation of Cellular Signal Pathways. *Future Neurology*, *14*(1), FNL9. DOI: 10.2217/fnl-2018-0028

Panche, A. N., Diwan, A. D., & Chandra, S. R. (2016). Flavonoids: An overview. *Journal of Nutritional Science*, *5*, e47. DOI: 10.1017/jns.2016.41

Papi, S., Ahmadizar, F., & Hasanvand, A. (2019). The role of nitric oxide in inflammation and oxidative stress. *Immunopathol Persa*, *5*(1), e08–e08. DOI: 10.15171/ipp.2019.08

Parr, A. J., & Bolwell, G. P. (2000). Phenols in the plant and in man. The potential for possible nutritional enhancement of the diet by modifying the phenols content or profile. *Journal of the Science of Food and Agriculture*, *80*(7), 985–1012. DOI: 10.1002/(SICI)1097-0010(20000515)80:7<985:AID-JSFA572>3.0.CO;2-7

Patil, S. P., Jain, P. D., Sancheti, J. S., Ghumatkar, P. J., Tambe, R., & Sathaye, S. (2022). Retraction notice to "Neuroprotective and neurotrophic effects of Apigenin and Luteolin in MPTP induced Parkinsonism in mice" [Neuropharmacology, 86 (2014) 192-202]. *Neuropharmacology*, *202*, 108876. DOI: 10.1016/j.neuropharm.2021.108876

Pinheiro Fernandes, F. D., Fontenele Menezes, A. P., de Sousa Neves, J. C., Fonteles, A. A., da Silva, A. T., de Araújo Rodrigues, P., & de Andrade, G. M. (2014). Caffeic acid protects mice from memory deficits induced by focal cerebral ischemia. *Behavioural Pharmacology*, *25*(7), 637–647. DOI: 10.1097/fbp.0000000000000076

Qian, Y., Guan, T., Huang, M., Cao, L., Li, Y., Cheng, H., & Yu, D. (2012). Neuroprotection by the soy isoflavone, genistein, via inhibition of mitochondria-dependent apoptosis pathways and reactive oxygen induced-NF-κB activation in a cerebral ischemia mouse model. *Neurochemistry International*, *60*(8), 759–767. DOI: 10.1016/j.neuint.2012.03.011

Rana, A., Samtiya, M., Dhewa, T., Mishra, V., & Aluko, R. E. (2022). Health benefits of polyphenols: A concise review. *Journal of Food Biochemistry*, *46*(10), e14264. DOI: 10.1111/jfbc.14264

Rehman, M. U., Wali, A. F., Ahmad, A., Shakeel, S., Rasool, S., Ali, R., & Khan, R. (2019). Neuroprotective Strategies for Neurological Disorders by Natural Products: An update. *Current Neuropharmacology, 17*(3), 247–267. DOI: 10.2174/1570159x16666180911124605

Renaud, J., & Martinoli, M. G. (2019). Considerations for the Use of Polyphenols as Therapies in Neurodegenerative Diseases. *International Journal of Molecular Sciences, 20*(8). Advance online publication. DOI: 10.3390/ijms20081883

Rendeiro, C., Rhodes, J. S., & Spencer, J. P. (2015). The mechanisms of action of flavonoids in the brain: Direct versus indirect effects. *Neurochemistry International, 89*, 126–139. DOI: 10.1016/j.neuint.2015.08.002

Ricciotti, E., & FitzGerald, G. A. (2011). Prostaglandins and inflammation. *Arteriosclerosis, Thrombosis, and Vascular Biology, 31*(5), 986–1000. DOI: 10.1161/atvbaha.110.207449

Rothwell, J. A., Perez-Jimenez, J., Neveu, V., Medina-Remón, A., M'Hiri, N., García-Lobato, P., & Scalbert, A. (2013). Phenol-Explorer 3.0: A major update of the Phenol-Explorer database to incorporate data on the effects of food processing on polyphenol content. *Database : The Journal of Biological Databases and Curation, 2013*, bat070. Advance online publication. DOI: 10.1093/database/bat070

Ruotolo, R., Minato, I., La Vitola, P., Artioli, L., Curti, C., Franceschi, V., & Ottonello, S. (2020). Flavonoid-Derived Human Phenyl-γ-Valerolactone Metabolites Selectively Detoxify Amyloid-β Oligomers and Prevent Memory Impairment in a Mouse Model of Alzheimer's Disease. *Molecular Nutrition & Food Research, 64*(5), e1900890. DOI: 10.1002/mnfr.201900890

Sharma, S. K., & Priya, S. (2017). Expanding role of molecular chaperones in regulating α-synuclein misfolding; implications in Parkinson's disease. *Cellular and Molecular Life Sciences : CMLS, 74*(4), 617–629. DOI: 10.1007/s00018-016-2340-9

Simunkova, M., Barbicrikova, Z., Jomova, K., Hudecova, L., Lauro, P., Alwasel, S. H., & Valko, M. (2021). Antioxidant vs. Prooxidant Properties of the Flavonoid, Kaempferol, in the Presence of Cu(II) Ions: A ROS-Scavenging Activity, Fenton Reaction and DNA Damage Study. *International Journal of Molecular Sciences, 22*(4). Advance online publication. DOI: 10.3390/ijms22041619

Singh, B., & Semwal, B. C. (2024). A Compressive Review on Source, Toxicity and Biological Activity of Flavonoid. *Current Topics in Medicinal Chemistry, 24*(24), 2093–2116. DOI: 10.2174/0115680266316032240718050055

Singh, N. A., Mandal, A. K., & Khan, Z. A. (2016). Potential neuroprotective properties of epigallocatechin-3-gallate (EGCG). *Nutrition Journal, 15*(1), 60. DOI: 10.1186/s12937-016-0179-4

Socci, V., Tempesta, D., Desideri, G., De Gennaro, L., & Ferrara, M. (2017). Enhancing Human Cognition with Cocoa Flavonoids. *Frontiers in Nutrition, 4*, 19. DOI: 10.3389/fnut.2017.00019

Sochocka, M., Diniz, B. S., & Leszek, J. (2017). Inflammatory Response in the CNS: Friend or Foe? *Molecular Neurobiology, 54*(10), 8071–8089. DOI: 10.1007/s12035-016-0297-1

Sochocka, M., Donskow-Łysoniewska, K., Diniz, B. S., Kurpas, D., Brzozowska, E., & Leszek, J. (2019). The Gut Microbiome Alterations and Inflammation-Driven Pathogenesis of Alzheimer's Disease-a Critical Review. *Molecular Neurobiology, 56*(3), 1841–1851. DOI: 10.1007/s12035-018-1188-4

Sontag, E. M., Samant, R. S., & Frydman, J. (2017). Mechanisms and Functions of Spatial Protein Quality Control. *Annual Review of Biochemistry, 86*, 97–122. DOI: 10.1146/annurev-biochem-060815-014616

Sorrenti, V., Ali, S., Mancin, L., Davinelli, S., Paoli, A., & Scapagnini, G. (2020). Cocoa Polyphenols and Gut Microbiota Interplay: Bioavailability, Prebiotic Effect, and Impact on Human Health. *Nutrients, 12*(7). Advance online publication. DOI: 10.3390/nu12071908

Strathearn, K. E., Yousef, G. G., Grace, M. H., Roy, S. L., Tambe, M. A., Ferruzzi, M. G., & Rochet, J. C. (2014). Neuroprotective effects of anthocyanin- and proanthocyanidin-rich extracts in cellular models of Parkinson s disease. *Brain Research, 1555*, 60–77. DOI: 10.1016/j.brainres.2014.01.047

Tan, K. W., Li, Y., Paxton, J. W., Birch, N. P., & Scheepens, A. (2013). Identification of novel dietary phytochemicals inhibiting the efflux transporter breast cancer resistance protein (BCRP/ABCG2). *Food Chemistry, 138*(4), 2267–2274. DOI: 10.1016/j.foodchem.2012.12.021

Tan, X., Liang, Z., Li, Y., Zhi, Y., Yi, L., Bai, S., & Li, Q. X. (2021). Isoorientin, a GSK-3β inhibitor, rescues synaptic dysfunction, spatial memory deficits and attenuates pathological progression in APP/PS1 model mice. *Behavioural Brain Research, 398*, 112968. DOI: 10.1016/j.bbr.2020.112968

Terao, J. (2017). Factors modulating bioavailability of quercetin-related flavonoids and the consequences of their vascular function. *Biochemical Pharmacology, 139*, 15–23. DOI: 10.1016/j.bcp.2017.03.021

Tönnies, E., & Trushina, E. (2017). Oxidative Stress, Synaptic Dysfunction, and Alzheimer's Disease. *Journal of Alzheimer's Disease, 57*(4), 1105–1121. DOI: 10.3233/jad-161088

Torello, C. O., Alvarez, M. C., & Olalla Saad, S. T. (2021). Polyphenolic Flavonoid Compound Quercetin Effects in the Treatment of Acute Myeloid Leukemia and Myelodysplastic Syndromes. *Molecules (Basel, Switzerland), 26*(19). Advance online publication. DOI: 10.3390/molecules26195781

Uddin, M. S., Tewari, D., Sharma, G., Kabir, M. T., Barreto, G. E., Bin-Jumah, M. N., & Ashraf, G. M. (2020). Molecular Mechanisms of ER Stress and UPR in the Pathogenesis of Alzheimer's Disease. *Molecular Neurobiology, 57*(7), 2902–2919. DOI: 10.1007/s12035-020-01929-y

Vafeiadou, K., Vauzour, D., Lee, H. Y., Rodriguez-Mateos, A., Williams, R. J., & Spencer, J. P. (2009). The citrus flavanone naringenin inhibits inflammatory signalling in glial cells and protects against neuroinflammatory injury. *Archives of Biochemistry and Biophysics, 484*(1), 100–109. DOI: 10.1016/j.abb.2009.01.016

Vicente, O., & Boscaiu, M. (2018). Flavonoids: Antioxidant Compounds for Plant Defence... and for a Healthy Human Diet. *Notulae Botanicae Horti Agrobotanici Cluj-Napoca, 46*(1), 14–21. DOI: 10.15835/nbha46110992

Wang, Q., Xu, J., Rottinghaus, G. E., Simonyi, A., Lubahn, D., Sun, G. Y., & Sun, A. Y. (2002). Resveratrol protects against global cerebral ischemic injury in gerbils. *Brain Research, 958*(2), 439–447. DOI: 10.1016/s0006-8993(02)03543-6

Wang, T. H., Wang, S. Y., Wang, X. D., Jiang, H. Q., Yang, Y. Q., Wang, Y., & Feng, H. L. (2018). Fisetin Exerts Antioxidant and Neuroprotective Effects in Multiple Mutant hSOD1 Models of Amyotrophic Lateral Sclerosis by Activating ERK. *Neuroscience*, *379*, 152–166. DOI: 10.1016/j.neuroscience.2018.03.008

Wu, Z., Huang, S., Li, T., Li, N., Han, D., Zhang, B., & Wang, J. (2021). Gut microbiota from green tea polyphenol-dosed mice improves intestinal epithelial homeostasis and ameliorates experimental colitis. *Microbiome*, *9*(1), 184. DOI: 10.1186/s40168-021-01115-9

Xie, J., Song, W., Liang, X., Zhang, Q., Shi, Y., Liu, W., & Shi, X. (2020). Protective effect of quercetin on streptozotocin-induced diabetic peripheral neuropathy rats through modulating gut microbiota and reactive oxygen species level. *Biomedicine and Pharmacotherapy*, *127*, 110147. DOI: 10.1016/j.biopha.2020.110147

Yang, T., Zhao, X., Zhang, Y., Xie, J., & Zhou, A. (2020). 6‴-Feruloylspinosin alleviated beta-amyloid induced toxicity by promoting mitophagy in Caenorhabditis elegans (GMC101) and PC12 cells. *The Science of the Total Environment*, *715*, 136953. DOI: 10.1016/j.scitotenv.2020.136953

Youdim, K. A., Dobbie, M. S., Kuhnle, G., Proteggente, A. R., Abbott, N. J., & Rice-Evans, C. (2003). Interaction between flavonoids and the blood-brain barrier: In vitro studies. *Journal of Neurochemistry*, *85*(1), 180–192. DOI: 10.1046/j.1471-4159.2003.01652.x

Youdim, K. A., Qaiser, M. Z., Begley, D. J., Rice-Evans, C. A., & Abbott, N. J. (2004). Flavonoid permeability across an in situ model of the blood-brain barrier. *Free Radical Biology & Medicine*, *36*(5), 592–604. DOI: 10.1016/j.freeradbiomed.2003.11.023

Zakaryan, H., Arabyan, E., Oo, A., & Zandi, K. (2017). Flavonoids: Promising natural compounds against viral infections. *Archives of Virology*, *162*(9), 2539–2551. DOI: 10.1007/s00705-017-3417-y

Zeng, N., Zhang, G., Hu, X., Pan, J., & Gong, D. (2019). Mechanism of fisetin suppressing superoxide anion and xanthine oxidase activity. *Journal of Functional Foods*, *58*, 1–10. DOI: 10.1016/j.jff.2019.04.044

Zhang, D., Hu, X., Qian, L., Wilson, B., Lee, C., Flood, P., & Hong, J. S. (2009). Prostaglandin E2 released from activated microglia enhances astrocyte proliferation in vitro. *Toxicology and Applied Pharmacology*, *238*(1), 64–70. DOI: 10.1016/j.taap.2009.04.015

Zhang, K., Dai, Z., Zhang, W., Gao, Q., Dai, Y., Xia, F., & Zhang, X. (2021). EDTA-based adsorbents for the removal of metal ions in wastewater. *Coordination Chemistry Reviews*, *434*, 213809. DOI: 10.1016/j.ccr.2021.213809

Zhang, Z., Tan, X., Sun, X., Wei, J., Li, Q. X., & Wu, Z. (2022). Isoorientin Affects Markers of Alzheimer's Disease via Effects on the Oral and Gut Microbiota in APP/PS1 Mice. *The Journal of Nutrition*, *152*(1), 140–152. DOI: 10.1093/jn/nxab328

Zhou, P., Zhao, X. N., Ma, Y. Y., Tang, T. J., Wang, S. S., Wang, L., & Huang, J. L. (2022). Virtual screening analysis of natural flavonoids as trimethylamine (TMA)-lyase inhibitors for coronary heart disease. *Journal of Food Biochemistry*, *46*(12), e14376. DOI: 10.1111/jfbc.14376

Ziberna, L., Fornasaro, S., Čvorović, J., Tramer, F., & Passamonti, S. (2014). Bioavailability of Flavonoids: The Role of Cell Membrane Transporters. In Watson, R. R., Preedy, V. R., & Zibadi, S. (Eds.), *Polyphenols in Human Health and Disease* (pp. 489–511). Academic Press.

Zuo, A. R., Dong, H. H., Yu, Y. Y., Shu, Q. L., Zheng, L. X., Yu, X. Y., & Cao, S. W. (2018). The antityrosinase and antioxidant activities of flavonoids dominated by the number and location of phenolic hydroxyl groups. *Chinese Medicine*, *13*, 51. DOI: 10.1186/s13020-018-0206-9

Chapter 15
Neuroprotective Potential of Bioactive Compounds

Priyanka Gupta
https://orcid.org/0009-0002-7668-9169
CMP College, University of Allahabad, Prayagraj, India

Poornima Yadav
https://orcid.org/0009-0007-6942-8346
CMP College, University of Allahabad, Prayagraj, India

Sudhi Shrivastava
https://orcid.org/0009-0008-5474-0852
CMP College, University of Allahabad, Prayagraj, India

Ratnesh Kumar Soni
https://orcid.org/0000-0002-5774-8664
Government K.N. Government P.G. College, India

ABSTRACT

Neurodegenerative disorders (NDs), notably Parkinson's disease (PD), Alzheimer's disease (AD), Huntington's disease (HD), Epilepsy, etc., are marked by progressive neuronal dysfunction, cognitive decline, and sensory-motor impairments. All these NDs result in clinical and biochemical changes, caused by the accumulation of misfolded proteins, such as Amyloid-β and tau in Alzheimer's disease and α-synuclein in PD. There are currently no cures for most of the NDs, and treatments are primarily symptomatic or aimed at slowing progression, highlighting the need for more research and innovation. Polyphenols and bioactive Compounds, including flavonoids like Naringenin in citrus fruit, alkaloids like Withania somnifera (Ashwagandha), terpenoids like Celastrus paniculatus (Malkangani), Azadirachta indica (Neem), Ginkgo biloba (Bal kowari), Seabuck thorn (Hippophae rhamnoides, etc., have gained significant attention for their neuroprotective potential. In addition, Polyphenols and bioactive compounds exhibit antioxidant activity, and anti-inflammatory effects.

DOI: 10.4018/979-8-3373-4139-2.ch015

1) INTRODUCTION

Neurodegenerative diseases are a group of disorders characterised by the progressive degeneration or death of neurons in the brain and spinal cord (Hippocampus in Alzheimer's, Substantia Nigra in Parkinson's) (Meo et al., 2020). Neurodegeneration involves the abnormal accumulation of misfolded proteins, such as amyloid-beta and α-synuclein, as well as toxic clumps, which disrupt cellular functions. Motor nerve fibres do not regenerate effectively, making damage permanent and progressive. NDs result from a combination of genetic, environmental, and age-related factors that interact in ways still not well understood. Drug-induced neurodegeneration is common. These conditions pose a unique challenge by blurring the line between mental and physical health, affecting memory, mood, slurred speech, behaviour, and movement simultaneously. Often, these diseases begin years or decades before symptoms appear, making early diagnosis difficult (Dar & Ahmad, 2020).

Biomarkers and imaging are being developed to detect them in this silent stage. However, natural solutions such as polyphenols and bioactive compounds, including *Resveratrol (*found in red grapes and blueberries*), Naringenin* (from lemon and oranges), *Bacopa monnieri* (Brahmi), *Withania somnifera* (Ashwagandha*), Azadirachta indica* (Neem*), Ginkgo biloba* (maidenhair tree), and *Seabuckthorn* (orange-yellow berries), offer promising aid in curing these neurodegenerative diseases due to their antioxidant, anti-inflammatory, and other neuroprotective properties. These compounds can significantly help slow disease progression and enhance cognitive function. Continued research, especially through clinical trials and the development of advanced delivery systems, is vital to harness their potential in preventing and treating neurodegenerative disorders fully. Thus, nature provides a valuable reservoir of compounds that may shape the future of neurodegenerative disease therapy. Neuroinflammation plays a crucial role in the development of these diseases, as demonstrated by the activation of microglia, the primary immune cells in the CNS (Adamu et al., 2024). Neurodegenerative disorders such as AD (Alzheimer's disease), PD (Parkinson's disease), and HD (Huntington's disease) present significant challenges to the healthcare system worldwide due to their progressive nature and the lack of effective treatments. (Giri et al., 2024) These conditions are often characterised by prolonged microglial activation and sustained inflammation (Thomas et al., 2024).

Figure 1. Etiology, impact, and therapeutic approaches in neurodegeneration

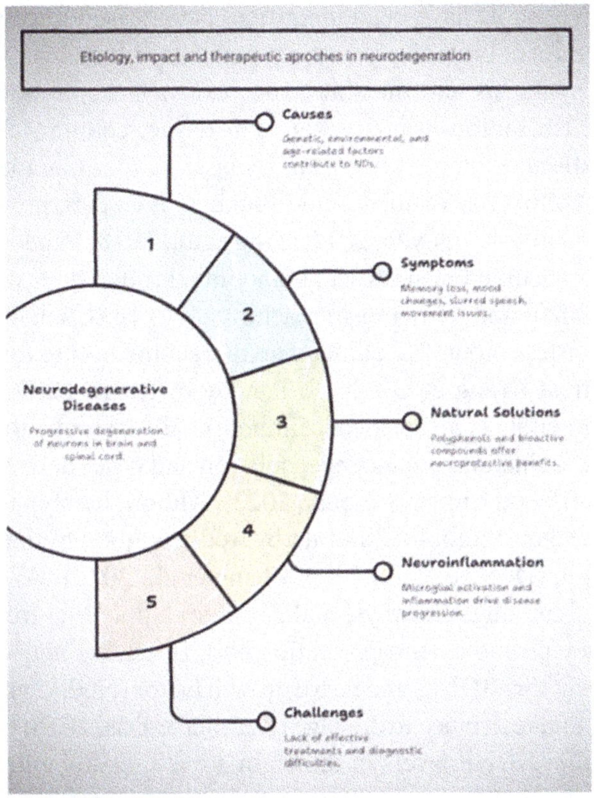

Neurodegeneration is triggered by the persistent cycle of inflammation and neuronal loss. Explaining the microglia's dual role is key to developing therapies that reduce inflammation and promote neuroprotection in neurodegeneration. (Han et al., 2024) Polyphenols, a broad class of naturally occurring bioactive compounds, have drawn significant attention due to their potential benefits in protecting neuronal health (Shi et al., 2024). Polyphenols are abundant in plants, which act as secondary metabolites for defence mechanisms and pigmentation. They are classified as flavonoids, phenolic acids, lignans, and stilbenes, characterised by specific structural properties (Sejbuk et al., 2024). Fruits, vegetables, tea, and wine are abundant sources of flavonoids, specifically flavones, flavanols, and isoflavones (Billowria et al., 2024). Phenolic acids like caffeic acid and ferulic acid are rich in coffee and grain products (Kumar & Konar, 2023). Polyphenols are studied for preventing age-related disorders, including neurodegenerative diseases, and facilitating efforts for healthy ageing through natural methods (Song et al., 2024). In our study, we investigate the current understanding of polyphenols and their effects on neuroinflammation and neurodegenerative disease.

Explaining the whole group of NDs mechanism and its relationship to the target's therapeutic Potential. Polyphenols are being studied for their promising benefits in slowing neurodegenerative disease progression and improving cognitive health because of their antioxidant, anti-inflammatory, and neuroprotective properties. It is recognised for treating neurodegenerative diseases. We are focusing on their mechanisms and therapeutic potential. Due to their antioxidant properties, polyphenols are capable of removing deleterious free radicals and boosting the activity of natural antioxidant enzymes, including superoxide dismutase, catalase, and glutathione peroxidase (Jomova et al., 2024; Dzah et al., 2024).

In plants, over 8000 polyphenols have been identified, categorised into subclasses based on factors like sources, chemical structure, or biological function, with varying inclusion criteria. Plants synthesise polyphenols through photosynthesis and the shikimate pathway, a common metabolic route (Orozco et al., 2019). Polyphenols distribution in plants varies at tissue, cellular, and subcellular levels, with insoluble phenols in cell walls and soluble phenols in vacuoles. Outer plant layers generally have higher polyphenol concentrations; factors like maturity, environment, plant part, processing, and storage impact polyphenol levels (Gomez-Pinilla & Nguyen, 2012; Yang et al., 2018; Pandey & Rizvi, 2009; Dueñas et al., 2015). Polyphenols are consumed in significant amounts through diet, with servings often containing over 1mg (Silva et al., 2020) and daily intake reaching up to 1g (Chahar et al., 2011; Leclerc et al., 2021). They impact food's taste, colour, flavour, and texture, contributing to bitterness and astringency. protect against oxidative stress (Yang et al., 2018; Pandey & Rizvi, 2009), UV radiation, and pests. The bioavailability of polyphenols is a significant limitation. To improve this, various pharmaceutical approaches are being explored, including nanoencapsulation and other delivery systems, which may enhance their neuroprotective effects (Kumar & Konar, 2023). Although polyphenols exert neuroprotective effects indirectly by altering gut microbiota and its byproducts, thereby influencing neurotransmitter production and brain function (Di Meo et al., 2020; Zhang et al., 2022). Polyphenols support neuronal survival by regulating key signalling pathways, making them a potential treatment for central nervous system (CNS) diseases. They promote neuroprotection by influencing neuroprotective factors, such as brain–derived neurotrophic factor (BDNF) and nerve growth factor (NGF), which are crucial for neuronal development, regeneration, and resistance to damage (Bathina & Das, 2015; Binder & Scharfman, 2004; Różańska et al., 2020). Higher BDNF levels in the brain are associated with better cognitive function, while decreased BDNF production may contribute to the development of brain diseases and disorders (Bathina & Das, 2015; Różańska et al., 2020; Moosavi et al., 2015).

Polyphenols may help treat neuroinflammatory diseases like multiple sclerosis by boosting BDNF production, which offers neuroprotection through immunomodulation. This suggests that polyphenols could be a valuable therapeutic approach for preventing and managing inflammatory neurological conditions (Baek et al., 2019; Binder & Scharfman, 2004). Studies show that gut microbes play a role in producing key brain chemicals like BDNF, dopamine, and serotonin. A healthy gut microbiome is linked to increased neuroprotection. One of the neuroprotective roles of polyphenols is to mitigate oxidative damage in the brain. Polyphenols exhibit anti-neuroinflammatory effects by blocking pro-inflammatory cytokines and regulating key pathways, such as NF-κB and AP-1.

Polyphenols promote neuronal health by boosting neurotrophic factors like BDNF, NGF, NT3, and NT4, and interacting with TRK receptors to activate key signalling pathways. This enhances neuronal survival, growth, and antiapoptotic effects by regulating proteins like Bcl-2 and BAX. (Moosavi et al., 2015). Since polyphenols can cross the blood-brain barrier, they can target brain lesions and support neuropathy. It also inhibits protein aggregation through multiple mechanisms, including antioxidant effects, promoting non-amyloidogenic protein degradation, binding to proteins to prevent hyperphosphorylation, and reducing amyloidogenic effects, neuronal loss, and apoptosis, thereby protecting the brain. Polyphenols have been shown to boost neurotransmitters like dopamine, serotonin, and norepinephrine in brain regions linked to mood regulation, potentially alleviating depressive and anxious symptoms. (Uddin et al., 2021).

Research has been found to improve behavioural deficits, prevent dopamine neuron loss, and regulate Cdk5 activity in MPTP-induced mice, suggesting potential therapeutic benefits for Parkinson's disease. Further investigation revealed that Korean red ginseng restored normal protein profiles in the striatum of MPTP-treated models, supporting its therapeutic potential (Kim et al., 2018). These alterations were linked to energy metabolism, oxidative phosphorylation, and neurodegenerative pathways. Moreover, the same research group discovered that Korean red ginseng also modulated protein expression in the substantia nigra, specifically those involved in neuronal development, energy metabolism, and neuroprotection, thereby promoting neuronal survival (Kim et al., 2018).

Panax ginseng extract has shown potential in improving cognitive function in Alzheimer's disease models by enhancing learning and memory, reducing oxidative stress, and modulating RAGE and NF-κB signalling pathways (Tan et al., 2015). Furthermore, fermented Panax ginseng extract has demonstrated benefits in alleviating memory impairment and reducing Aβ42 protein accumulation in both scopolamine-induced and transgenic mouse models (Kim et al., 2018). Withania somnifera root extracts have been found to positively impact motor function and muscle strength, with notable improvements observed in locomotor activity, rotarod performance, and limb withdrawal tests. The neuroprotective effects of *Withania somnifera* may be attributed to its antioxidant properties, including free radical scavenging, mitochondrial ATP preservation, and subsequent improvement in motor function. *Uncaria rhynchophylla* has demonstrated neuroprotective effects in Parkinson's disease models. According to Shim et al. (2009), it protected neuronal cells from 6-OHDA-induced toxicity by reducing cell death, ROS generation, and caspase-3 activity, while restoring GSH levels. Additionally, it prevented dopaminergic neuron loss in the substantia nigra. *Uncaria rhynchophylla* has shown neuroprotective effects in Parkinson's disease models. Specifically, it reduced neuronal cell death, ROS generation, and caspase-3 activity, while restoring GSH levels in 6-OHDA-treated PC12 cells. Additionally, it protected dopaminergic neurons in the substantia nigra of rats induced with 6-OHDA (Shim et al., 2009). According to research, Uncaria rhynchophylla protects dopaminergic neurons by promoting cell survival, reducing apoptosis, and inducing autophagy through regulation of MAPKs and Akt signalling pathways (Lan et al., 2018).

Research suggests that *Uncaria rhynchophylla* (a Chinese herb) is similar to Indian Ayurvedic plants such as Brahmi (*Bacopa monnieri*), Shankhpushpi (*Convolvulus),* and Jatamansi (*Nardostachys Jatamansi).* Their neuroprotective properties stem from the combined effects of their bioactive compounds, which target multiple pathways to mitigate neuronal damage. These compounds exhibit multifaceted benefits, including antioxidant effects (reducing ROS and boosting antioxidant defences), anti-inflammatory effects (suppressing inflammatory mediators), anti-apoptotic effects (regulating apoptosis related events and inhibiting caspase activation), and cognitive-enhancing effects (Chen et al., 2020). The study found that seaweed extracts exhibited protective effects by enhancing cell viability, mitigating oxidative stress, preserving mitochondrial membrane potential, and decreasing caspase-3 activity (Mohd Sairazi & Sirajudeen, 2020).

2) NEUROPROTECTIVE POTENTIAL OF PLANT-DERIVED ANTI-NEURODEGENERATIVE PHYTOCHEMICALS

Plants contain phytochemicals and bioactive compounds, which are secondary metabolites produced by plants. Phytochemicals are naturally occurring compounds that often contribute to colour, flavour, and resistance (Mohd Sairazi & Sirajudeen, 2020). These plant-derived natural compounds are not directly involved in growth or reproduction but play vital roles in defence, signalling, and environmental interaction. Phytochemicals enhance the efficacy of herbal remedies and are environmentally friendly (Mohamed et al., 2021). Many plant-derived secondary metabolites, also known as nutraceuticals, have demonstrated neuroprotective potential for preventing and treating neurodegenerative disorders such as Parkinson's disease (PD), Alzheimer's disease (AD), Huntington's disease (HD), and amyotrophic lateral sclerosis (ALS). Secondary metabolites, also called bioactive defenders, are found in foods like curcumin (turmeric), resveratrol (grapes and berries), quercetin (onions, apples), and Phytophenols in pomegranate, sea buckthorn, and other plants, highlighting the potential of plant-based compounds for treating neurodegenerative diseases (Di Paolo et al., 2019). Classes of secondary metabolites include alkaloids, flavonoids, terpenoids, phenolic acids, saponins, polyphenols, and lignans, which originate from sources such as fruits, vegetables, tea, wine, turmeric (spices), berries, coffee, whole grains, flaxseed, sesame seeds, and more, all contributing to neurodegenerative disease treatment. Curcumin, derived from the rhizome of Curcuma longa (turmeric), is a major bioactive compound in the polyphenol class with notable neuroprotective potential. Pomegranate (Punica granatum), rich in polyphenolic secondary metabolites like ellagic acids and urolithins, improves mitochondrial function and reduces dopaminergic neuron loss in Parkinson's disease. Sea buckthorn (Hippophae rhamnoides) contains flavonoids, vitamins, fatty acids, and phytosterols, enhancing cognitive function and making it a promising candidate for preventing and treating neurodegenerative diseases such as Alzheimer's and Parkinson's. Ginkgo biloba, a well-known herbal plant, exhibits neuroprotective potential, although its exact mechanisms in cognitive impairment are not fully understood. Resveratrol, found in grapes and berries, is a polyphenolic compound with anti-inflammatory properties. Quercetin, found in apples and onions, is a naturally occurring flavonoid especially useful for early intervention in neurodegenerative diseases (Cui et al., 2022).

In the time of Ancient Ayurveda and Traditional Chinese Medicine (TCM), it has been proven that all are related to the modern Neuroprotective potential of plant-derived natural compounds. Plants like *Withania somnifera* (Ashwagandha), *Bacopa monnieri* (Brahmi), *Curcuma longa* (Turmeric), and *Ginkgo biloba* have been used for centuries to enhance memory and improve cognitive function. Beginning scientific interest in medicinal plants grew during the colonial and pharmacognosy eras. Reserpine (*Rauwolfia serpentina*) was among the first plant-based CNS (Central nervous system) drugs. (Halder et al., 2021)

Modern neuropharmacology has begun to validate these uses via mechanistic studies demonstrating antioxidant, anti-amyloid, and neurogenesis-promoting activities. (Rahman et al., 2021).

Figure 2. Comprehensive overview of neurodegenerative diseases and neuroprotective properties of phytochemicals

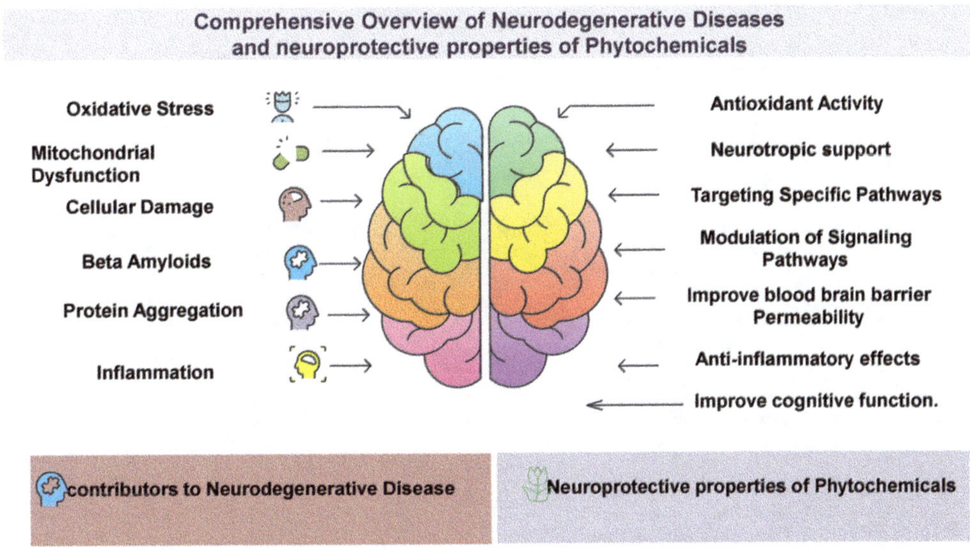

2.1. Advantages of plant-derived neuroprotective agents

- Multi-targeted action (antioxidant, anti-inflammatory, anti-apoptotic).
- Safer pharmacological profiles and lower toxicity.
- Potential for synergistic effects with conventional therapies.
- Easily accessible and culturally accepted in traditional medicine.
- Modulation of cellular signalling and epigenetics.
- Preventive and therapeutic potential.
- Enhancement of neuroplasticity and neurogenesis.

Table 1. Plant-derived compounds structure and their inspired neuroprotective strategies for cognitive preservation

	Common name	Scientific name of natural products	Nutrients	Chemical composition	Disorders	Neuroprotective effects	References
1)	Grapes and berries	Resveratol	Phytochemical (vitamins and minerals) rich.	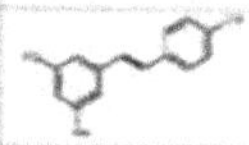	PD. AD, HD etc.	Prevent NDs.	Jiang et al., 2024.
2)	Apple and onion	Quercetin	Havcnoids (Omega fatty acids)	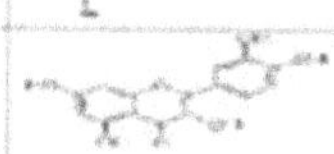	AD, HD, PD etc.	Enhance bioavailability.	Chiang et al., 2023.
3)	Pomegranate	Punica granatum	Bioactive compound (vit.,min., and fibres)	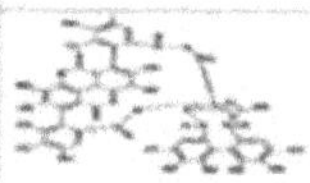	PD. AD	Enhance oxidative metabolism.	Ullah et al., 2023.
4)	Seabuck thorns	Hippophae rhamnoides	Flavonoids (vitamins, fatty acids).	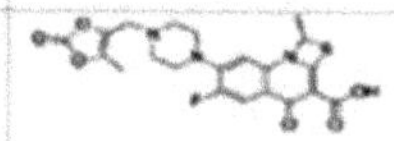	AD, HD, PD etc.	Boost the immune system.	Ren et al., 2025.
5)	Coffee	Coffee arabica	Ribunavin (vitamins rich).	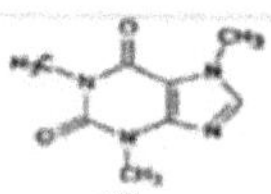		Reduced anti-amoebic response.	Wasim et al., 2020.
6)	Flaxseed	Linum ugitatis&im um	Phytochemical (Omega fatty acids rich).	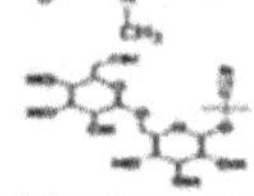	AD, PD, HD etc.	Reverse nerve cell damage.	Sanchez et al., 2023.
7)	Sesame seeds	Sesamu m indicum	B- vitamins rich	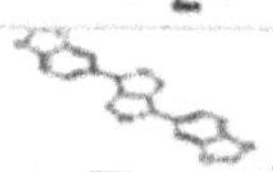	PO, AD.	Prevent NDs.	Mohamed et al.,
8)	Maidenhair	Gingko biloba	Bioactive (Terpenoids containing).	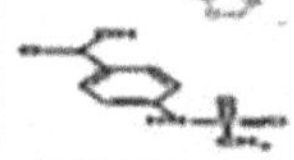	AD, PD, HD.	Neuroprotective potential.	Singh et al.,2019.
9)	Ashwagandha	Withania somnifer a	Bioactive compound (Protein, fibres rich)	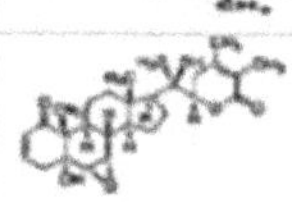	AD, PD	Reverse nerve cell damage	Mikulsa et al., 2023.
10)	Brahmi	Bacopa monnieri	Herbal (protein, minerals-rich).	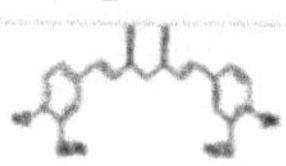	PD. AD.	Boosting memory capacity improves cognitive function.	Manap et al., 2019.
11)	Turmeric	Curcuma	Vitamins and minerals are rich.	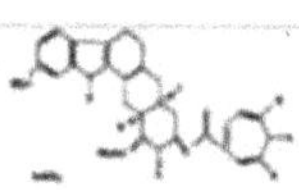	PD, HD, AD etc.	Improve information, prevent anxiety,	Bassoli et al., 2023.
12)	Reserpine	Rauwolfia serpentine	Alkaloids	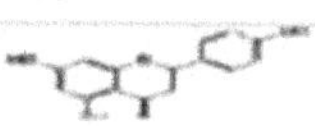	PD. AD.	Improve brain function.	Li et al., 2022.
13)	Lemons, oranges.	Naringerin	Havcnoids (vitamins rich)	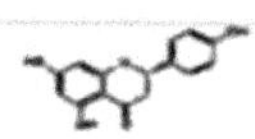	AD	Improve anti-inflammatory response.	Oyeniran et al., 2024.

3) ROLE OF NEUROMODULATING PHYTOPHENOL AND BIOACTIVE COMPOUNDS IN COGNITIVE EFFICACIES

Polyphenols and bioactive compounds can protect the brain and enhance mental performance. Phytochemicals like flavonoids and phenolic acids neutralise free radicals and reduce oxidative damage to Neurons. Neuroinflammation is a major contributor to cognitive decline in diseases like PD, AD, and

HD. Some bioactive compounds modulate key neurotransmitters like dopamine, serotonin, GABA, acetylcholine, improving mood, Memory, improving endothelial function, focus, increasing cerebral blood pressure, crucial for attention and cognition (Goyal et al., 2022) Neuromodulation is the process by which neuromodulators influence neuron function, not by directly exciting or inhibiting them like neurotransmitters, but by modifying how neurons respond to other signals. Neuromodulators include dopamine, serotonin, acetylcholine, Norepinephrine, and histamine. They act on metabotropic receptors and these neuromodulating actions are especially relevant to cognitive efficacies such as mood, attention, reward, and learning, and executive function, making Phytophenols promising candidates in the prevention and treatment of Neurodegenerative and cognitive disease (Meo et al., 2020) Phytophenols and Bioactive compounds can modulate brain function through several pathways, like neurotransmitter regulation, Antioxidant and Anti-inflammatory effects, Neurotrophic factors modulation (BDNF), Modulation of cell signalling pathways, and Epigenetics modulation. Their Potential in preventing or managing cognitive disorders such as Parkinson's, Alzheimer's, and age-related cognitive decline. They can delay disease progression by targeting multiple pathological hallmarks -beta-amyloid aggregation, tau hyperphosphorylation, synaptic dysfunction, and Neuroinflammation (Kumar & Konar, 2023).

Polyphenols protect the brain by neutralising reactive oxygen species (ROS), which can damage brain cells. It also reduces neuroinflammation by supporting cytokine release (e.g., IL-1β, TNF-α), inhibiting NF-kB expression, and modulating AP-1 activity. Polyphenols promote neuronal health by boosting neurotrophic factors like BDNF, NGF, NT3, and NT4, and interacting with TRK receptors to activate key signalling pathways. This enhances neuronal survival, growth, and antiapoptotic effects by regulating proteins like Bcl-2 and BAX. Since polyphenols can cross the blood-brain barrier, they can target brain lesions and support neuropathy treatment. Polyphenols inhibit protein aggregation through multiple mechanisms, including antioxidant effects, promoting non-amyloidogenic protein degradation, binding to proteins to prevent fibrillation, and reducing tau aggregation and hyperphosphorylation, ultimately protecting the CNS from Aβ plaque accumulation (Mursal et al.,2024).

Phenolic compounds can bind to amyloid-beta and alpha-synuclein, disrupting their structure and preventing toxicity. This interaction reduces amyloid plaque formation, protects dopaminergic neurons, and inhibits apoptosis. Research suggests that gut microbes influence brain chemistry by regulating neurotransmitters like BDNF, dopamine, serotonin, and GABA. A healthy gut microbiome may boost these molecules, promoting neuroprotection. Polyphenols have been shown to boost neurotransmitters like dopamine, serotonin, and norepinephrine in brain regions linked to mood regulation, potentially alleviating depressive and anxious symptoms. Polyphenols, characterised by hydroxyl groups on aromatic rings, are diverse plant compounds. Understanding their health benefits poses a significant challenge in modern medicine (Moosavi et al., 2015).

Flavonoids, a major class of phenolic compounds, are divided into subclasses such as flavones, anthocyanins, proanthocyanidins, flavanols, flavanones, and isoflavones, each with distinct examples like quercetin, catechin, and genistein. Natural phenolic compounds like quercetin and fisetin may help treat Parkinson's disease by fighting oxidative stress. A study explored their neuroprotective effects in an experimental model. In rats with 6-OHDA-induced dopamine depletion, curcumin and naringenin significantly preserved tyrosine hydroxylase-positive cells and dopamine levels in the brain. The high antioxidant capacity of naringenin and curcumin may enable them to cross the blood-brain barrier, providing neuroprotective effects. Research suggests plant phenolic extracts, like *Ginkgo biloba*, may help treat or prevent Amyotrophic Lateral Sclerosis (ALS), with some extracts showing sex-specific neuroprotective effects in animal models (Mutha et al., 2022).

A study found that ginseng extract improved outcomes in ALS mice by mitigating motor deficits and prolonging survival. Antioxidant effects, nerve growth factor modulation, and NO level changes may underlie these benefits. Quercetin has been found to provide neuroprotection in rat models by preserving tyrosine hydroxylase and supporting dopamine synthesis, potentially benefiting conditions like Parkinson's disease (Grabska-Kobyłecka et al., 2023).

4) MOLECULAR MECHANISM OF PHYTOPHENOL AND BIOACTIVE COMPOUNDS

Neurodegenerative diseases (NDs) such as Parkinson's disease (PD), Alzheimer's disease (AD), Huntington's disease (HD) are chronic, progressive conditions characterised by the loss of Neuronal structure and function. These diseases are driven by common pathological mechanisms, including oxidative stress, inflammation, mitochondrial dysregulation, protein misfolding, and neurotransmitter imbalance. In recent years, considerable attention has been focused on natural Polyphenols and plant-derived bioactive compounds for their ability to counteract neurodegeneration through multiple cellular and molecular mechanisms. (Stromsnes et al., 2021). Phytophenol and bioactive compounds from medicinal plants act through modulating neurotransmitter synthesis, signalling, receptor activity, and degradation, thereby unlocking multi-targeted neuroprotection. A central mechanism by which polyphenols confer neuroprotection is through their antioxidant activity. Serotonin receptors (5-HT1A, 5-HT2A) modulated by kaempferol, quercetin, and resveratrol, affecting mood and cognition. Chronic Neuroinflammation mediated by overactivated microglia and astrocytes contributes to neuronal injury in various neurodegenerative diseases. Catechins regulate dopamine transporters (DAT), preserving dopaminergic tone in parkinson's models. Several Bioactive compounds exert Neuroprotective effects by regulating neurotransmission. (Araujo et al., 2021).

Resveratrol: It is a naturally occurring polyphenolic compound belonging to the stilbenoid group. It consists of two phenol rings connected by a double bond with three hydroxyl groups that confer antioxidant activity and exists in two isomeric forms: trans-resveratrol and cis-resveratrol. Resveratrol is known to activate SIRT1 and PGC1a, an NAD+ dependent deacetylase that regulates ageing and stress resistance. Contributing to neuroprotection and cognitive preservation. Resveratrol has demonstrated the ability to bind directly to misfolded proteins, prevent plaque deposition, and enhance autophagy-mediated degradation. Resveratrol interferes with aggregation pathways, preventing fibril formation and promoting the formation of non-toxic oligomers (Yuan et al., 2013). Resveratrol monoamine oxidase-B(MAO-B) reduces dopamine breakdown and enhances dopamine transporter (DAT) function. Improve memory and learning in animal AD and PD Models. Resveratrol shows enhanced effects when combined with curcumin, quercetin, and omega-3 fatty acids. Advanced delivery systems and combination therapies may unlock its full therapeutic potential in combating neurodegenerative disorders. (Komorowska et al., 2020).

Naringenin: Naringenin belongs to the flavanone family of polyphenols, and it is a natural flavonoid. It has three hydroxyl OH groups at positions 5, 7, and 4 with a heterocyclic pyran ring. *Naringenin* influences dopaminergic pathways by protecting dopaminergic Neurons and regulating dopamine metabolism. Reduce lipid peroxidation and DNA fragmentation in brain tissue. CREB signalling improves cognitive function and memory through long-term plasticity. It maintains mitochondrial integrity by Peroxisome proliferator-activated receptor -gamma coactivator1-alpha (PGC1a). modulates Bcl-2/Bax ration and supresses caspase-3 activation. (Nouri et al., 2019). They also regulate the AMPK-SIRT1 axis, which is essential for cellular energy balance and mitochondrial health. *Naringenin* reduces amyloid-beta aggregation. Promotes neuronal differentiation, growth, and plasticity. Its ability to protect neurons, support synaptic function, and inhibit pathological protein aggregation positions it as a valuable natural candidate for preventing and managing neurodegenerative diseases (Goyal et al., 2022).

***Bacopa monnieri*:** It is rich in sources of Phytophenol, bioactive compounds, and antioxidants, and it is also called brahmi, renowned for its cognitive and neuroprotective benefits. Bacopa monnieri inhibits Acetylcholinesterase (AChE), thereby increasing acetylcholine availability and enhancing synaptic function. These effects are especially beneficial in the early stages of neurodegenerative conditions. Lipid peroxidation prevention preserves neuronal membrane integrity. Reduces chronic inflammation in brain ageing and neurodegeneration. Enhances mitochondrial respiration and ATP production. Bacosides influence gene expression related to neuronal plasticity, apoptosis, and oxidative stress resistance, etc. Neuroendocrine regulation stabilises cortisol levels and HPA axis response (Manap et al., 2019).

***Withania somnifera*:** It has adaptogenic activities that enhance dendritic branching, synaptogenesis, memory consolidation, reduce oxidative stress, and protect neuronal structure. It is upregulated by neurotrophic factors such as Brain-derived neurotrophic factor (BDNF) and Nerve growth factor (NGF). They balanced glutamatergic and GABAergic activity, which helps prevent excitotoxicity. Withanolides reduce glial activation and lower the production of cytokines, thereby dampening the inflammatory cascade. Its prebiotic actions influence neuroinflammation via gut microbiota (Dar & Ahmad, 2020).

Azadirachta indica: The key compounds of *Azadirachta indica* (Neem) are *Nimbi Salannin, kaempferol,* rutin, ferulic acid, margolone, etc. Gallic acid and ferulic acid restore mitochondrial membrane potential. Nimbin and azadirachtin downregulate pro-inflammatory cytokines (like TNF-α, IL-6, IL-1β). It inhibits NF-kB signalling and reduces COX2 and Inos expression. This has potential relevance for Alzheimer's disease, where cholinergic deficit is a hallmark. (Batra et al., 2022). Indica prevents cytochrome c release and caspase activation, which are involved in programmed neuronal death. Flavonoids regulate gene expression via histone acetylation and microRNA modulation in neurons. In animal studies, neem showed potential to reduce amyloid-beta aggregation and tau hyperphosphorylation, both central to Alzheimer's pathology. It shows improved memory, reduced oxidative markers, and preserved neuron density. Its Phytophenols and triterpenoids offer a blueprint for multi-targeted brain therapeutics, relevant for Alzheimer's, Parkinson's, stress-related cognitive dysfunction, and beyond. (Ishabiyi et al., 2023).

***Gingko biloba*:** It is the oldest living tree species in the environment, which has extraordinary neurotherapeutic properties. Its leaf extract, known as EGb 761, is a clinically validated nootropic and neuroprotectant. These compounds scavenge free radicals, protecting neurons from oxidative stress, a key factor in neuronal oxidation. Mediated by NO (Nitric oxide) pathways and reduction in platelet-activating factor (PAF). Ginkgo inhibits caspase-3, the Bax/Bcl-2 ratio, and other pro-apoptotic factors. Modulates the expression of neuroprotective genes through epigenetic regulation. Improves cerebral blood flow and oxygen delivery to neurons. It modifies gut microbiota, improving short-chain fatty acids production. (Singh et al., 2020).

***Seabuck thorns*:** It is also called holy fruits of the Himalayas, and its berries, leaves, and seeds are packed with neuroactive phytochemicals that confer powerful neuroprotective properties. Seabuck thorn extract has been found to upregulate BDNF, promote neurogenesis, synaptic plasticity, and learning and memory formation. Carotenoids shield neurons from oxidative DNA and mitochondrial damage. (Ren et al., 2025). It inhibits NF-kB signalling and Cox-2 expression, which are central to chronic brain inflammation. This is crucial for preventing neurodegeneration caused by mitochondrial dysfunction, especially in PD and AD. It prevents tau hyperphosphorylation, preserving cytoskeletal integrity. (Cho et al., 2017).

Table 2. Neuromodulating Role of Phytophenol and Bioactive Compounds

S. NO.	Phytochemical and Bioactive Compounds	Neuromodulating Role	Reference
1)	Quercetin (Apple, onion).	Crosses the blood-brain barrier, reduces ROS, modulates glutamatergic transmission, and protects against B-amyloid toxicity	Chiang et al., 2023
2)	Resveratol (grapes, berries)	Activates SIRTI Pathways, improves mitochondrial function, Neuroinflammation, and supports dopaminergic neurons.	Chiang et al., 2023
3)	Punica granatum (pomegranate)	Improving memory and learning contributes to mood stabilization and cognitive alertness, reducing Neuroinflammation.	Ullah et al., 2023.
4)	Hippophae rhamnoides (Seabuck thorns)	Enhancing mood, memory, and cognitive clarity, reducing neuronal damage and microglial activation.	Ren et al., 2025
5)	Linum usitatissimum (Flaxseed)	Suppressed pro-inflammatory cytokines improve learning and memory, protect dopaminergic neurons.	Sanchez et al., 2023.
6)	Coffee arabica (coffee)	Modulates cholinergic and dopaminergic pathways, reduces Amyloid plaque burden, and (coffee)cognitive flexibility.	Wasim et al., 2020
7)	Gingko biloba (maidenhair)	Modulates multiple neurotransmitter systems, supporting mood regulation and cognitive flexibility	Singh et al.,2019.
8)	Withania somnifera (Ashwagandha).	Improved memory and attention protect the Hippocampus from atrophy.	Mikulsa et al., 2023.
9)	Sesamum indicum	Reducing the release of pro-inflammatory cytokines such as TNF-alpha and 11-6	Mohamed et al.,
10)	Bacopa Monnieri (Brahmi).	Protecting neurons from free radical damage improves short-term and long-term memory.	Manap et al., 2019.
11)	Curcuma longa (Turmeric)	Enhances BDNF, modulates serotonin and dopamine pathways, inhibits Neuroinflammation and oxidative stress	Bassoli et al., 2023.
12)	Rauwolfia serpentina (Reserpine)	Reduced blood pressure, decreased heart rate, improved sleep	Li et al., 2022.
13)	Naringenin (lemons, Oranges)	Enhances dopaminergic transmission, influences serotonergic pathways, and enhances GABAergic activity.	Oyeniran et al., 2024.
14)	Azadirachta indica (Neem)	Enhancing memory and cognitive performance, reducing excitotoxic stress.	Batra et al., 2022

5) THERAPEUTIC INSIGHTS OF BIOACTIVE COMPOUNDS TO ALLEVIATE NEURODEGENERATION

Plants derived bioactive compounds from various sources like fruits, vegetables, roots, seeds, and edible flowers have shown potential anti-ageing properties (Gao et al.,2017). Mushrooms contain bioactive compounds, such as beta-glucans and chitin, which not only support fungal structure but also exhibit immunomodulatory and antioxidant effects, making them a focus of medical research (Luo et al.,2024). High levels of oxidative stress markers, such as malondialdehyde, and decreased antioxidant enzymes like superoxide dismutase and glutathione peroxidase can indicate the extent of oxidative damage and imbalance in the body (Liu et al., 2014).

Blueberries contain high levels of polyphenols with strong antioxidant effects. Studies indicate blueberry extracts may help protect brain cells, reduce inflammation, and delay dementia (Debnath et al., 2020). *Acacia dealbata* flower extracts exhibited antioxidant and anti-dementia properties by inhibiting acetylcholinesterase (AChE) (Paula et al., 2022). Similarly, *Sophora* species leaf extracts showed neuroprotective effects, modulating neurotransmitters and antioxidants, and potentially benefiting dementia treatment (Aly et al., 2021). *Bauhinia coccinea's* bioactive compounds, identified through HPLC-PDA, showed antioxidant and neuroprotective effects, including inhibition of amyloid-β aggregation and acetylcholinesterase activity, and protection against oxidative stress in neuronal cells (Kim et al., 2018). Research identified African plant extracts with strong AChE inhibitory activity, including *Ammocharis coranica* and *Scadoxus puniceus* (Masondo et al., 2019). A blend of Cyperus rotundus and Zingiber officinale also demonstrated neuroprotective effects, enhancing memory and reducing oxidative stress in a rat model of dementia (Sutalangka & Wattanathorn, 2017). *Cyperus rotundus* rhizome extract (CRE) exhibited potential neuroprotective benefits by mitigating oxidative damage and promoting cell survival in SH-SY5Y cells, suggesting therapeutic applications (Kumar & Konar, 2023).

Research found that *Ocimum basilicum* leaf extract improved cognitive function in scopolamine-induced mice by decreasing caspase-3, TNF-α, and AChE activity, while increasing glutathione and anti-inflammatory IL-10 levels (Singh et al., 2025). *Ampelopsin* A, derived from *Vitis vinifera*, exhibited neuroprotective properties by enhancing synaptic plasticity and cognitive function in mice, likely through its effects on BDNF/CREB signalling pathways (Hong et al., 2021). Asiaticoside, a compound primarily sourced from *Centella asiatica*, may offer therapeutic benefits for vascular dementia by addressing associated behavioural and cognitive changes.

Studies demonstrate the therapeutic potential of tilianin and kaempferol in neurodegenerative diseases, with *tilianin* improving cognitive function and reducing inflammation in vascular dementia, and kaempferol showing promise in mitigating oxidative stress, neuroinflammation, and cognitive impairment in Alzheimer's disease models (Shoaib et al., 2023). *Curcumin* and resveratrol showed potential in mitigating neurodegenerative effects by reducing oxidative stress, tau protein hyperphosphorylation, and amyloid-beta-induced cell damage (Yu et al., 2022). Caffeine, commonly found in coffee and tea, was studied for its potential to enhance the effects of donepezil, a medication for Alzheimer's disease. When combined, caffeine and donepezil showed increased inhibition of acetylcholinesterase (AChE) and antioxidant activity, suggesting a potential synergistic effect (Oboh et al., 2017). Harmine, derived from *Peganum harmala,* exhibits promising effects by crossing the blood-brain barrier and inhibiting AChE, which could have implications for treating neurodegenerative diseases. Ginkgo biloba extract (EGb7) and memantine didn't show effectiveness compared to other treatments like galantamine and donepezil for cognitive issues (Thancharoen et al., 2019).

5.1. In vitro studies:

In vitro models simulate disease-specific Neurodegenerative mechanisms using cultured neuronal cell lines or primary neurons. Some models, like SH-SY5Y human neuroblastoma cells used for PD and AD studies, PC12 rat pheochromocytoma cells responsive to nerve growth factor used for AD and oxidative stress studies, Primary cortical or hippocampal neurons mimic physiological neuronal behaviour, and Induced Pluripotent stem cells (iPSCs) generate patient-specific neurons. These models are exposed to neurotoxins (e.g., ALPHA BETA Peptides, 6-OHDA, MPTP, H202) to induce Neurodegeneration (Kumar & Konar, 2023).

There are several advantages of conducting in vitro studies in neurodegeneration research:

- Controlled environment for manipulation of variables to understand mechanisms.
- High throughput screening of numerous bioactive compounds.
- Identifies molecular pathways.

In vitro studies are essential in the early stages of drug discovery to understand the mechanism and assess the therapeutic potential of these compounds. Continued innovation in in vitro technologies and integration with omics approaches will accelerate the translation of bioactive compounds from bench to bedside in treating neurodegenerative diseases (Singh et al., 2020).

Table 3. Summarises the observed effects of various natural compounds and their neuroprotective potentials

S.NO.	COMPOUND	MODEL	EFFECT OBSERVED
1.	Curcumin	Alpha-beta treated SH-SY5Y.	Reduced ROS, improved cell viability.
2.	Resveratrol	Mpp+ treated SH-SY5Y.	Protected mitochondria, reduced oxidative stress.
3.	Berberine	6-OHDA- treated PC12.	Inhibited apoptosis, modulated autophagy.
4.	EGCG	H_2O_2-treated neurons.	Reduced oxidative stress and DNA damage.
5.	Quercetin	LPS activated microglia.	Reduced inflammation, suppressed cytokines.
6.	Withaferin A	Alpha-beta induced neuronal loss	Reduced neuronal apoptosis and Akt signaling.

6) GUT-BRAIN AXIS AND EFFICACY OF BIOACTIVE COMPOUNDS

The gut-brain axis is a complex, bidirectional communication system between the gastrointestinal tract (GI) and the central nervous system (CNS). The gut is not just a digestive tube- it is a second brain with the power to heal, protect, and enhance your first brain. This axis involves neural, hormonal, and immune signalling pathways and plays a crucial role in maintaining homeostasis, regulating mood, cognition, and overall brain health. Bioactive compounds, especially those derived from diet, are known to modulate the gut-brain axis and offer significant neuroprotective and cognitive benefits. Neem's polyphenols restore gut microbiota, which influences neuroinflammation and neurotransmitter synthesis. (Benameur et al., 2023). A network of neurons in the gut that communicates directly with the brain via the vagus nerve and influences brain function by producing neurotransmitters, metabolites, and modulating the immune response. Bioactive compounds can positively impact the gut microbiota, reduce inflammation, and promote neuronal function. (Singh et al., 2025). Enhance the growth of beneficial bacteria like *Lactobacillus* and *Bifidobacteria*. Reduce oxidative stress and neuroinflammation by inhibiting pro-inflammatory cytokines. Curcumin from turmeric modulates gut microbiota and reduces amyloid-beta aggregation in Alzheimer's. Flavonoids improve gut microbial diversity. *Ginsenosides* improve learning and memory via gut-brain interactions. Non-digestible compounds promote beneficial gut bacteria. Modulation of gut microbiota supports the growth of beneficial bacteria and suppresses pathogens. Suppress gut and brain inflammation by inhibiting NF-kB and reducing cytokine release. Strengthen the intestinal lining and blood-brain barrier, preventing toxin leakage. Bacterial products like short-chain fatty acids (SC-FAs), tryptophan metabolites, and secondary bile acids modulate mood, stress, and Neurodegeneration. Tryptophan, an essential amino acid, is a precursor of serotonin, 90% of which is produced in the gut (Loh et al., 2024). Bioactives reduce amyloid-beta accumulation by altering gut flora and reducing systemic inflammation in Alzheimer's disease. Gut microbiota alterations precede motor symptoms; polyphenols improve dopamine signalling and reduce GI inflammation in Parkinson's. Disruption of the gut-brain axis is associated with several neurological disorders (Parkinson's, Alzheimer's disease, etc). Herbs like Ashwagandha, Brahmi, and turmeric have dual roles as gut restorers and neurocalmers. Bioactive compounds serve as epigenetic diplomats between two kingdoms, microbes and neurons. Bioactive compounds act as ''neural gardeners'' shaping the growth and pruning of mental pathways from the soil of the microbiome. Emerging research suggests bioactives can rewire brain circuits via gut modulation; they influence Neuroplasticity, Neurogenesis, and Myelination. The gut-brain axis is a vital communication network that plays a significant role in brain health and disease. Bioactive compounds, especially from diet and medicinal plants, offer a natural, multi-targeted approach to modulating this axis. Through their antioxidant, anti-inflammatory, and microbiota-modulating properties, these compounds hold promise in preventing and managing neurodegenerative and neuropsychiatric disorders (Cassotta et al., 2025). Gut-brain axis optimisation, prebiotic polysaccharides in berries modulate gut microbiota, reduce systemic inflammation, and support cognitive health. Gut microbes modulate the hypothalamic-pituitary-adrenal (HPA) axis, reducing cortisol-induced damage to the hippocampus. Flavonoids metabolised by gut bacteria yield phenyl-Y-valerolactone, which crosses the BBB and exerts antioxidant and anti-inflammatory effects directly on neurons.

Figure 3. Relation of gut-brain axis with natural compounds

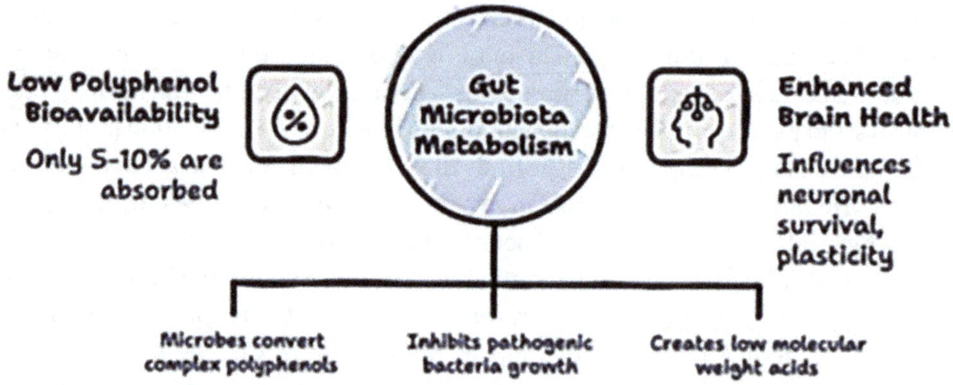

6.1. Absorption and Metabolism of antimicrobial activities of polyphenols in gut microbiota.

Polyphenols and bioactive compounds influence brain health through the gut-brain axis. their neuroprotective potential is closely linked to how they are absorbed, metabolised, and interact with gut microbes. Most polyphenols in the gut have low bioavailability in their native form; only 5-10% are absorbed in the small intestine, largely in the form of aglycones or simple conjugates. The majority, 90-95% reach the colon unaltered, where they are metabolised by gut microbiota. Gut microbiota plays a key role in breaking down complex polyphenols into simpler, bioactive metabolites (Marin et al., 2017). Antimicrobials inhibit pathogenic bacteria (e.g., *Clostridium perfringens, E.coli, Salmonella).* Modulation maintains a healthy gut microbiota balance, reducing gut inflammation and leaky gut, both of which are implicated in neurodegenerative disorders. Antimicrobial activity disrupts bacterial cell walls, inhibits the bacterial enzyme system, and interferes with microbial adhesion and biofilm formation. Microbes convert them into low molecular weight phenolic acids, which cross the blood-brain barrier. These metabolites mimic neurotransmitters, modulate inflammation, and influence neuronal survival and plasticity. The antimicrobial effect of polyphenols becomes a neurofilter, refining which bacterial products enter circulation and influence brain function. Polyphenols are seeds; the gut microbiota is soil; the fruits are brain health, so unlocking a new era of diet-driven neuroprotection (Bešlo et al., 2023).

Figure 4. Antimicrobial activity in gut microbiota

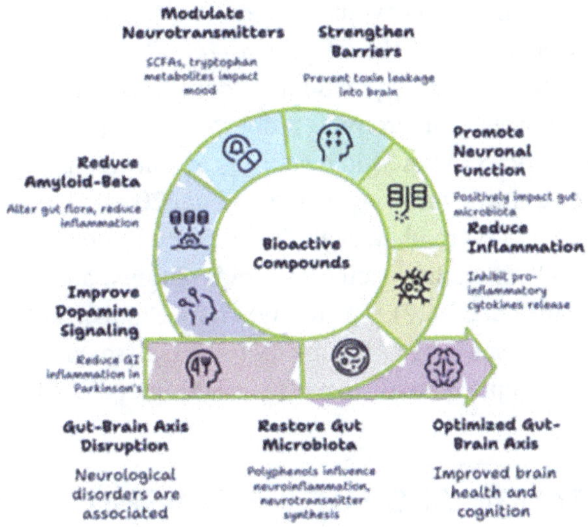

7) DELETERIOUS EFFECTS OF BIOACTIVE COMPOUNDS

The expanding use of herbal products has led to concerns about adverse effects from factors like intrinsic toxicity, adulteration, and interactions with other natural or synthetic substances. (Jordan et al., 2010). Certain plant compounds, such as alkaloids, can cause neurotoxic effects by disrupting sodium channels in neurons, which may lead to impaired brain function and damage (Spochacz et al., 2018). Attributing adverse events to individual ingredients in herbal products can be problematic due to their complex nature, highlighting potential risks associated with natural remedies. Some natural compounds, such as aristolochic acid and Glycyrrhiza spp., pose a health risk. Aristolochic acid is associated with kidney damage, and Glycyrrhiza spp. can lead to hypertension and fluid retention due to its chemical properties. The popular herb *Panax ginseng*, used to address fatigue and poor circulation, has been linked to adverse effects, including insomnia and irritability (Zhou et al., 2021).

A survey of natural remedy users reported common adverse events, including gastrointestinal issues, allergic reactions, and dizziness. Trigonella arabica has been found to have blood sugar-lowering effects when used systemically (Puri et al., 2006). While flavonoids are popular as dietary supplements, excessive intake may pose risks, such as DNA damage, free radical production, and disruption of hormone-regulation enzymes (Skibola & Smith, 2000). Combining herbal remedies with conventional medications can increase the risk of adverse interactions. (Izzo & Ernst, 2001). Herbal supplements like *Salvia miltiorrhiza and Ginkgo biloba* may enhance the effects of certain medications, such as warfarin, or increase the risk of severe bleeding when combined with ibuprofen. (Meisel et al., 2003).

Aconitum species contain significant amounts of diterpene alkaloids and flavonoids (Shyaula,2012). Aconitine interferes with cellular signalling, contributing to its neurotoxic effects, and can also affect drug metabolism by inducing p-glycoprotein expression (Wu et al.2016). *Atropa belladonna* (Deadly

Nightshade) has an alkaloid composition, particularly atropine and scopolamine, which is associated with its anticholinergic effects (Dani & Sharma, 2024). These alkaloids, including hyoscyamine, have dual effects on the central and peripheral nervous systems, offering therapeutic benefits, while also posing potential toxicity risks (Qadri et al., 2014). The *Atropa belladonna* plant's fruits, leaves, and roots contain tropane alkaloids, which can lead to poisoning if consumed. Hemlock water–dropwort (*Oenanthe crocata*) is known for its powerful neurotoxic properties. The plant's toxic compounds, including *Oenanthotoxin* and *cicutoxin*, block GABAergic responses, resulting in symptoms such as facial muscle paralysis and a characteristic strained smile (Appendino et al., 2009). The plant's chemical constituents are key factors in its neurotoxicity (Wyrembek et al., 2012). *Ricinus communis* (castor oil plant) contains a range of bioactive compounds, including ricinoleic acid, tannins, and alkaloids, which contribute to its neurotoxic effects (Chouhan et al., 2022). The unknown mechanisms of interactions between natural products and medications increase safety risks for unsupervised self-medication. Strengthening education and regulatory oversight can help prevent adverse effects (Radu et al., 2024).

8) CHALLENGE, LIMITATION, AND FUTURE PERSPECTIVES FOR RESEARCH

Earth's diverse ecosystems offer a vast reservoir of bioactive molecules with potential therapeutic applications (Rivai & Umar 2023), which could lead to significant advancements in medical interventions for neurodegenerative diseases. However, there is a lack of research and validation of the neuroprotective potential of Phytophenols and bioactive compounds. Very little data on standardisation is available; slow action, intellectual property issues, ethical concerns, and sustainability concerns are additional challenges. Phytophenol and bioactive products are regaining attention worldwide for advancements in extraction and formulation technologies, integration into preventive and holistic healthcare, and global recognition of traditional knowledge. Incorporating polyphenols into dietary recommendations or supplements could be a key strategy for slowing or preventing neuroinflammatory disorders, promoting healthier ageing, and improving quality of life. Future research should focus on identifying polyphenols, their molecular targets, enhancing their bioavailability, and elucidating their role in gut-brain signalling. Polyphenols may open new avenues for treating neurodegenerative diseases, showing promise for future medical breakthroughs.

More research is needed to elucidate how natural products interact with cellular pathways to achieve neuroprotection, which is crucial for their clinical application (Kim et al., 2018). The bioavailability of a treatment plays a crucial role in determining its clinical effectiveness, particularly for neurodegenerative diseases (Huang et al., 2024). The diverse and complex nature of natural products makes it challenging to achieve consistent results, as differences in extraction methods and plant sources can affect chemical profiles and study outcomes. In vitro studies have yielded encouraging results, but the lack of extensive in vivo and clinical research impedes the progression to practical applications. Many natural compounds, like curcumin and resveratrol, have limited bioavailability, which can undermine their therapeutic effectiveness in clinical settings despite promising preclinical results. (Esmaealzadeh et al., 2024) Differences in study design, such as dosages, models, and protocols, hinder the comparison of results and the drawing of conclusive findings. (Fernandes et al., 2021; Jiang et al., 2023).

Advances and future Perspectives of research are that 3D cell cultures and organoids are enhancing the physiological relevance of in vitro models. Co-culture systems introduced neuron glia interactions for studying neuroinflammation, and patient cells offer personalised medicine approaches that provide deeper insight into compound activity (Nahar et al, 2025).

CONCLUSION

The rising global burden of neurodegenerative diseases (NDs), such as Parkinson's disease (PD), Alzheimer's disease (AD), Huntington's disease (HD), etc., demands novel, safer, and more effective therapeutic strategies. In recent years, extensive scientific attention has been directed toward polyphenols and bioactive compounds derived from natural sources due to their promising neuroprotective potential. These naturally occurring compounds offer a multifaceted approach to combat the complex and multifactorial pathology of neurodegeneration. Many of these compounds exert modulatory effects on the neurotransmitter system for preventing the onset of neurodegeneration but also for slowing disease progression and enhancing cognitive function. One of the key strengths of polyphenols and bioactive compound is their ability to target multiple pathogenic processes simultaneously, a quality especially valuable in the treatment of multifactorial disorders. Furthermore, many of these compounds influence the gut-brain axis, suggesting a potential link between dietary habits, gut microbiota composition, and neurodegeneration. However, despite the overwhelming evidence from in vitro and animal studies, the clinical translation of polyphenols and bioactive compounds faces significant challenges. Moreover, standardised extraction, purification, and dosage protocols are necessary to ensure consistent and reproducible outcomes in clinical applications. There is also an increasing interest in exploring synergistic effects between different polyphenols and with conventional therapeutics. Additionally, the integration of omics technologies and computational approaches can further illuminate the molecular targets and optimise the pharmacokinetics of these compounds. Their antioxidant, anti-inflammatory, neurogenic, and signalling modulating properties offer a comprehensive neuroprotective shield. While the road from bench to bedside is complex and requires rigorous translational research, the evidence thus far strongly supports the integration of natural compounds into preventive and therapeutic frameworks. Continued interdisciplinary efforts in phytochemistry, neurosciences, pharmacology, and biotechnology will be crucial for unlocking the full potential of these compounds and ultimately improving brain health and cognitive longevity across populations.

REFERENCES

Adamu, A., Li, S., Gao, F., & Xue, G. (2024). The role of neuroinflammation in neurodegenerative diseases: Current understanding and future therapeutic targets. *Frontiers in Aging Neuroscience*, *16*, 1347987.

Aly, S. H., Elissawy, A. M., Fayez, A. M., Eldahshan, O. A., Elshanawany, M. A., & Singab, A. N. B. (2021). Neuroprotective effects of Sophora secundiflora, Sophora tomentosa leaves and formononetin on scopolamine-induced dementia. *Natural Product Research*, *35*(24), 5848–5852.

Baek, S. C., Park, M. H., Ryu, H. W., Lee, J. P., Kang, M. G., Park, D., Park, C. M., Oh, S. R., & Kim, H. (2019). Rhamnocitrin isolated from *Prunus padus* var. *seoulensis*: A potent and selective reversible inhibitor of human monoamine oxidase A. *Bioorganic Chemistry*, *83*, 317–325. 10.1016/j.bioorg.2018.10.051

Bathina, S., & Das, U. N. (2015). Brain-derived neurotrophic factor and its clinical implications. *Archives of Medical Science: AMS*, *11*(6), 1164–1178. 10.5114/aoms.2015.56342

Batra, N., Easwar Kumar, V., Nambiar, R., De Souza, C., Yuen, A., Le, U., Verma, R., Ghosh, P. M., & Vinall, R. L. (2022). Exploring the therapeutic potential of neem (*Azadirachta indica*) for the treatment of prostate cancer: A literature review. *Annals of Translational Medicine*, *10*(16), 894. Advance online publication. 10.21037/atm-22-94

Benameur, T., Porro, C., Twfieg, M.-E., Benameur, N., Panaro, M. A., Filannino, F. M., & Hasan, A. (2023). Emerging paradigms in inflammatory disease management: Exploring bioactive compounds and the gut microbiota. *Brain Sciences*, *13*(8), 1226. Advance online publication. 10.3390/brainsci13081226

Bešlo, D., Golubić, N., Rastija, V., Agić, D., Karnaš, M., Šubarić, D., & Lučić, B. (2023). Antioxidant activity, metabolism, and bioavailability of polyphenols in the diet of animals. *Antioxidants*, *12*(6), 1141. Advance online publication. 10.3390/antiox12061141

Bešlo, D., Golubić, N., Rastija, V., Agić, D., Karnaš, M., Šubarić, D., & Lučić, B. (2023). Antioxidant activity, metabolism, and bioavailability of polyphenols in the diet of animals. *Antioxidants*, *12*(6), 1141.

Billowria, K., Ali, R., Rangra, N. K., Kumar, R., & Chawla, P. A. (2024). Bioactive flavonoids: A comprehensive review on pharmacokinetics and analytical aspects. *Critical Reviews in Food Science and Nutrition*. Advance online publication. 10.1080/10408347.2022.2105641

Binder, D. K., & Scharfman, H. E. (2004). Brain-derived neurotrophic factor. *Growth Factors (Chur, Switzerland)*, *22*(3), 123–131. 10.1080/08977190410001723308

Cassotta, M., Armas Diaz, Y., Qi, Z., Yang, B., Grosso, G., Quiles, J. L., Battino, M., Godos, J., & Giampieri, F. (2025). Modernizing gut-brain axis research in nutritional science. *Trends in Food Science & Technology*. Advance online publication.

Chahar, M. K., Sharma, N., Dobhal, M. P., & Joshi, Y. C. (2011). Flavonoids: A versatile source of anticancer drugs. *Pharmacognosy Reviews*, *5*(9), 1–12. 10.4103/0973-7847.79093

Chen, L., Liu, Y., & Xie, J. (2020). The beneficial pharmacological effects of *Uncaria rhynchophylla* in neurodegenerative diseases: Focus on alkaloids. *Frontiers in Pharmacology*, *15*, 1436481. Advance online publication. 10.3389/fphar.2024.1436481

Cho, C. H., Jang, H., Lee, M., Kang, H., Heo, H. J., & Kim, D.-O. (2017). Sea buckthorn (*Hippophae rhamnoides* L.) leaf extracts protect neuronal PC-12 cells from oxidative stress. *Journal of Microbiology and Biotechnology*, *27*(7), 1257–1264. 10.4014/jmb.1704.04033

Cui, D., Chen, Y., Ye, B., Guo, W., & Wang, D. (2022). Natural products for the treatment of neurodegenerative diseases. *Phytomedicine*, *109*, 155101. Advance online publication. 10.1016/j.phymed.2023.155101

Dar, N. J., & Ahmad, M. (2020). Neurodegenerative diseases and *Withania somnifera* (L.): An update. *Journal of Ethnopharmacology*, *256*, 112769. Advance online publication. 10.1016/j.jep.2020.112769

de Araujo, F. F., de Paulo Farias, D., Neri-Numa, I. A., & Pastore, G. M. (2021). Polyphenols and their applications: An approach in food chemistry and innovation potential. *Food Chemistry*, *338*, 127535. Advance online publication. 10.1016/j.foodchem.2020.127535

Debnath, B., Majumdar, M., Bhowmik, M., Bhowmik, K. L., Debnath, A., & Roy, D. N. (2020). The effective adsorption of tetracycline onto zirconia nanoparticles synthesized by novel microbial green technology. *Journal of Environmental Management*, *261*, 110235.

Di Meo, F., Valentino, A., Petillo, O., Peluso, G., Filosa, S., & Crispi, S. (2020). Bioactive polyphenols and neuromodulation: Molecular mechanisms in neurodegeneration. *International Journal of Molecular Sciences*, *21*(7), 2564. Advance online publication. 10.3390/ijms21072564

Di Paolo, M., Papi, L., Gori, F., & Turillazzi, E. (2019). Natural products in neurodegenerative diseases: A great promise but an ethical challenge. *International Journal of Molecular Sciences*, *20*(20), 5170.

Dueñas, M., Martínez-Villaluenga, C., Limón, R. I., Peñas, E., & Frias, J. (2015). Effect of germination and elicitation on phenolic composition and bioactivity of kidney beans. *Food Research International*, *70*, 55–63.

Dzah, C. S., Zhang, H., Gobe, V., Asante-Donyinah, D., & Duan, Y. (2024). Anti-and pro-oxidant properties of polyphenols and their role in modulating glutathione synthesis, activity and cellular redox potential: Potential synergies for disease management. *Advances in Redox Research: an Official Journal of the Society for Redox Biology and Medicine and the Society for Free Radical Research-Europe*, *11*, 100099.

Gao, Y., Wei, Y., Wang, Y., Gao, F., & Chen, Z. (2017). *Lycium barbarum*: A traditional Chinese herb and a promising anti-aging agent. *Aging and Disease*, *8*(6), 778–791. 10.14336/AD.2017.0725

Giri, S., Choudhury, A., Praharaj, D. L., Singh, A., Vaidya, A., Harindranath, S., & Shukla, A. (2024). Changing etiological spectrum of hepatocellular carcinoma in India—A systematic review and meta-analysis. *Journal of Clinical and Experimental Hepatology*, *14*(5), 101391.

Gómez-Pinilla, F., & Nguyen, T. T. (2012). Natural mood foods: The actions of polyphenols against psychiatric and cognitive disorders. *Nutritional Neuroscience*, *15*(3), 127–133. 10.1179/1476830511Y.0000000035

Gomez-Pinilla, F., & Nguyen, T. T. (2012). Natural mood foods: The actions of polyphenols against psychiatric and cognitive disorders. *Nutritional Neuroscience*, *15*(3), 127–133.

Goyal, A., Verma, A., Dubey, N., Raghav, J., & Agrawal, A. (2022). Naringenin: A prospective therapeutic agent for Alzheimer's and Parkinson's disease. *Journal of Food Biochemistry*, *46*(1), e14415. Advance online publication. 10.1111/jfbc.14415

Grabska-Kobyłecka, I., Szpakowski, P., Król, A., Książek-Winiarek, D., Kobyłecki, A., Głąbiński, A., & Nowak, D. (2023). Polyphenols and their impact on the prevention of neurodegenerative diseases and development. *Nutrients*, *15*(15), 3454. Advance online publication. 10.3390/nu15153454

Halder, S., Anand, U., Nandy, S., Oleksak, P., Qusti, S., Alshammari, E. M., Batiha, G. E.-S., Koshy, E. P., & Dey, A. (2021). Herbal drugs and natural bioactive products as potential therapeutics: A review on pro-cognitives and brain boosters perspective. *Saudi Pharmaceutical Journal: SPJ: the Official Publication of the Saudi Pharmaceutical Society*, *29*(8), 879–907. 10.1016/j.jsps.2021.07.009

Han, K., Wang, J.-Y., Wang, P.-Y., & Peng, Y.-C.-H. (2024). Therapeutic potential of cannabidiol (CBD) in anxiety disorders: A systematic review and meta-analysis. *Psychiatry Research*, *338*, 116049. Advance online publication. 10.1016/j.psychres.2024.116049

Hong, W., Zhi, F. X., Kun, T. H., Hua, F. J., Ling, L. H., Fang, F., & Yang, L. C. (2021). 6-Gingerol attenuates ventilator-induced lung injury via anti-inflammation and antioxidative stress by modulating the PPARγ/NF-κB signalling pathway in rats. *International Immunopharmacology*, *92*, 107367.

Ishabiyi, F. O., Ogidi, J. O., Olukade, B. A., Amorha, C. C., El-Sharkawy, L. Y., Okolo, C. C., Adeniyi, T. M., Atasie, N. H., Ibrahim, A., & Balogun, T. A.. (2023). Computational evaluation of *Azadirachta indica*-derived bioactive compounds as potential inhibitors of NLRP3 in the treatment of Alzheimer's disease. *Journal of Alzheimer's Disease*, *92*(3), 979–998. 10.3233/JAD-221020

Jomova, K., Alomar, S. Y., Alwasel, S. H., Nepovimova, E., Kuca, K., & Valko, M. (2024). Several lines of antioxidant defense against oxidative stress: Antioxidant enzymes, nanomaterials with multiple enzyme-mimicking activities, and low-molecular-weight antioxidants. *Archives of Toxicology*, *98*(5), 1323–1367.

Jordan, P. A., Kayser-Bricker, K. J., & Miller, S. J. (2010). Asymmetric phosphorylation through catalytic P (III) phosphoramidite transfer: Enantioselective synthesis of D-myo-inositol-6-phosphate. *Proceedings of the National Academy of Sciences of the United States of America*, *107*(48), 20620–20624.

Kim, B., Kang, B., Vales, T. P., Yang, S. K., Lee, J., & Kim, H. J. (2018). Polyphenol-functionalized hydrogels using an interpenetrating chitosan network and investigation of their antioxidant activity. *Macromolecular Research*, *26*(1), 35–39.

Komorowska, J., Wątroba, M., & Szukiewicz, D. (2020). Review of the beneficial effects of resveratrol in neurodegenerative diseases such as Alzheimer's disease. *Advances in Medical Sciences*, *65*(2), 401–410. 10.1016/j.advms.2020.08.002

Kumar, A., & Konar, A. (2023). Neuromodulating bioactive compounds as potential cognitive therapeutics. *Frontiers in Aging Neuroscience*, *15*, 9936231. Advance online publication. 10.3389/fnagi.2023.9936231

Lan, D., Xu, N., Sun, J., Li, Z., Liao, R., Zhang, H., & Yi, W. (2018). Electroacupuncture mitigates endothelial dysfunction via effects on the PI3K/Akt signalling pathway in high fat diet-induced insulin-resistant rats. *Acupuncture in Medicine*, *36*(3), 162–169.

Leclerc, M., Dudonné, S., & Calon, F. (2021). Can natural products exert neuroprotection without crossing the blood–brain barrier? *International Journal of Molecular Sciences*, *22*(7), 3356. Advance online publication. 10.3390/ijms22073356

Loh, J. S., Mak, W. Q., & Khaw, K. Y. (2024). Microbiota–gut–brain axis and its therapeutic applications in neurodegenerative diseases. *Signal Transduction and Targeted Therapy*, *9*, 37. Advance online publication. 10.1038/s41392-024-01743-1

Luo, J., Ganesan, K., & Xu, B. (2024). Unlocking the power: New insights into the anti-aging properties of mushrooms. *Journal of Fungi (Basel, Switzerland)*, *10*(3), 215. Advance online publication. 10.3390/jof10030215

Manap, A. S. A., Vijayabalan, S., Madhavan, P., Chia, Y. Y., Arya, A., Wong, E. H., Rizwan, F., Bindal, U., & Koshy, S. (2019). Bacopa monnieri, a neuroprotective lead in Alzheimer's disease: A review on its properties, mechanisms of action, and preclinical and clinical studies. *Drug Target Insights*, *13*, 1177392819866412. 10.1177/1177392819866412

Masondo, N. A., Stafford, G. I., Aremu, A. O., & Makunga, N. P. (2019). Acetylcholinesterase inhibitors from southern African plants: An overview of ethnobotanical, pharmacological potential and phytochemical research including and beyond Alzheimer's disease treatment. *South African Journal of Botany*, *120*, 39–64.

Mohamed, E. A., Ahmed, H. I., Zaky, H. S., & Badr, A. M. (2021). Sesame oil mitigates memory impairment, oxidative stress, and neurodegeneration in a rat model of Alzheimer's disease: A pivotal role of NF-κB/p38MAPK/BDNF/PPAR-γ pathways. *Journal of Ethnopharmacology*, *280*, 114423. Advance online publication. 10.1016/j.jep.2021.114423

Mohd Sairazi, N. S., & Sirajudeen, K. N. S. (2020). Natural products and their bioactive compounds: Neuroprotective potentials against neurodegenerative diseases. *Evidence-Based Complementary and Alternative Medicine: eCAM*, *2020*(1), 6565396.

Moosavi, F., Hosseini, R., Saso, L., & Firuzi, O. (2015). Modulation of neurotrophic signaling pathways by polyphenols. *Drug Design, Development and Therapy*, *10*, 23–42. 10.2147/DDDT.S96936

Mutha, R. E., Tatiya, A. U., & Surana, S. J. (2022). Flavonoids as natural phenolic compounds and their role in therapeutics: An overview. *Phytomedicine Plus: International Journal of Phytotherapy and Phytopharmacology*, *2*(1), 100191. Advance online publication. 10.1016/j.phyplu.2021.100191

Nahar, L., Charoensup, R., Kalieva, K., Habibi, E., Guo, M., Wang, D., Kvasnica, M., Onder, A., & Sarker, S. D. (2025). Natural products in neurodegenerative diseases: Recent advances and future outlook. *Frontiers in Pharmacology*, *16*, 1529194. Advance online publication. 10.3389/fphar.2025.1529194

Nouri, Z., Fakhri, S., El-Senduny, F. F., Sanadgol, N., Abd-ElGhani, G. E., Farzaei, M. H., & Chen, J.-T. (2019). On the neuroprotective effects of naringenin: Pharmacological targets, signalling pathways, molecular mechanisms, and clinical perspective. *Biomolecules*, *9*(11), 690. Advance online publication. 10.3390/biom9110690

Orozco, M. F., Vázquez-Hernández, A., & Fenton-Navarro, B. (2019). Active compounds of medicinal plants, the mechanism for antioxidant and beneficial effects. *Phyton*, *88*(1), 1–10. 10.32604/phyton.2019.04525

Pandey, K. B., & Rizvi, S. I. (2009). Current understanding of dietary polyphenols and their role in health and disease. *Current Nutrition and Food Science*, *5*(4), 249–263.

Paula, H. S. C., Santiago, S. B., Araújo, L. A., Pedroso, C. F., Marinho, T. A., Goncalves, I. A. J., & Batista, K. A. (2022). An overview on the current available treatment for COVID-19 and the impact of antibiotic administration during the pandemic. *Brazilian Journal of Medical and Biological Research = Revista Brasileira de Pesquisas Médicas e Biológicas*, *55*, e11631.

Radu, A., Tit, D. M., Endres, L. M., Radu, A.-F., Vesa, C. M., & Bungau, S. G. (2024). Naturally derived bioactive compounds as precision modulators of immune and inflammatory mechanisms in psoriatic conditions. *Inflammopharmacology*, *32*(1), 527–549. 10.1007/s10787-024-01602-z

Rahman, M. H., Bajgai, J., Fadriquela, A., Sharma, S., Trinh, T. T., Akter, R., Jeong, Y. J., Goh, S. H., Kim, C.-S., & Lee, K.-J. (2021). Therapeutic potential of natural products in treating neurodegenerative disorders and their future prospects and challenges. *Molecules (Basel, Switzerland)*, *26*(17), 5327. Advance online publication. 10.3390/molecules26175327

Ren, R., Ma, J., Zhao, Y., Zhao, L., Zhang, G., Zheng, Y., & Zhang, Y. (2025). Nebulized seabuckthorn seed oil inhalation attenuates Alzheimer's disease progression in APP/PS1 mice. *Scientific Reports*, *15*, 6368. Advance online publication. 10.1038/s41598-025-56368-9

Rivai, B., & Umar, A. K. (2023). Neuroprotective compounds from marine invertebrates. *Journal of Basic and Applied Scientific Research*, *12*(1), 71. 10.1186/s43088-023-00407-3

Różańska, A., Russo, M., Cacciola, F., Salafia, F., Polkowska, Ż., Dugo, P., & Mondello, L. (2020). Concentration of potentially bioactive compounds in Italian extra virgin olive oils from various sources by using LC-MS and multivariate data analysis. *Foods*, *9*(8), 1120.

Sejbuk, M., Mirończuk-Chodakowska, I., Karav, S., & Witkowska, A. M. (2024). Dietary polyphenols, food processing and gut microbiome: Recent findings on bioavailability, bioactivity, and gut microbiome interplay. *Nutrients*, *16*(20), 3465. Advance online publication. 10.3390/nu16203465

Shi, W., Xie, H., Ouyang, K., Wang, S., Xiong, H., Woo, M. W., & Zhao, Q. (2024). The effect of rice protein-polyphenols covalent and non-covalent interactions on the structure, functionality and in vitro digestion properties of rice protein. *Food Chemistry*, *450*, 139241.

Shim, J. H., Greenblatt, M. B., Xie, M., Schneider, M. D., Zou, W., Zhai, B., & Glimcher, L. H. (2009). TAK1 is an essential regulator of BMP signalling in cartilage. *The EMBO Journal*, *28*(14), 2028.

Shoaib, S., Ansari, M. A., Fatease, A. A., Safhi, A. Y., Hani, U., Jahan, R., Alomary, M. N., Ansari, M. N., Ahmed, N., & Wahab, S.. (2023). Plant-derived bioactive compounds in the management of neurodegenerative disorders: Challenges, future directions, and molecular mechanisms involved in neuroprotection. *Pharmaceutics*, *15*(3), 749. Advance online publication. 10.3390/pharmaceutics15030749

Silva, F. L., Nascimento, G. O., Lopes, G. S., Matos, W. O., Cunha, R. L., Malta, M. R., & Trevisan, M. T. S. (2020). The concentration of polyphenolic compounds and trace elements in the Coffea arabica leaves: Potential chemometric pattern recognition of coffee leaf rust resistance. *Food Research International*, *134*, 109221.

Singh, A., Kaur, P., Kumar, M., Shafi, S., Kumar Upadhyaya, P., Tiwari, A., Tiwari, V., Rangra, N. K., Thirunavukarasu, V., Kumari, S., Roy, D., Ghosh, M., Arora, N., Sharma, N., & Garg, Y.. (2025). The role of phytochemicals in modulating the gut microbiota: Implications for health and disease. *Medicine in Microecology*, *19*, 100094. Advance online publication. 10.1016/j.medmic.2024.100094

Singh, S. K., Srivastav, S., Castellani, R. J., Plascencia-Villa, G., & Perry, G. (2019). Neuroprotective and antioxidant effect of *Ginkgo biloba* extract against AD and other neurological disorders. *Neurotherapeutics; the Journal of the American Society for Experimental NeuroTherapeutics*, *16*(3), 666–674. 10.1007/s13311-019-00767-8

Song, X. Y., Fan, C. X., Choudhary, M. I., & Wang, X. P. (2024). Neuro-regeneration or repair: Cell therapy of neurological disorders as a way forward. *Current Neuropharmacology*, *22*(14), 2272–2283.

Spochacz, M., Chowański, S., Walkowiak-Nowicka, K., Szymczak, M., & Adamski, Z. (2018). Plant-derived substances used against beetles–pests of stored crops and food–and their mode of action: A review. *Comprehensive Reviews in Food Science and Food Safety*, *17*(5), 1339–1366.

Stromsnes, K., Lagzdina, R., Olaso-González, G., Gimeno-Mallench, L., & Gambini, J. (2021). Pharmacological properties of polyphenols: Bioavailability, mechanisms of action, and biological effects in in vitro studies, animal models, and humans. *Biomedicines*, *9*(8), 1074. Advance online publication. 10.3390/biomedicines9081074

Sutalangka, C., & Wattanathorn, J. (2017). Neuroprotective and cognitive-enhancing effects of the combined extract of Cyperus rotundus and Zingiber officinale. *BMC Complementary and Alternative Medicine*, *17*(1), 135.

Tan, J. J. Y., Lim, Y. Y., Siow, L. F., & Tan, J. B. L. (2015). Effects of Drying on Polyphenol Oxidase and Antioxidant Activity of M orus alba Leaves. *Journal of Food Processing and Preservation*, *39*(6), 2811–2819.

Thancharoen, O., Limwattananon, C., Waleekhachonloet, O., Rattanachotphanit, T., Limwattananon, P., & Limpawattana, P. (2019). Ginkgo biloba extract (EGb761), cholinesterase inhibitors, and memantine for the treatment of mild-to-moderate Alzheimer's disease: A network meta-analysis. *Drugs & Aging*, *36*(5), 435–452.

Thomas, T., Friedrich, M., Rich-Griffin, C., Pohin, M., Agarwal, D., Pakpoor, J., & Buckley, C. D. (2024). A longitudinal single-cell atlas of anti-tumour necrosis factor treatment in inflammatory bowel disease. *Nature Immunology*, *25*(11), 2152–2165.

Uddin, M. S., Al Mamun, A., Rahman, M. M., Jeandet, P., Alexiou, A., Behl, T., Sarwar, M. S., Sobarzo-Sánchez, E., Ashraf, G. M., Sayed, A. A., Albadrani, G. M., Peluso, I., & Abdel-Daim, M. M. (2021). Natural products for neurodegeneration: Regulating neurotrophic signals. *Oxidative Medicine and Cellular Longevity*, *2021*, 8820406. Advance online publication. 10.1155/2021/8820406

Yang, L., Wen, K.-S., Ruan, X., Zhao, Y.-X., Wei, F., & Wang, Q. (2018). Response of the plant secondary metabolites to environmental factors. *Molecules (Basel, Switzerland)*, *23*(4), 762. Advance online publication. 10.3390/molecules23040762

Yuan, Y., Yuan, T., Wang, D., Tang, J., & Zhou, S. (2013). Sewage sludge biochar as an efficient catalyst for oxygen reduction reaction in an microbial fuel cell. *Bioresource Technology*, *144*, 115–120.

Zhang, Y., Yu, W., Zhang, L., Wang, M., & Chang, W. (2022). The interaction of polyphenols and the gut microbiota in neurodegenerative diseases. *Nutrients*, *14*(24), 5373. Advance online publication. 10.3390/nu14245373

Compilation of References

Aarts, M. M., & Tymianski, M. (2005). TRPMs and neuronal cell death. *Pflügers Archiv*, *451*(1), 243–249. DOI: 10.1007/s00424-005-1439-x PMID: 16044308

Aatif, M. (2023). Current understanding of polyphenols to enhance bioavailability for better therapies. *Biomedicines*, *11*(7), 2078. DOI: 10.3390/biomedicines11072078 PMID: 37509717

Abbasi, E., Aval, S. F., Akbarzadeh, A., Milani, M., Nasrabadi, H. T., Joo, S. W., Hanifehpour, Y., Nejati-Koshki, K., & Pashaei-Asl, R. (2014). Dendrimers: Synthesis, applications, and properties. *Advanced Drug Delivery Reviews*, *71*, 58–78. PMID: 24994950

Abbott, N. J., Patabendige, A. A. K., Dolman, D. E. M., Yusof, S. R., & Begley, D. J. (2009). Structure and function of the blood-brain barrier. *Neurobiology of Disease*, *37*(1), 13–25. DOI: 10.1016/j.nbd.2009.07.030 PMID: 19664713

Abbott, N. J., Rönnbäck, L., & Hansson, E. (2006). Astrocyte-endothelial interactions at the blood-brain barrier. *Nature Reviews. Neuroscience*, *7*(1), 41–53. DOI: 10.1038/nrn1824

Abdi Syahputra, R., Dalimunthe, A., Utari, Z. D., Halim, P., Sukarno, M. A., Zainalabidin, S., Salim, E., Gunawan, M., Nurkolis, F., Park, M. N., Luckanagul, J. A., Bangun, H., Kim, B., & Harahap, U. (2024). Nanotechnology and flavonoids: Current research and future perspectives on cardiovascular health. *Journal of Functional Foods*, *120*, 106355. DOI: 10.1016/j.jff.2024.106355

Abdullah, A., & Ravanan, P. (2018). Kaempferol mitigates Endoplasmic Reticulum Stress Induced Cell Death by targeting caspase 3/7. *Scientific Reports*, *8*(1), 2189. DOI: 10.1038/s41598-018-20499-7

Abraham, W. C., Jones, O. D., & Glanzman, D. L. (2019). Is plasticity of synapses the mechanism of long-term memory storage? *NPJ Science of Learning*, *4*(1), 9. DOI: 10.1038/s41539-019-0048-y PMID: 31285847

Ackery, A., Tator, C., & Krassioukov, A. (2004). A global perspective on spinal cord injury epidemiology. *Journal of Neurotrauma*, *21*(10), 1355–1370. DOI: 10.1089/neu.2004.21.1355 PMID: 15672627

Adake, P., Chandrashekar, R., & Rao, S. N. (2013). *THE PHARMA INNOVATION - JOURNAL Preclinical evaluation of antidepressant activity of Boswellia serrata by Tail Suspension Test.* 2(5), 13–19.

Adams, S. M., Aksenova, M. V., Aksenov, M. Y., Mactutus, C. F., & Booze, R. M. (2012). Soy isoflavones genistein and daidzein exert anti-apoptotic actions via a selective ER-mediated mechanism in neurons following HIV-1 Tat(1-86) exposure. *PLoS One*, *7*(5), e37540. DOI: 10.1371/journal.pone.0037540

Adamu, A., Li, S., Gao, F., & Xue, G. (2024). The role of neuroinflammation in neurodegenerative diseases: Current understanding and future therapeutic targets. *Frontiers in Aging Neuroscience*, *16*(April), 1–16. DOI: 10.3389/fnagi.2024.1347987 PMID: 38681666

Agus, A., Planchais, J., & Sokol, H. (2018). Gut microbiota regulation of tryptophan metabolism in health and disease. *Cell Host & Microbe*, *23*(6), 716–724. DOI: 10.1016/j.chom.2018.05.003 PMID: 29902437

Aguzzi, A., & O'Connor, T. (2010). Protein aggregation diseases: Pathogenicity and therapeutic perspectives. In *Nature Reviews Drug Discovery* (Vol. 9, Issue 3, pp. 237–248). Nat Rev Drug Discov. DOI: 10.1038/nrd3050

Agyei, D., & Danquah, M. K. (2011). Industrial-scale manufacturing of pharmaceutical-grade bioactive peptides from food proteins: Challenges & strategies. *Biotechnology Advances*, *29*(3), 272–277. DOI: 10.1016/j.biotechadv.2011.01.001 PMID: 21238564

Ahmad, A., Ali, T., Park, H. Y., Badshah, H., Rehman, S. U., & Kim, M. O. (2017). Neuroprotective effect of fisetin against amyloid-beta-induced cognitive/synaptic dysfunction, neuroinflammation, and neurodegeneration in adult mice. *Molecular Neurobiology*, *54*(3), 2269–2285. DOI: 10.1007/s12035-016-9795-4 PMID: 26944285

Ahmad, I., Pal, S., Singh, R., Ahmad, K., Dey, N., Srivastava, A., Ahmad, R., Suliman, M., Alshahrani, M. Y., Barkat, M. A., & Siddiqui, S. (2023). Antimicrobial peptide moricin induces ROS mediated caspase-dependent apoptosis in human triple-negative breast cancer via suppression of notch pathway. *Cancer Cell International*, *23*(1), 121. DOI: 10.1186/s12935-023-02958-y PMID: 37344820

Ahmad, S. R., Zeyaullah, M., AlShahrani, A. M., Dawria, A., Ali, H., Mohieldin, A., Altijani, A. A. G., Razi, U., Mehdi, M., Akram, S., & Hussain, E. R. (2024). Deciphering the enigma of neuron-glial interactions in neurological disorders. *Frontiers in Bioscience (Landmark Edition)*, *29*(4), 142. DOI: 10.31083/j.fbl2904142 PMID: 38682185

Ahmad, S. R., Zeyaullah, M., Khan, M. S., AlShahrani, A. M., Altijani, A. A. G., Ali, H., Dawria, A., Mohieldin, A., Alam, M. S., & Mohamed, A. O. A. (2024). Pharmacogenomics for neurodegenerative disorders - a focused review. *Frontiers in Pharmacology*, *15*(December), 1–11. DOI: 10.3389/fphar.2024.1478964 PMID: 39759457

Ahmad, S., Jo, M. H., Ikram, M., Khan, A., & Kim, M. O. (2021). Deciphering the potential neuroprotective effects of luteolin against Aβ1–42-Induced alzheimer's disease. *International Journal of Molecular Sciences*, *22*(17), 9583. DOI: 10.3390/ijms22179583 PMID: 34502488

Ahmed, M. (2025). Targeting aging pathways with natural compounds: a review of curcumin, epigallocatechin gallate, thymoquinone, and resveratrol. *Immunity & Ageing: I & A, 22*, 28.

Aileen Funke, S., & Willbold, D. (2012). Peptides for therapy and diagnosis of Alzheimer's disease. *Current Pharmaceutical Design*, *18*(6), 755–767. DOI: 10.2174/138161212799277752 PMID: 22236121

Akıncıoğlu, H., & Gülçin, İ. (2020). Potent acetylcholinesterase inhibitors: Potential drugs for Alzheimer's disease. *Mini-Reviews in Medicinal Chemistry*, *20*(8), 703–715. DOI: 10.2174/1389557520666200103100521 PMID: 31902355

Akram, R., Anwar, H., Javed, M. S., Imran, A., Rasul, A., Malik, S. A., Manzoor, M., Islam, F., Khan, I. U., Sajid, F., Iman, T., Shah, M. A., Sun, T., Hussain, G., & Shah, M. A. (2023). Natural molecules as promising players against diabetic peripheral neuropathy: an emerging nutraceutical approach. Https://Doi.Org/10.1080/10942912.2023.2189569, *26*(1), 894–914. DOI: 10.1080/10942912.2023.2189569

Akram, R., Muzaffar, H., Anwar, H., Malik, S. A., Sajid, F., Iman, T., Ahsan, A., Manzoor, M., Khan, I. U., & Hussain, G. (2023). *Role of Endogenous and Dietary Antioxidants in Brain Disorders*. 171–214. DOI: 10.1007/978-3-031-41188-5_8

Al Mamun, A., Nsiah, N. Y., Hossain, M. A., Islam, M. R., Herrera-Calderon, O., Baghel, M., & Hafez, M. M. (2024). Polyphenols targeting NF-κB pathway in neurological disorders: What we know so far? *International Journal of Biological Sciences*, *20*(4), 1332–1355. DOI: 10.7150/ijbs.90982 PMID: 38385077

Al Omran, Y., & Aziz, Q. (2014). The brain-gut axis in health and disease. *Microbial endocrinology: the microbiota-gut-brain axis in health and disease*, 135-153.

Alecu, I., & Bennett, S. A. L. (2019). Dysregulated lipid metabolism and its role in α-synucleinopathy in Parkinson's disease. In *Frontiers in Neuroscience* (Vol. 13, Issue APR). Frontiers Media S.A. DOI: 10.3389/fnins.2019.00328

Alharbi, M. H., Lamport, D. J., Dodd, G. F., Saunders, C., Harkness, L., Butler, L. T., & Spencer, J. P. E. (2016). Flavonoid-rich orange juice is associated with acute improvements in cognitive function in healthy middle-aged males. *European Journal of Nutrition*, *55*(6), 2021–2029. DOI: 10.1007/s00394-015-1016-9 PMID: 26280945

Ali, T., Kim, M. J., Rehman, S. U., Ahmad, A., & Kim, M. O. (2017). Anthocyanin-Loaded PEG-Gold Nanoparticles Enhanced the Neuroprotection of Anthocyanins in an Aβ1–42 Mouse Model of Alzheimer's Disease. *Molecular Neurobiology*, *54*(8), 6490–6506. DOI: 10.1007/s12035-016-0136-4 PMID: 27730512

Al-Khayri, J. M., Sahana, G. R., Nagella, P., Joseph, B. V., Alessa, F. M., & Al-Mssallem, M. Q. (2022). Flavonoids as potential anti-inflammatory molecules: A review. *Molecules (Basel, Switzerland)*, *27*(9), 2901. DOI: 10.3390/molecules27092901 PMID: 35566252

Allen, T. M., & Cullis, P. R. (2013). Liposomal drug delivery systems: From concept to clinical applications. *Science*, *339*(6119), 1232625. PMID: 23036225

Almad, A., Sahinkaya, F. R., & McTigue, D. M. (2011). Oligodendrocyte Fate after Spinal Cord Injury. In *Neurotherapeutics* (Vol. 8, Issue 2, pp. 262–273). Neurotherapeutics. DOI: 10.1007/s13311-011-0033-5

Almeida, A. F., Borge, G. I. A., Piskula, M., Tudose, A., Tudoreanu, L., Valentová, K., Williamson, G., & Santos, C. N. (2018). Bioavailability of quercetin in humans with a focus on interindividual variation. *Comprehensive Reviews in Food Science and Food Safety*, *17*(3), 714–731. DOI: 10.1111/1541-4337.12342 PMID: 33350133

Al-Obaid, S., Al-Khafaji, Z., Al-Jebori, S., & et al. (2020). Gut microbiome of native Arab Kuwaitis reveals diet-dependent variation. [Note: Journal title, volume, and pages/DOI missing in original – please complete if available]

Al-Obaidi, M. M. J., & Desa, M. N. M. (2018). Mechanisms of blood brain barrier disruption by different types of bacteria, and bacterial–host interactions facilitate the bacterial pathogen invading the brain. *Cellular and Molecular Neurobiology*, *38*(7), 1349–1368. DOI: 10.1007/s10571-018-0609-2 PMID: 30117097

Alpino, G. C. Á., Pereira-Sol, G. A., Dias, M. M. e., Aguiar, A. S., & Peluzio, M. C. G. (2024). Beneficial effects of butyrate on brain functions: A view of epigenetic. *Critical Reviews in Food Science and Nutrition*, *64*(12), 3961–3970. DOI: 10.1080/10408398.2022.2137776 PMID: 36287024

Alqahtani, T., Deore, S. L., Kide, A. A., Shende, B. A., Sharma, R., Chakole, R. D., Nemade, L. S., Kale, N. K., Borah, S., & Deokar, S. S. (2023). Mitochondrial dysfunction and oxidative stress in Alzheimer's disease, and Parkinson's disease, Huntington's disease and amyotrophic lateral sclerosis-an updated review. *Mitochondrion*, *71*, 83–92. DOI: 10.1016/j.mito.2023.05.007 PMID: 37269968

Alvarez, A. I., Real, R., Pérez, M., Mendoza, G., Prieto, J. G., & Merino, G. (2010). Modulation of the activity of ABC transporters (P-glycoprotein, MRP2, BCRP) by flavonoids and drug response. *Journal of Pharmaceutical Sciences*, *99*(2), 598–617. DOI: 10.1002/jps.21851

Álvarez-Pérez, B., Homs, J., Bosch-Mola, M., Puig, T., Reina, F., Verdú, E., & Boadas-Vaello, P. (2016). Epigallocatechin-3-gallate treatment reduces thermal hyperalgesia after spinal cord injury by down-regulating RhoA expression in mice. *European Journal of Pain (London, England)*, *20*(3), 341–352. DOI: 10.1002/ejp.722 PMID: 25913854

Aly, S. H., Elissawy, A. M., Fayez, A. M., Eldahshan, O. A., Elshanawany, M. A., & Singab, A. N. B. (2021). Neuroprotective effects of Sophora secundiflora, Sophora tomentosa leaves and formononetin on scopolamine-induced dementia. *Natural Product Research*, *35*(24), 5848–5852.

Alzubide, A., & Alhalafi, M. (2024). Gut-brain axis modulation: The role of probiotics in neurological health. *Journal of Neurogastroenterology*, *12*(3), 45–59.

Ambalam, P., Raman, M., Purama, R. K., & Doble, M. (2016). Probiotics, prebiotics and colorectal cancer prevention. *Best Practice & Research. Clinical Gastroenterology*, *30*(1), 119–131. DOI: 10.1016/j.bpg.2016.02.009 PMID: 27048903

Amen Rallapalli, K. L., Komor, R. A., Hassan, Y. M., Essmat, R. A., Ahmed, R. H., Azab, M. M., Shehata, N. R., & El-Sayed, W. M. (2025). Harnessing the Microbiome: CRISPR-Based Gene Editing and Antimicrobial Peptides in Combating Antibiotic Resistance and Cancer. *Probiotics and Antimicrobial Proteins*, •••, 1–31. PMID: 40377870

Am, J., & a. (2023). Medicinal Properties of Plant-Derived Compounds and its Significance. *Medicinal & Aromatic Plants*, *12*(3), 1–2. DOI: 10.35248/2167-0412.23.12.455

Anderson, R. C., Cookson, A. L., McNabb, W. C., Kelly, W. J., & Roy, N. C. (2012). Lactobacillus plantarum DSM 2648 is a potential probiotic that enhances intestinal barrier function. *FEMS Microbiology Letters*, *309*(2), 144–152. PMID: 20618863

Andres-Lacueva, C., Shukitt-Hale, B., Galli, R. L., Jauregui, O., Lamuela-Raventós, R. M., & Joseph, J. A. (2005). Anthocyanins in aged blueberry-fed rats are found in the brain and may enhance memory. *Nutritional Neuroscience*, *8*(2), 111–120. DOI: 10.1080/10284150500078117 PMID: 16053243

Anhê, F. F.. (2015). A polyphenol-rich cranberry extract increases Akkermansia muciniphila in obese mice. *Gut, 64*(6), 872–883. DOI: 10.1136/gutjnl-2014-307142 PMID: 25080446

Anhe, F. F., Varin, T. V., Mithieux, G., & Desjardins, Y. (2015). The gut microbiota and the beneficial effects of dietary polyphenols on human health. *Current Opinion in Clinical Nutrition and Metabolic Care, 18*(6), 519–525. DOI: 10.1097/MCO.0000000000000201

Anis, E., Xie, A., Brundin, L., & Brundin, P. (2022). Digesting recent findings: Gut alpha-synuclein, microbiome changes in Parkinson's disease. *Trends in Endocrinology and Metabolism, 33*(2), 147–157. DOI: 10.1016/j.tem.2021.11.005 PMID: 34949514

Ansari, M. A., Abdul, H. M., Joshi, G., Opii, W. O., & Butterfield, D. A. (2009). Protective effect of quercetin in primary neurons against Aβ (1–42): Relevance to Alzheimer's disease. *The Journal of Nutritional Biochemistry, 20*(4), 269–275. DOI: 10.1016/j.jnutbio.2008.03.002 PMID: 18602817

Ansari, M. A., Roberts, K. N., & Scheff, S. W. (2008). Oxidative stress and modification of synaptic proteins in hippocampus after traumatic brain injury. *Free Radical Biology & Medicine, 45*(4), 443–452. DOI: 10.1016/j.freeradbiomed.2008.04.038 PMID: 18501200

Antunes, M. S., Goes, A. T. R., Boeira, S. P., Prigol, M., & Jesse, C. R. (2014). Protective effect of hesperidin in a model of Parkinson's disease induced by 6-hydroxydopamine in aged mice. *Nutrition (Burbank, Los Angeles County, Calif.), 30*(11–12), 1415–1422. DOI: 10.1016/j.nut.2014.03.024 PMID: 25280422

Anusha, C., Sumathi, T., & Joseph, L. D. (2017). Protective role of apigenin on rotenone induced rat model of Parkinson's disease: Suppression of neuroinflammation and oxidative stress mediated apoptosis. *Chemico-Biological Interactions, 269*, 67–79. DOI: 10.1016/j.cbi.2017.03.016 PMID: 28389404

Apostolopoulos, V., Bojarska, J., Chai, T. T., Elnagdy, S., Kaczmarek, K., Matsoukas, J., & Toth, I. (2021). Short peptides: Global overview, limitations, and perspectives. *Molecules (Basel, Switzerland), 26*(2), 430. DOI: 10.3390/molecules26020430 PMID: 33467522

Ara, I., Maqbool, M., & Gani, I. (2022). Neuroprotective Activity of Herbal Medicinal Products: A Review. *International Journal of Current Research in Physiology and Pharmacology*, 1-10.

Aras, A. B., Guven, M., Akman, T., Ozkan, A., Sen, H. M., Duz, U., & Cosar, M. (2015). Neuroprotective effects of daidzein on focal cerebral ischemia injury in rats. *Neural Regeneration Research, 10*(1), 146–152. DOI: 10.4103/1673-5374.150724

Ardeshir, R. A., & Moarefvand, K. (2025). Selective apoptosis induction by antimicrobial peptide TP4 in MCF-7 breast cancer cells. *Scientific Reports, 15*(1), 15061. DOI: 10.1038/s41598-025-97906-3 PMID: 40301390

Arias-Sánchez, R. A., Torner, L., & Fenton Navarro, B. (2023). Polyphenols and Neurodegenerative Diseases: Potential Effects and Mechanisms of Neuroprotection. *Molecules (Basel, Switzerland), 28*(14), 1–15. DOI: 10.3390/molecules28145415 PMID: 37513286

Armstrong, R. (2020). What causes neurodegenerative disease? *Folia Neuropathologica, 58*(2), 93–112. DOI: 10.5114/fn.2020.96707 PMID: 32729289

Arnison, P. G., Bibb, M. J., Bierbaum, G., Bowers, A. A., Bugni, T. S., Bulaj, G., Camarero, J. A., Campopiano, D. J., Challis, G. L., Clardy, J., Cotter, P. D., Craik, D. J., Dawson, M., Dittmann, E., Donadio, S., Dorrestein, P. C., Entian, K.-D., Fischbach, M. A., Garavelli, J. S., & van der Donk, W. A. (2013). Ribosomally synthesized and post-translationally modified peptide natural products: Overview and recommendations for a universal nomenclature. *Natural Product Reports*, *30*(1), 108–160. DOI: 10.1039/C2NP20085F PMID: 23165928

Aruna, R., Geetha, A., & Suguna, P. (2014). Rutin modulates ASC expression in NLRP3 inflammasome: A study in alcohol and cerulein-induced rat model of pancreatitis. *Molecular and Cellular Biochemistry*, *396*(1-2), 269–280. DOI: 10.1007/s11010-014-2162-8

Aryal, B., Raut, B. K., Bhattarai, S., Bhandari, S., Tandan, P., Gyawali, K., Sharma, K., Ranabhat, D., Thapa, R., Aryal, D., Ojha, A., Devkota, H. P., & Parajuli, N. (2022). Potential Therapeutic Applications of Plant-Derived Alkaloids against Inflammatory and Neurodegenerative Diseases. *Evidence-Based Complementary and Alternative Medicine : eCAM*, *7299778*, 1–18. Advance online publication. DOI: 10.1155/2022/7299778 PMID: 35310033

Aseervatham, G. S. B., Suryakala, U., Doulethunisha, , Sundaram, S., Bose, P. C., & Sivasudha, T. (2016). Expression pattern of NMDA receptors reveals antiepileptic potential of apigenin 8-C-glucoside and chlorogenic acid in pilocarpine induced epileptic mice. *Biomedicine and Pharmacotherapy*, *82*, 54–64. DOI: 10.1016/j.biopha.2016.04.066 PMID: 27470339

Ashton, N. J., Benedet, A. L., Pascoal, T. A., Karikari, T. K., Lantero-Rodriguez, J., Brum, W. S., Mathotaarachchi, S., Therriault, J., Savard, M., Chamoun, M., Stoops, E., Francois, C., Vanmechelen, E., Gauthier, S., Zimmer, E. R., Zetterberg, H., Blennow, K., & Rosa-Neto, P. (2022). Cerebrospinal fluid p-tau231 as an early indicator of emerging pathology in Alzheimer's disease. *EBioMedicine*, *76*, 103836. Advance online publication. DOI: 10.1016/j.ebiom.2022.103836 PMID: 35158308

Asl, Z. R., Sepehri, G., & Salami, M. (2019). Probiotic treatment improves the impaired spatial cognitive performance and restores synaptic plasticity in an animal model of Alzheimer's disease. *Behavioural Brain Research*, *376*, 112183. DOI: 10.1016/j.bbr.2019.112183 PMID: 31472194

Association, A. s. (2019). 2019 Alzheimer's disease facts and figures. *Alzheimer's & dementia, 15*(3), 321-387.

Aswar, M., & Patil, V. (2016). Ferulic acid ameliorates chronic constriction injury induced painful neuropathy in rats. *Inflammopharmacology*, *24*(4), 181–188. DOI: 10.1007/s10787-016-0272-5 PMID: 27372349

ATAlAY, T., Gülşen, İ., Çölçimen, N., Alp, H., SOSuNCu, E., AlACA, I., Ak, H., & Rağbetli, M. (2017). Resveratrol treatment prevents hippocampal neurodegeneration in a rodent model of traumatic brain injury. *Turkish Neurosurgery*, *27*(6). PMID: 27509464

Attwell, D., Buchan, A. M., Charpak, S., Lauritzen, M., MacVicar, B. A., & Newman, E. A. (2010). Glial and neuronal control of brain blood flow. In *Nature* (Vol. 468, Issue 7321, pp. 232–243). Nature Publishing Group. DOI: 10.1038/nature09613

Augusto-Oliveira, M., Arrifano, G. P., Lopes-Araújo, A., Santos-Sacramento, L., Takeda, P. Y., Anthony, D. C., Malva, J. O., & Crespo-Lopez, M. E. (2019). What do microglia really do in healthy adult brain? *Cells*, *8*(10), 1293. DOI: 10.3390/cells8101293 PMID: 31652490

Aura, A. M. (2008). Microbial metabolism of dietary phenolic compounds in the colon. *Phytochemistry Reviews*, *7*(3), 407–429. DOI: 10.1007/s11101-008-9095-3

Avena-Woods, C. (2017). Overview of atopic dermatitis. *The American Journal of Managed Care*, *23*(6, Suppl), S115–S123. PMID: 28978208

Awuchi, C. G. (2023). Plants, phytochemicals, and natural practices in complementary and alternative system of medicine for treatment of central nervous system disorders. In *International Journal of Food Properties* (Vol. 26, Issue 1, pp. 1190–1213). Taylor and Francis Ltd. DOI: 10.1080/10942912.2023.2205039

Ayissi, V. B. O., Ebrahimi, A., & Schluesenner, H. (2014). Epigenetic effects of natural polyphenols: A focus on SIRT1-mediated mechanisms. *Molecular Nutrition & Food Research*, *58*(1), 22–32. DOI: 10.1002/mnfr.201300195 PMID: 23881751

Azeem, M., Hanif, M., Mahmood, K., Ameer, N., Chughtai, F. R. S., & Abid, U. (2023). An insight into anticancer, antioxidant, antimicrobial, antidiabetic and anti-inflammatory effects of quercetin: A review. *Polymer Bulletin*, *80*(1), 241–262. DOI: 10.1007/s00289-022-04091-8 PMID: 35125574

Aziz, N., Ruzza, C., Falcicchia, C., Guarino, A., Soukupova, M., Asth, L., Aleotti, V., Bettegazzi, B., Simonato, M., & Zucchini, S. (2024). Lack of Direct Effects of Neurotrophic Factors in an In Vitro Model of Neuroinflammation. *International Journal of Molecular Sciences*, *25*(8), 4160. Advance online publication. DOI: 10.3390/ijms25084160 PMID: 38673746

Babazadeh, A., Ghanbarzadeh, B., & Hamishehkar, H. (2023). Natural bioactive molecules as neuromedicines for the treatment/prevention of neurodegenerative diseases. *ACS Omega*, *8*(4), 3667–3683. DOI: 10.1021/acsomega.2c06098 PMID: 36743024

Bacchi, A. D. (2025). Beyond the Neuron: The Integrated Role of Glia in Psychiatric Disorders. *Neuroglia (Basel, Switzerland)*, *6*(2), 15. Advance online publication. DOI: 10.3390/neuroglia6020015

Badjakov, I., Kaluza, M., Ksiazek-Winiarek, D., Szpakowski, P., Czpakowska, J., Fijalkowska, J., & Glabinski, A. (2025). Polyphenols in the Central Nervous System: Cellular Effects and Liposomal Delivery Approaches. *International Journal of Molecular Sciences 2025, Vol. 26, Page 6477*, *26*(13), 6477. DOI: 10.3390/ijms26136477

Baek, S. C., Park, M. H., Ryu, H. W., Lee, J. P., Kang, M. G., Park, D., Park, C. M., Oh, S. R., & Kim, H. (2019). Rhamnocitrin isolated from *Prunus padus* var. *seoulensis*: A potent and selective reversible inhibitor of human monoamine oxidase A. *Bioorganic Chemistry*, *83*, 317–325. 10.1016/j.bioorg.2018.10.051

Bagga, D., Reichert, J. L., Koschutnig, K., Aigner, C. S., Holzer, P., Koskinen, K., Moissl-Eichinger, C., & Schöpf, V. (2018). Probiotics drive gut microbiome triggering emotional brain signatures. *Gut Microbes*, *9*(6), 486–496. DOI: 10.1080/19490976.2018.1460015 PMID: 29723105

Bagher, Z., Ehterami, A., Nasrolahi, M., Azimi, M., & Salehi, M. (2021). Hesperidin promotes peripheral nerve regeneration based on tissue engineering strategy using alginate/chitosan hydrogel: In vitro and in vivo study. *International Journal of Polymeric Materials, 70*(5), 299–308. DOI: 10.1080/00914037.2020.1713781

Bagriyanik, H., Ersoy, N., Cetinkaya, C., Ikizoglu, E., Kutri, D., Ozcana, T., Kamanga, L., & Kiray, M. (2014). The effects of resveratrol on chronic constriction injury of sciatic nerve in rats. *Neuroscience Letters, 561*, 123–127. DOI: 10.1016/j.neulet.2013.12.056 PMID: 24394911

Bahramsoltani, R., Rahimi, R., & Farzaei, M. H. (2017). Pharmacokinetic interactions of curcuminoids with conventional drugs: A review. *Journal of Ethnopharmacology, 209*, 1–12. DOI: 10.1016/j.jep.2017.07.022 PMID: 28734960

Bai, X., Bian, Z., Zhang, M., & Li, J. (2023). Targeting the Nrf2 signaling pathway using phytochemical ingredients: A novel therapeutic road map to combat neurodegenerative diseases. *Phytomedicine, 109*, 154582. Advance online publication. DOI: 10.1016/j.phymed.2022.154582 PMID: 36610130

Banc, R., Rusu, M. E., Filip, L., & Popa, D. S. (2023). The Impact of Ellagitannins and Their Metabolites through Gut Microbiome on the Gut Health and Brain Wellness within the Gut–Brain Axis. *Foods, 12*(2), 270. Advance online publication. DOI: 10.3390/foods12020270 PMID: 36673365

Banerjee, S., Biehl, A., Gadina, M., Hasni, S., & Schwartz, D. M. (2017). JAK–STAT signaling as a target for inflammatory and autoimmune diseases: Current and future prospects. *Drugs, 77*(5), 521–546. DOI: 10.1007/s40265-017-0701-9 PMID: 28255960

Bao, G., Li, C., Qi, L., Wang, N., & He, B. (2016). Tetrandrine protects against oxygen-glucose-serum deprivation/reoxygenation-induced injury via PI3K/AKT/NF-κB signaling pathway in rat spinal cord astrocytes. *Biomedicine and Pharmacotherapy, 84*(277), 925–930. DOI: 10.1016/j.biopha.2016.10.007 PMID: 27764754

Baquero, F., Lanza, V. F., Baquero, M. R., Del Campo, R., & Bravo-Vázquez, D. A. (2019). Microcins in Enterobacteriaceae: Peptide antimicrobials in the eco-active intestinal chemosphere. *Frontiers in Microbiology, 10*, 2261. DOI: 10.3389/fmicb.2019.02261 PMID: 31649628

Bardaweel, S. K., Gul, M., Alzweiri, M., Ishaqat, A., ALSalamat, H. A., & Bashatwah, R. M. (2018). Reactive oxygen species: The dual role in physiological and pathological conditions of the human body. *The Eurasian Journal of Medicine, 50*(3), 193–201. DOI: 10.5152/eurasianjmed.2018.17397 PMID: 30515042

Barreca, D., Bellocco, E., D'Onofrio, G., Nabavi, S. F., Daglia, M., Rastrelli, L., & Nabavi, S. M. (2016). Neuroprotective Effects of Quercetin: From Chemistry to Medicine. *CNS & Neurological Disorders - Drug Targets, 15*(8), 964–975. DOI: 10.2174/1871527315666160813175406

Barreca, D., Mandalari, G., Calderaro, A., Smeriglio, A., Trombetta, D., Felice, M. R., & Gattuso, G. (2020). Citrus flavones: An update on sources, biological functions, and health promoting properties. *Plants, 9*(3), 288. DOI: 10.3390/plants9030288 PMID: 32110931

Bartel, D. P. (2018). Metazoan micrornas. *Cell, 173*(1), 20–51. DOI: 10.1016/j.cell.2018.03.006 PMID: 29570994

Barua, C. C., Roy, J. D., Buragohain, B., Barua, A. G., Borah, P., & Lahkar, M. (2009). Anxiolytic effect of hydroethanolic extract of Drymaria cordata L Willd. *Indian Journal of Experimental Biology*, *47*(12), 969–973. PMID: 20329700

Basu, A., & Lyons, T. J. (2012). Berries: Emerging impact on cardiovascular health. *Nutrition Reviews*, *70*(2), 168–177. DOI: 10.1111/j.1753-4887.2011.00405.x PMID: 20384847

Basu, A., & Rhone, M. (2012). Blueberries: The evidence for brain health and neuroprotection. *Nutrition Reviews*, *70*(5), 283–294. DOI: 10.1111/j.1753-4887.2012.00495.x

Basu, P., & Basu, A. (2020). In vitro and in vivo effects of flavonoids on peripheral neuropathic pain. *Molecules (Basel, Switzerland)*, *25*(5), 1171. DOI: 10.3390/molecules25051171 PMID: 32150953

Batarfi, W. A., Yunus, M. H. M., Hamid, A. A., Lee, Y. T., & Maarof, M. (2024). Hydroxytyrosol: A promising therapeutic agent for mitigating inflammation and apoptosis. *Pharmaceutics*, *16*(12), 1504. DOI: 10.3390/pharmaceutics16121504 PMID: 39771483

Bathina, S., & Das, U. N. (2015). Brain-derived neurotrophic factor and its clinical implications. *Archives of Medical Science: AMS*, *11*(6), 1164–1178. 10.5114/aoms.2015.56342

Batra, N., Easwar Kumar, V., Nambiar, R., De Souza, C., Yuen, A., Le, U., Verma, R., Ghosh, P. M., & Vinall, R. L. (2022). Exploring the therapeutic potential of neem (*Azadirachta indica*) for the treatment of prostate cancer: A literature review. *Annals of Translational Medicine*, *10*(16), 894. Advance online publication. 10.21037/atm-22-94

Baur, J. A., & Sinclair, D. A. (2006). Therapeutic potential of resveratrol: The in vivo evidence. *Nature Reviews. Drug Discovery*, *5*(6), 493–506. DOI: 10.1038/nrd2060 PMID: 16732220

Bazira, P. J. (2021). Clinically applied anatomy of the vertebral column. [oxford]. *Surgery*, *39*(6), 315–323.

Bechara, S., & Popat, K. C. (2013). Patterned nanowire surfaces promote directional adhesion and proliferation of neural progenitor cells. *Journal of Biomedical Nanotechnology*, *9*(10), 1698–1706. DOI: 10.1166/jbn.2013.1667 PMID: 24015499

Begh, M. Z. A., Khan, J., Zehravi, M., Sweilam, S. H., Dinesh Raja, A., Muthukumar, A., Akiful Haque, M., Kar, N. R., Singh, L. P., Dharani Priya, B., Alshehri, M. A., Ahmad, I., Kang, S., Moon, S., Park, M. N., & Emran, T. (2024). Targeting Neurological Disorders with Stilbenes: Bridging the Preclinical-Clinical Gap. *International Journal of Biological Sciences*, *20*(14), 5474–5494. DOI: 10.7150/ijbs.102032 PMID: 39494329

Bellanti, F., Coda, A. R. D., Trecca, M. I., Lo Buglio, A., Serviddio, G., & Vendemiale, G. (2025). Redox Imbalance in Inflammation: The Interplay of Oxidative and Reductive Stress. In *Antioxidants* (Vol. 14, Issue 6). Multidisciplinary Digital Publishing Institute (MDPI). DOI: 10.3390/antiox14060656

Bellavite, P. (2023). Neuroprotective Potentials of Flavonoids: Experimental Studies and Mechanisms of Action. *Antioxidants*, *12*(2), 280. DOI: 10.3390/antiox12020280 PMID: 36829840

Bell, L., Lamport, D. J., Butler, L. T., & Williams, C. M. (2021). A review of the cognitive effects observed in humans following acute anthocyanin administration. *Nutrients*, *13*(1), 169. DOI: 10.3390/nu13010169 PMID: 33429859

Bellomio, A., Vincent, P. A., de Arcuri, B. F., Farías, R. N., & Morero, R. D. (2007). Microcin J25 has dual and independent mechanisms of action in Escherichia coli: RNA polymerase inhibition and increased superoxide production. *Journal of Bacteriology, 189*(11), 4180–4186. DOI: 10.1128/JB.00206-07 PMID: 17400747

Belščak-Cvitanović, A., Durgo, K., Huđek, A., Bačun-Družina, V., & Komes, D. (2018). Overview of polyphenols and their properties. In *Polyphenols: Properties, recovery, and applications* (pp. 3–44). Elsevier.

Benameur, T., Porro, C., Twfieg, M.-E., Benameur, N., Panaro, M. A., Filannino, F. M., & Hasan, A. (2023). Emerging paradigms in inflammatory disease management: Exploring bioactive compounds and the gut microbiota. *Brain Sciences, 13*(8), 1226. Advance online publication. 10.3390/brainsci13081226

Benameur, T., Soleti, R., & Porro, C. (2021). The potential neuroprotective role of free and encapsulated quercetin mediated by miRNA against neurological diseases. *Nutrients, 13*(4), 1318. DOI: 10.3390/nu13041318 PMID: 33923599

Benarroch, E. E. (2005). Neuron-astrocyte interactions: Partnership for normal function and disease in the central nervous system. *Mayo Clinic Proceedings, 80*(10), 1326–1338. DOI: 10.4065/80.10.1326 PMID: 16212146

Bento-Silva, A., Koistinen, V. M., Mena, P., Bronze, M. R., Hanhineva, K., Sahlstrøm, S., Kitrytė, V., Moco, S., & Aura, A. M. (2019). Factors affecting intake, metabolism and health benefits of phenolic acids: do we understand individual variability? *European Journal of Nutrition 2019 59:4, 59*(4), 1275–1293. DOI: 10.1007/s00394-019-01987-6

Bereiter-Hahn, J., & Vöth, M. (1994). Dynamics of mitochondria in living cells: Shape changes, dislocations, fusion, and fission of mitochondria. *Microscopy Research and Technique, 27*(3), 198–219. DOI: 10.1002/jemt.1070270303 PMID: 8204911

Bermudez-Brito, M., Plaza-Díaz, J., Muñoz-Quezada, S., Gómez-Llorente, C., & Gil, A. (2012). Probiotic mechanisms of action. *Annals of Nutrition & Metabolism, 61*(2), 160–174. DOI: 10.1159/000342079 PMID: 23037511

Berndtsson, E., Nynäs, A.-L., Newson, W., Langton, M., Andersson, R., Johansson, E., & Olsson, M. (2019). 21. The underutilised side streams of broccoli and kale–valorisation via proteins and phenols. In *Sustainable governance and management of food systems* (pp. 153–159). Academic. DOI: 10.3920/978-90-8686-892-6_21

Bernier, L. P., York, E. M., Kamyabi, A., Choi, H. B., Weilinger, N. L., & MacVicar, B. A. (2020). Microglial metabolic flexibility supports immune surveillance of the brain parenchyma. *Nature Communications 2020 11:1, 11*(1), 1–17. DOI: 10.1038/s41467-020-15267-z

Berthoud, H. R., Blackshaw, L. A., Brookes, S. J., & Grundy, D. (2004). Neuroanatomy of extrinsic afferents supplying the gastrointestinal tract. *Neurogastroenterology and Motility, 16*(s1), 28–33. DOI: 10.1111/j.1743-3150.2004.00471.x PMID: 15066001

Besedovsky, H. O., & del Rey, A. (2011). Central and peripheral cytokines mediate immune-brain connectivity. *Neurochemical Research, 36*(1), 1–6. DOI: 10.1007/s11064-010-0252-x PMID: 20820913

Bešlo, D., Golubić, N., Rastija, V., Agić, D., Karnaš, M., Šubarić, D., & Lučić, B. (2023). Antioxidant activity, metabolism, and bioavailability of polyphenols in the diet of animals. *Antioxidants*, *12*(6), 1141. Advance online publication. 10.3390/antiox12061141

Bhaskara, V. K., Mittal, B., Mysorekar, V. V., Amaresh, N., & Simal-Gandara, J. (2020). Resveratrol, cancer and cancer stem cells: A review on past to future. *Current Research in Food Science*, *3*(August), 284–295. DOI: 10.1016/j.crfs.2020.10.004 PMID: 33305295

Bhatt, R., Singh, D., Prakash, A., & Mishra, N. (2015). Development, characterization and nasal delivery of rosmarinic acid-loaded solid lipid nanoparticles for the effective management of Huntingtons disease. *Drug Delivery*, *22*(7), 931–939. DOI: 10.3109/10717544.2014.880860 PMID: 24512295

Bhullar, K. S., & Rupasinghe, H. P. V. (2013). Polyphenols: Multipotent therapeutic agents in neurodegenerative diseases. *Oxidative Medicine and Cellular Longevity*, *2013*, 891748. Advance online publication. DOI: 10.1155/2013/891748 PMID: 23840922

Bigagli, E. (2015). Gene and microRNA expression profiles in models of brain aging: effects of bioactive phenolic compounds.

Billowria, K., Ali, R., Rangra, N. K., Kumar, R., & Chawla, P. A. (2024). Bioactive flavonoids: A comprehensive review on pharmacokinetics and analytical aspects. *Critical Reviews in Food Science and Nutrition*. Advance online publication. 10.1080/10408347.2022.2105641

Binder, D. K., & Scharfman, H. E. (2004). Brain-derived neurotrophic factor. *Growth Factors (Chur, Switzerland)*, *22*(3), 123–131. DOI: 10.1080/08977190410001723308 PMID: 15518235

Biswas, P., Dey, D., Biswas, P. K., Rahaman, T. I., Saha, S., Parvez, A., Khan, D. A., Lily, N. J., Saha, K., Sohel, M., Hasan, M. M., Al Azad, S., Bibi, S., Hasan, M. N., Rahmatullah, M., Chun, J., Rahman, M. A., & Kim, B. (2022). A Comprehensive Analysis and Anti-Cancer Activities of Quercetin in ROS-Mediated Cancer and Cancer Stem Cells. *International Journal of Molecular Sciences*, *23*(19), 11746. Advance online publication. DOI: 10.3390/ijms231911746 PMID: 36233051

Blackshaw, L. A., Brookes, S. J. H., Grundy, D., & Schemann, M. (2007). Sensory transmission in the gastrointestinal tract. *Neurogastroenterology and Motility*, *19*(s1), 1–19. DOI: 10.1111/j.1365-2982.2006.00871.x PMID: 17280582

Boccazzi, M., Van Steenwinckel, J., Schang, A. L., Faivre, V., Le Charpentier, T., Bokobza, C., Csaba, Z., Verderio, C., Fumagalli, M., Mani, S., & Gressens, P. (2021). The immune-inflammatory response of oligodendrocytes in a murine model of preterm white matter injury: The role of TLR3 activation. *Cell Death & Disease*, *12*(2), 166. Advance online publication. DOI: 10.1038/s41419-021-03446-9 PMID: 33558485

Boespflug, E. L., Eliassen, J. C., Dudley, J. A., Shidler, M. D., Kalt, W., Summer, S. S., & Krikorian, R. (2018). Enhanced neural activation with blueberry supplementation in mild cognitive impairment. *Nutritional Neuroscience*, *21*(4), 297–305. DOI: 10.1080/1028415X.2017.1287833 PMID: 28221821

Boleij, A., Hechenbleikner, E. M., Goodwin, A. C., Badani, R., Stein, E. M., Lazarev, M. G., Ellis, B., Carroll, K. C., Albesiano, E., Wick, E. C., Platz, E. A., Pardoll, D. M., & Sears, C. L. (2015). The *Bacteroides fragilis* toxin gene is prevalent in the colon mucosa of colorectal cancer patients. *Clinical Infectious Diseases*, *60*(2), 208–215. DOI: 10.1093/cid/ciu787 PMID: 25305284

Bond, S., Saxena, S., & Sierra-Delgado, J. A. (2025). Microglia in ALS: Insights into Mechanisms and Therapeutic Potential. *Cells 2025, Vol. 14, Page 421, 14*(6), 421. DOI: 10.3390/cells14060421

Bond, S., Driano, J. M., & Costa, A. S. H. (2020). The integrated stress response and phosphorylated eukaryotic initiation factor 2α in neurodegeneration. *Journal of Neuropathology and Experimental Neurology*, *79*(2), 123–143. DOI: 10.1093/jnen/nlz129 PMID: 31913484

Borst, K., Dumas, A. A., & Prinz, M. (2021). Microglia: Immune and non-immune functions. *Immunity*, *54*(10), 2194–2208. DOI: 10.1016/j.immuni.2021.09.014 PMID: 34644556

Bourassa, M. W., Alim, I., Bultman, S. J., & Ratan, R. R. (2022). Butyrate, neuroepigenetics and the gut microbiome: Can a high-fiber diet improve brain health? *Neuroscience Letters*, *742*, 135566. DOI: 10.1016/j.neulet.2020.135566 PMID: 26868600

Bravo, J. A., Forsythe, P., Chew, M. V., Escaravage, E., Savignac, H. M., Dinan, T. G., Bienenstock, J., & Cryan, J. F. (2011). Ingestion of Lactobacillus strain regulates emotional behavior and central GABA receptor expression in a mouse via the vagus nerve. *Proceedings of the National Academy of Sciences of the United States of America*, *108*(38), 16050–16055. DOI: 10.1073/pnas.1102999108 PMID: 21876150

Breit, S., Kupferberg, A., Rogler, G., & Hasler, G. (2018). Vagus nerve as modulator of the brain–gut axis in psychiatric and inflammatory disorders. *Frontiers in Psychiatry*, *9*, 44. DOI: 10.3389/fpsyt.2018.00044 PMID: 29593576

Breton, J., Tennoune, N., Lucas, N., Francois, M., Legrand, R., Jacquemot, J., & Fetissov, S. O. (2016). Gut commensal E. coli proteins activate host satiety pathways following nutrient-induced bacterial growth. *Cell Metabolism*, *23*(2), 324–334. DOI: 10.1016/j.cmet.2015.10.017 PMID: 26621107

Brglez Mojzer, E., Knez Hrnčič, M., Škerget, M., Knez, Ž., & Bren, U. (2016). Polyphenols: Extraction methods, antioxidative action, bioavailability and anticarcinogenic effects. *Molecules (Basel, Switzerland)*, *21*(7), 901. DOI: 10.3390/molecules21070901 PMID: 27409600

Brickman, A. M., Khan, U. A., Provenzano, F. A., Yeung, L. K., Suzuki, W., Schroeter, H., & Small, S. A. (2014). Enhancing dentate gyrus function with dietary flavanols improves cognition in older adults. *Nature Neuroscience*, *17*(12), 1798–1803. DOI: 10.1038/nn.3850

Broide, R. S., Redwine, J. M., Aftahi, N., Young, W., Bloom, F. E., & Winrow, C. J. (2007). Distribution of histone deacetylases 1–11 in the rat brain. *Journal of Molecular Neuroscience*, *31*(1), 47–58. DOI: 10.1007/BF02686117 PMID: 17416969

Brookes, S. J., Spencer, N. J., Costa, M., & Zagorodnyuk, V. P. (2013). Extrinsic primary afferent signalling in the gut. *Nature Reviews. Gastroenterology & Hepatology*, *10*(5), 286–296. DOI: 10.1038/nrgastro.2013.29 PMID: 23438947

Brouillard, R., & Delaporte, B. (1977). Anthocyanin molecular forms in relation to color. *Phytochemistry*, *16*(1), 1–10. DOI: 10.1016/S0031-9422(00)83602-3

Bruce-Keller, A. J., Salbaum, J. M., & Berthoud, H. R. (2018). Harnessing gut microbes for mental health: Getting from here to there. [DOI] [PMC free article] [PubMed] [Google Scholar]. *Biological Psychiatry*, *83*(3), 214–223. DOI: 10.1016/j.biopsych.2017.08.014 PMID: 29031410

Buccellato, F. R., D'Anca, M., Fenoglio, C., Scarpini, E., & Galimberti, D. (2021). Role of oxidative damage in Alzheimer's disease and neurodegeneration: From pathogenic mechanisms to biomarker discovery. *Antioxidants*, *10*(9), 1353. DOI: 10.3390/antiox10091353 PMID: 34572985

Budagova, K. R., Zhmaeva, S. V., Grigor'ev, A. N., Goncharova, A. Y., & Kabakov, A. E. (2003). Flavonoid dihydroquercetin, unlike quercetin, fails to inhibit expression of heat shock proteins under conditions of cellular stress. *Biochemistry (Moscow)*, *68*(9), 1055–1061. DOI: 10.1023/a:1026081016923

Budson, A. E., & Solomon, P. R. (2021). *Memory loss, Alzheimer's disease, and dementia-e-book: a practical guide for clinicians*. Elsevier Health Sciences.

Burton-Freeman, B., Brzeziński, M., Park, E., Sandhu, A., Xiao, D., & Edirisinghe, I. (2019). A selective role of dietary anthocyanins and flavan-3-ols in reducing the risk of type 2 diabetes mellitus: A review of recent evidence. *Nutrients*, *11*(4), 841. DOI: 10.3390/nu11040841 PMID: 31013914

Butovsky, O., Jedrychowski, M. P., Moore, C. S., Cialic, R., Lanser, A. J., Gabriely, G., Koeglsperger, T., Dake, B., Wu, P. M., Doykan, C. E., Fanek, Z., Liu, L., Chen, Z., Rothstein, J. D., Ransohoff, R. M., Gygi, S. P., Antel, J. P., & Weiner, H. L. (2014). Identification of a unique TGF-β-dependent molecular and functional signature in microglia. *Nature Neuroscience*, *17*(1), 131–143. DOI: 10.1038/nn.3599 PMID: 24316888

Cabrera, C., Giménez, R., & Martínez-López, S. (2021). Nanoencapsulation of anthocyanins: Strategies to improve stability and bioavailability. *Food Chemistry*, *341*, 128203. DOI: 10.1016/j.foodchem.2020.128203

Cai, W., Wu, J., Wang, X., Huang, J., Shen, Z., & Chen, X. (2019). Epimedium flavonoids mitigate proteotoxicity and extend healthspan via DAF-16 in C. elegans. *Traditional Medicine and Modern Medicine*, *02*(01), 19–25. DOI: 10.1142/s2575900019500046

Calderaro, A., Patanè, G. T., Tellone, E., Barreca, D., Ficarra, S., Misiti, F., & Laganà, G. (2022). The neuroprotective potentiality of flavonoids on Alzheimer's disease. *International Journal of Molecular Sciences*, *23*(23), 14835. DOI: 10.3390/ijms232314835 PMID: 36499159

Calis, Z., Mogulkoc, R., & Baltaci, A. K. (2020). The roles of flavonols/flavonoids in neurodegeneration and neuroinflammation. *Mini-Reviews in Medicinal Chemistry*, *20*(15), 1475–1488. DOI: 10.2174/1389557519666190617150051 PMID: 31288717

Calou, I., Bandeira, M. A., Aguiar-Galvão, W., Cerqueira, G., Siqueira, R., Neves, K. R., Brito, G. A., & Viana, G. (2014). Neuroprotective Properties of a Standardized Extract from Myracrodruon urundeuva Fr. All. (Aroeira-Do-Sertão), as Evaluated by a Parkinson's Disease Model in Rats. *Parkinson's Disease*, *2014*(1), 519615. DOI: 10.1155/2014/519615 PMID: 25061534

Calvo-Rodriguez, M., Kharitonova, E. K., & Bacskai, B. J. (2020). Therapeutic strategies to target calcium dysregulation in Alzheimer's disease. *Cells, 9*(11), 2513. DOI: 10.3390/cells9112513 PMID: 33233678

Câmara, J. S., Perestrelo, R., Ferreira, R., Berenguer, C. V., Pereira, J. A. M., & Castilho, P. C. (2024). Plant-Derived Terpenoids: A Plethora of Bioactive Compounds with Several Health Functions and Industrial Applications—A Comprehensive Overview. *Molecules (Basel, Switzerland), 29*(16), 3861. DOI: 10.3390/molecules29163861 PMID: 39202940

Camargo, LC, Campos, GAA, Galante, P., Biolchi, AM, & Goncalves, JC, Lopes, KS ve Mortari, MR. (. (2018). Alzheimer hastalığının tedavisi için yeni tedavi yöntemlerine yönelik bir platform olarak hayvan zehirinden izole edilen peptitler. *Nöropeptitler, 67*, 79–86. DOI: 10.1016/j.npep.2017.11.010 PMID: 29198480

Camberos-Barraza, J., Camacho-Zamora, A., Bátiz-Beltrán, J. C., Osuna-Ramos, J. F., Rábago-Monzón, Á. R., Valdez-Flores, M. A., & De la Herrán-Arita, A. K. (2024). Sleep, glial function, and the endocannabinoid system: Implications for neuroinflammation and sleep disorders. *International Journal of Molecular Sciences, 25*(6), 3160.

Cameron, E. G., Götz, M., & Goldman, S. A. (2024). A molecular switch for neuroprotective astrocyte reactivity. *Nature, 626*(7999), 574–582. DOI: 10.1038/s41586-023-06935-3 PMID: 38086421

Campbell, J. C., Anderson, J. C., McFadgion, A., Gill, J., Zink, E., Patch, M., Callwood, G., & Campbell, D. (2018). The effects of intimate partner violence and probable traumatic brain injury on central nervous system symptoms. *Journal of Women's Health, 27*(6), 761–767. DOI: 10.1089/jwh.2016.6311 PMID: 29083256

Campolo, N., Bartesaghi, S., & Radi, R. (2014). Metal-catalyzed protein tyrosine nitration in biological systems. *Redox Report : Communications in Free Radical Research, 19*(6), 221–231. DOI: 10.1179/1351000214Y.0000000099 PMID: 24977336

Campos-Esparza, M. R., Sánchez-Gómez, M. V., & Matute, C. (2009). Molecular mechanisms of neuroprotection by two natural antioxidant polyphenols. *Cell Calcium, 45*(4), 358–368. DOI: 10.1016/j.ceca.2008.12.007 PMID: 19201465

Cani, P. D., Bibiloni, R., Knauf, C., Waget, A., Neyrinck, A. M., Delzenne, N. M., & Burcelin, R. (2012). Changes in gut microbiota control metabolic endotoxemia-induced inflammation in high-fat diet–induced obesity and diabetes in mice. *Diabetes, 57*(6), 1470–1481. DOI: 10.2337/db07-1403 PMID: 18305141

Cao, S. Y., Zhao, C. N., Xu, X. Y., Tang, G. Y., Corke, H., Gan, R. Y., & Li, H. B. (2019). Dietary plants, gut microbiota, and obesity: Effects and mechanisms. *Trends in Food Science & Technology, 92*, 194–204. DOI: 10.1016/j.tifs.2019.08.004

Cao, Y., Li, G., Wang, Y., Fan, Z., Yu, D., Wang, Z., & Bi, Y. (2010). Neuroprotective effect of baicalin on compression spinal cord injury in rats. *Brain Research, 1357*, 115–123. DOI: 10.1016/j.brainres.2010.07.108 PMID: 20708606

Carabotti, M., Scirocco, A., Maselli, M. A., & Severi, C. (2015). The gut–brain axis: Interactions between enteric microbiota, central and enteric nervous systems. *Annals of Gastroenterology, 28*(2), 203–209. https://www.ncbi.nlm.nih.gov/pmc/articles/PMC4367209/ PMID: 25830558

Carballo-Villalobos, A., González-Trujano, M., Alvarado-Vázquez, N., & López-Muñoz, F. (2017). Pro-inflammatory cytokines involvement in the hesperidin antihyperalgesic effects at peripheral and central levels in a neuropathic pain model. *Inflammopharmacology, 25*(2), 265–269. DOI: 10.1007/s10787-017-0326-3 PMID: 28265836

Cardenas-Aguayo Mdel, C., Kazim, S. F., Grundke-Iqbal, I., & Iqbal, K. (2013). Neurogenic and neurotrophic effects of BDNF peptides in mouse hippocampal primary neuronal cell cultures. *PLoS One, 8*(1), e53596. DOI: 10.1371/journal.pone.0053596 PMID: 23320097

Cardona, F., Andrés-Lacueva, C., Tulipani, S., Tinahones, F. J., & Queipo-Ortuño, M. I. (2013). Benefits of polyphenols on gut microbiota and implications in human health. *The Journal of Nutritional Biochemistry, 24*(8), 1415–1422. DOI: 10.1016/j.jnutbio.2013.05.001 PMID: 23849454

Carecho, R., Carregosa, D., Ratilal, B. O., Angeles, M., & Loncarevic-vasiljkovic, N. (2023). *Dietary (Poly) phenols in Traumatic Brain Injury.*

Carroll, C. S., & Moore, M. M. (2018). Ironing out siderophore biosynthesis: A review of non-ribosomal peptide synthetase (NRPS)-independent siderophore synthetases. *Critical Reviews in Biochemistry and Molecular Biology, 53*(4), 356–381. DOI: 10.1080/10409238.2018.1476449 PMID: 29863423

Carter, P., Gray, L. J., Troughton, J., Khunti, K., & Davies, M. J. (2010). Fruit and vegetable intake and incidence of type 2 diabetes mellitus: Systematic review and meta-analysis. *BMJ (Clinical Research Ed.), 341*, c4229. DOI: 10.1136/bmj.c4229

Casaus, P., Nilsen, T., Cintas, L. M., Nes, I. F., Hernández, P. E., & Holo, H. (1997). Enterocin B, a new bacteriocin from Enterococcus faecium T136 which can act synergistically with enterocin A. *Microbiology (Reading, England), 143*(Pt 7), 2287–2294. DOI: 10.1099/00221287-143-7-2287 PMID: 9245817

Casella, C., Lipp, I., Rosser, A., Jones, D. K., & Metzler-Baddeley, C. (2020). A Critical Review of White Matter Changes in Huntington's Disease. In *Movement Disorders* (Vol. 35, Issue 8, pp. 1302–1311). John Wiley and Sons Inc. DOI: 10.1002/mds.28109

Cassani, J., Dorantes-Barrón, A. M., Novales, L. M., Real, G. A., & Estrada-Reyes, R. (2014). Anti-depressant-like effect of kaempferitrin isolated from justicia spicigera schltdl (Acanthaceae) in two behavior models in mice: Evidence for the involvement of the serotonergic system. *Molecules (Basel, Switzerland), 19*(12), 21442–21461. DOI: 10.3390/molecules191221442 PMID: 25532842

Cassidy, A., & Minihane, A. M. (2017). The role of metabolism (and the microbiome) in defining the clinical efficacy of dietary flavonoids. *The American Journal of Clinical Nutrition, 105*(1), 10–22. DOI: 10.3945/ajcn.116.136051 PMID: 27881391

Cassotta, M., Armas Diaz, Y., Qi, Z., Yang, B., Grosso, G., Quiles, J. L., Battino, M., Godos, J., & Giampieri, F. (2025). Modernizing gut-brain axis research in nutritional science. *Trends in Food Science & Technology*. Advance online publication.

Castaño, E. M., Maarouf, C. L., Wu, T., Leal, M. C., Whiteside, C. M., Lue, L. F., Kokjohn, T. A., Sabbagh, M. N., Beach, T. G., & Roher, A. E. (2013). Alzheimer disease periventricular white matter lesions exhibit specific proteomic profile alterations. *Neurochemistry International, 62*(2), 145–156. DOI: 10.1016/j.neuint.2012.12.001 PMID: 23231993

Castro, D. M., Dillon, C., Machnicki, G., & Allegri, R. F. (2010). The economic cost of Alzheimer's disease: Family or public-health burden? *Dementia & Neuropsychologia*, *4*(4), 262–267. DOI: 10.1590/S1980-57642010DN40400003 PMID: 29213697

Cattaneo, A., Cattane, N., Galluzzi, S., Provasi, S., Lopizzo, N., Festari, C., Ferrari, C., Guerra, U. P., Paghera, B., Muscio, C., Bianchetti, A., Volta, G. D., Turla, M., Cotelli, M. S., Gennuso, M., Prelle, A., Zanetti, O., Lussignoli, G., Mirabile, D., & Frisoni, G. B. (2017). Association of brain amyloidosis with pro-inflammatory gut bacterial taxa and peripheral inflammation markers in cognitively impaired elderly. *Neurobiology of Aging*, *49*, 60–68. DOI: 10.1016/j.neurobiolaging.2016.08.019 PMID: 27776263

Cawthon, C. R., & de La Serre, C. B. (2018). Gut bacteria interaction with vagal afferents. *Brain Research, 1693*(Pt B), 134–139. https://doi.org/DOI: 10.1016/j.brainres.2018.03.021

Çevik, Ö., Erşahin, M., Şener, T. E., Tinay, İ., Tarcan, T., Çetinel, Ş., Şener, A., Toklu, H. Z., & Şener, G. (2013). Beneficial effects of quercetin on rat urinary bladder after spinal cord injury. *The Journal of Surgical Research*, *183*(2), 695–703. DOI: 10.1016/j.jss.2013.02.016 PMID: 23490140

Chagas, M. S. S., Behrens, M. D., Moragas-Tellis, C. J., Penedo, G. X., Silva, A. R., & Gonçalves-de-Albuquerque, C. F. (2022). Flavonols and flavones as potential anti-inflammatory, antioxidant, and antibacterial compounds. *Oxidative Medicine and Cellular Longevity*, *2022*(1), 9966750. DOI: 10.1155/2022/9966750 PMID: 36111166

Chahar, M. K., Sharma, N., Dobhal, M. P., & Joshi, Y. C. (2011). Flavonoids: A versatile source of anticancer drugs. *Pharmacognosy Reviews*, *5*(9), 1–12. 10.4103/0973-7847.79093

Cháirez-Ramírez, M. H., de la Cruz-López, K. G., & García-Carrancá, A. (2021). Polyphenols as antitumor agents targeting key players in cancer-driving signaling pathways. *Frontiers in Pharmacology*, *12*, 710304. DOI: 10.3389/fphar.2021.710304 PMID: 34744708

Chakrabarti, A., Geurts, L., Hoyles, L., Iozzo, P., Kraneveld, A. D., La Fata, G., & Shortt, C. (2022). The microbiota–gut–brain axis: Pathways to better brain health. Perspectives on what we know, what we need to investigate and how to put knowledge into practice. *Cellular and Molecular Life Sciences : CMLS*, *79*(2), 80. DOI: 10.1007/s00018-021-04060-w PMID: 35044528

Chakrabarti, S., Guha, S., & Majumder, K. (2018). Food-Derived Bioactive Peptides in Human Health: Challenges and Opportunities. *Nutrients*, *10*(11), 1738. DOI: 10.3390/nu10111738 PMID: 30424533

Chamoun, R., Suki, D., Gopinath, S. P., Goodman, J. C., & Robertson, C. (2010). Role of extracellular glutamate measured by cerebral microdialysis in severe traumatic brain injury. *Journal of Neurosurgery*, *113*(3), 564–570. DOI: 10.3171/2009.12.JNS09689 PMID: 20113156

Chandimali, N., Bak, S. G., Park, E. H., Lim, H. J., Won, Y. S., Kim, E. K., Park, S. I., & Lee, S. J. (2025). Free radicals and their impact on health and antioxidant defenses: a review. In *Cell Death Discovery* (Vol. 11, Issue 1). Springer Nature. DOI: 10.1038/s41420-024-02278-8

Chandrasekhar, Y., Phani Kumar, G., Navya, K., Ramya, E. M., & Anilakumar, K. R. (2018). Tannins from Terminalia chebula fruits attenuates GABA antagonist-induced anxiety-like behaviour via modulation of neurotransmitters. *The Journal of Pharmacy and Pharmacology*, *70*(12), 1662–1674. DOI: 10.1111/jphp.13007 PMID: 30198561

Chang, J., Qian, Z., Wang, B., Cao, J., Zhang, S., Jiang, F., Kong, R., Yu, X., Cao, X., Yang, L., & Chen, H. (2023). Transplantation of A2 type astrocytes promotes neural repair and remyelination after spinal cord injury. *Cell Communication and Signaling : CCS, 21*(1), 1–20. DOI: 10.1186/s12964-022-01036-6 PMID: 36797790

Chang, P. V., Hao, L., Offermanns, S., & Medzhitov, R. (2014). The microbial metabolite butyrate regulates intestinal macrophage function via histone deacetylase inhibition. *Proceedings of the National Academy of Sciences of the United States of America, 111*(6), 2247–2252. DOI: 10.1073/pnas.1322269111 PMID: 24390544

Charoenkwan, P., Chumnanpuen, P., Schaduangrat, N., Lio', P., Moni, M. A., & Shoombuatong, W. (2022). Improved prediction and characterization of blood-brain barrier penetrating peptides using estimated propensity scores of dipeptides. *Journal of Computer-Aided Molecular Design, 36*(11), 781–799. DOI: 10.1007/s10822-022-00476-z PMID: 36284036

Chatterjee, A., Kumar, S., Roy Sarkar, S., Halder, R., Kumari, R., Banerjee, S., & Sarkar, B. (2024). Dietary polyphenols represent a phytotherapeutic alternative for gut dysbiosis associated neurodegeneration: A systematic review. *The Journal of Nutritional Biochemistry, 129*, 109622. DOI: 10.1016/j.jnutbio.2024.109622 PMID: 38490348

Che Ramli, M. D., Sumari, N. A., Selamat, N. W., Muhammad, H., & Sanusi, J. (2020). Effectiveness of Flaxseed Oil on Peripheral Nerve Regeneration Following Crush Injury of Sciatic Nerve in Rat: Behavioural Analysis and an Electron Microscopic Study. *Malaysian Journal of Medicine & Health Sciences, 16*.

Che, F., Wang, G., Yu, J., Wang, X., Lu, Y., Fu, Q., Su, Q., Jiang, J., & Du, Y. (2017). Effects of epigallocatechin-3-gallate on iron metabolismin spinal cord motor neurons. *Molecular Medicine Reports, 16*(3), 3010–3014. DOI: 10.3892/mmr.2017.6919 PMID: 28677731

Chen, B., Zhao, J., Zhang, R., Zhang, L., Zhang, Q., Yang, H., & An, J. (2022). Neuroprotective effects of natural compounds on neurotoxin-induced oxidative stress and cell apoptosis. *Nutritional Neuroscience, 25*(5), 1078–1099. DOI: 10.1080/1028415X.2020.1840035 PMID: 33164705

Chen, C. G., Hung, T. H., Lee, C. Y., Wang, L. F., Wu, C. H., Ke, C. H., & Chen, S. F. (2014). Berberine protects against neuronal damage via suppression of glia-mediated inflammation in traumatic brain injury. *PLoS One, 9*(12), 1–30. DOI: 10.1371/journal.pone.0115694 PMID: 25546475

Cheng, T., Wang, W., Li, Q., Han, X., Xing, J., Qi, C., Lan, X., Wan, J., Potts, A., Guan, F., & Wang, J. (2016). Cerebroprotection of flavanol (-)-epicatechin after traumatic brain injury via Nrf2-dependent and-independent pathways. *Free Radical Biology & Medicine, 92*, 15–28. DOI: 10.1016/j.freeradbiomed.2015.12.027 PMID: 26724590

Chen, J., Lin, D., Zhang, C., Li, G., Zhang, N., Ruan, L., Yan, Q., Li, J., Yu, X., Xie, X., Pang, C., Cao, L., Pan, J., & Xu, Y. (2015). Antidepressant-like effects of ferulic acid: Involvement of serotonergic and norepinergic systems. *Metabolic Brain Disease, 30*(1), 129–136. DOI: 10.1007/s11011-014-9635-z PMID: 25483788

Chen, J., Yang, J., Ma, L., Li, J., Shahzad, N., & Kim, C. K. (2020). Structure-antioxidant activity relationship of methoxy, phenolic hydroxyl, and carboxylic acid groups of phenolic acids. *Scientific Reports*, *10*(1), 2611. DOI: 10.1038/s41598-020-59451-z PMID: 32054964

Chen, L., Cao, H., Huang, Q., Xiao, J., & Teng, H. (2022). Absorption, metabolism and bioavailability of flavonoids: A review. *Critical Reviews in Food Science and Nutrition*, *62*(28), 7730–7742. DOI: 10.1080/10408398.2021.1917508 PMID: 34078189

Chen, L., Liu, Y., & Xie, J. (2020). The beneficial pharmacological effects of *Uncaria rhynchophylla* in neurodegenerative diseases: Focus on alkaloids. *Frontiers in Pharmacology*, *15*, 1436481. Advance online publication. 10.3389/fphar.2024.1436481

Chen, M., Gu, H., Ye, Y., Lin, B., Sun, L., Deng, W., Zhang, J., & Liu, J. (2010). Protective effects of hesperidin against oxidative stress of tert-butyl hydroperoxide in human hepatocytes. *Food and Chemical Toxicology*, *48*(10), 2980–2987. DOI: 10.1016/j.fct.2010.07.037 PMID: 20678535

Chen, M.-M., Qin, J., Chen, S.-J., Yao, L.-M., Zhang, L.-Y., Yin, Z.-Q., & Liao, H. (2017). Quercetin promotes motor and sensory function recovery following sciatic nerve-crush injury in C57BL/6J mice. *The Journal of Nutritional Biochemistry*, *46*, 57–67. DOI: 10.1016/j.jnutbio.2017.04.006 PMID: 28458138

Chen, S. F., Hsu, C. W., Huang, W. H., & Wang, J. Y. (2008). Post-injury baicalein improves histological and functional outcomes and reduces inflammatory cytokines after experimental traumatic brain injury. *British Journal of Pharmacology*, *155*(8), 1279–1296. DOI: 10.1038/bjp.2008.345 PMID: 18776918

Chen, S. Y., Zheng, K., & Wang, Z. Q. (2016). Neuroprotective effects of ellagic acid on neonatal hypoxic brain injury via inhibition of inflammatory mediators and down-regulation of JNK/p38 MAPK activation. *Tropical Journal of Pharmaceutical Research*, *15*(2), 241–251. DOI: 10.4314/tjpr.v15i2.4

Chen, S., Wang, X., Cheng, Y., Gao, H., & Chen, X. (2023). A review of classification, biosynthesis, biological activities and potential applications of flavonoids. *Molecules (Basel, Switzerland)*, *28*(13), 4982. DOI: 10.3390/molecules28134982 PMID: 37446644

Chen, S.-Y., Gao, Y., Sun, J.-Y., Meng, X.-L., Yang, D., Fan, L.-H., Xiang, L., & Wang, P. (2020). Traditional Chinese medicine: Role in reducing β-amyloid, apoptosis, autophagy, neuroinflammation, oxidative stress, and mitochondrial dysfunction of Alzheimer's disease. *Frontiers in Pharmacology*, *11*, 497. DOI: 10.3389/fphar.2020.00497 PMID: 32390843

Chen, T. J., Feng, Y., Liu, T., Wu, T. T., Chen, Y. J., Li, X., & Wu, Y. C. (2020). Fisetin Regulates Gut Microbiota and Exerts Neuroprotective Effect on Mouse Model of Parkinson's Disease. *Frontiers in Neuroscience*, *14*, 549037. DOI: 10.3389/fnins.2020.549037

Chen, W. Q., Zhao, X. L., Hou, Y., Li, S. T., Hong, Y., Wang, D. L., & Cheng, Y. Y. (2009). Protective effects of green tea polyphenols on cognitive impairments induced by psychological stress in rats. *Behavioural Brain Research*, *202*(1), 71–76. DOI: 10.1016/j.bbr.2009.03.017 PMID: 19447283

Chen, Y., Liu, Z., & Gong, Y. (2024). Neuron-immunity communication: Mechanism of neuroprotective effects in EGCG. *Critical Reviews in Food Science and Nutrition*, *64*(25), 9333–9352. DOI: 10.1080/10408398.2023.2212069 PMID: 37216484

Chen, Y., Qi, Z., Qiao, B., Lv, Z., Hao, Y., & Li, H. (2019). Oxymatrine can attenuate pathological deficits of Alzheimer's disease mice through regulation of neuroinflammation. *Journal of Neuroimmunology*, *334*, 576978. DOI: 10.1016/j.jneuroim.2019.576978 PMID: 31177033

Chen, Y., Xu, J., & Chen, Y. (2021). Regulation of neurotransmitters by the gut microbiota and effects on cognition in neurological disorders. *Nutrients*, *13*(6), 2099. DOI: 10.3390/nu13062099 PMID: 34205336

Chen, Y., Zhou, J., & Wang, L. (2021). Role and mechanism of gut microbiota in human disease. * linical and Infection Microbiology, 11*. *Frontiers in Cellular and Infection Microbiology*, *625913*, 625913. Advance online publication. DOI: 10.3389/fcimb.2021.625913 PMID: 33816335

Chen, Z.-R., Huang, J.-B., Yang, S.-L., & Hong, F.-F. (2022). Role of cholinergic signaling in Alzheimer's disease. *Molecules (Basel, Switzerland)*, *27*(6), 1816. DOI: 10.3390/molecules27061816 PMID: 35335180

Cherry, J. D., Olschowka, J. A., & O'Banion, M. K. (2014). Neuroinflammation and M2 microglia: The good, the bad, and the inflamed. *Journal of Neuroinflammation*, *11*(1), 98. DOI: 10.1186/1742-2094-11-98 PMID: 24889886

Chesworth, R., Gamage, R., Ullah, F., Sonego, S., Millington, C., Fernandez, A., Liang, H., Karl, T., Münch, G., Niedermayer, G., & Gyengesi, E. (2021). Spatial Memory and Microglia Activation in a Mouse Model of Chronic Neuroinflammation and the Anti-inflammatory Effects of Apigenin. *Frontiers in Neuroscience*, *15*(July), 1–14. DOI: 10.3389/fnins.2021.699329 PMID: 34393713

Chiang, C. K., Hsu, S. P., Wu, C. T., Huang, J. W., Cheng, H. T., Chang, Y. W., & Liu, S. H. (2011). Endoplasmic reticulum stress implicated in the development of renal fibrosis. *Molecular Medicine (Cambridge, Mass.)*, *17*(11-12), 1295–1305. DOI: 10.2119/molmed.2011.00131

Chico, L., Ienco, E. C., Bisordi, C., Lo Gerfo, A., Petrozzi, L., Petrucci, A., Mancuso, M., & Siciliano, G. (2018). Amyotrophic lateral sclerosis and oxidative stress: A double-blind therapeutic trial after curcumin supplementation. *CNS & Neurological Disorders - Drug Targets*, *17*(10), 767–779. DOI: 10 .2174/1871527317666180720162029 PMID: 30033879

Chi, H., Chang, H.-Y., & Sang, T.-K. (2018). Neuronal cell death mechanisms in major neurodegenerative diseases. *International Journal of Molecular Sciences*, *19*(10), 3082. DOI: 10.3390/ijms19103082 PMID: 30304824

Chimento, A., De Luca, A., D'Amico, M., De Amicis, F., & Pezzi, V. (2023). The involvement of natural polyphenols in molecular mechanisms inducing apoptosis in tumor cells: A promising adjuvant in cancer therapy. *International Journal of Molecular Sciences*, *24*(2), 1680. DOI: 10.3390/ijms24021680 PMID: 36675194

Chiou, B., & Piao, X. (2021). Linking adhesion GPCRs to glial cell development and function. *Factors Affecting Neurodevelopment: Genetics, Neurology, Behavior, and Diet*, 93–103. DOI: 10.1016/B978-0-12-817986-4.00009-2

Chithra, Y., Dey, G., Ghose, V., Chandramohan, V., Gowthami, N., Vasudev, V., & Srinivas Bharath, M. M. (2023). Mitochondrial Complex I Inhibition in Dopaminergic Neurons Causes Altered Protein Profile and Protein Oxidation: Implications for Parkinson's disease. *Neurochemical Research*, *48*(8), 2360–2389. DOI: 10.1007/s11064-023-03907-x PMID: 36964824

Chitturi, J., Santhakumar, V., & Kannurpatti, S. S. (2019). Beneficial effects of kaempferol after developmental traumatic brain injury is through protection of mitochondrial function, oxidative metabolism, and neural viability. *Journal of Neurotrauma*, *36*(8), 1264–1278. DOI: 10.1089/neu.2018.6100 PMID: 30430900

Cho, C. H., Jang, H., Lee, M., Kang, H., Heo, H. J., & Kim, D.-O. (2017). Sea buckthorn (*Hippophae rhamnoides* L.) leaf extracts protect neuronal PC-12 cells from oxidative stress. *Journal of Microbiology and Biotechnology*, *27*(7), 1257–1264. 10.4014/jmb.1704.04033

Choi, G. Y., Kim, H. B., Hwang, E. S., Park, H. S., Cho, J. M., Ham, Y. K., Kim, J. H., Mun, M. K., Maeng, S., & Park, J. H. (2023). Naringin enhances long-term potentiation and recovers learning and memory deficits of amyloid-beta induced Alzheimer's disease-like behavioral rat model. *Neurotoxicology*, *95*, 35–45. DOI: 10.1016/j.neuro.2022.12.007 PMID: 36549596

Choi, H., & Oh, D. C. (2015). Considerations of the chemical biology of microbial natural products provide an effective drug discovery strategy. *Archives of Pharmacal Research*, *38*(9), 1591–1605. DOI: 10.1007/s12272-015-0639-y PMID: 26231248

Choi, S.-M., Kim, B. C., Cho, Y.-H., Choi, K.-H., Chang, J., Park, M.-S., Kim, M.-K., Cho, K.-H., & Kim, J.-K. (2014). Effects of Flavonoid Compounds on β-amyloid-peptide-induced Neuronal Death in Cultured Mouse Cortical Neurons. *Chonnam Medical Journal*, *50*(2), 45. DOI: 10.4068/cmj.2014.50.2.45 PMID: 25229015

Chong, J., Poutaraud, A., & Hugueney, P. (2009). Metabolism and roles of stilbenes in plants. *Plant Science*, *177*(3), 143–155. DOI: 10.1016/j.plantsci.2009.05.012

Chopade, P., Chopade, N., Zhao, Z., Mitragotri, S., Liao, R., & Chandran Suja, V. (2023). Alzheimer's and Parkinson's disease therapies in the clinic. *Bioengineering & Translational Medicine*, *8*(1), e10367. DOI: 10.1002/btm2.10367 PMID: 36684083

Chowdhury, T. (2024). Psychobiotics: Microbial interventions for mental health. *Advances in Microbial Physiology*, *84*, 112–145.

Christen, Y. (2000). Oxidative stress and Alzheimer disease. *The American Journal of Clinical Nutrition*, *71*(2), 621S–629S. DOI: 10.1093/ajcn/71.2.621s PMID: 10681270

Chu, A. J. (2022). Quarter-Century Explorations of Bioactive Polyphenols: Diverse Health Benefits. In *Frontiers in Bioscience - Landmark* (Vol. 27, Issue 4). Bioscience Research Institute. DOI: 10.31083/j.fbl2704134

Chung, I. R. S. (2010). Dietary polyphenols, deacetylases and chromatin remodeling in inflammation. *World Review of Nutrition and Dietetics*, *101*, 84–94. DOI: 10.1159/000314513 PMID: 20436255

Cichon, N., Saluk-Bijak, J., Gorniak, L., Przyslo, L., & Bijak, M. (2020). Flavonoids as a Natural Enhancer of Neuroplasticity—An Overview of the Mechanism of Neurorestorative Action. *Antioxidants*, *9*(11), 1035. DOI: 10.3390/antiox9111035 PMID: 33114058

Cintas, L. M., Casaus, P., Herranz, C., Håvarstein, L. S., Holo, H., & Nes, I. F. (2000). Biochemical and genetic evidence that Enterococcus faecium L50 produces enterocins L50A and L50B, the Sec-dependent enterocin P, and a novel bacteriocin secreted without an N-terminal extension termed enterocin Q. *Journal of Bacteriology*, *182*(23), 6806–6814. DOI: 10.1128/JB.182.23.6806-6814.2000 PMID: 11073927

Ciobanasu, C. (2021). Peptides-based therapy and diagnosis. Strategies for non-invasive therapies in cancer. *Journal of Drug Targeting*, *29*(10), 1063–1079. DOI: 10.1080/1061186X.2021.1906885 PMID: 33775187

Cirmi, S., Ferlazzo, N., Lombardo, G. E., Ventura-Spagnolo, E., Gangemi, S., Calapai, G., & Navarra, M. (2016). Neurodegenerative Diseases: Might Citrus Flavonoids Play a Protective Role? *Molecules (Basel, Switzerland)*, *21*(10). Advance online publication. DOI: 10.3390/molecules21101312

Ciupei, D., Colişar, A., Leopold, L., Stănilă, A., & Diaconeasa, Z. M. (2024). Polyphenols: From classification to therapeutic potential and bioavailability. *Foods*, *13*(24), 4131. DOI: 10.3390/foods13244131 PMID: 39767073

Číž, M., Dvořáková, A., Skočková, V., & Kubala, L. (2020). The role of dietary phenolic compounds in epigenetic modulation involved in inflammatory processes. *Antioxidants*, *9*(8), 691. DOI: 10.3390/antiox9080691 PMID: 32756302

Clarke, B. E., & Patani, R. (2020). The microglial component of amyotrophic lateral sclerosis. *Brain*, *143*(12), 3526–3539. DOI: 10.1093/brain/awaa309 PMID: 33427296

Clarke, G., Stilling, R. M., Kennedy, P. J., Stanton, C., Cryan, J. F., & Dinan, T. G. (2014). Minireview: Gut microbiota: the neglected endocrine organ. *Molecular Endocrinology (Baltimore, Md.)*, *28*(8), 1221–1238. DOI: 10.1210/me.2014-1108 PMID: 24892638

Clarke, L. E., Liddelow, S. A., Chakraborty, C., Münch, A. E., Heiman, M., & Barres, B. A. (2018). Normal aging induces A1-like astrocyte reactivity. *Proceedings of the National Academy of Sciences of the United States of America*, *115*(8), E1896–E1905. DOI: 10.1073/pnas.1800165115 PMID: 29437957

Clark, M. M., & Reguera, G. (2020). Biology and biotechnology of microbial pilus nanowires. *Journal of Industrial Microbiology & Biotechnology*, *47*(9–10), 897–907. DOI: 10.1007/s10295-020-02312-5 PMID: 33009965

Clemente, J. C., Ursell, L. K., Parfrey, L. W., & Knight, R. (2012). The impact of the gut microbiota on human health: An integrative view. *Cell*, *148*(6), 1258–1270. DOI: 10.1016/j.cell.2012.01.035 PMID: 22424233

Cobley, J. N., Fiorello, M. L., & Bailey, D. M. (2018). 13 reasons why the brain is susceptible to oxidative stress. *Redox Biology*, *15*, 490–503. DOI: 10.1016/j.redox.2018.01.008 PMID: 29413961

Coelho, V. R., Vieira, C. G., de Souza, L. P., da Silva, L. L., Pflüger, P., Regner, G. G., Papke, D. K. M., Picada, J. N., & Pereira, P. (2016). Behavioral and genotoxic evaluation of rosmarinic and caffeic acid in acute seizure models induced by pentylenetetrazole and pilocarpine in mice. *Naunyn-Schmiedeberg's Archives of Pharmacology*, *389*(11), 1195–1203. DOI: 10.1007/s00210-016-1281-z PMID: 27476160

Coelho, V. R., Vieira, C. G., De Souza, L. P., Moysés, F., Basso, C., Papke, D. K. M., Pires, T. R., Siqueira, I. R., Picada, J. N., & Pereira, P. (2015). Antiepileptogenic, antioxidant and genotoxic evaluation of rosmarinic acid and its metabolite caffeic acid in mice. *Life Sciences*, *122*, 65–71. DOI: 10.1016/j.lfs.2014.11.009 PMID: 25498895

Colton, C. A., & Gilbert, D. L. (1987). Production of superoxide anions by a CNS macrophage, the microglia. *FEBS Letters*, *223*(2), 284–288. DOI: 10.1016/0014-5793(87)80305-8 PMID: 2822487

Cong, P., Wang, T., Tong, C., Liu, Y., Shi, L., Mao, S., Shi, X., Jin, H., Liu, Y., & Hou, M. (2021). Resveratrol ameliorates thoracic blast exposure-induced inflammation, endoplasmic reticulum stress and apoptosis in the brain through the Nrf2/Keap1 and NF-κB signaling pathway. *Injury*, *52*(10), 2795–2802. DOI: 10.1016/j.injury.2021.08.019 PMID: 34454721

Cook, C., Stetler, C., & Petrucelli, L. (2012). Disruption of protein quality control in Parkinson's disease. *Cold Spring Harbor Perspectives in Medicine*, *2*(5), a009423. DOI: 10.1101/cshperspect.a009423

Corbetta, M., Ramsey, L., Callejas, A., Baldassarre, A., Hacker, C. D., Siegel, J. S., Astafiev, S. V., Rengachary, J., Zinn, K., Lang, C. E., Connor, L. T., Fucetola, R., Strube, M., Carter, A. R., & Shulman, G. L. (2015). Common behavioral clusters and subcortical anatomy in stroke. *Neuron*, *85*(5), 927–941. DOI: 10.1016/j.neuron.2015.02.027 PMID: 25741721

Coretti, L., Paparo, L., Riccio, M. P., Amato, F., Cuomo, M., Natale, A., Borrelli, L., Corrado, G., De Caro, C., Comegna, M., Buommino, E., Castaldo, G., Bravaccio, C., Chiariotti, L., Berni Canani, R., & Lembo, F. (2018). Gut microbiota features in young children with autism spectrum disorders. *Frontiers in Microbiology*, *9*, 3146. DOI: 10.3389/fmicb.2018.03146 PMID: 30619212

Corr, S. C., Li, Y., Riedel, C. U., O'Toole, P. W., Hill, C., & Gahan, C. G. M. (2007). Bacteriocin production as a mechanism for the antiinfective activity of Lactobacillus salivarius UCC118. *Proceedings of the National Academy of Sciences of the United States of America*, *104*(18), 7617–7621. DOI: 10.1073/pnas.0700440104 PMID: 17456596

Cosme, F., Aires, A., Pinto, T., Oliveira, I., Vilela, A., & Gonçalves, B. (2025). A Comprehensive Review of Bioactive Tannins in Foods and Beverages: Functional Properties, Health Benefits, and Sensory Qualities. In *Molecules* (Vol. 30, Issue 4, p. 800). Multidisciplinary Digital Publishing Institute. DOI: 10.3390/molecules30040800

Cragg, G. M., Grothaus, P. G., & Newman, D. J. (2009). Impact of natural products on developing new anti-cancer agents. *Chemical Reviews*, *109*(7), 3012–3043. DOI: 10.1021/cr900019j PMID: 19422222

Crawford, A. H., Tripathi, R. B., Foerster, S., McKenzie, I., Kougioumtzidou, E., Grist, M., Richardson, W. D., & Franklin, R. J. M. (2016). Pre-Existing Mature Oligodendrocytes Do Not Contribute to Remyelination following Toxin-Induced Spinal Cord Demyelination. *American Journal of Pathology*, *186*(3), 511–516. DOI: 10.1016/j.ajpath.2015.11.005 PMID: 26773350

Croall, I. D., Cowie, C. J. A., He, J., Peel, A., Wood, J., Aribisala, B. S., Mitchell, P., Mendelow, A. D., Smith, F. E., Millar, D., Kelly, T., & Blamire, A. M. (2014). White matter correlates of cognitive dysfunction after mild traumatic brain injury. *Neurology*, *83*(6), 494–501. DOI: 10.1212/WNL.0000000000000666 PMID: 25031282

Croft, K. D. (2016). Dietary polyphenols: Antioxidants or not? *Archives of Biochemistry and Biophysics*, *595*, 120–124. DOI: 10.1016/j.abb.2015.11.014 PMID: 27095227

Crozier, A., Clifford, M. N., & Ashihara, H. (2006). *Plant secondary metabolites. Occurrence, Structure and Role in the Human Diet*. Blackwell-Publishers. DOI: 10.1002/9780470988558

Cryan, J. F., & Dinan, T. G. (2012). Mind-altering microorganisms: The impact of the gut microbiota on brain and behaviour. *Nature Reviews. Neuroscience*, *13*(10), 701–712. DOI: 10.1038/nrn3346 PMID: 22968153

Cryan, J. F., O'Riordan, K. J., Cowan, C. S. M., Sandhu, K. V., Bastiaanssen, T. F. S., Boehme, M., Codagnone, M. G., Cussotto, S., Fulling, C., Golubeva, A. V., Guzzetta, K. E., Jaggar, M., Long-Smith, C. M., Lyte, J. M., Martin, J. A., Molinero-Perez, A., Moloney, G., Morelli, E., Morillas, E., & Dinan, T. G. (2019). The Microbiota-Gut-Brain Axis. *Physiological Reviews*, *99*(4), 1877–2013. DOI: 10.1152/physrev.00018.2018 PMID: 31460832

Cui, D., Chen, Y., Ye, B., Guo, W., & Wang, D. (2022). Natural products for the treatment of neurodegenerative diseases. *Phytomedicine*, *109*, 155101. Advance online publication. 10.1016/j.phymed.2023.155101

D'Alessandro-Gabazza, C. N., Kobayashi, T., Yasuma, T., Toda, M., Kim, H., Fujimoto, H., Hataji, O., Takeshita, A., Nishihama, K., Okano, T., Okano, Y., Nishii, Y., Tomaru, A., Fujiwara, K., D'Alessandro, V. F., Abdel-Hamid, A. M., Ren, Y., Pereira, G. V., Wright, C. L., & Gabazza, E. C. (2020). A Staphylococcus pro-apoptotic peptide induces acute exacerbation of pulmonary fibrosis. *Nature Communications*, *11*(1), 1539. DOI: 10.1038/s41467-020-15344-3 PMID: 32210242

D'Archivio, M., Filesi, C., Varì, R., Scazzocchio, B., & Masella, R. (2010). Bioavailability of the polyphenols: Status and controversies. *International Journal of Molecular Sciences*, *11*(4), 1321–1342. DOI: 10.3390/ijms11041321 PMID: 20480022

da Silva Carneiro, R., de Almeida Da Costa, M. H., Santos, J. Z. L. V., Uchôa, V. T., da Silva Lopes, L., & Mendes, A. N. (2025). Therapeutic actions of terpenes in neurodegenerative disorders and their correlations with the regulation and remodeling of the extracellular matrix. *Biocell*, *49*(1), 109–125. DOI: 10.32604/biocell.2024.058405

da Silva Pinto Neto, J., Serra, G. M., Xavier, L. P., & Santos, A. V. (2025). Chemodiversity and Biotechnological Potential of Microginins. *International Journal of Molecular Sciences*, *26*(13), 6117. DOI: 10.3390/ijms26136117 PMID: 40649895

Dąbek, J., Kułach, A., & Gąsior, Z. (2010). Nuclear factor kappa-light-chain-enhancer of activated B cells (NF-κB): A new potential therapeutic target in atherosclerosis? *Pharmacological Reports*, *62*(5), 778–783. DOI: 10.1016/S1734-1140(10)70338-8 PMID: 21098861

Dacosta, C., & Bao, Y. (2017). The role of MicroRNAs in the chemopreventive activity of sulforaphane from cruciferous vegetables. *Nutrients*, *9*(8), 902. DOI: 10.3390/nu9080902 PMID: 28825609

Dahan, A., Miller, J. M., & Amidon, G. L. (2009). Prediction of solubility and permeability class membership: Provisional BCS classification of the world's top oral drugs. *The AAPS Journal*, *11*(4), 740–746. DOI: 10.1208/s12248-009-9144-x PMID: 19876745

Dahiya, D., & Nigam, P. S. (2022). Probiotic biofilm formation in the gut: Mechanisms and implications. *Applied Microbiology and Biotechnology, 106*(5), 1985–2002. DOI: 10.1007/s00253-022-11821-5

Dai, W., Wang, H., Fang, J., Zhu, Y., Zhou, J., Wang, X., Zhou, Y., & Zhou, M. (2018). Curcumin provides neuroprotection in model of traumatic brain injury via the Nrf2-ARE signaling pathway. *Brain Research Bulletin, 140*, 65–71. DOI: 10.1016/j.brainresbull.2018.03.020 PMID: 29626606

Dal Bianco, A., Bradl, M., Frischer, J., Kutzelnigg, A., Jellinger, K., & Lassmann, H. (2008). Multiple sclerosis and Alzheimer's disease. *Annals of Neurology, 63*(2), 174–183. DOI: 10.1002/ana.21240 PMID: 17924575

Dalile, B., Van Oudenhove, L., Vervliet, B., & Verbeke, K. (2019). The role of short-chain fatty acids in microbiota–gut–brain communication. *Nature Reviews. Gastroenterology & Hepatology, 16*(8), 461–478. DOI: 10.1038/s41575-019-0157-3 PMID: 31123355

Daly, W., Yao, L., Zeugolis, D., Windebank, A., & Pandit, A. (2012). A biomaterials approach to peripheral nerve regeneration: Bridging the peripheral nerve gap and enhancing functional recovery. *Journal of the Royal Society, Interface, 9*(67), 202–221. DOI: 10.1098/rsif.2011.0438 PMID: 22090283

Dangles, O., & Fenger, J.-A. (2018). The chemical reactivity of anthocyanins and its consequences in food science and nutrition. *Molecules (Basel, Switzerland), 23*(8), 1970. DOI: 10.3390/molecules23081970 PMID: 30087225

Dao, M. C., Everard, A., Aron-Wisnewsky, J., Sokolovska, N., Prifti, E., Verger, E. O., Kayser, B. D., Levenez, F., Chilloux, J., Hoyles, L., Dumas, M.-E., Rizkalla, S. W., Doré, J., Cani, P. D., & Clément, K. (2024). Akkermansia muciniphila and improved metabolic health during a dietary intervention in obesity: Relationship with gut microbiome richness and ecology. *Gut Microbes, 16*(1), 1–17. DOI: 10.1080/19490976.2023.2297866

Dar, N. J., Bhat, J. A., John, U., & Bhat, S. A. (2024). Neuroglia in Neurodegeneration: Exploring Glial Dynamics in Brain Disorders. *Neuroglia 2024, Vol. 5, Pages 488-504, 5*(4), 488–504. DOI: 10.3390/neuroglia5040031

Darbandi, N., Ramezani, M., Khodagholi, F., & Noori, M. (2016). Kaempferol promotes memory retention and density of hippocampal CA1 neurons in intra-cerebroventricular STZ-induced experimental AD model in Wistar rats. *Biologija (Vilnius, Lithuania), 62*(3), 157–168. DOI: 10.6001/biologija.v62i3.3368

Dar, N. J., & Ahmad, M. (2020). Neurodegenerative diseases and *Withania somnifera* (L.): An update. *Journal of Ethnopharmacology, 256*, 112769. Advance online publication. 10.1016/j.jep.2020.112769

Dasgupta, B., & Milbrandt, J. (2007). Resveratrol stimulates AMP kinase activity in neurons. *Proceedings of the National Academy of Sciences of the United States of America, 104*(17), 7217–7222. DOI: 10.1073/pnas.0610068104 PMID: 17438283

Das, M., & Kasi, P. D. (2019). A Mini Review on the Protective Effect of Lignans for the Treatment of Neurodegenerative Disorders. *Journal of Nutrition. Food and Lipid Science, 2019*(01), 40–53. DOI: 10.33513/NFLS/1901-06

Daverey, A., & Agrawal, S. K. (2020). Curcumin protects against white matter injury through NF-κB and Nrf2 cross talk. *Journal of Neurotrauma*, *37*(10), 1255–1265. DOI: 10.1089/neu.2019.6749 PMID: 31914858

Davey, M. E., & O'Toole, G. A. (2000). Microbial biofilms: From ecology to molecular genetics. *Microbiology and Molecular Biology Reviews : MMBR*, *64*(4), 847–867. DOI: 10.1128/MMBR.64.4.847-867.2000 PMID: 11104821

Davis, D. S., & Carson, M. J. (2012). An introduction to CNS-resident microglia: definitions, assays, and functional roles in health and disease. In *Neural-immune interactions in brain function and alcohol related disorders* (pp. 3–29). Springer.

Day, J. J., & Sweatt, J. D. (2010). DNA methylation and memory formation. *Nature Neuroscience*, *13*(11), 1319–1323. DOI: 10.1038/nn.2666 PMID: 20975755

de Andrade Teles, R. B., Diniz, T. C., Costa Pinto, T. C., de Oliveira, R. G.Junior, Gama e Silva, M., de Lavor, É. M., & da Silva Almeida, J. R. G. (2018). Flavonoids as therapeutic agents in Alzheimer's and Parkinson's diseases: A systematic review of preclinical evidences. *Oxidative Medicine and Cellular Longevity*, *2018*(1), 7043213.

de Araujo, F. F., de Paulo Farias, D., Neri-Numa, I. A., & Pastore, G. M. (2021). Polyphenols and their applications: An approach in food chemistry and innovation potential. *Food Chemistry*, *338*, 127535. Advance online publication. 10.1016/j.foodchem.2020.127535

de Lima, E. P., Laurindo, L. F., Catharin, V. C. S., Direito, R., Tanaka, M., Jasmin Santos German, I., Lamas, C. B., Guiguer, E. L., Araújo, A. C., Fiorini, A. M. R., & Barbalho, S. M. (2025). Polyphenols, Alkaloids, and Terpenoids Against Neurodegeneration: Evaluating the Neuroprotective Effects of Phytocompounds Through a Comprehensive Review of the Current Evidence. *Metabolites*, *15*(2), 124. Advance online publication. DOI: 10.3390/metabo15020124 PMID: 39997749

De Strooper, B., & Karran, E. (2016). The Cellular Phase of Alzheimer's Disease. In *Cell* (Vol. 164, Issue 4, pp. 603–615). Cell Press. DOI: 10.1016/j.cell.2015.12.056

Debnath, B., Majumdar, M., Bhowmik, M., Bhowmik, K. L., Debnath, A., & Roy, D. N. (2020). The effective adsorption of tetracycline onto zirconia nanoparticles synthesized by novel microbial green technology. *Journal of Environmental Management*, *261*, 110235.

Decourt, B., D'Souza, G. X., Shi, J., Ritter, A., Suazo, J., & Sabbagh, M. N. (2022). The cause of Alzheimer's disease: The theory of multipathology convergence to chronic neuronal stress. *Aging and Disease*, *13*(1), 37. DOI: 10.14336/AD.2021.0529 PMID: 35111361

Deis, L., Quiroga, A. M., & De Rosas, M. I. (2021). Coloured compounds in fruits and vegetables and health. *Psychiatry and Neuroscience Update: From Epistemology to Clinical Psychiatry–Vol., IV*, 343–358. DOI: 10.1007/978-3-030-61721-9_25

Del Prado-Audelo, M. L., Cortés, H., Caballero-Florán, I. H., González-Torres, M., Escutia-Guadarrama, L., Bernal-Chávez, S. A., Giraldo-Gomez, D. M., Magaña, J. J., & Leyva-Gómez, G. (2021). Therapeutic Applications of Terpenes on Inflammatory Diseases. *Frontiers in Pharmacology*, *12*, 704197. DOI: 10.3389/fphar.2021.704197 PMID: 34483907

Del Rio, D., Rodriguez-Mateos, A., Spencer, J. P. E., Tognolini, M., & Borges, G. (2012). phenolics in human health: Structures, bioavailability, and evidence of protective effects against chronic diseases., 2013, 18. *DOI:* Https://Doi. Org/10.1089/Ars, 1818–1892.

Del Rio, D., Borges, G., & Crozier, A. (2010). Polyphenols and health: What compounds are involved? *Nutrition, Metabolism, and Cardiovascular Diseases*, *20*(1), 1–6. DOI: 10.1016/j.numecd.2009.05.015 PMID: 19713090

Del Rio, D., Rodriguez-Mateos, A., Spencer, J. P. E., Tognolini, M., Borges, G., & Crozier, A. (2010). Dietary (poly)phenolics in human health: Structures, bioavailability, and evidence of protective effects against chronic diseases. *Antioxidants & Redox Signaling*, *18*(14), 1818–1892. DOI: 10.1089/ars.2012.4581 PMID: 22794138

Del Rosario, B. R., O'Connell, J., Lynch, M. A., & Ryan, S. M. (2021). Blueberry supplementation improves anxiety-like behaviour and memory through modulation of the gut–brain axis in aged mice. *Nutrients*, *13*(2), 373. DOI: 10.3390/nu13020373 PMID: 33530406

Demir, E. A., Gergerlioglu, H. S., & Oz, M. (2015). Antidepressant-like effects of quercetin in diabetic rats are independent of hypothalamic-pituitary-adrenal axis. *Acta Neuropsychiatrica*, *28*(1), 23–30. DOI: 10.1017/neu.2015.45 PMID: 26234153

Deng, Q., Wu, C., Parker, E., Liu, T. C. Y., Duan, R., & Yang, L. (2024). Microglia and Astrocytes in Alzheimer's Disease: Significance and Summary of Recent Advances. In *Aging and Disease* (Vol. 15, Issue 4, pp. 1537–1564). International Society on Aging and Disease. DOI: 10.14336/AD.2023.0907

Deng, C., Yi, R., Fei, M., Li, T., Han, Y., & Wang, H. (2021). Naringenin attenuates endoplasmic reticulum stress, reduces apoptosis, and improves functional recovery in experimental traumatic brain injury. *Brain Research*, *1769*, 147591. DOI: 10.1016/j.brainres.2021.147591 PMID: 34324877

Deng, S., Gan, L., Liu, C., Xu, T., Zhou, S., Guo, Y., Zhang, Z., Yang, G. Y., Tian, H., & Tang, Y. (2023). Roles of Ependymal Cells in the Physiology and Pathology of the Central Nervous System. *Aging and Disease*, *14*(2), 468. DOI: 10.14336/AD.2022.0826-1 PMID: 37008045

Dent, K. A., Christie, K. J., Bye, N., Basrai, H. S., Turbic, A., Habgood, M., Cate, H. S., & Turnley, A. M. (2015). Oligodendrocyte birth and death following traumatic brain injury in adult mice. *PLoS One*, *10*(3), e0121541. DOI: 10.1371/journal.pone.0121541 PMID: 25798924

Depp, C. M., Nave, K., & Lab, K. N. (2022). Ageing-associated myelin dysfunction drives amyloid deposition in mouse models of Alzheimer's disease. *Alzheimer's & Dementia*, *18*(S4), e061183. Advance online publication. DOI: 10.1002/alz.061183

Depp, C., Sun, T., Sasmita, A. O., Spieth, L., Berghoff, S. A., Nazarenko, T., Overhoff, K., Steixner-Kumar, A. A., Subramanian, S., Arinrad, S., Ruhwedel, T., Möbius, W., Göbbels, S., Saher, G., Werner, H. B., Damkou, A., Zampar, S., Wirths, O., Thalmann, M., & Nave, K. A. (2023). Myelin dysfunction drives amyloid-β deposition in models of Alzheimer's disease. *Nature*, *618*(7964), 349–357. DOI: 10.1038/s41586-023-06120-6 PMID: 37258678

Deverman, B. E., & Patterson, P. H. (2009, October). Cytokines and CNS development. *Neuron*, *64*(1), 61–78. DOI: 10.1016/j.neuron.2009.09.002 PMID: 19840550

Devi, S., Kumar, V., Singh, S. K., Dubey, A. K., & Kim, J. J. (2021). Flavonoids: Potential Candidates for the Treatment of Neurodegenerative Disorders. *Biomedicines*, *9*(2). Advance online publication. DOI: 10.3390/biomedicines9020099

Devore, E. E., Kang, J. H., Breteler, M. M. B., & Grodstein, F. (2012). Dietary intakes of berries and flavonoids in relation to cognitive decline. *Annals of Neurology*, *72*(1), 135–143. DOI: 10.1002/ana.23594 PMID: 22535616

Dharshini, L. C. P., & Mandal, A. K. A. (2024). Regulation of gene expression by modulating microRNAs through Epigallocatechin-3-gallate in cancer. *Molecular Biology Reports*, *51*(1), 230. DOI: 10.1007/s11033-023-09145-2 PMID: 38281210

Dhingra, D., & Valecha, R. (2014). Punarnavine, an alkaloid isolated from ethanolic extract of Boerhaavia diffusa Linn. Reverses depression-like behaviour in mice subjected to chronic unpredictable mild stress. *Indian Journal of Experimental Biology*, *52*(8), 799–807. PMID: 25141543

Di Lorenzo, C., Colombo, F., Biella, S., Stockley, C., & Restani, P. (2021). Polyphenols and human health: The role of bioavailability. *Nutrients*, *13*(1), 273. DOI: 10.3390/nu13010273 PMID: 33477894

Di Meng, W.. (2016). Anti-inflammatory effects of Bifidobacterium longum subsp. infantis secretions on fetal human enterocytes are mediated by TLR-4 receptors. *American Journal of Physiology. Gastrointestinal and Liver Physiology*, *311*(4), G744–G753. DOI: 10.1152/ajpgi.00090.2016 PMID: 27562058

Di Meo, F., Valentino, A., Petillo, O., Peluso, G., Filosa, S., & Crispi, S. (2020). Bioactive polyphenols and neuromodulation: Molecular mechanisms in neurodegeneration. *International Journal of Molecular Sciences*, *21*(7), 2564. DOI: 10.3390/ijms21072564 PMID: 32272735

Di Paolo, M., Papi, L., Gori, F., & Turillazzi, E. (2019). Natural products in neurodegenerative diseases: A great promise but an ethical challenge. *International Journal of Molecular Sciences*, *20*(20), 5170.

Dias, M. C., Pinto, D. C., & Silva, A. M. (2021). Plant flavonoids: Chemical characteristics and biological activity. *Molecules (Basel, Switzerland)*, *26*(17), 5377. DOI: 10.3390/molecules26175377 PMID: 34500810

Dias, V., Junn, E., & Mouradian, M. M. (2013). The Role of Oxidative Stress in Parkinson's Disease. *Journal of Parkinson's Disease*, *3*(4), 461–491. DOI: 10.3233/jpd-130230

Diaz Heijtz, R., Wang, S., & Anuar, F.. (2011). Normal gut microbiota modulates brain development and behavior. [DOI] [PMC free article] [PubMed] [Google Scholar]. *Proceedings of the National Academy of Sciences of the United States of America*, *108*(7), 3047–3052. DOI: 10.1073/pnas.1010529108 PMID: 21282636

Dietrich, P., Shanmugasundaram, R., Shuyu, E., & Dragatsis, I. (2009). Congenital hydrocephalus associated with abnormal subcommissural organ in mice lacking huntingtin in Wnt1 cell lineages. *Human Molecular Genetics*, *18*(1), 142–150. DOI: 10.1093/hmg/ddn324 PMID: 18838463

Dinan, T. G., & Cryan, J. F. (2013). Melancholic microbes: A link between gut microbiota and depression? *Neurogastroenterology and Motility*, *25*(9), 713–719. DOI: 10.1111/nmo.12198 PMID: 23910373

Dinan, T. G., & Cryan, J. F. (2017). The microbiome-gut-brain axis in health and disease. *Gastroenterology Clinics of North America*, *46*(1), 77–89. DOI: 10.1016/j.gtc.2016.09.007 PMID: 28164854

Ding, H., Dolan, P. J., & Johnson, G. V. (2008). Histone deacetylase 6 interacts with the microtubule-associated protein tau. *Journal of Neurochemistry*, *106*(5), 2119–2130. DOI: 10.1111/j.1471-4159.2008.05564.x PMID: 18636984

Ding, L., Li, J., Song, B., Xiao, X., Zhang, B., Qi, M., Huang, W., Yang, L., & Wang, Z. (2016). Curcumin rescues high fat diet-induced obesity and insulin sensitivity in mice through regulating SREBP pathway. *Toxicology and Applied Pharmacology*, *304*, 99–109. DOI: 10.1016/j.taap.2016.05.011 PMID: 27208389

Ding, S., Chu, X.-L., Jin, Y.-X., Jiang, J.-J., Zhao, X., & Yu, M. (2023). Epigallocatechin gallate alleviates high-fat diet-induced hepatic lipotoxicity by targeting mitochondrial ROS-mediated ferroptosis. *Frontiers in Pharmacology*, *14*(March), 1–11. DOI: 10.3389/fphar.2023.1148814 PMID: 37025486

Ding, X., Yu, H., & Qiao, S. (2020). Lasso peptide microcin J25 effectively enhances gut barrier function and modulates inflammatory response in an enterotoxigenic Escherichia coli-challenged mouse model. *International Journal of Molecular Sciences*, *21*(18), 6500. DOI: 10.3390/ijms21186500 PMID: 32899529

Ding, Z., Cao, J., Shen, Y., Zou, Y., Yang, X., Zhou, W., Guo, Q., & Huang, C. (2018). Resveratrol promotes nerve regeneration via activation of p300 acetyltransferase-mediated VEGF signaling in a rat model of sciatic nerve crush injury. *Frontiers in Neuroscience*, *12*, 341. DOI: 10.3389/fnins.2018.00341 PMID: 29875625

Dominguez-Bello, M. G., De Jesus-Laboy, K. M., Shen, N., Cox, L. M., Amir, A., Gonzalez, A., Bokulich, N. A., Song, S. J., Hoashi, M., Rivera-Vinas, J. I., Mendez, K., Knight, R., & Clemente, J. C. (2016). Partial restoration of the microbiota of cesarean-born infants via vaginal microbial transfer. *Nature Medicine*, *22*(3), 250–253. DOI: 10.1038/nm.4039 PMID: 26828196

Dong, L., Yang, H., Wang, Z., Jiang, N., & Zhang, A. (2021). Antimicrobial peptide CC34 attenuates intestinal inflammation via downregulation of the NF-κB signaling pathway. 3 Biotech, 11(9), 397. DOI: 10.1007/s13205-021-02948-9

Dong, L., Li, Y., Zhang, Y., & Su, S. (2025). Cationic antimicrobial peptide CC34: Potential anticancer and apoptotic induction on cancer cells. *Amino Acids*, *57*(1), 1–13. DOI: 10.1007/s00726-025-03458-1 PMID: 40413361

Dong, Q., Yang, S., Liao, H., He, Q., & Xiao, J. (2023). Bioinformatics findings reveal the pharmacological properties of ferulic acid treating traumatic brain injury via targeting of ferroptosis. *International Journal of Food Properties*, *26*(1), 778–786. DOI: 10.1080/10942912.2023.2185178

Dong, W., Yang, B., Wang, L., Li, B., Guo, X., Zhang, M., Jiang, Z., Fu, J., Pi, J., Guan, D., & Zhao, R. (2018). Curcumin plays neuroprotective roles against traumatic brain injury partly via Nrf2 signaling. *Toxicology and Applied Pharmacology*, *346*, 28–36. DOI: 10.1016/j.taap.2018.03.020 PMID: 29571711

Dong, Y., Kalueff, A. V., & Song, C. (2017). N-methyl-d-aspartate receptor-mediated calcium overload and endoplasmic reticulum stress are involved in interleukin-1beta-induced neuronal apoptosis in rat hippocampus. *Journal of Neuroimmunology*, *307*, 7–13. https://doi.org/https://doi.org/10.1016/j.jneuroim.2017.03.005. DOI: 10.1016/j.jneuroim.2017.03.005 PMID: 28495142

Dong, Y., Yu, M., Wu, Y., Xia, T., Wang, L., Song, K., Zhang, C., Lu, K., & Rahimnejad, S. (2022). Hydroxytyrosol Promotes the Mitochondrial Function through Activating Mitophagy. *Antioxidants, 11*(5), 1–15. DOI: 10.3390/antiox11050893 PMID: 35624756

Dos Santos, C. N., Menezes, R., Carregosa, D., Valentova, K., Foito, A., McDougall, G., & Stewart, D. (2020). Flavonols and flavones. *Dietary Polyphenols: Their Metabolism and Health Effects*, 163-198.

Drljača, J., Milošević, N., Milanović, M., Abenavoli, L., & Milić, N. (2023). When the microbiome helps the brain-current evidence. *CNS Neuroscience & Therapeutics*, *29*(S1), 43–58. DOI: 10.1111/cns.14076 PMID: 36601680

Drossman, D. A., Tack, J., Ford, A. C., Szigethy, E., Törnblom, H., & Van Oudenhove, L. (2018). Neuromodulators for functional gastrointestinal disorders (disorders of gut– brain interaction): a Rome foundation working team report. *Gastroenterology, 154*(4), 1140-1171. e1141.

Dueñas, M., Martínez-Villaluenga, C., Limón, R. I., Peñas, E., & Frias, J. (2015). Effect of germination and elicitation on phenolic composition and bioactivity of kidney beans. *Food Research International, 70*, 55–63.

Du, G., Zhao, Z., Chen, Y., Li, Z., Tian, Y., Liu, Z., Liu, B., & Song, J. (2018). Quercetin protects rat cortical neurons against traumatic brain injury. *Molecular Medicine Reports*, *17*(6), 7859–7865. DOI: 10.3892/mmr.2018.8801 PMID: 29620218

Duncan, G. J., Manesh, S. B., Hilton, B. J., Assinck, P., Plemel, J. R., & Tetzlaff, W. (2020). The fate and function of oligodendrocyte progenitor cells after traumatic spinal cord injury. In *GLIA* (Vol. 68, Issue 2, pp. 227–245). John Wiley and Sons Inc. DOI: 10.1002/glia.23706

Duncan, I. D., & Radcliff, A. B. (2016). Inherited and acquired disorders of myelin: The underlying myelin pathology. In *Experimental Neurology* (Vol. 283, pp. 452–475). Academic Press., DOI: 10.1016/j.expneurol.2016.04.002

Duncan, S. H., Lobley, G. E., Holtrop, G., Ince, J., Johnstone, A. M., Louis, P., & Flint, H. J. (2007). Reduced dietary intake of carbohydrates by obese subjects results in decreased concentrations of butyrate and Faecalibacterium prausnitzii. *Applied and Environmental Microbiology*, *73*(4), 1073–1078. DOI: 10.1128/AEM.02340-06 PMID: 17189447

Dutt, Y., Pandey, R. P., Dutt, M., Gupta, A., Vibhuti, A., Raj, V. S., Chang, C.-M., & Priyadarshini, A. (2023). Liposomes and phytosomes: Nanocarrier systems and their applications for the delivery of phytoconstituents. *Coordination Chemistry Reviews*, *491*, 215251. DOI: 10.1016/j.ccr.2023.215251

Duval, A., & Avérous, L. (2016). Characterization and Physicochemical Properties of Condensed Tannins from Acacia catechu. *Journal of Agricultural and Food Chemistry*, *64*(8), 1751–1760. DOI: 10.1021/acs.jafc.5b05671 PMID: 26853705

Dwita, L. P., Iwo, M. I., Mauludin, R., & Elfahmi, . (2023). Neuroprotective potential of lignan-rich fraction of *Piper cubeba* L. by improving antioxidant capacity in the rat's brain. *Brazilian Journal of Biology*, *82*, e266573. DOI: 10.1590/1519-6984.266573 PMID: 36629543

Dzah, C. S., Zhang, H., Gobe, V., Asante-Donyinah, D., & Duan, Y. (2024). Anti-and pro-oxidant properties of polyphenols and their role in modulating glutathione synthesis, activity and cellular redox potential: Potential synergies for disease management. *Advances in Redox Research: an Official Journal of the Society for Redox Biology and Medicine and the Society for Free Radical Research-Europe*, *11*, 100099.

Echeverry, C., Arredondo, F., Abin-Carriquiry, J. A., Midiwo, J. O., Ochieng, C., Kerubo, L., & Dajas, F. (2010). Pretreatment with natural flavones and neuronal cell survival after oxidative stress: A structure-activity relationship study. *Journal of Agricultural and Food Chemistry*, *58*(4), 2111–2115. DOI: 10.1021/jf902951v

Edgar, N., & Sibille, E. (2012). A putative functional role for oligodendrocytes in mood regulation. *Translational Psychiatry*, *2*(5), e109–e109.

Edmondson, E. (2014). Hydrogen peroxide produced by mitochondrial monoamine oxidase catalysis: Biological implications. *Current Pharmaceutical Design*, *20*(2), 155–160. DOI: 10.2174/13816128113190990406 PMID: 23701542

Eijkelenboom, A., & Burgering, B. M. (2013). FOXOs: Signalling integrators for homeostasis maintenance. *Nature Reviews. Molecular Cell Biology*, *14*(2), 83–97. DOI: 10.1038/nrm3507 PMID: 23325358

El Fatimy, R., Li, S., Chen, Z., Mushannen, T., Gongala, S., Wei, Z., Balu, D. T., Rabinovsky, R., Cantlon, A., Elkhal, A., Selkoe, D. J., Sonntag, K. C., Walsh, D. M., & Krichevsky, A. M. (2018). MicroRNA-132 provides neuroprotection for tauopathies via multiple signaling pathways. *Acta Neuropathologica*, *136*(4), 537–555. DOI: 10.1007/s00401-018-1880-5 PMID: 29982852

El Gaamouch, F., Chen, F., Ho, L., Lin, H.-Y., Yuan, C., Wong, J., & Wang, J. (2022). Benefits of dietary polyphenols in Alzheimer's disease. *Frontiers in Aging Neuroscience*, *14*, 1019942. DOI: 10.3389/fnagi.2022.1019942 PMID: 36583187

Elahi, M. G., Hekmati, M., Esmaeili, D., Ziararti, P., & Yousefi, M. (2022). Determination of the antibacterial and apoptotic effects of Carvacrol and anti-cancer peptide. Jentashapir Journal of Cellular and Molecular Biology, 13(4).

Elkin, E. R., Higgins, C., Aung, M. T., & Bakulski, K. M. (2022). Metals exposures and DNA methylation: Current evidence and future directions. *Current Environmental Health Reports*, *9*(4), 673–696. DOI: 10.1007/s40572-022-00382-4 PMID: 36282474

Ellwardt, E., & Zipp, F. (2014). Molecular mechanisms linking neuroinflammation and neurodegeneration in MS. *Experimental Neurology*, *262*(Pt A), 8–17. https://doi.org/DOI: 10.1016/j.expneurol.2014.02.006

El-Saadony, M. T., Saad, A. M., Mohammed, D. M., Korma, S. A., Alshahrani, M. Y., Ahmed, A. E., Ibrahim, E. H., Salem, H. M., Alkafaas, S. S., Saif, A. M., Elkafas, S. S., Fahmy, M. A., Abd El-Mageed, T. A., Abady, M. M., Assal, H. Y., El-Tarabily, M. K., Mathew, B. T., AbuQamar, S. F., El-Tarabily, K. A., & Ibrahim, S. A. (2025). Medicinal plants: Bioactive compounds, biological activities, combating multidrug-resistant microorganisms, and human health benefits - a comprehensive review. *Frontiers in Immunology*, *16*(April), 1–37. DOI: 10.3389/fimmu.2025.1491777 PMID: 40375989

Eng, Q. Y., Thanikachalam, P. V., & Ramamurthy, S. (2018). Molecular understanding of Epigallocatechin gallate (EGCG) in cardiovascular and metabolic diseases. *Journal of Ethnopharmacology*, *210*, 296–310. DOI: 10.1016/j.jep.2017.08.035 PMID: 28864169

Engwa, G. A. (2018). Free Radicals and the Role of Plant Phytochemicals as Antioxidants Against Oxidative Stress-Related Diseases. In Asao, T., & Asaduzzaman, M. (Eds.), *Phytochemicals - Source of Antioxidants and Role in Disease Prevention*. IntechOpen.

Escartin, C., Galea, E., Lakatos, A., O'Callaghan, J. P., Petzold, G. C., Serrano-Pozo, A., Steinhäuser, C., Volterra, A., Carmignoto, G., Agarwal, A., Allen, N. J., Araque, A., Barbeito, L., Barzilai, A., Bergles, D. E., Bonvento, G., Butt, A. M., Chen, W. T., Cohen-Salmon, M., … Verkhratsky, A. (2021). Reactive astrocyte nomenclature, definitions, and future directions. *Nature Neuroscience 2021 24:3, 24*(3), 312–325. DOI: 10.1038/s41593-020-00783-4

Esmaeili, Y., Yarjanli, Z., Pakniyat, A., Aghajanzadeh, T., Namdar, A., & Kheiry, F. (2022). Targeting autophagy, oxidative stress, and ER stress for neurodegenerative disease treatment. *Journal of Controlled Release*, *345*, 147–175. DOI: 10.1016/j.jconrel.2022.03.001 PMID: 35248646

Esmeeta, A., Adhikary, S., Dharshnaa, V., Swarnamughi, P., Maqsummiya, Z. U., Banerjee, A., & Duttaroy, A. K. (2022). Plant-derived bioactive compounds in colon cancer treatment: An updated review. *Biomedicine and Pharmacotherapy*, *153*, 113384.

Esteras, N., Kundel, F., Amodeo, G. F., Pavlov, E. V., Klenerman, D., & Abramov, A. Y. (2017). Mitochondrial hyperpolarization in iPSC-derived neurons from patients of FTDP-17 with 10+16 MAPT mutation leads to oxidative stress and neurodegeneration. *Redox Biology*, *12*, 410–422. DOI: 10.1016/j.redox.2017.03.008 PMID: 28319892

Estruch, R., Ros, E., Salas-Salvadó, J., Covas, M. I., Corella, D., Arós, F., & Martínez-González, M. A. (2018). Primary Prevention of Cardiovascular Disease with a Mediterranean Diet Supplemented with Extra-Virgin Olive Oil or Nuts. *The New England Journal of Medicine*, *378*(25), e34. DOI: 10.1056/NEJMoa1800389

Estruel-Amades, S., González-Sarrías, A., Romo-Vaquero, M., García-Conesa, M. T., Selma, M. V., Espín, J. C., & Tomás-Barberán, F. A. (2019). Impact of anthocyanins on gut microbiota: Metabolites and health implications. *Nutrients*, *11*(8), 1812. DOI: 10.3390/nu11081812 PMID: 31390778

Fakhoury, M. (2018). Microglia and astrocytes in Alzheimer's disease: Implications for therapy. *Current Neuropharmacology*, *16*(5), 508–518.

Fakhri, S., Abbaszadeh, F., Moradi, S. Z., Cao, H., Khan, H., & Xiao, J. (2022). Effects of polyphenols on oxidative stress, inflammation, and interconnected pathways during spinal cord injury. *Oxidative Medicine and Cellular Longevity*, *2022*(1), 8100195. DOI: 10.1155/2022/8100195 PMID: 35035667

Fang, J., Pang, X., Yan, R., Lian, W., Li, C., Wang, Q., Liu, A.-L., & Du, G. H. (2016). Discovery of neuroprotective compounds by machine learning approaches. *RSC Advances*, *6*(12), 9857–9871. DOI: 10.1039/C5RA23035G

Fang, M., Chen, D., & Yang, C. S. (2007). Dietary polyphenols may affect DNA methylation123. *The Journal of Nutrition*, *137*(1), 223S–228S. DOI: 10.1093/jn/137.1.223S PMID: 17182830

Fan, H., Tang, H.-B., Shan, L.-Q., Liu, S.-C., Huang, D.-G., Chen, X., Chen, Z., Yang, M., Yin, X.-H., Yang, H., & Hao, D.-J. (2019). Quercetin prevents necroptosis of oligodendrocytes by inhibiting macrophages/microglia polarization to M1 phenotype after spinal cord injury in rats. *Journal of Neuroinflammation*, *16*(1), 206. DOI: 10.1186/s12974-019-1613-2 PMID: 31699098

Farbood, Y., Sarkaki, A., Dianat, M., Khodadadi, A., Haddad, M. K., & Mashhadizadeh, S. (2015). Ellagic acid prevents cognitive and hippocampal long-term potentiation deficits and brain inflammation in rat with traumatic brain injury. *Life Sciences*, *124*, 120–127. DOI: 10.1016/j.lfs.2015.01.013 PMID: 25637685

Farbood, Y., Sarkaki, A., Dolatshahi, M., Mansouri, S. M. T., & Khodadadi, A. (2015). Ellagic Acid Protects the Brain Against 6-Hydroxydopamine Induced Neuroinflammation in a Rat Model of Parkinson's Disease. *Basic and Clinical Neuroscience*, *6*(2), 83. https://pmc.ncbi.nlm.nih.gov/articles/PMC4636882/ PMID: 27307952

Farzaei, M. H., Bahramsoltani, R., Abbasabadi, Z., Braidy, N., & Nabavi, S. M. (2019). Role of green tea catechins in prevention of age-related cognitive decline: Pharmacological targets and clinical perspective. *Journal of Cellular Physiology*, *234*(3), 2447–2459. DOI: 10.1002/jcp.27289 PMID: 30187490

Fatuzzo, I., Niccolini, G. F., Zoccali, F., Cavalcanti, L., Bellizzi, M. G., Riccardi, G., & Minni, A. (2023). Neurons, nose, and neurodegenerative diseases: Olfactory function and cognitive impairment. *International Journal of Molecular Sciences*, *24*(3), 2117. DOI: 10.3390/ijms24032117 PMID: 36768440

Fazili, N. A., & Naeem, A. (2015). Anti-fibrillation potency of caffeic acid against an antidepressant induced fibrillogenesis of human α-synuclein: Implications for Parkinson's disease. *Biochimie*, *108*, 178–185. DOI: 10.1016/j.biochi.2014.11.011 PMID: 25461276

Feigin, V. L., Vos, T., Nichols, E., Owolabi, M. O., Carroll, W. M., Dichgans, M., Deuschl, G., Parmar, P., Brainin, M., & Murray, C. (2020). The global burden of neurological disorders: Translating evidence into policy. *Lancet Neurology*, *19*(3), 255–265. DOI: 10.1016/S1474-4422(19)30411-9 PMID: 31813850

Feng, J., Zhou, Y., Campbell, S. L., Le, T., Li, E., Sweatt, J. D., Silva, A. J., & Fan, G. (2010). Dnmt1 and Dnmt3a maintain DNA methylation and regulate synaptic function in adult forebrain neurons. *Nature Neuroscience*, *13*(4), 423–430. DOI: 10.1038/nn.2514 PMID: 20228804

Feng, L., Gwee, X., Kua, E. H., & Ng, T. P. (2010). Cognitive function and tea consumption in community dwelling older Chinese in Singapore. *The Journal of Nutrition, Health & Aging*, *14*(6), 433–438. DOI: 10.1007/s12603-010-0095-9

Feng, X., Yu, W., Li, X., Zhou, F., Zhang, W., Shen, Q., Li, J., Zhang, C., & Shen, P. (2017). Apigenin, a modulator of PPARγ, attenuates HFD-induced NAFLD by regulating hepatocyte lipid metabolism and oxidative stress via Nrf2 activation. *Biochemical Pharmacology*, *136*, 136–149. DOI: 10.1016/j.bcp.2017.04.014 PMID: 28414138

Feng, Y., Cui, Y., Gao, J. L., Li, M. H., Li, R., Jiang, X. H., Tian, Y. X., Wang, K. J., Cui, C. M., & Cui, J. Z. (2016). Resveratrol attenuates neuronal autophagy and inflammatoryinjury by inhibiting the TLR4/NF-κB signaling pathwayin experimental traumatic brain injury. *International Journal of Molecular Medicine*, *37*(4), 921–930. DOI: 10.3892/ijmm.2016.2495 PMID: 26936125

Fernandes, F. D. P., Menezes, A. P. F., de Sousa Neves, J. C., Fonteles, A. A., da Silva, A. T. A., de Araújo Rodrigues, P., do Carmo, M. R. S., de Souza, C. M., & de Andrade, G. M. (2014). Caffeic acid protects mice from memory deficits induced by focal cerebral ischemia. *Behavioural Pharmacology*, *25*(7), 637–647. DOI: 10.1097/FBP.0000000000000076 PMID: 25171077

Fernandez-Zafra, T., Codeluppi, S., & Uhlén, P. (2017). An ex vivo spinal cord injury model to study ependymal cells in adult mouse tissue. *Experimental Cell Research*, *357*(2), 236–242.

Ferrari Bardile, C., Sidik, H., Quek, R., Yusof, N. A. B. M., Garcia-Miralles, M., & Pouladi, M. A. (2021). Abnormal spinal cord myelination due to oligodendrocyte dysfunction in a model of Huntington's disease. *Journal of Huntington's Disease*, *10*(3), 377–384.

Ferreira, C., Vieira, P., Sá, H., Malva, J., Castelo-Branco, M., Reis, F., & Viana, S. (2024). Polyphenols: Immunonutrients tipping the balance of immunometabolism in chronic diseases. *Frontiers in Immunology*, *15*(March), 1–26. DOI: 10.3389/fimmu.2024.1360065 PMID: 38558823

Ferri, P., Angelino, D., Gennari, L., Benedetti, S., Ambrogini, P., Del Grande, P., & Ninfali, P. (2015). Enhancement of flavonoid ability to cross the blood-brain barrier of rats by co-administration with α-tocopherol. *Food & Function*, *6*(2), 394–400. DOI: 10.1039/c4fo00817k

Ferris, J. K., Greeley, B., Vavasour, I. M., Kraeutner, S. N., Rinat, S., Ramirez, J., & Boyd, L. A. (2022). In vivo myelin imaging and tissue microstructure in white matter hyperintensities and perilesional white matter. *Brain Communications*, *4*(3), fcac142.

Fetse, J., Kandel, S., Mamani, U. F., & Cheng, K. (2023). Recent developments in the development of therapeutic peptides. *Trends in Pharmacological Sciences*, *44*(7), 425–441. DOI: 10.1016/j.tips.2023.04.003 PMID: 37246037

Figueira, I., Garcia, G., Pimpão, R. C., Terrasso, A. P., Costa, I., Almeida, A. F., & Santos, C. N. (2017). Polyphenols journey through blood-brain barrier towards neuronal protection. *Scientific Reports*, *7*(1), 11456. DOI: 10.1038/s41598-017-11512-6

Figueira, I., Menezes, R., Macedo, D., Costa, I., & dos Santos, C. N. (2016). Polyphenols Beyond Barriers: A Glimpse into the Brain. *Current Neuropharmacology*, *15*(4), 562–594. DOI: 10.2174/1570159X14666161026151545 PMID: 27784225

Filosa, S., Di Meo, F., & Crispi, S. (2018). Polyphenols-gut microbiota interplay and brain neuromodulation. *Neural Regeneration Research*, *13*(12), 2055–2059. DOI: 10.4103/1673-5374.241429 PMID: 30323120

Finking, R., & Marahiel, MA. (2004). Ribozomalolmayan peptitlerin biyosentezi. *Annual Review of Microbiology*, *58*(1), 453–488. DOI: 10.1146/annurev.micro.58.030603.123615 PMID: 15487945

Fischbach, M. A., & Walsh, C. T. (2006). Assembly-line enzymology for polyketide and nonribosomal peptide antibiotics: Logic, machinery, and mechanisms. *Chemical Reviews*, *106*(8), 3468–3496. DOI: 10.1021/cr0503097 PMID: 16895337

Fisher, A. B. (2017). Peroxiredoxin 6 in the repair of peroxidized cell membranes and cell signaling. *Archives of Biochemistry and Biophysics*, *617*, 68–83. DOI: 10.1016/j.abb.2016.12.003 PMID: 27932289

Fleischer, J., Bumbalo, R., Bautze, V., Strotmann, J., & Breer, H. (2015). Expression of odorant receptor Olfr78 in enteroendocrine cells of the colon. *Cell and Tissue Research, 361*(3), 697–710. DOI: 10.1007/s00441-015-2165-0 PMID: 25813788

Fraga, C. G., Croft, K. D., Kennedy, D. O., & Tomás-Barberán, F. A. (2019). The effects of polyphenols and other bioactives on human health. *Food & Function, 10*(2), 514–528. DOI: 10.1039/C8FO01997E PMID: 30746536

Fraga-Corral, M., Otero, P., Echave, J., Garcia-Oliveira, P., Carpena, M., Jarboui, A., & Prieto, M. A. (2021). By-products of agri-food industry as tannin-rich sources: A review of tannins' biological activities and their potential for valorization. *Foods, 10*(1), 137.

Francisco, M., & Grau, R. (2025). Biofilm proficient Bacillus subtilis prevents neurodegeneration in Caenorhabditis elegans Parkinson's disease models via PMK-1/p38 MAPK and SKN-1/Nrf2 signaling. *Scientific Reports, 15*(1), 9864. DOI: 10.1038/s41598-025-93737-4 PMID: 40118903

Franzoni, F., Scarfò, G., Guidotti, S., Fusi, J., Asomov, M., & Pruneti, C. (2021). Oxidative Stress and Cognitive Decline: The Neuroprotective Role of Natural Antioxidants. In *Frontiers in Neuroscience* (Vol. 15). Frontiers Media S.A. DOI: 10.3389/fnins.2021.729757

Friedman, M. (2017). Nutritional and health benefits of blueberries. *Journal of Agricultural and Food Chemistry, 65*(35), 7653–7672. DOI: 10.1021/acs.jafc.7b01888 PMID: 28805053

Frost, G. R., & Li, Y. M. (2017). The role of astrocytes in amyloid production and Alzheimer's disease. *Open Biology, 7*(12), 170228.

Frost, G., Sleeth, M. L., Sahuri-Arisoylu, M., Lizarbe, B., Cerdan, S., Brody, L., Anastasovska, J., Ghourab, S., Hankir, M., Zhang, S., Carling, D., Swann, J. R., Gibson, G., Viardot, A., Morrison, D., Louise Thomas, E., & Bell, J. D. (2014). The short-chain fatty acid acetate reduces appetite via a central homeostatic mechanism. *Nature Communications, 5*(1), 3611. DOI: 10.1038/ncomms4611 PMID: 24781306

Frugier, T., Morganti-Kossmann, M. C., O'Reilly, D., & McLean, C. A. (2010). In situ detection of inflammatory mediators in post mortem human brain tissue after traumatic injury. *Journal of Neurotrauma, 27*(3), 497–507. DOI: 10.1089/neu.2009.1120 PMID: 20030565

Fruman, D. A., Chiu, H., Hopkins, B. D., Bagrodia, S., Cantley, L. C., & Abraham, R. T. (2017). The PI3K pathway in human disease. *Cell, 170*(4), 605–635. DOI: 10.1016/j.cell.2017.07.029 PMID: 28802037

Fu, J., Sun, H., Zhang, Y., Xu, W., Wang, C., Fang, Y., & Zhao, J. (2018). Neuroprotective Effects of Luteolin Against Spinal Cord Ischemia-Reperfusion Injury by Attenuation of Oxidative Stress, Inflammation, and Apoptosis. *Journal of Medicinal Food, 21*(1), 13–20. DOI: 10.1089/jmf.2017.4021 PMID: 28976796

Fülling, C., Dinan, T. G., & Cryan, J. F. (2019). Gut microbe to brain signaling: What happens in vagus…. *Neurogastroenterology and Motility, 31*(5), e13599. DOI: 10.1111/nmo.13599 PMID: 30897366

Fülling, C., Dinan, T. G., & Cryan, J. F. (2019). Gut microbe to brain signaling: What happens in vagus…. *Neuron, 101*(6), 998–1002. DOI: 10.1016/j.neuron.2019.02.008 PMID: 30897366

Furness, J. B., Rivera, L. R., Cho, H. J., Bravo, D. M., & Callaghan, B. (2013). The gut as a sensory organ. *Nature Reviews. Gastroenterology & Hepatology*, *10*(12), 729–740. DOI: 10.1038/nrgastro.2013.180 PMID: 24061204

Furtado, D., Björnmalm, M., Ayton, S., Bush, A. I., Kempe, K., & Caruso, F. (2018). Overcoming the blood–brain barrier: The role of nanomaterials in treating neurological diseases. *Advanced Materials*, *30*(46), 1801362. DOI: 10.1002/adma.201801362 PMID: 30066406

Fusco, W., Lorenzo, M. B., Cintoni, M., Porcari, S., Rinninella, E., Kaitsas, F., Lener, E., Mele, M. C., Gasbarrini, A., Collado, M. C., Cammarota, G., & Ianiro, G. (2023). Short-chain fatty-acid-producing bacteria: Key components of the human gut microbiota. *Nutrients*, *15*(9), 2211. DOI: 10.3390/nu15092211 PMID: 37432351

Fu, Y., Yang, J., Wang, X., Yang, P., Zhao, Y., Li, K., Chen, Y., & Xu, Y. (2018). Herbal Compounds Play a Role in Neuroprotection through the Inhibition of Microglial Activation. *Journal of Immunology Research*, *2018*(1), 9348046. DOI: 10.1155/2018/9348046 PMID: 29850641

Fymat, A. L. (2023). Multiple sclerosis: I. Symptomatology and etiology. *J Neurol Psychol Res*, *4*(1).

Gabathuler, R. (2010). Approaches to transport therapeutic drugs across the blood-brain barrier to treat brain diseases. *Neurobiology of Disease*, *37*(1), 48–57. DOI: 10.1016/j.nbd.2009.07.028 PMID: 19664710

Gahtani, R. M., Shoaib, S., Hani, U., Jayachithra, R., Alomary, M. N., Chauhan, W., Jahan, R., Tufail, S., & Ansari, M. A. (2024). Combating Parkinson's disease with plant-derived polyphenols: Targeting oxidative stress and neuroinflammation. *Neurochemistry International*, *178*, 105798. DOI: 10.1016/j.neuint.2024.105798 PMID: 38950626

Galiniak, S., Aebisher, D., & Bartusik-Aebisher, D. (2019). Health benefits of resveratrol administration. *Acta Biochimica Polonica*, *66*(1), 13–21. PMID: 30816367

Galli, R. L., Bielinski, D. F., & Shukitt-Hale, B. (2018). Blueberry supplemented diet improves memory and motor performance in aging mice. *Nutritional Neuroscience*, *21*(2), 103–110. DOI: 10.1080/1028415X.2016.1212985

Gandy, S. (2005). The role of cerebral amyloid β accumulation in common forms of Alzheimer disease. *The Journal of Clinical Investigation*, *115*(5), 1121–1129. PMID: 15864339

Gan, H., Ma, Q., Hao, W., Yang, N., Chen, Z. S., Deng, L., & Chen, J. (2024). Targeting autophagy to counteract neuroinflammation: A novel antidepressant strategy. In *Pharmacological Research* (Vol. 202). Academic Press., DOI: 10.1016/j.phrs.2024.107112

Gan, P., Ding, L., Hang, G., Xia, Q., Huang, Z., Ye, X., & Qian, X. (2020). Oxymatrine Attenuates Dopaminergic Neuronal Damage and Microglia-Mediated Neuroinflammation Through Cathepsin D-Dependent HMGB1/TLR4/NF-κB Pathway in Parkinson's Disease. *Frontiers in Pharmacology*, *11*(May), 1–13. DOI: 10.3389/fphar.2020.00776 PMID: 32528295

Gao, C., Jiang, J., Tan, Y., & Chen, S. (2023a). Microglia in neurodegenerative diseases: mechanism and potential therapeutic targets. *Signal Transduction and Targeted Therapy 2023 8:1*, *8*(1), 1–37. DOI: 10.1038/s41392-023-01588-0

Gao, L., Kuraji, R., Zhang, M. J., Martinez, A., Radaic, A., Kamarajan, P., Le, C., Zhan, L., Ye, C., Rangé, H., Sailani, M. R., & Kapila, Y. L. (2022). Nisin probiotic prevents inflammatory bone loss while promoting reparative proliferation and a healthy microbiome. *NPJ Biofilms and Microbiomes*, *8*(1), 45. DOI: 10.1038/s41522-022-00307-x PMID: 35672331

Gao, Y., Wei, Y., Wang, Y., Gao, F., & Chen, Z. (2017). *Lycium barbarum*: A traditional Chinese herb and a promising anti-aging agent. *Aging and Disease*, *8*(6), 778–791. 10.14336/AD.2017.0725

García-Conesa, M.-T., & Larrosa, M. (2020). *Polyphenol-rich foods for human health and disease* (Vol. 12). MDPI.

García-Mantrana, I., Selma-Royo, M., Alcántara, C., & et al. (2023). Review of the relationships between human gut microbiome, diet, and obesity. *Nutrients, 16*(23), Article 3996. https://doi.org/ (Note: Volume 16 is unusually high – double-check year/journal)DOI: 10.3390/nu16233996

Gatson, J. W., Liu, M.-M., Abdelfattah, K., Wigginton, J. G., Smith, S., Wolf, S., & Minei, J. P. (2013). Resveratrol decreases inflammation in the brain of mice with mild traumatic brain injury. *The Journal of Trauma and Acute Care Surgery*, *74*(2), 470–475. DOI: 10.1097/TA.0b013e31827e1f51 PMID: 23354240

Gautam, S., Karmakar, S., Batra, R., Sharma, P., Pradhan, P., Singh, J., Kundu, B., & Chowdhury, P. K. (2017). Polyphenols in combination with β-cyclodextrin can inhibit and disaggregate α-synuclein amyloids under cell mimicking conditions: A promising therapeutic alternative. *Biochimica et Biophysica Acta (BBA)-. Biochimica et Biophysica Acta. Proteins and Proteomics*, *1865*(5), 589–603. DOI: 10.1016/j.bbapap.2017.02.014 PMID: 28238838

Gauthier, S., Rosa-Neto, P., Morais, J. A., & Webster, C. (2018). World Alzheimer Report 2018: The state of the art of dementia research: New frontiers. Alzheimer's Disease International. https://www.alzint.org/resource/world-alzheimer-report-2018/

Gaynes, R. R. (2017). The discovery of penicillin – New insights after more than 75 years of clinical use. *Emerging Infectious Diseases*, *23*(5), 849–853. DOI: 10.3201/eid2305.161556

Geloso, M. C., Corvino, V., Marchese, E., Serrano, A., Michetti, F., & D'Ambrosi, N. (2017). The dual role of microglia in ALS: Mechanisms and therapeutic approaches. *Frontiers in Aging Neuroscience*, *9*, 242.

Gerhard, A., Pavese, N., Hotton, G., Turkheimer, F., Es, M., Hammers, A., & Brooks, D. J. (2006). In vivo imaging of microglial activation with [11C](R)-PK11195 PET in idiopathic Parkinson's disease. *Neurobiology of Disease*, *21*(2), 404–412.

Ge, X. H., Shao, L., & Zhu, G. J. (2018). Oxymatrine attenuates brain hypoxic-ischemic injury from apoptosis and oxidative stress: Role of p-Akt/GSK3β/HO-1/Nrf-2 signaling pathway. *Metabolic Brain Disease*, *33*(6), 1869–1875. DOI: 10.1007/s11011-018-0293-4 PMID: 30032345

Ghodsi, H., Rahimi, H. R., Aghili, S. M., Saberi, A., & Shoeibi, A. (2022). Evaluation of curcumin as add-on therapy in patients with Parkinson's disease: A pilot randomized, triple-blind, placebo-controlled trial. *Clinical Neurology and Neurosurgery*, *218*, 107300. DOI: 10.1016/j.clineuro.2022.107300 PMID: 35636380

Ghosh, N., Ghosh, R., Mandal, V., & Mandal, S. C. (2011). Recent advances in herbal medicine for treatment of liver diseases. *Pharmaceutical Biology*, *49*(9), 970–988. DOI: 10.3109/13880209.2011.558515 PMID: 21595500

Ghosh, P., & Saadat, A. (2023). Neurodegeneration and epigenetics: A review. [English Edition]. *Neurología : Publicación Oficial de la Sociedad Española de Neurología*, *38*(6), e62–e68. DOI: 10.1016/j.nrl.2021.01.016 PMID: 37344098

Ghosh, S., Whitley, C. S., Haribabu, B., & Jala, V. R. (2021). Regulation of intestinal barrier function by microbial metabolites. *Cellular and Molecular Gastroenterology and Hepatology*, *11*(5), 1463–1482. DOI: 10.1016/j.jcmgh.2021.02.007 PMID: 33610769

Giallongo, S., Lo Re, O., Lochmanová, G., Parca, L., Petrizzelli, F., Zdráhal, Z., Mazza, T., & Vinciguerra, M. (2021). Phosphorylation within intrinsic disordered region discriminates histone variant macroH2A1 splicing isoforms—macroH2A1. 1 and macroH2A1. 2. *Biology (Basel)*, *10*(7), 659. DOI: 10.3390/biology10070659 PMID: 34356514

Giallongo, S., Lo Re, O., & Vinciguerra, M. (2019). Macro histone variants: Emerging rheostats of gastrointestinal cancers. *Cancers (Basel)*, *11*(5), 676. DOI: 10.3390/cancers11050676 PMID: 31096699

Giallongo, S., Longhitano, L., Denaro, S., D'Aprile, S., Torrisi, F., La Spina, E., Giallongo, C., Mannino, G., Lo Furno, D., Zappalà, A., Giuffrida, R., Parenti, R., Li Volti, G., Tibullo, D., & Vicario, N. (2022). The role of epigenetics in neuroinflammatory-driven diseases. *International Journal of Molecular Sciences*, *23*(23), 15218. DOI: 10.3390/ijms232315218 PMID: 36499544

Giallongo, S., Řeháková, D., Biagini, T., Lo Re, O., Raina, P., Lochmanová, G., Zdráhal, Z., Resnick, I., Pata, P., Pata, I., Mistrík, M., de Magalhães, J. P., Mazza, T., Koutná, I., & Vinciguerra, M. (2022). Histone variant macroH2A1. 1 enhances nonhomologous end joining-dependent DNA double-strand-break repair and reprogramming efficiency of human iPSCs. *Stem Cells (Dayton, Ohio)*, *40*(1), 35–48. DOI: 10.1093/stmcls/sxab004 PMID: 35511867

Gibson, G. R., Hutkins, R., Sanders, M. E., Prescott, S. L., Reimer, R. A., Salminen, S. J., Scott, K., Stanton, C., Swanson, K. S., Cani, P. D., Verbeke, K., & Reid, G. (2017). Expert consensus document: The International Scientific Association for Probiotics and Prebiotics (ISAPP) consensus statement on the definition and scope of prebiotics. *Nature Reviews. Gastroenterology & Hepatology*, *14*(8), 491–502. DOI: 10.1038/nrgastro.2017.75 PMID: 28611480

Gil, E. S., & Hudson, S. M. (2004). Stimuli-reponsive polymers and their bioconjugates. *Progress in Polymer Science*, *29*(12), 1173–1222. DOI: 10.1016/j.progpolymsci.2004.08.003

Ginwala, R., Bhavsar, R., Chigbu, D. G. I., Jain, P., & Khan, Z. K. (2019). Potential role of flavonoids in treating chronic inflammatory diseases with a special focus on the anti-inflammatory activity of apigenin. *Antioxidants*, *8*(2), 35. DOI: 10.3390/antiox8020035 PMID: 30764536

Giorgetti, S., Greco, C., Tortora, P., & Aprile, F. A. (2018). Targeting amyloid aggregation: An overview of strategies and mechanisms. *International Journal of Molecular Sciences*, *19*(9), 2677. DOI: 10.3390/ijms19092677 PMID: 30205618

Giovanelli, G., & Dragoni, I. (2019). Anthocyanins: Functional ingredients for foods and beverages. *Journal of Functional Foods*, *59*, 43–56. DOI: 10.1016/j.jff.2019.04.011

Giraffa, G. (1995). Enterococcal bacteriocins: Their potential as anti-Listeria factors in dairy technology. *Food Microbiology*, *12*, 291–299. DOI: 10.1016/S0740-0020(95)80109-X

Giridharan, V. V., Thandavarayan, R. A., Sato, S., Ko, K. M., & Konishi, T. (2011). Prevention of scopolamine-induced memory deficits by schisandrin B, an antioxidant lignan from Schisandra chinensis in mice. *Free Radical Research*, *45*(8), 950–958. DOI: 10.3109/10715762.2011.571682 PMID: 21615274

Giri, S., Choudhury, A., Praharaj, D. L., Singh, A., Vaidya, A., Harindranath, S., & Shukla, A. (2024). Changing etiological spectrum of hepatocellular carcinoma in India—A systematic review and meta-analysis. *Journal of Clinical and Experimental Hepatology*, *14*(5), 101391.

Giuliano, C., Siani, F., Mus, L., Ghezzi, C., Cerri, S., Pacchetti, B., Bigogno, C., & Blandini, F. (2020). Neuroprotective effects of lignan 7-hydroxymatairesinol (HMR/lignan) in a rodent model of Parkinson's disease. *Nutrition (Burbank, Los Angeles County, Calif.)*, *69*, 110494. DOI: 10.1016/j.nut.2019.04.006 PMID: 31586482

Giusti, M. M., & Wrolstad, R. E. (2001). Characterization and measurement of anthocyanins by UV-visible spectroscopy. Current Protocols in Food Analytical Chemistry, F1.2.1–F1.2.13. https://doi.org/ DOI: 10.1002/0471142913.faf0102s00

Giusti, M. M., & Wrolstad, R. E. (2003). Acylated anthocyanins from edible sources and their applications in food systems. *Biochemical Engineering Journal*, *14*(3), 217–225. DOI: 10.1016/S1369-703X(02)00221-8

Gnasegaran, G. K., Agyei, D., Pan, S., Sarethy, I. P., Acquah, C., & Danquah, M. K. (2017). Process development for bioactive peptide production. Food Bioactives: Extraction and Biotechnology Applications, 91-110.

Godos, J., Caraci, F., Castellano, S., Currenti, W., Galvano, F., Ferri, R., & Grosso, G. (2020). Association Between Dietary Flavonoids Intake and Cognitive Function in an Italian Cohort. *Biomolecules 2020, Vol. 10, Page 1300, 10*(9), 1300. DOI: 10.3390/biom10091300

Godos, J., Caraci, F., Castellano, S., Currenti, W., Galvano, F., Ferri, R., & Grosso, G. (2020). Association between dietary flavonoids intake and cognitive function in older adults: A systematic review. *Frontiers in Nutrition*, *7*, 75. DOI: 10.3389/fnut.2020.00075

Godos, J., Pluchinotta, F. R., Caraci, F., Castellano, S., Galvano, F., & Grosso, G. (2020). Dietary polyphenols and cognitive decline: A systematic review. *Antioxidants*, *9*(12), 1236. DOI: 10.3390/antiox9121236 PMID: 33291467

Godyń, J., Jończyk, M., Panek, D., & Malawska, B. (2016). Therapeutic strategies for Alzheimer's disease in clinical trials. *Pharmacological Reports*, *68*(5), 947–958. DOI: 10.1016/j.pharep.2015.07.006 PMID: 26721364

Goldman, J. G., Volpe, D., Ellis, T. D., Hirsch, M. A., Johnson, J., Wood, J., & Wallock, K. (2024). Delivering multidisciplinary rehabilitation care in Parkinson's disease: An international consensus statement. *Journal of Parkinson's Disease*, *14*(1), 135–166.

Goldsteins, G., Hakosalo, V., Jaronen, M., Keuters, M. H., Lehtonen, Š., & Koistinaho, J. (2022). CNS Redox Homeostasis and Dysfunction in Neurodegenerative Diseases. In *Antioxidants* (Vol. 11, Issue 2). DOI: 10.3390/antiox11020405

Gomez de Cedron, M., & Ramírez de Molina, A. (2023). Metabolo-epigenetic interplay provides targeted nutritional interventions in chronic diseases and ageing. *Frontiers in Oncology*, *13*, 1169168–1169168. DOI: 10.3389/fonc.2023.1169168 PMID: 37404756

Gómez-Guzmán, M., Toral, M., Romero, M., Jiménez, R., Galindo, P., Sánchez, M., Zarzuelo, M. J., Olivares, M., Gálvez, J., & Duarte, J. (2015). Antihypertensive effects of probiotics *Lactobacillus* strains in spontaneously hypertensive rats. *Molecular Nutrition & Food Research*, *59*(11), 2326–2336. DOI: 10.1002/mnfr.201500290 PMID: 26255877

Gómez-Pinilla, F. (2008). Brain foods: The effects of nutrients on brain function. *Nature Reviews. Neuroscience*, *9*(7), 568–578. DOI: 10.1038/nrn2421 PMID: 18568016

Gomez-Pinilla, F., & Nguyen, T. T. (2012). Natural mood foods: The actions of polyphenols against psychiatric and cognitive disorders. *Nutritional Neuroscience*, *15*(3), 127–133.

Gong, M., Han, B., Wang, S., Liang, S., & Zou, Z. (2016). Icariin reverses corticosterone-induced depression-like behavior, decrease in hippocampal brain-derived neurotrophic factor (BDNF) and metabolic network disturbances revealed by NMR-based metabonomics in rats. *Journal of Pharmaceutical and Biomedical Analysis*, *123*, 63–73. DOI: 10.1016/j.jpba.2016.02.001 PMID: 26874256

Gong, Z., Ba, L., & Zhang, M. (2022). Dysfunction of the oligodendrocytes in amyotrophic lateral sclerosis. *Journal of Biomedical Research*, *36*(5), 336–342. DOI: 10.7555/JBR.36.20220009 PMID: 36106501

Gou, H.-Z., Zhang, Y.-L., Ren, L.-F., Li, Z.-J., & Zhang, L. (2022). How do intestinal probiotics restore the intestinal barrier? *Frontiers in Microbiology*, *13*, 929346. DOI: 10.3389/fmicb.2022.929346 PMID: 35910620

Goyal, A., Verma, A., Dubey, N., Raghav, J., & Agrawal, A. (2022). Naringenin: A prospective therapeutic agent for Alzheimer's and Parkinson's disease. *Journal of Food Biochemistry*, *46*(1), e14415. Advance online publication. 10.1111/jfbc.14415

Grabska-Kobylecka, I., Kaczmarek-Bak, J., Figlus, M., Prymont-Przyminska, A., Zwolinska, A., Sarniak, A., & Nowak, D. (2020). The Presence of Caffeic Acid in Cerebrospinal Fluid: Evidence That Dietary Polyphenols Can Cross the Blood-Brain Barrier in Humans. *Nutrients*, *12*(5). Advance online publication. DOI: 10.3390/nu12051531

Grabska-Kobyłecka, I., Szpakowski, P., Król, A., Książek-Winiarek, D., Kobyłecki, A., Głąbiński, A., & Nowak, D. (2023). Polyphenols and Their Impact on the Prevention of Neurodegenerative Diseases and Development. *Nutrients*, *15*(15), 1–34. DOI: 10.3390/nu15153454 PMID: 37571391

Grabska-Kobyłecka, I., Szpakowski, P., Król, A., Książek-Winiarek, D., Kobyłecki, A., Głąbiński, A., & Nowak, D. (2023). Polyphenols and their impact on the prevention of neurodegenerative diseases and development. *Nutrients*, *15*(15), 3454. Advance online publication. 10.3390/nu15153454

Gregory, P. D., Wagner, K., & Hörz, W. (2001). Histone acetylation and chromatin remodeling. *Experimental Cell Research, 265*(2), 195–202. DOI: 10.1006/excr.2001.5187 PMID: 11302684

Grenier, D., Marcoux, E., Azelmat, J., Ben Lagha, A., & Gauthier, P. (2020). Biocompatible combinations of nisin and licorice polyphenols exert synergistic bactericidal effects against Enterococcus faecalis and inhibit NF-κB activation in monocytes. *AMB Express, 10*(1), 1–8. DOI: 10.1186/s13568-020-01056-w PMID: 32632823

Grewal, A. K., Singh, T. G., Sharma, D., Sharma, V., Singh, M., Rahman, M. H., & Albadrani, G. M. (2021). Mechanistic insights and perspectives involved in neuroprotective action of quercetin. *Biomedicine and Pharmacotherapy, 140*, 111729. DOI: 10.1016/j.biopha.2021.111729 PMID: 34044274

Grice, E. A., & Segre, J. A. (2012). The human microbiome: Our second genome. *Annual Review of Genomics and Human Genetics, 13*(1), 151–170. DOI: 10.1146/annurev-genom-090711-163814 PMID: 22703178

Grigoletto, J., de Oliveira, C. V., Grauncke, A. C. B., de Souza, T. L., Souto, N. S., de Freitas, M. L., Furian, A. F., Santos, A. R. S., & Oliveira, M. S. (2016). Rosmarinic acid is anticonvulsant against seizures induced by pentylenetetrazol and pilocarpine in mice. *Epilepsy & Behavior, 62*, 27–34. DOI: 10.1016/j.yebeh.2016.06.037 PMID: 27448240

Griñán-Ferré, C., Corpas, R., Puigoriol-Illamola, D., Palomera-Ávalos, V., Sanfeliu, C., & Pallàs, M. (2021). Synergistic neuroprotective effects of a natural product mixture against AD hallmarks and cognitive decline in *Caenorhabditis elegans* and an SAMP8 mice model. *Nutrients, 13*(7), 2411. Advance online publication. DOI: 10.3390/nu13072411 PMID: 34371921

Gryka-Marton, M., Grabowska, A. D., & Szukiewicz, D. (2025). Breaking the Barrier: The Role of Proinflammatory Cytokines in BBB Dysfunction. *International Journal of Molecular Sciences, 26*(8), 3532. Advance online publication. DOI: 10.3390/ijms26083532 PMID: 40331982

Guerreiro, Í., Ferreira-Pêgo, C., Carregosa, D., Santos, C. N., Menezes, R., Fernandes, A. S., & Costa, J. G. (2022). Polyphenols and their metabolites in renal diseases: An overview. *Foods, 11*(7), 1060. DOI: 10.3390/foods11071060 PMID: 35407148

Gugliandolo, E., Peritore, A. F., D'amico, R., Licata, P., & Crupi, R. (2020). Evaluation of neuroprotective effects of quercetin against aflatoxin b1-intoxicated mice. *Animals (Basel), 10*(5), 1–13. DOI: 10.3390/ani10050898 PMID: 32455780

Gu, J. H., Ge, J., Li, M., Wu, F., Zhang, W., & Qin, Z.-H. (2012). Inhibition of NF-κB activation is associated with anti-inflammatory and anti-apoptotic effects of Ginkgolide B in a mouse model of cerebral ischemia/reperfusion injury. *European Journal of Pharmaceutical Sciences, 47*(4), 652–660. DOI: 10.1016/j.ejps.2012.07.016 PMID: 22850444

Gul, Z., Demircan, C., Bagdas, D., & Buyukuysal, R. L. (2016). Protective Effects of Chlorogenic Acid and its Metabolites on Hydrogen Peroxide-Induced Alterations in Rat Brain Slices: A Comparative Study with Resveratrol. *Neurochemical Research, 41*(8), 2075–2085. DOI: 10.1007/s11064-016-1919-8 PMID: 27161374

Guo, F., Liu, X., Cai, H., & Le, W. (2018). Autophagy in neurodegenerative diseases: pathogenesis and therapy. In *Brain Pathology* (Vol. 28, Issue 1, pp. 3–13). Blackwell Publishing Ltd. DOI: 10.1111/bpa.12545

Guo, C. Y., Sun, L., Chen, X. P., & Zhang, D. S. (2013). Oxidative stress, mitochondrial damage and neurodegenerative diseases. *Neural Regeneration Research*, *8*(21), 2003–2014. DOI: 10.3969/j.issn.1673-5374.2013.21.009 PMID: 25206509

Guo, D.-J., Li, F., Yu, P. H.-F., & Chan, S.-W. (2013). Neuroprotective effects of luteolin against apoptosis induced by 6-hydroxydopamine on rat pheochromocytoma PC12 cells. *Pharmaceutical Biology*, *51*(2), 190–196. DOI: 10.3109/13880209.2012.716852 PMID: 23035972

Guo, S., Wang, H., & Yin, Y. (2022). Microglia polarization from M1 to M2 in neurodegenerative diseases. *Frontiers in Aging Neuroscience*, *14*, 815347. DOI: 10.3389/fnagi.2022.815347 PMID: 35250543

Guo, W., Stoklund Dittlau, K., & Van Den Bosch, L. (2020). Axonal transport defects and neurodegeneration: Molecular mechanisms and therapeutic implications. *Seminars in Cell & Developmental Biology*, *99*, 133–150. DOI: 10.1016/j.semcdb.2019.07.010 PMID: 31542222

Gupta, S., & Dasmahapatra, A. K. (2020). Destabilization potential of phenolics on Aβ fibrils: Mechanistic insights from molecular dynamics simulation. *Physical Chemistry Chemical Physics*, *22*(35), 19643–19658. DOI: 10.1039/D0CP02459G PMID: 32830209

Gurcan, O., Gurcay, A. G., Kazanci, A., Senturk, S., Bodur, E., Karaca, E. U., Turkoglu, O. F., & Bavbek, M. (2017). Effect of asiatic acid on the treatment of spinal cord injury: An experimental study in rats. *Turkish Neurosurgery*, *27*(2), 259–264. DOI: 10.5137/1019-5149.JTN.15747-15.2 PMID: 27593768

Guven, H., Arici, A., & Simsek, O. (2019). Flavonoids in our foods: A short review. *Journal of Basic and Clinical Health Sciences*, *3*(2), 96–106. DOI: 10.30621/jbachs.2019.555

Guven, M., Bozkurt Aras, A., Akman, T., Murat Sen, H., Ozkan, A., Salis, O., Sehitoglu, I., Kalkan, Y., Silan, C., Deniz, M., & Cosar, M. (2015). Neuroprotective effect of p-coumaric acid in rat model of embolic cerebral ischemia. *Iranian Journal of Basic Medical Sciences.*, *18*(4), 356. https://pmc.ncbi.nlm.nih.gov/articles/PMC4439450/ PMID: 26019798

Guven, M., Sehitoglu, M. H., Yuksel, Y., Tokmak, M., Aras, A. B., Akman, T., Golge, U. H., Karavelioglu, E., Bal, E., & Cosar, M. (2015). The Neuroprotective Effect of Coumaric Acid on Spinal Cord Ischemia/Reperfusion Injury in Rats. *Inflammation*, *38*(5), 1986–1995. DOI: 10.1007/s10753-015-0179-0 PMID: 25943038

Guven, M., Yuksel, Y., Sehitoglu, M. H., Tokmak, M., Aras, A. B., Akman, T., Golge, U. H., Goksel, F., Karavelioglu, E., & Cosar, M. (2015). The Effect of Coumaric Acid on Ischemia–Reperfusion Injury of Sciatic Nerve in Rats. *Inflammation*, *38*(6), 2124–2132. DOI: 10.1007/s10753-015-0195-0 PMID: 26049410

Hajimoradi, M., Fazilati, M., Gharib-Naseri, M. K., & Sarkaki, A. (2015). Gallic acid and exercise training improve motor function, nerve conduction velocity but not pain sense reflex after experimental sciatic nerve crush in male rats. *Avicenna Journal of Phytomedicine*, *5*(4), 288. /pmc/articles/PMC4587608/

Halder, S., Anand, U., Nandy, S., Oleksak, P., Qusti, S., Alshammari, E. M., Batiha, G. E.-S., Koshy, E. P., & Dey, A. (2021). Herbal drugs and natural bioactive products as potential therapeutics: A review on pro-cognitives and brain boosters perspective. *Saudi Pharmaceutical Journal: SPJ: the Official Publication of the Saudi Pharmaceutical Society*, *29*(8), 879–907. 10.1016/j.jsps.2021.07.009

Hamman, J. H., Enslin, G. M., & Kotzé, A. F. (2005). Oral delivery of peptide drugs: Barriers and developments. *BioDrugs*, *19*, 165–177. DOI: 10.2165/00063030-200519030-00003 PMID: 15984901

Han, K. J., Lee, J. E., Lee, N. K., & Paik, H. D. (2020). Antioxidant and Anti-Inflammatory Effect of Probiotic Lactobacillus plantarum KU15149 Derived from Korean Homemade

Hanáková, Z., Hošek, J., Kutil, Z., Temml, V., Landa, P., Vaněk, T., & Šmejkal, K. (2017). Anti-inflammatory Activity of Natural Geranylated Flavonoids: Cyclooxygenase and Lipoxygenase Inhibitory Properties and Proteomic Analysis. *Journal of Natural Products*, *80*(4), 999–1006. DOI: 10.1021/acs.jnatprod.6b01011

Han, D., Dong, X., Zheng, D., & Nao, J. (2020). MiR-124 and the underlying therapeutic promise of neurodegenerative disorders. *Frontiers in Pharmacology*, *10*, 1555. DOI: 10.3389/fphar.2019.01555 PMID: 32009959

Han, K., Wang, J.-Y., Wang, P.-Y., & Peng, Y.-C.-H. (2024). Therapeutic potential of cannabidiol (CBD) in anxiety disorders: A systematic review and meta-analysis. *Psychiatry Research*, *338*, 116049. Advance online publication. 10.1016/j.psychres.2024.116049

Han, N., Wen, Y., Liu, Z., Zhai, J., Li, S., & Yin, J. (2022). Advances in the roles and mechanisms of lignans against Alzheimer's disease. *Frontiers in Pharmacology*, *13*, 960112. DOI: 10.3389/fphar.2022.960112 PMID: 36313287

Hano, C., & Tungmunnithum, D. (2020). *Plant polyphenols, more than just simple natural antioxidants: Oxidative stress, aging and age-related diseases* (Vol. 7). MDPI.

Han, S., Gim, Y., Jang, E. H., & Hur, E. M. (2022). Functions and dysfunctions of oligodendrocytes in neurodegenerative diseases. In *Frontiers in Cellular Neuroscience* (Vol. 16, p. 1083159). Frontiers Media S.A., DOI: 10.3389/fncel.2022.1083159

Hansen, C. H. F., Krych, L., Nielsen, D. S., Vogensen, F. K., Hansen, L. H., Sørensen, S. J., Buschard, K., & Hansen, A. K. (2012). Early life treatment with vancomycin propagates *Akkermansia muciniphila* and reduces diabetes incidence in the NOD mouse. *Diabetologia*, *55*(8), 2285–2294. DOI: 10.1007/s00125-012-2564-7 PMID: 22572803

Hanumanthappa, R., Parthasarathy, A., Heggannavar, G. B., Pc, K., Nanjaiah, H., Kumbhar, R., & Devaraju, K. S. (2024). Recent advances in therapeutic strategies for Alzheimer's and Parkinson's disease using protein/peptide co-modified polymer nanoparticles. *Neuroprotection*, *2*(4), 255–275. DOI: 10.1002/nep3.60

Hao, B., Yang, Z., Liu, H., Liu, Y., & Wang, S. (2024). Advances in Flavonoid Research: Sources, Biological Activities, and Developmental Prospectives. *Current Issues in Molecular Biology 2024, Vol. 46, Pages 2884-2925, 46*(4), 2884–2925. DOI: 10.3390/cimb46040181

Hao, J., Shen, W., Yu, G., Jia, H., Li, X., Feng, Z., Wang, Y., Weber, P., Wertz, K., Sharman, E., & Liu, J. (2010). Hydroxytyrosol promotes mitochondrial biogenesis and mitochondrial function in 3T3-L1 adipocytes. *The Journal of Nutritional Biochemistry*, *21*(7), 634–644. DOI: 10.1016/j.jnutbio.2009.03.012 PMID: 19576748

Hao, T., Du, X., Yang, S., Zhang, Y., & Liang, F. (2020). Astrocytes-induced neuronal inhibition contributes to depressive-like behaviors during chronic stress. *Life Sciences*, *258*, 118099. DOI: 10.1016/j.lfs.2020.118099 PMID: 32682917

Hardy, T. M., & Tollefsbol, T. O. (2011). Epigenetic diet: Impact on the epigenome and cancer. *Epigenomics*, *3*(4), 503–518. DOI: 10.2217/epi.11.71 PMID: 22022340

Harkey, H. L.III, White, E. A.IV, Tibbs, R. E.Jr, & Haines, D. E. (2003). A clinician's view of spinal cord injury. *Anatomical Record. Part B, New Anatomist*, *271*(1), 41–48. DOI: 10.1002/ar.b.10012 PMID: 12619085

Harris, J. J., Jolivet, R., & Attwell, D. (2012). Synaptic energy use and supply. *Neuron*, *75*(5), 762–777. DOI: 10.1016/j.neuron.2012.08.019 PMID: 22958818

Hartmann, R., & Meisel, H. (2007). Food-derived peptides with biological activity: From research to food applications. *Current Opinion in Biotechnology*, *18*(2), 163–169. DOI: 10.1016/j.copbio.2007.01.013 PMID: 17292602

Hasan, S., Khatri, N., Rahman, Z. N., Menezes, A. A., Martini, J., Shehjar, F., Mujeeb, N., & Shah, Z. A. (2023). Neuroprotective Potential of Flavonoids in Brain Disorders. *Brain Sciences*, *13*(9), 1258. DOI: 10.3390/brainsci13091258 PMID: 37759859

Hasegawa, Y., Suzuki, H., Sozen, T., Altay, O., & Zhang, J. H. (2011). Apoptotic mechanisms for neuronal cells in early brain injury after subarachnoid hemorrhage. *Early Brain Injury or Cerebral Vasospasm: Volume 1. Pathophysiology : the Official Journal of the International Society for Pathophysiology*, •••, 43–48.

Hassanein, E. H. M., Althagafy, H. S., Baraka, M. A., Abd-alhameed, E. K., Ibrahim, I. M., Abd El-Maksoud, M. S., Mohamed, N. M., & Ross, S. A. (2024). The promising antioxidant effects of lignans: Nrf2 activation comes into view. *Naunyn-Schmiedeberg's Archives of Pharmacology*, *397*(9), 6439–6458. DOI: 10.1007/s00210-024-03102-x PMID: 38695909

Hassanpour, S. H., & Doroudi, A. (2023). Review of the antioxidant potential of flavonoids as a subgroup of polyphenols and partial substitute for synthetic antioxidants. *Avicenna Journal of Phytomedicine*, *13*(4), 354. PMID: 37663389

Ha, T., Kim, M. K., Park, K.-S., Jung, W., Choo, H., & Chong, Y. (2018). Structural Modification of (-)-Epigallocatechin Gallate (EGCG) Shows Significant Enhancement in Mitochondrial Biogenesis. *Journal of Agricultural and Food Chemistry*, *66*(15), 3850–3859. DOI: 10.1021/acs.jafc.8b00364 PMID: 29514455

Hatrock, D., Caporicci-Dinucci, N., & Stratton, J. (2020). Ependymal cells and multiple sclerosis: Proposing a relationship. In *Neural Regeneration Research* (Vol. 15, Issue 2, pp. 263–264). Wolters Kluwer Medknow Publications. DOI: 10.4103/1673-5374.265551

Hayashi, A., Sato, T., Kamada, N., Mikami, Y., Matsuoka, K., Hisamatsu, T., Hibi, T., Roers, A., Yagita, H., Ohteki, T., Yoshimura, A., & Kanai, T. (2013). A single strain of Clostridium butyricum induces intestinal IL-10-producing macrophages to suppress acute experimental colitis in mice. *Cell Host & Microbe*, *13*(6), 711–722. DOI: 10.1016/j.chom.2013.05.013 PMID: 23768495

Hedreen, J. C., Berretta, S., & White, C. L.III. (2024). Postmortem neuropathology in early Huntington disease. *Journal of Neuropathology and Experimental Neurology*, *83*(5), 294–306. DOI: 10.1093/jnen/nlae022 PMID: 38553027

He, F. J., Nowson, C. A., & MacGregor, G. A. (2006). Fruit and vegetable consumption and stroke: Meta-analysis of cohort studies. *Lancet*, *367*(9507), 320–326. DOI: 10.1016/s0140-6736(06)68069-0

He, F. Q., Qiu, B. Y., Zhang, X. H., Li, T. K., Xie, Q., Cui, D. J., Huang, X. L., & Gan, H. T. (2011). Tetrandrine attenuates spatial memory impairment and hippocampal neuroinflammation via inhibiting NF-κB activation in a rat model of Alzheimer's disease induced by amyloid-β(1-42). *Brain Research*, *1384*, 89–96. DOI: 10.1016/j.brainres.2011.01.103 PMID: 21300035

Heidary Moghaddam, R., Samimi, Z., Moradi, S. Z., Little, P. J., Xu, S., & Farzaei, M. H. (2020). Naringenin and naringin in cardiovascular disease prevention: A preclinical review. *European Journal of Pharmacology*, *887*, 173535. DOI: 10.1016/j.ejphar.2020.173535

He, J., & Giusti, M. M. (2010). Anthocyanins: Natural colorants with health-promoting properties. *Annual Review of Food Science and Technology*, *1*(1), 163–187. DOI: 10.1146/annurev.food.080708.100754 PMID: 22129334

Heneka, M. T., Kummer, M. P., & Latz, E. (2014). Innate immune activation in neurodegenerative disease. *Nature Reviews. Immunology*, *14*(7), 463–477. DOI: 10.1038/nri3705 PMID: 24962261

Henrich, M. T., Oertel, W. H., Surmeier, D. J., & Geibl, F. F. (2023). Mitochondrial dysfunction in Parkinson's disease–a key disease hallmark with therapeutic potential. *Molecular Neurodegeneration*, *18*(1), 83. DOI: 10.1186/s13024-023-00676-7 PMID: 37951933

Heo, S.-D., Kim, J., Choi, Y., Ekanayake, P., Ahn, M., & Shin, T. (2020). Hesperidin improves motor disability in rat spinal cord injury through anti-inflammatory and antioxidant mechanism via Nrf-2/HO-1 pathway. *Neuroscience Letters*, *715*, 134619. DOI: 10.1016/j.neulet.2019.134619 PMID: 31715292

He, Q., Li, Z., Wang, Y., Hou, Y., Li, L., & Zhao, J. (2017). Resveratrol alleviates cerebral ischemia/reperfusion injury in rats by inhibiting NLRP3 inflammasome activation through Sirt1-dependent autophagy induction. *International Immunopharmacology*, *50*, 208–215. DOI: 10.1016/j.intimp.2017.06.029 PMID: 28683365

He, X., Liu, X., Yang, J., Du, H., Chai, N., Sha, Z., Geng, M., Zhou, X., & He, C. (2020). Tannic acid-reinforced methacrylated chitosan/methacrylated silk fibroin hydrogels with multifunctionality for accelerating wound healing. *Carbohydrate Polymers*, *247*, 116689. DOI: 10.1016/j.carbpol.2020.116689 PMID: 32829817

Higham, J. P., Hidalgo, A., & Gistelinck, M. (2019). Restoration of olfactory memory in *Drosophila* overexpressing human Alzheimer's disease associated tau by manipulation of L-type Ca^{2+} channels. *Frontiers in Cellular Neuroscience*, *13*, 409. Advance online publication. DOI: 10.3389/fncel.2019.00409 PMID: 31551716

Hill, C., Guarner, F., Reid, G., Gibson, G. R., Merenstein, D. J., Pot, B., Morelli, L., Canani, R. B., Flint, H. J., Salminen, S., Calder, P. C., & Sanders, M. E. (2014). Expert consensus document: The International Scientific Association for Probiotics and Prebiotics consensus statement on the scope and appropriate use of the term probiotic. *Nature Reviews. Gastroenterology & Hepatology*, *11*(8), 506–514. DOI: 10.1038/nrgastro.2014.66 PMID: 24912386

Hirsch, E. C., & Hunot, S. (2009). Neuroinflammation in Parkinson's disease: A target for neuroprotection? *Lancet Neurology*, *8*(4), 382–397. DOI: 10.1016/s1474-4422(09)70062-6

Ho, E., Beaver, L. M., Williams, D. E., & Dashwood, R. H. (2011). Dietary factors and epigenetic regulation for prostate cancer prevention. *Advances in Nutrition*, *2*(6), 497–510. DOI: 10.3945/an.111.001032 PMID: 22332092

Holbrook, J. A., Jarosz-Griffiths, H. H., Caseley, E., Lara-Reyna, S., Poulter, J. A., Williams-Gray, C. H., Peckham, D., & McDermott, M. F. (2021). Neurodegenerative disease and the NLRP3 inflammasome. *Frontiers in Pharmacology*, *12*, 643254. DOI: 10.3389/fphar.2021.643254 PMID: 33776778

Hölscher, C. (2014). Central effects of GLP-1: New opportunities for treatments of neurodegenerative diseases. *The Journal of Endocrinology*, *221*(1), T31–T41. DOI: 10.1530/JOE-13-0221 PMID: 23999914

Hong, B., Ohtake, Y., Itokazu, T., & Yamashita, T. (2024). Glial senescence enhances α-synuclein pathology owing to its insufficient clearance caused by autophagy dysfunction. *Cell Death Discovery*, *10*(1), 50. Advance online publication. DOI: 10.1038/s41420-024-01816-8 PMID: 38272865

Hong, W., Zhi, F. X., Kun, T. H., Hua, F. J., Ling, L. H., Fang, F., & Yang, L. C. (2021). 6-Gingerol attenuates ventilator-induced lung injury via anti-inflammation and antioxidative stress by modulating the PPARγ/NF-κB signalling pathway in rats. *International Immunopharmacology*, *92*, 107367.

Hooper, L. V., Littman, D. R., and Macpherson, A. J. (2012). Interactions between the microbiota and the immune system. *science*, *336*(6086), 1268-1273.

Hornedo-Ortega, R., Cerezo, A. B., De Pablos, R. M., Krisa, S., Richard, T., García-Parrilla, M. C., & Troncoso, A. M. (2018). Phenolic compounds characteristic of the mediterranean diet in mitigating microglia-mediated neuroinflammation. *Frontiers in Cellular Neuroscience*, *12*, 373. DOI: 10.3389/fncel.2018.00373 PMID: 30405355

Hossain, M. T., Furhatun-Noor, A., Matin, A., Tabassum, F., & Ar Rashid, H. (2021). A review study on the pharmacological effects and mechanism of action of tannins. *European Journal Pharmaceutical and Medical Research*, (August), 5–10.

Hou, K., Pan, H., Shahpasand-Kroner, H., Hu, C., Abskharon, R., Seidler, P., Mekkittikul, M., Balbirnie, M., Lantz, C., Sawaya, M. R., Dolinsky, J. L., Jones, M., Zuo, X., Loo, J. A., Frautschy, S., Cole, G., & Eisenberg, D. S. (2024). D-peptide-magnetic nanoparticles fragment tau fibrils and rescue behavioral deficits in a mouse model of Alzheimer's disease. *Science Advances, 10*(18), eadl2991. Advance online publication. DOI: 10.1126/sciadv.adl2991 PMID: 38691615

Hsu, C.-C., Huang, H.-C., Wu, P.-T., Tai, T.-W., & Jou, I.-M. (2016). Sesame oil improves functional recovery by attenuating nerve oxidative stress in a mouse model of acute peripheral nerve injury: Role of Nrf-2. *The Journal of Nutritional Biochemistry, 38*, 102–106. DOI: 10.1016/j.jnutbio.2016.09.003 PMID: 27732910

Hua, K. F., Chao, A. C., Lin, T. Y., Chen, W. T., Lee, Y. C., Hsu, W. H., Lee, S. L., Wang, H. M., Yang, D. I., & Ju, T. C. (2022). Ginsenoside compound K reduces the progression of Huntington's disease via the inhibition of oxidative stress and overactivation of the ATM/AMPK pathway. *Journal of Ginseng Research, 46*(4), 572–584. DOI: 10.1016/j.jgr.2021.11.003 PMID: 35818427

Huang, J., Chen, B., Wang, H., Hu, S., Yu, X., Reilly, J., He, Z., You, Y., & Shu, X. (2022). *Dihydromyricetin Attenuates Depressive-like Behaviors in Mice by Inhibiting the AGE-RAGE Signaling Pathway.*

Huang, F., Teng, K., Liu, Y., Wang, T., Xia, T., Yun, F., & Zhong, J. (2022). Nisin Z attenuates lipopolysaccharide-induced mastitis by inhibiting the ERK1/2 and p38 mitogen-activated protein kinase signaling pathways. *Journal of Dairy Science, 105*(4), 3530–3543. DOI: 10.3168/jds.2021-21356 PMID: 35181137

Huang, Q., Liao, C., Ge, F., Ao, J., & Liu, T. (2022). Acetylcholine bidirectionally regulates learning and memory. *Journal of Neurorestoratology, 10*(2), 100002. DOI: 10.1016/j.jnrt.2022.100002

Huang, Q., Wu, L. J., Tashiro, S., Gao, H. Y., Onodera, S., & Ikejima, T. (2005). (+)-Catechin, an ingredient of green tea, protects murine microglia from oxidative stress-induced DNA damage and cell cycle arrest. *Journal of Pharmacological Sciences, 98*(1), 16–24. DOI: 10.1254/jphs.fpj04053x

Huang, W. J., Zhang, X., & Chen, W. W. (2016). Role of oxidative stress in Alzheimer's disease. *Biomedical Reports, 4*(5), 519–522. DOI: 10.3892/br.2016.630 PMID: 27123241

Huang, W., Yan, J., Shen, B., & Zhong, Z. (2016). Anthocyanins protect against stress-induced neuroinflammation and cognitive deficits via gut–brain pathway modulation. *The Journal of Nutritional Biochemistry, 32*, 1–12. DOI: 10.1016/j.jnutbio.2016.01.001

Huang, Y., Chen, S., Luo, Y., & Han, Z. (2020). Crosstalk between inflammation and the BBB in stroke. *Current Neuropharmacology, 18*(12), 1227–1236. DOI: 10.2174/1570159X18666200620230321 PMID: 32562523

Huber, K., & Superti-Furga, G. (2011). After the grape rush: Sirtuins as epigenetic drug targets in neurodegenerative disorders. *Bioorganic & Medicinal Chemistry, 19*(12), 3616–3624. DOI: 10.1016/j.bmc.2011.01.018 PMID: 21306906

Huddleston, J. E. (2011). Metabolic regulation by ERK. *Nature Reviews. Molecular Cell Biology, 12*(9), 546. DOI: 10.1038/nrm3179 PMID: 21860387

Hu, G., Hong, D., Zhang, T., Duan, H., Wei, P., Guo, X., & Mu, X. (2018). Cynatratoside-C from Cynanchum atratum displays anti-inflammatory effect via suppressing TLR4 mediated NF-κB and MAPK signaling pathways in LPS-induced mastitis in mice. *Chemico-Biological Interactions*, *279*, 187–195. DOI: 10.1016/j.cbi.2017.10.017 PMID: 29113806

Hu, L. W., Yen, J. H., Shen, Y. T., Wu, K. Y., & Wu, M. J. (2014). Luteolin modulates 6-hydroxydopamine-induced transcriptional changes of stress response pathways in PC12 cells. *PLoS One*, *9*(5), e97880. DOI: 10.1371/journal.pone.0097880

Hung, C.-C., Lin, C.-H., Chang, H., Wang, C.-Y., Lin, S.-H., Hsu, P.-C., Sun, Y.-Y., Lin, T.-N., Shie, F.-S., Kao, L.-S., Chou, C.-M., & Lee, Y.-H. (2016). Astrocytic GAP43 induced by the TLR4/NF-κB/STAT3 axis attenuates astrogliosis-mediated microglial activation and neurotoxicity. *The Journal of Neuroscience : The Official Journal of the Society for Neuroscience*, *36*(6), 2027–2043. DOI: 10.1523/JNEUROSCI.3457-15.2016 PMID: 26865625

Hu, Q., Zhang, R., Dong, X., Yang, D., Du, Q., & Yu, W. (2024). Corrigendum to: Huperzine A ameliorates neurological deficits after spontaneous subarachnoid hemorrhage through endothelial cell pyroptosis inhibition. *Acta Biochimica et Biophysica Sinica*, *56*(5), 824. DOI: 10.3724/abbs.2024067 PMID: 38712432

Hussain, G., Huang, J., Rasul, A., Anwar, H., Imran, A., Maqbool, J., Razzaq, A., Aziz, N., Makhdoom, E., Konuk, M., & Sun, T. (2019). Putative Roles of Plant-Derived Tannins in Neurodegenerative and Neuropsychiatry Disorders: An Updated Review. *Molecules (Basel, Switzerland)*, *24*(12), 2213. Advance online publication. DOI: 10.3390/molecules24122213 PMID: 31200495

Hussain, G., Wang, J., Rasul, A., Anwar, H., Qasim, M., Zafar, S., Aziz, N., Razzaq, A., Hussain, R., De Aguilar, J.-L. G., & Sun, T. (2020). Current status of therapeutic approaches against peripheral nerve injuries: A detailed story from injury to recovery. *International Journal of Biological Sciences*, *16*(1), 116–134. DOI: 10.7150/ijbs.35653 PMID: 31892850

Hussain, G., Zhang, L., Rasul, A., Anwar, H., Sohail, M. U., Razzaq, A., Aziz, N., Shabbir, A., Ali, M., & Sun, T. (2018). Role of plant-derived flavonoids and their mechanism in attenuation of Alzheimer's and Parkinson's diseases: An update of recent data. *Molecules (Basel, Switzerland)*, *23*(4), 814. DOI: 10.3390/molecules23040814 PMID: 29614843

Hu, X., Leak, R. K., Shi, Y., Suenaga, J., Gao, Y., Zheng, P., & Chen, J. (2015). Microglial and macrophage polarization—New prospects for brain repair. *Nature Reviews. Neurology*, *11*(1), 56–64. DOI: 10.1038/nrneurol.2014.207 PMID: 25385337

Hwang, J.-Y., Aromolaran, K. A., & Zukin, R. S. (2017). The emerging field of epigenetics in neurodegeneration and neuroprotection. *Nature Reviews. Neuroscience*, *18*(6), 347–361. DOI: 10.1038/nrn.2017.46 PMID: 28515491

Hyronimus, B., Le Marrec, C., & Urdaci, M. C. (1998). Coagulin, a bacteriocin-like inhibitory substance produced by Bacillus coagulans I4. *Journal of Applied Microbiology*, *85*(1), 42–50. DOI: 10.1046/j.1365-2672.1998.00466.x PMID: 9721655

Ibero-Baraibar, I., Perez-Cornago, A., Ramirez, M. J., Martínez, J. A., & Zulet, M. A. (2016). An increase in plasma homovanillic acid with cocoa extract consumption is associated with the alleviation of depressive symptoms in overweight or obese adults on an energy restricted diet in a randomized controlled trial. *The Journal of Nutrition, 146*(4), 897S–904S. DOI: 10.3945/jn.115.222828 PMID: 26962189

Ibrahim, A. M., Pottoo, F. H., Dahiya, E. S., Khan, F. A., & Kumar, J. S. (2020). Neuron-glia interactions: Molecular basis of alzheimer's disease and applications of neuroproteomics. *The European Journal of Neuroscience*, *52*(2), 2931–2943.

Ignat, I., Volf, I., & Popa, V. I. (2011). A critical review of methods for characterisation of polyphenolic compounds in fruits and vegetables. *Food Chemistry*, *126*(4), 1821–1835. DOI: 10.1016/j.foodchem.2010.12.026 PMID: 25213963

Imamura, K., Hishikawa, N., Sawada, M., Nagatsu, T., Yoshida, M., & Hashizume, Y. (2003). Distribution of major histocompatibility complex class II-positive microglia and cytokine profile of Parkinson's disease brains. *Acta Neuropathologica*, *106*(6), 518–526. DOI: 10.1007/s00401-003-0766-2 PMID: 14513261

Imededdine, K., Yahya, M. F. Z. R., Bouzina, M. A., & Touil, A. (2023). Molecular interactions, binding stability, and synergistic inhibition on acetylcholinesterase activity of safranin O in combination with quercetin and gallic acid: In vitro and in silico study. *Journal of Molecular Structure*, *1286*, 135562. Advance online publication. DOI: 10.1016/j.molstruc.2023.135562

Imran, A., Maqbool, J., Rasul, A., Anwar, H., Malik, S. A., Akram, R., Sajid, F., Tariq, A. R., & Arshad, M. U. (2021). *Hussain, and G.* Neuroprotective Effects of Polyphenols., DOI: 10.2174/97816810893791210800006

Ishabiyi, F. O., Ogidi, J. O., Olukade, B. A., Amorha, C. C., El-Sharkawy, L. Y., Okolo, C. C., Adeniyi, T. M., Atasie, N. H., Ibrahim, A., & Balogun, T. A.. (2023). Computational evaluation of *Azadirachta indica*-derived bioactive compounds as potential inhibitors of NLRP3 in the treatment of Alzheimer's disease. *Journal of Alzheimer's Disease*, *92*(3), 979–998. 10.3233/JAD-221020

Ishisaka, M., Kakefuda, K., Yamauchi, M., Tsuruma, K., Shimazawa, M., Tsuruta, A., & Hara, H. (2011). Luteolin shows an antidepressant-like effect via suppressing endoplasmic reticulum stress. *Biological & Pharmaceutical Bulletin*, *34*(9), 1481–1486. DOI: 10.1248/bpb.34.1481 PMID: 21881237

Ishizawa, K., Komori, T., Sasaki, S., Arai, N., Mizutani, T., & Hirose, T. (2004). Microglial Activation Parallels System Degeneration in Multiple System Atrophy. *Journal of Neuropathology and Experimental Neurology*, *63*(1), 43–52. DOI: 10.1093/jnen/63.1.43 PMID: 14748560

Iskusnykh, I. Y., Zakharova, A. A., & Pathak, D. (2022). Glutathione in Brain Disorders and Aging. In *Molecules* (Vol. 27, Issue 1). MDPI. DOI: 10.3390/molecules27010324

Islam, F., Bepary, S., Nafady, M. H., Islam, M. R., Emran, T. B., Sultana, S., Huq, M. A., Mitra, S., Chopra, H., Sharma, R., Sweilam, S. H., Khandaker, M. U., & Idris, A. M. (2022). Polyphenols targeting oxidative stress in spinal cord injury: Current status and future vision. *Oxidative Medicine and Cellular Longevity*, *2022*(1), 8741787. DOI: 10.1155/2022/8741787 PMID: 36046682

Islam, F., Khan, M. M., Raza, S. S., Javed, H., Ahmad, A., Khan, A., Islam, F., & Safhi, M. M. (2012). Rutin protects dopaminergic neurons from oxidative stress in an animal model of parkinson's disease. *Neurotoxicity Research*, *22*(1), 1–15. DOI: 10.1007/s12640-011-9295-2 PMID: 22194158

Islam, F., Roy, S., Zehravi, M., Paul, S., Sutradhar, H., Yaidikar, L., & Nainu, F. (2024). Polyphenols targeting MAP kinase signaling pathway in neurological diseases: Understanding molecular mechanisms and therapeutic targets. *Molecular Neurobiology*, *61*(5), 2686–2706. DOI: 10.1007/s12035-023-03706-z PMID: 37922063

Itoh, T., Imano, M., Nishida, S., Tsubaki, M., Hashimoto, S., Ito, A., & Satou, T. (2011). (-)-Epigallocatechin-3-gallate protects against neuronal cell death and improves cerebral function after traumatic brain injury in rats. *Neuromolecular Medicine*, *13*(4), 300–309. DOI: 10.1007/s12017-011-8162-x PMID: 22038400

Itoh, T., Imano, M., Nishida, S., Tsubaki, M., Mizuguchi, N., Hashimoto, S., Ito, A., & Satou, T. (2012). (−)-Epigallocatechin-3-gallate increases the number of neural stem cells around the damaged area after rat traumatic brain injury. *Journal of Neural Transmission (Vienna, Austria)*, *119*(8), 877–890. DOI: 10.1007/s00702-011-0764-9 PMID: 22212485

Itoh, T., Tabuchi, M., Mizuguchi, N., Imano, M., Tsubaki, M., Nishida, S., Hashimoto, S., Matsuo, K., Nakayama, T., Ito, A., Munakata, H., & Satou, T. (2013). Neuroprotective effect of (–)-epigallocatechin-3-gallate in rats when administered pre-or post-traumatic brain injury. *Journal of Neural Transmission (Vienna, Austria)*, *120*(5), 767–783. DOI: 10.1007/s00702-012-0918-4 PMID: 23180302

Jaberi, K. R., Alamdari-Palangi, V., Savardashtaki, A., Vatankhah, P., Jamialahmadi, T., Tajbakhsh, A., & Sahebkar, A. (2024). Modulatory effects of phytochemicals on gut–brain axis: Therapeutic implication. *Current Developments in Nutrition*, *8*(6), 103785. DOI: 10.1016/j.cdnut.2024.103785 PMID: 38939650

Jabir, N. R., Khan, F. R., & Tabrez, S. (2018). Cholinesterase targeting by polyphenols: A therapeutic approach for the treatment of Alzheimer's disease. *CNS Neuroscience & Therapeutics*, *24*(9), 753–762. DOI: 10.1111/cns.12971 PMID: 29770579

Jacques, P. (2011). Surfactin and other lipopeptides from Bacillus spp. Biosurfactants: from genes to applications, 57-91.

Jaeger, B. N., Parylak, S. L., & Gage, F. H. (2018). Mechanisms of dietary flavonoid action in neuronal function and neuroinflammation. *Molecular Aspects of Medicine*, *61*, 50–62. DOI: 10.1016/j.mam.2017.11.003 PMID: 29117513

Jain, V., Pareek, A., Bhardwaj, Y. R., & Singh, N. (2013). Attenuating effect of standardized fruit extract of Punica granatum L in rat model of tibial and sural nerve transection induced neuropathic pain. *BMC Complementary and Alternative Medicine*, *13*(1), 274. DOI: 10.1186/1472-6882-13-274 PMID: 24499201

Jäkel, S., Agirre, E., Mendanha Falcão, A., van Bruggen, D., Lee, K. W., Knuesel, I., Malhotra, D., ffrench-Constant, C., Williams, A., & Castelo-Branco, G. (2019). Altered human oligodendrocyte heterogeneity in multiple sclerosis. *Nature 2019 566:7745*, *566*(7745), 543–547. DOI: 10.1038/s41586-019-0903-2

Jakobek, L. (2015). Interactions of polyphenols with carbohydrates, lipids and proteins. *Food Chemistry*, *175*, 556–567. DOI: 10.1016/j.foodchem.2014.12.013 PMID: 25577120

Jalouli, M., Rahman, M. A., Biswas, P., Rahman, H., Harrath, A. H., Lee, I. S., Kang, S., Choi, J., Park, M. N., & Kim, B. (2025). Targeting natural antioxidant polyphenols to protect neuroinflammation and neurodegenerative diseases: a comprehensive review. In *Frontiers in Pharmacology* (Vol. 16, p. 1492517). Frontiers Media SA. DOI: 10.3389/fphar.2025.1492517

Jamme, I., Petit, E., Divoux, D., Gerbi, A., Maixent, J.-M., & Nouvelot, A. (1995). Modulation of mouse cerebral Na+, K+-ATPase activity by oxygen free radicals. *Neuroreport*, *7*(1), 333–337. PMID: 8742483

Janda, E., Boi, L., & Carta, A. R. (2018). Microglial phagocytosis and its regulation: A therapeutic target in parkinson's disease? In *Frontiers in Molecular Neuroscience* (Vol. 11). Frontiers Media S.A., DOI: 10.3389/fnmol.2018.00144

Janle, E. M., Lila, M. A., Grannan, M., Wood, L., Higgins, A., Yousef, G. G., & Weaver, C. M. (2010). Pharmacokinetics and tissue distribution of 14C-labeled grape polyphenols in the periphery and the central nervous system following oral administration. *Journal of Medicinal Food*, *13*(4), 926–933. DOI: 10.1089/jmf.2009.0157

Javed, H., Khan, M. M., Ahmad, A., Vaibhav, K., Ahmad, M. E., Khan, A., Ashafaq, M., Islam, F., Siddiqui, M. S., Safhi, M. M., & Islam, F. (2012). Rutin prevents cognitive impairments by ameliorating oxidative stress and neuroinflammation in rat model of sporadic dementia of Alzheimer type. *Neuroscience*, *210*, 340–352. DOI: 10.1016/j.neuroscience.2012.02.046 PMID: 22441036

Javid, H., Oryani, M. A., Rezagholinejad, N., Esparham, A., Tajaldini, M., & Karimi-Shahri, M. (2024). RGD peptide in cancer targeting: Benefits, challenges, solutions, and potential integrin-RGD interactions. *Cancer Medicine*, *13*(2), e6800. DOI: 10.1002/cam4.6800 PMID: 38349028

Jaye, K., Li, C. G., Chang, D., & Bhuyan, D. J. (2022). The role of key gut microbial metabolites in the development and treatment of cancer. *Gut Microbes*, *14*(1), 2038865. DOI: 10.1080/19490976.2022.2038865 PMID: 35220885

Jenab, A., Roghanian, R., & Emtiazi, G. (2020). Bacterial Natural Compounds with Anti-Inflammatory and Immunomodulatory Properties (Mini Review). *Drug Design, Development and Therapy*, *14*, 3787–3801. DOI: 10.2147/DDDT.S261283 PMID: 32982183

Jeon, Y. M., Kwon, Y., Jo, M., Lee, S., Kim, S., & Kim, H. J. (2020). The Role of Glial Mitochondria in α-Synuclein Toxicity. In *Frontiers in Cell and Developmental Biology* (Vol. 8). Frontiers Media S.A., DOI: 10.3389/fcell.2020.548283

Jiang, E., Tang, Z., Yu, C., Yu, C., & Zhu, W. (2016). Protective effect of schisandrin B on cerebral ischemia reperfusion injury of rats and its mechanisms. *Journal of Jilin University Medicine Edition*, *42*(5), 860–865. DOI: 10.13481/J.1671-587X.20160504

Jiang, H., Zhang, X., Yu, Z., Zhang, Z., Deng, M., Zhao, J., & Ruan, B. (2018). Altered gut microbiota profile in patients with generalized anxiety disorder. *Journal of Psychiatric Research*, *104*, 130–136. DOI: 10.1016/j.jpsychires.2018.07.007 PMID: 30029052

Jiang, W., Huang, Y., Han, N., He, F., Li, M., Bian, Z., Liu, J., Sun, T., & Zhu, L. (2016). Quercetin suppresses NLRP3 inflammasome activation and attenuates histopathology in a rat model of spinal cord injury. *Spinal Cord*, *54*(8), 592–596. DOI: 10.1038/sc.2015.227 PMID: 26754474

Jiang, Y., Li, S., Liu, F., & Chen, X. (2020). Liposome-based delivery systems for polyphenols: Enhancing stability and bioavailability. *Journal of Functional Foods*, *70*, 103991. DOI: 10.1016/j.jff.2020.103991

Jiang, Y., Sun, J., Chandrapala, J., Majzoobi, M., Brennan, C., Zeng, X. A., & Sun, B. (2024). Recent progress of food-derived bioactive peptides: Extraction, purification, function, and encapsulation. *Food Frontiers*, *5*(3), 1240–1264. DOI: 10.1002/fft2.383

Jiang, Z., Zhang, J., Cai, Y., Huang, J., & You, L. (2017). Catechin attenuates traumatic brain injury-induced blood–brain barrier damage and improves longer-term neurological outcomes in rats. *Experimental Physiology*, *102*(10), 1269–1277. DOI: 10.1113/EP086520 PMID: 28678393

Jia, X., Mao, D., Guo, J., Ke, J., Zhu, Y., Zhao, X., Luo, Z., Liu, X., Tang, R., Hou, R., Lan, H., & zheng, Q. (2024). Epigallocatechin gallate attenuated high glucose-induced pancreatic beta cell dysfunction by modulating DRP1-mediated mitochondrial apoptosis pathways. *Scientific Reports*, *14*(1), 16809. DOI: 10.1038/s41598-024-67867-0 PMID: 39039202

Jie, Z., Xia, H., Zhong, S. L., Feng, Q., Li, S., Liang, S., Zhong, H., Liu, Z., Gao, Y., Zhao, H., Zhang, D., Su, Z., Fang, Z., Lan, Z., Li, J., Xiao, L., Li, J., Li, R., Li, X., & Kristiansen, K. (2017). The gut microbiome in atherosclerotic cardiovascular disease. *Nature Communications*, *8*(1), 845. Advance online publication. DOI: 10.1038/s41467-017-00900-1 PMID: 29018189

Jin, T., Zhang, Y., Botchway, B. O., Zhang, J., Fan, R., Zhang, Y., & Liu, X. (2022). Curcumin can improve Parkinson's disease via activating BDNF/PI3k/Akt signaling pathways. *Food and Chemical Toxicology*, *164*, 113091. DOI: 10.1016/j.fct.2022.113091 PMID: 35526734

Jin, W., Wang, J., Zhu, T., Yuan, B., Ni, H., Jiang, J., Wang, H., & Liang, W. (2014). Anti-inflammatory effects of curcumin in experimental spinal cord injury in rats. *Inflammation Research*, *63*(5), 381–387. DOI: 10.1007/s00011-014-0710-z PMID: 24468890

Jomova, K., Alomar, S. Y., Alwasel, S. H., Nepovimova, E., Kuca, K., & Valko, M. (2024). Several lines of antioxidant defense against oxidative stress: Antioxidant enzymes, nanomaterials with multiple enzyme-mimicking activities, and low-molecular-weight antioxidants. *Archives of Toxicology*, *98*(5), 1323–1367.

Jones, M. P., Dilley, J. B., Drossman, D., & Crowell, M. D. (2006, February). Brain-gut connections in functional GI disorders: Anatomic and physiologic relationships. *Neurogastroenterology and Motility*, *18*(2), 91–103. DOI: 10.1111/j.1365-2982.2005.00730.x PMID: 16420287

Jordan, P. A., Kayser-Bricker, K. J., & Miller, S. J. (2010). Asymmetric phosphorylation through catalytic P (III) phosphoramidite transfer: Enantioselective synthesis of D-myo-inositol-6-phosphate. *Proceedings of the National Academy of Sciences of the United States of America*, *107*(48), 20620–20624.

Joseph, J. A., Shukitt-Hale, B., Denisova, N. A., Bielinski, D., Martin, A., McEwen, J. J., & Bickford, P. C. (1999). Reversals of age-related declines in neuronal signaling and cognitive behavior with blueberry, spinach, or strawberry dietary supplementation. *The Journal of Neuroscience : The Official Journal of the Society for Neuroscience*, *19*(18), 8114–8121. DOI: 10.1523/JNEUROSCI.19-18-08114.1999 PMID: 10479711

Joseph, J. A., Shukitt-Hale, B., & Lau, F. C. (2007). Fruit polyphenols and their effects on neuronal signaling and behavior in senescence. *Annals of the New York Academy of Sciences*, *1100*(1), 470–485. DOI: 10.1196/annals.1395.052 PMID: 17460212

Joseph, S. V., Edirisinghe, I., & Burton-Freeman, B. M. (2016). Fruit polyphenols: A review of anti-inflammatory effects in humans. *Critical Reviews in Food Science and Nutrition*, *56*(3), 419–444. DOI: 10.1080/10408398.2013.767221 PMID: 25616409

Joshi, V., Mishra, R., Upadhyay, A., Amanullah, A., Poluri, K. M., Singh, S., & Mishra, A. (2019). Polyphenolic flavonoid (Myricetin) upregulated proteasomal degradation mechanisms: Eliminates neurodegenerative proteins aggregation. *Journal of Cellular Physiology*, *234*(11), 20900–20914. DOI: 10.1002/jcp.28695

Jucaite, A., Cselényi, Z., Kreisl, W. C., Rabiner, E. A., Varrone, A., Carson, R. E., Rinne, J. O., Savage, A., Schou, M., Johnström, P., Svenningsson, P., Rascol, O., Meissner, W. G., Barone, P., Seppi, K., Kaufmann, H., Wenning, G. K., Poewe, W., & Farde, L. (2022). Glia Imaging Differentiates Multiple System Atrophy from Parkinson's Disease: A Positron Emission Tomography Study with [11C]PBR28 and Machine Learning Analysis. *Movement Disorders*, *37*(1), 119–129. DOI: 10.1002/mds.28814 PMID: 34609758

Ju, D.-T., Kuo, W.-W., Ho, T.-J., Paul, C. R., Kuo, C.-H., Viswanadha, V. P., Lin, C.-C., Chen, Y.-S., Chang, Y.-M., & Huang, C.-Y. (2015). Protocatechuic acid from Alpinia oxyphylla induces Schwann cell migration via ERK1/2, JNK and p38 activation. *The American Journal of Chinese Medicine*, *43*(04), 653–665. DOI: 10.1142/S0192415X15500408 PMID: 26119854

Ju, D.-T., Liao, H.-E., Shibu, M. A., Ho, T.-J., Padma, V. V., Tsai, F.-J., Chung, L.-C., Day, C. H., Lin, C.-C., & Huang, C.-Y. (2015). Nerve regeneration potential of protocatechuic acid in RSC96 Schwann cells by induction of cellular proliferation and migration through IGF-IR-PI3K-Akt signaling. *The Chinese Journal of Physiology*, *58*(6), 412–419. DOI: 10.4077/CJP.2015.BAD340 PMID: 26717920

Jurcau, A. (2021). Insights into the pathogenesis of neurodegenerative diseases: Focus on mitochondrial dysfunction and oxidative stress. *International Journal of Molecular Sciences*, *22*(21), 11847. DOI: 10.3390/ijms222111847 PMID: 34769277

Juźwik, C. A., Drake, S. S., Zhang, Y., Paradis-Isler, N., Sylvester, C., Amar-Zifkin, A., Douglas, C., Morquette, B., Moore, C. S., & Fournier, A. E. (2019). microRNA dysregulation in neurodegenerative diseases: A systematic review. *Progress in Neurobiology*, *182*, 101664. Advance online publication. DOI: 10.1016/j.pneurobio.2019.101664 PMID: 31356849

Kaji, S., Maki, T., Ishimoto, T., Yamakado, H., & Takahashi, R. (2020). Insights into the pathogenesis of multiple system atrophy: Focus on glial cytoplasmic inclusions. In *Translational Neurodegeneration* (Vol. 9, Issue 1, pp. 1–15). BioMed Central. DOI: 10.1186/s40035-020-0185-5

Kale, J., Osterlund, E. J., & Andrews, D. W. (2018). BCL-2 family proteins: Changing partners in the dance towards death. *Cell Death and Differentiation*, *25*(1), 65–80. DOI: 10.1038/cdd.2017.186 PMID: 29149100

Kaliszewski, M., Knott, A. B., & Bossy-Wetzel, E. (2015). Primary cilia and autophagic dysfunction in Huntington's disease. *Cell Death and Differentiation*, *22*(9), 1413–1424.

Kalkan, Ç., Soykan, I., & Soydal, Ç. (2018). The role of probiotics in gut barrier function. *The Turkish Journal of Gastroenterology : the Official Journal of Turkish Society of Gastroenterology, 29*(2), 131–140. DOI: 10.5152/tjg.2018.18232

Kalt, W., Cassidy, A., Howard, L. R., Krikorian, R., Stull, A. J., Tremblay, F., & Rochefort, L. (2008). Recent research on the health benefits of blueberries and their anthocyanins. *Advances in Nutrition, 1*(1), 60–87. DOI: 10.3945/an.108.000304 PMID: 31329250

Kamada, N., Chen, G. Y., Inohara, N., & Núñez, G. (2013). Control of pathogens and pathobionts by the gut microbiota. *Nature Immunology, 14*(7), 685–690. DOI: 10.1038/ni.2608 PMID: 23778796

Kamada, N., Seo, S.-U., Chen, G. Y., & Núñez, G. (2013). Role of the gut microbiota in immunity and inflammatory disease. *Nature Reviews. Immunology, 13*(5), 321–335. DOI: 10.1038/nri3430 PMID: 23618829

Kanatani, K., Oshimura, M., & Sano, K. (1995). Isolation and characterization of acidocin A and cloning of the bacteriocin gene from Lactobacillus acidophilus. *Applied and Environmental Microbiology, 61*(3), 1061–1067. DOI: 10.1128/aem.61.3.1061-1067.1995 PMID: 7793908

Kang, K. S., Wen, Y., Yamabe, N., Fukui, M., Bishop, S. C., & Zhu, B. T. (2010). Dual beneficial effects of (-)-epigallocatechin-3-gallate on levodopa methylation and hippocampal neurodegeneration: In vitro and in vivo studies. *PLoS One, 5*(8), e11951. Advance online publication. DOI: 10.1371/journal.pone.0011951 PMID: 20700524

Kang, K., Park, Y. S., Park, M., Jang, M. J., Kim, S. M., Lee, J., Choi, J. Y., Jung, D. H., Chang, Y.-T., Yoon, M.-H., Lee, J. S., Nam, Y., & Choi, I. S. (2016). Axon-first neuritogenesis on vertical nanowires. *Nano Letters, 16*(1), 675–680. DOI: 10.1021/acs.nanolett.5b04458 PMID: 26645112

Kang, Y., & Cai, Y. (2018). Gut microbiota and hypertension: From pathogenesis to new therapeutic strategies. *Clinics and Research in Hepatology and Gastroenterology, 42*(2), 110–117. DOI: 10.1016/j.clinre.2017.09.006 PMID: 29102544

Kang, Y., Lee, J. H., Seo, Y. H., Jang, J. H., Jeong, C. H., Lee, S., & Park, B. (2019). Epicatechin Prevents Methamphetamine-Induced Neuronal Cell Death via Inhibition of ER Stress. *Biomolecules & Therapeutics, 27*(2), 145–151. DOI: 10.4062/biomolther.2018.092

Karak, P. (2019). Biological activities of flavonoids: An overview. *International Journal of Pharmaceutical Sciences and Research, 10*(4), 1567–1574.

Karczewski, J., Troost, F. J., Konings, I., Dekker, J., Kleerebezem, M., Brummer, R. J., & Wells, J. M. (2010). Regulation of human epithelial tight junction proteins by Lactobacillus plantarum in vivo and protective effects on the epithelial barrier. *American Journal of Physiology. Gastrointestinal and Liver Physiology, 298*(6), G851–G859. DOI: 10.1152/ajpgi.00327.2009 PMID: 20224007

Karioti, A., Carta, F., & Supuran, C. T. (2016). Phenols and polyphenols as carbonic anhydrase inhibitors. *Molecules (Basel, Switzerland), 21*(12), 1649. DOI: 10.3390/molecules21121649 PMID: 27918439

Karuppagounder, S., Madathil, S., Pandey, M., Haobam, R., Rajamma, U., & Mohanakumar, K. (2013). Quercetin up-regulates mitochondrial complex-I activity to protect against programmed cell death in rotenone model of Parkinson's disease in rats. *Neuroscience, 236*, 136–148. DOI: 10.1016/j.neuroscience.2013.01.032 PMID: 23357119

Kaspar, J. W., Niture, S. K., & Jaiswal, A. K. (2009). Nrf2:INrf2 (Keap1) signaling in oxidative stress. *Free Radical Biology & Medicine, 47*(9), 1304–1309. DOI: 10.1016/j.freeradbiomed.2009.07.035 PMID: 19666107

Kaur, P., & Sharma, S. (2018). Recent advances in pathophysiology of traumatic brain injury. *Current Neuropharmacology, 16*(8), 1224–1238. DOI: 10.2174/1570159X15666170613083606 PMID: 28606040

Kaur, S., Rana, A., Gangwani, S., & Sharma, R. (2012). Punica granatum attenuates sciatic nerve ligation induced-neuropathic pain. *International Journal of Pharmaceutical Sciences and Research, 3*(2), 509.

Kay, C. D., Mazza, G., Holub, B. J., & Wang, J. (2009). Anthocyanins and their metabolites in human health and disease. *Journal of Agricultural and Food Chemistry, 57*(13), 5823–5831. DOI: 10.1021/jf803548b PMID: 19579294

Kay, C. D., Pereira-Caro, G., Ludwig, I. A., Clifford, M. N., & Crozier, A. (2017). Anthocyanins and flavanones are poorly absorbed in humans but increase colonic metabolites. *Molecular Nutrition & Food Research, 61*(6), 1600911. DOI: 10.1002/mnfr.201600911

Kazim, S. F., Blanchard, J., Dai, C. L., Tung, Y. C., LaFerla, F. M., Iqbal, I. G., & Iqbal, K. (2014). Disease modifying effect of chronic oral treatment with a neurotrophic peptidergic compound in a triple transgenic mouse model of Alzheimer's disease. *Neurobiology of Disease, 71*, 110–130. DOI: 10.1016/j.nbd.2014.07.001 PMID: 25046994

Kejík, Z., Kaplánek, R., Masařík, M., Babula, P., Matkowski, A., Filipenský, P., & Jakubek, M. (2021). Iron Complexes of Flavonoids-Antioxidant Capacity and Beyond. *International Journal of Molecular Sciences, 22*(2). Advance online publication. DOI: 10.3390/ijms22020646

Kelly, J. R., Allen, A. P., Temko, A., Hutch, W., Kennedy, P. J., Farid, N., Murphy, E., Boylan, G., Bienenstock, J., Cryan, J. F., Clarke, G., & Dinan, T. G. (2017). Lost in translation? The potential psychobiotic *Lactobacillus rhamnosus* (JB-1) fails to modulate stress or cognitive performance in healthy male subjects. *Brain, Behavior, and Immunity, 61*, 50–59. DOI: 10.1016/j.bbi.2016.11.018 PMID: 27865949

Kelsey, N. A., Wilkins, H. M., & Linseman, D. A. (2010). Nutritionally derived anthocyanins as potential neuronal survival agents. *Molecular Nutrition & Food Research, 54*(3), 377–395. DOI: 10.1002/mnfr.200900430

Kent, K., Charlton, K., Roodenrys, S., Batterham, M., Potter, J., & Traynor, V. (2017). Consumption of anthocyanin-rich cherry juice improves sleep quality in adults with insomnia. *European Journal of Nutrition, 56*(8), 3303–3312. DOI: 10.1007/s00394-016-1300-7

Kerman, M., Kanter, M., Coşkun, K. K., Erboga, M., & Gurel, A. (2012). Neuroprotective effects of caffeic acid phenethyl ester on experimental traumatic brain injury in rats. *Journal of Molecular Histology, 43*(1), 49–57. DOI: 10.1007/s10735-011-9376-9 PMID: 22124729

Khamse, S., Sadr, S. S., Roghani, M., Hasanzadeh, G., & Mohammadian, M. (2015). Rosmarinic acid exerts a neuroprotective effect in the kainate rat model of temporal lobe epilepsy: Underlying mechanisms. *Pharmaceutical Biology*, *53*(12), 1818–1825. DOI: 10.3109/13880209.2015.1010738 PMID: 25874386

Khan, A., Ali, T., Rehman, S. U., Waheed, A., & Kim, M. O. (2020). Neuroprotection: Targeting multiple pathways by naturally occurring phytochemicals. *Biomedicines*, *8*(8), 284. Advance online publication. DOI: 10.3390/biomedicines8080284 PMID: 32806490

Khan, A., Ikram, M., Hahm, J. R., & Kim, M. O. (2020). Antioxidant and Anti-Inflammatory Effects of Citrus Flavonoid Hesperetin: Special Focus on Neurological Disorders. *Antioxidants*, *9*(7). Advance online publication. DOI: 10.3390/antiox9070609

Khan, N., Pelletier, D., McAlear, T. S., Croteau, N., Veyron, S., Bayne, A. N., Black, C., Ichikawa, M., Khalifa, A. A. Z., Chaaban, S., Kurinov, I., Brouhard, G., Bechstedt, S., Bui, K. H., & Trempe, J. F. (2021). Crystal structure of human PACRG in complex with MEIG1 reveals roles in axoneme formation and tubulin binding. *Structure (London, England)*, *29*(6), 572–586.e6. DOI: 10.1016/j.str.2021.01.001 PMID: 33529594

Khanna, S., Kumar, S., Sharma, P., Daksh, R., Nandakumar, K., & Shenoy, R. R. (2025). Flavonoids regulating NLRP3 inflammasome: A promising approach in alleviating diabetic peripheral neuropathy. *Inflammopharmacology*, *2025*(5), 1–32. DOI: 10.1007/s10787-025-01729-7 PMID: 40205269

Khan, S., Ullah, M. W., Siddique, R., Nabi, G., Manan, S., Yousaf, M., & Hou, H. (2016). Role of recombinant DNA technology to improve life. *International Journal of Genomics*, *2016*(1), 2405954. DOI: 10.1155/2016/2405954 PMID: 28053975

Khatri, D., & Juvekar, A. (2016). Abrogation of locomotor impairment in a rotenone-induced Drosophila melanogaster and zebrafish model of Parkinson's disease by ellagic acid and curcumin. *International Journal of Nutrition, Pharmacology, Neurological Diseases*, *6*(2), 90–96. DOI: 10.4103/2231-0738.179969

Khelfaoui, H., Ibaceta-Gonzalez, C., & Angulo, M. C. (2024). Functional myelin in cognition and neurodevelopmental disorders. *Cellular and Molecular Life Sciences 2024 81:1*, *81*(1), 1–14. DOI: 10.1007/s00018-024-05222-2

Khoo, H. E., Azlan, A., Tang, S. T., & Lim, S. M. (2017). Anthocyanidins and anthocyanins: Colored pigments as food, pharmaceutical ingredients, and the potential health benefits. *Food & Nutrition Research*, *61*(1), 1361779. DOI: 10.1080/16546628.2017.1361779 PMID: 28970777

Khorshidian, N., Khanniri, E., Mohammadi, M., Mortazavian, A. M., & Yousefi, M. (2021). Antibacterial activity of pediocin and pediocin-producing bacteria against Listeria monocytogenes in meat products. *Frontiers in Microbiology*, *12*, 709959. DOI: 10.3389/fmicb.2021.709959 PMID: 34603234

Kian, K., Khalatbary, A. R., Ahmadvand, H., Karimpour Malekshah, A., & Shams, Z. (2019). Neuroprotective effects of (−)-epigallocatechin-3-gallate (EGCG) against peripheral nerve transection-induced apoptosis. *Nutritional Neuroscience*, *22*(8), 578–586. DOI: 10.1080/1028415X.2017.1419542 PMID: 29292676

Kim, B., Kang, B., Vales, T. P., Yang, S. K., Lee, J., & Kim, H. J. (2018). Polyphenol-functionalized hydrogels using an interpenetrating chitosan network and investigation of their antioxidant activity. *Macromolecular Research*, *26*(1), 35–39.

Kim, D. H., Hung, T. M., Bae, K. H., Jung, J. W., Lee, S., Yoon, B. H., Cheong, J. H., Ko, K. H., & Ryu, J. H. (2006). Gomisin A improves scopolamine-induced memory impairment in mice. *European Journal of Pharmacology*, *542*(1–3), 129–135. DOI: 10.1016/j.ejphar.2006.06.015 PMID: 16824513

Kim, D., Nguyen, M. D., Dobbin, M. M., Fischer, A., Sananbenesi, F., Rodgers, J. T., Delalle, I., Baur, J. A., Sui, G., Armour, S. M., Puigserver, P., Sinclair, D. A., & Tsai, L. H. (2007). SIRT1 deacetylase protects against neurodegeneration in models for Alzheimer's disease and amyotrophic lateral sclerosis. *The EMBO Journal*, *26*(13), 3169–3179. DOI: 10.1038/sj.emboj.7601758 PMID: 17581637

Kim, H., Kim, Y. S., Kim, S. Y., & Suk, K. (2001). The plant flavonoid wogonin suppresses death of activated C6 rat glial cells by inhibiting nitric oxide production. *Neuroscience Letters*, *309*(1), 67–71. DOI: 10.1016/s0304-3940(01)02028-6

Kim, I. W., Choi, R. Y., Lee, J. H., Seo, M., Lee, H. J., Kim, M. A., Kim, S. H., Kim, I., & Hwang, J. S. (2021). Anticancer activity of periplanetasin-5, an antimicrobial peptide from the cockroach Periplaneta americana. *Journal of Microbiology and Biotechnology*, *31*(10), 1343–1349. DOI: 10.4014/jmb.2104.04040 PMID: 34409948

Kim, J.-E., Kim, T.-H., & Kang, T.-C. (2023). EGCG Attenuates CA1 Neuronal Death by Regulating GPx1, NF-κB S536 Phosphorylation and Mitochondrial Dynamics in the Rat Hippocampus following Status Epilepticus. *Antioxidants*, *12*(4), 966. Advance online publication. DOI: 10.3390/antiox12040966 PMID: 37107343

Kim, M., Jung, J., Jeong, N. Y., & Chung, H.-J. (2019). The natural plant flavonoid apigenin is a strong antioxidant that effectively delays peripheral neurodegenerative processes. *Anatomical Science International*, *94*(4), 285–294. DOI: 10.1007/s12565-019-00486-2 PMID: 30949912

Kim, S. W., Hooker, J. M., Otto, N., Win, K., Muench, L., Shea, C., & Fowler, J. S. (2013). Whole-body pharmacokinetics of HDAC inhibitor drugs, butyric acid, valproic acid and 4-phenylbutyric acid measured with carbon-11 labeled analogs by PET. *Nuclear Medicine and Biology*, *40*(7), 912–918. DOI: 10.1016/j.nucmedbio.2013.06.007 PMID: 23906667

Kim, S., Chun, H., Kim, Y., Kim, Y., Park, U., Chu, J., Bhalla, M., Choi, S. H., Yousefian-Jazi, A., Kim, S., Hyeon, S. J., Kim, S., Kim, Y., Ju, Y. H., Lee, S. E., Lee, H., Lee, K., Oh, S. J., Hwang, E. M., & Ryu, H. (2024). Astrocytic autophagy plasticity modulates Aβ clearance and cognitive function in Alzheimer's disease. *Molecular Neurodegeneration*, *19*(1), 1–26. DOI: 10.1186/s13024-024-00740-w PMID: 39044253

Kim, S.-R., Seong, K.-J., Kim, W.-J., & Jung, J.-Y. (2022). Epigallocatechin Gallate Protects against Hypoxia-Induced Inflammation in Microglia via NF-κB Suppression and Nrf-2/HO-1 Activation. *International Journal of Molecular Sciences*, *23*(7), 4004. Advance online publication. DOI: 10.3390/ijms23074004 PMID: 35409364

Kirschweng, B., Tátraaljai, D., Földes, E., & Pukánszky, B. (2017). Natural antioxidants as stabilizers for polymers. *Polymer Degradation & Stability*, *145*, 25–40. DOI: 10.1016/j.polymdegradstab.2017.07.012

Kishimoto, Y., Tani, M., & Kondo, K. (2013). Pleiotropic preventive effects of dietary polyphenols in cardiovascular diseases. *European Journal of Clinical Nutrition*, *67*(5), 532–535. DOI: 10.1038/ejcn.2013.29 PMID: 23403879

Kister, A., & Kister, I. (2023). Overview of myelin, major myelin lipids, and myelin-associated proteins. *Frontiers in Chemistry*, *10*, 1041961.

Klein, C., & Westenberger, A. (2012). Genetics of Parkinson's disease. *Cold Spring Harbor Perspectives in Medicine*, *2*(1), a008888. DOI: 10.1101/cshperspect.a008888 PMID: 22315721

Klemmensen, M. M., Borrowman, S. H., Pearce, C., Pyles, B., & Chandra, B. (2024). Mitochondrial dysfunction in neurodegenerative disorders. *Neurotherapeutics; the Journal of the American Society for Experimental NeuroTherapeutics*, *21*(1), e00292. DOI: 10.1016/j.neurot.2023.10.002 PMID: 38241161

Köbe, T., Witte, A. V., Schnelle, A., Tesky, V. A., Pantel, J., Schuchardt, J.-P., Hahn, A., Bohlken, J., Grittner, U., & Flöel, A. (2017). Impact of resveratrol on glucose control, hippocampal structure and connectivity, and memory performance in patients with mild cognitive impairment. *Frontiers in Neuroscience*, *11*, 105. DOI: 10.3389/fnins.2017.00105 PMID: 28326010

Kolida, S., & Gibson, G. R. (2007). Prebiotic capacity of inulin-type fructans. *The Journal of Nutrition*, *137*(11), 2503S–2506S. DOI: 10.1093/jn/137.11.2503S PMID: 17951493

Komorowska, J., Wątroba, M., & Szukiewicz, D. (2020). Review of the beneficial effects of resveratrol in neurodegenerative diseases such as Alzheimer's disease. *Advances in Medical Sciences*, *65*(2), 401–410. 10.1016/j.advms.2020.08.002

Kondo, S., El Omri, A., Han, J., & Isoda, H. (2015). Antidepressant-like effects of rosmarinic acid through mitogen-activated protein kinase phosphatase-1 and brain-derived neurotrophic factor modulation. *Journal of Functional Foods*, *14*, 758–766. DOI: 10.1016/j.jff.2015.03.001

Kong, Y. R., Tay, K. C., Su, Y. X., Khaw, K. Y., Wong, C. K., & Tan, W. N. (2021). Potential of Naturally Derived Alkaloids as Multi-Targeted Therapeutic Agents for Neurodegenerative Diseases. *Molecules 2021, Vol. 26, Page 728*, *26*(3), 728. DOI: 10.3390/molecules26030728

Kooshki, L., Ghasemi, N., & Azadbakht, M. (2023). The pivotal role of JAK/STAT and IRS/PI3K signaling pathways in neurodegenerative diseases: Mechanistic approaches to polyphenols and alkaloids. *Phytomedicine*, *112*, 154686. Advance online publication. DOI: 10.1016/j.phymed.2023.154686 PMID: 36804755

Korhonen, H., & Pihlanto, A. (2006). Bioactive peptides: Production and functionality. *International Dairy Journal*, *16*(9), 945–960. DOI: 10.1016/j.idairyj.2005.10.012

Kormas, P., & Moutzouri, A. (2023). Current psychological approaches in neurodegenerative diseases. In *Handbook of computational neurodegeneration* (pp. 261–289). Springer. DOI: 10.1007/978-3-319-75922-7_10

Koudoufio, M., Desjardins, Y., Feldman, F., Spahis, S., Delvin, E., & Levy, E. (2020). Insight into Polyphenol and Gut Microbiota Crosstalk: Are Their Metabolites the Key to Understand Protective Effects against Metabolic Disorders? *Antioxidants*, *9*(10). Advance online publication. DOI: 10.3390/antiox9100982

Kovacs, G. G. (2018). Concepts and classification of neurodegenerative diseases. [): Elsevier.]. *Handbook of Clinical Neurology*, *145*, 301–307. DOI: 10.1016/B978-0-12-802395-2.00021-3 PMID: 28987178

Krebs, B. (2016). Prebiotic and synbiotic treatment before colorectal surgery—Randomised double-blind trial. *Collegium Antropologicum*, *40*(1), 35–40. PMID: 27301235

Krikorian, R., Shidler, M. D., Nash, T. A., Kalt, W., Vinqvist-Tymchuk, M. R., Shukitt-Hale, B., & Joseph, J. A. (2010). Blueberry supplementation improves memory in older adults. *Journal of Agricultural and Food Chemistry*, *58*(7), 3996–4000. DOI: 10.1021/jf9029332 PMID: 20047325

Krishna, G., Ying, Z., & Gomez-Pinilla, F. (2019). Blueberry supplementation mitigates altered brain plasticity and behavior after traumatic brain injury in rats. *Molecular Nutrition & Food Research*, *63*(15), 1801055. DOI: 10.1002/mnfr.201801055 PMID: 31115168

Kropat, C., Mueller, D., Boettler, U., & Marko, D. (2013). Anthocyanins from blueberries inhibit inflammation in primary microglia. *Journal of Agricultural and Food Chemistry*, *61*(26), 6224–6230. DOI: 10.1021/jf401090k PMID: 23799644

Kruis, W., Frič, P., Pokrotnieks, J., Lukáš, M., Fixa, B., Kaščák, M., Kamm, M. A., Weismueller, J., Beglinger, C., Stolte, M., Wolff, C., & Schulze, J. (2004). Maintaining remission of ulcerative colitis with the probiotic Escherichia coli Nissle 1917 is as effective as with standard mesalazine. *Gut*, *53*(11), 1617–1623. DOI: 10.1136/gut.2003.037747 PMID: 15479682

Kruszka, J., Martyński, J., Szewczyk-Golec, K., Woźniak, A., & Nuszkiewicz, J. (2025). The Role of Selected Flavonoids in Modulating Neuroinflammation in Alzheimer's Disease: Mechanisms and Therapeutic Potential. *Brain Sciences*, *15*(5), 485. DOI: 10.3390/brainsci15050485 PMID: 40426656

Kübler, S., Soutschek, A., & Schubert, T. (2019). The causal role of the lateral prefrontal cortex for task-order coordination in dual-task situations: A study with transcranial magnetic stimulation. *Journal of Cognitive Neuroscience*, *31*(12), 1840–1856.

Kujawska, M., & Jodynis-Liebert, J. (2018). Polyphenols in Parkinson's disease: A systematic review of in vivo studies. *Nutrients*, *10*(5), 642. DOI: 10.3390/nu10050642 PMID: 29783725

Kulaszyńska, M., Kwiatkowski, S., & Skonieczna-Żydecka, K. (2024). The iron metabolism with a specific focus on the functioning of the nervous system. *Biomedicines*, *12*(3), 595.

Kulkarni, S. K., Bhutani, M. K., & Bishnoi, M. (2008). Antidepressant activity of curcumin: Involvement of serotonin and dopamine system. *Psychopharmacology*, *201*(3), 435–442. DOI: 10.1007/s00213-008-1300-y PMID: 18766332

Kulkarni, S. K., Dhir, A., & Akula, K. K. (2009). Potentials of curcumin as an antidepressant. *TheScientificWorldJournal*, *9*, 1233–1241. DOI: 10.1100/tsw.2009.137 PMID: 19882093

Kumar, G. P., Anilakumar, K. R., & Naveen, S. (2015). Phytochemicals having neuroprotective properties from dietary sources and medicinal herbs. In *Pharmacognosy Journal* (Vol. 7, Issue 1, pp. 1–17). DOI: 10.5530/pj.2015.1.1

Kumar, A., Chaudhary, T., & Mishra, J. (2013). Minocycline modulates neuroprotective effect of hesperidin against quinolinic acid induced Huntington's disease like symptoms in rats: Behavioral, biochemical, cellular and histological evidences. *European Journal of Pharmacology*, *720*(1–3), 16–28. DOI: 10.1016/j.ejphar.2013.10.057 PMID: 24211676

Kumar, A., & Konar, A. (2023). Neuromodulating bioactive compounds as potential cognitive therapeutics. *Frontiers in Aging Neuroscience*, *15*, 9936231. Advance online publication. 10.3389/fnagi.2023.9936231

Kumar, D., Ambasta, R. K., & Kumar, P. (2020). Ubiquitin biology in neurodegenerative disorders: From impairment to therapeutic strategies. *Ageing Research Reviews*, *61*, 101078. DOI: 10.1016/j.arr.2020.101078 PMID: 32407951

Kumari, A., Yadav, S. K., & Yadav, S. C. (2010). Biodegradable polymeric nanoparticles based drug delivery systems. *Colloids and Surfaces. B, Biointerfaces*, *75*(1), 1–18. DOI: 10.1016/j.colsurfb.2009.09.001 PMID: 19782542

Kumar, J., Park, K. C., Awasthi, A., & Prasad, B. (2015). Silymarin extends lifespan and reduces proteotoxicity in C. elegans Alzheimer's model. *CNS & Neurological Disorders - Drug Targets*, *14*(2), 295–302. DOI: 10.2174/1871527314666150116110212

Kumar, P., & Kumar, A. (2010). Protective effect of hesperidin and naringin against 3-nitropropionic acid induced Huntington's like symptoms in rats: Possible role of nitric oxide. *Behavioural Brain Research*, *206*(1), 38–46. DOI: 10.1016/j.bbr.2009.08.028 PMID: 19716383

Kumar, R., Kumar, R., Khurana, N., Singh, S. K., Khurana, S., Verma, S., Sharma, N., Vyas, M., Dua, K., Khursheed, R., Awasthi, A., & Vishwas, S. (2022). Improved neuroprotective activity of Fisetin through SNEDDS in ameliorating the behavioral alterations produced in rotenone-induced Parkinson's model. *Environmental Science and Pollution Research International*, *29*(33), 50488–50499. DOI: 10.1007/s11356-022-19428-z PMID: 35230633

Kumar, S., & Pandey, A. K. (2013). Chemistry and biological activities of flavonoids: An overview. *TheScientificWorldJournal*, *162750*. Advance online publication. DOI: 10.1155/2013/162750

Kumar, Y., Kaur, K., Shahi, A. K., Kairam, N., & Tyagi, S. K. (2017). Antilisterial, antimicrobial and antioxidant effects of pediocin and Murrayakoenigii berry extract in refrigerated goat meat emulsion. *Lebensmittel-Wissenschaft + Technologie*, *79*, 135–144. DOI: 10.1016/j.lwt.2017.01.028

Kung, H.-C., Lin, K.-J., Kung, C.-T., & Lin, T.-K. (2021). Oxidative stress, mitochondrial dysfunction, and neuroprotection of polyphenols with respect to resveratrol in Parkinson's disease. *Biomedicines*, *9*(8), 918. DOI: 10.3390/biomedicines9080918 PMID: 34440122

Küpeli Akkol, E., Karatoprak, G. Ş., Dumlupınar, B., Bahadır Acıkara, Ö., Arıcı, R., Yücel, Ç., Aynal, L. C., & Sobarzo Sánchez, E. (2025). Stilbenes Against Alzheimer's Disease: A Comprehensive Review of Preclinical Studies of Natural and Synthetic Compounds Combined with the Contributions of Developed Nanodrug Delivery Systems. *Molecules 2025, Vol. 30, Page 1982, 30*(9), 1982. DOI: 10.3390/molecules30091982

Kwon, H. S., & Koh, S.-H. (2020). Neuroinflammation in neurodegenerative disorders: The roles of microglia and astrocytes. *Translational Neurodegeneration, 9*(1), 42. DOI: 10.1186/s40035-020-00221-2 PMID: 33239064

Kyriakis, J. M., & Avruch, J. (2012). Mammalian MAPK signal transduction pathways activated by stress and inflammation: A 10-year update. *Physiological Reviews, 92*(2), 689–737. DOI: 10.1152/physrev.00028.2011 PMID: 22535895

Lacroix, S., Hamilton, L. K., Vaugeois, A., Beaudoin, S., Breault-Dugas, C., Pineau, I., Lévesque, S. A., Grégoire, C. A., & Fernandes, K. J. L. (2014). Central canal ependymal cells proliferate extensively in response to traumatic spinal cord injury but not demyelinating lesions. *PLoS One, 9*(1), e85916. Advance online publication. DOI: 10.1371/journal.pone.0085916 PMID: 24475059

Lagouge, M., Argmann, C., Gerhart-Hines, Z., Meziane, H., Lerin, C., Daussin, F., Messadeq, N., Milne, J., Lambert, P., Elliott, P., Geny, B., Laakso, M., Puigserver, P., & Auwerx, J. (2006). Resveratrol improves mitochondrial function and protects against metabolic disease by activating SIRT1 and PGC-1α. *Cell, 127*(6), 1109–1122. DOI: 10.1016/j.cell.2006.11.013 PMID: 17112576

Lahooti, B., Chhibber, T., Bagchi, S., Varahachalam, S. P., & Jayant, R. D. (2021). Therapeutic role of inflammasome inhibitors in neurodegenerative disorders. *Brain, Behavior, and Immunity, 91*, 771–783. DOI: 10.1016/j.bbi.2020.11.004 PMID: 33157255

Laird, M. D., Sukumari-Ramesh, S., Swift, A. E., Meiler, S. E., Vender, J. R., & Dhandapani, K. M. (2010). Curcumin attenuates cerebral edema following traumatic brain injury in mice: A possible role for aquaporin-4? *Journal of Neurochemistry, 113*(3), 637–648. DOI: 10.1111/j.1471-4159.2010.06630.x PMID: 20132469

Lam, P. Y., & Ko, K. M. (2012). Beneficial effect of (-)Schisandrin B against 3-nitropropionic acid-induced cell death in PC12 cells. *BioFactors (Oxford, England), 38*(3), 219–225. DOI: 10.1002/biof.1009 PMID: 22488872

Lamport, D. J., & Williams, C. M. (2021). Polyphenols and Cognition In Humans: An Overview of Current Evidence from Recent Systematic Reviews and Meta-Analyses. *Brain Plasticity, 6*(2), 139–153. DOI: 10.3233/bpl-200111

Lan, D., Xu, N., Sun, J., Li, Z., Liao, R., Zhang, H., & Yi, W. (2018). Electroacupuncture mitigates endothelial dysfunction via effects on the PI3K/Akt signalling pathway in high fat diet-induced insulin-resistant rats. *Acupuncture in Medicine, 36*(3), 162–169.

Landete, J. M. (2012). Plant and mammalian lignans: A review of source, intake, metabolism, intestinal bacteria and health. *Food Research International, 46*(1), 410–424. DOI: 10.1016/j.foodres.2011.12.023

Lao, C. D., Ruffin, M. T. t., Normolle, D., Heath, D. D., Murray, S. I., Bailey, J. M., & Brenner, D. E. (2006). Dose escalation of a curcuminoid formulation. *BMC Complementary and Alternative Medicine*, *6*, 10. DOI: 10.1186/1472-6882-6-10

Laricchiuta, D., Papi, M., Decandia, D., Panuccio, A., Cutuli, D., Peciccia, M., Mazzeschi, C., & Petrosini, L. (2024). The role of glial cells in mental illness: A systematic review on astroglia and microglia as potential players in schizophrenia and its cognitive and emotional aspects. *Frontiers in Cellular Neuroscience*, *18*(February), 1–26. DOI: 10.3389/fncel.2024.1358450 PMID: 38419655

Latha, R. C. R., & Daisy, P. (2011). Insulin-secretagogue, antihyperlipidemic and other protective effects of gallic acid isolated from Terminalia bellerica Roxb. in streptozotocin-induced diabetic rats. *Chemico-Biological Interactions*, *189*(1-2), 112–118. DOI: 10.1016/j.cbi.2010.11.005 PMID: 21078310

Lau, F. C., Bielinski, D. F., & Joseph, J. A. (2007). Inhibitory effects of blueberry extract on the production of inflammatory mediators in lipopolysaccharide-activated BV2 microglia. *Journal of Neuroscience Research*, *85*(5), 1010–1017. DOI: 10.1002/jnr.21205

Lau, J. L., & Dunn, M. K. (2018). Therapeutic peptides: Historical perspectives, current development trends, and future directions. *Bioorganic & Medicinal Chemistry*, *26*(10), 2700–2707. DOI: 10.1016/j.bmc.2017.06.052 PMID: 28720325

Le Marrec, C. H., Hyronimus, B., Bressollier, P., Verneuil, B., & Urdaci, M. C. (2000). Biochemical and genetic characterization of coagulin, a new antilisterial bacteriocin in the pediocin family of bacteriocins, produced by Bacillus coagulans I4. *Applied and Environmental Microbiology*, *66*(12), 5213–5220. DOI: 10.1128/AEM.66.12.5213-5220.2000 PMID: 11097892

le Tang, X., Wang, X., & Fang, G. (2021). Resveratrol ameliorates sevoflurane-induced cognitive impairment by activating the SIRT1/NF-κB pathway in neonatal mice. *The Journal of Nutritional Biochemistry*, *90*, 108579. DOI: 10.1016/j.jnutbio.2020.108579 PMID: 33388350

Leclerc, M., Dudonné, S., & Calon, F. (2021). Can natural products exert neuroprotection without crossing the blood–brain barrier? *International Journal of Molecular Sciences*, *22*(7), 3356. Advance online publication. 10.3390/ijms22073356

Leclerc, M., Dudonné, S., & Calon, F. (2021). Can Natural Products Exert Neuroprotection without Crossing the Blood-Brain Barrier? *International Journal of Molecular Sciences*, *22*(7). Advance online publication. DOI: 10.3390/ijms22073356

Lecomte, A., Casanova, A., Blatché, M. C., Gonzalès, J., Ajagli, A., Mazenq, L., . . . Larrieu, G. (2024, September). Nanowires vs neurons: Engineering network scale guidance and promoting biointerface through nanotopographical cues. In Micro Nano Engineering Conference 2024.

Lee, S., & Kovacs, G. G. (2024). The Irony of Iron: The Element with Diverse Influence on Neurodegenerative Diseases. In *International journal of molecular sciences* (Vol. 25, Issue 8, p. 4269). Multidisciplinary Digital Publishing Institute. DOI: 10.3390/ijms25084269

Lee, B., Sur, B., Park, J., Kim, S. H., Kwon, S., Yeom, M., Shim, I., Lee, H., & Hahm, D. H. (2013). Chronic administration of baicalein decreases depression-like behavior induced by repeated restraint stress in rats. *The Korean Journal of Physiology & Pharmacology; Official Journal of the Korean Physiological Society and the Korean Society of Pharmacology*, *17*(5), 393–403. DOI: 10.4196/kjpp.2013.17.5.393 PMID: 24227939

Lee, B., Sur, B., Yeom, M., Shim, I., Lee, H., & Hahm, D. H. (2012). Effect of berberine on depression- and anxiety-like behaviors and activation of the noradrenergic system induced by development of morphine dependence in rats. *The Korean Journal of Physiology & Pharmacology; Official Journal of the Korean Physiological Society and the Korean Society of Pharmacology*, *16*(6), 379–386. DOI: 10.4196/kjpp.2012.16.6.379 PMID: 23269899

Lee-Hilz, Y. Y., Boerboom, A. M., Westphal, A. H., Berkel, W. J., Aarts, J. M., & Rietjens, I. M. (2006). Pro-oxidant activity of flavonoids induces EpRE-mediated gene expression. *Chemical Research in Toxicology*, *19*(11), 1499–1505. DOI: 10.1021/tx060157q

Lee, J., Li, Y., Jang, J. H., Park, S., Lee, H. Y., & Jeong, Y. (2021). Microbial transformation of bile acids: Effects on host physiology and intestinal integrity. *International Journal of Molecular Sciences*, *22*(14), 7613. DOI: 10.3390/ijms22147613 PMID: 34299233

Lee, K. H., Cha, M., & Lee, B. H. (2020). Neuroprotective effect of antioxidants in the brain. *International Journal of Molecular Sciences*, *21*(19), 7152. DOI: 10.3390/ijms21197152 PMID: 32998277

Lee, S. K., & Wolfe, S. W. (2000). Peripheral nerve injury and repair. *The Journal of the American Academy of Orthopaedic Surgeons*, *8*(4), 243–252. DOI: 10.5435/00124635-200007000-00005 PMID: 10951113

Lee, S., Xie, J., & Chen, X. (2010). Peptides and peptide hormones for molecular imaging and disease diagnosis. *Chemical Reviews*, *110*(5), 3087–3111. DOI: 10.1021/cr900361p PMID: 20225899

Lee, W. Y., Savage, J. R., Zhang, J., Jia, W., Oottamasathien, S., & Prestwich, G. D. (2013). Prevention of anti-microbial peptide LL-37-induced apoptosis and ATP release in the urinary bladder by a modified glycosaminoglycan. *PLoS One*, *8*(10), e77854. DOI: 10.1371/journal.pone.0077854 PMID: 24204996

Lee, Y. K., Kwon, O., & Lee, H. (2019). Anthocyanins and gut microbiota: Modulation and health implications. *Nutrients*, *11*(8), 1773. DOI: 10.3390/nu11081773 PMID: 31374898

Leng, F., & Edison, P. (2021). Neuroinflammation and microglial activation in Alzheimer disease: Where do we go from here? *Nature Reviews. Neurology*, *17*(3), 157–172. DOI: 10.1038/s41582-020-00435-y PMID: 33318676

Lenzi, J., Rodrigues, A. F., Rós, A. de S., de Castro, B. B., de Lima, D. D., Magro, D. D. D., & Zeni, A. L. B. (2015). Ferulic acid chronic treatment exerts antidepressant-like effect: Role of antioxidant defense system. *Metabolic Brain Disease*, *30*(6), 1453–1463. DOI: 10.1007/s11011-015-9725-6 PMID: 26340979

Leonard, B. E. (2018). Inflammation and depression: A causal or coincidental link to the pathophysiology? *Acta Neuropsychiatrica*, *30*(1), 1–16. DOI: 10.1017/neu.2016.69 PMID: 28112061

Lesjak, M., Beara, I., Simin, N., Pintać, D., Majkić, T., Bekvalac, K., & Mimica-Dukić, N. (2018). Antioxidant and anti-inflammatory activities of quercetin and its derivatives. *Journal of Functional Foods, 40*, 68–75. DOI: 10.1016/j.jff.2017.10.047

Levy, G., Fishman, J. E., Xu, D., Chandler, B. T., Feketova, E., Dong, W., Qin, Y., Alli, V., Ulloa, L., & Deitch, E. A. (2013). Parasympathetic stimulation via the vagus nerve prevents systemic organ dysfunction by abrogating gut injury and lymph toxicity in trauma and hemorrhagic shock. *Shock (Augusta, Ga.), 39*(1), 39–44. DOI: 10.1097/SHK.0b013e31827b450d PMID: 23247120

Leyns, C. E. G., Ulrich, J. D., Finn, M. B., Stewart, F. R., Koscal, L. J., Serrano, J. R., Robinson, G. O., Anderson, E., Colonna, M., & Holtzman, D. M. (2017). TREM2 deficiency attenuates neuroinflammation and protects against neurodegeneration in a mouse model of tauopathy. *Proceedings of the National Academy of Sciences of the United States of America, 114*(43), 11524–11529. DOI: 10.1073/pnas.1710311114 PMID: 29073081

Li, H., Liu, Y., Sun, Y., Guo, H., Lv, S., Guo, W., Ren, J., Wang, Y., Zu, J., Yan, J., & Wang, N. (2024). Targeting astrocytes polarization after spinal cord injury: a promising direction. In *Frontiers in Cellular Neuroscience* (Vol. 18, p. 1478741). Frontiers Media SA. DOI: 10.3389/fncel.2024.1478741

Li, K., Li, J., Zheng, J., & Qin, S. (2019). Reactive astrocytes in neurodegenerative diseases. In *Aging and Disease*. DOI: 10.14336/AD.2018.0720

Li, C., Zug, C., Qu, H., Schluesener, H., & Zhang, Z. (2015). Hesperidin ameliorates behavioral impairments and neuropathology of transgenic APP/PS1 mice. *Behavioural Brain Research, 281*, 32–42. DOI: 10.1016/j.bbr.2014.12.012 PMID: 25510196

Liddelow, S. A., Guttenplan, K. A., Clarke, L. E., Bennett, F. C., Bohlen, C. J., Schirmer, L., Bennett, M. L., Münch, A. E., Chung, W. S., Peterson, T. C., Wilton, D. K., Frouin, A., Napier, B. A., Panicker, N., Kumar, M., Buckwalter, M. S., Rowitch, D. H., Dawson, V. L., Dawson, T. M., & Barres, B. A. (2017). Neurotoxic reactive astrocytes are induced by activated microglia. *Nature, 541*(7638), 481–487. DOI: 10.1038/nature21029 PMID: 28099414

Li, H., Chen, F.-J., Yang, W.-L., Qiao, H.-Z., & Zhang, S.-J. (2021). Quercetin improves cognitive disorder in aging mice by inhibiting NLRP3 inflammasome activation. *Food & Function, 12*(2), 717–725. DOI: 10.1039/D0FO01900C PMID: 33338087

Li, H., Wu, X., & Zhou, J. (2023). Gut microbiota dysbiosis and immune imbalance in autoimmune neurological diseases: A systematic review. *Neuroimmunomodulation, 30*(3), 123–136. DOI: 10.1159/000530341

Li, J., Zhang, L., Chu, Y., Namaka, M., Deng, B., Kong, J., & Bi, X. (2016). Astrocytes in oligodendrocyte lineage development and white matter pathology. *Frontiers in Cellular Neuroscience, 10*(MAY), 1–13. DOI: 10.3389/fncel.2016.00119 PMID: 27242432

Limanaqi, F., Biagioni, F., Busceti, C. L., Ryskalin, L., & Fornai, F. (2020). Merging the multi-target effects of phytochemicals in neurodegeneration: From oxidative stress to protein aggregation and inflammation. *Antioxidants, 9*(10), 1022. Advance online publication. DOI: 10.3390/antiox9101022 PMID: 33092300

Lin, C.-H., Chen, C.-C., Chiang, H.-L., Liou, J.-M., Chang, C.-M., Lu, T.-P., & Lin, H.-Y. (2019). Altered gut microbiota and inflammatory cytokine responses in patients with Parkinson's disease. *Journal of Neuroinflammation, 16*(1), 1–9. DOI: 10.1186/s12974-019-1528-y PMID: 31248424

Lin, K., Li, Y., Du Toit, E., Wendt, L., & Sun, J. (2021). Effects of Polyphenol Supplementations on Improving Depression, Anxiety, and Quality of Life in Patients With Depression. *Frontiers in Psychiatry, 12*(November), 1–13. DOI: 10.3389/fpsyt.2021.765485 PMID: 34819888

Lin, L., Li, C., Zhang, L., Zhang, Y., Gao, L., Li, T., Jin, L., Shen, Y., & Ren, D. (2023). Multi-dimensional deep learning drives efficient discovery of novel neuroprotective peptides from walnut protein isolates. *Food & Function, 14*(15), 6969–6984. DOI: 10.1039/D3FO01602A PMID: 37435725

Lin, L., Li, C., Zhang, T., Xia, C., Bai, Q., Jin, L., & Shen, Y. (2024). An in silico scheme for optimizing the enzymatic acquisition of natural biologically active peptides based on machine learning and virtual digestion. *Analytica Chimica Acta, 1298*, 342419. DOI: 10.1016/j.aca.2024.342419 PMID: 38462343

Li, R., Wang, X., Qin, T., Qu, R., & Ma, S. (2016). Apigenin ameliorates chronic mild stress-induced depressive behavior by inhibiting interleukin-1β production and NLRP3 inflammasome activation in the rat brain. *Behavioural Brain Research, 296*, 318–325. DOI: 10.1016/j.bbr.2015.09.031 PMID: 26416673

Li, R., Zhou, Y., Zhang, S., Li, J., Zheng, Y., & Fan, X. (2022). The natural (poly) phenols as modulators of microglia polarization via TLR4/NF-κB pathway exert anti-inflammatory activity in ischemic stroke. *European Journal of Pharmacology, 914*, 174660. DOI: 10.1016/j.ejphar.2021.174660 PMID: 34863710

Lisanti, C. J., Asbach, P., & Bradley, W. G. (2005). The ependymal "Dot-Dash" sign: An MR imaging finding of early multiple sclerosis. *AJNR. American Journal of Neuroradiology, 26*(8), 2033–2036. DOI: 10.1016/S0098-1672(08)70484-3 PMID: 16155155

Liss, J., Seleri Assunção, S., Cummings, J., Atri, A., Geldmacher, D., Candela, S., & Mintzer, J. (2021). Practical recommendations for timely, accurate diagnosis of symptomatic Alzheimer's disease (MCI and dementia) in primary care: A review and synthesis. *Journal of Internal Medicine, 290*(2), 310–334. DOI: 10.1111/joim.13244 PMID: 33458891

Liu, B., & Hong, J. S. (2003). Role of microglia in inflammation-mediated neurodegenerative diseases: Mechanisms and strategies for therapeutic intervention. In *Journal of Pharmacology and Experimental Therapeutics* (Vol. 304, Issue 1, pp. 1–7). J Pharmacol Exp Ther. DOI: 10.1124/jpet.102.035048

Liu, B., Xu, C., Wu, X., Liu, F., Du, Y., Sun, J., Tao, J., & Dong, J. (2015). Icariin exerts an antidepressant effect in an unpredictable chronic mild stress model of depression in rats and is associated with the regulation of hippocampal neuroinflammation. *Neuroscience, 294*(March), 193–205. DOI: 10.1016/j.neuroscience.2015.02.053 PMID: 25791226

Liu, G., Ren, G., Zhao, L., Cheng, L., Wang, C., & Sun, B. (2017). Antibacterial activity and mechanism of bifidocin A against Listeria monocytogenes. *Food Control, 73*, 854–861. DOI: 10.1016/j.foodcont.2016.09.036

Liu, H., Wang, J., & Chen, Y. (2024). Peptide and peptide-mimetic supplementation improve oxidative stress in neurodegenerative disorders. *Neuropharmacology, 214*, 109952. DOI: 10.1016/j.neuropharm.2024.109952

Liu, H., Yang, J., Li, L., Shi, W., Yuan, X., & Wu, L. (2016). The Natural Occurring Compounds Targeting Endoplasmic Reticulum Stress. *Evidence-Based Complementary and Alternative Medicine : eCAM, 7831282*. Advance online publication. DOI: 10.1155/2016/7831282

Liu, J., Zhou, H., Fan, W., & Dong, W. (2023). New insights into molecular mechanisms underlying neurodegenerative disorders. *Journal of Integrative Neuroscience*, *22*(3), 58. Advance online publication. DOI: 10.31083/j.jin2203058 PMID: 37258447

Liu, L., Liu, J., Bao, J., Bai, Q., & Wang, G. (2020). Interaction of microglia and astrocytes in the neurovascular unit. *Frontiers in Immunology*, *11*, 1024. DOI: 10.3389/fimmu.2020.01024 PMID: 32733433

Liu, L., Qi, W., Zhang, N., Zhang, J., Liu, S., Wang, H., Jiang, L., & Sun, Y. (2025). Nutraceuticals for gut–brain axis health: A novel approach to combat malnutrition and future personalised nutraceutical interventions. *Nutrients*, *17*(9), 1551. DOI: 10.3390/nu17091551 PMID: 40362863

Liu, M. H., Lin, Y. S., Sheu, S. Y., & Sun, J. S. (2009). Anti-inflammatory effects of daidzein on primary astroglial cell culture. *Nutritional Neuroscience*, *12*(3), 123–134. DOI: 10.1179/147683009x423274

Liu, Q. S., Li, S. R., Li, K., Li, X., Yin, X., & Pang, Z. (2017). Ellagic acid improves endogenous neural stem cells proliferation and neurorestoration through Wnt/β-catenin signaling in vivo and in vitro. *Molecular Nutrition & Food Research*, *61*(3), 1600587. Advance online publication. DOI: 10.1002/mnfr.201600587 PMID: 27794200

Liu, Q., Chen, J., & Wang, L.. (2024). Neuroprotective effects of peptide supplementation in Alzheimer's disease patients. *Neuropharmacology*, *202*, 109952.

Liu, Q., Yu, Z., Tian, F., Zhao, J., Zhang, H., Zhai, Q., & Chen, W. (2020). Surface components and metabolites of probiotics for regulation of intestinal epithelial barrier. *Microbial Cell Factories*, *19*(1), 23. Advance online publication. DOI: 10.1186/s12934-020-1289-4 PMID: 32024520

Liu, T., Li, X., Wang, Y., Luo, C., & Yin, Y. (2022). Dietary fiber supplementation improves gut barrier function and clinical outcomes in critically ill patients: A randomized clinical trial. *Clinical Nutrition (Edinburgh, Lothian)*, *41*(5), 1234–1241. DOI: 10.1016/j.clnu.2021.09.022

Liu, X., Tang, M., He, T.-Y., Zhao, S., Li, H.-Z., Li, Z., Guo, Y.-X., & Wang, X.-L. (2023). Resveratrol Improves Paclitaxel-Induced Cognitive Impairment in Mice by Activating SIRT1/PGC-1α Pathway to Regulate Neuronal State and Microglia Cell Polarization. *Drug Design, Development and Therapy*, *17*, 1125–1138. DOI: 10.2147/DDDT.S400936 PMID: 37077409

Liu, X., Yang, Y., & Urban, M. W. (2017). Stimuli-responsive polymeric nanoparticles. *Macromolecular Rapid Communications*, *38*(13), 1700030. DOI: 10.1002/marc.201700030 PMID: 28497535

Liu, Y. M., Hu, C. Y., Shen, J. D., Wu, S. H., Li, Y. C., & Yi, L. T. (2017). Elevation of synaptic protein is associated with the antidepressant-like effects of ferulic acid in a chronic model of depression. *Physiology & Behavior*, *169*, 184–188. DOI: 10.1016/j.physbeh.2016.12.003 PMID: 27940143

Liu, Y., Li, X., Zhou, X., Wang, J., & Ao, X. (2022). FADD as a key molecular player in cancer progression. *Molecular Medicine (Cambridge, Mass.)*, *28*(1), 132. DOI: 10.1186/s10020-022-00560-y

Liu, Y., Li, Y., Du, C., Kuang, S., Zhou, X., Zhang, J., & Ao, X. (2022). Underlying mechanisms of epithelial splicing regulatory proteins in cancer progression. *Journal of Molecular Medicine (Berlin, Germany)*, *100*(11), 1539–1556. DOI: 10.1007/s00109-022-02257-5

Liu, Y., Zhang, Y., Zheng, X., Fang, H., & Li, H. (2018). The synergistic neuroprotective effects of combined rosuvastatin and resveratrol pretreatment against cerebral ischemia/reperfusion injury. *Journal of Stroke and Cerebrovascular Diseases*, *27*(6), 1697–1704. DOI: 10.1016/j.jstrokecerebrovasdis.2018.01.033 PMID: 29525080

Liu, Z., Ren, Z., Zhang, J., Chuang, C.-C., Kandaswamy, E., Zhou, T., & Zuo, L. (2018). Role of ROS and nutritional antioxidants in human diseases. *Frontiers in Physiology*, *9*, 360203. DOI: 10.3389/fphys.2018.00477 PMID: 29867535

Liu, Z., Yao, X., Sun, B., Jiang, W., Liao, C., Dai, X., Chen, Y., Chen, J., & Ding, R. (2021). Pretreatment with kaempferol attenuates microglia-mediate neuroinflammation by inhibiting MAPKs–NF–κB signaling pathway and pyroptosis after secondary spinal cord injury. *Free Radical Biology & Medicine*, *168*, 142–154. DOI: 10.1016/j.freeradbiomed.2021.03.037 PMID: 33823244

Li, X., Wang, H., Li, J., & Li, Y. (2022). Synergistic neuroprotective effect of saikosaponin A and albiflorin on corticosterone-induced apoptosis in PC12 cells via regulation of metabolic disorders and neuroinflammation. *Molecular Biology Reports*, *49*(9), 8801–8813. DOI: 10.1007/s11033-022-07730-5 PMID: 36002654

Li, X., Wang, H., Wen, G., Li, L., Gao, Y., Zhuang, Z., Zhou, M., Mao, L., & Fan, Y. (2018). Neuroprotection by quercetin via mitochondrial function adaptation in traumatic brain injury: PGC-1α pathway as a potential mechanism. *Journal of Cellular and Molecular Medicine*, *22*(2), 883–891. DOI: 10.1111/jcmm.13313 PMID: 29205806

Li, X., Wang, X., & Zhang, Q. (2020). Advances in nanocarrier systems for anthocyanins: Implications for brain health. *Trends in Food Science & Technology*, *102*, 123–136. DOI: 10.1016/j.tifs.2020.05.007

Li, X., Zhang, Y., Yang, Z., Zhang, S., & Zhang, L. (2024). The Inhibition Effect of Epigallocatechin-3-Gallate on the Co-Aggregation of Amyloid-β and Human Islet Amyloid Polypeptide Revealed by Replica Exchange Molecular Dynamics Simulations. *International Journal of Molecular Sciences*, *25*(3), 1636. DOI: 10.3390/ijms25031636 PMID: 38338914

Li, X., Zhao, X., Xu, X., Mao, X., Liu, Z., Li, H., Guo, L., Bi, K., & Jia, Y. (2014). Schisantherin A recovers Aβ-induced neurodegeneration with cognitive decline in mice. *Physiology & Behavior*, *132*, 10–16. DOI: 10.1016/j.physbeh.2014.04.046 PMID: 24813830

Li, Y., Chen, M., Yan, J., Zhou, W., Gao, S., Liu, S., Li, Q., Zheng, Y., Cheng, Y., & Guo, Q. (2021). Tannic acid/Sr2+-coated silk/graphene oxide-based meniscus scaffold with anti-inflammatory and anti-ROS functions for cartilage protection and delaying osteoarthritis. *Acta Biomaterialia*, *126*, 119–131. DOI: 10.1016/j.actbio.2021.02.046 PMID: 33684536

Li, Z., Yi, C. X., Katiraei, S., Kooijman, S., Zhou, E., Chung, C. K., & Wang, Y. (2018). Butyrate reduces appetite and activates brown adipose tissue via the gut–brain neural circuit. *Gut*, *67*(7), 1269–1279. DOI: 10.1136/gutjnl-2017-314050 PMID: 29101261

Loane, D. J., & Kumar, A. (2016). Microglia in the TBI brain: The good, the bad, and the dysregulated. In *Experimental Neurology* (Vol. 275, Issue 0 3, pp. 316–327). Academic Press Inc. DOI: 10.1016/j.expneurol.2015.08.018

Loane, D. J., & Byrnes, K. R. (2010). Role of Microglia in Neurotrauma. *Neurotherapeutics; the Journal of the American Society for Experimental NeuroTherapeutics*, *7*(4), 366–377. DOI: 10.1016/j.nurt.2010.07.002 PMID: 20880501

Loh, J. S., Mak, W. Q., Tan, L. K. S., Ng, C. X., Chan, H. H., Yeow, S. H., Foo, J. B., Ong, Y. S., How, C. W., & Khaw, K. Y. (2024). Microbiota–gut–brain axis and its therapeutic applications in neurodegenerative diseases. *Signal Transduction and Targeted Therapy*, *9*(1), 37. DOI: 10.1038/s41392-024-01743-1 PMID: 38360862

López-Lázaro, M. (2009). Distribution and biological activities of the flavonoid luteolin. *Mini-Reviews in Medicinal Chemistry*, *9*(1), 31–59. DOI: 10.2174/138955709787001712

López, M. (2017). EJE PRIZE 2017: Hypothalamic AMPK: A golden target against obesity? *European Journal of Endocrinology*, *176*(6), R235–R246. DOI: 10.1530/EJE-16-0927 PMID: 28232370

Lorente Pons, A., Higginbottom, A., Cooper-Knock, J., Alrafiah, A., Alofi, E., Kirby, J., Shaw, P. J., Wood, J. D., & Highley, J. R. (2020). Oligodendrocyte pathology exceeds axonal pathology in white matter in human amyotrophic lateral sclerosis. *The Journal of Pathology*, *251*(3), 262–271. DOI: 10.1002/path.5455 PMID: 32391572

Loria, M. J., Ali, Z., Abe, N., Sufka, K. J., & Khan, I. A. (2014). Effects of Sceletium tortuosum in rats. *Journal of Ethnopharmacology*, *155*(1), 731–735. DOI: 10.1016/j.jep.2014.06.007 PMID: 24930358

Lou, H., Jing, X., Wei, X., Shi, H., Ren, D., & Zhang, X. (2014). Naringenin protects against 6-OHDA-induced neurotoxicity via activation of the Nrf2/ARE signaling pathway. *Neuropharmacology*, *79*(December), 380–388. DOI: 10.1016/j.neuropharm.2013.11.026 PMID: 24333330

Lou, M., Ji, S., Wu, R., Zhu, Y., Wu, J., & Zhang, J. (2025). Microbial production systems and optimization strategies of antimicrobial peptides: A review. *World Journal of Microbiology & Biotechnology*, *41*(2), 1–36. DOI: 10.1007/s11274-025-04278-x PMID: 39920500

Lovell, M. A., & Markesbery, W. R. (2007). Oxidative DNA damage in mild cognitive impairment and late-stage Alzheimer's disease. *Nucleic Acids Research*, *35*(22), 7497–7504. DOI: 10.1093/nar/gkm821 PMID: 17947327

Lu, Y., Wang, K., & Hu, L. (2025). Advancements in delivery systems for dietary polyphenols in enhancing radioprotection effects: challenges and opportunities. *npj Science of Food, 9*(1), 51.

Lu, B., Nagappan, G., Guan, X., Nathan, P. J., & Wren, P. (2013). BDNF-based synaptic repair as a disease-modifying strategy for neurodegenerative diseases. *Nature Reviews. Neuroscience*, *14*(6), 401–416. DOI: 10.1038/nrn3505 PMID: 23674053

Luca, A., Calandra, C., & Luca, M. (2018). Molecular bases of Alzheimer's disease and neurodegeneration: The role of neuroglia. In *Aging and Disease* (Vol. 9, Issue 6, pp. 1134–1152). International Society on Aging and Disease. DOI: 10.14336/AD.2018.0201

Lu, H., Qi, Y., Zhao, Y., & Jin, N. (2020). Effects of hydroxyl group on the interaction of carboxylated flavonoid derivatives with S. Cerevisiae α-glucosidase. *Current Computer-Aided Drug Design, 16*(1), 31–44. DOI: 10.2174/1573409914666181022142553 PMID: 30345924

Lu, K. T., Ko, M. C., Chen, B. Y., Huang, J. C., Hsieh, C. W., Lee, M. C., Chiou, R. Y. Y., Wung, B. S., Peng, C. H., & Yang, Y. L. (2008). Neuroprotective effects of resveratrol on MPTP-induced neuron loss mediated by free radical scavenging. *Journal of Agricultural and Food Chemistry, 56*(16), 6910–6913. DOI: 10.1021/jf8007212 PMID: 18616261

Luo, J., Ganesan, K., & Xu, B. (2024). Unlocking the power: New insights into the anti-aging properties of mushrooms. *Journal of Fungi (Basel, Switzerland), 10*(3), 215. Advance online publication. 10.3390/jof10030215

Luo, J., Si, H., Jia, Z., & Liu, D. (2021). Dietary anti-aging polyphenols and potential mechanisms. *Antioxidants, 10*(2), 283. DOI: 10.3390/antiox10020283 PMID: 33668479

Luo, Y., & Song, Y. (2021). Mechanism of Antimicrobial Peptides: Antimicrobial, Anti-Inflammatory and Antibiofilm Activities. *International Journal of Molecular Sciences, 22*(21), 11401. DOI: 10.3390/ijms222111401 PMID: 34768832

Lu, Q., Cai, X., Zhang, X., Li, S., Song, Y., Du, D., Dutta, P., & Lin, Y. (2018). Synthetic polymer nanoparticles functionalized with different ligands for receptor-mediated transcytosis across the blood–brain barrier. *ACS Applied Bio Materials, 1*(5), 1687–1694. DOI: 10.1021/acsabm.8b00502 PMID: 31815251

Luu, M., Steinhoff, U., & Visekruna, A. (2017). Functional heterogeneity of gut-resident regulatory T cells. *Clinical & Translational Immunology, 6*(9), e156. DOI: 10.1038/cti.2017.39 PMID: 28983404

Lymperopoulos, D., Dedemadi, A. G., Voulgari, M. L., Georgiou, E., Dafnis, I., Mountaki, C., Panagopoulou, E. A., Karvelas, M., Chiou, A., Karathanos, V. T., & Chroni, A. (2024). Corinthian Currants Promote the Expression of Paraoxonase-1 and Enhance the Antioxidant Status in Serum and Brain of 5xFAD Mouse Model of Alzheimer's Disease. *Biomolecules, 14*(4), 426. Advance online publication. DOI: 10.3390/biom14040426 PMID: 38672443

Macready, A. L., Kennedy, O. B., Ellis, J. A., Williams, C. M., Spencer, J. P., & Butler, L. T. (2009). Flavonoids and cognitive function: A review of human randomized controlled trial studies and recommendations for future studies. *Genes & Nutrition, 4*(4), 227–242. DOI: 10.1007/s12263-009-0135-4

Madsen, P. M., Desu, H. L., Vaccari, J. P. de R., Florimon, Y., Ellman, D. G., Keane, R. W., Clausen, B. H., Lambertsen, K. L., & Brambilla, R. (2020). Oligodendrocytes modulate the immune-inflammatory response in EAE via TNFR2 signaling. *Brain, Behavior, and Immunity, 84*, 132–146. DOI: 10.1016/j.bbi.2019.11.017 PMID: 31785393

Ma, G., & Chen, Y. (2020). Polyphenol supplementation benefits human health via gut microbiota: A systematic review via meta-analysis. *Journal of Functional Foods, 66*, 103829. DOI: 10.1016/j.jff.2020.103829

Maher, P. (2019). The potential of flavonoids for the treatment of neurodegenerative diseases. *International Journal of Molecular Sciences, 20*(12), 3056. DOI: 10.3390/ijms20123056 PMID: 31234550

Maher, P., Dargusch, R., Bodai, L., Gerard, P. E., Purcell, J. M., & Lawrence Marsh, J. (2011). Erk activation by the polyphenols fisetin and resveratrol provides neuroprotection in multiple models of Huntington's disease. *Human Molecular Genetics*, *20*(2), 261–270. DOI: 10.1093/hmg/ddq460 PMID: 20952447

Mahmoud, A. M., Mohammed, H. M., Khadrawy, S. M., & Galaly, S. R. (2017). Hesperidin protects against chemically induced hepatocarcinogenesis via modulation of Nrf2/ARE/HO-1, PPARγ and TGF-β1/Smad3 signaling, and amelioration of oxidative stress and inflammation. *Chemico-Biological Interactions*, *277*, 146–158. DOI: 10.1016/j.cbi.2017.09.015 PMID: 28935427

Ma, J., Li, Z., Zhang, W., Zhang, C., & Zhang, Y. (2021). Role of short-chain fatty acid–G-protein-coupled receptors in intestinal barrier function regulation. *Animal Nutrition*, *7*(1), 259–265. DOI: 10.1016/j.aninu.2020.08.005 PMID: 33997349

Makdissi, S., Parsons, B. D., & Di Cara, F. (2023). Towards early detection of neurodegenerative diseases: A gut feeling. *Frontiers in Cell and Developmental Biology*, *11*, 1087091. DOI: 10.3389/fcell.2023.1087091 PMID: 36824371

Makris, A. P., Karianaki, M., Tsamis, K. I., & Paschou, S. A. (2021). The role of the gut–brain axis in depression: Endocrine, neural, and immune pathways. *Hormones (Athens, Greece)*, *20*(1), 1–12. DOI: 10.1007/s42000-020-00236-4 PMID: 32827123

Maleki, S. J., Crespo, J. F., & Cabanillas, B. (2019). Anti-inflammatory effects of flavonoids. *Food Chemistry*, *299*, 125124. DOI: 10.1016/j.foodchem.2019.125124

Malin, D. H. (2016). Neuroprotective effects of blueberry polyphenols in Alzheimer's disease models. *Phytotherapy Research*, *30*(5), 726–735. DOI: 10.1002/ptr.5592

Malmierca, M. G., Izquierdo-Bueno, I., McCormick, S. P., Cardoza, R. E., Alexander, N. J., Barua, J., Lindo, L., Casquero, P. A., Collado, I. G., Monte, E., & Gutiérrez, S. (2016). Trichothecenes and aspinolides produced by Trichoderma arundinaceum regulate expression of Botrytis cinerea genes involved in virulence and growth. *Environmental Microbiology*, *18*(11), 3991–4004. DOI: 10.1111/1462-2920.13410 PMID: 27312485

Manach, C., Scalbert, A., Morand, C., Rémésy, C., & Jiménez, L. (2004). Polyphenols: Food sources and bioavailability. *The American Journal of Clinical Nutrition*, *79*(5), 727–747. DOI: 10.1093/ajcn/79.5.727 PMID: 15113710

Manach, C., Williamson, G., Morand, C., Scalbert, A., & Rémésy, C. (2005). Bioavailability and bioefficacy of polyphenols in humans. I. Review of 97 bioavailability studies. *The American Journal of Clinical Nutrition*, *81*(1, Suppl), 230S–242S. DOI: 10.1093/ajcn/81.1.230S PMID: 15640486

Manap, A. S. A., Vijayabalan, S., Madhavan, P., Chia, Y. Y., Arya, A., Wong, E. H., Rizwan, F., Bindal, U., & Koshy, S. (2019). Bacopa monnieri, a neuroprotective lead in Alzheimer's disease: A review on its properties, mechanisms of action, and preclinical and clinical studies. *Drug Target Insights*, *13*, 1177392819866412. 10.1177/1177392819866412

Mancuso, C., & Santangelo, R. (2018). Alzheimer's disease and gut microbiota modifications: The long way between preclinical studies and clinical evidence. *Pharmacological Research, 129*, 329–336. DOI: 10.1016/j.phrs.2017.12.009 PMID: 29233677

Mandel, S., Weinreb, O., Amit, T., & Youdim, M. B. H. (2005). Cell signaling pathways in the neuroprotective actions of flavonoids: Implications for COMT genotype and brain health. *Journal of Neural Transmission (Vienna, Austria), 112*(2), 293–300. DOI: 10.1007/s00702-004-0165-7

Mander, P. K., Jekabsone, A., & Brown, G. C. (2006). Microglia proliferation is regulated by hydrogen peroxide from NADPH oxidase. *The Journal of Immunology : Official Journal of the American Association of Immunologists, 176*(2), 1046–1052. DOI: 10.4049/jimmunol.176.2.1046 PMID: 16393992

Mangrulkar, S. V., Wankhede, N. L., Kale, M. B., Upaganlawar, A. B., Taksande, B. G., Umekar, M. J., Anwer, M. K., Dailah, H. G., Mohan, S., & Behl, T. (2023). Mitochondrial dysfunction as a signaling target for therapeutic intervention in major neurodegenerative disease. *Neurotoxicity Research, 41*(6), 708–729. DOI: 10.1007/s12640-023-00647-2 PMID: 37162686

Mansouri, M. T., Naghizadeh, B., Ghorbanzadeh, B., Farbood, Y., Sarkaki, A., & Bavarsad, K. (2013). Gallic acid prevents memory deficits and oxidative stress induced by intracerebroventricular injection of streptozotocin in rats. *Pharmacology, Biochemistry, and Behavior, 111*, 90–96. DOI: 10.1016/j.pbb.2013.09.002 PMID: 24036471

Maqbool, M., & Zehravi, M. (2021). Neuroprotective role of polyphenols in treatment of neurological disorders: A review. *Interventional Pain Medicine and Neuromodulation, 1*(1).

Marambaud, P., Zhao, H., & Davies, P. (2005). Resveratrol promotes clearance of Alzheimer's disease amyloid-β peptides. *The Journal of Biological Chemistry, 280*(45), 37377–37382. DOI: 10.1074/jbc.M508246200 PMID: 16162502

Markowiak, P., & Śliżewska, K. (2017). Effects of probiotics, prebiotics, and synbiotics on human health. *Nutrients, 9*(9), 1021. DOI: 10.3390/nu9091021 PMID: 28914794

Marmolejo-Martínez-Artesero, S., Aparicio, M., & Navarro, A. (2021). Endogenous mechanisms of neuroprotection: To boost or not to be. *Cells, 10*(2), 370. Advance online publication. DOI: 10.3390/cells10020370 PMID: 33578870

Marques, D., Moura-Louro, D., Silva, I. P., Matos, S., Dos Santos, C. N., & Figueira, I. (2024). Unlocking the potential of low-molecular-weight (Poly)phenol metabolites: Protectors at the blood-brain barrier frontier. *Neurochemistry International, 179*, 105836. DOI: 10.1016/j.neuint.2024.105836 PMID: 39151552

Marqus, S., Pirogova, E., & Piva, T. J. (2017). Evaluation of the use of therapeutic peptides for cancer treatment. *Journal of Biomedical Science, 24*(1), 21. DOI: 10.1186/s12929-017-0328-x PMID: 28320393

Martemucci, G., Costagliola, C., Mariano, M., D'andrea, L., Napolitano, P., & D'Alessandro, A. G. (2022). Free Radical Properties, Source and Targets, Antioxidant Consumption and Health. In *Oxygen* (Vol. 2, Issue 2, pp. 48–78). Multidisciplinary Digital Publishing Institute (MDPI). DOI: 10.3390/oxygen2020006

Martínez, G., Mijares, M. R., & De Sanctis, J. B. (2019). Effects of flavonoids and its derivatives on immune cell responses. *Recent Patents on Inflammation & Allergy Drug Discovery*, *13*(2), 84–104. DOI: 10.2174/1872213X13666190426164124 PMID: 31814545

Martín, R., Miquel, S., Benevides, L., Bridonneau, C., Robert, V., Hudault, S., Chain, F., Berteau, O., Azevedo, V., Chatel, J. M., Sokol, H., Bermúdez-Humarán, L. G., Thomas, M., & Langella, P. (2017). Functional characterization of novel Faecalibacterium prausnitzii strains isolated from healthy volunteers: A step forward in the use of F. prausnitzii as a next-generation probiotic. *Frontiers in Microbiology*, *8*, 1226. DOI: 10.3389/fmicb.2017.01226 PMID: 28713353

Martucciello, S., Masullo, M., Cerulli, A., & Piacente, S. (2020). Natural Products Targeting ER Stress, and the Functional Link to Mitochondria. *International Journal of Molecular Sciences*, *21*(6). Advance online publication. DOI: 10.3390/ijms21061905

Mary Ellen, S., Francisco, G., & Richard, G.. (2013). An update on the use and investigation of probiotics in health and disease. *Gut*, *62*(5), 787–796. DOI: 10.1136/gutjnl-2012-302504 PMID: 23474420

Masala, G., Assedi, M., Bendinelli, B., Ermini, I., Sieri, S., Grioni, S., & Palli, D. (2012). Fruit and vegetables consumption and breast cancer risk: The EPIC Italy study. *Breast Cancer Research and Treatment*, *132*(3), 1127–1136. DOI: 10.1007/s10549-011-1939-7

Mas-Bargues, C., Escrivá, C., Dromant, M., Borrás, C., & Viña, J. (2021). Lipid peroxidation as measured by chromatographic determination of malondialdehyde. Human plasma reference values in health and disease. In *Archives of Biochemistry and Biophysics* (Vol. 709). Academic Press Inc., DOI: 10.1016/j.abb.2021.108941

Mashhadizadeh, S., Farbood, Y., Dianat, M., Khodadadi, A., & Sarkaki, A. (2017). Therapeutic effects of ellagic acid on memory, hippocampus electrophysiology deficits, and elevated TNF-α level in brain due to experimental traumatic brain injury. *Iranian Journal of Basic Medical Sciences.*, *20*(4), 399. PMID: 28804609

Masondo, N. A., Stafford, G. I., Aremu, A. O., & Makunga, N. P. (2019). Acetylcholinesterase inhibitors from southern African plants: An overview of ethnobotanical, pharmacological potential and phytochemical research including and beyond Alzheimer's disease treatment. *South African Journal of Botany*, *120*, 39–64.

Matejuk, A., & Ransohoff, R. M. (2020). Crosstalk Between Astrocytes and Microglia: An Overview. In *Frontiers in Immunology* (Vol. 11, p. 507878). Frontiers Media S.A. DOI: 10.3389/fimmu.2020.01416

Matijašić, M., Meštrović, T., Čipčić Paljetak, H., Perić, M., Barešić, A., & Verbanac, D. (2020). Gut microbiota beyond bacteria—Mycobiome, virome, archaeome, and eukaryotic parasites in IBD. *International Journal of Molecular Sciences*, *21*(8), 2668. DOI: 10.3390/ijms21082668 PMID: 32290414

Matsuda, M., Huh, Y., & Ji, R.-R. (2019). Roles of inflammation, neurogenic inflammation, and neuroinflammation in pain. *Journal of Anesthesia*, *33*(1), 131–139. DOI: 10.1007/s00540-018-2579-4 PMID: 30448975

Mattsson, N., Insel, P. S., & Palmqvist, S.. (2015). Revolutionizing Alzheimer's disease and clinical trials through biomarkers. *Alzheimer's & Dementia*, *11*(5), 575–582. DOI: 10.1016/j.dadm.2015.09.001 PMID: 27239522

Mavroeidi, P., & Xilouri, M. (2021). Neurons and glia interplay in α-synucleinopathies. In *International Journal of Molecular Sciences* (Vol. 22, Issue 9, p. 4994). Multidisciplinary Digital Publishing Institute. DOI: 10.3390/ijms22094994

Ma, X., Liu, H., Li, J., Wang, Y., Ding, S., Zhang, B., & Wang, R. (2020). ADSCs-derived extracellular vesicles alleviate neuronal damage, promote neurogenesis and rescue memory loss in mice with Alzheimer's disease. *Journal of Controlled Release*, *327*, 688–702. DOI: 10.1016/j.jconrel.2020.09.019 PMID: 32931898

Ma, X., Li, Y., & Zhang, Q. (2022). Peptide-based interventions reduce cognitive impairment in Alzheimer's disease: A meta-analysis. *Neuroscience Letters*, *783*, 136699. DOI: 10.1016/j.neulet.2022.136699

Mayer, E. A. (2011). Gut feelings: The emerging biology of gut–brain communication. [A comprehensive recent review of the underlying neurobiology and bidirectional nature of the gut brain axis.]. *Nature Reviews. Neuroscience*, *12*(8), 453–466. DOI: 10.1038/nrn3071 PMID: 21750565

Mazmanian, S. K., Round, J. L., & Kasper, D. L. (2008). A microbial symbiosis factor prevents intestinal inflammatory disease. *Nature*, *453*(7195), 620–625. DOI: 10.1038/nature07008 PMID: 18509436

McDonald, J. W., & Sadowsky, C. (2002). Spinal-cord injury. *Lancet*, *359*(9304), 417–425. DOI: 10.1016/S0140-6736(02)07603-1 PMID: 11844532

McGeer, P. L., Itagaki, S., Boyes, B. E., & McGeer, E. G. (1988). Reactive microglia are positive for HLA-DR in the: Substantia nigra of Parkinson's and Alzheimer's disease brains. *Neurology*, *38*(8), 1285–1291. DOI: 10.1212/WNL.38.8.1285 PMID: 3399080

McGhie, T. K., & Walton, M. C. (2007). The bioavailability and absorption of anthocyanins: Towards a better understanding. *Molecular Nutrition & Food Research*, *51*(6), 702–713. DOI: 10.1002/mnfr.200700092 PMID: 17533653

Meher, A. K., Acharya, B., & Sahu, P. K. (2024). Probiotics: Bridging the interplay of a healthy gut and psychoneurological well-being. *Food Bioengineering*, *3*(1), 126–147. DOI: 10.1002/fbe2.12081

Mehnert, W., & Mäder, K. (2012). Solid lipid nanoparticles: Production, characterization, and applications. *Advanced Drug Delivery Reviews*, *64*, 83–101. DOI: 10.1016/j.addr.2012.09.021 PMID: 11311991

Meisel, H. (1997). Biochemical properties of bioactive peptides derived from milk proteins. *International Dairy Journal*, *7*(1), 21–32.

Mejía-Pitta, A., Broset, E., & de la Fuente-Nunez, C. (2021). Probiotic engineering strategies for the heterologous production of antimicrobial peptides. *Advanced Drug Delivery Reviews*, *176*, 113863. DOI: 10.1016/j.addr.2021.113863 PMID: 34273423

Meletis, K., Barnabé-Heider, F., Carlén, M., Evergren, E., Tomilin, N., Shupliakov, O., & Frisén, J. (2008). Spinal Cord Injury Reveals Multilineage Differentiation of Ependymal Cells. *PLoS Biology*, *6*(7), e182. DOI: 10.1371/journal.pbio.0060182 PMID: 18651793

Melrose, J. (2023). The potential of flavonoids and flavonoid metabolites in the treatment of neuro-degenerative pathology in disorders of cognitive decline. *Antioxidants, 12*(3), 663. DOI: 10.3390/antiox12030663 PMID: 36978911

Mendonça, E. L. S. S., Xavier, J. A., Fragoso, M. B. T., Silva, M. O., Escodro, P. B., Oliveira, A. C. M., Tucci, P., Saso, L., & Goulart, M. O. F. (2024). E-Stilbenes: General Chemical and Biological Aspects, Potential Pharmacological Activity Based on the Nrf2 Pathway. *Pharmaceuticals 2024, Vol. 17, Page 232, 17*(2), 232. DOI: 10.3390/ph17020232

Meng, H. Y., Shao, D. C., Li, H., Huang, X. D., Yang, G., Xu, B., & Niu, H. Y. (2018). Resveratrol improves neurological outcome and neuroinflammation following spinal cord injury through enhancing autophagy involving the AMPK/mTOR pathway. *Molecular Medicine Reports, 18*(2), 2237–2244. DOI: 10.3892/mmr.2018.9194 PMID: 29956767

Meng, K., Jia, H., Hou, X., Zhu, Z., Lu, Y., Feng, Y., Feng, J., Xia, Y., Tan, R., Cui, F., & Yuan, J. (2025). Mitochondrial Dysfunction in Neurodegenerative Diseases: Mechanisms and Corresponding Therapeutic Strategies. *Biomedicines, 13*(2), 1–30. DOI: 10.3390/biomedicines13020327 PMID: 40002740

Meng, T., Xiao, D., Muhammed, A., Deng, J., Chen, L., & He, J. (2021). Anti-Inflammatory Action and Mechanisms of Resveratrol. *Molecules (Basel, Switzerland), 26*(1), 229. Advance online publication. DOI: 10.3390/molecules26010229 PMID: 33466247

Merighi, S., Travagli, A., Nigro, M., Pasquini, S., Cappello, M., Contri, C., Varani, K., Vincenzi, F., Borea, P. A., & Gessi, S. (2023). Caffeine for Prevention of Alzheimer's Disease: Is the A2A Adenosine Receptor Its Target? *Biomolecules, 13*(6), 1–15. DOI: 10.3390/biom13060967 PMID: 37371547

Meyer, E., Mori, M. A., Campos, A. C., Andreatini, R., Guimarães, F. S., Milani, H., & de Oliveira, R. M. W. (2017). Myricitrin induces antidepressant-like effects and facilitates adult neurogenesis in mice. *Behavioural Brain Research, 316*, 59–65. DOI: 10.1016/j.bbr.2016.08.048 PMID: 27569185

Mietto, B. S., Mostacada, K., & Martinez, A. M. B. (2015). Neurotrauma and inflammation: CNS and PNS responses. *Mediators of Inflammation, 2015*(1), 251204. DOI: 10.1155/2015/251204 PMID: 25918475

Mignet, N., Seguin, J., & Chabot, G. G. (2013). Bioavailability of polyphenol liposomes: A challenge ahead. *Pharmaceutics, 5*(3), 457–471. DOI: 10.3390/pharmaceutics5030457 PMID: 24300518

Mihaylova, D., Dimitrova-Dimova, M., & Popova, A. (2024). Dietary Phenolic Compounds—Wellbeing and Perspective Applications. *International Journal of Molecular Sciences, 25*(9), 4769. DOI: 10.3390/ijms25094769 PMID: 38731987

Milinčić, D. D., Popović, D. A., Lević, S. M., Kostić, A. Ž., Tešić, Ž. L., Nedović, V. A., & Pešić, M. B. (2019). Application of polyphenol-loaded nanoparticles in food industry. *Nanomaterials (Basel, Switzerland), 9*(11), 1629. DOI: 10.3390/nano9111629 PMID: 31744091

Miller, M. G., Hamilton, D. A., Joseph, J. A., Shukitt-Hale, B. (2018). Dietary blueberry improves cognition among older adults in a randomized, double-blind, placebo-controlled trial. The Journals of Gerontology: Series A, 73(3), 399–406. https://doi.org/DOI: 10.1093/gerona/glx161

Miller, W. L. (1979). Use of recombinant DNA technology for the production of polypeptides. Cell Substrates: Their Use in the Production of Vaccines and Other Biologicals, 153-174.

Miller, I. (2018). The gut–brain axis: Historical reflections. *Microbial Ecology in Health and Disease*, *29*(2), 1542921. DOI: 10.1080/16512235.2018.1542921 PMID: 30425612

Miller, M. G., Thangthaeng, N., & Shukitt-Hale, B. (2018). Effects of blueberry supplementation on cognitive performance in older adults. *Journal of Gerontology: Series A*, *73*(3), 349–355. DOI: 10.1093/gerona/glx184

Minarovits, J., Banati, F., Szenthe, K., & Niller, H. H. (2015). Epigenetic regulation. *Patho-Epigenetics of Infectious Disease*, 1-25.

Miranda, M., Morici, J. F., Zanoni, M. B., & Bekinschtein, P. (2019). Brain-derived neurotrophic factor: A key molecule for memory in the healthy and the pathological brain. *Frontiers in Cellular Neuroscience*, *13*, 472800. DOI: 10.3389/fncel.2019.00363 PMID: 31440144

Mirshekar, M. A., Sarkaki, A., Farbood, Y., Naseri, M. K. G., Badavi, M., Mansouri, M. T., & Haghparast, A. (2018). Neuroprotective effects of gallic acid in a rat model of traumatic brain injury: Behavioral, electrophysiological, and molecular studies. *Iranian Journal of Basic Medical Sciences.*, *21*(10), 1056. PMID: 30524680

Mishra, A., Bandopadhyay, R., Singh, P. K., Mishra, P. S., Sharma, N., & Khurana, N. (2021). Neuroinflammation in neurological disorders: Pharmacotherapeutic targets from bench to bedside. *Metabolic Brain Disease*, *36*(7), 1591–1626. DOI: 10.1007/s11011-021-00806-4 PMID: 34387831

Mishra, S., Wang, S., Nagpal, R., Miller, B., Singh, R., Taraphder, S., & Yadav, H. (2023). Probiotics and prebiotics for mental health: Focus on depression and anxiety disorders. *Current Nutrition Reports*, *12*(1), 66–81. DOI: 10.1007/s13668-023-00457-0 PMID: 36790719

Missiego-Beltrán, J., & Beltrán-Velasco, A. I. (2024). The Role of Microbial Metabolites in the Progression of Neurodegenerative Diseases—Therapeutic Approaches: A Comprehensive Review. *International Journal of Molecular Sciences*, *25*(18), 10041. DOI: 10.3390/ijms251810041 PMID: 39337526

Mitrano, F., Tettamanti, M., & Ballabio, A. (2022). Artificial intelligence and personalized nutrition: Toward precision brain health. *Frontiers in Nutrition*, *9*, 823456. DOI: 10.3389/fnut.2022.823456

Mittal, P., & Agrawal, N. (2024). An Overview of Neurodegenerative Disorders. *Altered Metabolism: A Major Contributor of Comorbidities in Neurodegenerative Diseases*, 1-27.

Moghadas, M., Edalatmanesh, M. A., & Robati, R. (2016). Histopathological Analysis from Gallic Acid Administration on Hippocampal Cell Density, Depression, and Anxiety Related Behaviors in A Trimethyltin Intoxication Model. *Cell Journal*, *17*(4), 659–667. DOI: 10.22074/CELLJ.2016.3838 PMID: 26862525

Mohamed, E. A., Ahmed, H. I., Zaky, H. S., & Badr, A. M. (2021). Sesame oil mitigates memory impairment, oxidative stress, and neurodegeneration in a rat model of Alzheimer's disease: A pivotal role of NF-κB/p38MAPK/BDNF/PPAR-γ pathways. *Journal of Ethnopharmacology*, *280*, 114423. Advance online publication. 10.1016/j.jep.2021.114423

Mohammad, Z. B., Yudin, S. C. Y., Goldberg, B. J., Serra, K. L., & Klegeris, A. (2025). Exploring neuroglial signaling: Diversity of molecules implicated in microglia-to-astrocyte neuroimmune communication. *Reviews in the Neurosciences*, *36*(1), 91–117. DOI: 10.1515/revneuro-2024-0081 PMID: 39240134

Mohan Latha Kumari, A., Biju Latha, C., Padmanabhan, R. A., Vijayakumar, N., & Ramachandran, R. (2025). Antibacterial peptide nisin activates p53 gene expression and triggers apoptosis in human colon cancer cells. *Peptide Science (Hoboken, N.J.)*, *117*(1), e24379. DOI: 10.1002/pep2.24379

Mohd Sairazi, N. S., & Sirajudeen, K. N. S. (2020). Natural products and their bioactive compounds: Neuroprotective potentials against neurodegenerative diseases. *Evidence-Based Complementary and Alternative Medicine : eCAM*, *2020*(1), 6565396. DOI: 10.1155/2020/6565396 PMID: 32148547

Mohd Sairazi, N. S., & Sirajudeen, K. N. S. (2020). Natural products and their bioactive compounds: Neuroprotective potentials against neurodegenerative diseases. *Evidence-Based Complementary and Alternative Medicine: eCAM*, *2020*(1), 6565396.

Mokhtari, H., Yaghmaei, B., Sirati-Sabet, M., Jafari, N., Mardomi, A., Abediankenari, S., & Mahrooz, A. (2020). Epigallocatechin-3-gallate enhances the efficacy of MicroRNA-34a mimic and MiR-93 inhibitor co-transfection in prostate cancer cell line. *Iranian Journal of Allergy, Asthma, and Immunology*. Advance online publication. DOI: 10.18502/ijaai.v19i6.4930 PMID: 33463130

Mokra, D., Joskova, M., & Mokry, J. (2023). Therapeutic Effects of Green Tea Polyphenol (–)-Epigallocatechin-3-Gallate (EGCG) in Relation to Molecular Pathways Controlling Inflammation, Oxidative Stress, and Apoptosis. *International Journal of Molecular Sciences*, *24*(1), 340. Advance online publication. DOI: 10.3390/ijms24010340 PMID: 36613784

Molska, M., & Reguła, J. (2019). Potential mechanisms of probiotic action in the prevention and treatment of colorectal cancer. *Nutrients*, *11*(10), 2453. DOI: 10.3390/nu11102453 PMID: 31615096

Mongalo, N. I., McGaw, L. J., Segapelo, T. V., Finnie, J. F., & Van Staden, J. (2016). Ethnobotany, phytochemistry, toxicology and pharmacological properties of Terminalia sericea Burch. ex DC. (Combretaceae)–A review. *Journal of Ethnopharmacology*, *194*, 789–802. DOI: 10.1016/j.jep.2016.10.072 PMID: 27989875

Monkkonen, T., & Debnath, J. (2018). Inflammatory signaling cascades and autophagy in cancer. In *Autophagy*. DOI: 10.1080/15548627.2017.1345412

Montalto, M., Curigliano, V., Santoro, L., Vastola, M., Cammarota, G., Manna, R., Gasbarrini, A., & Gasbarrini, G. (2002). Management and treatment of lactose malabsorption. *World Journal of Gastroenterology*, *12*(2), 187–191. DOI: 10.3748/wjg.v12.i2.187 PMID: 16482616

Monteiro, L. S., & Paiva-Martins, F. (2022). Amino acids, amino acid derivatives and peptides as antioxidants. In *Lipid Oxidation in Food and Biological Systems: A Physical Chemistry Perspective* (pp. 381–404). Springer. DOI: 10.1007/978-3-030-87222-9_17

Mony, T. J., Elahi, F., Choi, J. W., & Park, S. J. (2022). Neuropharmacological Effects of Terpenoids on Preclinical Animal Models of Psychiatric Disorders: A Review. *Antioxidants*, *11*(9), 1834. DOI: 10.3390/antiox11091834 PMID: 36139909

Moosavi, F., Hosseini, R., Saso, L., & Firuzi, O. (2015). Modulation of neurotrophic signaling pathways by polyphenols. *Drug Design, Development and Therapy*, ●●●, 23–42. PMID: 26730179

Moradi Vastegani, S., Nasrolahi, A., Ghaderi, S., Tavassoly, I., & Ghasemi, A. (2023). Mitochondrial dysfunction and Parkinson's disease: Pathogenesis and therapeutic strategies. *Neurochemical Research, 48*(8), 2285–2308. DOI: 10.1007/s11064-023-03904-0 PMID: 36943668

Moradi, S. Z., Jalili, F., Farhadian, N., Joshi, T., Wang, M., Zou, L., & Xiao, J. (2022). Polyphenols and neurodegenerative diseases: Focus on neuronal regeneration. *Critical Reviews in Food Science and Nutrition*, *62*(13), 3421–3436. DOI: 10.1080/10408398.2020.1865870

Moradi, S. Z., Momtaz, S., Bayrami, Z., Farzaei, M. H., & Abdollahi, M. (2020). Nanoformulations of Herbal Extracts in Treatment of Neurodegenerative Disorders. *Frontiers in Bioengineering and Biotechnology*, *8*, 238. DOI: 10.3389/fbioe.2020.00238

Morgan, M. J., & Liu, Z. G. (2011). Crosstalk of reactive oxygen species and NF-κB signaling. *Cell Research*, *21*(1), 103–115. DOI: 10.1038/cr.2010.178 PMID: 21187859

Morris, G., Maes, M., Berk, M., Carvalho, A. F., & Puri, B. K. (2020). Nutritional ketosis as an intervention to relieve astrogliosis: Possible therapeutic applications in the treatment of neurodegenerative and neuroprogressive disorders. *European Psychiatry : the Journal of the Association of European Psychiatrists*, *63*(1), e8. Advance online publication. DOI: 10.1192/j.eurpsy.2019.13 PMID: 32093791

Morshedi, M., Hashemi, R., Moazzen, S., Sahebkar, A., & Hosseinifard, E.-S. (2019). Immunomodulatory and anti-inflammatory effects of probiotics in multiple sclerosis: A systematic review. *Journal of Neuroinflammation*, *16*(1), 1–11. DOI: 10.1186/s12974-019-1611-4 PMID: 31752913

Moser, I. (2008). Making Alzheimer's disease matter. Enacting, interfering and doing politics of nature. *Geoforum*, *39*(1), 98–110. DOI: 10.1016/j.geoforum.2006.12.007

Mouria, M., Gukovskaya, A. S., Jung, Y., Buechler, P., Hines, O. J., Reber, H. A., & Pandol, S. J. (2002). Food-derived polyphenols inhibit pancreatic cancer growth through mitochondrial cytochrome C release and apoptosis. *International Journal of Cancer*, *98*(5), 761–769. DOI: 10.1002/ijc.10202 PMID: 11920648

Moussa, C., Hebron, M., Huang, X., Ahn, J., Rissman, R. A., Aisen, P. S., & Turner, R. S. (2017). Resveratrol regulates neuro-inflammation and induces adaptive immunity in Alzheimer's disease. *Journal of Neuroinflammation*, *14*(1), 1. DOI: 10.1186/s12974-016-0779-0 PMID: 28086917

Muhammad, M. (2019). Tumor Necrosis Factor Alpha: A Major Cytokine of Brain Neuroinflammation. In Behzadi, P. (Ed.), *Cytokines*. IntechOpen., DOI: 10.5772/intechopen.85476

Mukherjee, N., Kumar, A. P., & Ghosh, R. (2015). DNA methylation and flavonoids in genitourinary cancers. *Current Pharmacology Reports*, *1*(2), 112–120. DOI: 10.1007/s40495-014-0004-8 PMID: 26005633

Mulica, P., Grünewald, A., & Pereira, S. L. (2021). Astrocyte-Neuron Metabolic Crosstalk in Neurodegeneration: A Mitochondrial Perspective. *Frontiers in Endocrinology*, *12*, 668517. DOI: 10.3389/fendo.2021.668517 PMID: 34025580

Müller, E., & Radler, F. (1993). Caseicin, a bacteriocin from Lactobacillus casei strain B80. *Folia Microbiologica, 38*(6), 441–446. DOI: 10.1007/BF02814392 PMID: 8150392

Mura, S., Nicolas, J., & Couvreur, P. (2013). Stimuli-responsive nanocarriers for drug delivery. *Nature Materials, 12*(11), 991–1003. DOI: 10.1038/nmat3776 PMID: 24150417

Murphy, C. E., Walker, A. K., & Weickert, C. S. (2021). Neuroinflammation in schizophrenia: The role of nuclear factor kappa B. *Translational Psychiatry, 11*(1), 1–13. DOI: 10.1038/s41398-021-01607-0 PMID: 34650030

Mustafa, A. G., & Alshboul, O. A. (2013). Pathophysiology of traumatic brain injury. *Neuroscience Journal, 18*(3), 222–235. PMID: 23887212

Mutha, R. E., Tatiya, A. U., & Surana, S. J. (2021). Flavonoids as natural phenolic compounds and their role in therapeutics: an overview. *Future Journal of Pharmaceutical Sciences 2021 7:1, 7*(1), 1–13. DOI: 10.1186/s43094-020-00161-8

Mutha, R. E., Tatiya, A. U., & Surana, S. J. (2022). Flavonoids as natural phenolic compounds and their role in therapeutics: An overview. *Phytomedicine Plus: International Journal of Phytotherapy and Phytopharmacology, 2*(1), 100191. Advance online publication. 10.1016/j.phyplu.2021.100191

Nagaoka, I., Suzuki, K., Niyonsaba, F., Tamura, H., & Hirata, M. (2012). Modulation of neutrophil apoptosis by antimicrobial peptides. *International Scholarly Research Notices, 2012*(1), 345791. PMID: 23724322

Nagarajan, S., Chellappan, D. R., Chinnaswamy, P., & Thulasingam, S. (2015). Ferulic acid pretreatment mitigates MPTP-induced motor impairment and histopathological alterations in C57Bl/6 mice. *Pharmaceutical Biology, 53*(11), 1591–1601. DOI: 10.3109/13880209.2014.993041 PMID: 25857436

Nagele, R. G., Wegiel, J., Venkataraman, V., Imaki, H., Wang, K. C., & Wegiel, J. (2004). Contribution of glial cells to the development of amyloid plaques in Alzheimer's disease. *Neurobiology of Aging, 25*(5), 663–674. DOI: 10.1016/j.neurobiolaging.2004.01.007 PMID: 15172746

Naghizadeh, B., & Mansouri, M. T. (2014). Protective effects of gallic acid against streptozotocin-induced oxidative damage in rat striatum. *Drug Research, 65*(10), 515–520. DOI: 10.1055/s-0034-1377012 PMID: 24941087

Nahar, L., Charoensup, R., Kalieva, K., Habibi, E., Guo, M., Wang, D., Kvasnica, M., Onder, A., & Sarker, S. D. (2025). Natural products in neurodegenerative diseases: Recent advances and future outlook. *Frontiers in Pharmacology, 16*, 1529194. Advance online publication. 10.3389/fphar.2025.1529194

Naik, B., Nirwane, A., & Majumdar, A. (2017). Pterostilbene ameliorates intracerebroventricular streptozotocin induced memory decline in rats. *Cognitive Neurodynamics, 11*(1), 35–49. DOI: 10.1007/s11571-016-9413-1 PMID: 28174611

Naoi, M., Maruyama, W., & Akao, Y. (2017). Neurotrophic function of phytochemicals for neuroprotection in aging and neurodegenerative disorders: Modulation of intracellular signaling and gene expression. *Journal of Neural Transmission (Vienna, Austria), 124*(12), 1515–1527. DOI: 10.1007/s00702-017-1797-5 PMID: 29030688

Naoi, M., Shamoto-Nagai, M., & Maruyama, W. (2019). Neuroprotection of Multifunctional Phytochemicals as Novel Therapeutic Strategy for Neurodegenerative Disorders: Antiapoptotic and Antiamyloidogenic Activities by Modulation of Cellular Signal Pathways. *Future Neurology*, *14*(1), FNL9. DOI: 10.2217/fnl-2018-0028

Naoi, M., Wu, Y., Shamoto-Nagai, M., & Maruyama, W. (2019). Mitochondria in Neuroprotection by Phytochemicals: Bioactive Polyphenols Modulate Mitochondrial Apoptosis System, Function and Structure. *International Journal of Molecular Sciences*, *20*(10), 2451. Advance online publication. DOI: 10.3390/ijms20102451 PMID: 31108962

Naomi, R., Yazid, M. D., Teoh, S. H., Balan, S. S., Shariff, H., Kumar, J., Bahari, H., & Embong, H. (2023). Dietary polyphenols as a protection against cognitive decline: Evidence from animal experiments; mechanisms and limitations. *Antioxidants*, *12*(5), 1054. DOI: 10.3390/antiox12051054 PMID: 37237920

Nasrabady, S. E., Rizvi, B., Goldman, J. E., & Brickman, A. M. (2018). White matter changes in Alzheimer's disease: a focus on myelin and oligodendrocytes. In *Acta neuropathologica communications* (Vol. 6, Issue 1, p. 22). BioMed Central. DOI: 10.1186/s40478-018-0515-3

Nath, J., Roy, R., Sathyamoorthy, Y. K., Paul, S., Goswami, S., Chakravarty, H., Paul, R., & Borah, A. (2022). Resveratrol as a therapeutic choice for traumatic brain injury: An insight into its molecular mechanism of action. *Brain Disorders (Amsterdam, Netherlands)*, *6*, 100038. DOI: 10.1016/j.dscb.2022.100038

National Library of Medicine (US). (2007, April 19). *The Effects of Green Tea on Parkinson's Disease* (Clinical Trial Registration No. NCT00461942). ClinicalTrials.gov. https://clinicaltrials.gov/ct2/show/NCT00461942

National Library of Medicine (US). (2014, February 12). *Grape Seed Extract in People With Memory Decline* (Clinical Trial Registration No. NCT02033941). ClinicalTrials.gov. https://clinicaltrials.gov/ct2/show/NCT02033941

Navabi, S. P., Badreh, F., Khombi Shooshtari, M., Hajipour, S., Moradi Vastegani, S., & Khoshnam, S. E. (2024). Microglia-induced neuroinflammation in hippocampal neurogenesis following traumatic brain injury. In *Heliyon* (Vol. 10, Issue 16, p. e35869). Elsevier. DOI: 10.1016/j.heliyon.2024.e35869

Navas, R. E., Cuppens, F., Cuppens, N. B., Toutain, L., & Papadopoulos, G. Z. (2020). MTD, where art thou? A systematic review of moving target defense techniques for IoT. *IEEE Internet of Things Journal*, *8*(10), 7818–7832. DOI: 10.1109/JIOT.2020.3040358

Nave, K. A., Asadollahi, E., & Sasmita, A. (2023). Expanding the function of oligodendrocytes to brain energy metabolism. In *Current Opinion in Neurobiology* (Vol. 83). Elsevier Ltd., DOI: 10.1016/j.conb.2023.102782

Nawaz, L., Grieve, D. J., Muzaffar, H., Iftikhar, A., & Anwar, H. (2024). Methanolic Extract of Phoenix Dactylifera Confers Protection against Experimental Diabetic Cardiomyopathy through Modulation of Glucolipid Metabolism and Cardiac Remodeling. *Cells*, *13*(14), 1196. DOI: 10.3390/cells13141196 PMID: 39056777

Neef, J., van Dijl, J. M., & Buist, G. (2021). Recombinant protein secretion by Bacillus subtilis and Lactococcus lactis: Pathways, applications, and innovation potential. *Essays in Biochemistry*, *65*(2), 187–195. DOI: 10.1042/EBC20200171 PMID: 33955475

Neider, D., Lindström, L. H., & Bodén, R. (2016). Risk factors for suicide among patients with schizophrenia: A cohort study focused on cerebrospinal fluid levels of homovanillic acid and 5-hydroxyindoleacetic acid. *Neuropsychiatric Disease and Treatment*, *12*, 1711–1714. DOI: 10.2147/NDT.S107178 PMID: 27468235

Nemzer, B. V., Al-Taher, F., Kalita, D., Yashin, A. Y., & Yashin, Y. I. (2025). Health-improving effects of polyphenols on the human intestinal microbiota: A review. *International Journal of Molecular Sciences*, *26*(3), 1335. DOI: 10.3390/ijms26031335 PMID: 39941107

Nesterova, Y. V., Povetieva, T. N., Suslov, N. I., Semenov, A. A., & Pushkarskiy, S. V. (2011). Antidepressant activity of diterpene alkaloids of aconitum baicalense turcz. *Bulletin of Experimental Biology and Medicine*, *151*(4), 425–428. DOI: 10.1007/s10517-011-1347-3 PMID: 22448357

Ng, Y. P., Or, T. C. T., & Ip, N. Y. (2015). Plant alkaloids as drug leads for Alzheimer's disease. In *Neurochemistry International* (Vol. 89, pp. 260–270). Elsevier Ltd., DOI: 10.1016/j.neuint.2015.07.018

Nichols, E., Steinmetz, J. D., Vollset, S. E., Fukutaki, K., Chalek, J., Abd-Allah, F., & Akram, T. T. (2022). Estimation of the global prevalence of dementia in 2019 and forecasted prevalence in 2050: An analysis for the Global Burden of Disease Study 2019. *The Lancet. Public Health*, *7*(2), e105–e125. DOI: 10.1016/S2468-2667(21)00249-8 PMID: 34998485

Nichols, E., Szoeke, C. E. I., Vollset, S. E., Abbasi, N., Abd-Allah, F., Abdela, J., Aichour, M. T. E., Akinyemi, R. O., Alahdab, F., Asgedom, S. W., Awasthi, A., Barker-Collo, S. L., Baune, B. T., Béjot, Y., Belachew, A. B., Bennett, D. A., Biadgo, B., Bijani, A., Bin Sayeed, M. S., & Murray, C. J. L. (2019). Global, regional, and national burden of Alzheimer's disease and other dementias, 1990–2016: A systematic analysis for the Global Burden of Disease Study 2016. *Lancet Neurology*, *18*(1), 88–106. DOI: 10.1016/S1474-4422(18)30403-4 PMID: 30497964

Niedzielska, E., Smaga, I., Gawlik, M., Moniczewski, A., Stankowicz, P., Pera, J., & Filip, M. (2016). Oxidative Stress in Neurodegenerative Diseases. In *Molecular Neurobiology* (Vol. 53, Issue 6, pp. 4094–4125). Humana Press Inc. DOI: 10.1007/s12035-015-9337-5

Nisha, I. J., & Elora, E. J. (2025). Natural bioactives and microbiome-based precision therapeutics for cancer, chronic, and neurodegenerative diseases. *PrimeAsia*, *6*(1), 1–8. DOI: 10.25163/primeasia.6110264

Njapdounke, S. K. J., Nkantchoua, N. G., Moto Okomolo, F. C., Taiwe, S. G., Neteydji, S., Pale, S., Ayissi Mbomo, E. R., & Ngo Bum, E. (2016). Anxiolytic - like properties of Hallea ciliata in mice. *African Journal of Traditional, Complementary, and Alternative Medicines : AJTCAM*, *13*(4), 1–7. DOI: 10.21010/ajtcam.v13i4.1 PMID: 28852713

Nongonierma, A. B., & FitzGerald, R. J. (2015). Bioactive properties of food-derived peptides: A review. *Journal of Functional Foods*, *17*, 640–656. DOI: 10.1016/j.jff.2015.06.021

Notter, T. (2021). Astrocytes in schizophrenia. *Brain and Neuroscience Advances*, *5*, 23982128211009148. Advance online publication. DOI: 10.1177/23982128211009148 PMID: 33997293

Nouri, Z., Fakhri, S., El-Senduny, F. F., Sanadgol, N., Abd-ElGhani, G. E., Farzaei, M. H., & Chen, J.-T. (2019). On the neuroprotective effects of naringenin: Pharmacological targets, signalling pathways, molecular mechanisms, and clinical perspective. *Biomolecules*, *9*(11), 690. Advance online publication. 10.3390/biom9110690

Novikov, V. N., Badaeva, A. V., Danilov, A. B., & Vorobyeva, Y. D. (2023). *Precision Neuronutrition: Personalized Approaches for Optimizing Brain Health.* Paper presented at the Biology and Life Sciences Forum. DOI: 10.3390/IECN2023-15524

Nowak, A., Kojder, K., Zielonka-Brzezicka, J., Wróbel, J., Bosiacki, M., Fabiańska, M., Wróbel, M., Sołek-Pastuszka, J., & Klimowicz, A. (2021). The use of Ginkgo biloba L. as a neuroprotective agent in the Alzheimer's disease. *Frontiers in Pharmacology*, *12*, 775034. DOI: 10.3389/fphar.2021.775034 PMID: 34803717

Nykjær, C. H., Brudek, T., Salvesen, L., & Pakkenberg, B. (2017). Changes in the cell population in brain white matter in multiple system atrophy. *Movement Disorders*, *32*(7), 1074–1082. DOI: 10.1002/mds.26979 PMID: 28394027

O'Riordan, K. J., Collins, M. K., Moloney, G. M., Knox, E. G., Aburto, M. R., Fülling, C., Morley, S. J., Clarke, G., Schellekens, H., & Cryan, J. F. (2022). Short chain fatty acids: Microbial metabolites for gut-brain axis signalling. *Molecular and Cellular Endocrinology*, *546*, 111572. DOI: 10.1016/j.mce.2022.111572 PMID: 35066114

O'Shea, J. J., & Plenge, R. (2012). JAK and STAT signaling molecules in immunoregulation and immune-mediated disease. *Immunity*, *36*(4), 542–550. DOI: 10.1016/j.immuni.2012.03.014 PMID: 22520847

Oberheim, N. A., Goldman, S. A., & Nedergaard, M. (2012). Heterogeneity of astrocytic form and function. *Methods in Molecular Biology (Clifton, N.J.)*, *814*, 23–45. DOI: 10.1007/978-1-61779-452-0_3 PMID: 22144298

Ocal, O., Borcek, A. O., Pasaoglu, O., Gundogdu, A. C., Kaplanoglu, G. T., & Baykaner, M. K. (2019). Can quercetin be an option for treatment of spinal cord injury? An experimental study. *Turkish Neurosurgery*, *29*(2), 247–253. DOI: 10.5137/1019-5149.JTN.23799-18.1 PMID: 30649798

Ohara, T., Yoshino, K., & Kitajima, M. (2010). Possibility of preventing colorectal carcinogenesis with probiotics. *Hepato-Gastroenterology*, *57*(101), 1411–1415. PMID: 21443095

Oirdi, E. (2024). Harnessing the Power of Polyphenols: A New Frontier in Disease Prevention and Therapy. *Pharmaceuticals (Basel, Switzerland)*, *17*(6), 692. https://www.mdpi.com/1424-8247/17/6/692#B197-pharmaceuticals-17-00692. DOI: 10.3390/ph17060692 PMID: 38931359

Ojha, S. (2021). *Containing Natural Products*. 1–26.

Ojha, S., Javed, H., Azimullah, S., Khair, S. B. A., & Haque, M. E. (2015). Neuroprotective potential of ferulic acid in the rotenone model of Parkinson's disease. *Drug Design, Development and Therapy*, *9*, 5499–5510. DOI: 10.2147/DDDT.S90616 PMID: 26504373

Oliveira, M. A., Heimfarth, L., Passos, F. R. S., Miguel-dos-Santos, R., Mingori, M. R., Moreira, J. C. F., Lauton, S. S., Barreto, R. S., Araújo, A. A., Oliveira, A. P., Oliveira, J. T., Baptista, A. F., Martinez, A. M. B., Quintans-Júnior, L. J., & Quintans, J. S. S. (2020). Naringenin complexed with hydroxypropyl-β-cyclodextrin improves the sciatic nerve regeneration through inhibition of p75NTR and JNK pathway. *Life Sciences*, *241*, 117102. DOI: 10.1016/j.lfs.2019.117102 PMID: 31790691

Olufunmilayo, E. O., Gerke-Duncan, M. B., & Holsinger, R. M. D. (2023). Oxidative stress and antioxidants in neurodegenerative disorders. *Antioxidants*, *12*(2), 517. DOI: 10.3390/antiox12020517 PMID: 36830075

Oluwole, O., Fernando, W. B., Lumanlan, J., Ademuyiwa, O., & Jayasena, V. (2022). Role of phenolic acid, tannins, stilbenes, lignans and flavonoids in human health – a review. *International Journal of Food Science & Technology*, *57*(10), 6326–6335. DOI: 10.1111/ijfs.15936

Orning, P., & Lien, E. (2021). Multiple roles of caspase-8 in cell death, inflammation, and innate immunity. *Journal of Leukocyte Biology*, *109*(1), 121–141. DOI: 10.1002/JLB.3MR0420-305R PMID: 32531842

Orozco, M. F., Vázquez-Hernández, A., & Fenton-Navarro, B. (2019). Active compounds of medicinal plants, the mechanism for antioxidant and beneficial effects. *Phyton*, *88*(1), 1–10. 10.32604/phyton.2019.04525

Ortiz, G. G., Flores Alvarado, L. J., Pacheco Moisés, F. P., Mireles Ramírez, M. A., González Renovato, E. D., Sánchez López, A. L., ... & Sánchez González, V. J. (2016). Cross-talk between glial cells and neurons: Relationship in Multiple Sclerosis.

Ou, Z., Deng, L., Lu, Z., Wu, F., Liu, W., Huang, D., & Peng, Y. (2020). Protective effects of Akkermansia muciniphila on cognitive deficits and amyloid pathology in a mouse model of Alzheimer's disease. *Nutrition & Diabetes*, *10*(1), 12. DOI: 10.1038/s41387-020-0115-8 PMID: 32321934

Oyagbemi, A. A., Omobowale, T. O., Saba, A. B., Olowu, E. R., Dada, R. O., & Akinrinde, A. S. (2016). Gallic acid ameliorates cyclophosphamide-induced neurotoxicity in Wistar rats through free radical scavenging activity and improvement in antioxidant defense system. *Journal of Dietary Supplements*, *13*(4), 402–419. DOI: 10.3109/19390211.2015.1103827 PMID: 26716793

Oyinloye, B. E., Adenowo, A. F., & Kappo, A. P. (2015). Reactive oxygen species, apoptosis, antimicrobial peptides and human inflammatory diseases. *Pharmaceuticals (Basel, Switzerland)*, *8*(2), 151–175. DOI: 10.3390/ph8020151 PMID: 25850012

Özbek, Z., Aydin, H., Kocman, A., Özkara, E., Şahin, E., Bektur, E., Vural, M., Köse, A., Arslantaş, A., & Baycu, C. (2017). Neuroprotective effect of genistein in peripheral nerve injury. *Turkish Neurosurgery*, *27*(5). PMID: 27759874

Özben, T., & Özben, S. (2019). Neuro-inflammation and anti-inflammatory treatment options for Alzheimer's disease. *Clinical Biochemistry*, *68*, 15–25. DOI: 10.1016/j.clinbiochem.2019.04.001 PMID: 30954437

Özben, T., & Özben, S. (2019b). Clinical biochemical markers in Alzheimer's disease and the effect of peptide interventions. *Clinical Biochemistry*, *71*, 20–28. DOI: 10.1016/j.clinbiochem.2019.04.015

Özben, T., & Özben, S. (2019c). Peptide-based antioxidant strategies in neurodegeneration. *Clinical Biochemistry*, *71*, 29–35. DOI: 10.1016/j.clinbiochem.2019.04.011

Ozdal, T., Sela, D. A., Xiao, J., Boyacioglu, D., Chen, F., & Capanoglu, E. (2016). The reciprocal interactions between polyphenols and gut microbiota and effects on bioaccessibility. *Nutrients*, *8*(2), 78. DOI: 10.3390/nu8020078 PMID: 26861391

Ozogul, Y., Ucar, Y., Tadesse, E. E., Rathod, N., Kulawik, P., Trif, M., Esatbeyoglu, T., & Ozogul, F. (2025). Tannins for food preservation and human health: A review of current knowledge. *Applied Food Research*, *5*(1), 100738. DOI: 10.1016/j.afres.2025.100738

Panche, A. N., Diwan, A. D., & Chandra, S. R. (2016). Flavonoids: An overview. *Journal of Nutritional Science*, *5*, e47. DOI: 10.1017/jns.2016.41 PMID: 28620474

Pandey, K. B., & Rizvi, S. I. (2009). Current understanding of dietary polyphenols and their role in health and disease. *Current Nutrition and Food Science*, *5*(4), 249–263.

Pandey, K. B., & Rizvi, S. I. (2009). Plant polyphenols as dietary antioxidants in human health and disease. *Oxidative Medicine and Cellular Longevity*, *2*(5), 270–278. DOI: 10.4161/oxim.2.5.9498 PMID: 20716914

Papi, S., Ahmadizar, F., & Hasanvand, A. (2019). The role of nitric oxide in inflammation and oxidative stress. *Immunopathol Persa*, *5*(1), e08–e08. DOI: 10.15171/ipp.2019.08

Pardridge, W. M. (2005). The blood-brain barrier: Bottleneck in brain drug development. *NeuroRx : the Journal of the American Society for Experimental NeuroTherapeutics*, *2*(1), 3–14. DOI: 10.1602/neurorx.2.1.3 PMID: 15717053

Pardridge, W. M. (2012). Drug transport across the blood–brain barrier. *Journal of Cerebral Blood Flow and Metabolism*, *32*(11), 1959–1972. DOI: 10.1038/jcbfm.2012.126 PMID: 22929442

Parent, M., Chitturi, J., Santhakumar, V., Hyder, F., Sanganahalli, B. G., & Kannurpatti, S. S. (2020). Kaempferol treatment after traumatic brain injury during early development mitigates brain parenchymal microstructure and neural functional connectivity deterioration at adolescence. *Journal of Neurotrauma*, *37*(7), 966–974. DOI: 10.1089/neu.2019.6486 PMID: 31830867

Parker, A., Fonseca, S., & Carding, S. R. (2020). Gut microbes and metabolites as modulators of blood-brain barrier integrity and brain health. *Gut Microbes*, *11*(2), 135–157. DOI: 10.1080/19490976.2019.1638722 PMID: 31368397

Parker, J. K., & Davies, B. W. (2022). Microcins reveal natural mechanisms of bacterial manipulation to inform therapeutic development. *Microbiology*, *168*(4), 001175. DOI: 10.1099/mic.0.001175 PMID: 35438625

Park, H., & Poo, M. M. (2013). Neurotrophin regulation of neural circuit development and function. *Nature Reviews. Neuroscience*, *14*(1), 7–23. DOI: 10.1038/nrn3379 PMID: 23254191

Parkhurst, C. N., Yang, G., Ninan, I., Savas, J. N., Yates, J. R.III, Lafaille, J. J., Hempstead, B. L., Littman, D. R., & Gan, W. B. (2013). Microglia promote learning-dependent synapse formation through brain-derived neurotrophic factor. *Cell*, *155*(7), 1596–1609. DOI: 10.1016/j.cell.2013.11.030 PMID: 24360280

Parr, A. J., & Bolwell, G. P. (2000). Phenols in the plant and in man. The potential for possible nutritional enhancement of the diet by modifying the phenols content or profile. *Journal of the Science of Food and Agriculture, 80*(7), 985–1012. DOI: 10.1002/(SICI)1097-0010(20000515)80:7<985:AID-JSFA572>3.0.CO;2-7

Parra, I., Martínez, I., Vásquez-Celaya, L., Gongora-Alfaro, J. L., Tizabi, Y., & Mendieta, L. (2023). Neuroprotective and immunomodulatory effects of probiotics in a rat model of Parkinson's disease. *Neurotoxicity Research, 41*(2), 187–200. DOI: 10.1007/s12640-022-00627-y PMID: 36662412

Parveen, A., Parveen, B., Parveen, R., & Ahmad, S. (2015). Challenges and guidelines for clinical trial of herbal drugs. *Journal of Pharmacy & Bioallied Sciences, 7*(4), 329–333. DOI: 10.4103/0975-7406.168035 PMID: 26681895

Parvez, S., Kaushik, M., Ali, M., Alam, M. M., Ali, J., Tabassum, H., & Kaushik, P. (2022). Dodging blood brain barrier with "nano" warriors: Novel strategy against ischemic stroke. *Theranostics, 12*(2), 689–719. DOI: 10.7150/thno.64806 PMID: 34976208

Patel, S., Keating, B. A., & Dale, R. C. (2023). Anti-inflammatory properties of commonly used psychiatric drugs. *Frontiers in Neuroscience, 16*, 1039379. Advance online publication. DOI: 10.3389/fnins.2022.1039379 PMID: 36704001

Patel, Y. C. (1999). Somatostatin and its receptor family. *Frontiers in Neuroendocrinology, 20*(3), 157–198. DOI: 10.1006/frne.1999.0183 PMID: 10433861

Pathak, D., & Sriram, K. (2023). Neuron-astrocyte omnidirectional signaling in neurological health and disease. *Frontiers in Molecular Neuroscience, 16*, 1169320.

Patil, K. R., Mahajan, U. B., Unger, B. S., Goyal, S. N., Belemkar, S., Surana, S. J., Ojha, S., & Patil, C. R. (2019). Animal models of inflammation for screening of anti-inflammatory drugs: Implications for the discovery and development of phytopharmaceuticals. *International Journal of Molecular Sciences, 20*(18), 4367. DOI: 10.3390/ijms20184367 PMID: 31491986

Patil, S. P., Jain, P. D., Sancheti, J. S., Ghumatkar, P. J., Tambe, R., & Sathaye, S. (2022). Retraction notice to "Neuroprotective and neurotrophic effects of Apigenin and Luteolin in MPTP induced Parkinsonism in mice" [Neuropharmacology, 86 (2014) 192-202]. *Neuropharmacology, 202*, 108876. DOI: 10.1016/j.neuropharm.2021.108876

Patil, S., Dandekar, P., & Pawar, S. (2021). Nanoemulsions as delivery vehicles for anthocyanins: Enhancing stability and bioactivity. *Food Research International, 143*, 110243. DOI: 10.1016/j.foodres.2021.110243

Patten, J. P. (1996). Trauma and the nervous system. In *Neurological Differential Diagnosis* (pp. 399–417). Springer. DOI: 10.1007/978-3-642-58981-2_23

Patterson, E., Cryan, J. F., Fitzgerald, G. F., Ross, R. P., Dinan, T. G., & Stanton, C. (2014). Gut microbiota, the pharmabiotics they produce and host health. [DOI] [PubMed] [Google Scholar]. *The Proceedings of the Nutrition Society, 73*(4), 477–489. DOI: 10.1017/S0029665114001426 PMID: 25196939

Paula, H. S. C., Santiago, S. B., Araújo, L. A., Pedroso, C. F., Marinho, T. A., Goncalves, I. A. J., & Batista, K. A. (2022). An overview on the current available treatment for COVID-19 and the impact of antibiotic administration during the pandemic. *Brazilian Journal of Medical and Biological Research = Revista Brasileira de Pesquisas Médicas e Biológicas*, *55*, e11631.

Pavithra, K., & Manimaran, V. (2024). A Review of Safety, Quality, Regulation, and Delivery Approaches for Phytopharmaceuticals. *Jordan Journal of Pharmaceutical Sciences*, *17*(2), 316–332. DOI: 10.35516/jjps.v17i2.1768

Payne, A. G. (2009). Experimental regimen targeting the ependyma slows disease progression in four patients with amyotrophic lateral sclerosis. *Medical Hypotheses*, *72*(5), 548–550. DOI: 10.1016/j.mehy.2008.12.032 PMID: 19200662

Payne, A., Nahashon, S., Taka, E., Adinew, G. M., & Soliman, K. F. (2022). Epigallocatechin-3-Gallate (EGCG): New therapeutic perspectives for neuroprotection, aging, and neuroinflammation for the modern age. *Biomolecules*, *12*(3), 371. DOI: 10.3390/biom12030371 PMID: 35327563

Peggion, C., Calì, T., & Brini, M. (2024). Mitochondria dysfunction and neuroinflammation in neurodegeneration: Who comes first? *Antioxidants*, *13*(2), 240. DOI: 10.3390/antiox13020240 PMID: 38397838

Pellegrini, M., & Strasser, A. (2013). Caspases, Bcl-2 family proteins and other components of the death machinery: their role in the regulation of the immune response. In *Madame Curie Bioscience Database* [Internet]. Landes Bioscience.

Peng, J., Yang, Z., Li, H., Hao, B., Cui, D., Shang, R., Lv, Y., Liu, Y., Pu, W., Zhang, H., He, J., Wang, X., & Wang, S. (2023). Quercetin Reprograms Immunometabolism of Macrophages via the SIRT1/PGC-1α Signaling Pathway to Ameliorate Lipopolysaccharide-Induced Oxidative Damage. *International Journal of Molecular Sciences*, *24*(6), 5542. Advance online publication. DOI: 10.3390/ijms24065542 PMID: 36982615

Peng, L., Li, Z.-R., Green, R. S., Holzman, I. R., & Lin, J. (2022). Butyrate enhances the intestinal barrier by facilitating tight junction assembly via activation of AMP-activated protein kinase in Caco-2 cell monolayers. *The Journal of Nutrition*, *152*(7), 1877–1885. DOI: 10.1093/jn/nxac069 PMID: 19625695

Peng, L., Parpura, V., & Verkhratsky, A. (2014). EDITORIAL Neuroglia as a Central Element of Neurological Diseases: An Underappreciated Target for Therapeutic Intervention. *Current Neuropharmacology*, *12*(4), 303–307. DOI: 10.2174/1570159X12999140829152550 PMID: 25342938

Peng, L., Verkhratsky, A., Gu, L., & Li, B. (2015). Targeting astrocytes in major depression. *Expert Review of Neurotherapeutics*, *15*(11), 1299–1306. DOI: 10.1586/14737175.2015.1095094 PMID: 26471936

Pérez-Jiménez, J., Neveu, V., Vos, F., & Scalbert, A. (2010). Identification of the 100 richest dietary sources of polyphenols: An application of the Phenol-Explorer database. *European Journal of Clinical Nutrition*, *64*(S3, Suppl 3), S112–S120. DOI: 10.1038/ejcn.2010.221 PMID: 21045839

Perluigi, M., Di Domenico, F., & Butterfield, D. A. (2024). OXIDATIVE DAMAGE IN NEURODEGENERATION: ROLES IN THE PATHOGENESIS AND PROGRESSION OF ALZHEIMER DISEASE. In *Physiological Reviews* (Vol. 104, Issue 1, pp. 103–197). American Physiological Society. DOI: 10.1152/physrev.00030.2022

Peterson, J., Dwyer, J., Adlercreutz, H., Scalbert, A., Jacques, P., & McCullough, M. L. (2010). Dietary lignans: Physiology and potential for cardiovascular disease risk reduction. *Nutrition Reviews*, *68*(10), 571–603. DOI: 10.1111/j.1753-4887.2010.00319.x PMID: 20883417

Peterson, J., Garges, S., Giovanni, M., McInnes, P., Wang, L., Schloss, J. A., Bonazzi, V., McEwen, J. E., Wetterstrand, K. A., Deal, C., Baker, C. C., Di Francesco, V., Howcroft, T. K., Karp, R. W., Lunsford, R. D., Wellington, C. R., Belachew, T., Wright, M., Giblin, C., & Guyer, M. (2009). The NIH Human Microbiome Project. *Genome Research*, *19*(12), 2317–2323. DOI: 10.1101/gr.096651.109 PMID: 19819907

Petrella, C., Farioli-Vecchioli, S., Cisale, G. Y., Strimpakos, G., Borg, J. J., Ceccanti, M., Fiore, M., Monteleone, G., & Nisticò, R. (2021). A healthy gut for a healthy brain: Preclinical, clinical and regulatory aspects. *Current Neuropharmacology*, *19*(5), 610–628. DOI: 10.2174/1570159X18666200730111528 PMID: 32744976

Picon, C., Jayaraman, A., Kirkwood, R., Pradeepan, S., Amor, S., & Reynolds, R. (2021). Neuron-specific activation of necroptosis signaling in multiple sclerosis cortical grey matter. *Acta Neuropathologica*, *141*(4), 585–604. DOI: 10.1007/s00401-021-02274-7 PMID: 33569629

Piers, K. L., Brown, M. H., & Hancock, R. E. (1993). Recombinant DNA procedures for producing small antimicrobial cationic peptides in bacteria. *Gene*, *134*(1), 7–13. DOI: 10.1016/0378-1119(93)90168-3 PMID: 8244033

Pislyagin, E. A., Tarbeeva, D. V., Yurchenko, E. A., Menchinskaya, E. S., Gorpenchenko, T. Y., Pokhilo, N. D., Kalinovskiy, A. I., Aminin, D. L., & Fedoreyev, S. A. (2025). Neuroprotective Activity of Oligomeric Stilbenes from Alpha Grape Stems in In Vitro Models of Parkinson's Disease. *International Journal of Molecular Sciences*, *26*(6), 2411. DOI: 10.3390/ijms26062411 PMID: 40141054

Pittenger, C., & Duman, R. S. (2008). Stress, depression, and neuroplasticity: A convergence of mechanisms. *Neuropsychopharmacology : Official Publication of the American College of Neuropsychopharmacology*, *33*(1), 88–109. DOI: 10.1038/sj.npp.1301574 PMID: 17851537

Pizarroso, N. A., Fuciños, P., Gonçalves, C., Pastrana, L., & Amado, I. (2021). A review on the role of food-derived bioactive molecules and the microbiota–gut–brain axis in satiety regulation. *Nutrients*, *13*(2), 632. DOI: 10.3390/nu13020632 PMID: 33669189

Pluta, R., Kiś, J., Januszewski, S., Jabłoński, M., & Czuczwar, S. J. (2022). Cross-talk between amyloid, tau protein and free radicals in post-ischemic brain neurodegeneration in the form of Alzheimer's disease proteinopathy. *Antioxidants*, *11*(1), 146. DOI: 10.3390/antiox11010146 PMID: 35052650

Pluznick, J. L. (2017). Microbial short-chain fatty acids and blood pressure regulation. *Current Hypertension Reports*, *19*(4), 1–5. DOI: 10.1007/s11906-017-0722-5 PMID: 28315048

Pluznick, J. L., Protzko, R. J., Gevorgyan, H., Peterlin, Z., Sipos, A., Han, J., Brunet, I., Wan, L.-X., Rey, F., Wang, T., Firestein, S. J., Yanagisawa, M., Gordon, J. I., Eichmann, A., Peti-Peterdi, J., & Caplan, M. J. (2013). Olfactory receptor responding to gut microbiota-derived signals plays a role in renin secretion and blood pressure regulation. *Proceedings of the National Academy of Sciences of the United States of America*, *110*(11), 4410–4415. DOI: 10.1073/pnas.1215927110 PMID: 23401498

Prince, M., Wimo, A., Guerchet, M., Ali, G. C., Wu, Y. T., & Prina, M. (2016). World Alzheimer Report 2015: The global impact of dementia. Alzheimer's Disease International. https://www.alzint.org/resource/world-alzheimer-report-2015/

Prinz, M., Masuda, T., Wheeler, M. A., & Quintana, F. J. (2021). Microglia and Central Nervous System-Associated Macrophages mdash from Origin to Disease Modulation. In *Annual Review of Immunology* (Vol. 39, pp. 251–277). Annual Reviews Inc., DOI: 10.1146/annurev-immunol-093019-110159

Prior, R. L., Wu, X., & Schaich, K. (2007). Standardized methods for the determination of antioxidant capacity and phenolics in foods and dietary supplements. *Journal of Agricultural and Food Chemistry*, *55*(4), 1081–1094. DOI: 10.1021/jf062583y PMID: 15884874

Proestos, C., & Varzakas, T. (2017). Aromatic plants: Antioxidant capacity and polyphenol characterisation. *Foods*, *6*(4), 28. DOI: 10.3390/foods6040028 PMID: 28375185

Puentes-Orozco, M., Albarracin, S. L., & Velásquez, M. M. (2024). Neuroinflammation and major depressive disorder: Astrocytes at the crossroads. *Frontiers in Cellular Neuroscience*, *18*(November), 1–8. DOI: 10.3389/fncel.2024.1504555 PMID: 39650796

Qian, Y., Guan, T., Huang, M., Cao, L., Li, Y., Cheng, H., & Yu, D. (2012). Neuroprotection by the soy isoflavone, genistein, via inhibition of mitochondria-dependent apoptosis pathways and reactive oxygen induced-NF-κB activation in a cerebral ischemia mouse model. *Neurochemistry International*, *60*(8), 759–767. DOI: 10.1016/j.neuint.2012.03.011

Qin, J., Li, R., Raes, J., Arumugam, M., Burgdorf, K. S., Manichanh, C., Nielsen, T., Pons, N., Levenez, F., Yamada, T., Mende, D. R., Li, J., Xu, J., Li, S., Li, D., Cao, J., Wang, B., Liang, H., Zheng, H., & Wang, J. (2010). A human gut microbial gene catalogue established by metagenomic sequencing. *Nature*, *464*(7285), 59–65. DOI: 10.1038/nature08821 PMID: 20203603

Quan, L., Uyeda, A., & Muramatsu, R. (2022). Central nervous system regeneration: the roles of glial cells in the potential molecular mechanism underlying remyelination. *Inflammation and Regeneration 2022 42:1*, *42*(1), 1–12. DOI: 10.1186/s41232-022-00193-y

Quesada-Vázquez, S., Eseberri, I., Les, F., Pérez-Matute, P., Herranz-López, M., Atgié, C., Lopez-Yus, M., Aranaz, P., Oteo, J. A., Escoté, X., Lorente-Cebrian, S., Roche, E., Courtois, A., López, V., Portillo, M. P., Milagro, F. I., & Carpéné, C. (2024). Polyphenols and metabolism: From present knowledge to future challenges. *Journal of Physiology and Biochemistry*, *80*(3), 603–625. DOI: 10.1007/s13105-024-01046-7 PMID: 39377969

Quideau, S., Deffieux, D., Douat-Casassus, C., & Pouységu, L. (2011). Plant polyphenols: Chemical properties, biological activities, and synthesis. *Angewandte Chemie International Edition*, *50*(3), 586–621. DOI: 10.1002/anie.201000044 PMID: 21226137

Quincozes-Santos, A., Santos, C. L., de Souza Almeida, R. R., da Silva, A., Thomaz, N. K., Costa, N. L. F., Weber, F. B., Schmitz, I., Medeiros, L. S., Medeiros, L., Dotto, B. S., Dias, F. R. P., Sovrani, V., & Bobermin, L. D. (2021). Gliotoxicity and Glioprotection: The Dual Role of Glial Cells. *Molecular Neurobiology*, *58*(12), 6577–6592. DOI: 10.1007/s12035-021-02574-9 PMID: 34581988

Qureshi, M., Al-Suhaimi, E. A., Wahid, F., Shehzad, O., & Shehzad, A. (2018). Therapeutic potential of curcumin for multiple sclerosis. *Neurological Sciences*, *39*(2), 207–214. DOI: 10.1007/s10072-017-3149-5 PMID: 29079885

Qu, Z., Zhang, J., Yang, H., Huo, L., Gao, J., Chen, H., & Gao, W. (2016). Protective effect of tetrahydropalmatine against d-galactose induced memory impairment in rat. *Physiology & Behavior*, *154*, 114–125. DOI: 10.1016/j.physbeh.2015.11.016 PMID: 26592138

Raabe, F. J., Slapakova, L., Rossner, M. J., Cantuti-Castelvetri, L., Simons, M., Falkai, P. G., & Schmitt, A. (2019). Oligodendrocytes as a new therapeutic target in schizophrenia: From histopathological findings to neuron-oligodendrocyte interaction. *Cells*, *8*(12), 1496. Advance online publication. DOI: 10.3390/cells8121496 PMID: 31771166

Radu, A., Tit, D. M., Endres, L. M., Radu, A.-F., Vesa, C. M., & Bungau, S. G. (2024). Naturally derived bioactive compounds as precision modulators of immune and inflammatory mechanisms in psoriatic conditions. *Inflammopharmacology*, *32*(1), 527–549. 10.1007/s10787-024-01602-z

Raees, S., Shah, S. A., Karim, A., Raees, S., Shah, S. A., & Karim, A. (2025). *Stilbenes: Emerging Applications in Health*. Agriculture, and Industry., DOI: 10.5772/intechopen.1010264

Raffaele, S., Boccazzi, M., & Fumagalli, M. (2021). Oligodendrocyte dysfunction in amyotrophic lateral sclerosis: Mechanisms and therapeutic perspectives. In *Cells* (Vol. 10, Issue 3, pp. 1–32). Multidisciplinary Digital Publishing Institute. DOI: 10.3390/cells10030565

Raheem, A., Liang, L., Zhang, G., & Cui, S. (2021). Modulatory Effects of Probiotics During Pathogenic Infections With Emphasis on Immune Regulation. *Frontiers in Immunology*, *12*, 616713. DOI: 10.3389/fimmu.2021.616713 PMID: 33897683

Rahman, M. H., Bajgai, J., Fadriquela, A., Sharma, S., Trinh, T. T., Akter, R., Jeong, Y. J., Goh, S. H., Kim, C.-S., & Lee, K.-J. (2021). Therapeutic potential of natural products in treating neurodegenerative disorders and their future prospects and challenges. *Molecules (Basel, Switzerland)*, *26*(17), 5327. Advance online publication. 10.3390/molecules26175327

Rahman, N., Khan, H., Zia, A., Khan, A., Fakhri, S., Aschner, M., Gul, K., & Saso, L. (2021). Bcl-2 modulation in p53 signaling pathway by flavonoids: A potential strategy towards the treatment of cancer. *International Journal of Molecular Sciences*, *22*(21), 11315. DOI: 10.3390/ijms222111315 PMID: 34768743

Rajman, M., & Schratt, G. (2017). MicroRNAs in neural development: From master regulators to fine-tuners. *Development (Cambridge, England)*, *144*(13), 2310–2322. DOI: 10.1242/dev.144337 PMID: 28676566

Rallapalli, K. L., & Komor, A. C. (2023). The design and application of DNA-editing enzymes as base editors. *Annual Review of Biochemistry*, *92*(1), 43–79. DOI: 10.1146/annurev-biochem-052521-013938 PMID: 37018843

Ramanathan, B., Wu, H., Ross, C. R., & Blecha, F. (2004). PR-39, a porcine antimicrobial peptide, inhibits apoptosis: Involvement of caspase-3. *Developmental and Comparative Immunology*, *28*(2), 163–169. DOI: 10.1016/S0145-305X(03)00135-6 PMID: 12969801

Rana, A., Samtiya, M., Dhewa, T., Mishra, V., & Aluko, R. E. (2022). Health benefits of polyphenols: A concise review. *Journal of Food Biochemistry*, *46*(10), e14264. DOI: 10.1111/jfbc.14264 PMID: 35694805

Ransohoff, R. M. (2016). How neuroinflammation contributes to neurodegeneration. *Science*, *353*(6301), 777–783. DOI: 10.1126/science.aag2590 PMID: 27540165

Ransom, B. R., & Ransom, C. B. (2011). Astrocytes: multitalented stars of the central nervous system. Astrocytes: Methods and Protocols, 3-7.

Rao, A. V., & Balasubramanian, M. (2021). Public health campaigns promoting flavonoid-rich diets for cognitive health. *Frontiers in Nutrition*, *8*, 667123. DOI: 10.3389/fnut.2021.667123

Rao, P. N., Mainkar, O., Bansal, N., Rakesh, N., Haffey, P., Urits, I., Orhurhu, V., Kaye, A. D., Urman, R. D., Gulati, A., & Jones, M. (2021). Flavonoids in the treatment of neuropathic pain. *Current Pain and Headache Reports*, *25*(7), 43. DOI: 10.1007/s11916-021-00959-y PMID: 33961144

Rashno, M., Sarkaki, A., Farbood, Y., Rashno, M., Khorsandi, L., Naseri, M. K. G., & Dianat, M. (2019). *RETRACTED: Therapeutic effects of chrysin in a rat model of traumatic brain injury: A behavioral, biochemical, and histological study*. Elsevier. DOI: 10.1016/j.lfs.2019.05.007

Rashno, M., Sarkaki, A., Farbood, Y., Rashno, M., Khorsandi, L., Naseri, M. K. G., & Dianat, M. (2023). Possible mechanisms involved in the neuroprotective effects of chrysin against mild traumatic brain injury-induced spatial cognitive decline: An in vivo study in a rat model. *Brain Research Bulletin*, *204*(October), 110779. DOI: 10.1016/j.brainresbull.2023.110779 PMID: 37827266

Ratan, Y., Rajput, A., & Sharma, A. (2023). An insight into cellular and molecular mechanisms underlying the pathogenesis of neurodegeneration in Alzheimer's disease. *Biomedicines*, *11*(5), 1398. Advance online publication. DOI: 10.3390/biomedicines11051398 PMID: 37239068

Rawani, N. S., Chan, A. W., Todd, K. G., Baker, G. B., & Dursun, S. M. (2025). The Role of Neuroglia in the Development and Progression of Schizophrenia. *Biomolecules*, *15*(1), 10. Advance online publication. DOI: 10.3390/biom15010010 PMID: 39858403

Ray, S., Dixon, C., & Banik, N. (2002). Molecular mechanisms in the pathogenesis of traumatic brain injury. *Histology and Histopathology*. PMID: 12371142

Rebas, E., Rzajew, J., Radzik, T., & Zylinska, L. (2020). Neuroprotective polyphenols: A modulatory action on neurotransmitter pathways. *Current Neuropharmacology*, *18*(5), 431–445. DOI: 10.2174/1570159X18666200106155127 PMID: 31903883

Reguera, G., McCarthy, K. D., Mehta, T., Nicoll, J. S., Tuominen, M. T., & Lovley, D. R. (2005). Extracellular electron transfer via microbial nanowires. *Nature*, *435*(7045), 1098–1101. DOI: 10.1038/nature03661 PMID: 15973408

Rehfeld, J. F. (2017). Cholecystokinin—From local gut hormone to ubiquitous messenger. *Frontiers in Endocrinology*, *8*, 47. DOI: 10.3389/fendo.2017.00047 PMID: 28450850

Rehman, M. U., Wali, A. F., Ahmad, A., Shakeel, S., Rasool, S., Ali, R., & Khan, R. (2019). Neuroprotective Strategies for Neurological Disorders by Natural Products: An update. *Current Neuropharmacology*, *17*(3), 247–267. DOI: 10.2174/1570159x16666180911124605

Reinisalo, M., Kårlund, A., Koskela, A., Kaarniranta, K., & Karjalainen, R. O. (2015). Polyphenol Stilbenes: Molecular Mechanisms of Defence against Oxidative Stress and Aging-Related Diseases. *Oxidative Medicine and Cellular Longevity*, *340520*, 1–24. Advance online publication. DOI: 10.1155/2015/340520 PMID: 26180583

Rejhová, A., Opattová, A., Čumová, A., Slíva, D., & Vodička, P. (2018). Natural compounds and combination therapy in colorectal cancer treatment. *European Journal of Medicinal Chemistry*, *144*, 582–594. DOI: 10.1016/j.ejmech.2017.12.039 PMID: 29289883

Renaud, J., & Martinoli, M.-G. (2019). Considerations for the use of polyphenols as therapies in neurodegenerative diseases. *International Journal of Molecular Sciences*, *20*(8), 1883. DOI: 10.3390/ijms20081883 PMID: 30995776

Rendeiro, C., Rhodes, J. S., & Spencer, J. P. (2015). The mechanisms of action of flavonoids in the brain: Direct versus indirect effects. *Neurochemistry International*, *89*, 126–139. DOI: 10.1016/j.neuint.2015.08.002

Renno, W. M., Al-Maghrebi, M., AlShammari, A., & George, P. (2013). (−)-Epigallocatechin-3-gallate (EGCG) attenuates peripheral nerve degeneration in rat sciatic nerve crush injury. *Neurochemistry International*, *62*(3), 221–231. DOI: 10.1016/j.neuint.2012.12.018 PMID: 23313191

Ren, R., Ma, J., Zhao, Y., Zhao, L., Zhang, G., Zheng, Y., & Zhang, Y. (2025). Nebulized seabuckthorn seed oil inhalation attenuates Alzheimer's disease progression in APP/PS1 mice. *Scientific Reports*, *15*, 6368. Advance online publication. 10.1038/s41598-025-56368-9

Renye, J. A.Jr, Needleman, D. S., Somkuti, G. A., & Steinberg, D. H. (2017). Complete genome sequence of Streptococcus thermophilus strain B59671, which naturally produces the broad-spectrum bacteriocin thermophilin 110. *Genome Announcements*, *5*(45), e01213–e01217. DOI: 10.1128/genomeA.01213-17 PMID: 29122869

Ren, Z., Wang, L., Cui, J., Huoc, Z., Xue, J., Cui, H., Mao, Q., & Yang, R. (2013). Resveratrol inhibits NF-kB signaling through suppression of p65 and IkappaB kinase activities. *Die Pharmazie*, *68*(8), 689–694. PMID: 24020126

Ren, Z., Zhang, R., Li, Y., Li, Y., Yang, Z., & Yang, H. (2017). Ferulic acid exerts neuroprotective effects against cerebral ischemia/reperfusion-induced injury via antioxidant and anti-apoptotic mechanisms in vitro and in vivo. *International Journal of Molecular Medicine*, *40*(5), 1444–1456. DOI: 10.3892/ijmm.2017.3127 PMID: 28901374

Réus, G. Z., Fries, G. R., Stertz, L., Badawy, M., Passos, I., Barichello, T., Kapczinski, F., & Quevedo, J. (2015). The role of inflammation and microglial activation in the pathophysiology of psychiatric disorders. *Neuroscience*, *300*, 141–154. DOI: 10.1016/j.neuroscience.2015.05.018 PMID: 25981208

Reynolds, C. F.III, Jeste, D. V., Sachdev, P. S., & Blazer, D. G. (2022). Mental health care for older adults: Recent advances and new directions in clinical practice and research. *World Psychiatry; Official Journal of the World Psychiatric Association (WPA)*, *21*(3), 336–363. DOI: 10.1002/wps.20996 PMID: 36073714

Riaz, T., Akram, M., & Laila, U. RidaZainab, Muhammad Talha Khalil, Momina Iftikhar, FethiAhmet Ozdemir, Gaweł Sołowski, Marcos Altable, AdonisSfera, Hamza Khalifa Ibrahim, P. P. (2023). *(PDF) Therapeutic applications of glycosides obtained from medicinal plants*. https://www.researchgate.net/publication/373272302_Therapeutic_applications_of_glycosides_obtained_from_medicinal_plants

Ribeiro, R., Alves, C., & Silva, C. (2022). Probiotic and polyphenol-enriched dairy products: Effects on gut microbiota and bioactive compound metabolism. *Food Research International*, *156*, 111224. DOI: 10.1016/j.foodres.2022.111224

Ricciotti, E., & FitzGerald, G. A. (2011). Prostaglandins and inflammation. *Arteriosclerosis, Thrombosis, and Vascular Biology*, *31*(5), 986–1000. DOI: 10.1161/atvbaha.110.207449

Richards, P., Parker, H. E., Adriaenssens, A. E., Hodgson, J. M., Cork, S. C., Trapp, S., & Reimann, F. (2014). Identification and characterization of GLP-1 receptor–expressing cells using a new transgenic mouse model. *Diabetes*, *63*(4), 1224–1233. DOI: 10.2337/db13-1440 PMID: 24296712

Riva, A., Ronchi, M., Petrangolini, G., Bosisio, S., & Allegrini, P. (2019). Improved oral absorption of quercetin from quercetin phytosome®, a new delivery system based on food grade lecithin. *European Journal of Drug Metabolism and Pharmacokinetics*, *44*(2), 169–177. DOI: 10.1007/s13318-018-0517-3 PMID: 30328058

Rivai, B., & Umar, A. K. (2023). Neuroprotective compounds from marine invertebrates. *Journal of Basic and Applied Scientific Research*, *12*(1), 71. 10.1186/s43088-023-00407-3

Roberfroid, M., Gibson, G. R., Hoyles, L., McCartney, A. L., Rastall, R., Rowland, I., Wolvers, D., Watzl, B., Szajewska, H., Stahl, B., Guarner, F., Respondek, F., Whelan, K., Coxam, V., Davicco, M.-J., Léotoing, L., Wittrant, Y., Delzenne, N. M., Cani, P. D., & Meheust, A. (2010). Prebiotic effects: Metabolic and health benefits. *British Journal of Nutrition*, *104*(S2), S1–S63. DOI: 10.1017/S0007114510003363 PMID: 20920376

Roca Pinilla, R. (2020). Development of a new generation of antimicrobial proteins based on a versatile nanoparticulated format and multidomain structure.

Rocca, W. A. (2018). The burden of Parkinson's disease: A worldwide perspective. *Lancet Neurology*, *17*(11), 928–929. DOI: 10.1016/S1474-4422(18)30355-7 PMID: 30287052

Rodrigues, A. M. G., Marcilio, F. D. S., Frazão Muzitano, M., & Giraldi-Guimarães, A. (2013). Therapeutic potential of treatment with the flavonoid rutin after cortical focal ischemia in rats. *Brain Research*, *1503*, 53–61. DOI: 10.1016/j.brainres.2013.01.039 PMID: 23370003

Rodríguez-García, C., Sánchez-Quesada, C., Toledo, E., Delgado-Rodríguez, M., & Gaforio, J. J. (2019). Naturally lignan-rich foods: A dietary tool for health promotion? *Molecules (Basel, Switzerland)*, *24*(5), 917. DOI: 10.3390/molecules24050917 PMID: 30845651

Rodríguez-Gómez, J. A., Kavanagh, E., Engskog-Vlachos, P., Engskog, M. K. R., Herrera, A. J., Espinosa-Oliva, A. M., Joseph, B., Hajji, N., Venero, J. L., & Burguillos, M. A. (2020). Microglia: Agents of the CNS Pro-Inflammatory Response. In *Cells* (Vol. 9, Issue 7). NLM (Medline). DOI: 10.3390/cells9071717

Rodríguez, J. M., Martínez, M. I., & Kok, J. (2002). Pediocin PA-1, a wide-spectrum bacteriocin from lactic acid bacteria. *Critical Reviews in Food Science and Nutrition*, *42*(2), 91–121. DOI: 10.1080/10408690290825475 PMID: 11934133

Rodriguez-Mateos, A., Cifuentes-Gomez, T., George, T. W., Heiss, C., Spencer, J. P., & Kroon, P. A. (2014). Impact of polyphenols on human health: Focus on gut microbiota. *Trends in Food Science & Technology*, *35*(2), 77–87. DOI: 10.1016/j.tifs.2014.04.007

Rodríguez-Vera, D., Abad-García, A., Vargas-Mendoza, N., Pinto-Almazán, R., Farfán-García, E. D., Morales-González, J. A., & Soriano-Ursúa, M. A. (2022). Polyphenols as potential enhancers of stem cell therapy against neurodegeneration. *Neural Regeneration Research*, *17*(10), 2093–2101. DOI: 10.4103/1673-5374.335826 PMID: 35259814

Rojas-García, A., Fernández-Ochoa, Á., Cádiz-Gurrea, M. de la L., Arráez-Román, D., & Segura-Carretero, A. (2023). Neuroprotective Effects of Agri-Food By-Products Rich in Phenolic Compounds. *Nutrients*, *15*(2), 449. Advance online publication. DOI: 10.3390/nu15020449 PMID: 36678322

Rolls, A., Shechter, R., & Schwartz, M. (2009). The bright side of the glial scar in CNS repair. *Nature Reviews. Neuroscience*, *10*(3), 235–241. DOI: 10.1038/nrn2591 PMID: 19229242

Roman, G. C., Jackson, R., Gadhia, R., Román, A., & Reis, J. (2019). Mediterranean diet: The role of long-chain ω-3 fatty acids in fish; polyphenols in fruits, vegetables, cereals, coffee, tea, cacao and wine; probiotics and vitamins in prevention of stroke, age-related cognitive decline, and Alzheimer disease. *Revue Neurologique*, *175*(10), 724–741. DOI: 10.1016/j.neurol.2019.08.005 PMID: 31521398

Rom, S., Zuluaga-Ramirez, V., Reichenbach, N. L., Erickson, M. A., Winfield, M., Gajghate, S., Christofidou-Solomidou, M., Jordan-Sciutto, K. L., & Persidsky, Y. (2018). Secoisolariciresinol diglucoside is a blood-brain barrier protective and anti-inflammatory agent: Implications for neuroinflammation. *Journal of Neuroinflammation*, *15*(1), 1–10. DOI: 10.1186/s12974-018-1065-0 PMID: 29373982

Rong, H., Liang, Y., & Niu, Y. (2018). Rosmarinic acid attenuates β-amyloid-induced oxidative stress via Akt/GSK-3β/Fyn-mediated Nrf2 activation in PC12 cells. *Free Radical Biology & Medicine*, *120*, 114–123. DOI: 10.1016/j.freeradbiomed.2018.03.028 PMID: 29555592

Rose, E. C., Odle, J., Blikslager, A. T., & Ziegler, A. L. (2021). Probiotics, prebiotics and epithelial tight junctions: A promising approach to modulate intestinal barrier function. *International Journal of Molecular Sciences*, *22*(13), 6729. DOI: 10.3390/ijms22136729 PMID: 34201613

Roseiro, L. B., Rauter, A. P., & Serralheiro, M. L. M. (2012). Polyphenols as acetylcholinesterase inhibitors: Structural specificity and impact on human disease. *Nutrition and Aging (Amsterdam, Netherlands)*, *1*(2), 99–111. DOI: 10.3233/NUA-2012-0006

Rossi, D., & Gaidano, G. (2003). Messengers of cell death: Apoptotic signaling in health and disease. *Haematologica*, *88*(2), 212–218. PMID: 12604411

Rossi, M., Corradini, C., Amaretti, A., Nicolini, M., Pompei, A., Zanoni, S., & Matteuzzi, D. (2005). Fermentation of fructooligosaccharides and inulin by bifidobacteria: A comparative study of pure and fecal cultures. *Applied and Environmental Microbiology*, *71*(10), 6150–6158. DOI: 10.1128/AEM.71.10.6150-6158.2005 PMID: 16204533

Rothwell, J. A., Perez-Jimenez, J., Neveu, V., Medina-Remón, A., M'Hiri, N., García-Lobato, P., & Scalbert, A. (2013). Phenol-Explorer 3.0: A major update of the Phenol-Explorer database to incorporate data on the effects of food processing on polyphenol content. *Database : The Journal of Biological Databases and Curation*, *2013*, bat070. Advance online publication. DOI: 10.1093/database/bat070

Round, J. L., & Mazmanian, S. K. (2010). Inducible Foxp3+ regulatory T-cell development by a commensal bacterium of the intestinal microbiota. *Cell Host & Microbe*, *8*(5), 377–387. DOI: 10.1016/j.chom.2010.10.006 PMID: 20566854

Rowland, I., Gibson, G., Heinken, A., Scott, K., Swann, J., Thiele, I., & Tuohy, K. (2018). Gut microbiota functions: Metabolism of nutrients and other food components. *European Journal of Nutrition*, *57*(1), 1–24. DOI: 10.1007/s00394-017-1445-8 PMID: 28393285

Royston, K. J. (2018). *Combinatorial Withaferin A and Sulforaphane and Their Potential Roles in Breast Cancer Prevention and Therapy*. The University of Alabama at Birmingham.

Różańska, A., Russo, M., Cacciola, F., Salafia, F., Polkowska, Ż., Dugo, P., & Mondello, L. (2020). Concentration of potentially bioactive compounds in Italian extra virgin olive oils from various sources by using LC-MS and multivariate data analysis. *Foods*, *9*(8), 1120.

Rudrapal, M., Khairnar, S. J., Khan, J., Dukhyil, A. B., Ansari, M. A., Alomary, M. N., Alshabrmi, F. M., Palai, S., Deb, P. K., & Devi, R. (2022). Dietary polyphenols and their role in oxidative stress-induced human diseases: Insights into protective effects, antioxidant potentials and mechanism (s) of action. *Frontiers in Pharmacology*, *13*, 806470. DOI: 10.3389/fphar.2022.806470 PMID: 35237163

Rudrapal, M., Rakshit, G., Singh, R. P., Garse, S., Khan, J., & Chakraborty, S. (2024). Dietary Polyphenols: Review on Chemistry/Sources, Bioavailability/Metabolism, Antioxidant Effects, and Their Role in Disease Management. *Antioxidants*, *13*(4), 429. Advance online publication. DOI: 10.3390/antiox13040429 PMID: 38671877

Ruff, R. L., McKerracher, L., & Selzer, M. E. (2008). Repair and neurorehabilitation strategies for spinal cord injury. *Annals of the New York Academy of Sciences*, *1142*(1), 1–20. DOI: 10.1196/annals.1444.004 PMID: 18990118

Ruiz Puentes, P., Henao, M. C., Cifuentes, J., Muñoz-Camargo, C., Reyes, L. H., Cruz, J. C., & Arbeláez, P. (2022). Rational discovery of antimicrobial peptides by means of artificial intelligence. *Membranes*, *12*(7), 708. DOI: 10.3390/membranes12070708 PMID: 35877911

Rummel, N. G., & Butterfield, D. A. (2022). Altered metabolism in Alzheimer disease brain: Role of oxidative stress. *Antioxidants & Redox Signaling*, *36*(16–18), 1289–1305. DOI: 10.1089/ars.2021.0177 PMID: 34416829

Ruotolo, R., Minato, I., La Vitola, P., Artioli, L., Curti, C., Franceschi, V., & Ottonello, S. (2020). Flavonoid-Derived Human Phenyl-γ-Valerolactone Metabolites Selectively Detoxify Amyloid-β Oligomers and Prevent Memory Impairment in a Mouse Model of Alzheimer's Disease. *Molecular Nutrition & Food Research*, *64*(5), e1900890. DOI: 10.1002/mnfr.201900890

Rutsch, A., Kantsjö, J. B., & Ronchi, F. (2020). The gut-brain axis: How microbiota and host inflammasome influence brain physiology and pathology. *Frontiers in Immunology, 11*, 604179. DOI: 10.3389/fimmu.2020.604179 PMID: 33362788

Sadiq, I. Z. (2023). Free radicals and oxidative stress: Signaling mechanisms, redox basis for human diseases, and cell cycle regulation. *Current Molecular Medicine, 23*(1), 13–35. DOI: 10.2174/1566524022666211222161637 PMID: 34951363

Safe, S., Jayaraman, A., Chapkin, R. S., Howard, M., Mohankumar, K., & Shrestha, R. (2021). Flavonoids: Structure–function and mechanisms of action and opportunities for drug development. *Toxicological Research, 37*(2), 147–162. DOI: 10.1007/s43188-020-00080-z PMID: 33868973

Sahadevan, R., Binoy, A., Shajan, I., & Sadhukhan, S. (2023). Mitochondria-targeting EGCG derivatives protect H9c2 cardiomyocytes from H2O2-induced apoptosis: Design, synthesis and biological evaluation. *RSC Advances, 13*(42), 29477–29488. DOI: 10.1039/D3RA04527G PMID: 37818277

Salami, M., & Soheili, M. (2022). Neuroprotective effects of probiotics via gut-brain axis signaling. *Journal of Functional Foods, 88*, 104887. DOI: 10.1016/j.jff.2021.104887

Saleem, A., Qurat-ul-Ain, , & Akhtar, M. F. (2022). Alternative Therapy of Psychosis: Potential Phytochemicals and Drug Targets in the Management of Schizophrenia. *Frontiers in Pharmacology, 13*, 895668. DOI: 10.3389/fphar.2022.895668 PMID: 35656298

Salini, F., Iacumin, L., Comi, G., & Dicks, L. M. T. (2023). Thermophilin 13: In silico analysis provides new insight in genes involved in bacteriocin production. *Microorganisms, 11*(3), 611. DOI: 10.3390/microorganisms11030611 PMID: 36985185

Salla, M., Karaki, N., El Kaderi, B., Ayoub, A. J., Younes, S., Abou Chahla, M. N., Baksh, S., & El Khatib, S. (2024). Enhancing the Bioavailability of Resveratrol: Combine It, Derivatize It, or Encapsulate It? *Pharmaceutics, 16*(4), 569. DOI: 10.3390/pharmaceutics16040569 PMID: 38675230

Salman, A., Koparde, R., & Zand, M. S. (2018). Determining the quantitative principles of T cell response to antigenic disparity in stem cell transplantation. *Frontiers in Immunology, 9*, 2284. Advance online publication. DOI: 10.3389/fimmu.2018.02284 PMID: 30364159

Salminen, S., Collado, M. C., Endo, A., Hill, C., Lebeer, S., & Mättö, J.. (2021). The International Scientific Association of Probiotics and Prebiotics (ISAPP) consensus statement on the definition and scope of postbiotics. *Nature Reviews. Gastroenterology & Hepatology, 18*(9), 649–667. DOI: 10.1038/s41575-021-00440-6 PMID: 33948025

Salter, M. W., & Stevens, B. (2017). Microglia emerge as central players in brain disease. *Nature Medicine, 23*(9), 1018–1027. DOI: 10.1038/nm.4397 PMID: 28886007

Salvesen, L., Winge, K., Brudek, T., Agander, T. K., Løkkegaard, A., & Pakkenberg, B. (2017). Neocortical Neuronal Loss in Patients with Multiple System Atrophy: A Stereological Study. *Cerebral Cortex (New York, N.Y.), 27*(1), 400–410. DOI: 10.1093/cercor/bhv228 PMID: 26464477

Sampson, T. R., Debelius, J. W., Thron, T., Janssen, S., Shastri, G. G., Ilhan, Z. E., Challis, C., Schretter, C. E., Rocha, S., Gradinaru, V., Chesselet, M.-F., Keshavarzian, A., Shannon, K. M., Krajmalnik-Brown, R., Wittung-Stafshede, P., Knight, R., & Mazmanian, S. K. (2016). Gut microbiota regulate motor deficits and neuroinflammation in a model of Parkinson's disease. *Cell, 167*(6), 1469–1480. DOI: 10.1016/j.cell.2016.11.018 PMID: 27912057

Sánchez, A., & Vázquez, A. (2017). Bioactive peptides: A review. *Food Quality and Safety, 1*(1), 29–46. DOI: 10.1093/fqs/fyx006

Sánchez-Martínez, J. D., Martínez-Gardeazabal, J., & Fernández-Villamarín, M. (2022). Blood–brain barrier permeability study of potential neuroprotective compounds recovered from plants and agri-food by-products. *Frontiers in Nutrition, 9*, 924596. Advance online publication. DOI: 10.3389/fnut.2022.924596 PMID: 35782945

Sanders, M. E., Heimbach, J. T., Pot, B., Tancredi, D. J., Lenoir-Wijnkoop, I., Lähteenmäki-Uutela, A., Gueimonde, M., & Bañares, S. (2011). Health claims substantiation for probiotic and prebiotic products. *Gut Microbes, 2*(3), 127–133. DOI: 10.4161/gmic.2.3.16174 PMID: 21646865

Sandhir, R., & Mehrotra, A. (2013). Quercetin supplementation is effective in improving mitochondrial dysfunctions induced by 3-nitropropionic acid: Implications in Huntington's disease. *Biochimica et Biophysica Acta. Molecular Basis of Disease, 1832*(3), 421–430. DOI: 10.1016/j.bbadis.2012.11.018 PMID: 23220257

SangHwon, L. S., Choi BoYoung, C. B., Lee SongHee, L. S., Kho, A., Jeong JeongHyun, J. J., Hong DaeKi, H. D., & Suh SangWon, S. S. (2017). Administration of protocatechuic acid reduces traumatic brain injury-induced neuronal death.

Santos-Beneit, F. (2024). What is the role of microbial biotechnology and genetic engineering in medicine? *MicrobiologyOpen, 13*(2), e1406. DOI: 10.1002/mbo3.1406 PMID: 38556942

Saraiva, C., Praça, C., Ferreira, R., Santos, T., Ferreira, L., & Bernardino, L. (2016). Nanoparticle-mediated brain drug delivery: Overcoming blood–brain barrier to treat neurodegenerative diseases. *Journal of Controlled Release, 235*, 34–47. DOI: 10.1016/j.jconrel.2016.05.044 PMID: 27208862

Sarao, L., Kaur, S., Malik, T., & Singh, A. (2022). Genistein and daidzein. In *Nutraceuticals and health care* (pp. 331–341). Elsevier. DOI: 10.1016/B978-0-323-89779-2.00016-8

Sarkaki, A., Farbood, Y., Gharib-Naseri, M. K., Badavi, M., Mansouri, M. T., Haghparast, A., & Mirshekar, M. A. (2015). Gallic acid improved behavior, brain electrophysiology, and inflammation in a rat model of traumatic brain injury. *Canadian Journal of Physiology and Pharmacology, 93*(8), 687–694. DOI: 10.1139/cjpp-2014-0546 PMID: 26222320

Sarkar, B., Ma, X., Agas, A., Siddiqui, Z., Iglesias-Montoro, P., Nguyen, P. K., Kim, K. K., Haorah, J., & Kumar, V. A. (2021). In vivo neuroprotective effect of a self-assembled peptide hydrogel. *Chemical Engineering Journal, 408*, 127295. DOI: 10.1016/j.cej.2020.127295 PMID: 37842134

Sarris, J. (2018). Herbal medicines in the treatment of psychiatric disorders: 10-year updated review. *Phytotherapy Research, 32*(7), 1147–1162. DOI: 10.1002/ptr.6055 PMID: 29575228

Sastre, M., Klockgether, T., & Heneka, M. T. (2006). Contribution of inflammatory processes to Alzheimer's disease: Molecular mechanisms. *International Journal of Developmental Neuroscience*, *24*(2–3), 167–176. DOI: 10.1016/j.ijdevneu.2005.11.014 PMID: 16472958

Satapathy, T., & Kumar, D. (2024). A comprehensive review and recent advancement in the application of tannins for treating Parkinson disease. *Pharmacological Research. Modern Chinese Medicine*, *12*, 100499. DOI: 10.1016/j.prmcm.2024.100499

Sawmiller, D., Li, S., Shahaduzzaman, M., Smith, A. J., Obregon, D., Giunta, B., Borlongan, C. V., Sanberg, P. R., & Tan, J. (2014). Luteolin reduces Alzheimer's disease pathologies induced by traumatic brain injury. *International Journal of Molecular Sciences*, *15*(1), 895–904. DOI: 10.3390/ijms15010895 PMID: 24413756

Saxton, R. A., & Sabatini, D. M. (2017). mTOR signaling in growth, metabolism, and disease. *Cell*, *168*(6), 960–976. DOI: 10.1016/j.cell.2017.02.004 PMID: 28283069

Scalbert, A., Johnson, I. T., & Saltmarsh, M. (2005). Polyphenols: Antioxidants and beyond. *The American Journal of Clinical Nutrition*, *81*(1), 215S–217S. DOI: 10.1093/ajcn/81.1.215S PMID: 15640483

Scalbert, A., & Williamson, G. (2000). Dietary intake and bioavailability of polyphenols. *The Journal of Nutrition*, *130*(8), 2073S–2085S. DOI: 10.1093/jn/130.8.2073S PMID: 10917926

Scearce-Levie, K., Sanchez, P. E., & Lewcock, J. W. (2020). Leveraging preclinical models for the development of Alzheimer disease therapeutics. *Nature Reviews. Drug Discovery*, *19*(7), 447–462. DOI: 10.1038/s41573-020-0065-9 PMID: 32612262

Schaffer, S., & Halliwell, B. (2012). Do polyphenols enter the brain and does it matter? Some theoretical and practical considerations. *Genes & Nutrition*, *7*(2), 99–109. DOI: 10.1007/s12263-011-0255-5 PMID: 22012276

Schmacht, M., Lorenz, E., & Senz, M. (2017). Microbial production of glutathione. *World Journal of Microbiology & Biotechnology*, *33*(6), 106. DOI: 10.1007/s11274-017-2277-7 PMID: 28466304

Schönfeld, P., & Reiser, G. (2021). How the brain fights fatty acids' toxicity. *Neurochemistry International*, *148*, 105050. DOI: 10.1016/j.neuint.2021.105050 PMID: 33945834

Schubert, J. J., Veronese, M., Marchitelli, L., Bodini, B., Tonietto, M., Stankoff, B., Brooks, D. J., Bertoldo, A., Edison, P., & Turkheimer, F. E. (2019). Dynamic 11C-PIB PET shows cerebrospinal fluid flow alterations in Alzheimer disease and multiple sclerosis. *Journal of Nuclear Medicine*, *60*(10), 1452–1460. DOI: 10.2967/jnumed.118.223834 PMID: 30850505

Schültke, E., Kamencic, H., Zhao, M., Tian, G. F., Baker, A. J., Griebel, R. W., & Juurlink, B. H. J. (2005). Neuroprotection following fluid percussion brain trauma: A pilot study using quercetin. *Journal of Neurotrauma*, *22*(12), 1475–1484. DOI: 10.1089/neu.2005.22.1475 PMID: 16379584

Schwarcz, R., Bruno, J. P., Muchowski, P. J., & Wu, H. Q. (2012). Kynurenines in the mammalian brain: When physiology meets pathology. *Nature Reviews. Neuroscience*, *13*(7), 465–477. DOI: 10.1038/nrn3257 PMID: 22678511

Sciences, N. A. o., Medicine, Policy, Affairs, G., Data, B. o. R., Information, . . . Statistics, T. (2019). *Reproducibility and replicability in science*: National Academies Press.

Scuderi, C., Verkhratsky, A., Parpura, V., & Li, B. (2021). Neuroglia in Psychiatric Disorders. *Advances in Neurobiology*, *26*, 3–19. DOI: 10.1007/978-3-030-77375-5_1 PMID: 34888828

Secher, T., Kassem, S., Benamar, M., Bernard, I., Boury, M., Barreau, F., Oswald, E., & Saoudi, A. (2017). Oral administration of the probiotic strain *Escherichia coli* Nissle 1917 reduces susceptibility to neuroinflammation and repairs experimental autoimmune encephalomyelitis-induced intestinal barrier dysfunction. *Frontiers in Immunology*, *8*, 1096. Advance online publication. DOI: 10.3389/fimmu.2017.01096 PMID: 28959254

Segers ME, Lebeer S. Towards a better understanding of Lactobacillus rhamnosus GG--host interactions. Microb Cell Fact. 2014;13 Suppl 1(Suppl 1):S7. .DOI: 10.1186/1475-2859-13-S1-S7

Sehar, U., Rawat, P., Reddy, A. P., Kopel, J., & Reddy, P. H. (2022). Amyloid beta in aging and Alzheimer's disease. *International Journal of Molecular Sciences*, *23*(21), 12924. DOI: 10.3390/ijms232112924 PMID: 36361714

Sejbuk, M., Mirończuk-Chodakowska, I., Karav, S., & Witkowska, A. M. (2024). Dietary polyphenols, food processing and gut microbiome: Recent findings on bioavailability, bioactivity, and gut microbiome interplay. *Nutrients*, *16*(20), 3465. Advance online publication. 10.3390/nu16203465

Sekhon, L. H., & Fehlings, M. G. (2001). Epidemiology, demographics, and pathophysiology of acute spinal cord injury. *Spine*, *26*(24S), S2–S12. DOI: 10.1097/00007632-200112151-00002 PMID: 11805601

Selma, M. V., Espín, J. C., & Tomás-Barberán, F. A. (2009). Interaction between phenolics and gut microbiota: Role in human health. *Journal of Agricultural and Food Chemistry*, *57*(15), 6485–6501. DOI: 10.1021/jf902107d PMID: 19580283

Selvaraj, N. R., Nandan, D., Nair, B. G., Nair, V. A., Venugopal, P., & Aradhya, R. (2025). Oxidative Stress and Redox Imbalance: Common Mechanisms in Cancer Stem Cells and Neurodegenerative Diseases. In *Cells* (Vol. 14, Issue 7). Multidisciplinary Digital Publishing Institute (MDPI). DOI: 10.3390/cells14070511

Sena, L. A., & Chandel, N. S. (2012). Physiological roles of mitochondrial reactive oxygen species. *Molecular Cell*, *48*(2), 158–167. DOI: 10.1016/j.molcel.2012.09.025 PMID: 23102266

Sengupta, U., & Kayed, R. (2022). Amyloid β, Tau, and α-Synuclein aggregates in the pathogenesis, prognosis, and therapeutics for neurodegenerative diseases. *Progress in Neurobiology*, *214*, 102270. DOI: 10.1016/j.pneurobio.2022.102270 PMID: 35447272

Sen, I., Nalini, A., Joshi, N. B., & Joshi, P. G. (2005). Cerebrospinal fluid from amyotrophic lateral sclerosis patients preferentially elevates intracellular calcium and toxicity in motor neurons via AMPA/kainate receptor. *Journal of the Neurological Sciences*, *235*(1–2), 45–54. DOI: 10.1016/j.jns.2005.03.049 PMID: 15936037

Şentürk, S., Yaman, M. E., Aydın, H. E., Güney, G., Bozkurt, I., Paksoy, K., & Abdioğlu, A. A. (2018). Effects of resveratrol on inflammation and apoptosis after experimental spinal cord injury.

Seo, J. S., Mantas, I., Svenningsson, P., & Greengard, P. (2021). Ependymal cells-CSF flow regulates stress-induced depression. *Molecular Psychiatry, 26*(12), 7308–7315. DOI: 10.1038/s41380-021-01202-1 PMID: 34234280

Serot, J. M., Zmudka, J., & Jouanny, P. (2012). A possible role for CSF turnover and choroid plexus in the pathogenesis of late onset alzheimer's disease. In *Journal of Alzheimer's Disease* (Vol. 30, Issue 1, pp. 17–26). IOS Press. DOI: 10.3233/JAD-2012-111964

Shaban, N. Z., Abdelrahman, S. A., & El-Kersh, M. A. (2021). Synergistic protective effect of *Beta vulgaris* with meso-2,3-dimercaptosuccinic acid against lead-induced neurotoxicity in male rats. *Scientific Reports, 11*(1), 252. Advance online publication. DOI: 10.1038/s41598-020-80669-4 PMID: 33420282

Shabbir, U., Tyagi, A., Elahi, F., Aloo, S. O., & Oh, D.-H. (2021). The potential role of polyphenols in oxidative stress and inflammation induced by gut microbiota in alzheimer's disease. *Antioxidants, 10*(9), 1370. DOI: 10.3390/antiox10091370 PMID: 34573002

Shahidi, F., & Pan, Y. (2022). Influence of food matrix and food processing on the chemical interaction and bioaccessibility of dietary phytochemicals: A review. *Critical Reviews in Food Science and Nutrition, 62*(23), 6421–6445. DOI: 10.1080/10408398.2021.1901650 PMID: 33787422

Shah, P. M., Priya, V. V., & Gayathri, R. (2016). Quercetin-a flavonoid: A systematic review. *Journal of Pharmaceutical Sciences and Research, 8*(8), 878.

Shaito, A., Al-Mansoob, M., Ahmad, S. M. S., Haider, M. Z., Eid, A. H., Posadino, A. M., Pintus, G., & Giordo, R. (2022). Resveratrol-Mediated Regulation of Mitochondria Biogenesis-associated Pathways in Neurodegenerative Diseases: Molecular Insights and Potential Therapeutic Applications. *Current Neuropharmacology, 21*(5), 1184–1201. DOI: 10.2174/1570159X20666221012122855 PMID: 36237161

Shang, L., Yang, F., Chen, Q., Dai, Z., Yang, G., Zeng, X., Qiao, S., & Yu, H. (2024). Bacteriocin Microcin J25's antibacterial infection effects and novel non-microbial regulatory mechanisms: Differential regulation of dopaminergic receptors. *Journal of Animal Science and Biotechnology, 15*(1), 156. DOI: 10.1186/s40104-024-01115-3 PMID: 39533384

Shareena, G., & Kumar, D. (2024). Deciphering the deep learning and machine learning tactics in advancement of neuroprotection by phytochemicals. In *NeuroPhytomedicine* (pp. 205–220). CRC Press.

Sharkey, K. A., & Mawe, G. M. (2023). The enteric nervous system. *Physiological Reviews, 103*(2), 1487–1564. DOI: 10.1152/physrev.00018.2022 PMID: 36521049

Sharma, K. K., Singh, D., & Rawat, S. (2018). Molecular dynamics simulation studies suggests unconventional roles of non-secretary laccases from enteropathogenic gut bacteria and Cryptococcus neoformans serotype D. *Computational Biology and Chemistry, 73*, 41–48. DOI: 10.1016/j.compbiolchem.2018.01.010 PMID: 29425935

Sharma, M., & Rajawat, N. K. (2025). Neurotoxicity study of copper oxide nanoparticles and the protective role of a probiotic (Lactobacillus acidophilus) in Swiss albino mice. *Toxicology and Industrial Health, 41*(7), 07482337251350165. DOI: 10.1177/07482337251350165 PMID: 40492926

Sharma, N., & Nehru, B. (2018). Curcumin affords neuroprotection and inhibits α-synuclein aggregation in lipopolysaccharide-induced Parkinson's disease model. *Inflammopharmacology*, *26*(2), 349–360. DOI: 10.1007/s10787-017-0402-8 PMID: 29027056

Sharma, R., Bhaskar, B., Sanodiya, B. S., Thakur, G. S., Jaiswal, P., Yadav, N., Sharma, A., & Bisen, P. S. (2014). Probiotic efficacy and potential of *Streptococcus thermophilus* modulating human health: A synoptic review. *IOSR Journal of Pharmacy and Biological Sciences*, *9*(3), 52–58. DOI: 10.9790/3008-09325258

Sharma, S. H., Thulasingam, S., & Nagarajan, S. (2017). Terpenoids as anti-colon cancer agents – A comprehensive review on its mechanistic perspectives. *European Journal of Pharmacology*, *795*, 169–178. DOI: 10.1016/j.ejphar.2016.12.008 PMID: 27940056

Sharma, S. K., & Priya, S. (2017). Expanding role of molecular chaperones in regulating α-synuclein misfolding; implications in Parkinson's disease. *Cellular and Molecular Life Sciences : CMLS*, *74*(4), 617–629. DOI: 10.1007/s00018-016-2340-9

Sharma, S., Singh, R. L., & Kakkar, P. (2011). Modulation of Bax/Bcl-2 and caspases by probiotics during acetaminophen induced apoptosis in primary hepatocytes. *Food and Chemical Toxicology*, *49*(4), 770–779. DOI: 10.1016/j.fct.2010.11.041 PMID: 21130831

Shen, Y., Zhang, Y., & Wu, H. (2020). Advances in enteric-coated and sustained-release delivery systems for polyphenols: Improving stability and bioavailability. *Journal of Functional Foods*, *70*, 103998. DOI: 10.1016/j.jff.2020.103998

Shim, J. H., Greenblatt, M. B., Xie, M., Schneider, M. D., Zou, W., Zhai, B., & Glimcher, L. H. (2009). TAK1 is an essential regulator of BMP signalling in cartilage. *The EMBO Journal*, *28*(14), 2028.

Shi, W., Xie, H., Ouyang, K., Wang, S., Xiong, H., Woo, M. W., & Zhao, Q. (2024). The effect of rice protein-polyphenols covalent and non-covalent interactions on the structure, functionality and in vitro digestion properties of rice protein. *Food Chemistry*, *450*, 139241.

Shi, Y., Manis, M., Long, J., Wang, K., Sullivan, P. M., Serrano, J. R., Hoyle, R., & Holtzman, D. M. (2019). Microglia drive APOE-dependent neurodegeneration in a tauopathy mouse model. *The Journal of Experimental Medicine*, *216*(11), 2546–2561. DOI: 10.1084/jem.20190980 PMID: 31601677

Shi, Z., Qiu, W., Xiao, G., Cheng, J., & Zhang, N. (2018). Resveratrol attenuates cognitive deficits of traumatic brain injury by activating p38 signaling in the brain. *Medical Science Monitor*, *24*, 1097–1103. DOI: 10.12659/MSM.909042 PMID: 29467361

Shoaib, M., Shehzad, A., Omar, M., Rakha, A., Raza, H., Sharif, H. R., Shakeel, A., Ansari, A., & Niazi, S. (2016). Inulin: Properties, health benefits and food applications. *Carbohydrate Polymers*, *147*, 444–454. DOI: 10.1016/j.carbpol.2016.04.020 PMID: 27178951

Shoaib, S., Ansari, M. A., Fatease, A. A., Safhi, A. Y., Hani, U., Jahan, R., Alomary, M. N., Ansari, M. N., Ahmed, N., & Wahab, S.. (2023). Plant-derived bioactive compounds in the management of neurodegenerative disorders: Challenges, future directions, and molecular mechanisms involved in neuroprotection. *Pharmaceutics*, *15*(3), 749. Advance online publication. 10.3390/pharmaceutics15030749

Shuai, K., & Liu, B. (2003). Regulation of JAK–STAT signalling in the immune system. *Nature Reviews. Immunology*, *3*(11), 900–911. DOI: 10.1038/nri1226 PMID: 14668806

Shukitt-Hale, B., Carey, A., Simon, L., Mark, D. A., & Joseph, J. A. (2008). Effects of blueberry supplementation on cognitive performance and oxidative stress in Alzheimer's disease mouse models. *Nutrition (Burbank, Los Angeles County, Calif.)*, *24*(3), 298–301. DOI: 10.1016/j.nut.2007.11.006

Shukitt-Hale, B., Cheng, V., & Joseph, J. A. (2015). Effects of blackberries on motor and cognitive function in aged rats. *Nutrition and Aging (Amsterdam, Netherlands)*, *3*(1–2), 55–65. DOI: 10.3233/NUA-150054

Siddarth, P., Li, Z., Miller, K. J., Ercoli, L. M., Merril, D. A., Henning, S. M., Heber, D., & Small, G. W. (2020). Randomized placebo-controlled study of the memory effects of pomegranate juice in middle-aged and older adults. *The American Journal of Clinical Nutrition*, *111*(1), 170–177. DOI: 10.1093/ajcn/nqz241 PMID: 31711104

Siddiqui, T., Khan, M. U., Sharma, V., & Gupta, K. (2024). Terpenoids in essential oils: Chemistry, classification, and potential impact on human health and industry. *Phytomedicine Plus : International Journal of Phytotherapy and Phytopharmacology*, *4*(2), 100549. DOI: 10.1016/j.phyplu.2024.100549

Sierra, A., Encinas, J. M., Deudero, J. J., Chancey, J. H., Enikolopov, G., Overstreet-Wadiche, L. S., Tsirka, S. E., & Maletic-Savatic, M. (2010). Microglia shape adult hippocampal neurogenesis through apoptosis-coupled phagocytosis. *Cell Stem Cell*, *7*(4), 483–495. DOI: 10.1016/j.stem.2010.08.014 PMID: 20887954

Silva dos Santos, J., Goncalves Cirino, J. P., de Oliveira Carvalho, P., & Ortega, M. M. (2021). The pharmacological action of kaempferol in central nervous system diseases: A review. *Frontiers in Pharmacology*, *11*, 565700. DOI: 10.3389/fphar.2020.565700 PMID: 33519431

Silva, F. C., Rodrigues, S., & Pintado, M. (2020). Encapsulation strategies for anthocyanins in functional beverages. *Food Hydrocolloids*, *103*, 105637. DOI: 10.1016/j.foodhyd.2020.105637

Silva, F. L., Nascimento, G. O., Lopes, G. S., Matos, W. O., Cunha, R. L., Malta, M. R., & Trevisan, M. T. S. (2020). The concentration of polyphenolic compounds and trace elements in the Coffea arabica leaves: Potential chemometric pattern recognition of coffee leaf rust resistance. *Food Research International*, *134*, 109221.

Silva, M. H., Batista, L. L., Malta, S. M., Santos, A. C., Mendes-Silva, A. P., Bonetti, A. M., Ueira-Vieira, C., & Dos Santos, A. R. (2024). Unveiling the Brazilian kefir microbiome: Discovery of a novel Lactobacillus kefiranofaciens (LkefirU) genome and in silico prospection of bioactive peptides with potential anti-Alzheimer properties. *BMC Genomics*, *25*(1), 884. DOI: 10.1186/s12864-024-10695-3 PMID: 39304820

Silva, N. A., Sousa, N., Reis, R. L., & Salgado, A. J. (2014). From basics to clinical: A comprehensive review on spinal cord injury. *Progress in Neurobiology*, *114*, 25–57. DOI: 10.1016/j.pneurobio.2013.11.002 PMID: 24269804

Silva, R. F., & Pogačnik, L. (2020). Polyphenols from food and natural products: Neuroprotection and safety. *Antioxidants*, *9*(1), 61. DOI: 10.3390/antiox9010061 PMID: 31936711

Silva, T., Fonseca, R., & Santos, J. (2020). Wearable sensors and mobile applications for personalized nutrition and cognitive monitoring. *Sensors (Basel), 20*(24), 7251. DOI: 10.3390/s20247251 PMID: 33348823

Silva, Y. P., Bernardi, A., & Frozza, R. L. (2020). The role of short-chain fatty acids from gut microbiota in gut-brain communication. *Frontiers in Endocrinology, 11*, 508738. DOI: 10.3389/fendo.2020.00025 PMID: 32082260

Simms, C. L., & Zaher, H. S. (2016). Quality control of chemically damaged RNA. *Cellular and Molecular Life Sciences : CMLS, 73*(19), 3639–3653. DOI: 10.1007/s00018-016-2261-7 PMID: 27155660

Simunkova, M., Alwasel, S. H., Alhazza, I. M., Jomova, K., Kollar, V., Rusko, M., & Valko, M. (2019). Management of oxidative stress and other pathologies in Alzheimer's disease. *Archives of Toxicology, 93*(9), 2491–2513. DOI: 10.1007/s00204-019-02538-y PMID: 31440798

Simunkova, M., Barbierikova, Z., Jomova, K., Hudecova, L., Lauro, P., Alwasel, S. H., & Valko, M. (2021). Antioxidant vs. Prooxidant Properties of the Flavonoid, Kaempferol, in the Presence of Cu(II) Ions: A ROS-Scavenging Activity, Fenton Reaction and DNA Damage Study. *International Journal of Molecular Sciences, 22*(4). Advance online publication. DOI: 10.3390/ijms22041619

Sinclair, A., Begg, D., Mathai, M., & Weisinger, R. (2007). *Omega 3 fatty acids and the brain : review of studies in depression.*

Singh, A., Kaur, P., Kumar, M., Shafi, S., Kumar Upadhyaya, P., Tiwari, A., Tiwari, V., Rangra, N. K., Thirunavukarasu, V., Kumari, S., Roy, D., Ghosh, M., Arora, N., Sharma, N., & Garg, Y.. (2025). The role of phytochemicals in modulating the gut microbiota: Implications for health and disease. *Medicine in Microecology, 19*, 100094. Advance online publication. 10.1016/j.medmic.2024.100094

Singh, A., Kukreti, R., Saso, L., & Kukreti, S. (2019). Oxidative stress: A key modulator in neurodegenerative diseases. *Molecules (Basel, Switzerland), 24*(8), 1583. DOI: 10.3390/molecules24081583 PMID: 31013638

Singhal, G., & Baune, B. T. (2017). Microglia: An interface between the loss of neuroplasticity and depression. *Frontiers in Cellular Neuroscience, 11*(September), 1–16. DOI: 10.3389/fncel.2017.00270 PMID: 28943841

Singh, B., & Semwal, B. C. (2024). A Compressive Review on Source, Toxicity and Biological Activity of Flavonoid. *Current Topics in Medicinal Chemistry, 24*(24), 2093–2116. DOI: 10.2174/0115680266316032240718050055

Singh, D., Agrawal, A., Singal, C. M. S., Pandey, H. S., Seth, P., & Sharma, S. K. (2020). Sinomenine inhibits amyloid beta-induced astrocyte activation and protects neurons against indirect toxicity. *Molecular Brain, 13*(1), 1–10. DOI: 10.1186/s13041-020-00569-6 PMID: 32127013

Singh, G., Kaur, J., Kaur, M., Singh, P., & Bhatti, R. (2020). Anti-nociceptive and anti-inflammatory effect of imperatorin: Evidences for involvement of COX-2, iNOS, NFκB and inflammatory cytokines. *The International Journal of Neuroscience, 130*(2), 176–185. DOI: 10.1080/00207454.2019.1667789 PMID: 31524564

Singh, N. A., Mandal, A. K., & Khan, Z. A. (2016). Potential neuroprotective properties of epigallocatechin-3-gallate (EGCG). *Nutrition Journal, 15*(1), 60. DOI: 10.1186/s12937-016-0179-4

Singhrao, S. K., Harding, A., Simmons, T., Robinson, S., Kesavalu, L., & Crean, S. (2014). Oral inflammation, tooth loss, risk factors, and association with progression of Alzheimer's disease. *Journal of Alzheimer's Disease, 42*(3), 723–737. DOI: 10.3233/JAD-140387 PMID: 24946875

Singh, S. K., Srivastav, S., Castellani, R. J., Plascencia-Villa, G., & Perry, G. (2019). Neuroprotective and antioxidant effect of *Ginkgo biloba* extract against AD and other neurological disorders. *Neurotherapeutics; the Journal of the American Society for Experimental NeuroTherapeutics, 16*(3), 666–674. 10.1007/s13311-019-00767-8

Singh, S., & Singh, T. G. (2020). Role of nuclear factor kappa B (NF-κB) signalling in neurodegenerative diseases: An mechanistic approach. *Current Neuropharmacology, 18*(10), 918–935. DOI: 10.2174/1570159X18666200207120949 PMID: 32031074

Sinha, K., & Mukhopadhyay, C. D. (2021). Immunomodulatory Effect of Plant-Based Extracts on Neurodegeneration. In Advanced Studies in Experimental and Clinical Medicine (pp. 31-52). Apple Academic Press. DOI: 10.1201/9781003057451-3

Sirangelo, I., Liccardo, M., & Iannuzzi, C. (2022). Hydroxytyrosol Prevents Doxorubicin-Induced Oxidative Stress and Apoptosis in Cardiomyocytes. *Antioxidants, 11*(6), 1087. Advance online publication. DOI: 10.3390/antiox11061087 PMID: 35739984

Sittipo, P., Choi, J., Lee, S., & Lee, Y. K. (2022). The function of gut microbiota in immune-related neurological disorders: A review. *Journal of Neuroinflammation, 19*(1), 154. DOI: 10.1186/s12974-022-02510-1 PMID: 35706008

Slavin, J. (2013). Fiber and prebiotics: Mechanisms and health benefits. *Nutrients, 5*(4), 1417–1435. DOI: 10.3390/nu5041417 PMID: 23609775

Smeriglio, A., Barreca, D., Bellocco, E., & Trombetta, D. (2017). Proanthocyanidins and hydrolysable tannins: Occurrence, dietary intake and pharmacological effects. *British Journal of Pharmacology, 174*(11), 1244–1262. DOI: 10.1111/bph.13630 PMID: 27646690

Śmiałek, J., Bzowska, M., Hinz, A., Mężyk-Kopeć, R., Sołtys, K., & Mak, P. (2022). Bacteriocin BacSp222 and its succinylated forms exhibit proinflammatory activities toward innate immune cells. *Journal of Inflammation Research, 15*, 4601–4621. DOI: 10.2147/JIR.S362066 PMID: 35982757

Śmiałek, J., Nowakowski, M., Bzowska, M., Bocheńska, O., Wlizło, A., Kozik, A., Dubin, G., & Mak, P. (2021). Structure, biosynthesis, and biological activity of succinylated forms of bacteriocin BacSp222. *International Journal of Molecular Sciences, 22*(12), 6256. DOI: 10.3390/ijms22126256 PMID: 34200765

Smith, D. H., Chen, X.-H., Pierce, J. E., Wolf, J. A., Trojanowski, J. Q., Graham, D. I., & Mcintosh, T. K. (1997). Progressive atrophy and neuron death for one year following brain trauma in the rat. *Journal of Neurotrauma, 14*(10), 715–727. DOI: 10.1089/neu.1997.14.715 PMID: 9383090

Smith, J. A., Das, A., Ray, S. K., & Banik, N. L. (2012). Role of pro-inflammatory cytokines released from microglia in neurodegenerative diseases. *Brain Research Bulletin*, *87*(1), 10–20. DOI: 10.1016/j.brainresbull.2011.10.004 PMID: 22024597

Sneath, P. H. A. (1966). Relations between chemical structure and biological activity in peptides. *Journal of Theoretical Biology*, *12*(2), 157–195. DOI: 10.1016/0022-5193(66)90112-3 PMID: 4291386

Socci, V., Tempesta, D., Desideri, G., De Gennaro, L., & Ferrara, M. (2017). Enhancing Human Cognition with Cocoa Flavonoids. *Frontiers in Nutrition*, *4*, 19. DOI: 10.3389/fnut.2017.00019

Sochocka, M., Diniz, B. S., & Leszek, J. (2017). Inflammatory Response in the CNS: Friend or Foe? *Molecular Neurobiology*, *54*(10), 8071–8089. DOI: 10.1007/s12035-016-0297-1

Sochocka, M., Donskow-Łysoniewska, K., Diniz, B. S., Kurpas, D., Brzozowska, E., & Leszek, J. (2019). The gut microbiome alterations and inflammation-driven pathogenesis of Alzheimer's disease—A critical review. *Molecular Neurobiology*, *56*(3), 1841–1851. DOI: 10.1007/s12035-018-1188-4 PMID: 29936690

Sofroniew, M. V. (2020). Astrocyte Reactivity: Subtypes, States, and Functions in CNS Innate Immunity. *Trends in Immunology*, *41*(9), 758–770. DOI: 10.1016/j.it.2020.07.004 PMID: 32819810

Soleymani, S., Habtemariam, S., Rahimi, R., & Nabavi, S. M. (2020). The what and who of dietary lignans in human health: Special focus on prooxidant and antioxidant effects. *Trends in Food Science & Technology*, *106*, 382–390. DOI: 10.1016/j.tifs.2020.10.015

Soltani, Z., Khaksari, M., Jafari, E., Iranpour, M., & Shahrokhi, N. (2015). Is genistein neuroprotective in traumatic brain injury? *Physiology & Behavior*, *152*, 26–31. DOI: 10.1016/j.physbeh.2015.08.037 PMID: 26367454

Song, H., Zhang, X., Wang, W., Liu, R., Zhao, K., Liu, M.-y., Gong, W., & Ning, B. (2018). Neuroprotective mechanisms of rutin for spinal cord injury through anti-oxidation and anti-inflammation and inhibition of p38 mitogen activated protein kinase pathway. *Neural Regeneration Research*, *13*(1), 128–134. DOI: 10.4103/1673-5374.217349 PMID: 29451217

Song, J., Du, G., Wu, H., Gao, X., Yang, Z., Liu, B., & Cui, S. (2021). Protective effects of quercetin on traumatic brain injury induced inflammation and oxidative stress in cortex through activating Nrf2/HO-1 pathway. *Restorative Neurology and Neuroscience*, *39*(1), 73–84. DOI: 10.3233/RNN-201119 PMID: 33612499

Song, X. Y., Fan, C. X., Choudhary, M. I., & Wang, X. P. (2024). Neuro-regeneration or repair: Cell therapy of neurological disorders as a way forward. *Current Neuropharmacology*, *22*(14), 2272–2283.

Sontag, E. M., Samant, R. S., & Frydman, J. (2017). Mechanisms and Functions of Spatial Protein Quality Control. *Annual Review of Biochemistry*, *86*, 97–122. DOI: 10.1146/annurev-biochem-060815-014616

Son, Y., Cheong, Y. K., Kim, N. H., Chung, H. T., Kang, D. G., & Pae, H. O. (2011). Mitogen-activated protein kinases and reactive oxygen species: How can ROS activate MAPK pathways? *Journal of Signal Transduction*, *792639*, 1–6. Advance online publication. DOI: 10.1155/2011/792639 PMID: 21637379

Sorond, F. A., Hurwitz, S., Salat, D. H., Greve, D. N., & Fisher, N. D. (2013). Neurovascular coupling, cerebral white matter integrity, and response to cocoa in older people. *Neurology*, *81*(10), 904–909. DOI: 10.1212/WNL.0b013e3182a351aa PMID: 23925758

Sorrenti, V., Ali, S., Mancin, L., Davinelli, S., Paoli, A., & Scapagnini, G. (2020). Cocoa Polyphenols and Gut Microbiota Interplay: Bioavailability, Prebiotic Effect, and Impact on Human Health. *Nutrients*, *12*(7). Advance online publication. DOI: 10.3390/nu12071908

Souza, L. C., de Gomes, M. G., Goes, A. T. R., Del Fabbro, L., Filho, C. B., Boeira, S. P., & Jesse, C. R. (2013). Evidence for the involvement of the serotonergic 5-HT1A receptors in the antidepressant-like effect caused by hesperidin in mice. *Progress in Neuro-Psychopharmacology & Biological Psychiatry*, *40*(1), 103–109. DOI: 10.1016/j.pnpbp.2012.09.003 PMID: 22996046

Sowers, A., Wang, G., Xing, M., & Li, B. (2023). Advances in antimicrobial peptide discovery via machine learning and delivery via nanotechnology. *Microorganisms*, *11*(5), 1129. DOI: 10.3390/microorganisms11051129 PMID: 37317103

Spaas, J., van Veggel, L., Schepers, M., Tiane, A., van Horssen, J., Wilson, D. M., Moya, P. R., Piccart, E., Hellings, N., Eijnde, B. O., Derave, W., Schreiber, R., & Vanmierlo, T. (2021). Oxidative stress and impaired oligodendrocyte precursor cell differentiation in neurological disorders. In *Cellular and Molecular Life Sciences* (Vol. 78, Issue 10, pp. 4615–4637). Springer. DOI: 10.1007/s00018-021-03802-0

Spagnuolo, C., Moccia, S., & Russo, G. L. (2018). Anti-inflammatory effects of flavonoids in neurodegenerative disorders. *European Journal of Medicinal Chemistry*, *153*, 105–115. DOI: 10.1016/j.ejmech.2017.09.001 PMID: 28923363

Spagnuolo, C., Napolitano, M., Tedesco, I., Moccia, S., Milito, A., & Luigi Russo, G. (2016). Neuroprotective role of natural polyphenols. *Current Topics in Medicinal Chemistry*, *16*(17), 1943–1950. DOI: 10.2174/1568026616666160204122449 PMID: 26845551

Speciale, A., Cimino, F., Saija, A., & Fazzari, M. (2014). Cyanidin-3-O-glucoside protects microglial cells against LPS-induced inflammation by inhibiting NF-κB and iNOS activation. *Food and Chemical Toxicology*, *65*, 1–6. DOI: 10.1016/j.fct.2013.12.016

Spencer, J. P. E. (2010). The impact of fruit flavonoids on memory and cognition. *British Journal of Nutrition*, *104*(S3), S40–S47. DOI: 10.1017/S0007114510003934 PMID: 20955649

Spiegel, M., Andruniów, T., & Sroka, Z. (2020). Flavones' and flavonols' antiradical structure–activity relationship—A quantum chemical study. *Antioxidants*, *9*(6), 461. DOI: 10.3390/antiox9060461 PMID: 32471289

Spochacz, M., Chowański, S., Walkowiak-Nowicka, K., Szymczak, M., & Adamski, Z. (2018). Plant-derived substances used against beetles–pests of stored crops and food–and their mode of action: A review. *Comprehensive Reviews in Food Science and Food Safety*, *17*(5), 1339–1366.

Sriganesh, K., Shanthanna, H., & Busse, J. W. (2016). A brief overview of systematic reviews and meta-analyses. *Indian Journal of Anaesthesia*, *60*(9), 689–694. DOI: 10.4103/0019-5049.190628 PMID: 27729699

Stacchiotti, A., & Corsetti, G. (2020). Natural compounds and autophagy: Allies against neurodegeneration. *Frontiers in Cell and Developmental Biology*, *8*, 555409. Advance online publication. DOI: 10.3389/fcell.2020.555409 PMID: 33072744

Staniszewski, A., & Kordowska-Wiater, M. (2021). Probiotic and anticancer properties of Lactobacillus spp. *Applied Sciences (Basel, Switzerland)*, *11*(8), 3702. DOI: 10.3390/app11083702

Stilling, R. M., Dinan, T. G., & Cryan, J. F. (2016). Microbial genes, brain & behaviour – epigenetic regulation of the gut–brain axis. *Genes Brain & Behavior*, *15*(1), 13–27. DOI: 10.1111/gbb.12271

Stith, S. S., Diviant, J. P., Brock, C., & Vigil, J. (2018). Patient-reported symptom relief following medical cannabis consumption. *Frontiers in Pharmacology*, *9*, 916. Advance online publication. DOI: 10.3389/fphar.2018.00916 PMID: 30210337

Stoker, T. B., Mason, S. L., Greenland, J. C., Holden, S. T., Santini, H., & Barker, R. A. (2022). Huntington's disease: Diagnosis and management. *Practical Neurology*, *22*(1), 32–41. DOI: 10.1136/practneurol-2021-003074 PMID: 34413240

Strandwitz, P. (2018). Neurotransmitter modulation by the gut microbiota. *Brain Research*, *1693*, 128–133. DOI: 10.1016/j.brainres.2018.03.015 PMID: 29903615

Strathearn, K. E., Yousef, G. G., Grace, M. H., Roy, S. L., Tambe, M. A., Ferruzzi, M. G., & Rochet, J. C. (2014). Neuroprotective effects of anthocyanin- and proanthocyanidin-rich extracts in cellular models of Parkinson s disease. *Brain Research*, *1555*, 60–77. DOI: 10.1016/j.brainres.2014.01.047

Strieker, M., Tanović, A., & Marahiel, M. A. (2010). Nonribosomal peptide synthetases: Structures and dynamics. *Current Opinion in Structural Biology*, *20*(2), 234–240. DOI: 10.1016/j.sbi.2010.01.009 PMID: 20153164

Stromsnes, K., Lagzdina, R., Olaso-González, G., Gimeno-Mallench, L., & Gambini, J. (2021). Pharmacological properties of polyphenols: Bioavailability, mechanisms of action, and biological effects in in vitro studies, animal models, and humans. *Biomedicines*, *9*(8), 1074. Advance online publication. 10.3390/biomedicines9081074

Subaraja, M., Arokiyaraj, S., & Mathew, P. (2024). Neuroprotective effect of huperzine-A against cadmium chloride-induced Huntington's disease in Drosophila melanogaster model. *Journal of King Saud University. Science*, *36*(8), 103319. DOI: 10.1016/j.jksus.2024.103319

Subedi, L., & Yumnam, S. (2021). Terpenoids from Abies holophylla Attenuate LPS-Induced Neuroinflammation in Microglial Cells by Suppressing the JNK-Related Signaling Pathway. *International Journal of Molecular Sciences 2021, Vol. 22, Page 965*, *22*(2), 965. DOI: 10.3390/ijms22020965

Suganya, S. N., & Sumathi, T. (2017). Effect of rutin against a mitochondrial toxin, 3-nitropropionicacid induced biochemical, behavioral and histological alterations-a pilot study on Huntington's disease model in rats. *Metabolic Brain Disease*, *32*(2), 471–481. DOI: 10.1007/s11011-016-9929-4 PMID: 27928694

Sun, Y., Tong, H., Yang, T., Liu, L., Li, X. J., & Li, S. (2022). Insights into White Matter Defect in Huntington's Disease. In *Cells* (Vol. 11, Issue 21, p. 3381). Multidisciplinary Digital Publishing Institute. DOI: 10.3390/cells11213381

Sun, G., Miao, Z., Ye, Y., Zhao, P., Fan, L., Bao, Z., Tu, Y., Li, C., Chao, H., Xu, X., & Ji, J. (2020). Curcumin alleviates neuroinflammation, enhances hippocampal neurogenesis, and improves spatial memory after traumatic brain injury. *Brain Research Bulletin*, *162*, 84–93. DOI: 10.1016/j.brainresbull.2020.05.009 PMID: 32502596

Sunhe, Y.-X., Zhang, Y.-H., Fu, R.-J., Xu, D.-Q., & Tang, Y.-P. (2024). Neuroprotective effect and preparation methods of berberine. *Frontiers in Pharmacology*, *15*(September), 1–23. DOI: 10.3389/fphar.2024.1429050 PMID: 39309003

Sun, J., Xu, J., Yang, B., Chen, K., Kong, Y., Fang, N., Gong, T., Wang, F., Ling, Z., & Liu, J. (2020). Effect of Clostridium butyricum against Microglia-Mediated Neuroinflammation in Alzheimer's Disease via Regulating Gut Microbiota and Metabolites Butyrate. *Molecular Nutrition & Food Research*, *64*(2), e1900636. DOI: 10.1002/mnfr.201900636 PMID: 31835282

Sun, J., Yang, J., Miao, X., Loh, H. H., Pei, D., & Zheng, H. (2021). Proteins in DNA methylation and their role in neural stem cell proliferation and differentiation. *Cell Regeneration (London, England)*, *10*(1), 1–12. DOI: 10.1186/s13619-020-00070-4 PMID: 33649938

Sun, Y., Luo, Z.-M., Guo, X.-M., Su, D.-F., & Liu, X. (2015). An updated role of microRNA-124 in central nervous system disorders: A review. *Frontiers in Cellular Neuroscience*, *9*, 193. DOI: 10.3389/fncel.2015.00193 PMID: 26041995

Suraweera, T. L., Vasantha Rupasinghe, H. P., Dellaire, G., & Xu, Z. (2020). Regulation of Nrf2/are pathway by dietary flavonoids: A friend or foe for cancer management? In *Antioxidants* (Vol. 9, Issue 10, pp. 1–44). MDPI. DOI: 10.3390/antiox9100973

Susin, S. A., Zamzami, N., & Kroemer, G. (1998). Mitochondria as regulators of apoptosis: Doubt no more. *Biochimica et Biophysica Acta (BBA)-. Biochimica et Biophysica Acta. Bioenergetics*, *1366*(1-2), 151–165. DOI: 10.1016/S0005-2728(98)00110-8

Sutalangka, C., & Wattanathorn, J. (2017). Neuroprotective and cognitive enhancing effects of the combined extract of Cyperus rotundus and Zingiber officinale. *BMC Complementary and Alternative Medicine*, *17*(1), 135.

Su, Y., Wang, D., Liu, N., Yang, J., Sun, R., & Zhang, Z. (2023). Clostridium butyricum improves cognitive dysfunction in ICV-STZ-induced Alzheimer's disease mice via suppressing TLR4 signaling pathway through the gut-brain axis. *PLoS One*, *18*(6), e0286086. DOI: 10.1371/journal.pone.0286086 PMID: 37267300

Swanson, K. V., Deng, M., & Ting, J. P.-Y. (2019). The NLRP3 inflammasome: Molecular activation and regulation to therapeutics. *Nature Reviews. Immunology*, *19*(8), 477–489. DOI: 10.1038/s41577-019-0165-0 PMID: 31036962

Szwajgier, D., Borowiec, K., & Pustelniak, K. (2017). The Neuroprotective Effects of Phenolic Acids: Molecular Mechanism of Action. *Nutrients*, *9*(5), 477. DOI: 10.3390/nu9050477 PMID: 28489058

Szymczak, P., & Szczurek, E. (2023). Artificial intelligence-driven antimicrobial peptide discovery. *Current Opinion in Structural Biology*, *83*, 102733. DOI: 10.1016/j.sbi.2023.102733 PMID: 37992451

Szymczak, P., Zarzecki, W., Wang, J., Duan, Y., Wang, J., Coelho, L. P., de la Fuente-Nunez, C., & Szczurek, E. (2025). AI-driven antimicrobial peptide discovery: Mining and generation. *Accounts of Chemical Research*, *58*(12), 1831–1846. Advance online publication. DOI: 10.1021/acs.accounts.0c00594 PMID: 40459283

Tabatabaee, M. (2024). Microbiome-Glia Crosstalk: Bridging the Communication Divide in the Central Nervous System. In *Neuroglia* (Vol. 5, Issue 2, pp. 89–104). Multidisciplinary Digital Publishing Institute. DOI: 10.3390/neuroglia5020007

Taché, Y., Bonaz, B., & Million, M. (2018). Brain-gut axis and its role in gastrointestinal motility and functional bowel disorders. *Gastroenterology Clinics of North America*, *47*(2), 321–339. DOI: 10.1016/j.gtc.2018.01.007

Taib, T., Leconte, C., Van Steenwinckel, J., Cho, A. H., Palmier, B., Torsello, E., Kuen, R. L., Onyeomah, S., Ecomard, K., Benedetto, C., Coqueran, B., Novak, A. C., Deou, E., Plotkine, M., Gressens, P., Marchand-Leroux, C., & Besson, V. C. (2017). Neuroinflammation, myelin and behavior: Temporal patterns following mild traumatic brain injury in mice. *PLoS One*, *12*(9), e0184811. Advance online publication. DOI: 10.1371/journal.pone.0184811 PMID: 28910378

Tajbakhsh, M., Karimi, A., Fallah, F., & Akhavan, M. M. (2017). Overview of ribosomal and non-ribosomal antimicrobial peptides produced by Gram positive bacteria. *Cellular and Molecular Biology*, *63*(10), 20–32. DOI: 10.14715/cmb/2017.63.10.4 PMID: 29096754

Takano, T., & Soderling, S. H. (2021). Tripartite synaptomics: Cell-surface proximity labeling in vivo. *Neuroscience Research*, *173*, 14–21. DOI: 10.1016/j.neures.2021.05.002 PMID: 34019951

Takano, T., Tian, G. F., Peng, W., Lou, N., Libionka, W., Han, X., & Nedergaard, M. (2006). Astrocyte-mediated control of cerebral blood flow. *Nature Neuroscience*, *9*(2), 260–267. DOI: 10.1038/nn1623 PMID: 16388306

Talarico, T. L., Casas, I. A., Chung, T. C., & Dobrogosz, W. J. (1988). Production and isolation of reuterin, a growth inhibitor produced by Lactobacillus reuteri. *Antimicrobial Agents and Chemotherapy*, *32*(12), 1854–1858. DOI: 10.1128/AAC.32.12.1854 PMID: 3245697

Tamagno, E., Guglielmotto, M., Vasciaveo, V., & Tabaton, M. (2021). Oxidative stress and beta amyloid in Alzheimer's disease. Which comes first: The chicken or the egg? *Antioxidants*, *10*(9), 1479. DOI: 10.3390/antiox10091479 PMID: 34573112

Tamberi, L., Belloni, A., Pugnaloni, A., Rippo, M. R., Olivieri, F., Procopio, A. D., & Bronte, G. (2024). The Influence of Myeloid-Derived Suppressor Cell Expansion in Neuroinflammation and Neurodegenerative Diseases. In *Cells* (Vol. 13, Issue 7). Multidisciplinary Digital Publishing Institute (MDPI). DOI: 10.3390/cells13070643

Tan, Y., Zheng, Y., Xu, D., Sun, Z., Yang, H., & Yin, Q. (2021). Galectin-3: a key player in microglia-mediated neuroinflammation and Alzheimer's disease. In *Cell and Bioscience* (Vol. 11, Issue 1, pp. 1–13). BioMed Central. DOI: 10.1186/s13578-021-00592-7

Tanaka, M., Chock, P. B., & Stadtman, E. R. (2007). Oxidized messenger RNA induces translation errors. *Proceedings of the National Academy of Sciences of the United States of America, 104*(1), 66–71. DOI: 10.1073/pnas.0609737104 PMID: 17190801

Tandon, A., Singh, S. J., & Chaturvedi, R. K. (2018). Stem cells as potential targets of polyphenols in multiple sclerosis and Alzheimer's disease. *BioMed Research International, 2018*(1), 1483791. DOI: 10.1155/2018/1483791 PMID: 30112360

Tang, Y., & Le, W. (2016). Differential Roles of M1 and M2 Microglia in Neurodegenerative Diseases. In *Molecular Neurobiology* (Vol. 53, Issue 2, pp. 1181–1194). Humana Press Inc. DOI: 10.1007/s12035-014-9070-5

Tang, L., Zeng, Z., Zhou, Y., Wang, B., Zou, P., Wang, Q., Ying, J., Wang, F., Li, X., Xu, S., Zhao, P., & Li, W. (2021). Bacillus amyloliquefaciens SC06 Induced AKT–FOXO signaling pathway-mediated autophagy to alleviate oxidative stress in IPEC-J2 Cells. *Antioxidants, 10*(10), 1545. DOI: 10.3390/antiox10101545 PMID: 34679680

Tan, J. J. Y., Lim, Y. Y., Siow, L. F., & Tan, J. B. L. (2015). Effects of Drying on Polyphenol Oxidase and Antioxidant Activity of M orus alba Leaves. *Journal of Food Processing and Preservation, 39*(6), 2811–2819.

Tan, K. W., Li, Y., Paxton, J. W., Birch, N. P., & Scheepens, A. (2013). Identification of novel dietary phytochemicals inhibiting the efflux transporter breast cancer resistance protein (BCRP/ABCG2). *Food Chemistry, 138*(4), 2267–2274. DOI: 10.1016/j.foodchem.2012.12.021

Tan, L. Y., Yeo, X. Y., Bae, H.-G., Lee, D. P. S., Ho, R. C., Kim, J. E., Jo, D.-G., & Jung, S. (2021). Association of gut microbiome dysbiosis with neurodegeneration: Can gut microbe-modifying diet prevent or alleviate the symptoms of neurodegenerative diseases? *Life (Chicago, Ill.), 11*(7), 698. DOI: 10.3390/life11070698 PMID: 34357070

Tan, S. H., Karri, V., Tay, N. W. R., Chang, K. H., & Wong, K. H. (2019). Emerging pathways to neurodegeneration: Dissecting the critical molecular mechanisms in Alzheimer's disease, Parkinson's disease. *Biomedicine and Pharmacotherapy, 111*, 765–777. DOI: 10.1016/j.biopha.2018.12.101 PMID: 30612001

Tan, X., Liang, Z., Li, Y., Zhi, Y., Yi, L., Bai, S., & Li, Q. X. (2021). Isoorientin, a GSK-3β inhibitor, rescues synaptic dysfunction, spatial memory deficits and attenuates pathological progression in APP/PS1 model mice. *Behavioural Brain Research, 398*, 112968. DOI: 10.1016/j.bbr.2020.112968

Tauffenberger, A., & Magistretti, P. J. (2021). Reactive oxygen species: Beyond their reactive behavior. *Neurochemical Research, 46*(1), 77–87. DOI: 10.1007/s11064-020-03208-7 PMID: 33439432

Tavan, M., Hanachi, P., de la Luz Cádiz-Gurrea, M., Segura Carretero, A., & Mirjalili, M. H. (2024). Natural Phenolic Compounds with Neuroprotective Effects. *Neurochemical Research, 49*(2), 306–326. DOI: 10.1007/s11064-023-04046-z PMID: 37940760

Teixeira, M., Souza, C., Menezes, A., Carmo, M., Fonteles, A., Gurgel, J., Lima, F., Viana, G., & Andrade, G. (2013). Catechin attenuates behavioral neurotoxicity induced by 6-OHDA in rats. *Pharmacology, Biochemistry, and Behavior, 110*, 1–7. DOI: 10.1016/j.pbb.2013.05.012 PMID: 23714698

Teka, T., Zhang, L., Ge, X., Li, Y., Han, L., & Yan, X. (2022). Stilbenes: Source plants, chemistry, biosynthesis, pharmacology, application and problems related to their clinical Application-A comprehensive review. *Phytochemistry*, *197*, 113128. DOI: 10.1016/j.phytochem.2022.113128 PMID: 35183567

Teleanu, D. M., Niculescu, A.-G., Lungu, I. I., Radu, C. I., Vladâcenco, O., Roza, E., Costăchescu, B., Grumezescu, A. M., & Teleanu, R. I. (2022). An overview of oxidative stress, neuroinflammation, and neurodegenerative diseases. *International Journal of Molecular Sciences*, *23*(11), 5938. DOI: 10.3390/ijms23115938 PMID: 35682615

Teleanu, R. I., Niculescu, A.-G., Roza, E., Vladâcenco, O., Grumezescu, A. M., & Teleanu, D. M. (2022). Neurotransmitters—Key factors in neurological and neurodegenerative disorders of the central nervous system. *International Journal of Molecular Sciences*, *23*(11), 5954. DOI: 10.3390/ijms23115954 PMID: 35682631

Tenland, E., Krishnan, N., Rönnholm, A., Kalsum, S., Puthia, M., Mörgelin, M., Davoudi, M., Otrocka, M., Alaridah, N., Glegola-Madejska, I., Sturegård, E., Schmidtchen, A., Lerm, M., Robertson, B. D., & Godaly, G. (2018). A novel derivative of the fungal antimicrobial peptide plectasin is active against Mycobacterium tuberculosis. *Tuberculosis (Edinburgh, Scotland)*, *113*, 231–238. DOI: 10.1016/j.tube.2018.10.008 PMID: 30514507

Terao, J. (2017). Factors modulating bioavailability of quercetin-related flavonoids and the consequences of their vascular function. *Biochemical Pharmacology*, *139*, 15–23. DOI: 10.1016/j.bcp.2017.03.021

Thakare, V. N., Aswar, M. K., Kulkarni, Y. P., Patil, R. R., & Patel, B. M. (2017). Silymarin ameliorates experimentally induced depressive like behavior in rats: Involvement of hippocampal BDNF signaling, inflammatory cytokines and oxidative stress response. *Physiology & Behavior*, *179*, 401–410. DOI: 10.1016/j.physbeh.2017.07.010 PMID: 28711395

Thancharoen, O., Limwattananon, C., Waleekhachonloet, O., Rattanachotphanit, T., Limwattananon, P., & Limpawattana, P. (2019). Ginkgo biloba extract (EGb761), cholinesterase inhibitors, and memantine for the treatment of mild-to-moderate Alzheimer's disease: A network meta-analysis. *Drugs & Aging*, *36*(5), 435–452.

Thapa, A., Jett, S. D., & Chi, E. Y. (2016). Curcumin attenuates amyloid-β aggregate toxicity and modulates amyloid-β aggregation pathway. *ACS Chemical Neuroscience*, *7*(1), 56–68. DOI: 10.1021/acschemneuro.5b00214 PMID: 26529184

Thau, L., Reddy, V., & Singh, P. (2022). Anatomy, Central Nervous System. *BMJ (Clinical Research Ed.)*, *1*(4293), 478–478. DOI: 10.1136/bmj.1.4293.478 PMID: 31194336

Thiruvengadam, M., Venkidasamy, B., Subramanian, U., Samynathan, R., Ali Shariati, M., Rebezov, M., & Fedoseeva, N. (2021). Bioactive compounds in oxidative stress-mediated diseases: Targeting the NRF2/ARE signaling pathway and epigenetic regulation. *Antioxidants*, *10*(12), 1859. DOI: 10.3390/antiox10121859 PMID: 34942962

Thoda, C., & Touraki, M. (2023). Probiotic-derived bioactive compounds in colorectal cancer treatment. *Microorganisms*, *11*(8), 1898. DOI: 10.3390/microorganisms11081898 PMID: 37630458

Thomas, T., Friedrich, M., Rich-Griffin, C., Pohin, M., Agarwal, D., Pakpoor, J., & Buckley, C. D. (2024). A longitudinal single-cell atlas of anti-tumour necrosis factor treatment in inflammatory bowel disease. *Nature Immunology*, *25*(11), 2152–2165.

Thota, R. N., Rosato, J. I., Dias, C. B., Burrows, T. L., Martins, R. N., & Garg, M. L. (2020). Dietary supplementation with curcumin reduce circulating levels of glycogen synthase kinase-3β and islet amyloid polypeptide in adults with high risk of type 2 diabetes and Alzheimer's disease. *Nutrients*, *12*(4), 1032. DOI: 10.3390/nu12041032 PMID: 32283762

Thyab Gddoa Al-sahlany, S., Altemimi, A., Al-Manhel, A., Niamah, A., Lakhssassi, N., & Ibrahim, S.ThyabGddoa Al-sahlany. (2020). Purification of bioactive peptide with antimicrobial properties produced by Saccharomyces cerevisiae. *Foods*, *9*(3), 324. DOI: 10.3390/foods9030324 PMID: 32168785

Tian, F., Xiao, Y., Li, X., Zhai, Q., Wang, G., Zhang, Q., Zhang, H., & Chen, W. (2015). Protective effects of Lactobacillus plantarum CCFM8246 against copper toxicity in mice. *PLoS One*, *10*(11), e0143318. DOI: 10.1371/journal.pone.0143318 PMID: 26605944

Tian, T., Xie, W., Liu, L., Fan, S., Zhang, H., Qin, Z., & Yang, C. (2023). Industrial application of antimicrobial peptides based on their biological activity and structure-activity relationship. *Critical Reviews in Food Science and Nutrition*, *63*(21), 5430–5445. DOI: 10.1080/10408398.2021.2019673 PMID: 34955061

Tian, W., Han, X.-G., Liu, Y.-J., Tang, G.-Q., Liu, B., Wang, Y.-Q., Xiao, B., & Xu, Y.-F. (2013). Intrathecal epigallocatechin gallate treatment improves functional recovery after spinal cord injury by upregulating the expression of BDNF and GDNF. *Neurochemical Research*, *38*(4), 772–779. DOI: 10.1007/s11064-013-0976-5 PMID: 23344852

Timewell, M., Butikofer, L., & Garcia-Larsen, V. (2023). Polyphenols as modulators of gut barrier function: Caffeoyltyramine and feruloyltyramine restore TNF-α–induced epithelial damage. *Food & Function*, *14*(4), 7613–7624.

Tlaskalová-Hogenová, H., Štěpánková, R., Kozáková, H., Hudcovic, T., Vannucci, L., Tučková, L., Rossmann, P., Bártová, J., Sokol, D., Funda, D. P., Borovská, D., Zákostelská, Z., & Hrncir, T. (2011). The role of gut microbiota (commensal bacteria) and the mucosal barrier in the pathogenesis of inflammatory and autoimmune diseases and cancer: Contribution of germ-free and gnotobiotic animal models of human diseases. *Cellular & Molecular Immunology*, *8*(2), 110–120. DOI: 10.1038/cmi.2010.67 PMID: 21278760

Tönnies, E., & Trushina, E. (2017). Oxidative Stress, Synaptic Dysfunction, and Alzheimer's Disease. *Journal of Alzheimer's Disease*, *57*(4), 1105–1121. DOI: 10.3233/jad-161088

Tooley, K. L. (2020). Effects of the human gut microbiota on cognitive performance, brain structure and function: A narrative review. *Nutrients*, *12*(10), 3009. Advance online publication. DOI: 10.3390/nu12103009 PMID: 33007941

Torello, C. O., Alvarez, M. C., & Olalla Saad, S. T. (2021). Polyphenolic Flavonoid Compound Quercetin Effects in the Treatment of Acute Myeloid Leukemia and Myelodysplastic Syndromes. *Molecules (Basel, Switzerland)*, *26*(19). Advance online publication. DOI: 10.3390/molecules26195781

Tortorella, S. M., Royce, S. G., Licciardi, P. V., & Karagiannis, T. C. (2015). Dietary sulforaphane in cancer chemoprevention: The role of epigenetic regulation and HDAC inhibition. *Antioxidants & Redox Signaling*, *22*(16), 1382–1424. DOI: 10.1089/ars.2014.6097 PMID: 25364882

Traiffort, E., Morisset-Lopez, S., Moussaed, M., & Zahaf, A. (2021). Defective oligodendroglial lineage and demyelination in amyotrophic lateral sclerosis. In *International Journal of Molecular Sciences* (Vol. 22, Issue 7, p. 3426). Multidisciplinary Digital Publishing Institute. DOI: 10.3390/ijms22073426

Travica, N., D'Cunha, N. M., Naumovski, N., Kent, K., Beck, K. L., Lane, K. E., Firth, J., Sarris, J., & Young, J. M. (2020). Polyphenols and cognitive function: A systematic review and meta-analysis. *Nutrients*, *12*(9), 2618. DOI: 10.3390/nu12092618 PMID: 32867278

Travica, N., D'Cunha, N. M., Naumovski, N., Kent, K., Mellor, D. D., Firth, J., & Georgousopoulou, E. N. (2020). The effect of anthocyanins on cognitive performance: A systematic review and meta-analysis of randomized controlled trials. *Brain, Behavior, and Immunity*, *87*, 629–637. DOI: 10.1016/j.bbi.2020.01.016

Travica, N., D'Cunha, N. M., Naumovski, N., Kent, K., Mellor, D., & McKune, A. (2020). Anthocyanin-rich interventions reduce depressive symptoms in adults: A systematic review and meta-analysis. *Advances in Nutrition*, *11*(3), 458–472. DOI: 10.1093/advances/nmz113

Trépanier, M. O., Hopperton, K. E., Mizrahi, R., Mechawar, N., & Bazinet, R. P. (2016). Postmortem evidence of cerebral inflammation in schizophrenia: A systematic review. *Molecular Psychiatry*, *21*(8), 1009–1026. DOI: 10.1038/mp.2016.90 PMID: 27271499

Triantafillidis, J. K., Nasioulas, G., & Kosmidis, P. A. (2009). Colorectal cancer and inflammatory bowel disease: Epidemiology, risk factors, mechanisms of carcinogenesis and prevention strategies. *Anticancer Research*, *29*(7), 2727–2737. PMID: 19596953

Tsimogiannis, D., & Oreopoulou, V. (2019). Classification of phenolic compounds in plants. In *Polyphenols in plants* (pp. 263–284). Elsevier. DOI: 10.1016/B978-0-12-813768-0.00026-8

Tsoi, P. S., Quan, M. D., & Choi, M. L. (2023). Aggregation of disordered proteins associated with neurodegeneration. *International Journal of Molecular Sciences*, *24*(4), 3380. Advance online publication. DOI: 10.3390/ijms24043380 PMID: 36834792

Tsuda, T. (2012). Dietary anthocyanin-rich plants: Biochemical basis and recent progress in health benefits studies. *Molecular Nutrition & Food Research*, *56*(1), 159–170. DOI: 10.1002/mnfr.201100526 PMID: 22102523

Tsuda, T. (2016). Possible health effects of dietary anthocyanins: Antioxidant and anti-inflammatory actions. *Journal of Agricultural and Food Chemistry*, *64*(12), 2273–2279. DOI: 10.1021/acs.jafc.5b01035

Tsujimoto, Y., & Shimizu, S. (2007). Role of the mitochondrial membrane permeability transition in cell death. *Apoptosis*, *12*(5), 835–840. DOI: 10.1007/s10495-006-0525-7 PMID: 17136322

Tsunemi, T., Ishiguro, Y., Yoroisaka, A., Valdez, C., Miyamoto, K., Ishikawa, K., Saiki, S., Akamatsu, W., Hattori, N., & Krainc, D. (2020). Astrocytes protect human dopaminergic neurons from α-synuclein accumulation and propagation. *The Journal of Neuroscience : The Official Journal of the Society for Neuroscience*, *40*(45), 8618–8628. DOI: 10.1523/JNEUROSCI.0954-20.2020 PMID: 33046546

Tuohy, K. M., Probert, H. M., Smejkal, C. W., & Gibson, G. R. (2003). Using probiotics and prebiotics to improve gut health. *Digestive and Liver Disease*, *35*(5), 282–289. DOI: 10.1016/S1590-8658(03)00088-0 PMID: 12927512

Turner, R. S., Thomas, R. G., Craft, S., Van Dyck, C. H., Mintzer, J., Reynolds, B. A., Brewer, J. B., Rissman, R. A., Raman, R., Aisen, P. S., Mintzer, J., Reynolds, B. A., Karlawish, J., Galasko, D., Heidebrink, J., Aggarwal, N., Graff-Radford, N., Sano, M., Petersen, R., & Alexopoulis, G. (2015). A randomized, double-blind, placebo-controlled trial of resveratrol for Alzheimer disease. *Neurology*, *85*(16), 1383–1391. DOI: 10.1212/WNL.0000000000002035 PMID: 26362286

Tzounis, X., Rodriguez-Mateos, A., Vulevic, J., Gibson, G. R., Kwik-Uribe, C., & Spencer, J. P. E. (2011). Prebiotic evaluation of cocoa-derived flavanols in healthy humans. *The American Journal of Clinical Nutrition*, *93*(1), 62–72. DOI: 10.3945/ajcn.110.000075 PMID: 21068351

Uddin, M. S., Al Mamun, A., Rahman, M. M., Jeandet, P., Alexiou, A., Behl, T., Sarwar, M. S., Sobarzo-Sánchez, E., Ashraf, G. M., Sayed, A. A., Albadrani, G. M., Peluso, I., & Abdel-Daim, M. M. (2021). Natural products for neurodegeneration: Regulating neurotrophic signals. *Oxidative Medicine and Cellular Longevity*, *2021*, 8820406. Advance online publication. 10.1155/2021/8820406

Uddin, M. S., Kabir, M. T., Rahman, M. S., Alim, M. A., Rahman, M. M., & Rauf, A. (2020). Neuroprotective role of polyphenols against oxidative stress-mediated neurodegeneration. *European Journal of Pharmacology*, *886*, 173412. Advance online publication. DOI: 10.1016/j.ejphar.2020.173412 PMID: 32771668

Uddin, M. S., Tewari, D., Sharma, G., Kabir, M. T., Barreto, G. E., Bin-Jumah, M. N., & Ashraf, G. M. (2020). Molecular Mechanisms of ER Stress and UPR in the Pathogenesis of Alzheimer's Disease. *Molecular Neurobiology*, *57*(7), 2902–2919. DOI: 10.1007/s12035-020-01929-y

Udenigwe, C. C., & Aluko, R. E. (2012). Food protein-derived bioactive peptides: Production, processing, and potential health benefits. *Journal of Food Science*, *77*(1), R11–R24. DOI: 10.1111/j.1750-3841.2011.02455.x PMID: 22260122

Ullah, H., De Filippis, A., Santarcangelo, C., & Daglia, M. (2020). Epigenetic regulation by polyphenols in diabetes and related complications. *Mediterranean Journal of Nutrition and Metabolism*, *13*(4), 289–310. DOI: 10.3233/MNM-200489

Urdzikova, L. M., Ruzicka, J., Karova, K., Kloudova, A., Svobodova, B., Amin, A., Dubisova, J., Schmidt, M., Kubinova, S., & Jhanwar-Uniyal, M. (2017). A green tea polyphenol epigallocatechin-3-gallate enhances neuroregeneration after spinal cord injury by altering levels of inflammatory cytokines. *Neuropharmacology*, *126*, 213–223. DOI: 10.1016/j.neuropharm.2017.09.006 PMID: 28899730

Utpal, B. K., Roy, S. C., Zehravi, M., Sweilam, S. H., Raja, A. D., Haque, M. A., Nayak, C., Balakrishnan, S., Singh, L. P., Panigrahi, S., Alshehri, M. A., Rab, S. O., Minhaj, N. S., & Emran, T. B. (2025). Polyphenols as Wnt/β-catenin pathway modulators: A promising strategy in clinical neurodegeneration. *Animal Models and Experimental Medicine*, *8*(2), 266–286. DOI: 10.1002/ame2.12525 PMID: 39808166

Uzar, E., Alp, H., Cevik, M. U., Fırat, U., Evliyaoglu, O., Tufek, A., & Altun, Y. (2012). Ellagic acid attenuates oxidative stress on brain and sciatic nerve and improves histopathology of brain in streptozotocin-induced diabetic rats. *Neurological Sciences*, *33*(3), 567–574. DOI: 10.1007/s10072-011-0775-1 PMID: 21922312

Vaessen, M. J., Saj, A., Lovblad, K. O., Gschwind, M., & Vuilleumier, P. (2016). Structural white-matter connections mediating distinct behavioral components of spatial neglect in right brain-damaged patients. *Cortex*, *77*, 54–68. DOI: 10.1016/j.cortex.2015.12.008 PMID: 26922504

Vafeiadou, K., Vauzour, D., Lee, H. Y., Rodriguez-Mateos, A., Williams, R. J., & Spencer, J. P. (2009). The citrus flavanone naringenin inhibits inflammatory signalling in glial cells and protects against neuroinflammatory injury. *Archives of Biochemistry and Biophysics*, *484*(1), 100–109. DOI: 10.1016/j.abb.2009.01.016

Vaillant-Beuchot, L., Mary, A., Pardossi-Piquard, R., Bourgeois, A., Lauritzen, I., Eysert, F., Kinoshita, P. F., Cazareth, J., Badot, C., Fragaki, K., Bussiere, R., Martin, C., Mary, R., Bauer, C., Pagnotta, S., Paquis-Flucklinger, V., Buée-Scherrer, V., Buée, L., Lacas-Gervais, S., & Chami, M. (2021). Accumulation of amyloid precursor protein C-terminal fragments triggers mitochondrial structure, function, and mitophagy defects in Alzheimer's disease models and human brains. *Acta Neuropathologica*, *141*(1), 39–65. DOI: 10.1007/s00401-020-02234-7 PMID: 33079262

Vaiserman, A., & Lushchak, O. (2021). DNA methylation changes induced by prenatal toxic metal exposure: An overview of epidemiological evidence. *Environmental Epigenetics*, *7*(1), dvab007. PMID: 34631153

Vaissière, T., Sawan, C., & Herceg, Z. (2008). Epigenetic interplay between histone modifications and DNA methylation in gene silencing. *Mutation Research/Reviews in Mutation Research*, *659*(1-2), 40–48. DOI: 10.1016/j.mrrev.2008.02.004 PMID: 18407786

Van De Kerkhof, N. W. A., Fekkes, D., Van Der Heijden, F. M. M. A., Egger, J. I. M., & Verhoeven, W. M. A. (2015). Relationship between Plasma Homovanillic Acid and Outcome in Patients with Psychosis Spectrum Disorders. *Neuropsychobiology*, *71*(4), 212–217. DOI: 10.1159/000431095 PMID: 26279280

van Dinteren, S. (2024). *Defining the window of opportunity of prenylated phenolics as natural antimicrobials: Extraction, antimicrobial activity, and impact on the intestinal epithelium.* Wageningen University and Research, Volpe, C. M. O., Villar-Delfino, P. H., Dos Anjos, P. M. F., & Nogueira-Machado, J. A. (2018). Cellular death, reactive oxygen species (ROS) and diabetic complications. *Cell Death & Disease*, *9*(2), 119. PMID: 38331884

Van Kesteren, C. F. M. G., Gremmels, H., De Witte, L. D., Hol, E. M., Van Gool, A. R., Falkai, P. G., Kahn, R. S., & Sommer, I. E. C. (2017). Immune involvement in the pathogenesis of schizophrenia: A meta-analysis on postmortem brain studies. *Translational Psychiatry*, *7*(3), e1075–e11. DOI: 10.1038/tp.2017.4 PMID: 28350400

Van Praag, H., Fleshner, M., Schwartz, M. W., & Mattson, M. P. (2014). Exercise, energy intake, glucose homeostasis, and the brain. *The Journal of Neuroscience : The Official Journal of the Society for Neuroscience*, *34*(46), 15139–15149. DOI: 10.1523/JNEUROSCI.2814-14.2014 PMID: 25392482

van Tilburg, A. Y., Cao, H., van der Meulen, S. B., Solopova, A., & Kuipers, O. P. (2019). Metabolic engineering and synthetic biology employing Lactococcus lactis and Bacillus subtilis cell factories. *Current Opinion in Biotechnology*, *59*, 1–7. DOI: 10.1016/j.copbio.2019.01.007 PMID: 30784872

Vandeputte, D., Falony, G., Vieira-Silva, S., Tito, R. Y., Joossens, M., & Raes, J. (2021). Prebiotic inulin-type fructans induce specific changes in the human gut microbiota. *Gut*, *66*(11), 1968–1974. DOI: 10.1136/gutjnl-2016-313271 PMID: 28213610

Vanes, L. D., Mouchlianitis, E., Barry, E., Patel, K., Wong, K., & Shergill, S. S. (2019). Cognitive correlates of abnormal myelination in psychosis. *Scientific Reports*, *9*(1), 1–9. DOI: 10.1038/s41598-019-41679-z PMID: 30914748

Vasile, F., Dossi, E., & Rouach, N. (2017). Human astrocytes: Structure and functions in the healthy brain. *Brain Structure & Function*, *222*(5), 2017–2029. DOI: 10.1007/s00429-017-1383-5 PMID: 28280934

Vassiliadis, G., Destoumieux-Garzón, D., Lombard, C., Rebuffat, S., & Peduzzi, J. (2010). Isolation and characterization of two members of the siderophore-microcin family, microcins M and H47. *Antimicrobial Agents and Chemotherapy*, *54*(1), 288–297. DOI: 10.1128/AAC.00744-09 PMID: 19884380

Vauzour, D. (2012). Dietary Polyphenols as Modulators of Brain Functions: Biological Actions and Molecular Mechanisms Underpinning Their Beneficial Effects. *Oxidative Medicine and Cellular Longevity*, *914273*, 1–16. Advance online publication. DOI: 10.1155/2012/914273 PMID: 22701758

Vauzour, D., Rodriguez-Mateos, A., Corona, G., Oruna-Concha, M. J., & Spencer, J. P. E. (2010). Polyphenols and human health: Prevention of disease and mechanisms of action. *Nutrients*, *2*(11), 1106–1131. DOI: 10.3390/nu2111106 PMID: 22254000

Vauzour, D., Vafeiadou, K., Rodriguez-Mateos, A., Rendeiro, C., & Spencer, J. P. E. (2008). The neuroprotective potential of flavonoids: A multiplicity of effects. *Genes & Nutrition*, *3*(3–4), 115–126. DOI: 10.1007/s12263-008-0091-4 PMID: 18937002

Vauzour, D., Vafeiadou, K., & Spencer, J. P. E. (2010). Anthocyanins and neuronal signaling: Implications for memory. *Molecular Nutrition & Food Research*, *54*(3), 313–329. DOI: 10.1002/mnfr.200900444

Vellucci, L., Mazza, B., Barone, A., Nasti, A., De Simone, G., Iasevoli, F., & de Bartolomeis, A. (2025). The Role of Astrocytes in the Molecular Pathophysiology of Schizophrenia: Between Neurodevelopment and Neurodegeneration. *Biomolecules*, *15*(5), 1–36. DOI: 10.3390/biom15050615 PMID: 40427508

Vendrame, S., Guglielmetti, S., Riso, P., Arioli, S., Klimis-Zacas, D., & Porrini, M. (2011). Six-week consumption of a wild blueberry powder drink increases bifidobacteria in the human gut. *Journal of Agricultural and Food Chemistry*, *59*(24), 12815–12820. DOI: 10.1021/jf2028686 PMID: 22060186

Verkhratsky, A., & Semyanov, A. (2025). *Physiology of neuroglia of the central nervous system. 209*, 69–91. DOI: 10.1016/B978-0-443-19104-6.00005-X

Verkhratsky, A., Hol, E. M., de Witte, L. D., & Aronica, E. (2025). Neuroglia in the healthy brain. *Handbook of Clinical Neurology*, *209*, 1–5. DOI: 10.1016/B978-0-443-19104-6.00008-5 PMID: 40122619

Verkhratsky, A., Matteoli, M., Parpura, V., Mothet, J., & Zorec, R. (2016). Astrocytes as secretory cells of the central nervous system: Idiosyncrasies of vesicular secretion. *The EMBO Journal*, *35*(3), 239–257. DOI: 10.15252/embj.201592705 PMID: 26758544

Verkhratsky, A., & Nedergaard, M. (2018). Physiology of astroglia. *Physiological Reviews*, *98*(1), 239–389. DOI: 10.1152/physrev.00042.2016 PMID: 29351512

Verkhratsky, A., Rodríguez, J. J., & Parpura, V. (2013). Astroglia in neurological diseases. *Future Neurology*, *8*(2), 149–158. DOI: 10.2217/fnl.12.90 PMID: 23658503

Vezzani, B., Carinci, M., Patergnani, S., Pasquin, M. P., Guarino, A., Aziz, N., Pinton, P., Simonato, M., & Giorgi, C. (2020). The dichotomous role of inflammation in the CNS: A mitochondrial point of view. *Biomolecules*, *10*(10), 1–36. DOI: 10.3390/biom10101437 PMID: 33066071

Vicente, O., & Boscaiu, M. (2018). Flavonoids: Antioxidant Compounds for Plant Defence... and for a Healthy Human Diet. *Notulae Botanicae Horti Agrobotanici Cluj-Napoca*, *46*(1), 14–21. DOI: 10.15835/nbha46110992

Vijayalakshmi, K., Alladi, P. A., Sathyaprabha, T. N., Subramaniam, J. R., Nalini, A., & Raju, T. R. (2009). Cerebrospinal Fluid from sporadic Amyotrophic Lateral Sclerosis patients induces degeneration of a cultured motor neuron cell line. *Brain Research*, *1263*, 122–133. DOI: 10.1016/j.brainres.2009.01.041 PMID: 19368830

Villarino, A. V., Kanno, Y., & O'Shea, J. J. (2015). Mechanisms of Jak/STAT signaling in immunity and disease. *The Journal of Immunology : Official Journal of the American Association of Immunologists*, *194*(1), 21–27. DOI: 10.4049/jimmunol.1401867 PMID: 25527793

Villemagne, V. L., Doré, V., Burnham, S. C., Masters, C. L., & Rowe, C. C. (2018). Imaging tau and amyloid-β proteinopathies in Alzheimer disease and other conditions. *Nature Reviews. Neurology*, *14*(4), 225–236. DOI: 10.1038/nrneurol.2018.9 PMID: 29449700

Viña, J., Escudero, J., Baquero, M., Cebrián, M., Carbonell-Asíns, J. A., Muñoz, J. E., Satorres, E., Meléndez, J. C., Ferrer-Rebolleda, J., Cózar-Santiago, M. P., Santabárbara-Gómez, J. M., Jové, M., Pamplona, R., Tarazona-Santabalbina, F. J., & Borrás, C. (2022). Genistein effect on cognition in prodromal Alzheimer's disease patients. The GENIAL clinical trial. *Alzheimer's Research & Therapy*, *14*(1), 164. DOI: 10.1186/s13195-022-01097-2 PMID: 36329553

Vinolo, M. A. R., Rodrigues, H. G., Nachbar, R. T., & Curi, R. (2011). Regulation of inflammation by short chain fatty acids. *Nutrients*, *3*(10), 858–876. DOI: 10.3390/nu3100858 PMID: 22254083

Vinters, H. V., & Kleinschmidt-DeMasters, B. (2018). General pathology of the central nervous system. In *Greenfield's Neuropathology-Two Volume Set* (pp. 25–82). CRC Press.

Visanji, N. P., Brooks, P. L., Hazrati, L. N., & Lang, A. E. (2014). The prion hypothesis in Parkinson's disease: Braak to the future. In *Acta Neuropathologica Communications* (Vol. 2, Issue 1). BioMed Central Ltd. DOI: 10.1186/2051-5960-1-2

Vitini, E., Alvarez, S., Medina, M., & Perdigón, G. (2000). Gut mucosal immunostimulation by lactic acid bacteria. *Biocell*, *24*(3), 223–232. PMID: 11201658

Vogt, L., Ramasamy, U., Meyer, D., Pullens, G., Venema, K., Faas, M. M., Schols, H. A., & de Vos, P. (2013). Immune modulation by different types of β2→1-fructans is toll-like receptor dependent. *PLoS One*, *8*(7), e68367. Advance online publication. DOI: 10.1371/journal.pone.0068367 PMID: 23861894

Vuolo, M. M., Lima, V. S., & Junior, M. R. M. (2019). Phenolic compounds: Structure, classification, and antioxidant power. In *Bioactive compounds* (pp. 33–50). Elsevier. DOI: 10.1016/B978-0-12-814774-0.00002-5

Walker, L. C. (2020). Aβ plaques. *Free Neuropathology*, *1*, 1–31. PMID: 33345256

Wallace, T. C. (2011). Anthocyanins in cardiovascular disease. *Advances in Nutrition*, *2*(1), 1–7. DOI: 10.3945/an.110.000042 PMID: 22211184

Wan, F., Wong, F., Collins, J. J., & de la Fuente-Nunez, C. (2024). Machine learning for antimicrobial peptide identification and design. *Nature Reviews Bioengineering*, *2*(5), 392–407. DOI: 10.1038/s44222-024-00152-x PMID: 39850516

Wang, J., Zhen, Y., Yang, J., Yang, S., & Zhu, G. (2024). Recognizing Alzheimer's disease from perspective of oligodendrocytes: Phenomena or pathogenesis? In *CNS Neuroscience and Therapeutics* (Vol. 30, Issue 3, p. e14688). John Wiley and Sons Inc. DOI: 10.1111/cns.14688

Wang, Y., Zhang, X., Liu, Z., & Chen, L. (2023). Probiotic-induced modulation of Th2 cells and IL-10 production alleviates experimental autoimmune encephalomyelitis

Wang, B., & Wang, X. M. (2009). Schisandrin B protects rat cortical neurons against Aβ1-42- induced neurotoxicity. *Die Pharmazie*, *64*(7), 450–454. DOI: 10.1691/PH.2009.9552 PMID: 19694182

Wang, F. B., & Powley, T. L. (2007). Vagal innervation of intestines: Afferent pathways mapped with new en bloc horseradish peroxidase adaptation. *Cell and Tissue Research*, *329*(2), 221–230. DOI: 10.1007/s00441-007-0413-7 PMID: 17453246

Wang, H., Zhang, C., Li, M., Liu, C., Wang, J., Ou, X., & Han, Y. (2022). Antimicrobial peptides mediate apoptosis by changing mitochondrial membrane permeability. *International Journal of Molecular Sciences*, *23*(21), 12732. DOI: 10.3390/ijms232112732 PMID: 36361521

Wang, J., Fan, S. M., & Zhang, J. (2019). Epigallocatechin-3-gallate ameliorates lipopolysaccharide-induced acute lung injury by suppression of TLR4/NF-κB signaling activation. *Brazilian Journal of Medical and Biological Research = Revista Brasileira de Pesquisas Médicas e Biológicas*, *52*(7), e8092. DOI: 10.1590/1414-431x20198092 PMID: 31241712

Wang, J., Liu, Y. M., Cao, W., Yao, K. W., Liu, Z. Q., & Guo, J. Y. (2012). Anti-inflammation and antioxidant effect of Cordymin, a peptide purified from the medicinal mushroom Cordyceps sinensis, in middle cerebral artery occlusion-induced focal cerebral ischemia in rats. *Metabolic Brain Disease*, *27*(2), 159–165. DOI: 10.1007/s11011-012-9282-1 PMID: 22327557

Wang, L., Wang, N., Zhang, W., Cheng, X., Yan, Z., Shao, G., Wang, X., Wang, R., & Fu, C. (2022). Therapeutic peptides: Current applications and future directions. *Signal Transduction and Targeted Therapy*, *7*(1), 48. DOI: 10.1038/s41392-022-00904-4 PMID: 35165272

Wang, M. L., Rivlin, M., Graham, J. G., & Beredjiklian, P. K. (2019). Peripheral nerve injury, scarring, and recovery. *Connective Tissue Research*, *60*(1), 3–9. DOI: 10.1080/03008207.2018.1489381 PMID: 30187777

Wang, P., Jiang, L., Zhou, N., Zhou, H., Liu, H., Zhao, W., Zhang, H., Zhang, X., & Hu, Z. (2018). Resveratrol ameliorates autophagic flux to promote functional recovery in rats after spinal cord injury. *Oncotarget*, *9*(9), 8427–8440. DOI: 10.18632/oncotarget.23877 PMID: 29492205

Wang, Q., Xue, H., Yue, Y., Hao, S., Huang, S.-H., & Zhang, Z. (2022). Role of mitophagy in the neurodegenerative diseases and its pharmacological advances: A review. *Frontiers in Molecular Neuroscience*, *15*, 1014251. DOI: 10.3389/fnmol.2022.1014251 PMID: 36267702

Wang, Q., Xu, J., Rottinghaus, G. E., Simonyi, A., Lubahn, D., Sun, G. Y., & Sun, A. Y. (2002). Resveratrol protects against global cerebral ischemic injury in gerbils. *Brain Research*, *958*(2), 439–447. DOI: 10.1016/S0006-8993(02)03543-6 PMID: 12470882

Wang, Q., Yang, S., Zhang, X., Zhang, S., Chen, L., Wang, W., Chen, N., & Yan, J. (2024). Inflammasomes in neurodegenerative diseases. *Translational Neurodegeneration*, *13*(1), 65. DOI: 10.1186/s40035-024-00459-0 PMID: 39710713

Wang, R., Zhang, Y., Li, J., & Zhang, C. (2017). Resveratrol ameliorates spatial learning memory impairment induced by Aβ1–42 in rats. *Neuroscience*, *344*, 39–47. DOI: 10.1016/j.neuroscience.2016.08.051 PMID: 27600946

Wang, T., Wang, S., Wang, X., Jiang, H., Yang, Y., Wang, Y., Cheng, J., Zhang, C., Liang, W., & Feng, H. (2018). Fisetin exerts antioxidant and neuroprotective effects in multiple mutant hSOD1 models of amyotrophic lateral sclerosis by activating ERK. *Neuroscience*, *379*, 152–166. DOI: 10.1016/j.neuroscience.2018.03.008 PMID: 29559385

Wang, T.-y., Li, Q., & Bi, K. (2018). Bioactive flavonoids in medicinal plants: Structure, activity and biological fate. *Asian Journal of Pharmaceutical Sciences*, *13*(1), 12–23. DOI: 10.1016/j.ajps.2017.08.004 PMID: 32104374

Wang, W.-Y., Tan, M.-S., Yu, J.-T., & Tan, L. (2015). Role of pro-inflammatory cytokines released from microglia in Alzheimer's disease. *Annals of Translational Medicine*, *3*(10), 136. PMID: 26207229

Wang, X., Hu, D., Zhang, L., Lian, G., Zhao, S., Wang, C., Yin, J., Wu, C., & Yang, J. (2014). Gomisin A inhibits lipopolysaccharide-induced inflammatory responses in N9 microglia via blocking the NF-κB/MAPKs pathway. *Food and Chemical Toxicology*, *63*, 119–127. DOI: 10.1016/j.fct.2013.10.048 PMID: 24211520

Wang, X., & Jia, J. (2023). Magnolol improves Alzheimer's disease-like pathologies and cognitive decline by promoting autophagy through activation of the AMPK/mTOR/ULK1 pathway. *Biomedicine and Pharmacotherapy*, *161*, 114473. Advance online publication. DOI: 10.1016/j.biopha.2023.114473 PMID: 36889111

Wang, X., Qi, Y., & Zheng, H. (2022). Dietary polyphenol, gut microbiota, and health benefits. *Antioxidants, 11*(6), 1212. DOI: 10.3390/antiox11061212 PMID: 35740109

Wang, X., Zhang, P., & Zhang, X. (2021). Probiotics regulate gut microbiota: An effective method to improve immunity. *Molecules (Basel, Switzerland), 26*(19), 6076. DOI: 10.3390/molecules26196076 PMID: 34641619

Wang, Y., Chen, L., Liang, L., & Guo, Y. (2015). Anthocyanins from blueberries mitigate Aβ-induced apoptosis via caspase-3 inhibition and mitochondrial protection in SH-SY5Y cells. *Neurochemistry International, 89*, 40–48. DOI: 10.1016/j.neuint.2015.07.007

Wang, Y., Chen, W. J., Han, Y. Y., Xu, X., Yang, A. X., Wei, J., Hong, D., Fang, X., & Chen, T. T. (2023). Neuroprotective effect of engineered Clostridium butyricum-pMTL007-GLP-1 on Parkinson's disease mice models via promoting mitophagy. *Bioengineering & Translational Medicine, 8*(3), e10505. DOI: 10.1002/btm2.10505 PMID: 37206220

Wang, Y., Wang, K., Yan, J., Zhou, Q., & Wang, X. (2022). Recent progress in research on mechanisms of action of natural products against Alzheimer's disease: Dietary plant polyphenols. *International Journal of Molecular Sciences, 23*(22), 13886. DOI: 10.3390/ijms232213886 PMID: 36430365

Wang, Z., Huang, J., Liu, C., Liu, L., Shen, Y., Shen, C., & Liu, C. (2019). BAF45D Downregulation in Spinal Cord Ependymal Cells Following Spinal Cord Injury in Adult Rats and Its Potential Role in the Development of Neuronal Lesions. *Frontiers in Neuroscience, 13*, 471535. DOI: 10.3389/fnins.2019.01151 PMID: 31736692

Wan, J., Li, Y., Chen, D., Yu, B., Zheng, P., Mao, X., Yu, J., & He, J. (2016). Expression of a tandemly arrayed plectasin gene from Pseudoplectania nigrella in Pichia pastoris and its antimicrobial activity. *Journal of Microbiology and Biotechnology, 26*(3), 461–468. DOI: 10.4014/jmb.1508.08091 PMID: 26643963

Wareham, L. K., Liddelow, S. A., Temple, S., Benowitz, L. I., Di Polo, A., Wellington, C., & Bu, G. (2022). Solving neurodegeneration: Common mechanisms and strategies for new treatments. *Molecular Neurodegeneration, 17*(1), 23. DOI: 10.1186/s13024-022-00524-0 PMID: 35313950

Wei, D. C., & Morrison, E. H. (2023). *Histology, Astrocytes*. StatPearls.

Wei, I.-H., Tu, H.-C., Huang, C.-C., Tsai, M.-H., Tseng, C.-Y., & Shieh, J.-Y. (2011). (-)-Epigallocatechin gallate attenuates NADPH-d/nNOS expression in motor neurons of rats following peripheral nerve injury. *BMC Neuroscience, 12*(1), 52. DOI: 10.1186/1471-2202-12-52 PMID: 21627848

Wenzel, T. J., Gates, E. J., Ranger, A. L., & Klein, J. R. (2022). Short-chain fatty acids reduce inflammation in LPS-stimulated human monocytes via histone deacetylase inhibition. *Journal of Immunology Research, 2022*, 8299834.

Werner, C., & Engelhard, K. (2007). Pathophysiology of traumatic brain injury. *British Journal of Anaesthesia, 99*(1), 4–9. DOI: 10.1093/bja/aem131 PMID: 17573392

What, H., Projects, N., Link, B., Plan, E., Management, E., & Identify, S. (2016). Festivals Toolkit. *Health & Safety : Risk Assessment & Management, 63*(8), 1429–1451. DOI: 10.1002/glia.22859.NG2-glia

White, R. E., & Jakeman, L. B. (2008). Don't fence me in: Harnessing the beneficial roles of astrocytes for spinal cord repair. *Restorative Neurology and Neuroscience*, *26*(2–3), 197–214. DOI: 10.3233/RNN-2008-00442 PMID: 18820411

Whitney, E. N., Rolfes, S. R., Crowe, T., & Walsh, A. (2019). *Understanding nutrition*. Cengage AU.

Whyte, A. R., Schafer, G., & Williams, C. M. (2016). Cognitive effects following acute wild blueberry supplementation in 7–10-year-old children. *European Journal of Nutrition*, *55*(6), 2151–2162. DOI: 10.1007/s00394-015-1029-4 PMID: 26437830

William Raja, T. R., Duraipandiyan, V., Ignacimuthu, S., Janakiraman, U., & Packiam, S. M. (2023). Role of polyphenols in alleviating alzheimer's disease: A review. *Current Medicinal Chemistry*, *30*(35), 4032–4047. DOI: 10.2174/0929867330666221202152540 PMID: 36476438

Williams, C. M., El Mohsen, M. A., Vauzour, D., Rendeiro, C., Butler, L. T., Ellis, J. A., & Spencer, J. P. E. (2008). Blueberry-induced cognitive improvements in aged rats are associated with enhanced brain regional activation and behavior. *Free Radical Biology & Medicine*, *45*(2), 295–305. DOI: 10.1016/j.freeradbiomed.2008.04.008 PMID: 18457678

Williamson, C., & Rajajee, V. (2021). Traumatic brain injury: Epidemiology, classification, and pathophysiology. *Waltham: UpToDate*.

Williamson, M. R., Kwon, W., Woo, J., Ko, Y., Maleki, E., Yu, K., Murali, S., Sardar, D., & Deneen, B. (2024). Learning-associated astrocyte ensembles regulate memory recall. *Nature 2024 637:8045*, *637*(8045), 478–486. DOI: 10.1038/s41586-024-08170-w

Williamson, G. (2025). Bioavailability of Food Polyphenols: Current State of Knowledge. *Annual Review of Food Science and Technology*, *16*(1), 315–332. DOI: 10.1146/annurev-food-060721-023817 PMID: 39899845

Williamson, G., & Clifford, M. N. (2017). Role of the small intestine, colon and microbiota in determining the metabolic fate of polyphenols. *Biochemical Pharmacology*, *139*, 24–39. DOI: 10.1016/j.bcp.2017.03.012 PMID: 28322745

Williams, R. J., Spencer, J. P. E., & Rice-Evans, C. (2009). Flavonoids: Antioxidants or signaling molecules? *Free Radical Biology & Medicine*, *36*(7), 838–849. DOI: 10.1016/j.freeradbiomed.2004.01.001 PMID: 15019969

Wilson, G. R., Tan, J. T., Brody, K. M., Taylor, J. M., Delatycki, M. B., & Lockhart, P. J. (2009). Expression and localization of the Parkin Co-Regulated Gene in mouse CNS suggests a role in ependymal cilia function. *Neuroscience Letters*, *460*(1), 97–101. DOI: 10.1016/j.neulet.2009.05.043 PMID: 19463890

Winiarska-Mieczan, A., Kwiecień, M., Jachimowicz-Rogowska, K., Donaldson, J., Tomaszewska, E., & Baranowska-Wójcik, E. (2023). Anti-Inflammatory, Antioxidant, and Neuroprotective Effects of Polyphenols-Polyphenols as an Element of Diet Therapy in Depressive Disorders. *International Journal of Molecular Sciences*, *24*(3), 2258. Advance online publication. DOI: 10.3390/ijms24032258 PMID: 36768580

Wink, M. (2000). Interference of alkaloids with neuroreceptors and ion channels. *Studies in Natural Products Chemistry, 21*(PART B), 3–122. DOI: 10.1016/S1572-5995(00)80004-6

Wright-Jin, E. C., & Gutmann, D. H. (2019). Microglia as Dynamic Cellular Mediators of Brain Function. *Trends in Molecular Medicine, 25*(11), 967–979. DOI: 10.1016/j.molmed.2019.08.013 PMID: 31597593

Wróbel-Biedrawa, D., Grabowska, K., Galanty, A., Sobolewska, D., & Podolak, I. (2022). A flavonoid on the brain: Quercetin as a potential therapeutic agent in central nervous system disorders. *Life (Chicago, Ill.), 12*(4), 591. DOI: 10.3390/life12040591 PMID: 35455082

Wu, A., Ying, Z., & Gomez-Pinilla, F. (2006). Dietary curcumin counteracts the outcome of traumatic brain injury on oxidative stress, synaptic plasticity, and cognition. *Experimental Neurology, 197*(2), 309–317. DOI: 10.1016/j.expneurol.2005.09.004 PMID: 16364299

Wu, J., Chen, H., Li, H., Tang, Y., Yang, L., Cao, S., & Qin, D. (2016). Antidepressant potential of chlorogenic acid-enriched extract from Eucommia ulmoides Oliver bark with neuron protection and promotion of serotonin release through enhancing synapsin I expression. *Molecules (Basel, Switzerland), 21*(3), 260. Advance online publication. DOI: 10.3390/molecules21030260 PMID: 26927040

Wu, J., Li, Q., Wang, X., Yu, S., Li, L., Wu, X., Chen, Y., Zhao, J., & Zhao, Y. (2013). Neuroprotection by curcumin in ischemic brain injury involves the Akt/Nrf2 pathway. *PLoS One, 8*(3), e59843. DOI: 10.1371/journal.pone.0059843 PMID: 23555802

Wu, J., Maoqiang, L., Fan, H., Zhenyu, B., Qifang, H., Xuepeng, W., & Liulong, Z. (2016). Rutin attenuates neuroinflammation in spinal cord injury rats. *The Journal of Surgical Research, 203*(2), 331–337. DOI: 10.1016/j.jss.2016.02.041 PMID: 27363641

Wu, S. K., Wang, L., Wang, F., & Zhang, J. (2024). Resveratrol improved mitochondrial biogenesis by activating SIRT1/PGC-1α signal pathway in SAP. *Scientific Reports, 14*(1), 1–15. DOI: 10.1038/s41598-024-76825-9 PMID: 38167627

Wu, X., Beecher, G. R., Holden, J. M., Haytowitz, D. B., Gebhardt, S. E., & Prior, R. L. (2006). Concentrations of anthocyanins in common foods in the United States and estimation of normal consumption. *Journal of Agricultural and Food Chemistry, 54*(11), 4069–4075. DOI: 10.1021/jf0603001 PMID: 16719536

Wu, X., Beecher, G. R., Holden, J. M., Haytowitz, D. B., Gebhardt, S. E., & Prior, R. L. (2021). Concentration of anthocyanins in common foods and beverages. *Journal of Agricultural and Food Chemistry, 69*(22), 6260–6272. DOI: 10.1021/acs.jafc.1c02012 PMID: 34043345

Wu, Y., Chen, Q., Wen, B., Wu, N., He, B., & Chen, J. (2021). Berberine Reduces Aβ42 Deposition and Tau Hyperphosphorylation via Ameliorating Endoplasmic Reticulum Stress. *Frontiers in Pharmacology, 12*(July), 1–14. DOI: 10.3389/fphar.2021.640758 PMID: 34349640

Wu, Y., & Cui, J. (2020). (-)-Epigallocatechin-3-gallate provides neuroprotection via AMPK activation against traumatic brain injury in a mouse model. *Naunyn-Schmiedeberg's Archives of Pharmacology, 393*(11), 2209–2220. DOI: 10.1007/s00210-020-01841-1 PMID: 32062732

Wu, Y., Zhong, L., Yu, Z., & Qi, J. (2019). Anti-neuroinflammatory effects of tannic acid against lipopolysaccharide-induced BV2 microglial cells via inhibition of NF-κB activation. *Drug Development Research, 80*(2), 262–268. DOI: 10.1002/ddr.21490 PMID: 30724376

Wu, Z., Huang, S., Li, T., Li, N., Han, D., Zhang, B., & Wang, J. (2021). Gut microbiota from green tea polyphenol-dosed mice improves intestinal epithelial homeostasis and ameliorates experimental colitis. *Microbiome, 9*(1), 184. DOI: 10.1186/s40168-021-01115-9

Xie, S., & Li, F. (2024). Ependymal cells: roles in central nervous system infections and therapeutic application. *Journal of Neuroinflammation 2024 21:1, 21*(1), 1–14. DOI: 10.1186/s12974-024-03240-2

Xie, J., Bi, Y., Zhang, H., Dong, S., Teng, L., Lee, R. J., & Yang, Z. (2020). Cell-penetrating peptides in diagnosis and treatment of human diseases: From preclinical research to clinical application. *Frontiers in Pharmacology, 11*, 697. DOI: 10.3389/fphar.2020.00697 PMID: 32508641

Xie, J., Song, W., Liang, X., Zhang, Q., Shi, Y., Liu, W., & Shi, X. (2020). Protective effect of quercetin on streptozotocin-induced diabetic peripheral neuropathy rats through modulating gut microbiota and reactive oxygen species level. *Biomedicine and Pharmacotherapy, 127*, 110147. DOI: 10.1016/j.biopha.2020.110147

Xing, L., Zhang, H., Qi, R., Tsao, R., & Mine, Y. (2019). Recent advances in the understanding of the health benefits and molecular mechanisms associated with green tea polyphenols. *Journal of Agricultural and Food Chemistry, 67*(4), 1029–1043. DOI: 10.1021/acs.jafc.8b06146 PMID: 30653316

Xu, B., Bai, L., Chen, L., Tong, R., Feng, Y., & Shi, J. (2022). Terpenoid natural products exert neuroprotection via the PI3K/Akt pathway. *Frontiers in Pharmacology, 13*, 1036506. DOI: 10.3389/fphar.2022.1036506 PMID: 36313360

Xue, J., Tao, K., Wang, W., & Wang, X. (2024). What Can Inflammation Tell Us about Therapeutic Strategies for Parkinson's Disease? *International Journal of Molecular Sciences, 25*(3), 1641. Advance online publication. DOI: 10.3390/ijms25031641 PMID: 38338925

Xu, H., Wang, E., Chen, F., Xiao, J., & Wang, M. (2021). Neuroprotective phytochemicals in experimental ischemic stroke: Mechanisms and potential clinical applications. *Oxidative Medicine and Cellular Longevity, 2021*(1), 6687386. DOI: 10.1155/2021/6687386 PMID: 34007405

Xu, J., Wang, H., Ding, K., Zhang, L., Wang, C., Li, T., Wei, W., & Lu, X. (2014). Luteolin provides neuroprotection in models of traumatic brain injury via the Nrf2–ARE pathway. *Free Radical Biology & Medicine, 71*, 186–195. DOI: 10.1016/j.freeradbiomed.2014.03.009 PMID: 24642087

Xu, J., Wang, H., Lu, X., Ding, K., Zhang, L., He, J., Wei, W., & Wu, Y. (2014). Posttraumatic administration of luteolin protects mice from traumatic brain injury: Implication of autophagy and inflammation. *Brain Research, 1582*, 237–246. DOI: 10.1016/j.brainres.2014.07.042 PMID: 25093609

Xu, L., Botchway, B. O., Zhang, S., Zhou, J., & Liu, X. (2018). Inhibition of NF-κB signaling pathway by resveratrol improves spinal cord injury. *Frontiers in Neuroscience, 12*, 690. DOI: 10.3389/fnins.2018.00690 PMID: 30337851

Xu, L., Wang, J., Ding, Y., Wang, L., & Zhu, Y. J. (2022). Current Knowledge of Microglia in Traumatic Spinal Cord Injury. In *Frontiers in Neurology* (Vol. 12, p. 796704). Frontiers Media S.A., DOI: 10.3389/fneur.2021.796704

Xu, P., Wang, S., Yu, X., Su, Y., Wang, T., Zhou, W., Zhang, H., Wang, Y., & Liu, R. (2014). Rutin improves spatial memory in Alzheimer's disease transgenic mice by reducing Aβ oligomer level and attenuating oxidative stress and neuroinflammation. *Behavioural Brain Research, 264*, 173–180. DOI: 10.1016/j.bbr.2014.02.002 PMID: 24512768

Xu, Q., Yi, L. T., Pan, Y., Wang, X., Li, Y. C., Li, J. M., Wang, C. P., & Kong, L. D. (2008). Antidepressant-like effects of the mixture of honokiol and magnolol from the barks of Magnolia officinalis in stressed rodents. *Progress in Neuro-Psychopharmacology & Biological Psychiatry*, *32*(3), 715–725. DOI: 10.1016/j.pnpbp.2007.11.020 PMID: 18093712

Xu, X.-Y., Wang, D.-Y., Li, Y.-P., Deyrup, S. T., & Zhang, H.-J. (2022). Plant-derived lignans as potential antiviral agents: A systematic review. *Phytochemistry Reviews*, *21*(1), 239–289. DOI: 10.1007/s11101-021-09758-0 PMID: 34093097

Xu, Y. Q., Jin, S. J., Liu, N., Li, Y. X., Zheng, J., Ma, L., Du, J., Zhou, R., Zhao, C. J., Niu, Y., Sun, T., & Yu, J. Q. (2014). Aloperine attenuated neuropathic pain induced by chronic constriction injury via anti-oxidation activity and suppression of the nuclear factor kappa B pathway. *Biochemical and Biophysical Research Communications*, *451*(4), 568–573. DOI: 10.1016/j.bbrc.2014.08.025 PMID: 25128276

Xu, Y., Lin, D., Yu, X., Xie, X., Wang, L., Lian, L., Fei, N., Chen, J., Zhu, N., Wang, G., Huang, X., & Pan, J. (2016). The antinociceptive effects of ferulic acid on neuropathic pain: Involvement of descending monoaminergic system and opioid receptors. *Oncotarget*, *7*(15), 20455–20468. DOI: 10.18632/oncotarget.7973 PMID: 26967251

Xylaki, M., Atzler, B., & Outeiro, T. F. (2019). Epigenetics of the Synapse in Neurodegeneration. *Current Neurology and Neuroscience Reports*, *19*(10), 1–10. DOI: 10.1007/s11910-019-0995-y PMID: 31440934

Yahfoufi, N., Alsadi, N., Jambi, M., & Matar, C. (2018). The immunomodulatory and anti-inflammatory role of polyphenols. *Nutrients*, *10*(11), 1618. DOI: 10.3390/nu10111618 PMID: 30400131

Yang, C., Fei, Y., Qin, Y., Luo, D., Yang, S., Kou, X., Zi, Y., Deng, T., & Jin, M. (2015). Bacterial flora changes in conjunctiva of rats with streptozotocin-induced type 1 diabetes. *PLoS One*, *10*(7), e0133021. Advance online publication. DOI: 10.1371/journal.pone.0133021 PMID: 26176548

Yang, J., Xiong, L. L., Wang, Y. C., He, X., Jiang, L., Fu, S. J., Han, X. F., Liu, J., & Wang, T. H. (2018). Oligodendrocyte precursor cell transplantation promotes functional recovery following contusive spinal cord injury in rats and is associated with altered MicroRNA expression. *Molecular Medicine Reports*, *17*(1), 771–782. DOI: 10.3892/mmr.2017.7957 PMID: 29115639

Yang, L., Wen, K.-S., Ruan, X., Zhao, Y.-X., Wei, F., & Wang, Q. (2018). Response of the plant secondary metabolites to environmental factors. *Molecules (Basel, Switzerland)*, *23*(4), 762. Advance online publication. 10.3390/molecules23040762

Yang, S.-G., Wang, X.-W., Qian, C., & Zhou, F.-Q. (2022). Reprogramming neurons for regeneration: The fountain of youth. *Progress in Neurobiology*, *214*, 102284. DOI: 10.1016/j.pneurobio.2022.102284 PMID: 35533809

Yang, T., Kong, B., Gu, J.-W., Kuang, Y.-Q., Cheng, L., Yang, W.-T., Xia, X., & Shu, H.-F. (2014). Anti-apoptotic and anti-oxidative roles of quercetin after traumatic brain injury. *Cellular and Molecular Neurobiology*, *34*(6), 797–804. DOI: 10.1007/s10571-014-0070-9 PMID: 24846663

Yang, T., Zhao, X., Zhang, Y., Xie, J., & Zhou, A. (2020). 6‴-Feruloylspinosin alleviated beta-amyloid induced toxicity by promoting mitophagy in Caenorhabditis elegans (GMC101) and PC12 cells. *The Science of the Total Environment*, *715*, 136953. DOI: 10.1016/j.scitotenv.2020.136953

Yang, W., Chen, X., Li, Y., Guo, S., Wang, Z., & Yu, X. (2020). Advances in Pharmacological Activities of Terpenoids. *Natural Product Communications*, *15*(3), 1934578X20903555. Advance online publication. DOI: 10.1177/1934578X20903555

Yang, X., Xu, S., Qian, Y., & Xiao, Q. (2017). Resveratrol regulates microglia M1/M2 polarization via PGC-1α in conditions of neuroinflammatory injury. *Brain, Behavior, and Immunity*, *64*, 162–172. DOI: 10.1016/j.bbi.2017.03.003 PMID: 28268115

Yang, Y. H., Wang, Z., Zheng, J., & Wang, R. (2015). Protective effects of gallic acid against spinal cord injury-induced oxidative stress. *Molecular Medicine Reports*, *12*(2), 3017–3024. DOI: 10.3892/mmr.2015.3738 PMID: 25955644

Yang, Y., Ye, Y., Kong, C., Su, X., Zhang, X., Bai, W., & He, X. (2019). MiR-124 enriched exosomes promoted the M2 polarization of microglia and enhanced hippocampus neurogenesis after traumatic brain injury by inhibiting TLR4 pathway. *Neurochemical Research*, *44*(4), 811–828. DOI: 10.1007/s11064-018-02714-z PMID: 30628018

Yan, L., Guo, M.-S., Zhang, Y., Yu, L., Wu, J.-M., Tang, Y., & Chen, Q. (2022). Dietary plant polyphenols as the potential drugs in neurodegenerative diseases: Current evidence, advances, and opportunities. *Oxidative Medicine and Cellular Longevity*, *2022*(1), 5288698. DOI: 10.1155/2022/5288698 PMID: 35237381

Yan, P., Bai, L., Lu, W., Gao, Y., Bi, Y., & Lv, G. (2017). Regulation of autophagy by AMP-activated protein kinase/sirtuin 1 pathway reduces spinal cord neurons damage. *Iranian Journal of Basic Medical Sciences.*, *20*(9), 1029. PMID: 29085598

Yan, S. X., Lang, J. L., Song, Y. Y., Wu, Y. Z., Lv, M. H., Zhao, X., Liu, Y. H., & Xu, C. Y. (2015). Studies on anti-depressant activity of four flavonoids isolated from Apocynum venetum linn (Apocynaceae) leaf in mice. *Tropical Journal of Pharmaceutical Research*, *14*(12), 2269–2277. DOI: 10.4314/tjpr.v14i12.17

Yan, Z., Zhong, Y., Duan, Y., Chen, Q., & Li, F. (2020). Antioxidant mechanism of tea polyphenols and its impact on health benefits. *Animal Nutrition*, *6*(2), 115–123. DOI: 10.1016/j.aninu.2020.01.001 PMID: 32542190

Yao, M., Zhou, X., & Zhang, L. (2022). Polymeric nanoparticles for improving anthocyanin bioavailability: A review. *Journal of Agricultural and Food Chemistry*, *70*(12), 3645–3657. DOI: 10.1021/acs.jafc.2c00112

Yao, S., Wang, L., Chen, Q., Lu, T., Pu, X., & Luo, C. (2021). The effect of mild hypothermia plus rutin on the treatment of spinal cord injury and inflammatory factors by repressing TGF-β/smad pathway. *Acta Cirurgica Brasileira*, *36*(3), e360307. DOI: 10.1590/acb360307 PMID: 33978063

Yao, Y., Li, F., & Li, Y. (2021). Stimuli-responsive delivery systems for anthocyanins: Targeted release and enhanced bioactivity. *Food Hydrocolloids*, *112*, 106363. DOI: 10.1016/j.foodhyd.2020.106363

Yavtushenko, I. V., Nazarenko, S. M., Katrushov, O. V., & Kostenko, V. O. (2020). Quercetin limits the progression of oxidative and nitrosative stress in the rats' tissues after experimental traumatic brain injury. *Wiadomosci Lekarskie (Warsaw, Poland)*, *73*(10), 2127–2132. DOI: 10.36740/WLek202010104 PMID: 33310934

Yi, L. T., Li, J., Li, H. C., Su, D. X., Quan, X. B., He, X. C., & Wang, X. H. (2012). Antidepressant-like behavioral, neurochemical and neuroendocrine effects of naringenin in the mouse repeated tail suspension test. *Progress in Neuro-Psychopharmacology & Biological Psychiatry*, *39*(1), 175–181. DOI: 10.1016/j.pnpbp.2012.06.009 PMID: 22709719

Yi, L. T., Xu, H. L., Feng, J., Zhan, X., Zhou, L. P., & Cui, C. C. (2011). Involvement of monoaminergic systems in the antidepressant-like effect of nobiletin. *Physiology & Behavior*, *102*(1), 1–6. DOI: 10.1016/j.physbeh.2010.10.008 PMID: 20951716

Yin, F. (2023). Lipid metabolism and Alzheimer's disease: clinical evidence, mechanistic link and therapeutic promise. In *FEBS Journal* (Vol. 290, Issue 6, pp. 1420–1453). John Wiley and Sons Inc. DOI: 10.1111/febs.16344

Yin, H., Guo, Q., Li, X., Tang, T., Li, C., Wang, H., Sun, Y., Feng, Q., Ma, C., Gao, C., Yi, F., & Peng, J. (2018). Curcumin Suppresses IL-1β Secretion and Prevents Inflammation through Inhibition of the NLRP3 Inflammasome. *Journal of Immunology (Baltimore, Md. : 1950)*, *200*(8), 2835–2846. DOI: 10.4049/jimmunol.1701495

Yin, X., Yin, Y., Le Cao, F., Chen, Y. F., Peng, Y., Hou, W. G., Sun, S. K., & Luo, Z. J. (2012). Tanshinone IIA attenuates the inflammatory response and apoptosis after traumatic injury of the spinal cord in adult rats. *PLoS One*, *7*(6), 1–10. DOI: 10.1371/journal.pone.0038381 PMID: 22675554

Yin, Z., Zheng, T., Ho, C.-T., Huang, Q., Wu, Q., & Zhang, M. (2022). Improving the stability and bioavailability of tea polyphenols by encapsulations: A review. *Food Science and Human Wellness*, *11*(3), 537–556. DOI: 10.1016/j.fshw.2021.12.011

Yi, S., Jiang, X., Tang, X., Li, Y., Xiao, C., Zhang, J., & Zhou, T. (2020). IL-4 and IL-10 promotes phagocytic activity of microglia by up-regulation of TREM2. *Cytotechnology*, *72*(4), 589–602. DOI: 10.1007/s10616-020-00409-4 PMID: 32623621

Yiu, G., & He, Z. (2006). Glial inhibition of CNS axon regeneration. *Nature Reviews. Neuroscience*, *7*(8), 617–627. DOI: 10.1038/nrn1956 PMID: 16858390

Yoo, J. Y., Groer, M., Dutra, S. V. O., Sarkar, A., & McSkimming, D. I. (2020). Gut microbiota and immune system interactions. *Microorganisms*, *8*(10), 1587. DOI: 10.3390/microorganisms8101587 PMID: 33076307

Yoon, C. S. (2025). Natural Products in the Treatment of Neuroinflammation at Microglia: Recent Trend and Features. *Cells 2025, Vol. 14, Page 571, 14*(8), 571. DOI: 10.3390/cells14080571

Youdim, K. A., Dobbie, M. S., Kuhnle, G., Proteggente, A. R., Abbott, N. J., & Rice-Evans, C. (2003). Interaction between flavonoids and the blood–brain barrier: In vitro studies. *Journal of Neurochemistry, 85*(1), 180–192. DOI: 10.1046/j.1471-4159.2003.01652.x PMID: 12641740

Youdim, K. A., Martin, A., & Joseph, J. A. (2000). Incorporation of anthocyanins into neuronal membranes protects against oxidative stress-mediated damage. *The Journal of Nutritional Biochemistry, 11*(6), 276–281. DOI: 10.1016/S0955-2863(00)00081-4

Youdim, K. A., Qaiser, M. Z., Begley, D. J., Rice-Evans, C. A., & Abbott, N. J. (2004). Flavonoid permeability across an in situ model of the blood-brain barrier. *Free Radical Biology & Medicine, 36*(5), 592–604. DOI: 10.1016/j.freeradbiomed.2003.11.023

Yuan, M., & Wu, H. (2022). Astrocytes in the Traumatic Brain Injury: the Good and the Bad. In *Experimental Neurology* (Vol. 348, p. 113943). Academic Press. DOI: 10.1016/j.expneurol.2021.113943

Yuan, Y., Yuan, T., Wang, D., Tang, J., & Zhou, S. (2013). Sewage sludge biochar as an efficient catalyst for oxygen reduction reaction in an microbial fuel cell. *Bioresource Technology, 144*, 115–120.

Yuan, Z., Syed, M. A., Panchal, D., Rogers, D., Joo, M., & Sadikot, R. T. (2012). Curcumin mediated epigenetic modulation inhibits TREM-1 expression in response to lipopolysaccharide. *The International Journal of Biochemistry & Cell Biology, 44*(11), 2032–2043. DOI: 10.1016/j.biocel.2012.08.001 PMID: 22890222

Yu, F., Wang, Y., Stetler, A. R., Leak, R. K., Hu, X., & Chen, J. (2022). Phagocytic microglia and macrophages in brain injury and repair. *CNS Neuroscience & Therapeutics, 28*(9), 1279–1293. DOI: 10.1111/cns.13899 PMID: 35751629

Yu, X., Jiang, X., Zhang, X., Chen, Z., Xu, L., Chen, L., Wang, G., & Pan, J. (2016). The effects of fisetin on lipopolysaccharide-induced depressive-like behavior in mice. *Metabolic Brain Disease, 31*(5), 1011–1021. DOI: 10.1007/s11011-016-9839-5 PMID: 27209403

Yu, Y., Liao, X., Xie, X., Li, Q., Chen, X., & Liu, R. (2024). The role of neuroglial cells communication in ischemic stroke. *Brain Research Bulletin, 209*, 110910. DOI: 10.1016/j.brainresbull.2024.110910 PMID: 38423190

Zaidi, F. K., & Deep, S. (2020). Scutellarin inhibits the uninduced and metal-induced aggregation of α-Synuclein and disaggregates preformed fibrils: Implications for Parkinson's disease. *The Biochemical Journal, 477*(3), 645–670. DOI: 10.1042/BCJ20190705 PMID: 31939603

Zakaryan, H., Arabyan, E., Oo, A., & Zandi, K. (2017). Flavonoids: Promising natural compounds against viral infections. *Archives of Virology, 162*(9), 2539–2551. DOI: 10.1007/s00705-017-3417-y

Zbieta, E., & Ebas, R. (2025). Role of Flavonoids in Protecting Against Neurodegenerative Diseases—Possible Mechanisms of Action. *International Journal of Molecular Sciences 2025, Vol. 26, Page 4763, 26*(10), 4763. DOI: 10.3390/ijms26104763

Zbinden, A., Pérez-Berlanga, M., De Rossi, P., & Polymenidou, M. (2020). Phase separation and neurodegenerative diseases: A disturbance in the force. *Developmental Cell*, *55*(1), 45–68. DOI: 10.1016/j.devcel.2020.09.014 PMID: 33049211

Zengin, G., Locatelli, M., Carradori, S., Mocan, A. M., & Aktumsek, A. (2016). Total Phenolics, Flavonoids, Condensed Tannins Content of Eight Centaurea Species and Their Broad Inhibitory Activities against Cholinesterase, Tyrosinase, α-Amylase and α-Glucosidase. *Notulae Botanicae Horti Agrobotanici Cluj-Napoca*, *44*(1), 195–200. DOI: 10.15835/nbha44110259

Zeng, K. W., Zhang, T., Fu, H., Liu, G. X., & Wang, X. M. (2012). Schisandrin B exerts anti-neuroinflammatory activity by inhibiting the Toll-like receptor 4-dependent MyD88/IKK/NF-κB signaling pathway in lipopolysaccharide-induced microglia. *European Journal of Pharmacology*, *692*(1–3), 29–37. DOI: 10.1016/j.ejphar.2012.05.030 PMID: 22698579

Zeng, N., Zhang, G., Hu, X., Pan, J., & Gong, D. (2019). Mechanism of fisetin suppressing superoxide anion and xanthine oxidase activity. *Journal of Functional Foods*, *58*, 1–10. DOI: 10.1016/j.jff.2019.04.044

Zeng, P., Fang, M., Zhao, H., & Guo, J. (2021). A network pharmacology approach to uncover the key ingredients in Ginkgo Folium and their anti-Alzheimer's disease mechanisms. *Aging (Albany NY)*, *13*(14), 18993–19012. DOI: 10.18632/aging.203348 PMID: 34315132

Zerbinati, C., & Iuliano, L. (2017). Cholesterol and related sterols autoxidation. *Free Radical Biology & Medicine*, *111*, 151–155. DOI: 10.1016/j.freeradbiomed.2017.04.013 PMID: 28428001

Zhai, X., Ding, Y., Wang, Q., Zhang, H., & Li, F. (2016). Rutin acid ameliorates neural apoptosis induced by traumatic brain injury via mitochondrial pathways in mice. *Neuroimmunomodulation*, *23*(3), 179–187. DOI: 10.1159/000448716 PMID: 27644033

Zhang, C., Qiu, M., & Fu, H. (2024). Oligodendrocytes in central nervous system diseases: the effect of cytokine regulation. In *Neural Regeneration Research* (Vol. 19, Issue 10, pp. 2132–2143). Wolters Kluwer Medknow Publications. DOI: 10.4103/1673-5374.392854

Zhang, H., Zhang, X., Chai, Y., Wang, Y., Zhang, J., & Chen, X. (2025). Astrocyte-mediated inflammatory responses in traumatic brain injury: mechanisms and potential interventions. In *Frontiers in Immunology* (Vol. 16, p. 1584577). Frontiers Media SA. DOI: 10.3389/fimmu.2025.1584577

Zhang, B., Wang, B., Cao, S., & Wang, Y. (2015). Epigallocatechin-3-gallate (EGCG) attenuates traumatic brain injury by inhibition of edema formation and oxidative stress. *The Korean Journal of Physiology & Pharmacology; Official Journal of the Korean Physiological Society and the Korean Society of Pharmacology*, *19*(6), 491. DOI: 10.4196/kjpp.2015.19.6.491 PMID: 26557015

Zhang, B., Wei, Y.-Z., Wang, G.-Q., Li, D.-D., Shi, J.-S., & Zhang, F. (2019). Targeting MAPK pathways by naringenin modulates microglia M1/M2 polarization in lipopolysaccharide-stimulated cultures. *Frontiers in Cellular Neuroscience*, *12*, 531. DOI: 10.3389/fncel.2018.00531 PMID: 30687017

Zhang, C., Zheng, X., Wan, X., Shao, X., Liu, Q., Zhang, Z., & Zhang, Q. (2014). The potential use of H102 peptide-loaded dual-functional nanoparticles in the treatment of Alzheimer's disease. *Journal of Controlled Release*, *192*, 317–324. DOI: 10.1016/j.jconrel.2014.07.050 PMID: 25102404

Zhang, D., Hu, X., Qian, L., Wilson, B., Lee, C., Flood, P., & Hong, J. S. (2009). Prostaglandin E2 released from activated microglia enhances astrocyte proliferation in vitro. *Toxicology and Applied Pharmacology*, *238*(1), 64–70. DOI: 10.1016/j.taap.2009.04.015

Zhang, F., Li, F., & Chen, G. (2014). Neuroprotective effect of apigenin in rats after contusive spinal cord injury. *Neurological Sciences*, *35*(4), 583–588. DOI: 10.1007/s10072-013-1566-7 PMID: 24166720

Zhang, F., Shi, J. S., Zhou, H., Wilson, B., Hong, J. S., & Gao, H. M. (2010). Resveratrol protects dopamine neurons against lipopolysaccharide-induced neurotoxicity through its antiinflammatory actions. *Molecular Pharmacology*, *78*(3), 466–477. DOI: 10.1124/mol.110.064535 PMID: 20554604

Zhang, G., Liu, Y., Xu, L., Sha, C., Zhang, H., & Xu, W. (2019). Resveratrol alleviates lipopolysaccharide-induced inflammation in PC-12 cells and in rat model. *BMC Biotechnology*, *19*(1), 10. DOI: 10.1186/s12896-019-0502-1 PMID: 30691440

Zhang, H., Chen, Y., Wang, Z., Xie, G., Liu, M., Yuan, B., Chai, H., Wang, W., & Cheng, P. (2022). Implications of gut microbiota in neurodegenerative diseases. *Frontiers in Immunology*, *13*, 785644. DOI: 10.3389/fimmu.2022.785644 PMID: 35237258

Zhang, H., Shao, Y., Yang, S., Xu, L., & Wang, Y. (2021). Protocatechuic acid and its role in the prevention and treatment of neurodegenerative diseases. *Nutrients*, *13*(4), 1147. DOI: 10.3390/nu13041147 PMID: 33807148

Zhang, K., Dai, Z., Zhang, W., Gao, Q., Dai, Y., Xia, F., & Zhang, X. (2021). EDTA-based adsorbents for the removal of metal ions in wastewater. *Coordination Chemistry Reviews*, *434*, 213809. DOI: 10.1016/j.ccr.2021.213809

Zhang, L. Q., Sa, F., Chong, C. M., Wang, Y., Zhou, Z. Y., Chang, R. C. C., Chan, S. W., Hoi, P. M., & Yuen Lee, S. M. (2015). Schisantherin A protects against 6-OHDA-induced dopaminergic neuron damage in zebrafish and cytotoxicity in SH-SY5Y cells through the ROS/NO and AKT/GSK3β pathways. *Journal of Ethnopharmacology*, *170*, 8–15. DOI: 10.1016/j.jep.2015.04.040 PMID: 25934514

Zhang, L., Wang, H., Zhou, Y., Zhu, Y., & Fei, M. (2018). Fisetin alleviates oxidative stress after traumatic brain injury via the Nrf2-ARE pathway. *Neurochemistry International*, *118*(February), 304–313. DOI: 10.1016/j.neuint.2018.05.011 PMID: 29792955

Zhang, P., & Ma, X. (2015). Effect of rutin on spinal cord injury through inhibition of the expression of MIP-2 and activation of MMP-9, and downregulation of Akt phosphorylation. *Molecular Medicine Reports*, *12*(5), 7554–7560. DOI: 10.3892/mmr.2015.4357 PMID: 26502930

Zhang, Q., Lenardo, M. J., & Baltimore, D. (2017). 30 years of NF-κB: A blossoming of relevance to human pathobiology. *Cell*, *168*(1-2), 37–57. DOI: 10.1016/j.cell.2016.12.012 PMID: 28086098

Zhang, S., Gai, Z., Gui, T., Chen, J., Chen, Q., & Li, Y. (2021). Antioxidant effects of protocatechuic acid and protocatechuic aldehyde: Old wine in a new bottle. *Evidence-Based Complementary and Alternative Medicine : eCAM*, *2021*(1), 6139308. DOI: 10.1155/2021/6139308 PMID: 34790246

Zhang, S., Liu, Y., Zhao, Z., & Xue, Y. (2010). Effects of green tea polyphenols on caveolin-1 of microvessel fragments in rats with cerebral ischemia. *Neurological Research*, *32*(9), 963–970. DOI: 10.1179/016164110X12700393823570 PMID: 20444327

Zhang, W., Xiao, D., Mao, Q., & Xia, H. (2023). Role of neuroinflammation in neurodegeneration development. *Signal Transduction and Targeted Therapy*, *8*(1), 267. DOI: 10.1038/s41392-023-01486-5 PMID: 37433768

Zhang, X., Du, L., Zhang, W., Yang, Y., Zhou, Q., & Du, G. (2017). Therapeutic effects of baicalein on rotenone-induced Parkinson's disease through protecting mitochondrial function and biogenesis. *Scientific Reports*, *7*(1), 1–14. DOI: 10.1038/s41598-017-07442-y PMID: 28855526

Zhang, Y., Cai, P., Cheng, G., & Zhang, Y. (2022). A Brief Review of Phenolic Compounds Identified from Plants: Their Extraction, Analysis, and Biological Activity. *Natural Product Communications*, *17*(1), 1934578X211069721. Advance online publication. DOI: 10.1177/1934578X211069721

Zhang, Y., Chen, L., & Zhao, Y. (2021). Digital health interventions in personalized brain nutrition and cognitive support. *Nutrients*, *13*(10), 3572. DOI: 10.3390/nu13103572 PMID: 34684573

Zhang, Y., Peiying, L., Qianqian, K., Shiyan, X., Xiaojing, L., & Limin, S. (2017). The contents of terpene trilactone and flavonoid in leaves of seedlings from ancient female ginkgo trees in China. *Horticultural Plant Journal*, *3*(4), 165–171. DOI: 10.1016/j.hpj.2017.06.002

Zhang, Y., Wang, Z., Peng, J., Gerner, S. T., Yin, S., & Jiang, Y. (2021). Gut microbiota-brain interaction: An emerging immunotherapy for traumatic brain *injury. Experimental Neurology*, *337*, 113585. DOI: 10.1016/j.expneurol.2020.113585 PMID: 33370556

Zhang, Y., Yu, W., Zhang, L., Wang, M., & Chang, W. (2022). The interaction of polyphenols and the gut microbiota in neurodegenerative diseases. *Nutrients*, *14*(24), 5373. DOI: 10.3390/nu14245373 PMID: 36558531

Zhang, Y., Zheng, T., Ma, D., Shi, P., Zhang, H., Li, J., & Sun, Z. (2023). Probiotics *Bifidobacterium lactis* M8 and *Lactobacillus rhamnosus* M9 prevent high blood pressure via modulating the gut microbiota composition and host metabolic products. *mSystems*, *8*(6), e0033123. Advance online publication. DOI: 10.1128/msystems.00331-23 PMID: 37855616

Zhang, Z., Hamada, H., & Gerk, P. M. (2019). Selectivity of dietary phenolics for inhibition of human monoamine oxidases A and B. *BioMed Research International*, *2019*, 1–12. Advance online publication. DOI: 10.1155/2019/8361858 PMID: 30809547

Zhang, Z., Tan, X., Sun, X., Wei, J., Li, Q. X., & Wu, Z. (2022). Isoorientin Affects Markers of Alzheimer's Disease via Effects on the Oral and Gut Microbiota in APP/PS1 Mice. *The Journal of Nutrition*, *152*(1), 140–152. DOI: 10.1093/jn/nxab328

Zhao, J. Y., Zhou, Y., Zhou, C. W., & Zhan, K. Bin, Yang, M., Wen, M., & Zhu, L. Q. (2025). Revisiting the critical roles of reactive microglia in traumatic brain injury. In *International journal of surgery (London, England)* (Vol. 111, Issue 6, pp. 3942–3978). Int J Surg. DOI: 10.1097/JS9.0000000000002420

Zhao, H., Chen, S., Gao, K., Zhou, Z., Wang, C., Shen, Z., Guo, Y., Li, Z., Wan, Z., Liu, C., & Mei, X. (2017). Resveratrol protects against spinal cord injury by activating autophagy and inhibiting apoptosis mediated by the SIRT1/AMPK signaling pathway. *Neuroscience*, *348*, 241–251. DOI: 10.1016/j.neuroscience.2017.02.027 PMID: 28238848

Zhao, J., Bi, W., Xiao, S., Lan, X., Cheng, X., Zhang, J., Lu, D., Wei, W., Wang, Y., Li, H., Fu, Y., & Zhu, L. (2019). Neuroinflammation induced by lipopolysaccharide causes cognitive impairment in mice. *Scientific Reports*, *9*(1), 5790. DOI: 10.1038/s41598-019-42286-8 PMID: 30962497

Zhao, J., Liang, Q., Sun, Q., Chen, C., Xu, L., Ding, Y., & Zhou, P. (2017). (−)-Epigallocatechin-3-gallate (EGCG) inhibits fibrillation, disaggregates amyloid fibrils of α-synuclein, and protects PC12 cells against α-synuclein-induced toxicity. *RSC Advances*, *7*(52), 32508–32517. DOI: 10.1039/C7RA03752J

Zhao, R., Wu, X., Bi, X.-Y., Yang, H., & Zhang, Q. (2022). Baicalin attenuates blood-spinal cord barrier disruption and apoptosis through PI3K/Akt signaling pathway after spinal cord injury. *Neural Regeneration Research*, *17*(5), 1080–1087. DOI: 10.4103/1673-5374.324857 PMID: 34558536

Zhao, Y.-N., Li, W.-F., Li, F., Zhang, Z., Dai, Y.-D., Xu, A.-L., Qi, C., Gao, J.-M., & Gao, J. (2013). Resveratrol improves learning and memory in normally aged mice through microRNA-CREB pathway. *Biochemical and Biophysical Research Communications*, *435*(4), 597–602. DOI: 10.1016/j.bbrc.2013.05.025 PMID: 23685142

Zhao, Z., Li, X., & Li, Q. (2017). Curcumin accelerates the repair of sciatic nerve injury in rats through reducing Schwann cells apoptosis and promoting myelinization. *Biomedicine and Pharmacotherapy*, *92*, 1103–1110. DOI: 10.1016/j.biopha.2017.05.099 PMID: 28622711

Zheng, Y., Sun, J., Luo, Z., Li, Y., & Huang, Y. (2024). Emerging mechanisms of lipid peroxidation in regulated cell death and its physiological implications. In *Cell Death and Disease* (Vol. 15, Issue 11). Springer Nature. DOI: 10.1038/s41419-024-07244-x

Zhou, B., Zhu, Z., Ransom, B. R., & Tong, X. (2020). Oligodendrocyte lineage cells and depression. *Molecular Psychiatry 2020 26:1, 26*(1), 103–117. DOI: 10.1038/s41380-020-00930-0

Zhou, Y., Shao, A., Yao, Y., Tu, S., Deng, Y., & Zhang, J. (2020). Dual roles of astrocytes in plasticity and reconstruction after traumatic brain injury. In *Cell Communication and Signaling* (Vol. 18, Issue 1, pp. 1–16). BioMed Central. DOI: 10.1186/s12964-020-00549-2

Zhou, J., Yang, Z., Shen, R., Zhong, W., Zheng, H., Chen, Z., Tang, J., & Zhu, J. (2021). Resveratrol Improves Mitochondrial Biogenesis Function and Activates PGC-1α Pathway in a Preclinical Model of Early Brain Injury Following Subarachnoid Hemorrhagc. *Frontiers in Molecular Biosciences*, *8*, 620683. DOI: 10.3389/fmolb.2021.620683 PMID: 33968980

Zhou, L., Fan, L., Yi, X., Zhou, Z., Liu, C., Fu, R., Dai, C., Wang, Z., Chen, X., Yu, P., Chen, D., Tan, G., Wang, Q., & Ning, C. (2018). Soft Conducting Polymer Hydrogels Cross-Linked and Doped by Tannic Acid for Spinal Cord Injury Repair. *ACS Nano*, *12*(11), 10957–10967. DOI: 10.1021/acsnano.8b04609 PMID: 30285411

Zhou, L., Wu, Z., Wang, G., Xiao, L., Wang, H., Sun, L., & Xie, Y. (2020). Long-term maternal separation potentiates depressive-like behaviours and neuroinflammation in adult male C57/BL6J mice. *Pharmacology, Biochemistry, and Behavior*, *196*, 172953. DOI: 10.1016/j.pbb.2020.172953 PMID: 32450088

Zhou, M., & Pang, X. e. (2025). Polyphenols and miRNA interplay: A novel approach to combat apoptosis and inflammation in Alzheimer's disease. *Frontiers in Aging Neuroscience*, *17*, 1571563. DOI: 10.3389/fnagi.2025.1571563 PMID: 40400913

Zhou, P., Zhao, X. N., Ma, Y. Y., Tang, T. J., Wang, S. S., Wang, L., & Huang, J. L. (2022). Virtual screening analysis of natural flavonoids as trimethylamine (TMA)-lyase inhibitors for coronary heart disease. *Journal of Food Biochemistry*, *46*(12), e14376. DOI: 10.1111/jfbc.14376

Zhou, S., Yu, L., Zhang, Z., & Zhang, H. (2016). Gut microbiota and polyphenols: From absorption to bioactivity. *Food Research International*, *89*, 253–261. DOI: 10.1016/j.foodres.2016.09.029

Zhou, Y., Dhaher, R., Parent, M., Hu, Q. X., Hassel, B., Yee, S. P., Hyder, F., Gruenbaum, S. E., Eid, T., & Danbolt, N. C. (2019). Selective deletion of glutamine synthetase in the mouse cerebral cortex induces glial dysfunction and vascular impairment that precede epilepsy and neurodegeneration. *Neurochemistry International*, *123*, 22–33. DOI: 10.1016/j.neuint.2018.07.009 PMID: 30053506

Zhou, Y.-F., Guo, B., Ye, M.-J., Liao, R.-F., & Li, S.-L. (2016). Protective effect of rutin against H2O2-induced oxidative stress and apoptosis in human lens epithelial cells. *Current Eye Research*, *41*(7), 933–942. DOI: 10.3109/02713683.2015.1082186 PMID: 26576853

Zhou, Z. D., Xie, S. P., Saw, W. T., Ho, P. G. H., Wang, H. Y., Zhou, L., Zhao, Y., & Tan, E. K. (2019). The therapeutic implications of tea polyphenols against dopamine (DA) neuron degeneration in Parkinson's disease (PD). *Cells*, *8*(8), 911. DOI: 10.3390/cells8080911 PMID: 31426448

Zhu, G., Wang, X., Chen, L., Lenahan, C., Fu, Z., Fang, Y., & Yu, W. (2022). Crosstalk between the oxidative stress and glia cells after stroke: From mechanism to therapies. *Frontiers in Immunology*, *13*, 852416. DOI: 10.3389/fimmu.2022.852416 PMID: 35281064

Zhu, H., Bian, C., Yuan, J., Chu, W., Xiang, X., Chen, F., Wang, C., Feng, H., & Lin, J. (2014). Curcumin attenuates acute inflammatory injury by inhibiting the TLR4/MyD88/NF-κB signaling pathway in experimental traumatic brain injury. *Journal of Neuroinflammation*, *11*(1), 59. DOI: 10.1186/1742-2094-11-59 PMID: 24669820

Zhuo, C., Tian, H., Song, X., Jiang, D., Chen, G., Cai, Z., Ping, J., Cheng, L., Zhou, C., & Chen, C. (2023). Microglia and cognitive impairment in schizophrenia: Translating scientific progress into novel therapeutic fiinterventions. *Schizophrenia*, *9*(1), 42. Advance online publication. DOI: 10.1038/s41537-023-00370-z PMID: 37429882

Zhu, X., Liu, Q., Wang, M., Liang, M., Yang, X., Xu, X., Zou, H., & Qiu, J. (2011). Activation of Sirt1 by resveratrol inhibits TNF-α induced inflammation in fibroblasts. *PLoS One*, *6*(11), e27081. DOI: 10.1371/journal.pone.0027081 PMID: 22069489

Zhu, Y.-L., Sun, M.-F., Jia, X.-B., Cheng, K., Xu, Y.-D., Zhou, Z.-L., Zhang, P.-H., Qiao, C.-M., Cui, C., Chen, X., Yang, X.-S., & Shen, Y.-Q. (2019). Neuroprotective effects of Astilbin on MPTP-induced Parkinson's disease mice: Glial reaction, α-synuclein expression and oxidative stress. *International Immunopharmacology*, *66*, 19–27. DOI: 10.1016/j.intimp.2018.11.004 PMID: 30419450

Ziberna, L., Fornasaro, S., Čvorović, J., Tramer, F., & Passamonti, S. (2014). Bioavailability of Flavonoids: The Role of Cell Membrane Transporters. In Watson, R. R., Preedy, V. R., & Zibadi, S. (Eds.), *Polyphenols in Human Health and Disease* (pp. 489–511). Academic Press.

Zmora, N., Zilberman-Schapira, G., Suez, J., Mor, U., Dori-Bachash, M., Bashiardes, S., Kotler, E., Zur, M., Regev-Lehavi, D., Brik, R. B.-Z., Federici, S., Cohen, Y., Linevsky, R., Rothschild, D., Moor, A. E., Ben-Moshe, S., Harmelin, A., Itzkovitz, S., Maharshak, N., & Elinav, E. (2018). Personalized gut mucosal colonization resistance to empiric probiotics is associated with unique host and microbiome features. *Cell*, *174*(6), 1388–1405. DOI: 10.1016/j.cell.2018.08.041 PMID: 30193112

Zong, Y., Li, H., Liao, P., Chen, L., Pan, Y., Zheng, Y., Zhang, C., Liu, D., Zheng, M., & Gao, J. (2024). Mitochondrial dysfunction: Mechanisms and advances in therapy. *Signal Transduction and Targeted Therapy*, *9*(1), 124. DOI: 10.1038/s41392-024-01839-8 PMID: 38744846

Zou, P., Wu, C., Liu, T. C. Y., Duan, R., & Yang, L. (2023). Oligodendrocyte progenitor cells in Alzheimer's disease: from physiology to pathology. In *Translational Neurodegeneration* (Vol. 12, Issue 1, pp. 1–20). BioMed Central. DOI: 10.1186/s40035-023-00385-7

Zou, P., Liu, X., Li, G., & Wang, Y. (2018). Resveratrol pretreatment attenuates traumatic brain injury in rats by suppressing NLRP3 inflammasome activation via SIRT1. *Molecular Medicine Reports*, *17*(2), 3212–3217. DOI: 10.3892/mmr.2017.8241 PMID: 29257276

Zughaibi, T. A., Suhail, M., Tarique, M., & Tabrez, S. (2021). Targeting PI3K/Akt/mTOR pathway by different flavonoids: A cancer chemopreventive approach. *International Journal of Molecular Sciences*, *22*(22), 12455. DOI: 10.3390/ijms222212455 PMID: 34830339

Zuo, A. R., Dong, H. H., Yu, Y. Y., Shu, Q. L., Zheng, L. X., Yu, X. Y., & Cao, S. W. (2018). The antityrosinase and antioxidant activities of flavonoids dominated by the number and location of phenolic hydroxyl groups. *Chinese Medicine*, *13*, 51. DOI: 10.1186/s13020-018-0206-9

About the Contributors

Ali Imran is an accomplished food scientist and academic with a distinguished career in nutritional research, functional foods, and metabolic health interventions. Currently serving as an Associate Professor in the Department of Food Science at Government College University Faisalabad (GCUF), Pakistan, he is also a Post-Doctoral Research Fellow at The University of Melbourne, Australia. Dr. Imran's expertise lies in the valorization of agro-industrial waste, the development of polyphenol-based functional foods, and dietary interventions for metabolic disorders, with a strong focus on neurodegenerative diseases. His research has significantly contributed to understanding the role of bioactive compounds—such as catechins, theaflavins, and flavonoids—in modulating cognitive function and brain health. With over 128 peer-reviewed publications in high-impact journals and multiple books and book chapters to his credit, Dr. Imran is an active contributor to scientific advancements in food and nutrition. He is also an editorial board member and reviewer for leading journals in the field. His extensive consultancy work with global organizations, including GAIN and Nutrition International, highlights his influence in policy-driven food fortification and public health nutrition. A recipient of numerous research productivity awards and international fellowships, Dr. Imran continues to drive innovation at the intersection of nutrition and neuroscience, making substantial contributions to cognitive health research.

Ghulam Hussain is an esteemed neuroscientist and academic with extensive expertise in neurodegenerative disorders, nerve regeneration, and the therapeutic potential of natural bioactive compounds. He currently serves as an Associate Professor at the Department of Physiology, Government College University Faisalabad (GCUF), Pakistan, where he also holds administrative roles, including Chairperson of the Department of Physiology. Dr. Hussain earned his Ph.D. in Neurosciences from the University of Strasbourg, France, with postdoctoral research in neurophysiology and neurochemistry. His research primarily focuses on the mechanisms of nerve regeneration, neurodegeneration, and the role of natural products in cognitive and neurological disorders. His contributions to neuroscience include molecular insights into neuroinflammatory pathways, drug discovery for neurological disorders, and the physiological impact of polyphenols on neuroprotection. With a strong publication record, Dr. Hussain has authored numerous high-impact journal articles, book chapters, and edited books, including works on neuroprotective effects of polyphenols, oxidative stress, and metabolic brain disorders. He is actively involved in various international scientific organizations, including the Federation of European Neuroscience Societies (FENS), the Society for Neuroscience (SfN), and the International Society for Neurochemistry (ISN). His commitment to advancing neuroscience is reflected in his role as an editor and reviewer for prestigious journals, international research collaborations, and contributions to neuroscience conferences worldwide. His research aims to bridge the gap between functional foods, nutraceuticals, and neurodegenerative disease management, making significant strides in natural interventions for cognitive health.

***

Qammar Abbas is currently serving as a faculty member in the Department of Emerging Allied Health Technologies at the University of Lahore, Pakistan. He is actively involved in teaching and research within the Faculty of Allied Health Sciences, with a focus on neurophysiology, epigenetics, and environmental health sciences. Mr. Abbas has contributed to the academic development of students through curriculum design, examination content development, and scholarly publications. His research interests include clinical neurophysiology, environmental risk assessment, and the integration of emerging technologies into allied health education. He is dedicated to promoting interdisciplinary research and enhancing academic standards in the health sciences domain

Moin Akhtar is perusing his Bachelors in Zoology from the Department of Zoology, Government College University Faisalabad, Pakistan.

Rao Mubashar Ali is a microbiologist at Capital University of Science and Technology working on probiotics from fermented food products and their metabolic by products like bacteriocins. I am passionate about microbiological research and aims to pursue a career in laboratory sciences and research.

Muhammad Awais is currently pursuing MPhil in Physiology from Government College University Faisalabad. His research focuses on investigating the cellular and molecular mechanisms underlying pancreatic β-cell regeneration, particularly emphasizing the role of the Notch/Ngn3 signaling axis and its modulation by Withania somnifera (Ashwagandha). Through this work, he would elucidate novel therapeutic strategies for diabetes by exploring how herbal bioactives can influence β-cell differentiation and functional restoration.

Nimra Aziz is an Assistant Professor and Head of the Department of Emerging Allied Health Technologies at the University of Lahore. She holds a PhD in Translational Neurosciences and Neurotechnologies from University of Ferrara, Italy and leads undergraduate programs in Neurophysiology, Cardiac Perfusion, Cardiovascular, Dermatology, and Renal & Dialysis Technologies. Her research interests include translational neuroscience, neurophysiology, neuroinflammation, and electrophysiological biomarkers, with experience in in vivo and in vitro models. Dr. Aziz is committed to innovative, student-centered teaching and academic leadership, focusing on advancing allied health education, research excellence, and clinical relevance in Pakistan.

Bushra Basharat is currently pursuing her Bachelors in Physiology from the Department of Physiology, Government College University Faisalabad, Pakistan.

Shazia Batool is an Academic and Program Coordinator in the Department of Emerging Allied Health Technologies at the University of Lahore. She holds a BS in Renal Dialysis Technology and an MPhil in Human Physiology, with distinction as a gold medalist. With over five years of clinical experience and expertise in physiology, she bring both academic and practical insights to teaching. She actively contributes to research, curriculum development, and academic coordination, while also serving on departmental committees.

Pankaj Singh Chandel serves as an Assistant Professor and Head in the School of Hospitality and Tourism Management at AAFT University of Media and Arts. Prior to this, he held academic positions at Dev Sanskriti Vishwavidyalaya, Haridwar, where he taught in the Department of Tourism Management from July 2014 to March 2025. Dr. Chandel holds a PhD in Tourism and a Post Graduate Diploma in Management (PGDM) in Tourism and Travel Services Management from the Indian Institute of Tourism and Travel Management. His primary research interests center on sustainable tourism practices and the development of homestay tourism in India. He is recognized for his expertise in homestay tourism and is noted as an accomplished speaker in the field. Dr. Chandel has published research articles, books, and book chapters in Scopus-indexed journals, and has contributed to academic publications with leading publishers such as IGI Global, Taylor & Francis, Springer Nature, IEEE, and Emerald, reflecting his commitment to advancing scholarship in tourism studies. Dr. Chandel is actively involved in professional organizations, including the Travel Agents Association of Uttarakhand and the Adventure Tour Operators Association of India (ATOAI), where he contributes to sustainability initiatives within the tourism sector. His engagement with these organizations enables him to bridge academic research and industry practice.

Emaan Fatima is a 22-year-old undergraduate student currently pursuing a Bachelor's degree in Microbiology at Capital University of Science and Technology, Islamabad. With a strong passion for scientific research and healthcare innovation, she is committed to exploring the microscopic world and its impact on human health. Emaan is dedicated to expanding her knowledge in areas such as infectious diseases, immunology, and biotechnology, aiming to contribute meaningfully to the field through academic excellence and hands-on laboratory experience.

Shah Gull is an experienced microbiologist and educator with over five years of teaching and research experience in Pakistan's leading academic institutions. She holds an M.Phil. in Microbiology (CGPA 3.91) and a B.S. (Hons.) in Applied Microbiology from the University of Veterinary and Animal Sciences, Lahore. Currently serving as a Lecturer at the Capital University of Science and Technology, Islamabad, Shahgull has previously worked at PMAS-Arid Agriculture University and UVAS, Lahore. Her research has focused on foodborne pathogens, antibiotic resistance, and bioremediation, with several publications and GenBank sequence submissions to her credit. Shahgull has taught a wide range of microbiology courses, designed student assessments, and coordinated academic and extracurricular activities. Her technical expertise includes PCR, ELISA, DNA/RNA extraction, microbial isolation, and biosafety practices in BSL-2/3 labs. She has also led an ORIC-funded bioentrepreneurial project on oyster mushroom cultivation. A proactive contributor to scientific workshops and conferences, Shahgull is a Global Outreach Member of the American Society for Microbiology and a lifelong UVAS alumni member.

Arslan Iftikhar is an Assistant Professor of Physiology at Government College University, Faisalabad, Pakistan, with a diverse academic background that spans from a Doctor of Veterinary Medicine (DVM) to a PhD in Life & Health Sciences from the prestigious Institut de Génétique et de Biologie Moléculaire et Cellulaire (IGBMC), University of Strasbourg, France. His PhD research explored the pivotal role of histone variants in DNA repair, laying the groundwork for his current focus on regenerative medicine and natural product-based therapies. As the head of the Nutritional Genomics Research Lab (NGRL) at GCUF, he leads cutting-edge research on the regenerative potential of natural products, targeting pancreatic β-cell repair, insulin secretion, and oxidative stress mitigation. He has successfully supervised 17 MPhil graduates and is currently mentoring 2 PhD and 6 MPhil scholars. His team has published extensively in leading Q1 journals, such as Nucleic Acid Research, Cells, Heliyon, and Molecules. An active member of global scientific societies—including The Physiological Society (UK), American Physiological Society (USA), and European Society of Endocrinology (UK)—he remains committed to staying at the forefront of emerging trends in physiology and biomedical sciences. More recently, he has integrated artificial intelligence into his research to bridge traditional physiology with advanced technologies, addressing contemporary healthcare challenges. He is passionate about mentoring the next generation of scientists, fostering collaboration, and driving impactful research to inspire transformative advancements in healthcare.

Nazmeen Ismail is an MS bioscience student with research intrest in microbiology and its application. She is currently contributing to a book chapter on the gut brain axis.

Dt. Khushi Kashyap is a Clinical Dietitian, Certified Lifestyle Coach, and Diabetes Educator currently serving as an Assistant Professor at the School of Wellness, AAFT University. With expertise in evidence-based nutrition and sustainable food systems, she focuses her research on clean-label entrepreneurship, ESG compliance in wellness ventures, and nutrition education innovation. She has actively contributed to health campaigns, corporate wellness programs, and student-driven initiatives that combine health science with responsible business practices. Her academic background is complemented by field experience in promoting nutrition-focused startups that align with ethical governance and environmental sustainability standards.

Asra Khan is a dedicated academic professional currently working as a Demonstrator in the Cardiac Perfusion Technology department at The University of Lahore. With over two years of teaching experience, she has actively contributed to the academic and clinical training of undergraduate students pursuing careers in cardiac sciences. Her role involves conducting lectures, facilitating laboratory sessions, and mentoring students in the fundamentals and advanced practices of cardiac perfusion. Asra is particularly passionate about bridging the gap between theoretical knowledge and hands-on clinical application, aiming to prepare her students for the dynamic demands of healthcare environments. She is continually engaged in academic growth and is keenly interested in expanding her expertise through research, professional development, and collaboration within the cardiovascular education community.

Hamna khan is a PhD scholar in neuroscience at department of physiology specializing in neurodegenerative and genetic disorders.

Saba Mehboob is currently serving as a Lecturer in the Department of Physiology at Government College University, Faisalabad. She is also pursuing her Ph.D. in Physiology at the same institution. Her doctoral research focuses on elucidating the molecular mechanisms underlying β-cell regeneration and the major signaling pathways involved in pancreatic function and metabolic regulation. With a strong interest in cellular physiology and molecular biology, she aims to contribute to the development of novel therapeutic strategies for diabetes and related metabolic disorders.

Maryam Mustafa is an M.Phil. scholar in the Department of Physiology at Government College University, Faisalabad. Her research focuses on exploring the therapeutic potential of various phytochemicals in the treatment of peripheral nerve injury. With a keen interest in neurophysiology and natural product-based interventions, she aims to contribute to the development of novel, safe, and effective strategies for nerve regeneration and functional recovery.

Humaira Muzaffar is serving as an Assistant Professor in the Department of Physiology, Government College University Faisalabad, Pakistan since October, 2020.

Muhammad Hannan is a dedicated researcher at the Neuroregeneration and Genetics Laboratory (NGL), where his work centers on peripheral nerve injury, particularly diabetic peripheral neuropathy, neurodegeneration, and therapeutic target discovery. His research expertise combines molecular docking, network pharmacology, and experimental neuroscience models to investigate novel treatment strategies for neurological complications of diabetes. At NGL, he contributes to projects involving in vitro and in vivo studies, neurotoxicology, and computational drug discovery approaches to identify and validate potential therapeutic compounds. With a strong interest in linking natural bioactive compounds and modern pharmacological tools, he aims to provide new insights into the management of diabetic peripheral nerve injury and related neurodegenerative disorders. His long-term vision is to advance NGL's role as a hub for multidisciplinary neuroscience research, bridging genetics, pharmacology, and neurobiology to design effective, evidence-based therapeutic interventions for patients.

Nasreen Noor is a graduate in Biosciences from Capital University of Science and Technology, Islamabad. Her academic focus lies in sustainable biotechnology and microbial metabolites, with a strong interest in applying biological research to address environmental and agricultural challenges.

Meenambiga S S Assistant Professor completed 6 years of teaching experience and 4 years of research experience in the field of Biotechnology. Passionate about teaching and have trained many young students in the field of Mycology, Natural Product Research, Nanoparticle synthesis, Drug Discovery from Plants and fungal sources and in-silico analysis of drug targets using modern Bio-informatics tools. Skilled in scientific writing covering the following aspects: Proposal submission, research articles suitable for national and international publications,drafting, proof reading and editing research thesis and expertise in critically reviewing the research manuscript for indexed journals. Member of Board of Studies for B.Tech Biotechnology curriculum in VISTAS. Guided more than 20 undergraduate students projects and served as Principal Investigator in the student's project funded by Tamilnadu State Council for Science and Technology. Published more than 30 research and review articles in national and international indexed journal and has volunteered as reviewer for many reputed journals. Life member of Mycological Society of India, BRSI, IARA, IAENG and many other professional organizations. Completed

Saher Sajjad is currently pursuing a Master's degree in Biosciences and holds a Bachelor of Science in Biotechnology from Capital University of Science and Technology, graduating with distinction (CGPA 3.84). Her academic and research interests lie in microbial biotechnology, molecular biology, and bioinformatics. She has conducted research on the termite gut microbiome for potential applications in biofuel production and has presented her findings at the 12th International Conference on Biological and Computational Sciences (C-BICS 2024). She has also contributed microbial sequence data to the NCBI database. Saher is actively involved in GENIX society and is passionate about advancing scientific research through collaboration and continuous learning.

Alishba Sarwat is a graduate of BS Biosciences from Capital University of Science and Technology (CUST), Islamabad. Her academic focus includes molecular biology, bioinformatics, and sustainable solutions in biotechnology. She has worked on transforming agricultural waste into bioplastics as her final year project. Her research interests lie in environmental biotechnology and sustainable development. She aspires to pursue higher education and research in molecular sciences.

Anushay Shahab A final year undergraduate student majoring in microbiology, research projects encompassing bioinformatics driven drug design, in siliso vaccine design and precision medicine for biological datasets.

Richa Sharma is a clinical nutritionist, educator, and wellness advocate with a strong background in therapeutic nutrition and preventive health. With over 5 years of experience in academic research and personalized diet planning, she specializes in translating complex nutritional science into practical, everyday guidance. Her work focuses on holistic well-being, brain health, and lifestyle-based interventions. She is the founder of Eat Healthy for Tomorrow and also leads a home-based initiative, The Wellness Box by Dt. Richa, offering mindful, nutritious meals. Passionate about functional foods, she continues to explore the connection between food, longevity, and quality of life through both teaching and writing.

Ankita Shrivastava is a dedicated Dietitian and Assistant Professor at the School of Wellness, blending clinical expertise with academic excellence. She holds a Master's degree in Food and Nutrition and is UGC-NET qualified, reflecting her deep understanding of the science behind nutrition and its impact on health and well-being. With a strong foundation in dietetics and a passion for teaching, Ankita is committed to nurturing future health professionals and promoting evidence-based nutrition practices. Her interests lie in holistic health, food science, wellness education, and lifestyle modification strategies that empower individuals to live healthier lives. Driven by purpose and compassion, Ankita combines theory and practice to bridge the gap between academic knowledge and real-world application in the field of wellness and nutrition.

Syeda Meryam Waseem is a passionate researcher at the Neuroregeneration and Genetics Laboratory (NGL), where she focuses on advancing the understanding of neurodegenerative diseases and peripheral nerve injuries. Her research integrates molecular neuroscience, genetics, and pharmacology to uncover novel therapeutic targets for conditions such as neurodegeneration, neuropathic pain, and nerve repair. At NGL, she is actively engaged in projects that utilize network pharmacology, molecular docking, in vitro and in vivo studies, and neurotoxicology models. With a strong interest in exploring bioactive compounds from medicinal plants, she aims to bridge traditional medicine with modern neuroscience. Her long-term vision is to contribute to the development of innovative, evidence-based therapeutic strategies while expanding NGL's impact as a center for multidisciplinary collaboration in neuroscience and genetics research.

Aqsa Qasim Zafar is currently pursuing bachelor's degree in Microbiology at the Capital University of Science and Technology in Islamabad, Pakistan. Her academic pursuits are driven by a strong interest in understanding the intricacies of microbial life and its applications and is passionate about exploring the role of microorganisms in food systems and delving into the molecular mechanisms that underpin their behavior.

Index

A

B

C

D

N

O

P

R

S

T

www.ingramcontent.com/pod-product-compliance
Lightning Source LLC
LaVergne TN
LVHW060252011026
849161LV00011B/2699